HANDBOOK OF
TECHNOLOGY
MANAGEMENT

Other McGraw-Hill Books of Interest

HANDBOOK OF TECHNOLOGY MANAGEMENT

Gerard H. "Gus" Gaynor, Editor in Chief

McGraw-HILL

New York   San Francisco   Washington, D.C.   Auckland   Bogotá
Caracas   Lisbon   London   Madrid   Mexico City   Milan
Montreal   New Delhi   San Juan   Singapore
Sydney   Tokyo   Toronto

Library of Congress Cataloging-in-Publication Data

Handbook of technology management / Gerard H. "Gus" Gaynor, editor in
   chief.
         p.   cm.
   Includes bibliographical references and index.
   ISBN 0-07-023619-4
   1. Technology—Management—Handbooks, manuals, etc. I. Gaynor,
Gerard H.
T49.H257   1996
658—dc20   96-11637
                                            CIP

McGraw-Hill

A Division of The McGraw·Hill Companies

1 2 3 4 5 6 7 8 9 0 DOC/DOC 9 0 1 0 9 8 7 6

ISBN 0-07-023619-4

*The sponsoring editor for this book was Harold B. Crawford, the editing
supervisor was Peggy Lamb, and the production supervisor was Pamela A.
Pelton. It was set in Times Roman by Terry Leaden of McGraw-Hill's Professional
Book Group composition unit.*

Printed and bound by R. R. Donnelley & Sons Company.

This book is printed on acid-free paper.

McGraw-Hill books are available at special quantity discounts to use as premiums
and sales promotions, or for use in corporate training programs. For more information, please write to the Director of Special Sales, McGraw-Hill, 11 West 19th
Street, New York, NY 10011. Or contact your local bookstore.

CONTENTS

Part 2   Methodologies, Tools, and Techniques

Chapter 11. Tools for Analyzing Organizational Impacts of New Technology   *Ann Majchrzak*   **11.3**

Chapter 12. Forecasting and Planning Technology   *Frederick Betz*   **12.1**

Chapter 13. Knowledge Mapping: A Tool for Management of Technology   *Karol I. Pelc*   **13.1**

Chapter 14. The Process of Developing an R&D Strategy   *Robert Szakonyi*   **14.1**

CONTRIBUTORS

Paul S. Adler is currently Associate Professor at the University of Southern California School of Business Administration. He began his education in Australia and moved to France in 1974, where he received his doctorate in Economics and Management while working as a Research Economist for the French government. Before coming to USC in 1991, Dr. Adler was a Visiting Scholar at the Brookings Institution and a Post-doctoral Research Fellow at the Harvard Business School. He also taught at Columbia and Stanford. His principal research interests are in the management of new technologies in production and engineering operations. Dr. Adler has published widely in academic and managerial journals both in the United States and overseas. Most recently he published two edited volumes, "Technology and the Future of Work," and "Usability: Turning Technologies into Tools." He is on the editorial boards of *IEEE Transactions in Engineering Management, Journal of Engineering and Technology Management*, and *International Journal of Human Factors in Manufacturing*, and currently serves as Senior Editor at Organization Science and Associate Editor at Management Science. (CHAP. 4)

Sarah Beckman has B.S., M.S., and Ph.D. degrees from the Department of Industrial Engineering and Engineering Management at Stanford University and an M.S. in Statistics from the same institution. Dr. Beckman is Co-Director of the Management of Technology Program at the University of California, Berkeley's Haas School of Business. Her primary research interests are in green design and manufacturing, product definition processes for software products, and the effects of vertical "disintegration." Prior to and concurrent with her involvement at the Haas School, Dr. Beckman worked for the Hewlett-Packard Company, most recently as Director of the Product Generation Change Management Team. (CHAP. 24)

Frederick Betz received a Ph.D. in physics from the University of California at Berkeley and later changed fields to management science and operations research. He taught in management schools for several years and is currently at the National Science Foundation, funding research in several areas of science and engineering. He is the author of several books about the management of technology, the most recent being *Strategic Technology Management* (McGraw-Hill, 1991). (CHAPS. 8 and 12)

Klaus K. Brockhoff is professor of Technology Management and director of the Institute for Research in Innovation Management at the University of Kiel, Germany. He was a visiting professor at institutions in Austria, Belgium, Sweden, and the United States. He acts as consultant to various governments and to industrial firms. (CHAP. 27)

Alok K. Chakrabarti is Dean of the School of Industrial Management, New Jersey Institute of Technology, Newark, N.J., and a distinguished professor for technology management at the same institution. He has published extensively in national and international journals and has taught management of technology courses as a visiting professor in Germany and India. (CHAP. 27)

Joseph P. Cory of the Cory Group is a Management and Educational Consultant in the Management of Technology. Currently, he is Visiting Associate Professor at Stevens Institute of Technology where he was responsible for developing the curriculum leading to an M.S. in the Management of Technology. A practitioner for more than 35 years across most industry sectors, he also wrote and reviewed for the *International Journal of Technology Management* and other journals. Recently, he left IBM where he was curriculum manager and developed and taught courses in Technology Management. He monitored Technology Management programs for IBM at MIT Sloan School and the National Technological University. He represented IBM at CIMS at Lehigh University and is a reviewer and writer of articles for the International Association in the Management of Technology. He held various positions at IBM in development, marketing, market research, plans, and controls. He also was a senior management consultant with The Diebold Group and associate engineer at David Sarnoff Research Center. (CHAP. 2)

Dov Dvir is a senior lecturer and the head of Management of Technology Department in the Holon Center for Technological Education. He is also a lecturer in Management of Technology at the Tel Aviv University, Faculty of Management, the Leon Recanati School of Business Administration. Previously, he was a senior officer in the Israeli Defense Force (IDF). He holds a B.S. in electrical engineering from the Technion—Israel Institute of Technology; an M.S. in operations research and an M.B.A. from Tel Aviv University; and a Ph.D. in management from Tel Aviv University. (CHAP. 32)

Lynn W. Ellis is scholar-in-residence at the University of New Haven (Connecticut) where he is a retired professor of management. He is a former research director at ITT, and the author of *The Financial Side of Industrial Research Management* (Wiley 1984) as well as many articles on issues related to the management of research and development. He has a doctorate of professional studies in management degree from Pace University. Dr. Ellis is a Life Fellow of the Institute of Electrical and Electronics Engineers, and a Fellow of the American Association for the Advancement of Science. (CHAP. 26)

Gerard H. (Gus) Gaynor brings more than 40 years of technical and management expertise in managing technology and innovation. After more than a decade in upper atmosphere research and development he began a 25-year career at 3M. Upon retirement from 3M, as a senior technology executive, he organized G. H. Gaynor and Associates, Inc., a consulting firm, focusing on managing technology from a systems perspective. McGraw-Hill has published two of his books: *Achieving the Competitive Edge Through Integrated Technology Management* and *Exploiting Cycle Time in Technology Management*. Since retirement he has been appointed twice as a Senior Fulbright Scholar. (CHAPS. 1 and 33)

Juergen Hauschildt is professor of organization and director of the Institute of Business Administration at the University of Kiel, Germany. He is the author of numerous books and papers, among them *Innovation Management*. He was awarded an honorary doctors degree from the University of Rostock. (CHAP. 27)

Anil B. Jambekar is Chairman and Professor of Operations and Engineering Management at Michigan Technological University. He also serves as a director of the graduate program in Operations Management and has been principal in continuing development of the program. He received D.Sc. in Operations Research from Washington University, St. Louis, and has worked as an Industrial Engineer at Olin, E. Alton, Ill. His current research is in the area of systems thinking, quality engineering, and management, and productivity issues with regard to technology management. He has more than 35 scholarly publications and served as a consultant to several private organizations. (CHAPS. 19 and 29)

John E. Juhasz, founder and president of the Micron Group International, a management consulting firm, has 26 years of combined technical-managerial responsibility, including product-system development and strategic planning. Industry experience spans process control, automotive electronics, equipment manufacturing, aerospace systems, and management services. John holds eleven patents and is the recipient of numerous awards including Rockwell International's Engineer of the Year. He is author-developer of a unique, computer-based systems methodology, published as

Structured Enterprise Modeling. At Baldwin Wallace College, Berea, Ohio, John enjoys teaching Management Information Science and Decision Models in the Executive M.B.A. program. He holds an M.S.E.E., with emphasis on Systems Engineering and a M.B.A. with an area of concentration in International Business. (CHAP. 16)

John Kanz is President of DMS OnLine Locator, Inc., Scottsdale, Arizona, a specialized international database service. His analyses of competitive forces, technology trends, and institutional factors in the global semiconductor industry have received wide attention. Prior research as Associate Professor and Chair, Management of Technology, The University of Calgary, combined with more than 25 years in management at major U.S. electronics firms while chairing U.S. and international electronics standards committees, gives a unique perspective to his frequent publications. He is a registered professional engineer and holds a B.S. in Physics (University of Washington), an M.S. in physics (University of Illinois), and an M.B.A. and Ph.D. in Management from the Peter Drucker Center at Clarement. (CHAP. 6)

Ronald Neil Kostoff received a Ph.D. in Aerospace and Mechanical Sciences from Princeton University in 1967. At Bell Labs, he performed technical studies in support of the NASA Office of Manned Space Flight and economic and financial studies in support of AT&T Headquarters. At the U.S. Department of Energy, he managed the Nuclear Applied Technology Development Division, the Fusion Systems Studies Program, and the Advanced Technology Program. At the Office of Naval Research, he was Director of Technical Assessment for many years. His present interests revolve around improved methods to assess the impact of research. (CHAP. 31)

Danny Lam is Research Fellow at the Economic Development Institute at Auburn University where he co-directs the Project on Institutions of Advanced Development. During 1992–94, he was a Fellow of the Pacific Basin Research Center-Center for Science and International Affairs, John F. Kennedy School of Government, Harvard University. He has been a fellow of the Pacific Cultural Foundation and a Visiting Scholar at the Chung Hwa Institute for Economic Research in Taiwan. Dr. Lam was a guest editor of the Special Issue on the Competitiveness Debate in Business and the Contemporary World. His publications have appeared in *Harvard International Law Journal, Journal of Northeast Asian Studies*, and *The Atlantic Monthly*. Dr. Lam completed his Ph.D. Thesis on the Development of Taiwan's Computer and Electronics Industries at Carleton University in 1992. (CHAP. 6)

Alvin P. Lehnerd holds a B.S.E.E. degree from Ohio University and an M.E.A. degree from George Washington University. Recently retired as vice president of Steelcase office furniture products, he has previous experience with DuPont, General Electric, Black & Decker, and Sunbeam. Mr. Lehnerd has led several well known developments and innovations resulting in new product successes including the Dustbuster, the first electronic and global iron, orthopedic surgical tools, the complete redesign of Black & Decker double insulated power tools and several Steelcase office furniture products. Currently a consultant to industry, he has expertise in a number of areas including product and process innovation, global competitiveness, strategic allocation of technology and technology resources, and new business–new venture development. He has lectured and presented papers at MIT's Sloan School of Management, University of California at Los Angeles, Cal Tech, National Academy of Engineering, and Harvard Business School and for many industry and government organizations. (CHAP. 25)

Kevin B. Lowe is a doctoral candidate at Florida International University. His work includes several scholarly articles which have appeared in journals such as *The Academy of Management Journal, The Leadership Quarterly*, and *Journal of Advances in International Comparative Management*. He is a member of the Academy of Management, the Society of Industrial and Organizational Psychology, and the Southern Management Association. His current research interests include the study of best practices for the management of human resources in international settings, the effects of multiculturalism on group processes, and the evolving role of leaders in the post-industrial organization. (CHAP. 20)

Gary S. Lynn is an Assistant Professor of Marketing at the University of Alabama in Huntsville. Dr. Lynn received a bachelors degree in mechanical engineering from Vanderbilt, a masters in business from the Kellogg School, and a Ph.D. in management and technology from Rensselaer Polytechnic Institute. Dr. Lynn's specialty is new product development and marketing under uncertainty. He has written and co-authored four books and several articles on new product development and marketing, including *From Concept to Market* (John Wiley & Sons). Dr. Lynn has licensed his own products and patents in the United States and abroad and has conducted countertrade negotiations with the Minister of Trade in the Philippines and the Minister of Economics in Taiwan. Dr. Lynn was a technical designer and marketing specialist at General Electric with a secret security clearance. He started, built, and sold three companies. He is listed in *Who's Who Leading American Business Executives* and was the editor of Probus Innovation Book Series. (CHAP. 34)

Ann Majchrzak is Associate Professor at the Institute of Safety and Systems Management (ISSM) with a joint appointment at the Industrial Systems and Engineering Department. Dr. Majchrzak has conducted research, teaching, and consulting on the integration of organizational and human factors in the design, implementation, and management of manufacturing technology. She has authored *The Human Side of Factory Automation: Managerial and Human Resource Strategies for Making Automation Succeed* and *HITOP: A Reference Manual for Highly-Integrated Technology, Organization and People*. She is co-director of the ACTION consortium, a $10 million investment of the Air Force Mantech Program, National Center for Manufacturing Sciences, DEC, HP, TI, GM & Hughes to create computer-based tools to facilitate integration. (CHAP. 11)

Richard McNichols earned his B.S. in Finance from the University of Illinois at Chicago in 1980, his CPA certificate in February 1981, and his M.B.A. from the University of Baltimore in 1989. He is the treasurer of NUM Corporation, a high tech manufacturer of CNC and Robotic Controls. He is also an active speaker at university and continuing professional education seminars and the author of articles on multinational financial management and factory automation. (CHAP. 35)

Paul A. Nelson is Chairman and Associate Professor of Business Administration and Management at Michigan Technological University. He received a Ph.D. degree in Industrial Organization from the University of Wisconsin, and served as an Army field artillery officer, operations research analyst, and comptroller staff officer before coming to Michigan Tech in 1972. His current research interest is in the area of justification and implementation of advanced manufacturing technology and implementation of change in organizations. He has co-authored two books on institutional impediments to the adoption of geothermal energy technology, and has served as a consultant to the U.S. Department of Energy and the Federal Energy Regulatory Commission on the effectiveness of pooling agreements among electric utilities. (CHAP. 29)

Marvin L. Patterson is founder and President of Innovation Resultants International (IRI), a San Diego firm dedicated to helping client companies become more competitive at new product innovation. Prior to founding IRI, Patterson enjoyed twenty years in new product development with Hewlett-Packard Company including five years as Director of HP's Corporate Engineering department, a group focused on improving HP's product development process. Marv is the author of *Accelerating Innovation: Improving the Process of Product Development* (Van Nostrand Reinhold, New York, 1993). (CHAP. 10)

Alan W. Pearson is the Director of the R&D Research Unit at Manchester Business school. He is past director of the M.B.A. program and past Dean of the Faculty of Manchester Business School. Currently he is the chairman of the management committee for the recently formed Manchester Federal School of Business and Management. He was awarded a honorary doctorate degree from the University of Kiel, a centennial medal from IEEE for contributions to engineering management, and—jointly with Klaus K. Brockhoff—the Max Planck Research Award. (CHAPS. 15 and 27)

Karol I. Pelc, M.S. Electrical Engineering (Technical University of Wroclaw, Poland), Ph.D. Electronics (Uppsala University, Sweden), Ph.D. Economics/Management (Technical University of Wroclaw), is a professor of engineering management at Michigan Technological University. He has worked as a designer of electronic instruments, a manager of manufacturing, and a manager of a research laboratory in the electric power industry. He has served as consultant to several companies and lectured at universities in Europe, India, Japan, and the United States. His current research interests include: R&D management, dynamics of technological innovation, interdisciplinary knowledge mapping, knowledge systems and AI for engineering/technology management, international technology management, and the Japanese innovation management system. Dr. Pelc serves on the editorial boards of these journals: *Technological Forecasting and Social Changes* (U.S.), *R&D Management* (U.K.), and *Transformations* (Poland). He is also a member of the Academy of Management, American Society for Engineering Management, IEEE, Engineering Management Society, International Association for the Management of Technology, and RADMA Association for R&D Management (U.K.). (CHAP. 13 and 19)

Aleda V. Roth, associate professor of operations management. Before joining the Kenan-Flagler faculty, Roth was an associate professor at Duke University's Fuqua School of Business. She also held faculty positions at Boston University and Ohio State University. Her research interests include empirical and theoretical investigation of operations strategy and world class performance. Roth is the principal investigator of the CME Global Business Process Reengineering Project and the Knowledge Factory Research Project, and a co-principal investigator of the Global Manufacturing Strategy and Technology Futures Project. She is a member of the U.S. Quality Council-II and Quality Management Center of the Conference Board and Board of Directors of the Operations Management Association. Roth received a Ph.D. from the College of Administrative Science at Ohio State University, an MSPH from the School of Public Health at UNC-Chapel Hill, and a B.S. from Ohio State University. (CHAP. 38)

Susan Walsh Sanderson is an associate professor at Rensselaer Polytechnic Institute where she teaches in the School of Management. She has previously held teaching and research positions at Carnegie-Mellon University, the Conference Board, and the Colegio de Mexico in Mexico City as well as visiting positions at Harvard Business School and MIT. She served on the advisory committee of the Design and Operations Management Program of the National Science Foundation. She is an author of the Design and Manufacturing Learning Environment and other interactive multimedia material for teaching management of design and manufacturing. She has published in *Research Policy, The Design Management Journal*, and other business an academic journals. (CHAP. 23)

Terri A. Scandura is an Associate Professor of Management at the University of Miami. She received her Ph.D. from the University of Cincinnati in 1988 and teaches Organizational Behavior at the undergraduate, M.B.A., and doctoral levels. She has published scholarly articles in numerous journals including the *Journal of Applied Psychology* and the *Academy of Management Journal*, and she is on the editorial board of the *Journal of Management*. She is a member of the Academy of Management, the Society of Industrial and Organizational Psychology (Division 14 of the American Psychological Association), and the Southern Management Association. Dr. Scandura's current research interests are in the development of international competitiveness through leadership, team-building, and innovation best practices. (CHAP. 20)

Theodore W. Schlie is an Associate Professor of Technology Management in the College of Business and Economics at Lehigh University where he teaches and does research in the fields of Innovation Management and Policy, Production/Operations Management, Business Policy and Strategy, and International Competitiveness. Dr. Schlie is also Associate Director for Research at Lehigh's Center for Innovation Management Studies where he is responsible for administering a nationwide research grants program in the area of innovation management funded by a number of industrial sponsors. He is co-director of Lehigh's Management of Technology graduate program and the Director of Field Research Projects for the National Technological University's Masters Degree Program in Management of Technology. He is a consultant to a number of industrial and governmental clients. Dr. Schlie received both his M.S. and Ph.D. degrees from the Department of Industrial Engineering and Management Sciences at Northwestern University. (CHAP. 7)

Aaron J. Shenhar, is the James J. Renier Visiting Professor in Technological Leadership at the Center for the Development of Technological Leadership, University of Minnesota. Dr. Shenhar holds five academic degrees in engineering, statistics, and economic systems—acquired in Israel and at Stanford University. Prior to academia, Dr. Shenhar accumulated extensive engineering, scientific, and management experience in a high-tech environment. Working for more than 20 years in the defense industry in Israel, he went through all phases of technical management, from functional and project manager, to vice president of human resources, and finally, to manager of a large systems development division. Dr. Shenhar's current teaching and research interests include management of technology and innovation, and project and product development management. He is a recognized speaker and consultant to various leading high-technology organizations. (CHAP. 4 and 32)

J. Daniel Sherman received a B.S. degree from the University of Iowa, an M.A. degree from Yale University, and a Ph.D. in organizational theory/organizational behavior from the University of Alabama. In 1989–90 he was a visiting scholar at the Stanford Center for Organizations Research at Stanford University. He currently serves as chairman of the Management and Marketing department at the University of Alabama. His research has been published in *Academy of Management Journal, Psychological Bulletin, Journal of Management, Personnel Psychology, IEEE Transactions on Engineering Management*, and other journals. His current research interests focus on the management of innovation. (CHAP. 28)

Jeffrey C. Shuman is an Associate Professor of Management at Bentley College in Waltham, Massachusetts. He received a bachelor's degree in Electrical Engineering from Lowell Technological Institute and a masters and doctorate in Management from Rensselaer Polytechnic Institute. Dr. Shuman has crafted a unique career as an entrepreneur, consultant and professor, and in active research or entrepreneurship. He has founded or been part of the founding team of four businesses, served as a consultant to dozens of entrepreneurs, and has developed and taught new courses in entrepreneurship that tap state-of-the-art knowledge about business creation. And he has contributed to this knowledge base by authoring dozens of articles and two books on entrepreneurship and the venture creation process. (CHAP. 21)

Clayton G. Smith (deceased) was Professor of Management at Oklahoma City University. Professor Smith held a B.B.A. degree from Pace University and M.S. and Ph.D. degrees from Purdue University. He was a member of the Academy of Management, the American Management Association, and the Strategic Management Society. Professor Smith's research interests and publications concerned the "interface" of Strategic Management and the Management of Technology: the specific interests concerned diversification into young industries (based on major product innovations), technological substitution, and strategies for building strong competitive positions in markets with established competitors. (CHAP. 22)

Wm. E. Souder holds the Alabama Eminent Scholar Endowed Chair in Management of Technology at the University of Alabama in Huntsville (UAH). At UAH, Dr. Souder is also the founder and Director of the Center for the Management of Science and Technology (CMOST), and he also holds positions as Professor of Engineering and Professor of Management Science. Dr. Souder received a B.S. from Purdue University, an M.B.A. from St. Louis University, and a Ph.D. in Management Science from St. Louis University. He has twelve years of varied industrial management experience, seven years of government laboratory experience and over twenty years of experience in academe. He is the author of over two hundred papers and six books. (CHAP. 28)

M. J. W. Stratford is currently a Visiting Fellow at the Manchester Business School. He was previously a senior consultant in the Corporate Management Services Group at ICI Plc where he had more than 20 years experience in modeling systems for R&D project and portfolio management. (CHAP. 15)

David J. Sumanth is Professor of Industrial Engineering and Founding Director of the UM Productivity Research Group (UM PRG) since 1979 at the University of Miami. He founded the International Conference series on Productivity and Quality Research (ICPQR's), and chaired the first five conferences from 1987 to 1995. He also founded The International Society for Productivity and Quality Research (ISPQR) in 1993, and serves as Chairman of its board. Dr. Sumanth is the author/co-author/editor of more than 20 books, monographs, and video courses, and more than 100 publications. His more than 60 honors, recognitions, and awards include the Outstanding Industrial Engineer of the Year Award (1983, 1984), the George Washington Honor Medal for Excellence in Economic Education (1987), and Fellow of the World Academy of Productivity Science (1989). (CHAPS. 3 and 17)

John J. Sumanth is an undergraduate research assistant in the UM Productivity Research Group. He is presently pursuing a degree in Biomedical Engineering at the University of Miami. His academic honors include: Provost's Honor Roll, Dean's List, induction into Alpha Lamda Delta Honor Society, and a listing in The National Dean's List. His research interests are in technology and "competitive economics." (CHAPS. 3 and 17)

Paul M. Swamidass is the Associate Director, Thomas Walter Center for Technology Management, Auburn University, Auburn, Alabama. His undergraduate degree is in mechanical engineering. He has a doctorate in production/operations management from the University of Washington, Seattle. His professional background includes seven years of experience as production manager for a large power station equipment manufacturer. His research and teaching interests include manufacturing strategy, manufacturing technology, and international manufacturing. His publications have appeared in leading academic as well as practitioner journals including *Management Science*, the *Academy of Management Review*, the *Journal of Management, IEEE Spectrum*, the *International Journal of Production Research, Long Range Planning*, and the *Journal of International Business Studies*. He is known for his report Technology on the Factory Floor II, which is a study of technology use in U.S. manufacturing plants. The study was sponsored by the National Association of Manufacturers and the National Science Foundation. (CHAP. 37)

Robert Szakonyi is Director, Center of Technology Management of IIT Center. Szakonyi has a B.A. from Dartmouth College and a Ph.D. from Columbia University. Previously he worked at SRI International (Stanford Research Institute), George Washington University, and American University. Dr. Szakonyi has consulted with or conducted research at over 500 companies in the United States, Canada, and Europe during the last 17 years. In 1983 he participated in the White House Conference Committee on Technology. He has authored seven books: *World Class R&D Management, Managing New Product Technology, How to Successfully Keep R&D Projects on Track, Technology Management: Case Studies in Innovation*, Volumes 1, 2, 3, and 4. (CHAP. 14)

Hans L. Thamhain is Associate Professor of Management at Bentley College in Waltham, Massachusetts. He received masters degrees in Engineering and Business Administration, and a doctorate in Management from Syracuse University. Dr. Thamhain has held engineering and management positions with GTE, General Electric, and Westinghouse, and is well known for his research on Engineering Team Building and Project Management. Dr. Thamhain is a frequent speaker at major conferences, has written more than 60 research papers and four books on engineering/project management, and is consulted in all phases of technology management. (CHAPS. 9 and 21)

Mustafa (Vic) Uzumeri is an assistant professor in operations management at Auburn University. He has worked in transportation engineering, international contracting, business development, test facility marketing, and software product management. His research interests include operations management, management of technology, product planning, and the strategic implications of emerging quality management standards such as ISO 9000. Dr. Uzumeri has published in *Design Management*

Journal, Research Policy, Manufacturing Engineer, and the *Southern Business and Economic Journal*. (CHAP. 23)

Rias J. van Wyk is a graduate of Harvard University, the University of Stellenbosch, and the University of Pretoria. At present he teaches management of technology at the Graduate School of Business, University of Cape Town, South Africa, and at the Center for the Development of Technological Leadership, University of Minnesota, Minneapolis, Minnesota, USA. He has taught or done research at various centers throughout the world. These include: MIT, Cambridge, Massachusetts; Case Western Reserve University, Cleveland, Ohio; IMD, Lausanne, Switzerland; IAE, Aix-en-Provence, France; Tel-Aviv University, Israel; and the Technical University, Karlsruhe, Germany. He conducts in-company courses in mapping and forecasting technology and offers board-room briefings on the changing technological frontier. He has served on a variety of corporate boards for two decades. (CHAP 5, SECS. 1 and 2)

Mary Ann Von Glinow is Professor of Management and International Business at Florida International University. She was President of the Academy of Management in 1995. Dr. Von Glinow's research interests have centered around the management of high technology and professional employees, and international HRM. She has authored more than 100 journal articles and six books. Her current work is on Best Practice Innovators, which looks at global best practice in the management of engineering professionals. (CHAP. 20)

Abdus Wadee is currently working with Eltec, a regional training and enterprise company, where he is involved in designing training programs and collaborative ventures with industry after undertaking research for a high degree at the Manchester Business School. (CHAP. 15)

Thomas G Whiston is a Senior Fellow and Director of Studies of the M.Sc. in Technology and Innovation Management at the Science Policy Research Unit, University of Sussex. His main areas of research relate to global forecasting and related policy analysis, higher education policy, and the influence of technical change upon organization structures. He has published several texts and approximately 150 articles, policy studies and monographs related to these areas. He has been a senior consultant to most UK research councils and several international agencies. During 1992/93 he was co-leader of a large EC program entitled Global Perspective 2010: Tasks for Science and Technology, a program involving about forty international researchers. The results have been published in twenty-three volumes. (CHAPS. 18 and 30)

Alan Wilkinson was Technical Director of a major division of a multinational chemical company before joining Manchester Business School as Administrative Director. Recently retired he played a major part in setting up the industrial contacts and supervising the early stages of the research. (CHAP. 15)

Edith Wilson has been with Hewlett-Packard for more than fifteen years and has had a variety of positions ranging from being an engineer at Hewlett-Packard Research Laboratories in 1980 to her current position as Manager of Product Definition and Prospecting for New Businesses at the Corporate level. Her background includes a Bachelor's Degree in Bio-Medical Engineering from Duke University and both a Master's and an Engineer's Degree in Mechanical Engineering from Stanford University. Ms. Wilson's interest in the marketability of excellent design is the thrust of her department's endeavors to improve product definitions throughout Hewlett-Packard. The department focuses on encouraging divisions to only pursue excellent opportunities and to redeploy or cancel the less attractive projects early in the development cycle. Currently, she also holds a position as Senior Lecturer at Stanford where she teaches graduate students the interrelationship of engineering design and marketing and business fundamentals. She lives in Palo Alto, California, with her husband and daughter. (CHAP. 36)

ABOUT THE EDITOR IN CHIEF

Gerard H. (Gus) Gaynor is the principal of G. H. Gaynor and Associates, Inc., a technology management consulting firm. A 24-year career at 3M (Minnesota Mining and Manufacturing) provided him with extensive executive-level experience in managing technology in a global context. A recent Fulbright Scholar in management and a sought-after lecturer, Mr. Gaynor is active in several professional societies. He also serves on the industrial advisory boards of several universities and professional journals and is president of the IEEE's Engineering Management Society. He is the author of two previous McGraw-Hill books: *Achieving the Competitive Edge Through Integrated Technology Management* and *Exploiting Cycle Time in Technology Management*. Gus Gaynor is based in Minneapolis, Minnesota.

PREFACE

Technology Management (TM) can be viewed from many different perspectives. The word *technology* itself is subject to many different interpretations—artifact, knowledge, and/or, whatever interpretation one wishes to use. Some only consider information as technology. Thus the approaches to technology management can be quite different depending on how users describe *technology, management*, and *technology management*. The authors of this handbook approach the topics from different experiences that are associated with many different environments. So, if in considering any particular issue, you find different or even contradictory approaches, recognize that context of TM will determine the description. There are not many individuals or corporations that consider TM from a holistic perspective and as an integrative activity.

This handbook looks at technology from a system perspective. It focuses attention on integration of all of the technology related issues. It should not be construed as a handbook of engineering, research, development, manufacturing, information, or some limited technology issue. The handbook takes a wide-angle view of TM. It is not single-issue oriented. It has no recipes for success. It raises the issues that are essential if organizations wish to improve their performance levels and the rewards from investing in technology.

The purpose is not only to provide information about TM and expand the scope and meaning of TM but to raise issues that force reconsideration of some fundamental management principles. As managers have recently learned, gurus are not the answer; going back to fundamentals provides greater assurance of success. Those fundamental management principles require an emphasis on integration of resources related to TM.

TM crosses all disciplines and all levels of an organization. Depending on the size and type of organization it may even be part of the CEO's domain. The handbook does not deal with technology but its management. The authors recognize that technology is a rich source of opportunity and not a threat and that managing technology is not simply a technical issue but a major business responsibility.

TM includes the decisions of scientists and engineers from the lowest to the highest levels, the decisions of science and engineering managers at all levels, and the decision processes of managers and executives from marketing, from sales, from human resources, and representatives of all the affiliated functional disciplines. Technology does not take precedence in the decision processes, but how that technology can be implemented within the constraints of the business, the marketplace, and the economy does.

The construct of the Handbook focuses on the two basic reasons why businesses invest in technology: to improve financial performance and enhance their competitive position. The Handbook brings together the latest thinking, research results, and practices related to TM. It considers TM as an integrating activity.

The Handbook is divided into seven parts:

- Perspectives on Technology Management
- Methodologies, Tools, and Techniques
- Education and Learning

- The New Product Process
- Managing Technology Management
- Case Histories and Studies
- Appendix

Part 1, Perspectives on Technology Management sets forth some broad fundamental issues. Chapter 1 describes TM as an integrative function and sets the stage for describing TM in a business context. Chapter 2 presents a business architecture for TM. Chapter 3 explores "Technology Cycle" as an approach to TM. Chapter 4 focuses on the organization's technology base. Chapter 5 reflects on the implications of corporate boards and executive committees understanding of technology analysis. Chapters 6 and 7 consider the issues of technology strategy, competitiveness, and competitive advantage from an institutional-managerial perspective and the contribution of technology to the business enterprise. Chapter 8 focuses on targeted basic research as related to industry-university partnerships. Chapters 9 and 10 provide two complementary approaches to managing the process of technology-based innovation.

Part 2, Methodologies, Tools, and Techniques considers approaches that should force organizations to rethink their current practices in managing technology. Chapter 11 suggests some tools for analyzing organizational impact of new technology. Chapter 12 considers issues in forecasting and planning technology. Chapter 13 reveals the benefits of using technology mapping as a tool for TM. Chapter 14 considers the process of developing an R&D strategy in the TM context. Chapter 15 looks at decision support systems in R&D project management through use of matrices. Chapter 16 brings to managers an enterprise engineering approach in the systems age. Chapter 17 develops the issues related to managing the "Technology Gradient" for global competitiveness.

Part 3, Education and Learning places these concepts in perspective. Chapter 18 recognizes the importance of learning far beyond what is proposed by the learning gurus and the business press. Chapter 19 lays down the principles of learning processes for TM. Chapter 20 draws attention to the importance of technological literacy. Chapter 21 looks at learning skill requirements in developing technology managers.

Part 4, The New Product Process discusses issues related to TM. Chapter 22 deals with MT in the product substitution context. Chapter 23 develops a framework and model for product family competition. Chapter 24 looks at the process for managing product definition in software product development. Chapter 25 provides some real life examples and processes for renewing product platforms.

Part 5, Managing Technology Management sets forth in eight chapters the issues of implementing a systems approach to TM. Chapter 26 develops selected models for managing the TM process. Chapter 27 brings together the real life situations involved in managing functional and other interfaces. Chapter 28 builds on previous work and identifies the factors that influence effective integration of technical organizations. Chapter 29 looks at the barriers to implementing an integrated TM approach. Chapter 30 focuses on the need for developing interdisciplinary relationships in order to improve functional performance. Chapter 31 looks at the effectiveness and impact on research performance in the context of TM. Chapter 32 organizes the long-term success dimensions in technology-based organizations. Chapter 33 recognizes the lack of adequate tools to measure the effectiveness and impact of TM and provides a listing of questions that must be asked in developing meaningful measurement systems.

Part 6, Case Histories and Studies provides information from the real world about the issues involved in TM. Chapter 34, the NutraSweet case involves just about every aspect of TM from the concept to commercialization cycle. Chapter 35 looks at communication as being at the crossroads for success in linking the technology resources of four countries. Chapter 36 raises the important issue of product definition as a key to successful product design and marketplace acceptance. Chapter 37, a manufacturing study, focuses on the issues related to benchmarking manufacturing. Chapter 38 lays down the principles for building strategic agility through a neo-operations strategy that links combinative competitive capabilities and advanced manufacturing technology strategies under the rubric of economies of knowledge.

Part 7, Appendix A provides the reader with a listing of universities, for reference only, that offer courses or advanced degree programs in TM.

ACKNOWLEDGMENTS

The list of the individuals who have affected my thinking regarding this subject of technology management is extensive and multidisciplinary. I am positive that many of them are not aware of the lessons they may have taught, either positive or negative. I obviously come with certain preconceived notions of TM because as a practicing engineer whose experience has spanned research, development, design, and manufacturing, as well as management in each of these functions in both a domestic and international setting, the word "integration" became a guiding force in my decision processes. How would my decision affect other functions, other design parameters, and so on, was always a question that needed a response. To me technology was and is a business issue and not solely a technology issue. As such, successful implementation of technology management (TM) requires systems thinking as contrasted to more conventional promotion of some single-issues.

I want to take this opportunity to thank all the authors who participated in this Handbook. They come from diverse experiences and with different perspectives that span the academic to practitioner continuum. TM requires this linkage of the academic and practitioner perspectives. While TM is essentially a practitioners discipline, knowledge and understanding of the issues can only come from research into the field of technology management. As editor of the Handbook, my sincere thanks to each of them for taking the time to pass on their knowledge and perspectives as related to this discipline of technology management.

My sincere thanks to McGraw-Hill's sponsoring editor Harold B. Crawford for his interest in promoting the concept of the *Handbook of Technology Management*, to Ms. Peggy Lamb, editing manager, and to Pamela A. Pelton, the production supervisor. My compliments to them and their staff for their high level of professionalism.

Finally, to my wife Shirley, my partner who made the decision with me to pursue this effort, my sincere thanks for her interest, collaboration, and cooperation.

Gerard H. (Gus) Gaynor
Editor in Chief

P · A · R · T · 1

PERSPECTIVES ON MANAGEMENT OF TECHNOLOGY

CHAPTER 1
MANAGEMENT OF TECHNOLOGY:
DESCRIPTION, SCOPE, AND IMPLICATIONS

Gerard H. (Gus) Gaynor

G. H. Gaynor and Associates, Inc.
Minneapolis, Minnesota

1.1 AN OVERVIEW

In a speech at the First International Conference on Engineering Management in 1986 Edward Roberts, David Sarnoff professor of management of technology at Massachusetts Institute of Technology, suggested that the failures of the automotive, office equipment, and electronics industries were not a result of trade, economic, or political policies but from the inability of industry to implement programs in technology management. He emphasized the need for

- Integrating technology into the firm's strategic objectives
- Taking a proactive stance in introducing new technologies, new products, and new processes with a greater emphasis on cycle time
- Increasing the productivity and performance of the firm's technical community
- Understanding the interdisciplinary needs in project management
- Analyzing the resources and infrastructure to effectively select the technical scope of the work effort

Dr. Roberts also focused on the need for increasing the manager's understanding of the issues related to the *management of technology* (MOT). His major concern was that universities could be accused of false advertising in their depiction of MOT courses. He said that there is seldom any teaching of either engineering management or MOT. There is a teaching of both engineering and management as separate disciplines, but no teaching of engineering and management combined. For some reason the "and" was omitted.

This situation is not unusual; academic institutions rarely recognize interdisciplinary study or research. Somehow it offends them. But MOT, with its need for integration, requires an interdisciplinary approach.

Since that time many books have been written on MOT and several thousand papers presented at conferences. The industry press has also focused attention on the many issues involved in MOT. Yet organizations fail to manage their technology effectively, efficiently, and with the economic use of resources. Projects continue to miss timely implementation, continue to fail to meet requirements and specifications, and continue to exceed allocated funding.

In the intervening years organizations have failed to focus any significant effort on managing technology as an integrated effort. While many organizations have been caught up in the frenzy of hiring the latest guru anointed by the business press, that effort has resulted in focusing attention on single issues which may afford some short-term benefit but provide no benefit for sustainable future performance. Single-issue management, as a principle of management, ignores the realities of the competitive forces that allow an organization to meet the expectations of the stakeholders. By now, most organization have discovered that they cannot Baldrige or ISO 9000 their way to economic success for either the short or long term.

The suggestions presented by Dr. Roberts continue to be valid not only for managing the technology-related issues but also for the total business enterprise. Management of technology (MOT) cannot be accomplished by some predetermined recipe or prescription. Guidelines provide direction and the caution lights, but the fact remains that MOT can be implemented effectively only within the context of a specific organization with consideration of its strengths and weaknesses and in relation to its available resources and infrastructure.

This introductory chapter, on the general characteristics, scope, and implications of technology management, includes

- An overview
- A general description of technology management
- Delineation of the scope of technology management
- A system model—resources, infrastructure, and activities
- An integrated and holistic model
- A brief introduction to strategic, operational, and management issues
- Classification of technologies
- A section on education in technology management
- Summary and conclusions

1.2 MANAGEMENT OF TECHNOLOGY: A DESCRIPTION

Is it necessary to ask the question: What is management of technology? The answer is a resounding "yes." You may argue that the answers are self-evident, but any discussion of MOT usually brings out a diversity of opinion—opinion but not knowledge or understanding. The two words *management* and *technology* carry the burden of many different meanings. The combination of the two words presents additional complexities. To many, MOT means managing engineering. To others it means managing information, managing research, managing development, managing manufacturing operations, managing the activities of engineers and scientists, or managing functional activities without concern for the total spectrum of activities that encompass the business concept to commercialization process. Those interrelated activities must be integrated into a technology management system. There is one key word that must be emphasized in the management of technology: integration. MOT involves managing the system; it also involves managing the pieces. Neither the system nor the pieces can be subordinated. MOT involves integrating the "pieces" into an acceptable "whole" by focusing attention on the interdependence of the pieces.

1.2.1 Management and Administration

To understand management of technology, it is necessary to clarify what we mean by *management* and what we mean by *technology* and then explain *management of technology*. The discussion is further complicated by the fact that technology can be viewed in many different ways. As an example, technology can be viewed as a

- Tool
- Physical manifestation
- Knowledge
- Applied science
- Academic discipline

We view technology as a tool. It is a means for accomplishing some end. Viewing technology as an applied science limits the scope and essentially focuses on the issues engaged by engineers and scientists. MOT as an academic discipline could provide a benefit at some future point in time when academia becomes oriented toward intensive multidisciplinary research that possesses some semblance of relevance and is directed toward problem solving and problem finding. The physical manifestation of technology surfaces in all our lives 24 hours a day and has no relevance to this discussion. To reiterate, technology will be considered as a tool throughout this chapter.

To understand just what MOT includes, it is necessary to clearly differentiate between management and administration. Management and administration are not the same; there is a clear distinction between the two, and that differentiation is important. Simply stated, *management* involves degrees of creativity, leadership, risk, and concern about future performance, while *administration* involves supervising the assigned activities or tasks that are essential to keep an organization afloat.

Management, or, more appropriately, managing, is a complex process—much more complex than administration. Such factors as managerial creativity and innovation and taking a proactive approach make special demands. Pushing the frontiers, regardless of discipline or function, requires people with not only special talent but also foresight of what is possible on the basis of available resources and infrastructure. Management of technology requires leadership. That leadership function must focus on the long-term as well as the short-term requirements in order to maintain the viability of the firm. That kind of leadership requires focus and discipline, moves the organization into the future, and takes place at all levels of the organization. The traditional approach that the leader is at the top is no longer viable, if ever it was. It is certainly not viable for managing technology where technology leadership is expected at all levels. Chief executive officers (CEOs) do not make the decisions that determine success or failure of technology.

1. Management. Management is not a science. As much as researchers try, they cannot and have not developed any consistent theories that guide human performance in a logical sequence from point *a* to point *b*. Not only are people different, but the same person acts and reacts differently in different situations. There are no mathematical equations that can be written and then applied to describe the interactions that take place between people. At best, management is an applied art that involves using the linkage of data, information, knowledge, and the social interaction between people in solving problems or pursuing opportunities.

2. Administration. Administration, on the other hand, implies following rules and regulations. It implies following predetermined processes and methods. Words such as *creativity, innovation,* and *risk* are not in the administrator's lexicon. This does not mean that creativity and innovation are not essential to the organization. It does imply that these activities will be more tightly controlled. It does not excuse these administration functions from exercising their creativity in a systematic way toward continuous improvement. In MOT administration, creative accounting on a daily basis would most likely be challenged. Creativity and innovation may not be desirable attributes for the payroll department. Paychecks must be issued in the correct amount and on time. Administration, however, must be differentiated from the general negative attitudes toward bureaucracy. Bureaucracies are essential to survival of any business, but those bureaucracies must be effective and efficient. In essence, *administration* means fulfilling the routine activities of the organization. These routine activities can neither be ignored nor dominate the actions of the organization.

This distinction between management and administration becomes of greater importance in considering the issues related to the management of technology. For many years managers involved in managing the technical activities of the business have tended to spend a greater percentage of their time on routine administrative matters. Managers in science and engineering in many situations have become paper pushers. Much of this evolution from management to administration came about as a result of some misguided human resource professionals and academic behavioral scientists—the manager was not to interfere and was to let scientists and engineers do their own thing and let every employee start at the bottom of the learning curve. In the process, the role of the manager as teacher was totally ignored.

Both management and administration are essential. The question that must be answered in the context of the organization's purposes and objectives is: What is the proper balance between management and administration? That question must be answered, and the answer depends on the many characteristics that define the organization—the resources, infrastructure, and the activities in which the organization engages. The proper balance may be quite different for an automotive company like Ford and a multiproduct innovative company like 3M. The balance will be quite different for an order-entry department and a research department. The best product cannot be introduced on time, within specifications and requirements, and at estimated cost without the effective performance of many routine activities.

1.2.2 What Is Technology?

The word *technology* usually conjures up many different images and generally refers to what has been described as the "high-tech" (high-technology) industries. Limiting technology to high-tech industries such as computers, chips, superconductivity, genetic engineering, robotics, and so on focuses excessive attention on what the media consider newsworthy. Limiting technology to science, engineering, and mathematics also loses sight of other supporting technologies. Technology includes more than machines, processes, and inventions. There are many different manifestations of technology;

some are very simple and others, very complex. A description of technology in the MOT context must go beyond the traditional.

Technology can be described in different ways:

1. Technology is the means for accomplishing a task—it includes whatever is needed to convert resources into products or services.
2. Technology includes the knowledge and resources that are required to achieve an objective.
3. Technology is the body of scientific and engineering knowledge which can be applied in the design of products and/or processes or in the search for new knowledge.

1.2.3   Management as a Technology

Within this context, is *management* a technology? The response can be a resounding "yes" or a resounding "no." The response depends on the limits placed on the description of *technology*.

Every management action requires a process—or at the least should follow a process. But that process must be accompanied by substance, action, and integrity. The "seat of the pants" and "play it by ear" approaches have outlived their usefulness. Gut reactions may be acceptable but should be validated from other perspectives.

It seems almost trite to say that all decisions should follow some predetermined process regardless of whether the decision involves a major financial investment or the introduction of some new human resource program. Both involve allocation of resources, so a fiscally responsible action must be guided by some process. To that extent management as a technology can be defined simply:

> Management as a technology can be described as the process of integrating the business unit resources and infrastructure in the fulfillment of its defined purposes, objectives, strategies, and operations.

This is a simple statement with significant implications for management of technology. If the broader descriptions in 1 and 2 (in the list at the end of Sec. 1.2.2) are accepted, then management definitely is a technology. If the restrictive approach of description 3 (technology as a body of scientific and engineering knowledge) is used, management would probably not be considered as a technology.

It could be argued that descriptions 1 and 2 are so broad that they encompass all of management and further that considering management as a technology is stretching the description of technology. It is true that the broad perspective is all-encompassing, but then technology in one form or another or to a greater or lesser extent drives most organizations—especially those that are concerned about the future. If it does not drive the product base, it does drive the distribution process from order entry to customer satisfaction. Technology cannot be restricted to the manufacturing industries. It encompasses not only the manufacturing sector but all industries—agriculture, airlines, banks, communication, entertainment, fast food, clothing, hospitals, insurance, investment, and so on—and determines future viability of the business unit as well as the industry.

There is no limit to the way in which organizations can describe technology. The important point is that organizations define what they mean by *technology*. This chapter uses the description of technology as the means for accomplishing a task; it includes whatever is needed to convert resources into products and services. This is a holistic approach and differentiates MOT from the single-issues approach—managing engineering, managing research, and so on.

1.3 MANAGEMENT OF TECHNOLOGY SCOPE

Figure 1.1 shows a five-phase continuum for managing technology. This is a simplified version but is presented here to show what MOT includes. At the present time it is difficult to find organizations that manage technology as an integrated function and holistically. The 5 phases are arbitrary because in reality there could be 20 or more different phases. In this scenario each function would represent a phase. Much of what is described as MOT is generally a phase 1 effort involving some overlap with phase 2. Managing research and development usually receives the greatest amount of attention in reference to management of technology. These two functions also overlap with what is often referred to as *managing engineering.* I purposely eliminate the use of "managing engineers and/or engineering" because engineers are not the only specialists involved in research and development—there are many scientists and science majors working in these two functions. Phase 1 does not involve managing technology. It is, pure and simple, management of research and management of development since these functions are often only loosely connected.

Phase 2 adds design and manufacturing to research and development and at least links the product side of the business with the process side. Little by little organizations have learned that designs must eventually be transformed into some form of tooling and eventually to a manufacturable product that meets certain quality and reliability standards. Phase 2 requires integration of these four functions into a cohesive group.

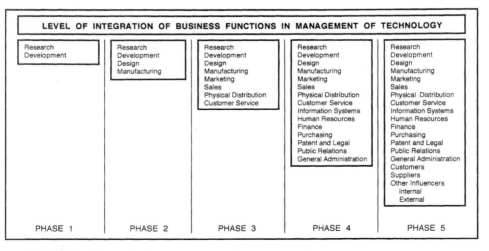

FIGURE 1.1   Phases of technology integration for management of technology.

1.3.1 Phase 3—The Beginning of MOT

Phase 3 adds the marketing, sales, physical distribution, and customer service activities. It is the minimum level at which an organization can claim to have a semblance of an MOT approach. Certain conditions preclude pursuing the requirements of phase 3. The primary constraint involves trying to integrate eight different major functions with a multiplicity of subfunctions into some form of cohesive system. But these are the functions that are directly responsible for new products and processes and must be integrated in the process of introducing new products.

The project approach provides the means for integrating these functions. Products and processes are generally developed through some form of project approach. But those activities are seldom integrated. While research and development may be working on a new product, the remaining functions in phase 3 may be sitting on the sidelines waiting for something to happen. They are part of the project in name only. Approaches such as concurrent engineering have attempted to resolve some of the issues, but with relatively little success.

Successful implementation of phase 3 requires a new way of thinking and a new model. It begins with a clearly defined approach as to what is required in project management. This type of project management has nothing to do with planning or programming systems. It has nothing to do with control of costs or schedules. The model deals with the up-front work that must be performed by all the functions. It involves inputs from all associated functions relative to an understanding of the resource capabilities, an evaluation of the infrastructure, a definition of the decision criteria, a well-developed statement of purpose, a project specification, and an evaluation and validation of assumptions. All these requirements must be integrated through manageable and enforceable feedback loops to take into account the possibilities and limitations as established by each function. What must emerge from the dialogue is a business unit plan that raises the issues as to the factors that define the success of the project. The issues for which no solution exists must be placed on the table. Contingency approaches must be clearly stated. The approach that says "we don't have the right people for this project, but we'll get it done somehow" does not work.

Implementing a phase 3-or-beyond approach to MOT requires certain specific operational characteristics:

- A level of integrity (call it honesty or ethics; it must be practiced) far above what is generally practiced.

- The ability to face up to the unresolved and potentially unresolved issues; problems cannot be hidden.

- Continuous sensitivity to issues that may in some way modify the initial assumptions; if original assumptions have changed, recognize those changes and act accordingly.

Phase 4 adds the administrative functions to the integration process. These include the major functions such as human resources, finance, purchasing, patent and legal, public relations, and general administration. Phase 5, the ultimate level of integration, adds customers, suppliers, and other internal and external influencers.

These five phases represent one approach for considering management of technology as a model for managing. Attempting to reach phase 5 is not an easy task when research shows that most organizations operate someplace between phase 1 and phase 2. In reality management of technology begins in phase 3, where integration of the concept to commercialization process begins.

1.4   SYSTEM MODEL: RESOURCES, INFRASTRUCTURE, AND ACTIVITIES

Discussing MOT as an abstract concept does not provide much insight into the complexity of practicing the basics of MOT. Management of technology is practitioner-based. In that sense MOT is complicated by the fact that it involves interaction of people with their strengths, weaknesses, foibles, biases, aspirations, and so on. But the difficulties in exploiting MOT come from a narrow description of the resources of an organization, a lack of consideration of the business infrastructure, and little, if any, consideration for the specific activities that are assigned the resources. Resources include more than people, plant and equipment, and money.

1.4.1   Resources

Figure 1.2 is a model relating business resources, infrastructure, and activities. It lists the primary elements related to each. This classification is one used by the author to expand the range of the elements involved when dealing in real-life situations.

The classification of available resources, while broad in scope, will consist of many subcategories depending on the particular business unit under consideration. People, plant and equipment, and finance are the traditional resources. These are inadequate in a technology environment. As an example, intellectual property, information, organizational characteristics, time, and customers and suppliers are seldom considered as resources. These resources are inside or outside the organization and all are interdependent. No single resource, by itself, provides any beneficial business result.

In the leftmost column of Fig. 1.2, technology is highlighted as one of 11

RESOURCES	**INFRASTRUCTURE**	**ACTIVITIES**
• People	• Purposes	• Business
• Intellectual Property	• Objectives	• Management
• Information	• Strategies	• System
• Organizational Characteristics	• Organizational Structure	• Project
• **Technology**	• Guiding Principles	• Functional
• Time	• Policies and Practices	• Group
• Customers	• Management Attitudes	• Individual
• Suppliers	• Management Expertise	• External
• Plant and Equipment	• Support for Innovation	
• Facilities	• Acceptance of Risk	
• Finance	• Communication	

FIGURE 1.2   Elements of resources, infrastructure, and activities in the context of management of technology.

resources; the successful use of technology depends on the availability of the other 10 resources. This is not a profound discovery, yet organizations ignore these relationships. As an example, intellectual property is a resource, but the *not-invented-here* (NIH) approach flourishes—start at the bottom of the learning curve, do not build on what is known—reinvent the wheel. Time is a vital resource. It cannot be replaced. It includes total time, cycle time duration, and timing. Information is a resource, but the sources and integrity of the information must be known. Organizational characteristics are a resource and include more than culture. Even from the limited perspective of culture, it is a resource. But when characteristics are described as those principles and practices that differentiate one organization from another, the importance as a resource is even more profound.

1.4.2 Infrastructure

The business unit infrastructure elements listed in Fig. 1.2 are equally important and determine the viability of the resources. Infrastructure plays a major role in business performance. The utilization and viability of resources depend on the supporting infrastructure. Purposes (mission), objectives, and strategies must flow down to the people who make things happen—those responsible for doing the work. Organizational structure—the real operative organizational structure, not the rectangular boxes on the organization chart—must meet the needs for a particular activity. Guiding principles, policies and practices, and management attitudes determine how people respond to the organization's purposes, objectives, and strategies. The breadth of management expertise and knowledge provides the underlying understanding for effective decision support and decision making. Managers who do not understand what they are managing and play the role of administrator make good candidates for the profit-prevention department. If the infrastructure does not support innovation and accept the associated risk, none will surface. And finally, the manner in which the organization communicates—not just from the top down, but from the bottom up—and laterally—determines how this infrastructure is perceived by those who are asked to respond.

1.4.3 Activities

Resources and infrastructure do not reside in a vacuum. They are applied to some type of activity and can be classified in many different ways. The point is that it is necessary to differentiate among them in consideration of the resources and infrastructure. Each of these activities will be impacted differently by the resources and the infrastructure. There is a continuum from almost totally ignoring the other relationships to being totally dependent on them. An individual person involved in an activity may ignore the other resources and infrastructure. This will depend on the individual's proactive qualities, independence of thought, past track record of accomplishment, self-confidence, and the other personal attributes that allow an individual to function as a creative and contributing maverick.

Activities that involve more than one person enter a totally different domain that must take into account the supporting resources and infrastructure. Consider, as an example, the specific activity related to a project. In this situation exploitation of all the resources is essential. The infrastructure must support the objectives of the project. The project must meet the requirements of the purposes, objectives, and strategies of the organization. An acceptable organizational structure must not impede the forward progress; the guiding principles and the policies and practices cannot unnecessarily restrict the freedom to act. Management attitudes and expertise will avoid micro managing yet allow for constructive involvement based on past knowledge and experience—perhaps even some wisdom. Without management's support for innovation and acceptance of the associated risk, the project will provide limited benefit. The use of teams or teamwork has been intentionally excluded from this listing of activities. All the categories listed under activities except "individual" involve some type of team effort coupled with individual proactive initiatives.

1.4.4   Technology as a Resource

The relationship of the elements of this model relating resources, infrastructure, and activities is complex. But then, there is no reason to think that managing is a simple process. It may appear so from a macro perspective or when quantified in a two by two matrix in two dimensions, but the real world is more complex.

In Fig. 1.2, technology is highlighted as one of the resources. But technology as a resource is effective only if it is applied to some specific activity and within the confines of a particular infrastructure. The same is true for every other element of the resources. People without technology, without available time, and without input from customers do not enhance performance. People without a supporting infrastructure do not enhance performance. People without assigned or self-generated activities do not enhance performance.

1.5   INTEGRATED AND HOLISTIC MODEL

Management of technology is explained with the aid of Fig. 1.3, which shows a model of what I describe as the tripartite organization. It includes the same 17 functions shown in Fig. 1.1, phase 5. It includes three main internal organizational units: *product genesis, distribution,* and *administration,* and a fourth unit, *supporting influences,* that is external to the organization.

1.5.1   Product Genesis

Product genesis provides the creativity and innovation for new products. It includes research, development, design, and manufacturing. The conventional wisdom of discussing R&D as a unified function leads management to some erroneous conclusions. For that reason I differentiate between research and development and add the additional function of design.

It may be good public relations to focus on R&D expenditures, but in most organizations the "R" is a relatively small part of the total—most of the expenditures are allocated to "development." Most organizations allocate a minimum of resources to what would be classified as research. Most of what is classified as research involves a search for a solution to some unknown requirement of a product or process development program—generally only some small segment of the total project. These projects usually deal with knowledge that is integrated in some new way or processed in some new way in order to meet a product specification. The process basically involves packaging existing knowledge in new ways. This is not to minimize the creativity required by design and development. On the contrary, that creativity is vital but seldom yields new discoveries.

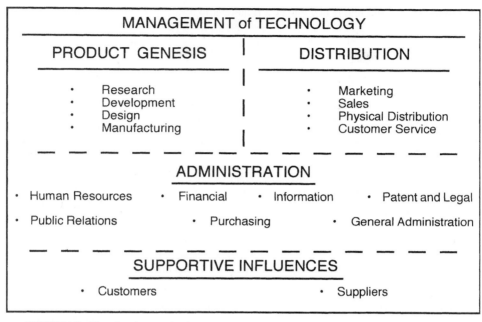

FIGURE 1.3 The tripartite organization.

The function of product design, which includes the aesthetic, functional, and operational considerations of a product, seldom receives the required attention. I have included design as a supporting entity because of its much-too-often overlooked but essential importance. Design involves the details as contrasted to the big picture. Yet, the execution of the details determines the success of a product. While it is true that the aesthetic design of an automobile will receive much greater attention from top management, the functional and operational factors determine the number of recalls. Automobile executives do not design drivetrains.

Manufacturing, as a part of product genesis, is often left out of the MOT loop. The approach that process development follows product development is not, and never was, a viable approach to effective business management. The spread of such unsubstantiated advice has led many industries to the brink of self-destruction—witness the past performance of the auto industry and the electronics industry and the long difficult climb to some acceptable level of performance. Manufacturing requires a process research focus that must be integrated with the other functions of the tripartite organizational model.

1.5.2 Distribution

Distribution comprises the functions of marketing, sales, physical distribution, and customer service. These four distribution functions represent those required to interact with the product genesis functions to agree on a product concept, to define the market, to find the customers, to negotiate the sale, to provide the customer with before-and-after service, and finally and the most forgotten: to bring the customer feedback to the organization in a timely manner. The distribution function also provides a rich source for new-product ideas.

1.5.3   Administration

Administration technologies are more difficult to identify than the technologies of the product genesis and distribution functions. When it comes to administration, many questions are raised at the suggestion that these functions utilize specific technologies. While science and engineering, in the strictest and classic sense, may not be involved in these administration functions, they have their own technologies. The use of information systems as technology in these administration functions is accepted, but skepticism arises when the word *technology* is attributed to the related administration functions. I suggest that each of these administrative functions has its own set of technologies, keeping in mind that technology is defined as "a means for accomplishing a task."

Human Resources.   Human resources technologies include personal appraisal forms, flexible scheduling, compensation based on performance, the behavioral science base, educational technology, and all the work processes used in meeting their objectives.

Financial.   Financial and accounting departments also use specific technologies. The processes used to justify investments in capital equipment, research, new-product development, and the miscellany of business operations are all technologies. Additional technologies come into play as financial departments attempt to make a clear distinction between project justification and project evaluation. Are balance sheets and profit/loss statements a result of processes and technology? How many decades did it take the financial and accounting functions to recognize that traditional accounting methods did not meet the needs of manufacturing organizations?

Information.   Information, as a function, is generally considered to involve technology. But this is primarily because of the extensive use of computer hardware and software. As in the other functions, process plays a major role in managing information— if not, it must. Just think about how much data you receive and how little information. The techniques by which information departments make decisions are technology. The eventual use of expert systems or artificial intelligence provides additional examples of information technologies.

Patent and Legal.   *Patent and legal,* which in recent years has expanded to protecting the intellectual property of the organization, uses processes which are technologies. Contracts, agreements, and other legal documents are a form of technology. Pareto's 80/20 rule (80 percent of the work performed in 20 percent of the time) is a technology. It is a "means for accomplishing a task" more effectively. It may substitute using the tools for evaluating situations to determine the degree of risk rather than dotting every "i" and crossing every "t" and still ending up in litigation.

Public Relations.   Public relations uses technologies to convey the interests of the organization. Those interests usually revolve around content and focus attention on process. The process by which private (individual) knowledge and public (community) knowledge are differentiated plays a major role, whether related to products, social

```
Audiovisual services
Benefits administration: pension, medical, etc.
Communication services: internal and external
Custodial services
Data processing
Economic studies
Education and training
Environmental monitoring
Fire and emergency evacuation
Food services: cafeterias, vending machines, etc.
Library services
Mail services: internal and external
Maintenance of office equipment
Medical: health and hygiene
Office maintenance and rearrangement
Supplies
Toxicology services
Transportation
Travel services and analysis
Waste and scrap disposal or recycling
Others depending on the organization
```

FIGURE 1.4   Subfunctions in general administration.

issues, catastrophes, and so on. The advanced technologies will play a major role in developing new communication models.

Purchasing.   Purchasing plays a major role in project success. Its processes determine whether projects are completed on time, within specifications, and within cost estimates. Restrictive policies and procedures, methods of selecting vendors, and limiting access to vendors solely by purchasing or with a purchasing representative create delays. These processes are a type of technology.

General Administration.   General administration functions comprise a group of subfunctions that are essential in support of the organization's objectives. Figure 1.4 provides examples of some of these administrative subfunctions where technologies exist but may not be self-evident.

1.5.4   Product and Service Organizations

This tripartite organizational structure of Fig. 1.3 applies to product as well as service industries. A close look at the differences between product and service businesses clearly shows that both are similar, if not identical.

An accounting firm, generally considered a service industry, sells a product. That product is information in the form of a report. It utilizes all the functions—research, development, design, and manufacturing—of the tripartite organization shown in Fig. 1.3. You may accept research, development, and design as part of an accounting firm's functions but probably question the use of the term *manufacturing* in this accounting example. In essence, the work effort to prepare the contents—the mental work—of that report falls into the classification of production. It is necessary to expand the use of the word *manufacturing* only to include information products and recognize the high level of knowledge, skill, and intellect required to produce them.

The fast-food industry, at the opposite end of the intellectual spectrum, is classified as a service industry, but it embodies all the elements of the tripartite organization. There is probably no dispute that the functions of research, design, and development apply, but the word *manufacturing* probably raises some concerns—however, every fast-food establishment has its own production facilities—the kitchen.

These two examples, with accounting at one end of the spectrum and fast-food establishments at the other end, require only a new view of what *manufacturing* includes. At the accounting end of the spectrum the product is information developed by highly educated and skilled professionals—at the fast-food establishment the product is food in one form or another provided by lower-level education and skills.

What are the appropriate limits for management of technology? The comprehensive and holistic approach described around Fig. 1.3, of the tripartite organizational structure, may be considered idealistic, but the gap between the present segmented approach to technology management and the comprehensive approach must be closed.

The past malpractice in technology management was prologue. The failure of projects to meet the specifications, the schedule, and the cost projections provides sufficient evidence that not only are the technologies related to science and engineering and information mismanaged but the supporting technologies in the many support functions are totally ignored.

The media more than adequately cover major and publicly oriented project failures. In recent times the Denver airport and the Chunnel (underwater Channel tunnel) connecting England and France are but two major examples. In both cases the time schedules and the costs far exceeded the original projections. But the thousands of projects in industry, academia, and government that miss their projections seldom make the headlines. In many cases managers rationalize their organization's mismanagement with a shrug of the shoulder and "we can't expect perfection." But what is the cost?

Probably less than 5 percent of all projects meet the projected specifications, schedules, and costs. For those of you who question this statement, look back at the work effort for the last 1 to 5 years in your own organization and determine how many projects met the original projections—no excuses for anything, no fudging—the projections were either met or not met. There are lessons to be learned from such an exercise, and that investigation will show the lack of managing technology—not just managing the science- and engineering-related aspects but all the technologies relevant to the business unit. As an example, think of the potential impact of human resource technology if creativity and innovation became a dominant factor. Human resource technologies need to be thought of in a new way. They require a shift in the basic human resource model.

The new model expects creativity and innovation from human resource professionals at all levels and in all assignments. That new way of thinking requires understanding processes and determining the impact of those processes before attempting implementation. At the very least, the model demands examination of the effectiveness and the efficiency of human resource activities. The model also requires determination of the levels of creativity and routine. What benefits arise from the multipage annual or semiannual appraisals? What purpose do they serve? Why use them? Would a blank sheet of paper with the individual's name at the top suffice? Do managers really know how to appraise performance? This new model demands dissatisfaction with the status quo. The same model applies to all business unit functions, including finance, marketing, and so on.

1.6 STRATEGIC, OPERATIONAL, AND MANAGEMENT ISSUES

The underlying elements of any organization include its purpose or purposes, its vision, its objectives, its strategies, its operations (doing the work to achieve its purposes, vision, and objectives), and its management of the process from purposes to customer satisfaction. A view of MOT from the perspective of strategy, operations, and management shows the extent to which MOT in reality is congruent with managing the enterprise.

1.6.1 Strategic Issues

The strategic issues of MOT require greater attention by managers involved in developing business unit strategy. Consider the following strategic issues:

- Understanding the scope of managing technology
- Managing technology—different levels
- Technology managers—who manages technology?
- Adding value with technology
- Developing a technology policy
- Bridging the gap between technology policy and results
- Precursors to technology strategy
- Including technology in business strategy
- Rationalizing strategy and operations
- Managing the decision-making processes
- Systems thinking—the imperative
- Negative impact of single-issue management
- The role of technology in achieving competitive advantage
- Managing technology in a dynamic environment

Figure 1.5 details the subelements of these issues involved in integrating technology.

During the strategic-planning craze, technology was essentially ignored. Elaborate strategic plans were developed without any strategy. Volumes were prepared but were seldom reviewed after approval. Strategic-planning processes yielded volumes of data instead of information, an interjection of operational detail but insufficient as an operational plan, and prepared on a basis of at least questionable, if not false, assumptions. Those assumptions included the assumptions of a static rather than a dynamic environment, were based on a questionable premise of an annual event, dealt with data rather than information, focused on analysis without comparable emphasis on synthesis, failed to translate the strategy in meaningful terms throughout the organization, and ignored technology that affects over 75 percent of the sales value of production.

The degree to which each item in Fig. 1.5 affects performance will depend on the organization's purpose(s), industry linkage, and current competitive position and the activities required to attain the objectives. One point is certain—technology cannot be ignored and cannot be given short shrift. That strategy begins with an understanding of the basics of MOT and the role of technology in the business enterprise.

- Understanding the role of technology in the business

  A business, not technical responsibility
  Integrating resources, infrastructure, and activities

- Including technology in business strategy

  Research
  Development
  Design
  Manufacturing
  Marketing
  Administration

- Managing technology—different levels

  Globally
  Corporate, sector, group, division, strategic business unit
  Research, development, design, manufacturing, marketing, sales, physical distribution, customer service, all administrative and support functions
  Managing technology at the project level—team level
  Managing technology by scientists, engineers, and other professionals

- Technology managers—who manages technology?

  At different levels—from executive to individual decision maker
  Description of role
  Required skills, experience, and knowledge
  Balancing technology and management competency

- Adding value with technology: leveraging available resources

  Internal and worldwide
  External and worldwide

- Developing a technology policy
- Bridging the gap between technology policy and results

  Transforming the thinkers into the thinkers and doers
  Transforming the doers into the doers and thinkers

- Precursors to technology strategy
- Rationalizing strategy and operations
- Managing the decision-making process—practices and models

  Consensus
  Quantitative and qualitative
  Intuitive
  Balanced

- Systems thinking
- Impact of single-issue management
- Role of technology in achieving competitive advantage
- Dynamics of technology management

FIGURE 1.5   Strategic issues in the management of technology.

This list of strategic issues may be considered too extensive and too difficult to consider or implement, or you may flip the coin to the other side and ask, "What's new?" But in that process you also must make a decision as to which elements are not important. Can decision processes be discounted? Can systems thinking be replaced with single-issue-management? Can rationalizing strategy with operations be eliminated? Can technology continue to be ignored in the process of developing a strategy? Is there a viable technology strategy? Is the process of including the technology issues in business strategy essential? Absolutely, yes.

1.6.2 Operational Issues

The operational issues of MOT present a similar vast array of topics that must be defined in the context of the business. It is even more extensive than the list of strategic issues. Consider the following major categories of operational issues that must be resolved:

- Idea and concept generation
- Forecasting
- Evaluating
- Justifying investments
- Planning management
- Managing the project management process
- Managing discontinuities
- Descriptions—how, where, and why
- Resolving problems and exploring opportunities
- System cycle time management
- Technological intelligence
- Innovation
- Entrepreneurship
- Technology transfer
- Information
- Functional integration
- Investing in research
- Organizing for effective product development
- Market-pull and/or technology-push
- Introducing new processes
- Introducing new products
- Selecting, monitoring, and terminating projects
- Integrating technology, products, and markets
- Linking purposes, objectives, and strategies
- Focusing on value-adding activities
- Resolving the information paradox
- Effectiveness, efficiency, and economic use of resources

- Analysis followed by synthesis determines results
- Differentiating the means and the ends
- Eliminating the barriers to effective management of technology
- Developing and using business unit technology plans
- Organizing and allocating resources
- Developing as-is profiles (the plus/minus analysis)
- Closing the gaps from competence to capability to competitive advantage
- Implementing activity-based management
- Implementing the project approach at all levels
- Auditing research, development, technologies, and potential new products and processes

Figures 1.6 and 1.7 list these major operational issues randomly with many subcategories but without attempting to prioritize their importance. The relative importance or priority of the topics depends on the particular business problem or opportunity. As with the strategic issues, which of these categories can be eliminated?

FIGURE 1.6   Part 1 operational issues.

- Idea and concept generation

  Forecasting
  Technologies
  Products
  Markets
  Productivity and performance

- Evaluating

  Infrastructure
  Resources
  Products
  Processes
  Markets
  Work practices
  Productivity and performance
  Organizational competencies and capabilities

- Justifying investments

  Resources
  Business functions
  Strategic and operational fit
  Risk analysis
  Value-added products and processes
  Justification or evaluation of alternatives

- Planning management

  From a dynamic or static perspective
  Up-front work
  Process—thought—process with thought
  Restrictions: business dynamics, change in requirements, new knowledge

FIGURE 1.6   (*Continued*) Part 1 operational issues.

- Managing the project management process

  Designing the process
  Planning
  Scheduling
  Designing and developing the end product or service
  Implementing, monitoring, controlling, terminating, etc.

- Managing discontinuities

  Descriptions—how, where, and why
  Technology—product—process—social, political, and economic

- Resolving problems and exploring opportunities

  Observing
  Synthesizing
  Describing
  Integrating data, information, knowledge, and wisdom

- System cycle time management

  Description—more than concurrent engineering
  Benefits and costs
  Sources for optimizing
  Types of cycle time: system, project, product, and process

- Technological intelligence

  Developing a process: internal and external
  Recognizing emerging technologies
  Using technological intelligence for technology leadership

- Innovation

  Description and scope
  Organizational environment
  Role of creativity
  In all business functions—not just science and engineering
  Innovating on schedule

(Continued)

1.6.3   Management Issues

The management issues include the fundamentals associated with managing any organization. The following list provides some broad categories:

- People-related
- Developing competent personnel
- Overcoming objections and resistance to change
- Competencies and capabilities
- Productivity and performance
- Specialization and segmentation
- Providing a balanced environment
- Educating the organization

FIGURE 1.7   Part 2 operational issues.

- Entrepreneurship

  Beyond the hype—the reality
  Organizational environment
  Successes and failures
  Internal and external

- Technology transfer

  Within the organization, business units, functions, individuals
  Academia and industry
  Government agencies and industry
  Global opportunities
  Consortium

- Information transfer

  Top-down, bottom-up, horizontally
  Sources of information beyond computers, beyond record keeping (libraries, periodicals, patents, other intellectual property)

- Functional integration

  Going beyond communication
  Processes for achieving

- Investing in research

  Corporate or operational
  Internal and external

- Organizing for effective product development

  Requirements—total system
  The process
  Limitations of formalized structure
  Need for flexibility
  Balancing freedom and constraint
  Balancing stability and change
  Creating an appropriate environment

- Market-pull and/or technology-push
- Introducing new processes in every business function

  Reasons for
  Benefits to be derived from
  Delineating the issues
  Making the transition

- Focusing the organization
- Integrating business functions
- Achieving gains from technology management
- Facing realities

Figures 1.8 and 1.9 provide a detailed listing of the management topics that must be considered. None can be ignored, and the listing provides a checklist for the practicing manager or professional. But MOT requires more than using the listing as a checklist. The topics listed under "people-related" are fundamental. Too often man-

FIGURE 1.7 (*Continued*) Part 2 operational issues.

- Introducing new products

  Classification, description, and assessment of new products
  The planning process from concept to commercialization (marketplace)
- Selecting, monitoring, and terminating projects
- Integrating technology, products, and markets
- Linking purposes, objectives, and strategies
- Focusing on value-adding activities
- Resolving the information paradox
- Effectiveness, efficiency, and economic use of resources
- Analysis followed by synthesis determines results
- Differentiating the means and the ends
- Eliminating the barriers to effective management of technology
- Developing and using business unit technology plans
- Organizing and allocating resources
- Developing "as-is" profiles (the plus/minus analysis)
- Closing the gaps from competence to capability to competitive advantage
- Implementing activity-based management—not activity-based accounting
- Implementing the project approach at all levels
- Auditing

  Research
  Design
  Development
  Manufacturing
  Marketing
  New products
  New processes
  Technologies

(*Continued*)

agers at all levels fail to consider these issues in some formal or at least conscious manner. As an example, "identifying the required knowledge" for a particular project cannot be ignored; yet too often, assumptions are made about the required knowledge that are inconsistent with the requirements. As a result, the wrong people are assigned.

Another example may consider the topic of "raising expectations." Certainly expectations can be debated, but even in today's competitive environment a general reluctance exists on the part of managers to establish meaningful stretch targets.

An additional example may serve to illustrate the point. Assume that manufacturing is proposing cost-reduction targets. The manager suggests 3 percent as a logical target. Some person in upper management states that 3 percent is insufficient and insists on a 10 percent reduction. The usual complaints arise from manufacturing that the targets are unrealistic. The year-end results show a reduction of 8 percent.

How does upper management view such a situation? Does management penalize manufacturing in some way for not meeting a 10 percent reduction target, or do they recognize the group's performance? An 8 percent cost reduction is certainly better than 3 percent. In the initial stage manufacturing played it safe. It is unfortunate that top management does not realize that there is a certain amount of game playing in all cost-reduction or improvement programs. Manufacturing knew that costs could be reduced by 3 percent without any great amount of extra effort—no new thinking, just more of the same. At the same time, the arbitrary 10 percent established by manage-

- People-related

  Identifying the required knowledge
  Selecting the appropriate participants
  Developing a critical mass
  Integrating the unique characteristics and talent of the participants
  Focusing on career emergence rather than career planning
  Raising expectations
  Stimulating as contrasted to motivating
  Identifying and rewarding the new heroes
  Role of personal characteristics and human behavior
  Managers as teachers

- Developing competent personnel

  Increasing flow of new ideas and concepts
  The business operating philosophy—the important fundamentals of managing, business practice, and managing technology
  Balancing depth with breadth
  Becoming multidisciplinary and multifunctional, leading toward interdisciplinarity
  Deveioping a high-performance organization
  Understanding the art and practice of managing
  Accepting change
  Managing transitions
  Balancing team and individual performance

- Overcoming resistance

  New ideas and concepts
  New thinking
  New technologies
  Anything new and different

- Competencies and capabilities

  Transition from individual to business unit
  Core (also platform) and support
  Individual competence to business competency

- Productivity and performance

  Expectations
  Measurement

- Specialization and segmentation

  For building organizational flexibility
  Realities of providing flexibility
  For whom, when, and how much
  Organizational focus and flexibility

FIGURE 1.8   Part 1 management issues.

- Providing a balanced environment

  Balancing creativity and routine
  Business discipline
  Balancing stability and change
  Freedom and constraint
  Financial success and stakeholder rewards
  Leading and following

- Educating the organization

  Emphasize "thinking" as a precursor and postrequirement
  Understand the fundamentals involved in going from illiteracy, to literacy, to acquiring information, to knowledge, to learning from application
  Continuous education: what type, how much, when, and for whom
  Applying the "learning" of the organization

- Focusing the organization
- Integrating and achieving the potential gains from

  Systems thinking
  Management by objectives
  Participative management
  Benchmarking
  Empowerment
  Exploiting organizational learning
  Introducing the right amount of operational discipline
  Work analysis
  Organizational structure
  Reconcepting the organization
  Focusing on the thinking/learning organization

- Achieving the gains from technology management

  Automation of work in all functions
  Computer-integrated manufacturing
  Computer-aided design, engineering, manufacturing, and others
  Concurrent or simultaneous engineering
  Design theory: all functions including administrative
  Information systems that provide information
  Using technology and business models and simulations
  Using other computer-aided technologies such as artificial intelligence, expert systems, chaos theory, catastrophe theory
  Understanding the limitations of computer technology
  Business-related processes

- Facing realities

  Abandoning the old myths
  Eliminating the organizational noise
  Dealing with issues rather than personalities
  Playing games
  Work from the basics

FIGURE 1.9 Part 2 management issues.

ment, while appearing totally unrealistic in the eyes of manufacturing, did provide the necessary impetus to obtain better results. Managers should not be averse to setting difficult targets provided they do not become overly concerned about some arbitrary targets but look for the actual realistic gains.

The issue of raising expectations requires serious attention. Consider the situation associated with raising the expectations of engineers and scientists. Only one question needs to be answered to demonstrate that these professionals spend probably less than 50 percent of their time in their professional endeavors. The other 50 percent involves routine work well below their level of expertise. Decisions regarding this topic as well as others must be made consciously. Assumptions must be validated and qualified. Information should not be accepted without questioning the validity and the integrity of the assumptions.

Now consider linking these operational issues with the strategic issues and subsequently with the management issues. What results is a continuum from strategic to operational to management, to strategic or operational, and so on in a continuous feedback loop. A change in one requires a change in the other. You may argue that strategy is a management issue. It may have been at one time, but people at low levels in the organization make strategic technology decisions. A relatively young engineer can make some major technology strategic decisions in the product or process development area. A young marketing or sales representative may provide creative input for future market development, new-product requirements, or needs of customers.

1.7   CLASSIFICATION OF TECHNOLOGIES

Technology is technology is technology, and so on. Not true. Managing technologies requires some delineation of the technologies into categories. There are no agreed-on approaches for classification. Technologies can be classified according to any of the following categories:

- *State-of-the-art-technologies:* those technologies that equal or surpass the competitors
- *Proprietary technologies:* those technologies protected by patents or secrecy agreements that provide a measurable competitive advantage
- *Known technologies:* those technologies that may be common to many organizations but are used in unique ways
- *Core technologies:* those technologies that are essential to maintain a competitive position
- *Leveraging technologies:* those technologies that support several products, product lines, or classes of products
- *Supporting technologies:* technologies that support the core technologies
- *Pacing technologies:* technologies whose rate of development controls the rate of product or process development
- *Emerging technologies:* technologies that are currently under consideration for future products or processes
- *Scouting technologies:* formal tracking of potential product and process technologies for future study or application
- *Idealized unknown basic technologies:* technologies that, if available, would provide a significant benefit in some aspect of life

It is not important that an organization use this classification. It is important, however, to develop some format for classifying technologies in a manner that is meaningful to the specific organization. Management seldom faces up to the realities of technology classification. Too often technologies used by competitors are considered as proprietary, the list of supporting technologies is limited because they are known, every technology seems to be a leveraging one, the list is too long, the list is too short, and so on. In reality, most organizations depend totally on known technologies and may only be combining them in creative ways. Some realism must be applied to this classification process. Every technology is not a proprietary technology. Every technology is not a core technology. Many commonly known technologies are not leveraging technologies. This classification of technologies must be rationalized in some way. As an example, is there a proper ratio for these technologies:

State-of-the-art

Proprietary

Known

Core

Leveraging

Supporting

Pacing

Emerging

Scouting

There is some ratio of these technology classes which determines where organizations invest their resources. Those percentages will vary significantly not only within organizations but also within organizations in the same industry. The purpose for identifying these classes is to recognize where those resources are being applied. For example, some resources must be applied to scouting new technologies, but a relatively small percentage. Yet many organizations fail to recognize the need to allocate some resources to this class.

The classification allows putting technologies into perspective. While the core technologies may be considered the most important, the other classes cannot be ignored. Competence in supporting technologies may be just as important as the core technologies. Work on emerging technologies must be planned on the basis of business unit strategy. Scouting technologies that flow into the emerging technologies and then into other categories represent future technological opportunities. The future demands consideration of the long-term technology needs. Technology classification can also be viewed from the perspective of proprietary technology that provides competitive advantage and state-of-the-art technology which may be available to competitors.

Technology classification will also be modified by the type of product or process under consideration. As an example, new products and processes can be classified as

- "Me, too" products or processes
- Minor product or process improvement
- Major product or process improvement
- New-to-the-market products or processes
- Breakthrough products or processes

Of course, each organization must describe just exactly what terms such as "me, too," "minor," "major," "new to the market," and "breakthrough" really mean. It is difficult for organizations to acknowledge that their products fall into the me, too category.

There are no generalized rules that describe the relation between types of technologies and their application to different types of products or processes. At one end of the technology continuum, a breakthrough product or process could involve state-of-the-art or proprietary technologies. But a similar breakthrough product or process could involve known technologies but combined in a unique way.

1.8 EDUCATION IN TECHNOLOGY MANAGEMENT

Some universities offer courses in the management of technology. These usually take the form similar to an executive M.B.A.—every other weekend on alternate weeks or some such part-time arrangement. Most of these programs lack substance and focus more attention on a graduate degree than on the content. Awarding a master's degree when the subject matter is often at the undergraduate level gives false impressions of the educational value. A review of many curricula reveals no clear understanding of purposes, objectives, and strategies. MOT courses are another product offered in the smorgasbord of master's-level programs.

There is no simple process for delineating the differences of the objectives of various MOT programs. One analysis, from a study by Kamm,* shows that course objectives show a great deal of diversity in thinking about what MOT should include in the curriculum. My analysis of Kamm's study shows that if the verbs used throughout the discussion of academic curricula in MOT are identified, some interesting conclusions can be drawn. The verbs have been classified in relation to "objectives for whom." Do these verbs relate to professors, students, or both? Here are the findings:

- Professors: clarify, design, develop, give, help, introduce, offer, present, promote, provide, utilize
- Students: acquire, come to understand, gain, overcome, study, understand
- Professors and students: consider, examine, explore, identify, recognize

Course objectives described by these verbs convey the idea of transmitting data and possibly information as contrasted to providing opportunities for gaining knowledge. There is a distinct difference between information and knowledge.

The objective-related verbs are weak—in essence, professors give, and students receive. The objectives of MOT courses must emphasize "doing something" of significance that will lead to knowledge rather than data or information. This approach of acquiring data and information ignores the basic requirements of MOT: creativity and innovation. It appears that those studying MOT need not be the idea generators and the innovators, but will manage the idea generators and innovators. This approach by the academic community in teaching MOT is unfortunate. It places us on the road to

---

*Judith Kamm, Bentley College, Waltham, Mass. Report on academic programs in the management of technology. Presented during a business meeting of the Technology and Innovation Management Division of the Academy of Management, August 1991.

developing another generation of technology managers who cannot provide the necessary technological leadership—that leadership requires making a creative and innovative contribution to the enterprise; thus we develop another generation of technology
administrators rather than technology managers. MOT, a multidisciplinary as well as
interdisciplinary discipline, is a practitioner's discipline. Academia must focus on relevance and at the same time not ignore the theoretical foundations where they apply.

Those contemplating attending a course in MOT must ask some questions. For
instance, is the purpose to obtain a degree or to gain an understanding of the issues in
management of technology? With a bias, I suggest that the purpose is to develop an
understanding of the principles of MOT. MOT involves managing with the systems
approach. Without hesitation, I say accounting or economics majors will not manage
technology. Nothing against accountants or economists. They play a major role in
business management, but they do not have the background to manage technology.
MOT is a hands-on profession. It demands a broad understanding of technologies
related to the specific business.

1.9   SUMMARY AND CONCLUSIONS

Management of technology provides firms with many opportunities for improving performance. But gaining that additional effectiveness and efficiency comes with a price.
The cost of entry requires a change in thinking, a new management model, a focus on
the integration of the pieces of the system, and ignoring the quick fixes promoted by
the latest management guru. MOT is not a quick-fix approach. Figure 1.10 summarizes the discussion of the description, scope, and implications of MOT.

The summary of MOT illustrated in Fig. 1.10 brings together the phases of MOT,
the primary elements, the business issues, and the tripartite organization. Each of these
integrators interacts with all others. At a minimum, at least 10 interact at any one time.
This is based on the fact that in the phases of MOT, only one phase would be involved
at any one time. The complexity of these interactions is further magnified by the number of subcategories involved. As an example, a phase 1 approach to MOT involves
resources, infrastructure, activities, management, strategy, operations, and the three
major segments of the tripartite organization. If a systematic management process is
practiced, the number of these interactions will not pose a major problem. At the same
time, those interactions cannot be ignored. They are real, and that is why MOT
requires their consideration as part of the system.

For a moment consider the cost in correcting the malpractice in management due to
not meeting project requirements and specifications, not completing projects on
schedule, and not meeting projected costs. When less than 5 percent of all projects are
completed within specification, on schedule, and on time, it is certainly time to reappraise the ways of managing. These corrections will not be made using the single-
issue management model, in which every function attempts to build walls around its
empire. That kind of activity is totally unproductive in a competitive economy. So to
introduce MOT to your organization, consider some of the following suggestions.

- Competitiveness begins inside the organization. More importantly, it begins with
  the individual. Compete with yourself, your past performance, and not with some
  external competitor.
- Recognize the role of process—not methods or methodology but process—process
  with substance, and not just going through the process.

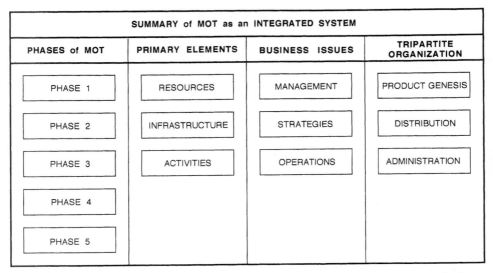

FIGURE 1.10 Summary of all the integrators involved in the management of technology. Each of these elements interacts with every other element, thus complicating the process.

- Reappraise strategic planning. Focus on strategy without all the associated bureaucracy. Strategy is the first step toward implementation of objectives. Separate strategy from planning. Planning responds to what should be done, why it should be done, who is going to do it, and when it will be done.

- Recognize that there is no single strategy of the firm. The strategies of all the functional organizations must be integrated into some form of cohesive strategy. The strategy of research cannot countermand the corporate strategy; it must support it. The same is true for all functions.

- Developing a strategy is an iterative process. It cannot be done on the annual 12-month cycle. Strategy must be reviewed daily to determine whether it continues to be viable. Strategy is not a book, but a paragraph on a sheet of paper.

- In the final analysis, strategy is not what is written on paper but what is in the minds of the participants.

- People play a major role in successful implementation of MOT. Do not allow human resource policies to reduce every person to some lowest common denominator. Teams are a means to an end, but ideas and concepts come from individual creativity and innovation. Managers cannot become tour directors or social directors, and people cannot mentally retire at age 35.

- Interfaces must be managed—one of the critical management issues. The competitive environment is dynamic; it cannot be treated as static.

- People are an organization's greatest asset. A cliché. Yes. Make it a reality not by vying for "best manager of the year" award but the manager who challenged the staff and provided opportunities for interesting and business-oriented effort.

- Recognize that all investments must be justified. That applies not only to product and process development and to capital investment but to all work in the adminis-trative functions. Technology can be justified if it provides a benefit. If it does not, it should not be implemented. That benefit, however, must be measured qualitative-ly as well as quantitatively. Clearly delineate the benefits from the investment.

- Emerging technologies—whether related to product, process, manufacturing, or other activities—should be treated as unknowns. Even scaleup of processes requires research.

- Establish criteria for investing in new technology. Objection by a single individual, regardless of management level, cannot be justified. Decisions to invest must be based on business needs and not only on financial analysis.

- Raise the levels of expectation for every person from the CEO to the lowest level. Emphasize activities that increase the value added for the effort expended.

- Technological literacy becomes an absolute must for MOT. Obviously the level depends on the need. But people involved in making technology-related decisions (at least those of significant scope) must understand exactly what the decisions involve.

- MOT is not a mystery. It requires following fundamentals of management. It requires taking a systems approach. It requires including more than the activities of scientists and engineers. MOT involves the complete organization.

- Managers at all levels and in all professions must recognize that people generally do not buy technology—they buy performance in some form or another.

CHAPTER 2

A BUSINESS ARCHITECTURE FOR TECHNOLOGY MANAGEMENT

Joseph P. Cory, Ph.D.

Cory Group
Rowayton, Connecticut
Stevens Institute of Technology
Hoboken, New Jersey

Man is limited not so much by his tools as by his vision.

PASCALE AND ATHOS
The Art of Japanese Management, 1981

2.1 TECHNOLOGY MANAGEMENT AS AN EMERGING DISCIPLINE

Technology management is rapidly emerging as a discipline combining the elements of business management and engineering. In support of the meaning of technology management one description views this discipline as

the research and education on how to:

- Manage the technology component of individual product life cycles,
- Capitalize on process technology to gain a competitive advantage, and
- Relate and integrate product and process technologies.

Technology management is applicable to every phase of technology-oriented businesses (in either application or development) such as marketing (services) and planning activities as well as R&D, product development, and manufacturing.[1] Major elements referred by the National Research Council Task Force Report on Management of Technology include

1. Research management

2. Product planning and development

3. Project management

4. Integrated manufacturing processes

5. Production control

6. Quality assurance

7. Information systems design and use

8. Software development

9. Product vendoring

10. Corporate technology

11. Integration of technical [disciplines] with business and financial decision making

"Formal knowledge of technology management is valuable, not only for managers of R&D, but also for manufacturing, marketing, financial, and general corporate management."[2] Therefore, it is important to view the management of technology not only across all disciplines and industries but also in a global environment in a rapidly shrinking society. Both manufacturing and services industries are in need of this knowledge today, although the latter is just beginning to be cited with frequency.

2.1.1 Goals and Strategies of a Firm

A firm's goals and strategies represent the aggregate of its products, technologies, and services. "For such goals to be credible they must be linked directly to the set of development projects the firm intends to undertake [or are in-process or in the marketplace]."[3] Therefore, a business architecture in technology management follows for new development projects and their functions. It identifies what business parameters to address to make a good business decision regarding key problems encountered in achieving the objectives and goals of the strategies of a new development project. These problems occur at different steps in the strategic and development processes.

2.1.2 Development Chain

The development of any product or service can be described as a chain of acceptances by management in its travels from idea conceptualization to the marketplace. Essentially, each acceptance step is a gate that allows one to proceed to the next major step. Concurrent engineering has allowed one to accelerate this process, but the overall business decisions related to acceptance are essentially the same. This chain of acceptance steps is described later, but is defined below to include

1. The research and development of an idea or invention
2. The competitive evaluation of the idea
3. The research required in technologies to develop the idea
4. The transfer or purchase of a selected technology for support of a process, a component of a product, or a further development of the technology for its own marketability
5. The acceptance of a proposal to develop the product
6. The acceptance of the design of the product and/or technology
7. The acceptance of the product's manufacturability, i.e., at the volumes and specific quality levels required
8. The acceptance by the customer of the product that is marketed

2.1.3   Managing the Chain

Managing these steps, effectively and efficiently, is often met with some difficulty. Problems emerge in implementing functions that are needed to achieve the acceptance criteria. The management practice is "sometimes ineffective resulting in lack of schedule integrity and resource drains,"[4] thus impacting the improvement of the total development process.

An assessment criterion that allows management to make functionally sound and less costly decisions often is lacking. The cost of lack of commitment to schedule and the drains on resources of the firm result in a poor "payout" to the development project.

2.1.4   Use of Technologies

Technologies can be used as a component, module of a product, or a part of a process. An example of a technology within a product is the automatic gear that is an intrinsic part of an automobile. An example of a technology that is part of a process is the laminating process in metallurgy to produce sheet metal. The ideas behind technologies such as these are often applicable to other uses and can be expressed in the form of derived demand.

2.1.5   Decision Problems in the Use of Technology

The decision as to whether to use a technology can be problematic for a variety of reasons. The technology within a product may not be responsive to customer needs in quality, function, or performance. Or the technology chosen for a process may be difficult to implement or lead to poor and costly results. Also, it is possible that an alternate technology is the right one rather than the technology selected. In addition, the technology may lack synchronization with the committed schedule, the engineering changes may be excessive, the wrong product may be produced (i.e., unresponsive to the customer's needs), the development resources may be displaced to other products for various reasons, or the technology transferred may be changed by product development to the degree that the technology no longer resembles the technology which it represents or for which it was originally intended. There are many other reasons why wrong decisions and actions can cause problems in managing technology.

2.1.6   General Development Steps

It is necessary to look at the general development steps (shown later in Figs. 2.2 and 2.3) and identify key checkpoints in decision making that relate to business parameters.

These steps have a business assessment criteria associated with them for which there must be a payout or payoff. Payoff, in economic theory, often is expressed in the form of a matrix and is "the interaction between arguments or conditions exemplified in terms of firms or people."[5] The payoff is the interaction between the accepted development steps and the required business parameters. In reality, these business parameters, aggregately, contribute to the business strategy of the firm. The action to respond to this payout can be a benefit to both firm and customer alike, or can be an improvement in the operational cost of the business. The effective managing of technology in the business decision-making process directly entails the evaluation, use, and consequences of technology. Therefore, technology management requires an architecture (a structure of interfaces) that contains a technical and business decision-making process recognizing these technological impacts.

2.1.7   Technical and Business Views

Technology management puts into perspective the ways of advancing technology for the benefit of its recipients. To survive, a business must be profitable. If the firm is profitable, it must be responding, positively, to the demand of its products and services.

Therefore, it is necessary to be cognizant of both the technical and business views. From a product development viewpoint, management is faced with many choices in technology. Technologies are researched, developed, or procured for the products intended in the marketplace. These technologies must be coupled to the payout of the business.

From a business viewpoint, the products result in profitability to the firm and benefits to the customer. Effective management of technology is achieved when the products that the firm markets are profitable and the processes that are developed to build them are cost-effective. Technology is the means of solving the customer's problems for two primary reasons: to help them become more efficient and to help them grow. Also, the firm's processes for supporting these products can produce cost-effectiveness and time savings through the improvement of productive yields.

2.2   INTRODUCTION TO THE BUSINESS ARCHITECTURE

The architecture that follows shows the major functional steps in research and development with their corresponding examples of business parameters. These steps are those used in the industrial sector and describe the technology and product strategy processes employed in the development process. Analogous steps exist in the service sectors but are not addressed here.

2.2.1   Definition of a Functional Step

Each critical step in the process, from research to the marketplace, is identified and mapped to corresponding business factor(s) that provide a return or payoff. The business architecture, consisting of a technology and product development chain, is mapped against the corresponding business factors; these business factors, in turn, form a chain which is defined. Each functional step invokes one or more processes and, if managed effectively, responds favorably to the business factor(s) of the enterprise. A functional step generates other steps for which business decisions must be made. These subordinate steps may be identified only to make a point.

2.2.2 Understanding the Architecture (Process)

To understand any process, it is necessary to identify its goals and the strategies used to achieve these goals. It is said that a goal "should focus on significant areas of organizational need."[6] A *strategy* is "a guide to action or a channel of thinking,"[7] i.e., a general direction employed to achieve specific goals. There can be more than one strategy to achieve the same goal. For example, let us say that a computer manufacturer is to develop information-handling products that will be responsive to customers' needs in the office environment. One solution is to automate the office using workstations, printers, and intermediate-size computers to respond to those needs. One manufacturing strategy may be to build these products using the company's resources. Another manufacturing strategy may be to use vendors in place of the firm's own manufacturing facilities. The goal is still the same. In the same fashion, a technology is developed with function, quality objectives, and performance as its goals. One strategy may use a particular innovation as the technological solution, yet another strategy may consider a completely different innovation as a solution.

2.2.3 An Example in Technology Research and Transfer

As an example, technology research and transfer is used in Fig. 2.1 to show the practice of technology management in responding to the business. The model shows a process that identifies major development steps beginning with technology research and technology transfer, followed by product development, and market segmentation leading to an ultimate customer payoff. Correspondingly, on the business side, there is also a process which shows projects funded on the basis of customer requirements (wants and needs), funding priorities, the strategy of the business, and the contribution that the projects must respond to achieve revenue and profit objectives. Successful correspondence between development steps and payout to business objectives through decision making gives the ultimate successful result.

2.2.4 Phases of the Architecture

The overall architecture is in two phases: the research-and-technology phase shown in Fig. 2.2 and the product development phase shown in Fig. 2.3. It is stated that "The product development process represents the crucial interface between applied research, development engineering, and the market groups."[8]

2.3 RESEARCH-AND-TECHNOLOGY PHASE

The research-and-technology process phase begins with research and closes with successful technology transfer to the product development phase. The goals of the technology strategy are represented often by quality objectives, performance objectives, and function(s). The steps include

- The research and development of the idea or concept
- An analysis of competition that relates to the idea
- The creativity and innovation to develop technology(ies) derived from the idea

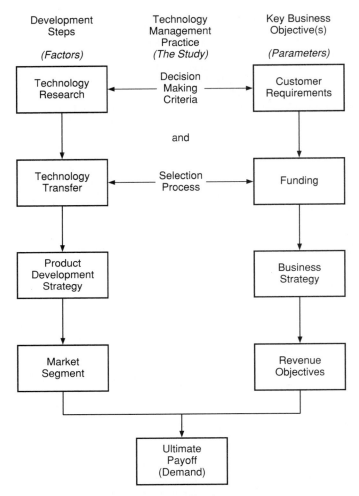

FIGURE 2.1   Technology management flowchart.

- The technology research, including technology forecasting, selection, and the transfer of technology to development

2.3.1   Research and Development (R&D)

One standard definition states that research is an undirected basic science or a directed or applied science.[9] Development in research can result in innovative products and processes from either of these sources. Research differs from development of a product in that it is an exploration of ideas rather than a tangible output to be marketed. There is no guarantee that a new idea will be successful. R&D must provide a high

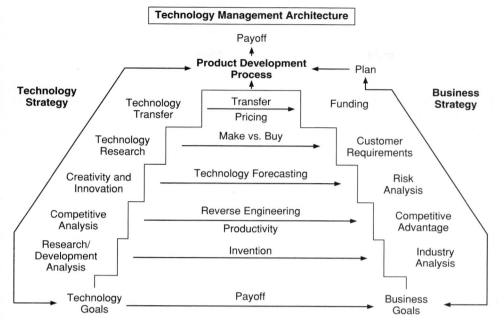

FIGURE 2.2   Research and Technology Phase.

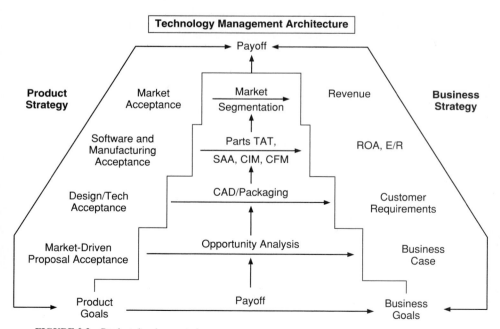

FIGURE 2.3   Product development phase.

degree of insurance so that future gaps can be closed to meet corporate goals, such as profitability. Therefore, many innovations are necessary and should be allowed to proliferate to assure the achievement of these goals. "Invention is an idea that must next be reduced to practice, i.e., one must show the technical idea is feasible and can be demonstrated."[10]

Industry analysis is one step used by business strategists to examine the external environment to assist in establishing business goals. Thus, a payout to industry analysis can be achieved through the practical results responding in invention. For example, the invention of semiconductors was well received and was a handsome payout to the industry providing a new generation of computers.

2.3.2 Competitive Analysis

Competitive analysis verifies that the innovation is valid, timely, and responsive to competition. This analysis is portrayed by asking four questions:

- "What are the implications of the interaction of the probable competitors' moves that have been identified?
- Are the firm's strategies converging and likely to clash?
- Do firm's have sustainable growth rates that match the industry's forecasted growth rate, or will a gap be created that will invite entry?
- Will probable moves combine to hold implications for industry structure?"[11]

It is fair to say that technology management is instrumental when applied through forecasting, assessment, and implementation of technologies for competitive advantage. *Competitive advantage* views technology as a way of improving a process, thus reducing costs, or providing customers with a best-of-breed benefit. In addition, for example, consider analyzing a manufacturing process. A competitive advantage against a competitor can be achieved by assessing a best-of-breed process of any firm regardless of whether the process is a competitor's or from another industry.

2.3.3 Creativity and Innovation

"Creativity by thought is invention and that inventiveness is a quality usually required and always desirable in all phases of the innovation process."[12] Formulation of "the creative process contains the following stages:

- The perception of the problem (coming from the R&D process) with its idea(s),
- Frustration of the inability to solve it,
- Relaxation or sleeping on the problem,
- Illumination or sudden inspiration, and
- Solution and verification."[13]

Innovation exploits opportunity to seek a return on investment. *Creativity* provides the forum for innovation by being one of the requirements for the successful entrepreneurs in their quest to innovate. Six stages are outlined in the innovation process:[14]

- Preproject stage, including inside and outside R&D communications of what may be of interest to the firm and networking ideas with peers
- Project possibilities: what could be useful to the customer
- Initiation of the project: managing the idea with the marketplace
- Execution: managing the innovative project
- Outcome evaluation: evaluating the development of the innovation
- Project transfer: transferring the development to the next point for further work on the project.

As simply as this is stated, the important point is that innovation as differentiated from creativity has a constant reference to the marketplace because that is where the opportunity (return on investment) and the risk for success or failure lie. Risk analysis assesses creative and innovative ventures in technology to determine technologies that are worth considering. Many risk analyses are common tools for the entrepreneur who exploits the opportunity to get the maximum return on investment. For example, Marconi, who did not invent the telegraph, exploited the opportunity with risk and received a handsome payoff from the return on investment.

2.3.4 Technology Research

It is stated that there is a technology paradox which concludes that all technologies are fated to be replaced, eventually; however, most attempts to replace them will fail. "Starting with the beginning step—a bright new idea—those that survive all the way to application are probably less than 1 per cent."[15] This statistic calls for the primary task of technology forecasting to be performed as a required part of technology research in this process.

Technology forecasting is defined "as a collection of formalized processes or methods of future technology evolution caused by developments in science and society, and the interactions between these developments." Note that

> Scientific phenomena are precursors to technology. All technologies are based on exploiting the possibilities of guiding the selection of the courses of natural events and states of a phenomenon....The discovery of a scientific phenomenon is a precondition to the utilization of that phenomenon for a technology—a precursor to technology forecasting.[16]

There are two major types of technology forecasting: exploratory and normative technology forecasting. *Exploratory forecasting* begins with today's knowledge and is oriented to the future. An often-cited example is the succession of techniques used for the function of lighting, in which the 1860s' paraffin candle progressed to Edison's first lamp, followed by the cellulose filament, tungsten filament, sodium lamp, mercury lamp, fluorescent lamp, and white light (and now the halogen lamp).[17] This is a deliberate evolution.

Normative technology forecasting first assesses future needs of society and market potential and analyzes these for their technological potential. An example of this type of forecasting is the automatic gear shift. To expand the demand for driving automobiles to a significantly greater number of users, it was necessary to satisfy a need for "ease of use" with a much simpler and less intimidating technology by embedding the gear shift within the product and providing "automatic" control to the driver.

Surveys find that "innovation shows clearly that normative technology forecasting has a [greater] potential for success than exploratory technology forecasting."[18] Still another approach to technology forecasting is viewed as a creative problem-solving process with techniques which can be used equally well to stimulate group creativity. An example of this was "the completion of the lunar mission with astronauts requires the completion of a number of inventive problem-solving steps that can often be approached in a number of different ways."[19] Customer requirements help identify technologies that are needed to assist in solutions for customer operations or to help customers grow.

Accordingly, "some people, especially economists outside the steel industry, already saw clearly the need for major technological change that would convert steel-making into a chemical flow process. This temperature-conserving flow process would have a significant impact on the economy and did."[20]

2.3.5   Technology Transfer

The technology transfer step results in the transfer of a technology from its development to product development and manufacturing. It can be considered as part of a component or module in a product, or a product itself for derived demand, or it may be used to support or implement a process.

There are two major concerns in technology transfer viewed in two dimensions: technology and people. The first dimension is "the problem of transferring information about physical phenomena, equipment, analytical and manipulative techniques, terminology, etc. associated with the technology."[21] The transferal of information can introduce ambiguities in specifications, misinterpretations of meaning, and lack of on-the-job training to understand the technology and its interdependencies and architecture. The second dimension "concerns the feelings and attitudes in both organizations [of] R&D and product development engineering [regarding] the two sets of people with different skills, values, and priorities to become successful in passing the baton from one to the other."[22] This is often a problem in management style and practice. Funding is needed in the technology transfer process to support the transfer of critical technology for inclusion in a product or process. For example, there is "an important property of technological innovations for full-scale production settings that remain to be considered and that is the impact of learning effects on unit production costs and pricing of an innovative product."[23]

2.3.6   The Payoff

The payoff of a timely technology can make a firm gain a market edge by lowering price, even with a strategy that places it below production cost. The expectation is that increased sales would ensure future cost reductions from increased volume of output due to increased sales.

2.4   PRODUCT DEVELOPMENT PROCESS PHASE

In most manufacturing firms the critical product development steps are similar. These functional steps include

- A product proposal
- Design of the product with its relevant technologies
- Capabilities of the manufacturing or software processes to build and support the product
- The marketing (and servicing) of the product to potential customers

These steps constitute the major gating factors for delivering products and services. These steps, like the ones in the research and technology phase described in Sec. 2.3, constitute the major checkpoints for delivering technologies to product development or for process implementation of the product.

2.4.1 Product or Technology Proposal

A *product proposal* is a common instrument used by management to review and consider new products or major enhancements in the product line. This proposal is best described with the following content:[24]

- An opportunity analysis which analyzes the characteristics of demand and competition
- The sociotechnical environment which assists in defining the opportunity
- The feasibility and capability requirements of the firm's engineering, manufacturing, and marketing resources which relate to the potential product solution
- A market structure and market segmentation evaluation of the competitive economics, and the design and communication alternatives
- The overall firm's line of business policies and strategies as they relate to the new-product opportunity

Using this as the background of the proposal, the outline of key parts to its contents must be a part of the business case and include

- The product's business objectives such as revenue and customer demand from benefits derived from the product or technology
- A description of the product or technology
- The market channel to be used
- The segment(s) of the marketplace being addressed
- Preliminary market tests with their results to determine the usefulness for the product, if possible
- A description of the market opportunity, e.g., what the market is and who the competitors are
- A description of the product opportunity from the firm's viewpoint
- The opportunity sizing, e.g., how large is the opportunity in the market sought after and how much is anticipated for this product
- The financial risk and summary of measurements such as return on investment, customer value, and return on assets
- Work schedules, resources required, and expected time for completion and delivery

On review and acceptance of the proposal the product is funded, dropped, or returned to the "drawing board." The positive response to the business case to the proposal is the payoff. If the analysis of the opportunity shows a potential for the proposed product, then a reasonable business case can be developed and a proposal accepted.

2.4.2 Product Design and Technology

In this step, the design specifications, which include the technologies to be incorporated, are developed. Design and product engineering reviews and modifies these specifications for manufacturing implementation. The acceptance of the design and technology is reached after the models of the product design and the technology are developed and the product is stress-tested for design limitations and in compliance with customer requirements. Shortcomings to that compliance must be assessed to determine the impact to anticipated demand. To arrive at the best design point, the design and technology of a product must respond to customer requirements. Managing the technology, for instance, in computer-assisted design or in the packaging of a product with human factors is the payout to customer wants and needs.

2.4.3 Manufacturing and Software Development

On completion of the verification for design, manufacturability must be assured through stress testing for capability of assuring function objectives, volumes, performance, and quality objectives. Software, as required, must be in place and fully tested. At this point, these actions must be capable of achieving the proper return on assets and intended minimal cost and capital expenditures compared to expected revenues.

The payoff from return on assets is not realized, for example, in managing technology in a computer-integrated manufacturing (CIM) environment unless it is cost-effective in relationship to the revenues that result.

2.4.4 Market Release, Marketing, and Servicing Products

The final step that leads the product to the marketplace is a market release process and the implementation of the marketing plan to market the product or technology. Final plans to market a new product through marketing programs and selected market channels are put into place. A marketing program, for example, can include advertising in the press, TV exposure, or promotions using trade journals. Market channels can range from mail order to selling the product directly to companies. The sales strategy for the product should be in place. A program to service the product through either direct installation or installation aids, as required, should be developed. In addition, a program to service and maintain the product may also be part of a firm's objectives. This is highly desirable in the case of complex and high-tech products. The ultimate payoff is customer delight and satisfaction, and the revenue derived through market acceptance and increased demand. This can be successful only with the right technology and product responding to the correct market segment.

2.5 SUMMARY

A business architecture for technology management is fundamental to understanding the relationship between functional disciplines and managing technology as the causal parameter for producing effective results. The morphology presented identifies business objectives as the practical considerations of the firm. Without that, significant costs can be experienced. What is demonstrated here are the examples of business parameters and their relationship to critical functional acceptance factors in the "value chain" of the enterprise from research to the marketplace. Technology is the lead factor and, to be managed successfully, requires a progression of decisions from a business perspective. Without that, it can become a costly venture. These numerous examples in this architecture demonstrate the need for technology management to realize the firm's profitability and customer satisfaction as the ultimate payout to the firm.

2.6 REFERENCES

1. *The Technology Management Field: Definition and Scope,* unpublished communication, MIT Sloan School, Cambridge, Mass., no date, pp. 1–2.
2. National Research Council, *Management of Technology: The Hidden Competitive Advantage,* National Academy Press, Washington, D.C., 1987, p. 13.
3. Kim B. Clark and Steven C. Wheelwright, *Managing New Product and Process Development,* Free Press, New York, 1993, p. 100.
4. Don Clausing, *Improved Total Development Process,* unpublished communication, MIT School of Engineering, no date, p. 3A.
5. Paul A. Samuelson and William D. Nordhaus, *Economics,* McGraw-Hill, New York, 1985, p. 554.
6. George A. Steiner, *Strategic Planning,* Free Press, New York, 1979, p. 280.
7. Ibid., pp. 348–349.
8. Modesto A. Maidique and Peter Patch, "Corporate Strategy and Technological Policy," in *Readings in the Management of Innovation,* Michael L. Tushman and William L. Moore, eds., Pitman, Boston, 1982, p. 284.
9. J. J. Verschuur, *Technologies and Markets,* Peter Peregrinus, Ltd., London, 1984, pp. 2–3.
10. Frederick Betz, *Strategic Technology Management,* McGraw-Hill, New York, 1993, p. 23.
11. Michael E. Porter, *Competitive Strategy,* Free Press, New York, 1980, p. 71.
12. Michael J. C. Martin, *Managing Technological Innovation and Entrepreneurship,* Reston, Reston, Va., 1984, p. 134.
13. Ibid., p. 137.
14. Edward B. Roberts and Alan R. Fusfeld, "Critical Functions: Needed Roles in the Innovation Process," in *Managing Professionals in Innovative Organizations,* Ralph Katz, ed., Ballinger, Cambridge, Mass., 1988, pp. 102–104.
15. Lowell W. Steele, *Managing Technology,* McGraw-Hill, New York, 1989, p. 44.
16. Verschuur, op. cit. (Ref. 9), p. 16.
17. Ibid., p. 18.
18. James M. Utterback, "Innovation in Industry and the Diffusion of Technology," *Science,* **183:** 622, 1974.
19. Martin, op. cit. (Ref. 12), p. 87.
20. Peter Drucker, *Innovation and Entrepreneurship,* Harper & Row, New York, 1985, p. 33.
21. Steele, op cit. (Ref. 15), p. 158.
22. Ibid.
23. Martin, op. cit., p. 2.
24. Edgar A. Pessemier, *Product Management, Strategy and Organization,* Krieger, Malabar, Fla., 1986, p. 343.

CHAPTER 3
THE "TECHNOLOGY CYCLE" APPROACH TO TECHNOLOGY MANAGEMENT

Dr. David J. Sumanth

Professor and Founding Director,
UM Productivity Research Group
Department of Industrial Engineering
University of Miami
Coral Gables, Florida

John J. Sumanth

Research Assistant
UM Productivity Research Group
University of Miami
Coral Gables, Florida

3.1 INTRODUCTION

Eight years ago, Peters (1987) warned us with some alarming statistics about what he called the "accelerating American decline":

- The U.S. per capita gross national product (GNP) slipped below Japan in 1986.

- The average wage for a 25- to 34-year-old white male declined by 26 percent from 1973 to 1983 in constant dollars, and the comparable figure for 35- to 44-year-olds declined by 14 percent.

- The average business productivity (output per worker-hour) barely crept along at 1 percent since 1973 (compared to the 2 percent during 1965–1973, and 3 percent during 1950–1965). Manufacturing productivity grew at only 2.5 percent per year from 1950 to 1985, contrasting with Japan's 8.4 percent, Germany and Italy's 5.5 percent, France's 5.3 percent, Canada's 3.5 percent, and Britain's 3.1 percent.

- As compared to 10 banks toppling in 1981, in 1986 188 banks failed—the largest number in one year since the Great Depression.

- As many as 30 million people have been dislocated by the "restructuring" in manufacturing during the last decade. Since 1980 the *Fortune 500* companies have shed a staggering 2.8 million jobs.
- In 1986, even the trade balance in high-technology goods went into the red (into debt).
- A formidable $41 billion positive trade balance in the services in 1981 has all but disappeared.

Other authors also subscribe to Peters' assessment of America's deteriorating economic condition, including manufacturing competitiveness:

- It is estimated that it takes 55 years for 90 percent of U.S. manufacturers to adopt a new technology, compared with 18 years in Japan (*Focus,* 1991).
- Japan's use of numerically controlled machine tools is estimated to be *one and a half times* the U.S. rate. Japanese manufacturers use industrial robots at *seven times* the rate of U.S. manufacturers (Shapira, 1990).

As these statistics show, the United States has still not responded well to these growing concerns, while being shackled with vast trade deficits, and an inability to be competitive in the foreign markets. In 1989, while the U.S. trade deficit was approximately $125 billion, Japan enjoyed a trade surplus of $58 billion, and Germany reached $71 billion (Tenner and DeToro, 1992). Many other authors had warned before of the necessity for improved U.S. competitiveness, including Eckstein et al. (1984), Scott and Lodge (1984), Hayes and Wheelwright (1984), and Thurow (1985). Clearly, the U.S. competitiveness has been challenged severely in recent years. Because of this lack of competitiveness (and possibly other political considerations), Japan has tried to keep American products from being imported into its markets, resulting in a bitter trade war between the two economic powers. Although the Japanese administration recently has agreed to liberalize the import policies, to reduce the perceived inequalities in the trade, the competitive issues for the United States still remain. Apart from the trillions of dollars in national debt and billions in trade deficits, there are many other reasons for the lack of competitiveness in the U.S. industry. However, since our objective here is to prescribe a technology management approach that enables companies to enhance their competitiveness, it is sufficient to say that we need some new, but relevant thinking, to make our enterprises more competitive. We now present one such approach in the context of managing technology.

3.2 BASIC DEFINITIONS

The terms *technology management* and *management of technology* (MOT) are used interchangeably here. For our discussion, we define *technology* in a broad sense.

Technology is the means by which a tangible or intangible product (or service) is produced (or offered) in the "market." Thus, the use of technology can be seen even as far back as the times of primitive humans, whose technologies included stones to create fires, bows, and arrows to kill animals for food, canoes carved out of tree trunks to cross lakes, etc. Although these technologies have now been cast aside in favor of more developed tools, the purpose for which they were used has not changed—simply the means by which results are achieved. Because of the incredible pace at which technologies are being introduced into the market nowadays, the technologies we designate as "high-tech" today might be primitive and antiquated from the standpoint of people just a few years or decades from now. Consider the fairly recently introduced 486 microprocessor personal computer technology and today's Pentium microchip. The latter is more high-tech in our perception although the former has been in the open market for only about 7 years. The term *market* indicates that there must be a demand or need. The terms *tangible* and *intangible* in this definition are purely relative in nature. The distinction between these terms takes on relevant meaning in appropriate context. A few years ago, a famous university on the U.S. West Coast used goats to graze the grass on its campus, as a cost-effective "technological substitute" for more expensive lawnmowers. Certainly, in this context, there was a need to mow the lawns, and goats were an alternative technology, producing a tangible product—a mowed lawn! The use of animals in today's highly technical, computerized world stands out as an aberrant and somewhat awkward occurrence. Yet, it should also kindle us into thinking, "Why limit technology to just inanimate objects?" For centuries, we have used animals to accomplish tasks that our size made us incapable of doing. Even today, in some Far Eastern countries, elephants are used to move huge logs of lumber, as substitutes for forklift trucks. Dolphins were even used by the U.S. Navy some years ago to detect and uncover potentially dangerous mines in the Persian Gulf. Aren't dolphins an example of a cost-effective substitute for sophisticated mine-detecting devices? As these examples show, technology comes in more forms than just microchips and hardware. With this argument in mind, we can be led into a broader thinking on the subject of technology management. In order to improve our competitive power, we need to better understand the complexity of managing work, technology, resources, and human relations (Stephens, 1977).

Two other terms, *innovation* and *invention,* must be clarified for the purpose of our discussion here. Although these two terms are normally used interchangeably, there is really a distinct difference in the two meanings. Schmookler (1966) pointed out that

> Every *invention* is (a) a new combination of (b) preexisting knowledge which (c) satisfies some want. When an enterprise produces a good or service or uses a method or input that is new to it, it makes a technical change. The first enterprise to make a given technical change is presumably an imitator and its action imitation.

On the other hand, *innovation* is seen by Schonberger and Knod (1991) as

> technological breakthroughs—new products, services, and techniques—when they occur, but is more often the result of modest, incremental, improvements to existing products, services, and operations (the "tinkerer's tool box").

According to Michiyuki Uenohara, research director at NEC Corporation, innovation is "the result of tiny improvements in a thousand places" (Gross, 1989). John P. McTague, vice president for research at Ford Motor Company, expounds on the point, saying, "The cumulation of a large number of small improvements is the surest path, in most industries, to increasing your competitive advantage" (Port, 1989). Utterback (1986) and Bienayme (1986) also discuss the dynamics of innovation in sufficient detail.

3.3 A SYSTEMS APPROACH TO TECHNOLOGY MANAGEMENT

Over the years, many concepts have been developed in managing technology. However, Malaska's work (1987) comes close to a systems view, through a conscious development of technology, by a principle he proposed:

> The use of natural forces, matter and space on terms set by man and directed by him, must be adapted to the ecological entity of which it is a part, and must be carried out according to the principles which ecological development has shown to be correct. The recycling of matter and the many-stage use of energy must be established into the technical way of life and requisite exchange of entropy with the ultimate environment must be safeguarded.

A systems view, idealistic as it may be, is a necessity today, particularly in view of the fact that today's businesses are forced to operate in an international arena in order to preserve their market shares and economic stability. The world market is not only going to be a forum for the flow of goods such as cars, computer chips, and corporate bonds, but also, as Johnson (1991) sees it, a world market for labor. In fact, those economies that do not "play" their business game in the international arena generally limit their domestic growth as well. The total systems approach challenges enterprises to be aware of *all* the available resources (including animals!) and to make the most of those available, while designing technological systems.

3.4 THE "TECHNOLOGY CYCLE" AND THE "TECHNOLOGY FLOW PROCESS"

David Sumanth, in his earlier work (1988), proposes a *total systems approach to technology management* (TSTM) through what he calls the *technology cycle* (TC) (Fig. 3.1). He contends that the management of technology in enterprises is not just a one-shot deal, but rather a continuous process, involving *five* distinctly different phases of technology: *awareness, acquisition, adaptation, advancement,* and *abandonment.*

1. Awareness phase. This is the first phase of the technology cycle, in which a company has a formal mechanism to become aware of emerging technologies relevant to the company's needs. Some companies form "think tanks" with engineers and scientists, who research from around the world, and gather information through computer bulletin services, journals, magazines, books, conferences, international product exhibitions, etc. This information is synthesized and put in short internal report form for the benefit of corporate strategic planners and technology policy makers.

2. Acquisition phase. This phase involves the actual acquisition of a particular technology. To go from the awareness phase to the acquisition phase, a company's technology group, in collaboration with the industrial engineering group, would do a technical feasibility study, as well as an economic feasibility study, before justifying and acquiring a new technology. Of course, companies which do not spend much time and effort in either the technical feasibility study or the economic feasibility study usually face serious repercussions down the road through a rapid technological obsolescence, or through the acquisition of an inappropriate technology for their needs. For

THE "TECHNOLOGY CYCLE"

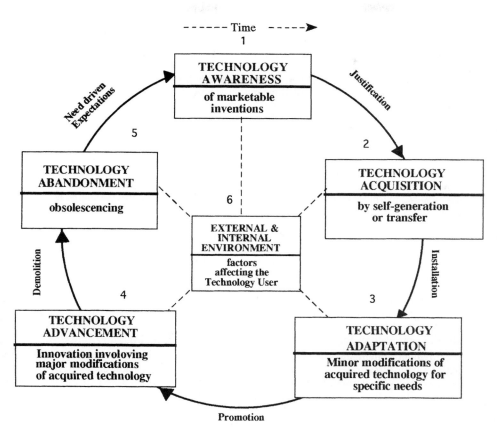

FIGURE 3.1 The *technology cycle* (TC), showing the five basic elements of technology management at any given level (product, service, function, work center, plant/division, corporation, industry, national, or international) applicable to deal with an existing or new technology. The dashed lines represent "analysis." (*Copyright Dr. David J. Sumanth, 1988. All rights reserved.*)

example, a major computer equipment manufacturer acquired an IBM 7535 robot in the early 1980s, assuming that it would replace an injection-molding operator. However, because of inadequate and inappropriate economic feasibility, the company found that the robot was costing more than the savings projected. After a few months, the company put aside the robot and brought back the human operator! Sometimes major plant relocation decisions are made, overriding both technical and economic feasibility recommendations. At times, these decisions have nothing to do with technical factors, but rather, are the result of someone's personal bias while making a policy decision in a boardroom.

3. Adaptation phase. Virtually every enterprise ends up adapting an acquired technology for its particular needs. Of course, if the homework is done correctly, the transition from acquisition to adaptation becomes much smoother and less expensive. Conversely, if sufficient time and effort have not gone into studying the relevance of a particular technology to a company's present needs, a great deal of rework and adaptation result. This not only frustrates the people acquiring the technology but also, and more importantly, slows down the assimilation rate, causes major productivity losses, and results in severe quality problems. Clearly, good planning and preparation before acquiring new technologies ensures the expected greater economic returns. This becomes a more dominant problem when companies import technologies from elsewhere. For example, a Far Eastern company once brought in a western fertilizer plant without first studying what *not* to bring. Because of its lack of preparation, the company did not know that the technical collaborator was using that part of the equipment because of the cold, snowy climate it was operating in. Thus, the company was installing equipment which was inappropriate for the humid tropical climate in which it operated.

4. Advancement phase. When capital is limited, as has become the case for many companies today, one cannot indiscriminately purchase and abandon technologies with scarce money. Therefore, it becomes imperative to improvise the acquired technologies for one's home needs. Companies like Lincoln Electric have taken this thinking to a new height. They are a world leader in electric arc-welding equipment, and generate most of their process technologies internally, eventually patenting them because they cannot find equipment out among the vendors. For the most part, it advances its technologies through the efforts of its design and development engineers. A company which buys stator-winding machines from a company like Lincoln Electric, within the legally permissible limits, may be able to improvise the feed rates, the winding patterns, and other such features in order to enhance the original technological capabilities of the equipment. Similarly, an automotive company, which might spend several billions of dollars to retool for new models, might have to create advancement features for its basic tooling in order to reduce the overall tooling costs.

5. Abandonment phase. This last phase of the technology cycle is probably one of the most critical ones, because this is where decisions are made concerning the obsolescence of a particular technology. With the rapid discarding of existing technologies (product-based, process-based, information-based, and management-based), timing for new technologies is critical for winning in the business game, let alone for survival. Posturing for new technologies involves many interdependent variables, including the competition's product entry timing, the customer's ability to absorb and invest in new technologies, the technical knowledge and skills needed, the spare-parts management program, and the marketing and advertising channels available. Bad timing in prematurely abandoning a product could result in lost revenues on one hand, but on the other hand, waiting too long to abandon might also result in lost revenues because a customer has found a better alternative in competition. There doesn't seem to be an easy formula to make the selection—it is still an art—but it can be done with greater input of information from different areas of the company such as research and development, marketing, and production.

The TC fits within the broader *technology flow process* (Fig. 3.2), in which all technologies that are abandoned after being marketed, or inventions not commercialized to start with, are archived for a certain period of time, ranging from a few days to possibly several decades.

Central to the TC concept is an analysis of the external and internal environment factors in each of its five phases. Figure 3.3 is a summary of 12 such factors.

THE "TECHNOLOGY FLOW PROCESS"

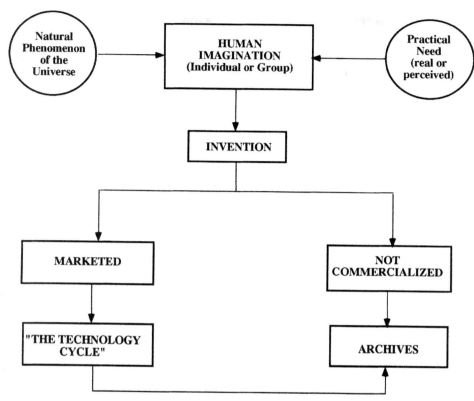

FIGURE 3.2   The *technology flow process* for any given technology. The technology cycle is shown in detail in Fig. 3.1. (*Copyright, Dr. D. J. Sumanth, 1988.*)

One or more of these factors may be more important than the others in each of the five phases, and for each type of technology under consideration. For example, in the acquisition phase of the TC, a developing-country enterprise, which has a limited foreign exchange to buy sophisticated equipment, will consider the economic factor to be relatively more important. Accordingly, the company may opt for *self-generation* rather than transfer of technology, by designing and building machinery of its own. In a multinational company, external factors, including the political and cultural factors, sometimes may become more dominant in relative importance than the economic and technical factors. For example, prior to the 1990s, some Japanese multinational companies were wise to offer their technology and know-how in India without demanding a 51 percent share, but rather settling for 45 percent, because of an understanding and appreciation for the political and cultural requirements prevalent and imposed by the Indian government. Thus, they gained entry to some vital markets well before others in the West did.

Factors affecting technology user	General meaning
Economic	Economic viability in the short term and/or long term
Social	Factors dictated by the social needs of people affected by the technology in question
Political	Factors imposed and/or controlled by the policies and directives of political system(s) in which the technology user functions
Ecological	Ecology-sustaining facets that emphasize balance between natural and synthetic or artificial systems
Nature-imposed	Factors strictly imposed and/or controlled by the natural systems of the universe in a predictable or unpredictable manner
Technical	Functionality at expected levels of performance, as defined by the user
Educational	Level of education and training
	Proficiency in skills acquired
Psychological	Emotional and mental states
	Behavioral characteristics
	Perceptual conditions
Personal health and safety–related	Factors directly affecting the health, safety, and general well-being of individuals who will be affected by the technology in question
Cultural	Overall, general impact of culture (national and corporate)
Moral, ethical, religious	Moral, ethical, or religious standards prevalence-mandated or informally accepted
Institutional	Factors created as a result of institutional bureaucracies, policies, and guidelines

FIGURE 3.3   External and internal environmental factors affecting the technology user.

All five phases of the TC are equally important. Depending on the type of product and technology, the magnitude of activities and analysis might vary with time. For example, in a multinational corporation that is operating in a highly oligopolistic market, where new product introduction is very common (such as in the personal computer market), the level of activity is usually high in the *advancement* and *abandonment* phases of the TC, on account of the need for rapid changes in technology, both at the product level and in the process stage. Technology management must be viewed as a broader umbrella for productivity and quality management. When total productivity is optimized, technology management takes on a more systemic meaning (Sumanth, 1984).

3.5   TEN BASIC TENETS FOR THE MANAGEMENT OF TECHNOLOGY (MOT)

A *tenet* is a principle based on observation, intuition, experience, and in some cases, empirical analysis. Ten (10) tenets are proposed next, as guiding principles for an enterprise to operate within a TC framework. They recognize that short-term treatments of any issue in general, and technology management in particular, are at best suboptimizations, and so, will not lead to more long-lasting solutions in adapting and advancing technology. Let us take some time now to discuss these principles in detail.

3.5.1 *Value diversification* is a poor substitute for MOT.

Value diversification refers to the improvement of stockholders' investments in a company through quick-fix solutions on paper, such as mergers, acquisitions, and other stock-enhancing strategies. Unfortunately, this traditional approach to value enhancement results in mostly short-term gains and long-term pains. Every company ought to identify core technologies and core competencies, and then hone them to get the most out of those for innovating products and/or services. When IBM acquired ROLM Corporation many years ago, IBM was trying to complement its core technologies in mainframe computers and personal computers with the core technology of ROLM, communication systems. Unfortunately, this did not work out very well, and IBM eventually sold ROLM. In the early 1980s, McGraw-Hill, whose core technologies are in publishing, books, journals, and related products, went into the personal computer business with Odyssey with a totally different core technology that didn't work, either. Another example of a questionable, huge investment was the Ford Motor Company's purchase in 1989 of the British automobile company Jaguar for $2.5 billion, rather than investing in next-generation technology (Luttwak, 1993). Because of Jaguar's old technology, poor workmanship, inferior quality, and outdated manufacturing processes, at least another $1.5 billion in investment has had to go into raising the company to only today's standards (Flynn and Treece, 1994). Acquisitions and mergers for the sake of financial leverage can backfire significantly if the core technologies involved are not inherently complementing to each other. MOT has a formal focus to nurture and cultivate core technologies and core competencies. As Prahalad and Hamel (1990) argue, "In the long run, competitiveness derives from an ability to build, at lower cost and more speedily than competitors, the core competencies that spawn unanticipated products." This is a painstaking, long-term process and not a quick-fix solution. A thoughtfully monitored technology management results in long-term gains, in terms of both financial considerations and product longevity.

3.5.2 Manufacturability must keep pace with inventiveness and marketability.

In industries in general, and manufacturing companies in particular, people in manufacturing functions often find themselves coping with increasingly aggressive marketing strategies and design strategies. Manufacturing in the United States is being troubled by intense competition from the Pacific Rim and European trading partners, who are developing new and better technologies and techniques to increase their advantage in product design and manufacturing process (Gold, 1994). Yet, in today's globally competitive marketplace, it is not only a necessity for manufacturability to be in step with marketing and design strategies but also a luxury, serving as an important weapon to chip away the market share of the competition. Timing in designing, manufacturing, and marketing products and/or services has become extremely crucial. This calls for what we call "modular technologies," which enable a company to have a tremendous flexibility to quickly package together innovative products and/or services to beat the competition, let alone to survive the fierce global competitive game.

3.5.3   Quality and total productivity are inseparable concepts in managing technology.

In the 1970s, here in the United States, productivity was of major concern, particularly after the 1973 Oil Embargo, and the ensuing Japanese "automobile attack." In the 1980s, after the famous NBC documentary, "If Japan Can, Why Can't We?," emphasis on quality reemerged with great ferocity and intensity. The *total quality management* (TQM) movement made quality a common prerequisite for ensuring competitiveness, even in the domestic markets. With the onset of the information superhighways in the late 1980s, and the rapidly changing global communication technology panorama, time has become a third crucial strategic variable in a company's drive to be competitive. The nontraditional *total productivity management* approach (Sumanth, 1981, 1984, 1995) to competitiveness forcefully argues that quality, price, and time are the three competitive dimensions which must be simultaneously created and monitored for companies to be long-lasting. Quality and total productivity are like two sides of the same coin or two rails of the same track. Companies that have excellent quality and competitive prices still cannot do well unless they can bring products of highest value to the marketplace in the least time possible. Information technologies have made it possible today to order products 24 hours a day from the luxury of one's home through the Home Shopping Network (HSN) and others, where the customer expects a rapid response rate. In fact, Thurow (1992) predicts that in the twenty-first century there will be high-tech and low-tech final products, but almost every product in every industry—from fast food to textiles—will be produced with high-tech processes. Therefore, product technologies and process technologies must be managed carefully to ensure that all these three dimensions of competitiveness are working together to enhance market share and profitability.

3.5.4   It is management's responsibility to bring about technological change and job security for long-term competitiveness.

We have seen, particularly in the last 5 years, that American management has gone on a downsizing binge in the name of streamlining and cost cutting. From 1987 through 1991, five million jobs in 85 percent of the *Fortune 1000* firms had been affected by this process (Cascio, 1993). Often, technological improvements have been associated with such downsizing. Unfortunately, this is a poor business strategy because it underestimates the employees' ability to manage not only existing technologies that their company has but also their creative capabilities to create and perfect new ones. Employees must feel that they have job security, particularly when they are responsible for suggesting and implementing new technologies. They feel betrayed after they spend hours of hard work designing a technologically advanced environment for greater competitiveness, only to find themselves victims of their own making. This need not be the case. Companies often spend millions of dollars trying to mitigate the negative effects of low morale, job dissatisfaction, and consequent low productivity, following a layoff or cost-cutting measure, right after a major technological change. A smart approach to managing technology is to look at the competitiveness challenge as a holistic one. The Japanese have been excellent in taking such a systemic view while managing all their basic technologies.

3.5.5 Technology must be the "servant," not the "master"; the "master" is still the human being.

Until recently, *we* used to be the masters of technology, our servant. *We* used to drive technology, but today we have become technology's servant, and technology is driving *us*. We believe that we have crossed a "technology threshold," whereby our response to technology has become one of catching up. Many companies are unable to cope with the dramatic changes taking place in the very nature of technologies. This, in turn, puts a company in a reactive posture, rather than a proactive one. Companies which are learning the art of managing new technologies have a better chance at being a technology master instead of a technology servant. The chaos that companies face today in responding to "rapid technologies" can be harnessed as a positive strategy to create opportunities for new products and/or services. The recent bill passed by the U.S. House of Representatives to deregulate the broadcasting and communications industry will no doubt spin some companies into utter chaos. On the other hand, the proactive ones will create technology strategies involving the creation of altogether radically new products and services. It will not be too long before we see integrated communication systems, combining technologies related to television sets, computers, VCRs, telephones, and fax (facsimile) machines. Cable companies will soon be in the computer business; and computer companies, in the telecommunication business. It is impossible to even conceive the extent of the technological integration revolution we will be facing even before we enter the twenty-first century. Our wristwatches might become microcomputer devices, working as remote-control units and information retrieval systems. We might see a series of technology thresholds bombarding us in the years to come, and every time we cross one of them, companies have an opportunity to convert technological chaos into economic opportunities.

3.5.6 The consequences of technology selection can be more serious than expected because of systemic effects.

This principle has major impact on the economic viability of the twenty-first-century organization, because we will be selecting multiple technologies with a rapidity that is hard to comprehend at this time. Product technologies will become obsolete in such short periods of time that they will resemble the toy industry, where the average shelf life of a product may be only one season, or sometimes only a month. We are already beginning to see personal computers fall into this category. In the early 1980s, when the personal computer was something new to all of us, the average shelf life was approximately 4 to 5 years for a model to become obsolete. Today, just 13 years later, the average shelf life has been whittled down to less than 1 year. By the time a company decides to update its PC technology to state of the art and acquire it, that technology has just about become obsolete. In anticipation of even greater obsolescences, companies will usually wait for newer models in both hardware and software. The rapid turnover in both of these categories of technologies makes it even more difficult to implement newer ones. However, the penalty for *not* updating can also be severe in terms of lost revenues. As the technology cycle demonstrated previously in this chapter, selection of any technologies must be carefully planned and executed, taking into consideration as many system variables as possible.

3.5.7 Continuous education and training in a constantly changing workplace is a necessity, not a luxury.

Companies have traditionally slashed the education and training budgets during times of economic downturn. Today, this would be a foolish strategy, because most employees know how to use process and information technologies to the fullest extent. Having spent millions, and sometimes billions of dollars in such technologies, it would be most uneconomical not to get the most out of these expensive technologies. Sometimes, a million-dollar piece of equipment has a 20 percent downtime, costing hundreds of thousands of dollars in lost revenues, simply because operators and engineers have not been trained in all aspects of its operation. The more the education and training for managing technologies, the greater the utilization rates would be, and hence, the greater the economic leverage. For example, in a multinational bank, such as Citibank, the technology group strives hard for customers to do worldwide banking in any one of 14 languages. The company spends much time and energy educating and training the personal bankers, the customer service representatives, and others, in an effort to offer their clients the ever-increasing portfolio of products and services in a competitive manner. Citibank creates its own technologies for almost everything, ranging from its ATMs (automatic teller machines) to its banking terminals. Because of the sophisticated nature of today's technologies, continuing education budgets keep rising in well-managed companies. In the years to come, we can only see greater rates of such continuing education, to enhance knowledge and skills in using technologies to their fullest extent possible.

3.5.8 *Technology gradient* is a dynamic component of the technology management process, to be monitored for strategic advantage.

The term *technology gradient* refers to the technical advantage an enterprise enjoys with respect to its licensees and its competitors. Normally, most sensible multinationals maintain a sufficiently high technology gradient with respect to their licensees, particularly if the latter are even remotely associated with a product line that competes anywhere in the world. This is particularly true when the technologies are radically new, for example, biotechnology, global networking technology, etc. Technology gradient, which is the subject of another chapter in this handbook, is a powerful concept for managing the technological advantage that the company enjoys with respect to its competition worldwide. Briefly, a company monitoring its technology gradient can be in one of four postures: technology leader, technology follower, technology yielder, or technology loser. Depending on the technology advantage a company wishes to enjoy, it must consciously position itself as one of these. Obviously, the company would want to be in one of these areas, depending on which phase of the technology life cycle it is in. For example, if a company is in the declining phase of a product life cycle, no matter how hard it tries to be a technology leader in that phase, the returns on its technological investments will be so marginal that it is better off being a technology yielder during that stage for short periods of time. Ideally, however, a company must overlap its technologies so as to minimize the technology yielding positions and maximize the technology leadership positions.

3.5.9 The *RTC factor* must be carefully analyzed and meticulously monitored for gaining the most out of any technology, particularly a new one.

The *RTC factor* refers to the magnitude and nature of resistance to change. Unfortunately, very little is known about the process of the RTC factor, and the rational means to minimize it. At the same time, however, we now know that a high RTC factor can lead to work slowdowns, poor employee morale, high maintenance costs, and even serious sabotages. Management has to recognize that a creative, lively workforce is better than stagnant, high-priced technology. Research shows that when new technologies are implemented, "total productivity" at first drops because of the *natural* response of employees—resistance to accept the new technologies as viable means—before it picks up again. Keys (1995) furthers this by stating that often

> The competitive threat of a new technology-based business paradigm (and its early implementation success) does often, unfortunately, prolong the last gasps of life in the old technology because it temporarily forces a really serious competitive pressure on the old technology. Under this real threat, it is amazing how much excess effort, costs, and inefficiencies can be wrung out of the last defense stand of the "old guard." It is usually a wasted effort, which delays the acceleration of the inevitable new "S" curve and prolongs the agony of the old.

However, as employees get used to the new technologies, their acceptance rate improves, their attitude to the technology becomes more matter-of-fact, and their proficiency and skill rate also return to normal levels. A proactive approach to minimizing the RTC factor is to explain the benefits of new technologies to both the company, and the employees themselves.

3.5.10 Information linkage must keep pace with technology growth.

As pointed out during previous discussion, information networks are evolving so rapidly that unless companies take advantage of linking up to such networks, they lose opportunities for new revenue streams. For example, companies that quickly capitalized on the accessibility to the Internet increased their market share through an exposure of their products and/or services to millions of people around the world. We barely understand the potential of the Internet through the World Wide Web (WWW). Within a company, it has become an absolute necessity to keep all the employees informed of the latest technological developments within their own company so that unnecessary duplication of costly effort is avoided, and product changes and client updates are offered on an on-line basis so that customer responsiveness can be in real time. Time lags can cause serious miscommunications, particularly with multinationals. For example, if a component is eliminated in a product and this information is not communicated to the company's worldwide spare-parts inventory system, retail clerks somewhere in Indonesia or Taiwan may still be carrying the part on shelves unnecessarily, increasing their inventory carrying costs. Companies like Caterpillar and International Harvester maintain global inventory management systems so efficiently

1. *Value diversification* is a poor substitute for the management of technology (MOT).
2. Manufacturability must keep pace with inventiveness and marketability.
3. Quality and total productivity are inseparable concepts in managing technology.
4. It is management's responsibility to bring about technological change and job security for long-term competitiveness.
5. Technology must be the "servant," not the "master." The "master" is still the human being.
6. The consequences of technology selection can be more serious than expected because of systemic effects.
7. Continuous education and training in a constantly changing workplace is a necessity, not a luxury.
8. The *technology gradient* is a dynamic component of the technology management process, to be monitored for strategic advantage.
9. The *RTC factor* must be carefully analyzed and meticulously monitored for gaining the most out of any technology, particularly a new one.
10. Information linkage must keep pace with technology growth.

FIGURE 3.4   The 10 basic tenets of technology management.

that within 24 hours they can have a part made available to any retailer around the world. In such situations, this tenet has even greater relevance and respect. These 10 tenets are summarized in Fig. 3.4, for a ready reference.

3.6 CONCLUSIONS

As companies are increasingly faced with decisions impacting not only their competitors, but the global market as well, a greater need will arise for swift, yet calculated decision making on the part of top management. With that added pressure and responsibility, economic enterprises will be thrust into the dilemma of choosing between two basic strategies: (1) achieving the "ideal" goal at a high sacrifice level or (2) going for the most practical solution at the least cost possible in terms of time and effort. Unfortunately, when the choice is for the latter strategy, the long-term outcomes often convince the enterprise that it should have gone with the first one, even if all of it could not be achieved. Companies might be wise to adopt a motto such as "Excellence is our starting point and perfection our ultimate goal." Even though it is not practical to define "perfection," the mental target to achieve it helps to develop a set of guiding philosophies that are well thought out, long-lasting, and consistent, even when uncertainty looms large.

This chapter has attempted to provide a technology management perspective that is centralized around a systems approach to management. Technology can no longer be considered as just fancy machines and computer chips. It is a vital, growing agglomeration of earthly resources fashioned by human ingenuity, and steered by a vital force, namely, a systems perspective. This is due to the rapidly changing, technological age we live in today, in which every technology that we adopt or discard has a systemic effect on the process of an organization. Great aims and preparations are needed as we

PROS

1. Domestic and international competitiveness can be significantly improved with improvements in quality, total productivity, and unit costs.

2. Organizational structures based on integrating technology requirements with sociohuman needs will have greater resilience to withstand generally unforeseen market forces.

3. Opportunities to treat causes rather than symptoms in a TSTM will prevent major diseases in an organic enterprise.

4. Uniformity in technology management at strategic and tactical levels will promote greater standardization and simplification with inherently greater potential for constant improvement.

5. When downward economic surges occur, the TSTM has a greater probability of minimizing the maximum "pain" of an enterprise.

CONS

1. Simultaneous practice of all the 10 basic tenets of technology management requires a bold, decisive, consistent direction on the part of top management.

2. The adaptation of the TSTM becomes a greater challenge in multinationals and autonomous companies than the nationally based, relatively more centralized firms.

3. Extreme uncertainty in one or more of the twelve external and internal environment factors (of Fig. 3.3) may cause top management to abruptly abandon the TSTM for a path of least resistance.

FIGURE 3.5   The pros and cons of the total systems approach to technology management.

enter the exciting, yet challenging twenty-first century, to avoid the temptations of suboptimization and its costly consequences. Although a systems approach to technology management is prescribed in this chapter, the effort needed to implement it, in terms of time, money, and patience, is great and often arduous. However, the benefits of a successful implementation of this approach greatly outweigh the costs, and make the time and effort spent well worth it. The advantages and disadvantages of this approach are highlighted in Fig. 3.5. It is hoped that enterprises would revisit the following basic points when practicing technology management:

1. Technology is not merely machinery and advanced electronics, but rather any means to accomplish a purpose in a broad, holistic sense.

2. The management of technology is not an exclusive practice, independent of other variables, but rather, is a complex, interconnected systemic process.

3. Suboptimal solutions are created when an enterprise focuses solely on the easiest, and most logical remedy to the problem, failing to see a broader proposal of worthy solutions from a systemic standpoint.

At a time when product innovation strategies in Japanese companies like Sony, Nippon Instrument Manufacturing, and Pioneer are becoming more and more integrative between technology and humans (Tanaka, 1987), it is increasingly evident that enterprises have no choice but to develop more systems-oriented thinking to manage technologies for maintaining and enhancing their competitive position in the ever-complex, ever-growing globalism.

3.7 SUMMARY

As we approach the twenty-first century, there are rapid changes taking place in social, political, economic, and intellectual atmospheres that are characterizing today's complex and sophisticated companies with much shorter product life cycles than ever before, international market oligopoly, and highly diversified, vast knowledge bases from which to draw. Multinational companies are, and foreseeably will be, the prime channels for free enterprise in the near future because of the global business dynamics. This chapter has proposed a conceptual framework for a total systems view of technology management for companies and organizations, especially those competing in the global market. The technology cycle concept was introduced within this framework, to clearly identify the basic elements of technology management. Also presented was a technology flow process, to give companies a panoramic view of technology management as they plan and develop their strategic and tactical goals for competitiveness in domestic and world markets. The "pros" and "cons" of this systems approach were argued on the basis of 10 tenets of technology management.

3.8 REFERENCES

Bienayme, A., "The Dynamics of Innovation," *Internatl. J. Technol. Management,* **1**: 133–160, 1986.

Cascio, W. F., "Downsizing: What Do We Know? What Have We Learned?," *Acad. Management Executive,* 95–104, 1993.

Eckstein, O., et al. *Report on U.S. Manufacturing Industries,* Data Resource and International, 1984.

Flynn, J., and J. B. Treece, "Is the Jinx Finally off Jaguar?" *Business Week,* p. 62, Oct. 10, 1994.

Focus "The 21st Century Challenge: Is America Ready for it?," National Center for Manufacturing Sciences, 1991.

Gold, D. M., "Overcoming the Competitive Barriers Facing Smaller Manufacturers: A Discussion of the Need for the Manufacturing Extension Partnership," *Management of Technology IV, Proceedings of the Fourth International Conference on Management of Technology,* Miami, Feb. 27–March 4, 1994.

Gross, N., "A Wave of Ideas, Drop by Drop," *Business Week Special Report on Innovation in America,* pp. 22–30, 1989.

Hayes, R. H., and S. C. Wheelwright, *Restoring Our Competitive Edge: Competing through Manufacturing,* Wiley, New York, 1984.

Johnson, W. B., "Global Work Force 2000: The New World Labor Market," *Harvard Business Review,* (2):115–127, 1991.

Keys, L. K., "Continuing Technology Management Challenges for the Automobile Industry," *Technol. Management,* **1**(6):233, 1995.

Luttwak, E. N., *The Endangered American Dream,* Simon and Schuster, New York, 1993, 248–255.

Malaska, P., "Environmental Problems of Modern Societies," *Internatl. J. Technol. Management,* **2**(2):263–278, 1987.

Peters, T., *Thriving on Chaos—Handbook for Management Revolution,* Knopf, New York, 1987.

Port, O., "Back to Basics," *Business Week Special Report on Innovation in America,* pp. 14–18, 1989.

Prahalad, C. K., and G. Hamel, "The Core Competence of the Corporation," *Harvard Business Review,* pp. 79–91, May–June, 1990.

Schmookler, J., *Invention and Economic Growth,* Harvard University Press, Cambridge, Mass., 1966.

Schonberger, R. J., and E. M. Knod, *Operations Management—Improving Customer Service,* Irwin, Boston, 1991.

Scott, B. R., and G. C. Lodge, *U.S. Competitiveness in the World Economy,* Harvard Business School Press, Cambridge, Mass., 1984.

Shapira, P. *"Modernizing Manufacturing: New Policies to Build Industrial Extension Services,"* Economic Policy Institute, 1990.

Stephens, J. C. *Managing Complexity: Work, Technology Resources and Human Relations,* Lomond Publications, Mount Airy, Md., 1977.

Sumanth, D. J., *Productivity Engineering and Management,* McGraw-Hill, New York, 1984, 1985, 1990, 1994.

Tanaka, Y., "A Proposal for a Product Innovation Method through Fusion Strategy in the Intelligent Computer Aided Management (ICAM) Age," *Technovation,* **6:**3–23, 1987.

Tenner, A. R., and I. J. DeToro, *Total Quality Management: Three Steps to Continuous Improvement,* Addison-Wesley, Reading, Mass., 1992.

Thurow, L. C., "The New Economics of High Technology," *Harper's Magazine,* **284**(1702):15–17, 1992.

Thurow, L. C., *The Zero Sum Solution,* Touchstone Books, New York, 1985.

Utterback, J. M., "Innovation and Corporate Strategy," *Internatl. J. Technol. Management, 2* (2): 263–278, 1986.

CHAPTER 4

THE TECHNOLOGICAL BASE OF THE COMPANY

Aaron J. Shenhar
University of Minnesota
Minneapolis, Minnesota

Paul S. Adler
University of Southern California
Los Angeles, California

4.1 INTRODUCTION

Technology-based organizations are constantly confronted with dynamic and unpredictable changes in markets, products, and technologies. How do they stay competitive in a rapidly changing environment? Obviously, such companies compete on the basis of their technological strength, but what is this strength? And what makes one company stronger than the other in terms of technological capabilities? The company's physical assets and its financial resources are obviously parts of the answer, but they are hardly sufficient. The success of a technology-based company depends crucially on what we call its *technological base,*[1] namely, its ability to exploit technology as a core competency, to invest in future technology, to incorporate better technology in its products and services, and to do so in a shorter time period, with lower costs, and with better performance than competitors.

What, therefore, exactly is the technological base of a company? In this chapter we define and discuss the technological base concept and its relevance to assessing the technology-based company's strength. Several examples of successful and unsuccessful firms illustrate the scope of our concept:

- When IBM finally decided, in the early 1980s, to enter into the personal computer (PC) business, it certainly had the skilled computer and electronics people. But a key ingredient was missing: IBM realized that it did not have the appropriate structure for this new product line. An effective PC business would need to be much more agile than most of IBM's core businesses. So IBM structured the new division for the PC business as an *independent business unit,* more autonomous than most of IBM's divisions from corporate control. This allowed the unit to make the essential quick decisions and develop its new product exceptionally fast. Nevertheless, IBM found it necessary to use external help, relying on a tiny company called Microsoft, to develop the complementary operating system software for its new computer.

- When Swiss watchmakers faced the introduction of digital watches by American and Japanese companies in the early 1970s, most discovered belatedly that they were not prepared for the new electronic era. Most Swiss watch companies knew very little about the technology of either integrated circuits or digital displays embedded in digital watches. Moreover, they had neither the right organizational structure for developing electronic devices nor the appropriate project management processes and decision-making procedures to deal with the much faster pace of technical, product, and market change. The world preeminence of the Swiss watch industry was destroyed in 10 years.

- More recently, following the demise of the Soviet Union and the end of the Cold War, the American defense industry was left with a vast surplus of engineering design, systems development, and production capabilities. Only a few companies were able to adapt quickly to this change; many others did not. Some companies stayed focused on defense activities, while undergoing considerable downsizing and turmoil. The companies that succeeded often found it necessary to establish joint ventures with commercial sector partners who could help them adapt their defense-related products and technologies to new types of customers. A notable example was GM-Hughes Electronics. The firm was able to exploit its satellite communication capabilities in a joint venture with Thompson-RCA and move into the consumer market of small-dish "direct TV."

As these examples show, whether a technology-based company is contemplating a strategic change or evaluating its existing strategy, it must critically examine its capacity to exploit change. The technological base concept is a framework for assessing this capacity. In the following sections we will define the elements of a company's technological base and identify some key managerial issues for evaluating and reshaping it.

4.2 COMPONENTS OF THE TECHNOLOGICAL BASE

The *technological base* of a company is the ability of an organization to develop an ongoing stream of new products which meet the market needs, to manufacture these products while maintaining appropriate levels of quality and cost, to develop or adapt new technology to meet future needs, and to respond promptly to unexpected competitive moves or unforeseen opportunities. This functional definition of the technological base implies that in addition to caring for its current products, processes, and projects, the organization must look into the future.[2] Among other things, it must consider whether it is technologically equipped to sustain a competitive advantage over a period of years. Is it able meet or initiate new opportunities and threats created by its environment and its competitors, and is it able to change and adapt quickly to new situations and circumstances?[3]

What kinds of questions should managers ask themselves as they attempt to assess their company's technological base? Our framework groups the elements of this assessment into five distinct but interacting components. Each contributes to the organization's technological capability in a different way (see Fig. 4.1):

- *Core technological assets.* This is the technical—and most obvious—part of the technological base. Such assets are the heart of the organization's core competencies.[4] They constitute the set of technologies embedded in products and processes that are key to the company's present and future competitiveness.

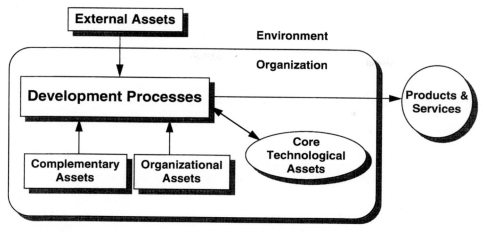

FIGURE 4.1   The technological base.

- *Organizational assets.* These are the factors that allow the firm to create and exploit new technologies. They include five elements: the skill profile of the employees and managers, the procedures for decision making and information sharing, the organizational structure, the strategies that guide action, and the culture that shapes shared assumptions and values.

- *External assets.* These are the links between the firm and its environment—a network of connections the organization establishes with the external world. They include relations with current and potential partners, rivals, suppliers, customers, professional associations, research and educational institutes, consultants, political actors, and local communities.

- *Development processes.* Two key processes[5] are critical to technology-based firms. First, the product and process development process generates the actual output of the company and creates value for customers. Second, and fueling the first, the technology development process creates the next generation of technologies which eventually become part of the core technological assets.

- *Complementary assets.* Even successful technology-based companies need more than technology to compete effectively. Complementary assets may be needed in marketing, distribution, manufacturing, field service, or information systems.

To illustrate these components and analyze the key managerial issues involved in each, we will refer to cases from companies who have successfully utilized elements of their technological base and achieved competitiveness in the face of dynamic change.

4.3   CORE TECHNOLOGICAL ASSETS

The company's core technological assets reflect the collective learning in specific technical areas and the resulting reservoir of technology know-how across various business units. They depend on the accumulated experience in developing technologies, as well as exploiting them in new products and processes.

Superior technological assets are based on a commitment to working across internal boundaries. Technological assets are not derived from a single product, nor are they normally confined to a single business. They should be nurtured and exploited across many functions, product lines, and business units.

One of the most successful companies in focusing and exploiting its core technological assets is Sharp Corporation of Japan. This company has defined itself—among other things—as an optoelectronics company, centered on optosemiconductors which act as converters between light and electricity. It developed the world's largest solar cells for a lighthouse in the early 1960s, and further research led to the development of solar cells for satellites in the 1970s. Sharp also developed the technology of electroluminescent (EL) displays and liquid crystal displays (LCDs).[6] LCDs were incorporated into what was then "the world's thinnest calculator" made by Sharp in the early 1970s. This development led to a stream of products, all based on the same core assets. Sharp introduced an alphabetical LCD for calculators in the late 1970s and a large monochrome LCD for personal computers and word processors in the 1980s. Using a new thin-film transistor (TFT) active-matrix technology, the company then developed a 3-in color LCD with faster response and a higher picture quality, and a 14-in color TFT LCD. On the basis of these LCDs, the company introduced in the early 1990s a number of first-in-the-world products, such as the 110-in color LCD video projector, a 8.6-in wall-mount LCD monitor, and the ViewCam camcorder with a 4-in color LCD monitor.

Canon and 3M also exemplify this concept. Canon has developed its position in miniature mechanics and fine optics for its camera business. It combined this capability together with a separate competence in microelectronics to develop a myriad of business lines and additional products, such as electronic cameras, videocameras, ink-jet printers, laser printers, faxes, calculators, and copiers. And the Minnesota-based 3M Company exploited its capabilities in coated abrasives, adhesives, and tapes, in a wide variety of products—well beyond its famous Post-it notes.

A detailed assessment of an organization's core technological assets should encompass both product and process technologies. The first step is to develop a list or "map" of relevant technologies.[7] In diversified businesses, such mapping may confront two challenges. First, the organization must find the appropriate aggregation level for the many discrete relevant technologies. Strategic planning cannot encompass more than a dozen or so major groups. Second, and more difficult, is the challenge of identifying the right dimensions along which to aggregate. The best mapping is rarely given by the academic disciplines (mechanical vs. electrical, etc.) or by the organizational chart (one business or function vs. another). It typically takes several iterations before an organization can develop a technology map that is neither too detailed nor too aggregated, and one that is neither too functional nor too product-oriented. But these iterations are extremely valuable not only for the map they produce but also for the common understanding and vocabulary they create between technologists and managers.[8]

One dimension of aggregation that has proved useful is the distinction between base, key, pacing, and emergent technologies. Base technologies are necessary for being in the game; however, they do not provide any competitive advantage. All industry players have equal access to these technologies. Key technologies are critical to competitive advantage since they offer the opportunity for meaningful product or process differentiation. Pacing technologies are not currently deployed in the industry but have the proven potential of becoming key technologies. And emerging technologies are on the horizon, as yet unproved, but potentially important.[9] Pacing and emerging technologies are extremely important. The most strategically significant technologies might be ones for which the organization is not currently generating any project proposals.[10]

The state of the firm's core technological assets can be evaluated along two dimensions. First, and most obviously, each business needs to assess its technological strengths and weaknesses relative to the external world—to its competitors and to the evolving technological frontiers. Here the technology life-cycle notion is very useful.[11] In some industries (such as chemicals), patents are a powerful competitive lever, and an assessment of the firm's relative patent position is needed. In other industries, other indicators of competitive technological standing need to be developed. Second, the organization needs to assess its ability to deal with the interdependencies between technologies and its current and projected product portfolio. Technologies developed should become the entire company's asset, and potentially contribute to every relevant product or process.

4.4   ORGANIZATIONAL ASSETS

Core technological assets are the tip of the iceberg; it is the firm's organizational assets that create these competencies and position the company to exploit them to competitive advantage. Organizational assets can be divided into five key elements: skills, procedures, structure, strategy, and culture.

4.4.1   Skills

To assess its skills, an organization should have a clear map of both the managerial and the technical skills of managers, engineers, and scientists, as well as those of technicians and nontechnical personnel.

Two dimensions of the technical skill mix need careful study:

1. What *types* of skills does the organization have? To which professional groups do the staff belong? What types of degrees do they possess?

2. What *levels* of skills can the firm tap in these domains? What is the mix of educational levels? Of experience levels? Of real expertise?

Types and levels are both difficult to assess. As with technological assets, the more challenging part is classifying the relevant types of skills. Typically this cannot be determined from the organizational chart or from the personnel classifications. These may serve as a first cut; ultimately, however, the organization needs to refer to its strategic direction and the external environment to know whether to classify engineers as mechanical versus electrical, or product versus process designers.

The next step in assessing technical skills is to determine what types and levels will be needed in the future, and then act on the current/needed gap. Texas Instruments, for example, one of the leading semiconductor companies today, did not invent the transistor, nor did it have any previous knowledge in semiconductors. However, during its early years, TI decided to move from being a geophysical service company to becoming a transistor manufacturer. As part of its new strategy, the company found it necessary to hire a group of experts, many of them from Bell Labs, where the transistor was originally invented. It also sent its executives and researchers to attend seminars on transistors offered by Western Electric as part of a licensing agreement.[12]

Of particular importance is an assessment of the management team's technical skills and conversely, the technical staff's managerial skills. Managers lacking the needed technical skills can remedy the situation by some individual knowledge building or by finding a technologist capable of translating between the business and technical worlds. The technical staff's managerial skills are critical to an organization's technical performance. Does the organization have sufficient people to provide leadership to its engineers and scientists? Does it have experienced project managers who have sound technical judgment and intuition, who can make the needed tradeoff decisions, and who can lead cross-functional teams and advanced complex projects? Does it have managers on the factory floor who understand how to implement new manufacturing automation? The company must also look for a balanced mixture of "critical functions" to the innovative process. Does it have enough idea generators, gatekeepers, champions, project leaders, and sponsors.[13] The lack of one or more of these functions may seriously reduce the chance of successful innovation.

Finally, an assessment of the skills area must consider the skill development capabilities of the company. Does the company provide employees with sufficient opportunities, guidance, time, and other resources to develop their skills and knowledge?

This assessment should be made for both the technical and managerial fields. Does the company help the professional staff to deepen their knowledge in their original skill area? Does it help them obtain advanced degrees? And does it encourage them to move into new areas? And for managers, does it help them develop the proper mix of technical, human, administrative, and strategic skills?[14]

4.4.2 Procedures

Procedures are the managerial routines and criteria guiding the way in which the organization is making decisions and disseminating information. Decision-making procedures most relevant to the technological base involve planning, control, and problem solving. They either support the organization's main business processes or are related to other internal service or administrative processes. Planning includes selecting technologies, products, and markets, as well as organizing, staffing, budgeting, and resource allocation. Control procedures include personnel evaluations, organizational performance criteria, and project control mechanisms. Problem-solving procedures include both routines for technical problem solving and mechanisms for resolving management conflicts. Information dissemination procedures govern the flow of technical and business information. Any assessment should encompass all these types of procedures.

The difference between good and badly designed procedures influences greatly the ability of the organization to develop and exploit technology. Take a ubiquitous example: meetings. People spend a lot of time in meetings, starting from the board and top-management meetings, and down to the working-level meetings. Effective meetings concentrate on the real issues, are efficient and productive, enhance motivation, empower people, and increase cohesiveness in the organization. Meetings will be perceived as a waste of time when nothing significant gets done or if things get done too slowly. The formal and informal procedures for preparing and conducting meetings can contribute greatly to any organization's effectiveness. Managers may consider whether the issues to be discussed are equivocal and ambiguous enough to warrant the use of "rich media" such as face-to-face meetings, or are perhaps straightforward, and even large amounts of information can be exchanged through less costly mechanisms such a periodic written reports.[15]

The key criterion for assessing procedures is whether they facilitate or impede organizational learning. A project selection procedure, for example, can be designed to encourage the right mix of creative bottom-up initiative and rigorous review, or alternatively it can become a bureaucratic deterrent, creating unnecessarily formalistic hurdles and politicized project promotion games. Texas Instrument's OST system is a case in point. Its *objective, strategies, and tactics* (OST) system is a systematic, hierarchical procedure to plan for the future and encourage innovation. It was intended to focus attention on the more important but less urgent activities that ensure an organization's future: long-term planning, new-product development, and new major strategic thrusts.[16] But in later years, it degenerated into a bureaucratic minefield. The degree of formality and detail was not an issue; the way it was used was.

4.4.3 Structure

Structure in the present context is the way businesses organize their activities into divisions, plants, and subunits. It is how they divide the work among functions, departments, and projects, as well as between local and overseas operations. The structure of the organization has a great impact on its ability to meet new challenges in technological development and project management.

The basic dilemma in organizational structure is between (1) the need to keep people who are focused on the same types of tasks or disciplines together, to ensure that they remain up-to-date in that functional field, and (2) the need for cohesiveness in key business processes that run across different functions, to ensure that projects do not suffer for lack of timely information or appropriate incentives. Central determinants of the appropriate structure are the rate of change of the functional knowledge base (faster change indicating great reliance on the functional dimension), subsystem interdependence in the projects, and the duration of the project assignments (with the last two indicating greater reliance on the product dimension).[17]

Organizations need to attend to both dimensions simultaneously. Formal forms for doing this are called *matrix* structures. There are many organizational and behavioral barriers to the effective operation of a matrix structure,[18] and many managers are therefore reluctant to adopt this form. However, whether formalized or not, matrix forms are unavoidable; and instead of backing away from the matrix, organizations should accompany its introduction with complementary adaptation. In fact, most cross-functional teams in product development projects are based on the matrix concept.[19] New management skills and procedures are often needed, as well as modified compensation and reward systems, and greater strategic consensus and cultural integration.

One could also include under structure a second component: geographic location and proximity of groups and people. The physical layout of the organization plays a key role in enhancing or impeding the informal flow of information between groups, both within and across functional units.[20] Indeed, some firms achieve the goals of the matrix by having an exclusively functional chart, but dispersing the functional people into co-located product groups. Here the informal communication created by co-location balances the formal communication channels of the reporting structure. An example is the Ford Motor Company, which decided in the early 1990s to combine its previously scattered product development efforts into multifunctional development centers. To enhance efficiency, each development center will be responsible for developing one type of car; this type will be sold by all Ford dealers across the globe.[21]

There is typically no one correct organizational design. Every design has its strengths and weaknesses. The key aspects of the influence of the structure on performance are information and incentives; namely, is the structure facilitating or impeding the communication flows, and is it creating useful or counterproductive incentives?

4.4.4 Strategy

Strategy is the way the organization attempts to outperform its competitors. It is those things at which it tries to excel. Different companies pursue different strategies, such as being first in the market, imitating successful leaders, product customization, or delivering the lowest cost. Each strategy requires a distinct alignment of other organizational assets and a long-term process of learning.

Effective strategies are hard to formulate and even harder to implement. The organizational processes of formulating and implementing strategy, and the substantive content of that strategy, are typically deeply embedded in the organizational fabric of the business. As a result, they are typically not amenable to particularly rapid change—top management's desires notwithstanding. It is true that some organizations put a premium on strategic flexibility, and in some environments such flexibility may be particularly valuable.[22] However, flexibility is only one criterion among others for assessing the strategic element of the technological base.

We can identify at least two other criteria. First, an effective strategy is one that has been effectively "deployed" down into the various functions and layers to ensure an operative fit between different units' priorities. Many firms have learned how to elaborate explicit business strategies; only a few of them have learned, however, how to formulate and implement strategies for specific functions such as engineering or manufacturing, and even fewer have learned the art of cross-departmental strategizing for managing critical assets such as technology and human resources. Obviously, the technology strategy subelement is of particular importance to the technological base, but it will be of little use unless it is well integrated with both the business strategy and the relevant functional strategies. Functions that are most relevant to defining and implementing the technology strategy are R&D, manufacturing, information systems, and marketing.[23]

Apart from flexibility and fit, a final assessment criterion is worth highlighting. Is the form of these strategies that of a detailed itinerary or a compass heading? The itinerary form can be an effective guide only if the environment is stationary and well known.[24] Fewer and fewer industries offer such easy environments. In a dynamic environment, characterized by a lot of uncertainty, flexibility may be very valuable; but the organization still needs to trace substantive lines of development for itself, and in a dynamic environment these lines of development can be specified only as an overall compass heading. This requires building real insight into the nature of the organization's current and projected capabilities and into its fit with the evolving market needs. The need for such insight explains the value of "strategic intent"[25] and business focus—a clear sense of what the organization needs to master and what it can afford to let others do for it—as opposed to unconstrained and unrelated diversification. Strategic focus and the quality of insight are key criteria for evaluating the strategy component of the technological base. The greater the focus and insight, the more effective is the strategic planning and the longer is its time horizon.[26]

Sharp Corporation's technology strategy is a good example of such insight. Their strategy was directed at investing in technologies which would serve as the "nucleus" of the company in the future. Such technologies should have powerful relevance to many products. For example, Sharp kept investing heavily in LCD factories in the 1990s, believing that this technology could be leveraged into several end products in the future. Conversely, Sharp avoided direct random-access memories (DRAMs), capacitors, and resistors as commodities readily available from competitive suppliers and focused on specialties such as masked read-only memories (ROMs). Once it chose to develop a technology, Sharp committed to it for the long run. It pursued, for example, LCD research throughout the 1970s, although the market for LCDs did not take off until the 1980s. Similarly, it continued research in gallium arsenide (GaAs) laser diodes long after most competitors had abandoned their research in this area. This strategy paid off with the big market growth for CD players in the 1980s. Sharp became the world leader in supplying the laser diode "optical needles" for the growing market of compact disks.[27]

4.4.5 Culture

Culture is usually the most difficult organizational asset to evaluate; but it is also the most powerful. *Culture* is the shared and relatively enduring pattern of basic values, beliefs, and assumptions in an organization.[28] It is "the way we do things around here" and the collection of norms, behaviors, and values that guide action without managerial intervention. Culture provides implicit guidance for desired behavior; it frames right and wrong, and highlights the things for which individuals are most admired and recognized.

The culture at 3M, for example, is one of continuous innovation. In this organization's environment, introducing new products is the most important criterion for success and accordingly, the most valued activity. The company worships stories of new-product successes and how new innovations came about. Stories such as the famous Post-it legend are diffused widely and shared from one generation to the next. The company conducts continuous training of its engineers and managers in the characteristics of innovation and how to enhance successful innovations. At 3M, even if managers tried to change strategy and deemphasize innovation, the culture would probably sustain innovative behavior for many years.

Culture can work either for or against competitive advantage. Culture was the main driving force behind the development—and failure—of Texas Instruments' home computer in the late 1970s. In this case, the strong "know it all" and "not invented here" (NIH) culture pushed the company into developing a computer around its failed 9900 microprocessor. This 16-bit microprocessor was introduced ahead of its time, when the market was just tasting the previous generation of 8-bit microprocessors. When Texas Instruments was unable to leap-frog the rest of the industry with this microprocessor, it launched the development of a home computer which incorporated the poor-selling microprocessor. Although many engineers advocated the use of a cheaper microprocessor, TI's culture insisted on using the more expensive, and not-yet-utilized 16-bit microprocessor. After enjoying some initial success in the early 1980s, TI was unable to stay competitive against the lower-priced VIC-20 computer, made by Commodore. This rival product was based on a lower-cost 8-bit microprocessor. The outcome for TI was painful and resulted in the greatest loss in the company's history.[29]

In assessing culture, Schein's approach is particularly useful.[30] He distinguishes three layers of culture. First, the visible and most tangible layer is made up of the artifacts of the company's culture such as pay scales and office space. Second, underlying these artifacts are the normative values of the organization. Third, beneath these values, and typically invisible, are shared assumptions about how the world works. An assessment of the cultural element of the technological base should explore all three layers.

In identifying and trying to influence culture, managers should be aware of the six areas in which culture can be recognized:

- *Vocabulary*—words and phrases typically expressing important values and ideas. Sony, for example, by emphasizing its leadership is using the phrase "BMW," meaning "beat Matsushita whatsoever."

- *Methodology*—the established norms under which an organization gets things done. For example, do we use internal task forces or external consultants?
- *Rules of conduct*—the unwritten do's and don'ts that guide day-to-day behavior, from the appropriate dress code, to situations in office protocol, to decision making and etiquette.
- *Values*—such as the belief in being the best, the belief in the importance of people as individuals, the belief in superior quality and service.
- *Rituals*—such as types of announcements, holiday parties, or the way people are introduced.
- *Myths and stories*—who are the heroes, and who is the butt of the jokes? What do people from the lab talk about after a few drinks together? Are the stars acknowledged to be those making positive contributions, or are they people beating the system? Is it good or foolish to work hard and long hours?

4.5 EXTERNAL ASSETS

External assets are a crucial component of the organization's technological base. The ability of the organization to find, build, and exploit technology depends on its network of partnerships, contracts, and business relationships. Equally important are the informal linkages established by employees with their peers and colleagues outside the organization. This includes networks of professionals, managers, and executives in local, national, and international associations.

A first category of external assets is composed of downstream links to customers. Companies should ask: How many customers do we know that can provide ideas for new products or new uses? And how well does the organization learn from these lead customers?[31]

A second category is composed of upstream links to materials and component suppliers, equipment vendors, and subcontractors. The organization should assess whether it has appropriate links with the best people at the best companies and whether these relations are sufficiently collaborative to sustain a long-term two-way commitment.

Links to competitors can also be critical. They are a source of knowledge, experience, and lessons to learn. They also fuel the organization's momentum by providing drive and excitement. Porter discussed the role of "good" competitors in improving competitive advantage, industry structure, market development, and entry barriers. Through these means, a firm's competitive base can be enhanced.[32] It is therefore important to assess the quality and configuration of competitors as well as the efficacy of the organization's ties to its competitors. Seeing competitors as assets rather than hindrances, companies in related businesses can create forums to share ideas and pursue benchmarking.[33]

Building and maintaining external links requires an appropriate set of internal organizational assets. Managing downstream linkages, for example, requires skills to interpret customers' comments, procedures to ensure the systematic collection and analysis of field information, organizational structures to ensure that results of this analysis flow to the appropriate people and that these people have some incentive to act on it, a strategic framework that focuses people's attention on learning from users, and a cultural context that avoids the NIH syndrome.

In some industries, regulations have a considerable impact on product innovation (e.g., FDA approval for new drugs) or on the organization's internal operations (e.g., EPA or OSHA regulations). Both types of regulation can have a considerable impact on the firm's processes and internal activities. This component of the technological base should therefore be assessed in terms of the appropriateness of internal compliance policies and the productivity of relations with the regulators.

Finally, in some industries, the political environment can play an important role in shaping the firm's technological base. Recent years have seen industry players mobilize to seek protection from foreign competition—including technology-based competition—and to seek government support for domestic technology development. Other types of societal issues, such as ecological concerns, can also influence the organization's technology agenda, and an assessment of relations with the relevant social movement is often necessary.[34]

4.6 DEVELOPMENT PROCESSES

For technology-based firms, the two key processes are the product and process-generation process and the technology-generation process. The latter generates the organization's future technological assets that are used as inputs into the former. Both processes span the entire innovation cycle, from idea to commercialization—including idea generation, formulating and screening proposals, conceptual and detail development, and internal or external sales. Like other business processes, development processes require the cooperation of different departments within the company.[35]

One way to structure the evaluation of this process is to divide it into three broad phases: preproject, project, and postproject. The *preproject phase* encompasses the assessment and mapping of technology and products[36] as well as setting the "structural context"[37] and the "strategic context" for innovative projects. These preproject activities focus the organization's attention on ideas and opportunities, and thus play a critical role in shaping the subsequent outcomes.

The key factors for effective idea generation and identifying opportunities are maintaining links to external knowledge sources and across internal boundaries. 3M, for example, in addition to continuously collecting customer feedback and ideas, has established a method of articulating radically new product proposals without explicitly getting customer requests. Often in cases of radical innovation customers simply don't know what they need, until they see or have it (and then, most likely, they "can't live without it").

The key tasks within the project phase are product definition, project definition, and project organization. *Product definition* involves positioning the product as a breakthrough, next generation, or derivative.[38] It also involves transforming customer requirements into product attributes and specifications. For this purpose companies employ techniques such as *quality function deployment* (QFD), as developed by Toyota,[39] or *conjoint analysis,* as used by Sunbeam.[40]

Once the product is defined and specified, the organization needs to define the project—its type, scope, schedule, and budget. Projects can be classified according to their level of technological uncertainty (from low to high, or superhigh)[41] and according to their system scope (assembly, system, or array).[42] Depending on the nature of the project, different organizational forms, team structures, and leadership approaches are needed.

Modern project management techniques are getting special attention as companies come under increasing pressure to shorten their time to market. Companies are creating a reservoir of project management tools, program applications, planning and control methods, and documentation formats, in an attempt to learn how to manage their projects more efficiently.

The *postproject phase* governs what the organization will learn from past projects for its subsequent activity. Does the organization systematically conduct postproject reviews? Does it collect data that enables it to compare this project's performance against that of comparable projects in the past? How objective is the assessment? The key factors for effective postproject activities are the expectations communicated by senior managers and the culture that rewards good decisions and not just good outcomes. The organization's development processes can truly become a part of the technological base if management commits the organization to "learning across projects."[43] Reengineering the company's project management processes has become an important learning thrust in organizations seeking to reduce cost and improve quality of service to their customers.[44]

4.7 COMPLEMENTARY ASSETS

Generating profits from technological innovation often requires a collection of supporting capabilities. Complementary assets, such as information technology, distribution, after-sales services, field support, and manufacturing, can create the difference between temporary and long-term success.

The most critical complementary assets are those which exhibit a strong dependence on the specific innovation. This was the case in Mazda's failed attempt to commercialize its rotary-engine automobiles. Specialized repair facilities to support the new type of engine were simply nonexistent. Similarly EMI, which invented and developed the first generation of *computerized tomography* (CT) scanners, was unable to sustain its market position since it lacked the required chain of service, support, and in-hospital sales access. Eventually, it lost its market share leadership, first to Technicare, and then to GE—both possessing the complementary capabilities.[45]

In certain cases, information technology is a critical complementary asset. American Airlines, a major transportation service provider, was able to achieve substantial advantage from its advanced automated ticketing, routing, and reservation system, SABRE. Sometimes a company may possess the main technology embedded in its innovation, but it is unable to commercialize the final product without an additional support technology. IBM's reliance on Microsoft's MS-DOS operating system is a typical example. Without this help IBM could not have developed its first PC in a record development time of one year. And when Cray Research markets its supercomputers, or Sun Microsystems its workstations, they essentially give away supporting software to enhance the sale and use of their advanced hardware.

Securing control of complementary assets may be a key factor in sustained success. Large companies are more likely to possess the relevant assets. Small firms may not have the time or resources to develop the required capability. They must then create alliances with owners of the specialized assets, thus creating potential future competitors. However, even large corporations often find it necessary to join forces with others to speed up development or reduce costs. When the personal computer industry began to move from desktop toward laptop computers, most U.S. firms found themselves sourcing critical technologies such as flat panel displays from Japanese competitors: Compaq from Citizen, TI and AST from Sharp, AT&T from Matsushita, and IBM from Toshiba. Another way to gain access to complementary assets is by acquisition. That was IBM's motivation in its move to acquire Lotus with its network GroupWare software. In considering an acquisition of this type, companies must assess, however, the short-long-term diversification test: Does the acquisition add true sustainable value?[46]

Special attention should be given to manufacturing capabilities. Often an innovator would produce the first and second generations of an innovative product, and then fully or partially outsource manufacturing to a low-cost manufacturer with reduced labor costs or overhead expenses. Although such a decision may make economic sense in the short term, it may result in surrendering key technological assets in the long run. This was clearly the case for color TVs and other consumer electronics products. Many of them were invented by American companies, but these firms later lost market share or entire businesses to off-shore manufacturers who started their involvement by joint manufacturing with American producers.[47]

Complementary assets are product-, industry-, and innovation-specific. When assessing a specific innovation, a company must identify all the necessary capabilities which are needed to successfully develop, produce, market, distribute, and maintain the product or service in the long run. Ownership of the relevant complementary asset is clearly an advantage, but it comes at a cost. Strategic partnerships are an increasingly common alternative. (Such partnership then becomes part of the company's external assets.) When GM-Hughes planned its move into the commercial TV reception business, it had sought external help in securing required complementary assets. Although it was relatively easy for the company to build a new generation of commercial satellites, making the consumer components—the home decoder boxes and reception dishes—was another story. For building this part, Hughes found it necessary to distance itself from its attitude as a long-time Pentagon contractor who is known for high quality—and even higher costs. Hughes was also looking for entries into a completely new market: the consumer arena. Although Hughes engineers were eager to build the electronic home components, the company chose RCA-Thompson as its partner. RCA assured Hughes it would build the consumer units while meeting the price target of $700, and its 11,000-dealer retail network guaranteed distribution.[48]

4.8 ADAPTING THE TECHNOLOGICAL BASE

Management may often initiate a change in the company. Such a change may be needed to better adapt the organization to its environment or to be able to carry forward the company's strategic decisions such as moving into a new group of products, a new technology, or a new type of market. The technological base provides a framework, not only for assessment but also for implementing change and for analyzing the difficulty involved. Of the five elements we have identified, it is usually the organizational assets that prove to be the limiting factor. Furthermore, if the organizational assets are effectively managed, one can positively control the other four elements. Among the elements of the organizational assets we can often find a hierarchy:

- *Skills.* The most direct determinant of the organization's ability to derive benefits from new technological opportunities is the skill base of the organization. Do the personnel have the skills required to effectively select, develop, operate, and maintain the technological capabilities?

- *Procedures.* Whether skills are effectively deployed will depend on prevailing procedures—in particular, the procedures for decision making, personnel selection, and human resources development.

- *Structure.* Whether these procedures—which prescribe certain roles—are effectively implemented or degenerate over time will depend on their congruence with the incentives and information flows created by the organizational structure. What specialized units have been established? To whom do they report? And how do they contribute to the development processes?

- *Strategy.* These structures, in turn, will evolve to reflect the priorities embodied in the organization's strategy. What are the competitive priorities of the firm? How are these formulated? How are they translated into resource allocation?

- *Culture.* Behind these priorities, we often find culture—the values and assumptions that bind the organization and give it continuity over time.

This hierarchical order helps us understand the dynamics of change in the organization. The five components can be seen as five levels of organizational learning in two senses. First, the greater the magnitude of change in technology or market that the organization seeks to effect, the higher in this hierarchy the organization needs to make adaptations. Simple technological or market changes typically require modest changes in skills and procedures. More radical changes, on the other hand, typically call for organizational changes not only in skills and procedures but also in structure and strategy. Revolutionary technological changes, such as those required by the defense industry in its conversion to commercialization, usually call for changes in all five levels—including culture.

These five elements also form a hierarchy in another sense: the lower levels of organizational learning are typically amenable to faster change than higher levels. The higher levels are more "viscous" and resistant to change. New skills can be recruited in a matter of weeks or months. New procedures typically take several months to develop and implement. New organization charts can be drafted overnight, but getting the organization to work effectively in the new framework usually takes 6 months to a year. New strategies can be decreed, but effectively mobilizing the organization to internalize and implement them typically requires a year, and often longer. And finally culture, if it is manageable at all, usually takes several years to change (see Fig. 4.2).

The remaining components—external assets, development processes, and complementary assets—may also be targets of change. They cannot be changed, however, without an appropriate adjustment in some of the organizational assets. External assets, for example, can be changed, but that will typically require new procedures for information gathering, and perhaps even a cultural change to open the organization to the external world. This was the case for General Electric, which directed itself to become a "global, borderless company." The process of change took more than 10 years, and even then some elements of the "old culture" were still evident throughout the company.[49] Similarly, development processes can be changed, but doing so typically requires changes to skills, procedures, and sometimes to structure and culture, too. Finally, complementary assets can sometimes be developed in-house, but that typically requires new skills, and sometimes new organizational units. If unable to build an in-house capability, the company must resort to external help, and must again take the time to build the strategic alliance with its new "external assets."

There are, of course, exceptions, but rapid change in the organizational fabric can only be effective in peculiar circumstances. In the mid-1980s, a large, New York–based financial services organization had been plagued by poor processing performance. The company realized the urgency of completely overhauling its back-office technological competencies and organizational assets and reassessing its new system development project capability. They could see that the magnitude of change they were seeking would necessitate not only equipment, skill, and procedural changes but also a major transformation of their structures, strategy, and culture. So they decided to replace the entire operations top-management team, nearly half of the other managers, and one-third of the employees. The change proved highly effective, but it still took 2 years to pay off. Even this 2-year time span was possible only because they

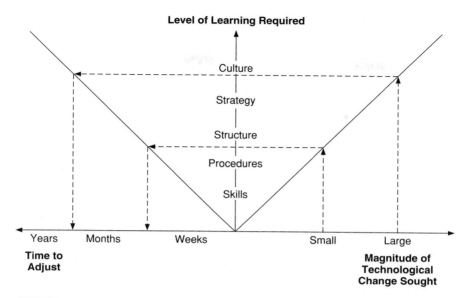

FIGURE 4.2 Dynamics of organizational change. (*Source: P. S. Adler and A. Shenhar, "Adapting Your Technological Base: The Organizational Challenge," Sloan Management Review,* **25**: *25–37, 1990.*)

were located in Manhattan, where there is a large pool of experienced financial industry experts.

This exception thus "proves the rule" that companies that wish to capitalize on technology's ability to make a contribution to their performance—rather than seeking merely to minimize technology change's negative impact—need to carefully assess the strengths and weaknesses of their technological base as well as the time it takes to remedy those weaknesses and build new strengths.

4.9 REFERENCES

1. P. S. Adler and A. Shenhar, "Adapting Your Technological Base: The Organizational Challenge," *Sloan Management Review,* **25**: 25–37, 1990.

2. C. K. Prahalad and G. Hamel, *Competing for the Future,* Harvard Business School Press, Boston, 1994.

3. E. B. Roberts and C. A. Berry, "Entering New Businesses," *Sloan Management Review,* pp. 3–17, Spring 1985.

4. C. K. Prahalad and G. Hamel, "The Core Competence of the Corporation," *Harvard Business Review,* pp. 79–91, May–June 1990.

5. M. Hammer and J. Champy, *Reengineering the Corporation,* Harper Business, New York, 1993.

6. Sharp Corporation, "Corporate Strategy," *Harvard Business School,* HBS-9-793-064, Jan. 1994.

7. J. D. Jasper, "Inventory Your Technology for Increased Awareness and Profit," *Research Management,* **28**(4): 16–20, 1980.

8. G. R. Mitchell, "New Approaches for the Strategic Management of Technology," in *Technology in the Modern Corporation,* M. Horwitch, ed., Pergamon Press, New York, 1986.

9. A. D. Little, "The Strategic Management of Technology," *Eur. Management Forum,* Booz-Allen and Hamilton, Cambridge, 1981.

10. P. A. Roussel, N. Saad, and T. J. Erickson, *Third Generation R&D,* Harvard Business School Press, Boston, 1991.

11. R. Foster, *Innovation—the Attacker's Advantage,* Summit Books, New York, 1986.

12. M. Jelinek and C. Schoonhoven, *The Innovation Marathon,* Jossey-Bass, San Francisco, 1993.

13. E. B. Roberts and A. Fusfeld, "Staffing the Technology-Based Organization," *Sloan Management Review,* **22**(3): 19–34, 1981.

14. A. J. Shenhar and H. Thamhain, "A New Mixture of Management Skills—Meeting the High-Technology Managerial Challenges," *Human Systems Management,* **13**(1): 27–40, 1994.

15. R. L. Daft and R. H. Lengel, "Organizational Information Requirements, Media Richness and Structural Design," *Management Sci.,* **32:** 554–571, 1986.

16. M. Jelinek and C. Schoonhoven, op cit. (Ref. 12).

17. T. J. Allen, "Organizational Structure, Information Technology, and R&D Productivity," *IEEE Transact. Engineering Management,* **4:** 212–217, 1986.

18. P. R. Lawrence, H. F. Kolodny, and S. M. Davis, "The Human Side of the Matrix," *Organizational Dynamics,* Summer 1977.

19. S. C. Wheelwright and K. B. Clark, *Revolutionizing Product Development,* Free Press, New York, 1992.

20. A. J. Allen, *Managing the Flow of Technology,* MIT Press, Cambridge, Mass., 1977.

21. *Business Week,* pp. 66–68, March 23, 1995.

22. A. Bhide, "Hustle as Strategy," *Harvard Business Review,* pp. 59–65, Sept.–Oct. 1986.

23. P. S. Adler, D. W. McDonald, and F. McDonald, "Strategic Management for Technical Functions," *Sloan Management Review,* **33**(2): 19–28, 1992.

24. R. H. Hayes, "Strategic Planning: Forward in Reverse?" *Harvard Business Review,* pp. 111–119, Nov.–Dec. 1985.

25. Prahalad and Hamel, op. cit. (Ref. 4).

26. R. A. Burgelman, "Strategy-Making as a Social Learning Process: The Case of Internal Corporate Venturing," *Interfaces,* **18**(3): 74–85, 1988.

27. Sharp Corporation, op. cit. (Ref. 6).

28. M. A. Von Glinow, *The New Professionals,* Ballinger, Cambridge, Mass., 1988.

29. J. Nocera, "Death of a Computer," *Infoworld,* pp. 59–63, 63–65, June 4, 11, 1984.

30. E. H. Schein, "Coming to an Awareness of Corporate Culture," *Sloan Management Review,* pp. 3–16, Winter 1984.

31. E. Von Hippel, *The Sources of Innovation,* Oxford University Press, New York, 1988.

32. M. E. Porter, *Competitive Advantage,* Free Press, New York, 1985.

33. F. G. Tucker, S. M. Zivan, and R. C. Camp, "How to Measure Yourself Against the Best," *Harvard Business Review,* Jan.–Feb. 1987.

34. I. H. Ansoff, *Implanting Strategic Management,* Prentice-Hall International, Englewood Cliffs, N.J., 1984.

35. Hammer and Champy, op. cit. (Ref. 5).

36. C. H. Willyard and C. W. McClees, "Motorola's Technology Road-map Process," *Research Management,* **30**(3): 13–19, 1987.

37. R. A. Burgelman, "A Process Model of Internal Corporate Venturing in the Diversified Major Firm," *Admin. Sci. Quart.,* 28: 223–244, 1983.

38. Wheelwright and Clark, op. cit. (Ref. 19).

39. J. R. Hauser and D. Clausing, "The House of Quality," *Harvard Business Review,* pp. 63–73, May–June, 1988.

40. A. L. Page and H. F. Rosenbaum, "Redesigning Product Lines with Conjoint Analysis: How Sunbeam Does It." *J. Product Innovation Management,* **4:** 120–137, 1987.

41. A. J. Shenhar, "From Low to High-Tech Project Management," *R&D Management,* **23**(3): 199–214, 1993.

42. A. J. Shenhar and D. Dvir, "Managing Technology Projects: A Contingent Exploratory Approach," *Proceedings of the 28th Annual Hawaii International Conference on System Sciences,* IEEE, Computer Society Press, Washington, D.C., 1995.

43. R. H. Hayes, S. C. Wheelwright, and K. B. Clark, *Dynamic Manufacturing,* Free Press, New York, 1988.

44. Hammer and Champy, op. cit.

45. D. J. Teece, "Profiting From Technological Innovations: Implications for Integration, Collaboration, Licensing and Public Policy," in *The Competitive Challenge: Strategies for Industrial Innovation and Renewal,* D. J. Teese, ed., School of Business Administration, University of California, Berkeley, 1987.

46. M. Porter, "From Competitive Advantage to Corporate Strategy," *Harvard Business Review,* pp. 43–59, May–June 1987.

47. R. A. Bettis, S. P. Bradley, and G. Hamel, "Outsourcing and Industrial Decline," *Acad. Management Executive,* **6**(1): 7–22, 1992.

48. *Business Week,* pp. 66–68, March 13, 1995.

49. N. M. Tichy and S. Sherman, *Control Your Destiny or Someone Else Will,* Currency-Doubleday, New York, 1993.

CHAPTER 5

SECTION 1 THE CORPORATE BOARD AND THE NEED FOR TECHNOLOGY ANALYSIS*

Rias Van Wyk

University of Capetown
South Africa
University of Minnesota
Center for Development of Technological Leadership
Minneapolis, Minnesota

5.1 INTRODUCTION

How should the corporate board be involved in management of technology (MOT)? Many executives would wonder about the relevance of this question, arguing that it is not the traditional role of the board to be involved in MOT at all. Many would agree with Steve Lohr of the *New York Times* in his description of the ceremonial role of the board.[1]

> It used to be so easy, so civilized, more a social ritual than a chore. The members of the corporate board would fly in the night before, dine with their old friend, the chief executive, sharing after-dinner brandy and camaraderie. Each director would receive a slender binder of briefing papers that, typically, got a cursory read before the formality of the next morning's board meeting. The chief executive called the tune, the board hummed along.

This is no longer the case. At present corporate boards are rethinking their roles. Financial reporters and academic authors are taking a hard look at the role and functions of the board and are detecting subtle shifts in expectations and perceptions. These could alter significantly the role of the board in many areas.[2-5]

Coincidentally with the changing role of the corporate board there is a growing interest in MOT. One of the strongest trends in evidence in society today is that of increasing technification. Technology is a large and growing part of most people's daily experience. Not only do we see the emergence of new materials, products, and

*Financial support by the Human Sciences Research Council is gratefully acknowledged. The Council did not initiate this research and does not necessarily agree with the findings. The role of the Honeywell Foundation in supporting the Center for the Development of Technological Leadership is acknowledged with appreciation.

information systems, but many activities that in the past were performed by people are being increasingly performed by technological entities. The technological landscape within which we all live is mutating and evolving continuously. This rapid evolution is the direct result of the massive investments in science and technology that all the economically advanced countries of the world are committed to. Each year 2 to 3 percent of global financial resources are allocated to research and development (R&D). At the leading edge technological change is measured at over 50 percent per annum. This poses particular challenges to the corporate board which carries the mandate to ensure the profitable survival of the corporation within this changing technological environment.

On the basis of preliminary evidence, this chapter suggests four propositions concerning the role of corporate boards in MOT:

- Corporate boards are taking on more active roles.
- Corporate boards are becoming increasingly involved in strategy formation.
- The strategic dimension of MOT is receiving increasing emphasis.
- Corporate boards will be increasingly involved in MOT.

While each proposition needs to be more fully researched, certain implications for boardroom procedure may already be discerned. In the following sections the grounds for each proposition will be set out and the implications for boardroom procedure investigated.

5.2 CORPORATE BOARDS TAKING MORE ACTIVE ROLES

May we detect a change in the dominant mode of operation of corporate boards? Do we note a shift in position? In a recent study Demb and Neubauer address this very question. The authors differentiate three modes of operation:[6]

- *The watchdog.* The role of the board is that of monitor. "This implies a post-factum assessment, primarily in terms of how successfully the organization conducts its business."
- *The trustee.* The board serves as the guardian of the assets. "Implicit in this role is the sense that the trustee is responsible for evaluating *what* the corporation defines as its business, as well as how well that business is conducted."
- *The pilot.* The board takes an active role in directing the business of the corporation. "A pilot board is active, gathers a great deal of information, and takes on the decision roles the other archetypes leave solely to management."

It would appear that corporate boards can occupy different positions on a broad spectrum of possible modes of operation. We may describe this spectrum as ranging all the way from a *review-and-react* (RR) mode that is somewhat uninvolved and nonparticipative, to an *envision-and-enact* (EE) mode that is highly involved and participative. What do Demb and Neubauer find? They come to the conclusion that, in the future, "The board will have more to handle, a bigger portfolio, and there will be a need to shift emphasis more toward the conduct arena."[7]

This provides us with the basis for the first proposition: *Corporate boards are taking on more active roles.*

5.3 CORPORATE BOARDS INCREASINGLY INVOLVED IN STRATEGY

But if boards are becoming more active and involved, what are they becoming involved in? In a series of interviews of corporate board members, many members saw strategy as their major responsibility. "Almost two out of three board members interviewed identified 'setting the strategic direction of the company' as one of the jobs of the board. The percentage is probably an understatement. When asked directly at another point in the interview, 'Are you involved in setting strategy for the company?' only one in five answered no."[8]

This impression is reinforced by the results of a formal survey of board members attending an executive course at the International Institute for Management Development (IMD), Lausanne, Switzerland. The results are outlined in Fig. 5.1.

Other activities included in the survey but not reported separately here include ensuring compliance with corporate law and regulations; providing a broad view; monitoring the environment; handling shareholder relationships; and setting overall culture, ethics, and image.[9]

Clearly, if there is one activity that board members regard as their legitimate domain, it is strategy. They feel they are involved in managing the destination of the corporation and in ensuring its survival as a viable entity.

This evidence gives rise to the second proposition put forward in this chapter: *Corporate boards are becoming increasingly involved in strategy formation.*

5.4 STRATEGIC DIMENSION OF MOT RECEIVES INCREASING EMPHASIS

As a consequence of the technification of society there is a growing interest in the formal management of technology (MOT). MOT is that part of management concerned with exploring the potential of new technologies and developing the technological base of the corporation to utilize this potential. New developments in product, process, and information technologies have to be monitored, evaluated, and—possibly—utilized. These decisions require a subtle understanding of the interplay between technology, economics, and the environment, as well as a sound understanding of the impact of technology on corporate functions.

Tasks and responsibilities	Percentage of respondents identifying this task
Set strategy, corporate policies, overall direction, mission, vision	75
Oversee and monitor top management and CEO	45
Handle succession, hiring and firing CEO and top management	26
Approve and review financial plans, budgets, and resource allocation	23
Serve as watchdog for shareholders, dividends	23
Make key financial decisions, handle mergers and acquisitions	21
Advise and support top management	21

FIGURE 5.1   The job of the board: responses of participants in IMD course.

This interest in MOT is of very recent origin. One indication of the relative youthfulness of the field is the short history of professional recognition. The Technology and Innovation Management Division (TIM) of the Academy of Management was formed only in the late 1980s. The European Technology Management Initiative (ETMI) was launched soon after. The International Association for the Management of Technology (IAMOT) was established in 1992. Most academic courses in MOT are very young, and only a handful of universities have track records dating back for more than a decade.

A particular feature of MOT is the way in which it permeates the entire business setting. It is relevant to many functional areas and to all levels in the corporate hierarchy. In a recent study of science and technology policy in the European Economic Community (EEC), MOT was characterized as the area "where long-term visions and short-term improvements meet."[10] MOT is concerned with both operational and strategic issues.

More specifically, MOT is concerned with:

• The level of nuts and bolts—i.e., the design, making, and maintenance of individual products
• The level of corporate functions—e.g., managing operations, marketing, R&D
• The level of corporate strategy—i.e., as part of managing the "destiny" of the corporation

5.4.1   MOT at the Functional Level

The concern with MOT at the functional level is widely understood. It is the area that naturally comes to mind when the term *management of technology* is used. The links with operations, manufacturing, R&D, material development, product design, process reengineering, TQM (total quality management), and productivity are actively pursued.

When viewed at this level technology forms a natural alliance with science and engineering. It reflects the area of the specialist. It logically extends the domain of influence of the engineering division or the laboratory.

5.4.2   MOT at the Strategic Level

The role of MOT at the strategic level is not that immediately apparent. Few management procedures exist involving MOT at this level. It is rarely addressed in management education.

While the literature concerned with MOT at the functional level has, in the past, constituted the largest part of the writing on MOT, recent years have seen a significant increase in the writings concerned with strategy. Together with the growing emphasis on the strategic dimension in MOT, there is also a growing belief that corporate strategy (CS) and MOT are inextricably linked and should be managed as such.[11,12]

Evidence has also been advanced that present approaches to strategy and technology are "fatally flawed" and in need of revision. In short, the frequently advanced view that the technological interests of an organization are dictated by its mission is seriously challenged. It is too limiting. It is the technological insight of the strategic managers that should determine and shape the corporate mission, not the other way around.[13]

No wonder a recent review article on MOT emphasizes anew the importance of integrating technology into corporate strategy,[14] and no wonder that ETMI has chosen as its theme for the Fifth International Forum on Technology Management: *Closing the gap between technology and corporate strategy.*[15] This evidence provides grounds for the third proposition: *The strategic dimension in MOT is receiving increasing emphasis.*

5.5 CORPORATE BOARDS INCREASINGLY INVOLVED IN MOT

If MOT is becoming increasingly strategic in nature, and if corporate boards are becoming increasingly involved in strategy, it follows, by way of inference, that corporate boards will be increasingly involved in MOT. Do we have any evidence that this is happening already? At this stage, not too much. There is some evidence that leading-edge companies are assigning technology to a specific board committee. One example is that of Medtronic Inc. in the United States. As early as 1974 while most boards had only an "audit" and a "compensation" committee, Medtronic established a "research committee." In 1977 the name of this committee was changed to "research and technology." In 1987, the name was changed again to "technology and quality," signifying the board's commitment to the quality concept. Furthermore, in addition to financial data, Medtronic's board receives reports on the strategic outlook of the company which covers indicators such as market share, customer surveys, and technological comparisons with competitors.[16]

Similarly, Motorola has seven committees of the Board of Directors, one committee of which is responsible for technology. This committee "identifies and assesses significant technological issues and needs affecting the company."[17]

This evidence provides grounds for the fourth proposition outlined in this chapter: *Corporate boards will be increasingly involved in MOT.* This proposition has far-reaching consequences for boardroom behavior and the responsibilities of individual members. Board members will have to develop an interest in and a familiarity with technological matters. Technology will have to be an explicit item on the boardroom agenda.

5.6 IMPLICATIONS FOR BOARDS

What does an increasing involvement in MOT mean for the corporate board? Of late much has been written about the responsibilities of boards and about guidelines for proper corporate governance. Underlying most of the discussions is the criterion of "due diligence." Board members are expected to apply their minds to the task at hand and inform themselves adequately of the issues they are dealing with. New arrangements are being experimented with in the boardroom to enable boards to perform the functions set for themselves. Examples include the greater use of nonexecutive directors, the allocation of areas of responsibility, and the appointment of dedicated committees.[17,18]

In the case of MOT it will be necessary for boards to access the necessary sources of *information,* to develop the relevant *literacy,* and to introduce appropriate *procedures* to discharge their responsibilities.

5.6.1 Information

Corporate boards should require adequate information on the changing technological landscape. Just as economists and other strategic analysts of the macro setting are invited into the boardroom to present economic outlooks, political perspectives, and other relevant backdrops, so competent technology analysts should be employed to give a *technological outlook.* To an extent this does happen already. But more often than not, discussion of the technological scene is limited to the predefined and immediate technological interests of the corporation, not the entire technological setting within which the organization will have to survive and prosper. Board members should insist on an *overview of the entire technological landscape* to provide them with an appropriate structure to view and comprehend technological change.

5.6.2 Expertise

In a recent article Lorsch emphasizes the importance of knowledge to enable directors to function properly. *"Knowledge* is the appropriate word here instead of the more frequently used *information,* because the director's real problem is not lack of information but its content and context."[19] To meet this objective, board members may wish to take steps to enhance their *technological expertise.* In this respect technological expertise does not necessarily mean an in-depth and detailed understanding of particular technologies. It goes far wider. It means a comfortable familiarity with the whole spectrum of technologies.

Where would board members turn to for training of this nature? Unfortunately, the field is not very well served at this stage. Many authors have suggested approaches to such a comprehensive technological field. Examples that come to mind include a systems theory of technology,[20] technometrics,[21,22] a functional approach to technology,[23] technology analysis,[24] technocology,[25] and a comprehensive theory of technology.[26]

There is a challenge to develop appropriate educational programs in this area. It is expected that business schools, centers for the development of technological leadership, and associations of corporate directors (such as the National Association of Corporate Directors in the U.S.A. and the Institute of Directors in the U.K.), as well as international associations of professionals (such as the International Association for the Management of Technology and the European Technology Management Initiative), will respond to this challenge. In this respect a recent survey by the American Assembly of Collegiate Schools of Business on resources available for courses in the management of innovation and technology provides a useful point of departure.[27]

In selecting appropriate courses directors should ensure that the following topics are covered:

- How to identify an appropriate unit of analysis when dealing with the phenomenon of technology
- How to understand different technologies by using a common analytical framework
- How to classify technologies for management purposes
- How to track technological trends
- How to observe and forecast technological fusion
- How to chart potential breakthrough zones
- How to evaluate technologies in terms of social preferences

An outline of an executive short course covering these aspects is dealt with in Sec. 2 of this chapter.

5.6.3 Procedures

The corporate board should clearly define its role in the *strategy formation* process. This is not easy. Strategic planning is undergoing a major revision at present, and previously accepted procedures are being probed and questioned. One approach is to involve board members in the *strategic-scanning* process. This can be done by means of an explicit exercise devoted to the surveillance of the corporate environment, including the technological landscape.[28]

Within the strategic-scanning approach an interactive procedure is followed involving a dialogue between the corporate board and technology analysts. Major trends are explored and a number of "landmark technologies" identified which reflect the most outstanding features of the technological landscape.[29]

CHAPTER 5

SECTION 2   TECHNOLOGY ANALYSIS: A FOUNDATION FOR TECHNOLOGICAL EXPERTISE

5.7   INTRODUCTION

This section introduces the field of technology analysis, a new field of inquiry seeking a comprehensive approach to technology as opposed to the more conventional specialized and differential approach. It seeks a common grammar for dealing with all technologies and concepts for understanding entire technological landscapes. Its roots are in the boardroom and the attempts to provide a strategic perspective on technology to senior executives.

5.7.1   Basics

As stated in Sec. 1 of this chapter, MOT focuses on three levels within the organization:

- Individual products and processes, i.e., management concerned with issues at the nuts-and-bolts level
- Functional areas (e.g., operations) and corporationwide concerns (e.g., quality and productivity)
- Strategy, i.e., selecting a corporate destiny and guiding the corporation through a turbulent technological landscape

To effectively manage technology, all corporate managers must have a certain level of technological literacy. This does not necessarily mean in-depth expertise but a comfortable familiarity with the whole gamut of technologies. Technology analysis provides the tools for thought for achieving such a comfortable familiarity.

Six basic tools cover the essence of technology analysis:

- A standard format for viewing and describing individual technologies
- A classification of technologies
- A table of technological interactions
- A cascade of trends describing technological change
- A chart of technological breakthrough zones
- A profile of social preferences with respect to technology

Technology analysis may be viewed as a "skyway theory" of technology. Just as a skyway is a passageway linking various buildings, allowing easy access to the main areas without attempting to cover every floor, so technology analysis allows easy access to the essential features of all technologies without attempting to grasp every detail.

5.7.2 Defining Technology

There are many definitions of technology. For the purposes of this handbook, the following one is convenient:

> Technology is a "set of means" created by people to facilitate human endeavor. In the briefest possible terms technology may be viewed as "created capability."

This definition is very cryptic and leaves many subtleties unsaid. Some elucidation is necessary:

- The emphasis on *means*. This is the essence of technology; it is not an end in itself.
- The use of the term *created*. Technology is not natural. It is made by people and is therefore artificial.
- The *size* of the "set of means." This can be limited or universal, depending on the focus of the analyst.
- *Facilitate* human endeavor. This is normally taken to mean to enhance human performance or enable tasks beyond human capability. However, in many cases technology can render human endeavor obsolete.

The boundaries of the definition have been drawn to include technology in all its manifestations, such as (1) emerging versus established technology and (2) high technology, conventional technology, intermediate technology, and subsistence technology. It excludes the social and environmental *impact* of technology, specifically, technology assessment.

5.7.3 Choosing a Unit of Analysis

To select a unit of analysis we need to answer a key question: Where does created capability reside? Where can we observe it and take its measure?

There are many possible answers to this question. For instance, technology could reside in the minds of people—they have the potential for creating goods and services. On a more mundane level, technology may be observed in the many artifacts used by society, such as materials, tools, machines, devices, and procedures. And there are more. What we are after is a concept for technology studies which corresponds to the concept of organism in biological studies—i.e., a recognizable unit that can become the focus of the analyst's attention, that can be subdivided into discernible parts that exist within a larger community. For the purpose of this chapter, the unit of analysis chosen is a *technological entity*—an uncommon name for a complex cluster of hardware, algorithm, and human skills. One or more such technological entities constitute the "set of means" referred to in the definition.

The entity concept is illustrated in Fig. 5.2.

In practice technological entities are embedded in an organizational framework, sometimes referred to as *orgware*. The notion of technological entity is very flexible. It can be large or small, simple or complex, concrete or abstract, old or new—its specifications will depend on the focus of the analyst.

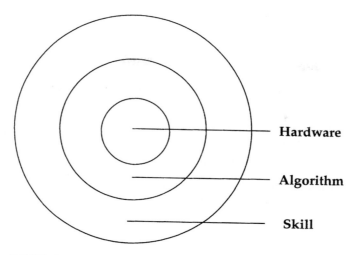

FIGURE 5.2   Technological entity.

5.8 DESCRIBING TECHNOLOGIES

5.8.1   The Notion of a Standard Format

Each technological specialty uses its own unique set of words and concepts. This complicates communication between technical and nontechnical people, and even between various technical specialists. The first tool of technology analysis is a *standard format* for describing all technologies.[30]
    The description responds to six questions:

- What does the technological entity do—what is its *function?*
- How does it do it—what is the *principle of operation?*
- How well does it do it—what is the level of *performance?*
- What does the technological entity look like—what is its *structure?*
- What is it made of—from what *material?*
- How big is it—what is its *size?*

This tool is the basic tool of technology analysis. It is used as an analytical tool in its own right, but also to devise others.

5.8.2   Function

While it is easy to describe the function of a technological entity, it is a little difficult to standardize this description. One way to help do this is to use a very simple format of one verb and one noun to describe function. The verb says what the technological entity does, and the noun indicates what it does it to. There are three possible *verbs* that describe what a technological entity can do:

- *Processing*—to receive inputs and to work with them and to produce outputs of another kind
- *Transporting*—to receive inputs and to send them a certain distance
- *Storing*—to receive inputs and to hold them for a longer or shorter duration

The *nouns* are a little more difficult to standardize. So many different things can be processed, transported, or stored. However, it is possible to define three broad categories:

- *Matter*—that which has mass or volume
- *Energy*—that which can cause work
- *Information*—meaning conveyed in structured signals

By using the three nouns and the three verbs outlined above, it is possible to introduce a large measure of standardization into our description of function.[20]

5.8.3   The Principle of Operation

Unfortunately, there is no standardized way of describing principles of operation. Individual technology analysts will have to use their judgment and intuition. One example may serve to illustrate the presence of different principles. It concerns signal processing. This can be done manually where a hand sign can indicate the presence or absence of a certain condition. It can be done mechanically, electromechanically, electronically, or photonically. Each of these ways of signaling employs a different principle of operation for an entity performing a given function.

5.8.4   Performance

There are probably as many measures of performance as there are technological entities. The technology analyst should seek measures which recur frequently. Four may be identified:

- *Efficiency*—measured as the ratio of output to input
- *Capacity*—in processing and transporting entities, measured as the output per unit of time; in storing entities, as the output stored over time
- *Density*—a measure of output in relation to the space required by the entity producing that output
- *Accuracy*—a measure of preciseness or resolution which reflects the clarity or exactness with which an output may be produced

5.8.5   Structure

There is no standard prescription for describing structure. Technology analysts usually refer to three aspects of structure:

- Shape
- Configuration
- Complexity

Complexity is a key dimension in dealing with structure. To help describe complexity it is useful to refer to the so-called technological hierarchy, which distinguishes between various levels of technological entity:[31]

- Supersystem
- Product group
- Product
- Component
- Part
- Material

The higher up in the technological hierarchy that an entity is located, the more complex it is in that it embraces more subsystems. The greater the linkage between subsystems, the greater overall complexity.

5.8.6 Material

In describing the material the technological entity is made of, it is useful to distinguish between functional and structural features.

The *functional* features of a material reflect its special attributes such as photovoltaic, superconducting, shock-absorbing, and elastic. The *structural* features normally refer to general attributes such as strength and rigidity. Sometimes the functional and structural features are referred to as primary and secondary attributes, respectively.[31]

5.9 CLASSIFYING TECHNOLOGY

5.9.1 The Need for Classification

Classification is at the heart of management. Each functional area employs a classification that we take for granted. In the case of accounting, we classify entries into meaningful accounts. In the case of marketing, we use the marketing mix to help us understand various possible marketing initiatives. In the case of technology, which is a new area of involvement for management, standard classifications have yet to emerge. This has not been an area of much research. The major ideas are summarized below.

5.9.2 Various Approaches to Classification

Horner recently undertook a major survey of various approaches to classification. These belong to two major families:[32] a bibliographic approach and a taxonomic approach.

A *bibliographic* approach looks at what has been created and attempts to put these creations into meaningful categories. When many items have been created in one category, that is a large category. If no items are created, no category is required. A *taxonomic* approach uses a more formal structure. It suggests a number of cells of equal value. Some of these cells may be crowded, some empty.

Teichmann suggests five categories, all taxonomic:[33]

- Characteristics of historical development
- Natural laws or scientific concepts, embodied
- Branch of production
- Function within a branch or process of production
- Principles of construction

There have also been a number of more pragmatic approaches to classification. The most recent is that by Farrell, who suggests seven "kingdoms" of technologic species: shelter, health, communication, tools, packaging, raw materials, and transport.[34]

5.9.3 Approach Adopted in Technology Analysis

In attempting to do a technology scan or a technology audit it is necessary to find a system of classification that meets the following criteria:

- It should start with a simple distinction such as a twofold or threefold one.
- It should be capable of further expansion.
- It should be intuitively appealing.
- It should fit on to a one- or two-page format.

After considering many options, this author finds a basic twofold categorization useful. It distinguishes between materials and technological entities.

The subclassification of materials involves the distinction between (1) basic materials and (2) composites. *Basic materials* can hardly be classified into a limited number of categories. The size of the periodic table of the elements bears witness to that. The following classification is based on Ashby's published theory and is suggested as a first approach:[35]

- Metals
- Polymers
- Ceramics
- Carbons (pure)

Composite materials are classified in terms of the type of composition:

- Matrices (weaves)
- Laminates or bonds
- Alloys

Technological entities can be classified in terms of many schemes. The five approaches of Teichmann would be one possibility, while any of the elements used in the description of individual technological entities would be a valid candidate.

In the absence of a generally accepted format the functional classification is suggested as a simple and practical basis classification. It can be used to categorize any technology, whether simple or complex, large or small, modern or ancient. This classification scheme is illustrated in Fig. 5.3.

When using the nine-cell table the technology analyst has to obtain clarity on

Manipulators of matter	Processing Transporting Storing	
Manipulators of energy	Processing Transporting Storing	
Manipulators of information	Processing Transporting Storing	

FIGURE 5.3   The nine-cell functional classification.

- The predominant output which best describes that technology
- The exact level on the technological hierarchy on which the analyst wished to focus, (e.g., product or product group)
- Whether the technological entity will be viewed independently (i.e., in terms of its immediate task) or with reference to higher levels (i.e., role within a defined system)

5.9.4   Refinements When Dealing with Matter or Energy-Handling Entities

In early economic texts the authors sometimes drew a distinction between three categories of *processing matter:*

- Mechanical processing
- Chemical processing
- Biological processing

Mechanical processing refers to all processing concerned with altering the external features, shape, or configuration of an object, such as planing, sawing, and joining. *Chemical* processing refers to all processing concerned with altering the composition of the materials the object is made of, i.e., chemical reaction. *Biological* processing refers to all processing involving living organisms, e.g., growing and fermenting. This distinction is sometimes used to subclassify material and energy processing in the nine-cell functional classification.

Miller, writing on living systems, introduced a further set of refinements which may be used to subclassify the nine-cell functional classification.[36] In the case of matter- and energy-processing entities, Miller suggests a distinction between converter and producer entities. *Conversion* refers to the breaking down of inputs into a more usable form, while *production* refers to the assembly of these forms into new "stable associations."

In the case of matter and energy transportation entities, Miller distinguished between:

- Ingestion
- Distribution
- Extrusion

Ingestion refers to transporting outputs from without to within, *distribution* refers to moving inputs internally, and *extrusion* refers to carrying outputs from within to without.

5.9.5 Refinements for Dealing with Information-Handling Entities

In the case of information-processing entities, Miller distinguished between associating and deciding. *Associating* is concerned with "forming enduring associations" in information received. *Deciding* is concerned with receiving inputs from other parts, manipulating them in terms of given criteria, and then sending information to certain parts.

Miller also introduces further subtleties with respect to information handling and describes the roles of

- Input transduction
- Output transduction
- Decoding
- Encoding

These are very useful concepts in certain specialized applications. They are not explored further in this text.

The author of this chapter identifies the following elements in a message to help in description and classification:

- *Content*—the meaning of the message
- *Code*—the format in which meaning is expressed
- *Carrier*—the physical elements (matter or energy) which bear the code
- *Channel*—the guideways along which the carriers move
- *Construct*—the overall structure of the message
- *Count*—the size of the message (usually expressed in bits)

5.9.6 Use of the Nine-Cell Functional Classification

The nine-cell functional classification is not in widespread use—in fact, there is no standard classification of technologies in general use. However, it is becoming increasingly popular as companies gain experience with it and discover its flaws and its strengths. Examples exist of four practical applications:

- To structure a technology audit, i.e., to classify core technologies within an organization
- To structure a technology scan, i.e., to provide a basis for reviewing emerging technologies in the global technological environment
- To study interactions between various technologies
- To provide an overview of the portfolio of projects of a research organization

5.10   IDENTIFYING TECHNOLOGICAL INTERACTIONS

5.10.1   The Value of Identifying Interaction

One of the strongest phenomena in technological evolution is the notion of technological fusion. Various technologies come together to produce a new complex technological system far more advanced than any of its predecessors. We need to find a simple way of studying patterns of interaction between various technologies.

5.10.2   Various Types of Technological Associations

Four types of technological association have been identified:[37]

- *Contingent*—the one technology is intimately dependent on the other
- *Supplementary*—changes in one technology affect the other in the same direction
- *Independent*—there is no link
- *Competitive*—the one technology competes with the other and could replace it

How does one go about reviewing an entire technological setting seeking out the types of associations mentioned above?

5.10.3   A "Tableau Technologique"

One approach that is being experimented with is a technology matrix. For a given technological setting a number of selected technologies are identified and represented in a matrix format. Patterns of interaction are then sought by searching for the following relationships:

- One technology is a component of another.
- There is a common material.
- There is a common principle.
- There is a common structure.

A technology matrix is then constructed, and for each cell the type of interaction is noted. In some cases analysts have experimented with the possibility of quantifying the interaction.[38]

5.11   TRACKING TECHNOLOGICAL CHANGE

5.11.1   The Need to Keep Abreast of Technological Change

One cannot think of technology without thinking of constant change. The technological landscape undergoes continuous evolution, fed by the vast resources devoted to research and development. Not many authors have attempted to give a panoramic view of technological change. We have to structure one for ourselves.[39]

5.11.2   The Cascade Approach to Viewing Technological Change

This approach suggests that technological change should be viewed as a cascade of trends.[40] Five levels may be distinguished as technological change cascades through the landscape (see Fig. 5.4):

At level 1, technological change is viewed as changing *material* characteristics. With technological advance, materials become progressively better both functionally and structurally.

At level 2, technological *change* is viewed as changes in the *size, structure, and principle of operation* of entities.

- *Size.* Some entities become larger, some smaller, and some exhibit an increasing size range.
- *Structure.* Some entities change shape and configuration; most become more complex.
- *Principle of operation.* There is a succession of dominant principles of operation.

At level 3, technological change is viewed as improved *performance features* of the technological entity concerned.

- Efficiency improvement as measured by the output/input ratio.
- Capacity improvement as measured by the output/time ratio.
- Functional density improvement as measured by the output/space ratio.
- Accuracy improvement as measured by the decrease in relative variation experienced.

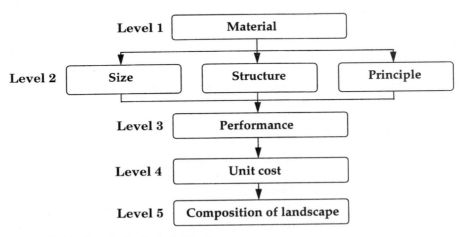

FIGURE 5.4   Cascade of technological trends.

At level 4, technological change is viewed as a *decrease in real unit cost per unit output* when employing the particular technological entity.

At level 5, technological change is viewed as an increasing share of the *technological landscape* occupied by the entity concerned.

5.11.3   Graphing Technological Change

It is possible to plot these changes by using three types of graphic presentations:

- Curves of technological parameters
- Cost curve
- Substitution and diffusion curve

Curves of technological parameters are used to depict changing material characteristics or changing performance levels. Mostly they exhibit a pattern of successive S curves (denoting successive generations of materials, or successive generations of technological entities). The S curve derives its shape from the pattern of increasing increments as technological development takes off, followed by a period of decreasing increments as the effect of some constraint makes itself felt. *Cost curves* are typically decay curves. Please note that they are expressed in real terms. *Substitution and diffusion curves* help describe the pattern of competition between existing technologies and new technologies, as well as the pattern of diffusion of a new technology. Frequently an orderly pattern emerges, with new technologies gaining an increasing share of the market, but at a decreasing rate as market saturation occurs.

5.12   CHARTING TECHNOLOGICAL BREAKTHROUGH ZONES

5.12.1   The Notion of Breakthrough Zones

Most technology managers would be interested in areas where breakthroughs are imminent. Can we identify and chart such breakthrough zones? To define a breakthrough zone we have to clarify the concept of technological constraints.[41–43]

5.12.2   Technological Constraints

For the purposes of technology analysis, constraints have been classified into three categories.

Constraints of the First Order.   These constraints are temporary barriers that may be transcended with the process of technological evolution. Barriers are caused by limitations set by particular material characteristics, by the nature of the technological principle employed, or by the structure or size of a technological entity. Barriers are usually specific to a particular technological entity or class of entity. A barrier is overcome when a new material, principle, or structure is successfully introduced into an existing technological entity.

Constraints of the Second Order.   These are theoretical limits defining boundaries beyond which technological entities cannot perform. They are defined by the laws of physics and cannot be overcome by innovative effort. Examples would be the speed of light in a vacuum, temperature of zero degrees Kelvin, and a complete vacuum. They are not specific to a particular technological entity or class of technological entity but are universal and would apply to an entire technological category.

Constraints of the Third Order.   These are admitted to allow for incomplete present knowledge. They lie in the range of metaphysics and are therefore beyond the range of rational analysis. For the purpose of technology analysis they are not considered here.

5.12.3   Identifying a Breakthrough Zone

A breakthrough zone may be identified as a situation in which

- A technological trend is approaching a barrier (constraint of the first order)
- This constraint is very much lower than a theoretical limit (constraint of the second order)
- A large part of uncharted territory remains for the technology to advance into

Such a breakthrough zone is illustrated in Fig. 5.5.

5.12.4   Charting Breakthrough Zones

There are various ways of charting breakthrough zones. One example of such a chart is given in Fig. 5.6. This chart is based on a combination of two analytical tools:

- The nine-cell functional classification
- Trends extracted from the cascade theory

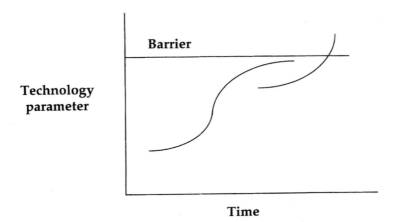

FIGURE 5.5   Breakthrough zone.

Classification of technological entities		Structural		Performance			
		Size	Complexity	Efficiency	Capacity	Density	Accuracy
Manipulators of matter	Processing Transporting Storing						
Manipulators of energy	Processing Transporting Storing						
Manipulators of information	Processing Transporting Storing						

FIGURE 5.6   Chart of breakthrough zones.

By combining the two, we can create a chart of various technological breakthrough zones. Although charts of this nature have not yet found much use in practice, they are useful for alerting management to possible breakthrough areas. They create a mentality likened to that of a cat waiting at a mouse hole for a whisker to appear.

5.13   SOCIAL PREFERENCES AND TECHNOLOGY

5.13.1   Interaction between Social and Technological Trends

It is well known that close interaction exists between technology and society. How can this be charted? De Vulpian recognizes four categories of interaction:[44]

- *Allergy*—denoting the rejection by society of a particular technology
- *Deviation*—denoting the partial acceptance of a given technology, but not without reaction and not without a significant restructuring
- *Enforced penetration*—denoting the emplacement of a given technology by a powerful agency
- *Synergy*—denoting the enthusiastic acceptance of a given technology by society

The interesting problem lies in finding which technologies will cause which reaction pattern.

5.13.2   Embedded Social Values

It is very difficult to identify social values that impact technology. One approach is to identify clusters of values that have some relevance to technology.[45] These values may be grouped under a number of headings:

- *Safety*—technology should enhance and not jeopardize human safety.
- *Health*—the use of the technology should enhance and not diminish human health.
- *Energistics*—the technology should employ renewable energy sources as far as possible.

- *Ecology*—the technology should contribute to long-term sustainability and habitability as far as possible.
- *Entropics*—the technology should contribute minimally to the entropy of the globe.
- *Economics*—the technology should obey economics requirements including the new set of socioeconomic values.

5.13.3   The Special Role of Environmentalism

Environmental concerns are becoming a dominant theme in technology analysis. Two areas need to be understood:

- The emergence of ultimate environmental criteria
- The emergence of environmentally based legislation and regulation

In the latter case technology analysts should be aware of the use of environmental regulation as a way of protecting domestic markets.

5.14   REFERENCES

1. S. Lohr, "Pulling Down the Corporate Clubhouse," *New York Times*, April 12, Section 3, p. 1, 1992.
2. G. Anders, "Barbarians in the Boardroom," *Harvard Business Review*, pp. 79–89, July–Aug. 1992.
3. Working Group on Corporate Governance, "A New Compact for Owners and Directors," *Harvard Business Review*, pp. 141–147, July–Aug. 1991.
4. J. H. Dobrzynski, "Activist Boards, Yes. Panicky Boards, No," *Business Week*, no. 3299, p. 40, Dec. 28, 1992.
5. J. H. Dobrzynski, "The Board Revolts Prove the System Works. Right? Wrong," *Business Week*, no. 3305, p. 35, Feb. 15, 1993.
6. A. Demb and F. F. Neubauer, *The Corporate Board*, Oxford University Press, New York, 1992, p. 55.
7. Ibid., p. 198.
8. Ibid., p. 43.
9. Ibid., p. 44.
10. B. Dankbaar, *Research and Technology Management in Enterprises: Issues for Community Policy*, Commission of the European Communities, Brussels, 1994, p. 14.
11. P. S. Adler, "Technology Strategy: A Guide to the Literature," in *Research on Technology Innovation and Policy*, R. A. Burgelman and R. S. Rosenbloom, eds., vol. 4, JAI Press, Greenwich, Conn., 1989, pp. 25–251.
12. J. P. Ulhoi, "Linking Technology Management to Strategic Management," in *Management of Technology III*, T. M. Khalil and B. Bayraktar, eds., Institute of Industrial Engineers, Norcross, Ga., 1992, pp. 195–204.
13. R. J. Van Wyk, "Technology Analysis and Corporate Governance," in *Design for Competitiveness, Proceedings of the International Conference on Technology Management*, J. D. Clark and W. O. Troxell, eds., Manufacturing Excellence Center, Denver, Colo., 1993.
14. M. M. J. Berry and J. H. Taggart, "Managing Technology and Innovation: A Review," *R&D Management*, **24**(4): pp. 341–353, 1994.

15. European Technology Management Initiative, "Closing the Gap between Technology and Corporate Strategy," *Brochure for 5th International Forum on Technology Management,* Espoo, Finland, 1995.

16. N. Spaulding, *Medtronic, Inc.,* Case N2-494-096, President and Fellows of Harvard College, Harvard Business School, Boston, 1994.

17. Motorola, Inc., *Notice of Annual Meeting of Stockholders,* Schaumberg, Ill., 1995.

18. R. I. Tricker, "The Board's Role in Strategy Formulation. Some Cross Cultural Comparisons," *Futures,* **26**(4): 403–415, 1994.

19. Jay W. Lorsch, "Empowering the Board," *Harvard Business Review,* pp. 107–117, Jan.–Feb. 1995.

20. G. Ropohl, *Eine systemtheorie der Technik,* Carl Hanser Verlag, Munich and Vienna, 1979.

21. H. Grupp and O. Hohmeyer, "A Technometric Model for the Assessment of Technological Standards and Their Application to Selected Technology Intensive Products," *Technol. Forecasting Social Change,* **30**(2): 123–138, 1986.

22. D. Sahal, "Foundations of Technometrics," *Technol. Forecasting Social Change,* **27**(1): 1–38, 1985.

23. H. Majer, "Technology Measurement: The Functional Approach," *Technol. Forecasting Social Change,* **27**(2/3): 335–351, 1985.

24. R. J. Van Wyk, "Technological Analysis: An Imperative for Modern Management," *Futures,* **19**(3): 347–349, 1987.

25. M. K. Badawy and A. M. Badawy, "Directions for Scholarly Research in Management of Technology—Editorial Commentary," *J. Engineering Technol. Management,* **10**(1/2): 1, 1993.

26. C. Farrell, "A Theory of Technological Progress," *Technol. Forecasting Social Change,* **44**: 161–178, 1993.

27. W. R. Boulton, *Resource Guide to the Management of Innovation and Technology,* American Assembly of Collegiate Schools of Business, St Louis, Mo., 1993.

28. R. J. Van Wyk, "When Eagle-Eyed Strategists Scan the Technological Landscape," Working Paper no. 1, *Center for the Development of Technological Leadership,* University of Minnesota, Minneapolis, 1994.

29. R. J. Van Wyk, "A Standard Framework for Product Protocols," in *Technology Management,* vol. I, T. Khalil, et al., eds., Interscience, Geneva, 1988, pp. 93–99.

30. G. De Wet, "Technology Space Maps for Technology Management and Audits," in *Management of Technology,* vol. III, T. M. Khalil and B. A. Bayraktar, eds., Institute of Industrial Engineers, Norcross, Ga., 1992, pp. 1235–1243.

31. N. Waterman, "Materials for the 80s and 90s," *Materials Design,* **5**: 121–125, June–July 1984.

32. D. S. Horner, "Frameworks for Technology Analysis and Classification," *J. Information Sci.,* **18**: 57–68, 1992.

33. D. Teichmann, "On the Classification of the Technological Sciences," in *Contributions to a Philosophy of Technology,* F. Rapp, ed., Reidel, Dordrecht, 1974.

34. Farrell, op cit. (Ref. 26).

35. M. F. Ashby, *Materials Selection in Mechanical Design,* App. C, Pergamon Press, Oxford, 1992.

36. J. G. Miller, *Living Systems,* McGraw-Hill, New York, 1978, pp. xxiii–xxix, 3.

37. V. Mahajan and Y. Wind, "Innovation Diffusion Models: A Reexamination," in *Innovation Diffusion Models of New Product Acceptance,* V. Mahajan and Y. Wind, eds., Ballinger, Cambridge, Mass., 1986, pp. 3–25.

38. R. J. Van Wyk, "Towards a Tableau Technologique: A Contribution to Technology Analysis," in *Management of Technology III, The Key to Global Competitiveness, Proceedings of the Third International Conference on Management of Technology*, T. M. Khalil and B. A. Bayraktar, eds., Industrial Engineering and Management Press, Norcross, Ga., 1992.

39. J. R. Bright, "Opportunity and Threat in Technological Change," *Harvard Business Review*, **41:** 76–86, Nov.–Dec. 1963.

40. R. J. Van Wyk, G. Haour, and S. Japp, "Permanent Magnets: A Technological Analysis," *R&D Management*, **21**(4): 301–308, 1991.

41. R. U. Ayres, *Technological Forecasting and Long Range Planning*, McGraw-Hill, New York, 1968, p. 106.

42. R. U. Ayres, "Barriers and Breakthroughs: An 'Expanding Frontier' Model of the Technology-Industry Life Cycle," in *Management of Technological Change: Context and Case Studies*, G. Rosegger, ed., Elsevier, New York, 1988.

43. R. J. Van Wyk, "The Notion of Technological Limits: An Aid to Forecasting," *Futures*, **17**(3): pp. 213–214, 1985.

44. A. De Vulpian, *New Directions for Innovation in Products and Services*, Congres d'Esomar de Rome, Paris, COFREMCA, 1984, p. 32.

45. R. J. Van Wyk, "Technological Change: A Macro Perspective," *Tech. Forecasting Social Change*, **15:** 281–296, 1979.

CHAPTER 6

TECHNOLOGY, STRATEGY, AND COMPETITIVENESS

AN INSTITUTIONAL-MANAGERIAL PERSPECTIVE

John Kanz, Ph.D./P.E.
DMS OnLine Locator, Inc.
Scottsdale, Arizona

Danny Lam, Ph.D./M.B.A.
Economic Development Institute
Auburn University
Auburn, Alabama

You can always tell who were the pioneers in this business. They're lying face-down in the mud with arrows in their backs. ANON.

6.1 INTRODUCTION

A cliché in many firms, this adage is merely the age-old advice to "play it safe; don't stick your neck out." But rapid technological change renders it meaningless for managers, especially those responsible for firm strategies. Most, not just those in "high-tech," now encounter serious technology-related challenges more frequently, perhaps when approving a complex new telephone system or new computer software or maybe when facing an innovative, cutting-edge competitor. Managements must effectively and efficiently execute such choices to meet the challenges of today's dominant, intertwined change agents: extreme competition, globalization, and technology. Virtually all managers acknowledge increased competition. Most are aware of the pressures of globalization. Fewer are conscious of the full scope and impact of technological change. Very few have developed real understanding of technical issues or familiarity with associated management tools.

That adage is also reflected in a related set of problems with conventional strategic management, whose failures are cited as a major reason for the dominance of financial management at so many firms.[1] Even more pernicious is a resultant tendency toward "denominator management"[2] in which lack of effective strategic direction leads firms toward shrinking the denominators (assets, costs, headcounts, etc.), rather than growing the numerators (revenues, profits, value-added, etc.), of those familiar financial formulas that measure firm performance. Both trends are barriers to the kind of technology-based pioneering that is central to U.S. competitive advantage.[3-5] When technological ignorance is combined with traditional risk aversion, competitiveness suffers.

This chapter suggests an alternative that is rooted in existing strategic management models and frameworks,* but utilizes an institutional perspective which permits simple, logical extensions of these familiar tools and places them in a realistic competitive context. We start with a broad definition of technology and then examine conventional strategic management approaches and their limited ability to encompass technological change. We then look at relationships between management and technology within U.S. and Japanese institutional contexts. With resource-based views of the firm as a general guide, we extend one well-established model to link technological change and strategic management to firm competitiveness. We conclude with a suggested list of technology management competencies.

6.2 TECHNOLOGY

6.2.1 Definition of Technology

We need clear definitions; *technology* is often confused with science or products. Fred Bucy, former chairman of Texas Instruments, argued:[6]

> Science is the systematic pursuit of knowledge, while technology is the application of that knowledge to the production of specific goods and services. Technology is the design and manufacturing know-how to produce goods.... Products...are the result of technology but are not themselves technology; and while science is almost always the basis of technology, it, also, is not technology.

Thus, technology is ultimately know-how. It is the ability to identify needs, to define products (goods or services) that meet those needs, and to marshal capabilities to build and market the products. A complementary but broader view of technology is provided by Shoshana Zuboff:[7]

> Technology represents intelligence systematically applied to the problem of the body. It functions to amplify and surpass the organic limits of the body; it compensates for the body's fragility and vulnerability.

Even if we treat technological development as esoteric and distinct, these definitions remind us that it is really ubiquitous. Lowell Steele called it "the system by which society satisfies its needs and desires."[8]

6.2.2 Invisible Technologies

Bucy and Zuboff also tell us that every firm uses technologies to help define its competencies and competitiveness, even if not all are easily identifiable as such. Also, acceptance of a technology, or its manifestations, can be very gradual. Fire, the

---

Frameworks are structured sets of guidelines, less rigid (and perhaps less precise) than models, but more focused than general principles.

masonry arch, or even the typewriter have been around long enough to no longer appear "technical," but rather "invisible."

The present fascination with "reengineering"[9] implicitly recognizes the quasi-technical nature of business operations, using an industrial engineering perspective from which such technologies are identified, improved, and organized. Some are simply "ways of doing things" or particular operational methods: sticking pins in a wall-mounted calendar for scheduling, clever document storage and filing systems, providing an extra invoice copy to accounting, etc. Such invisible technologies can be especially vulnerable and are frequently outside the purview of managements until used by some competitor. Retail and fast-food chains seemed comfortably "low-tech" until certain companies began effectively employing information (and other) technologies for competitive advantage.

"Visible" technologies like telecommunications, personal computers, electronic white boards, or computer-based management control systems almost always replace older invisible technologies—conventional PBX systems, typewriters, chalkboards, wall calendars, etc.—and will, in turn, become invisible as they become more familiar. These particular examples are part of the information technology (IT) explosion. But that depends on advances in silicon chip, display, software, materials, and other technologies which represent an even more dangerous type of invisible technology. Managers, including IT managers, tend to ignore these until too late, when they suddenly become visible well up on the technological food chain.

6.2.3 Technology Changes

There have always been areas of uncertainty, art, and craft in technology. Success has invariably depended on individual intelligence, insight, and skill. But even allowing for traditional "fuzziness," the frontiers of today's sciences are fundamentally different and more subtle, pervasive, and complex. From early probings with simple instruments and techniques by investigators like Maxwell and the Curies, our superior understanding of nature is now driven by esoteric approaches: from sunlight and prisms to massive particle accelerators, from crude dissections to gene splicing, from test tubes to simulating complex chemical reactions on supercomputers. In addition, technology drives the industrial metamorphosis from producing "things" to producing "knowledge"[10] with ever less tangible firm outputs.

6.2.4 Benefits and Threats of Technological Change

One cannot manage technology without understanding its socioeconomic impacts. On one hand, technology is widely viewed as a primary benefactor of humankind and a source of material, and even aesthetic or spiritual, advancement.[11] Vannevar Bush, who spanned government, industry, and academia with singular grace, called technology the cornerstone of American economic life and national security.[12] Perhaps the most cogent protechnology academic arguments, first effectively advanced by Joseph Schumpeter[13] and later reinforced by Solow[14] and others, view such development as the engine which drives economic progress. On the other hand, resistance to change is commonplace, and technological change is particularly feared; the dark side of technology is a popular theme.

There is a also resurgence of historical concerns about "ordinary" workers. Perhaps Marx[15] best articulated technology's potential to deskill the worker, to isolate him/her from the means of production, and to diminish worker power and influence. The concerns of nineteenth century manufacturing workers are now those of professionals and managers threatened by (partly technology-driven) reengineering, restructuring, and downsizing.[16,17] Ironically, resistance to technology introduction in the workplace may be growing,[18,19] even as we learn how to use it to provide greater employee empowerment and richer work environments. Meeting such challenges requires technologically informed management.

6.2.5   Management's Missing Dimension

Managing technology represents a missing dimension in management theory, education, training, and practice. This is not surprising since "professional management" is largely an American development, and U.S. technological dominance in the post-World War II period readily provided "spinoffs." Also, there was little serious foreign competition in technology-rich products like electronics and automobiles into the 1970s.

These factors, and relative U.S. institutional stability from about 1955 to 1980, gave rise to certain management precepts. First, a generalist orientation at business schools promoted the assumption that firm-level competitiveness could be ensured by well-trained managers who could manage anything with the help of specialists. Second, analytical management frameworks which relied on mainstream economic theory came into widespread use. Unfortunately, that underlying theory employs equilibrium concepts where technological change is essentially ignored. Thus, technology became an exogenous variable beyond primary top-management concern.

Also, logical positivism in much of social science, business, and economics research displaced an earlier structural-institutional perspective. As global competition ignited an increasingly acerbic debate on U.S. competitiveness,[20,21] this older and valuable framework was largely overlooked, deflecting and exacerbating arguments that now extend to the nature of the debate itself.[22]

6.3   STRATEGY AND MANAGEMENT

Mainstream management, especially strategic management frameworks and models, evolved largely during a more stable era. Rapid change along technical, economic, demographic, political, cultural, and other dimensions may well erode their legitimacy. However, those same forces make idiosyncratic, reactive, intuitive, or "seat of the pants" management less viable. Shifting markets, shorter product lifespans, capital intensification, and expansions of scale and scope amplify the likelihood and costs of uninformed management blunders. Successfully navigating a firm and its employees through such environmental turbulence demands a strategic viewpoint, effective change management, supervision transcending command and control, and at least reasonable understanding of the nature and impacts of technological change.

6.3.1   Strategic Management

Strategic management drives the firm; it sets the overall direction of the firm; it creates competitive advantage for the firm (or impairs that of its competition). Approaches range from formal and rational (synoptic) to reactive and adaptive (not quite including seat of the pants), but all are being questioned. As d'Aveni says, "Strategic concepts such as fit, sustainable advantage, barriers to entry, long-range planning, the use of financial goals to control strategy implementation, and SWOT [strengths, weaknesses, opportunities, threats] analysis all fall apart when the dynamics of competition are considered."[23]

In today's complex, shifting world, all actions are relative, all actions are really interactions, and each significant action requires evaluation of its long-term evolution and path.[24] Traditional concepts and tools focus attention on strategy, but can no longer fully encompass its formulation and implementation.

Traditional Strategy Setting. Traditional synoptic (rational) strategic management starts with "vision," something that is often dismissed as a soft-headed or idealistic concept or a mere surrogate for CEO ego. Perhaps "foresight"[2] is better, but under any label there must be a clear view of where the firm is headed and what it is to become. Lack of vision plagues public and private sector organizations from small shops to giant conglomerates, and CEOs from sole proprietors to U.S. presidents.

Strategic management then identifies and defines specific key factors such as finance, markets, channels, or competitor characteristics (sometimes including technological issues), usually through comprehensive environmental "scanning." This includes analyzing (1) relevant elements of the firm's sociopoliticoeconomic-technical environment; (2) structural characteristics of its industry; and (3) the firm's internal characteristics, strengths, weaknesses, and interrelationships. While time- and resource-intensive, there is no substitute for environmental vigilance.

Traditional Strategy: Fit vs. Stretch. A basic question is "fit" vs. "stretch."[2] Scanning results are contrasted with the firm's vision, strengths, and weaknesses and the threats and opportunities it faces. Conventional strategic management, perhaps reflecting traditional risk aversion, is biased toward strategic goals and objectives that fit within firm strengths and capabilities, subtly restricting its choices. Fit/stretch decisions are individualistic and situational. They involve subtle and complex risk-benefit tradeoffs. However, accepting an inherent "tilt" toward fit, while perhaps less risky in the short term, can prove fatal over the longer run.

Traditional Strategy Implementation. In the conventional hierarchical approach to strategic management (Fig. 6.1), the firm's vision is reflected in the strategies it formulates, in its strategic goals (long-term and general), and in its specific objectives (shorter-term, measurable, time-based). Strategies are implemented, and results monitored and controlled by familiar mechanisms such as policies, practices, procedures, budgets, schedules, audits, and reports, which provide guidance, information, and restraint mechanisms that keep the firm on course.

Influence is bidirectional along the structure. Vision drives strategies, which, in turn, drive all supporting elements and activities. At the same time, each underlying element must be consistent with and supportive of that above it. Thus, practices uphold policies which further objectives which are consistent with strategic goals, etc. These vertical relationships must also be consistent horizontally through the organization's structure.

However, this introduces dilemmas. First, environmental changes outside the firm (the economy, competitive or market shifts, political changes, new technologies, etc.) or within the firm (new skill sets, loss of key personnel, worsening financial performance, etc.) dictate virtually continuous monitoring, scanning, and (probably) adjustment. Second, rigid imposition of strategic conformity can be counterproductive; coherence can inhibit innovation, responsiveness, and needed change.[25] Thus, there must always be balances among consistency and adaptation, control and empowerment. Monitoring and control mechanisms must be structured and operated for responsiveness and flexibility, and strategies must incorporate controlled adaptation (strategic incrementalism).

FIGURE 6.1   Traditional strategic management hierarchical structure.

6.3.2   Limitations of Traditional Strategic Management

Particularly frustrating to management scholars are successes of firms which ignore established strategic management principles or firms cited for enlightened strategic management shortly before decline or oblivion.[26] Many observers doubt that conventional strategic management boosts firm performance or ultimate success,[27] although some surveys do show a positive correlation.[28]

Rigid, formalized strategic planning has fallen into disfavor,[29] and that now extends to more broadly based strategic management approaches. Reports show systematic environmental scanning and competitive factor trend analysis occupy a vanishingly small portion of top management time and attention.[2] That probably reflects (1) poor results from mechanistic long-range strategic planning amid rapid, multidimensional change; (2) intense competition requiring more "firefighting" from thinner management ranks, (3) lack of integration with firm operations, and (4) failure of traditional strategic management to encompass key change drivers, especially technology. This last point emphasizes a primary problem: the static nature of most strategic analysis frameworks and models[30] which inherently address existing competitive situations.[1,2] These provide only a snapshot of a moment in time and clues to pressures and forces. They are often based on economic theory and hypothetical equilibria which are increasingly divorced from business realities.

Even basic strategic management precepts like sustainable competitive advantage are being questioned; they easily become distractions.[1] Successful firms now destroy existing competitive advantages by introducing successor products before their competition, for example, Microsoft's frequent new software releases and Sun Technology's "burning" of current products with better and cheaper workstations. Contrast these with IBM's reluctance to introduce minicomputers for fear of cannibalizing mainframes, a mistake repeated with personal computers vs. minicomputers.

Today's remarkable vision can become tomorrow's accepted wisdom; firms must continually scan and adjust. The learning organization must selectively forget[2] because success reinforces persistence of management frames and resistance to new ideas. A few decades ago, "superstitious learning" of false cause-effect connections[31] taught many successful American managers that high-quality products and services were unimportant.

6.4 TECHNOLOGICAL CHANGE AND TRADITIONAL STRATEGIC MANAGEMENT

6.4.1 Strategic Analysis

Traditional strategic management tools typically attempt to answer:

What is the position of the firm in its industry?

How attractive is that industry?

They are helpful simplifications of more complex realities; the real world can never be reduced to simple matrices and two-dimensional models. Also, they were not intended to clarify technology-related issues; rather, they partially accommodate them. In spite of these and other shortcomings, they are still useful within certain limits. We examine three popular types.

Firm-Level: Strengths, Weaknesses, Opportunities, Threats (SWOT). Systematically analyzing a firm's strengths and weaknesses is basic. Results can then be played against (external or internal) environmental threats and specific opportunities (old, new, potential, in present markets, different markets, or entirely new competitive space). Technologies should always be part of SWOT. For example, a firm's proprietary technology can be a major strength, while a new competitor technology could represent a significant threat. Table 6.1 reviews some simple examples.

SWOT results can be very dependent on the knowledge, understanding, and frames of those performing it (a particular problem with technologies), can easily degenerate into generalities, and tend to overlook invisible technologies.

Firm and Infrastructure: Value-Chain Analysis. The *value chain,* popularized by Michael Porter,[32] is a schematic representation of the way a firm adds value to products (goods or services) for distribution channels or customers. Typically there are five primary value-adding activities (inbound logistics, operations, marketing, outbound logistics, and service), and these underpin support activities (e.g., administration, technology development, human resources, procurement). Complex linkages, among

TABLE 6.1   Technologies Redefine SWOT

Strengths	Key technology is proprietary to the firm
	Firm on cutting edge of technology in fast-moving industry
	Firm employs highly competent technologists
Weaknesses	Firm has aging technologies in fast-moving industry
	Key technical employees nearing retirement
Opportunities	Firm offered favorable terms on license for key technology
	Apparent spinoff opportunities for existing firm technology
Threats	Major competitor adopts new "killer" technology
	New low-cost device performs key firm function for customers

TABLE 6.2   Linkages of Direct Technology-Driven Changes

Technology development simplifies product design	*Operations* costs reduced, *procurement* requirements changed, *marketing* redefined, *service* simplified, etc.
Supplier introduces electronic order handling and shipping	*Procurement* procedures change, *operations* introduces JIT (just-in-time), *marketing* has to explain fewer stockouts to customers
Service introduces on-line customer feedback system	*Marketing* has new customer-oriented campaign, *operations* uses feedback for quality improvement—customer costs reduced, *technology development* incorporates customer ideas for product improvements
Outbound logistics introduces customer-designated shipping packages and lot sizes	*Marketing* has another new campaign—customer costs reduced, *service* has lower customer demands and costs

the firm's primary and support activities, with its channels or customers, and with its suppliers, largely determine the efficiency and effectiveness of its operations. Those, together with the competencies of individual activities, reflect management's ability to optimize, integrate, utilize, and leverage its resources to add value, satisfy customers, and demonstrate competitiveness.

Value-chain analysis requires a thorough understanding of the firm and its operations and external relationships. Technologies play roles (note technology development), and good analysis should find relevant linkages (see Table 6.2), but technological change is again accommodated, not emphasized. Also, such issues can be outside the purview of the analysts.

Industry: Five-Forces Model.   Again from Michael Porter,[33] the five-forces structural analysis framework deduces power relationships among five entities which make up an industry: *interfirm rivalry, suppliers, customers, new entrants,* and *substitutes* for the firm's products. Each of these exerts pressures on the firm, and each, in turn, is affected by the firm. Relative power relationships, such as influences the firm has over its suppliers (or vice versa) or the pressures felt by the firm from its customers (or again vice versa), can be made apparent in some detail.

There is obvious overlap between five-forces and SWOT analysis: new entrants and substitutes typically represent threats, compliant suppliers would indicate strengths, etc. But five forces can also expose barriers to new entrants or different market segments, help assess the sustainability of the firm's position, and provide clues as to the relevancy of its strategies. It can include technology issues, as shown in Table 6.3.

Porter and others have written about technological change shifting entry or mobility barriers and resulting changes in relative power positions. This type of analysis can reveal other technological issues, but again is a framework on which technological change components can be mounted without exposing their full effects.

6.4.2   Traditional Tools: A Summary

Traditional strategic management's analytical tools can address technological issues with some utility. However, in varying degrees they suffer from (1) a lack of differentiation of technological change from other types of change; (2) little insight into the

TABLE 6.3 Technological Change Restructures Industry

New *supplier* technology	Improved *supplier* differentiation or cost reduction, potentially higher *supplier* power, potentially greater cost competition
Technology lowers plant and equipment costs	Entry barriers lowered for *new entrants,* potentially greater cost competition
Technology drives higher plant and equipment costs	Entry barriers raised for *new entrants*—may increase motivation for *substitutes* or increase firm power
New technology-driven *substitutes*	Firm power endangered—entry barriers, new competitive space created
Firm's proprietary technology diffused	Reduced firm power—lower barriers for *new entrants,* potentially greater cost competition

dynamic processes involved in technological change; (3) a dependence on (frequently atechnical) analysts' knowledge, expertise, and understanding; and (4) some question about the credibility of underlying concepts. They lack the focus and emphasis required for contemporary strategic management analysis.

6.5   IN SEARCH OF MANAGEMENT PARADIGMS

There are examples of enlightened management of technologies, but underlying principles are poorly understood and approaches highly fragmented;[8] conventional strategic management has not kept up. Its frameworks scarcely provide reliable, long-term solutions, and that contributes to disillusion with "business-school wisdom" and its perceived arcane and irrelevant focus.[34,35] But business academics also complain of managements' search for a "magic bullet" solution to problems. This has produced a series of narrowly focused management "fads" with disappointing longer-term results[36,37] and at least three special management paradigms.

6.5.1   Special Case 1: High-Tech Management

Business schools have long included operations management courses, but those often seem closer to industrial engineering than to other management specializations. Neither is central to their respective curricula, but they are the bases for studies of management in high-tech firms (those which depend on technological progress, emphasize new designs and products, have consequential research and development, and/or employ significant numbers of technical specialists). Studies of research, new product development, innovation management, and technical marketing contributed to understanding the characteristics and dynamics of technological change, technology diffusion and implementation, and managing specialized personnel (technology can be regarded as the pace-setter for managing creative specialized employees[8]). For many observers, this largely constitutes "technology management."

There are at least two problems with that view: (1) this scarcely defines a comprehensive field of study and practice which covers most major issues[38] and (2) isolating the resulting knowledge and insight to the high-tech sector discounts their real value.

6.5.2 Special Case 2: Strategic Technology Management

An extension of this work in recent strategic management research, practice, and writing advocates integrating technology and business strategies.[8,12,39–42] These are critical steps toward a fully integrative approach to strategic management. They explicitly recognize and explicate the central role of technology in firm performances, with an associated view of technologies as primary supports for improved products and processes. They typically define strategies as composed of resource allocation plans, policies, procedures, programs, and projects for such objectives,[40] reflecting an implicit assumption of technologies as separable (although interacting), distinct, and not always central. While useful, they still lack a view of technologies as pervasive forces which must be integrated implicitly and explicitly into core strategic thinking and into every phase of activities at *all kinds of companies.*

6.5.3 Special Case 3: Information Technology Management

Information technologies (ITs) challenge generalist* managers in "nontechnical" activities with technology forecasting and selection, technology investment, innovation management, and management of specialized technical personnel—the sort of dilemmas faced by high-tech managers. A debilitating "management gap" (see below) between generalist management and the IT professionals on whom they must depend[42] has been noted. At the same time, distinctions between IT, other technologies, and firm operations are dissolving. The growing service component in modern economies and increased automation in manufacturing and service sectors mean that elements of IT are now pervasive throughout the value chains of most businesses. That makes any management gap a serious threat.

Information technology is a primary change driver for managerial and academic attitudes and practices. IT can reshape industries and firms, large and small, in everything from health care to auto manufacturing to home building to entertainment to insurance. IT reconfigures competitive advantages, redefines organizational structures, recasts strategies, reengineers processes, repatterns employee-management relationships, and will ultimately help reform definitions of management. But IT is only one dimension of technological change; a much broader strategic management view is needed.

6.5.4 The Value of Academic Paradigms

A word should be said in defense of academics. Within the cut and thrust of academic debate, frameworks and models are often attempts to bring out underlying operational characteristics through radical simplifications of reality; and *they are frequently labeled as such.* These are but a stage in a long process hopefully leading to better understanding how the world really works. Prominent scholars[30] have been very open about these limitations, but managers searching for a quick-fix solution often take the proffered models, frameworks, and tools much too literally.

---

*Throughout this chapter, the term *generalist manager* is used to denote a manager without special technical training or education.

6.6 DICHOTOMY: TECHNOLOGY AND MANAGEMENT

Fully integrating technological issues into strategic management represents revolutionary change for management frames.

6.6.1 Presumptions

The common management view is that managing technologies is something apart from "conventional" management. It is frequently delegated to a specialist "chief technology officer" (in many firms, the "chief information officer") whose career ceiling lies well below CEO level.[42] This "disconnect" between top management and technology managers is reflected in a recent world survey of 95 leading R&D-intensive companies. The survey found technical executives on the boards of directors of over 90 percent of the Japanese firms surveyed; the U.S. figure was less than 20 percent.[43]

Technology is sometimes "managed" by purchasing what is required from outside the firm as needs develop. Limited understanding of the "commodity" being purchased can lead to results reminiscent of the conglomerate craze of past decades (e.g., Exxon's foray into electronics, McDonnell-Douglas' in semiconductors). These two interrelated beliefs—(1) that technology is a factor "outside" normal management and (2) that it can be purchased when needed—help explain the lack of technology management skills in U.S. firms.

6.6.2 Implications

In his recent article on the theory of business, Peter Drucker observes[44]

> A theory of the business has three parts. First, there are assumptions about the environment of the organization: society and its structure, the market, the customer, and technology. Second, there are assumptions about the specific mission of the organization.... Third, there are assumptions about the core competencies needed to accomplish the organization's mission.... The assumptions about the environment define what an organization is paid for. The assumptions about mission define what an organization defines to be meaningful results; in other words, they point to how it envisions itself making a difference in the economy and in the society at large. Finally, the assumptions about core competencies define where an organization must excel in order to maintain leadership.

Thus, technological change undermines assumptions about a "theory of business" in all three areas simultaneously. Recognized or invisible, technologies are now key competitive factors, even in tradition-minded areas like education, insurance, banking, or publishing. Pervasive in every phase of business, commerce, government, education, health care, etc., they can transform virtually any activity. The impacts of failing to integrate technology and business are strikingly apparent in "natural" monopolies such as telephone, electricity, or cable television. All were mature industries where technologies seemed reasonably stable, firm roles were clear, and there was limited competition. Each is now being recreated by new technologies like wireless communications (cellular phones, direct broadcast satellites), energy cogeneration, or advanced control techniques (and resulting changes in regulatory regimes).

The belief that technology can be purchased assumes that (1) willing and capable vendors for needed technologies can be found readily; (2) technologies can be freely transferred, absorbed, and integrated into purchaser systems and structures; and (3) managing, developing, and elaborating acquired technologies require no special top management insight or expertise. It is strengthened by the (now largely discredited[45]) spinoff concept of university and government-funded research providing a reservoir of technologies.

In reality, many cutting-edge technologies are difficult to acquire. Where readily acquirable, they often cannot be effectively utilized by a management unfamiliar with them.[46] Technological success at American firms like 3M, Motorola, and Hewlett-Packard has depended on managers familiar with their company's core technologies and able to regularly leverage them into new products.

6.6.3   The Management Gap

This separation of management and technology has produced the aforementioned management gap. At least as far back as C. P. Snow's two cultures, marked differences have been noted in outlooks, value sets, and paradigms of generalist top management and mid- and upper-level technical managements.[25] Similar differences are apparent at various levels among other generalist managers, technical managers, and research-oriented technologists.[8,47–51] As most firms develop significant technical activities (some of which may involve invisible technologies), the barriers and obstacles that this cultural and organizational fissure creates are no longer tolerable. Technology pervasiveness now makes it a general issue.[52]

6.6.4   Failed Concepts and Newer Initiatives

As noted by David Teece, the economic models, underlying many management frameworks

> ...take technology as given, ignoring the fact that the options open to a manager almost always include some degree of innovative improvements in existing ways to do things.... Technology is [commonly assumed] to be uniformly available to all. Or, if technology is proprietary, then it is information that can be embedded in a "book of blueprints." [Rather than] know-how [which] is often tacit...[hence not transferable]...without demonstration and involvement.[53]

Economists now appear to be rediscovering earlier scholars[13] who pointed to technology as a major economic driver. There is more interest in strategic technology management (see above), and competitiveness analysis may be returning to earlier structural-institutional frameworks.[54] However, mainstream strategic management, although perhaps "ripe for a revolution," has yet to really change, largely for institutional reasons.

6.7 INSTITUTIONAL SOURCES OF TECHNOLOGY MANAGEMENT COMPETENCIES

6.7.1 Business Schools

Management research initiatives, values and areas of interest, and pedagogy are centered largely in established university-based management schools which function as "gatekeepers" for training America's managerial elite and tend to homogenize managerial outlook, culture, values, and episteme. Courses in operations management (nearly all schools), manufacturing management (some schools), and technology management (a few schools) can scarcely be central to curricula where technological issues are peripheral. That helps explain a profound ignorance, among many managers in many companies, about (1) linkages between technological change and firm competitiveness and (2) how to manage them.

An academic discipline which grew out of "general management" and evolved into "business policy and planning" finally became known as *strategic management*. Courses are usually integrative or "capstone" since they encompass the whole firm or industry. That seems a natural base for technology issues, but associated research and teaching, although varied and reasonably inclusive, seldom address technology issues in depth. Indeed, technological change represents concepts inherently antithetical to traditional strategic management. Technological change brings benefits and disruptions, both of which create uncertainties* difficult to address within conventional planning contexts.[1,25]

Thus, the dominant tradition of management education rests on a curriculum which is neither protechnological nor antitechnological, but atechnical. It implicitly resists systematic introduction of technological issues for a variety of reasons: the aforementioned tensions, the lack of accommodation of technological change in existing management frameworks, those frameworks' roots in mainstream economics models which treat technology as exogenous, and a general failure to fully gauge the pervasiveness of technologies in contemporary businesses. The education, values, and behavior of a large segment of professional management reflect this, along with an emphasis on "sticking to your knitting" (core businesses), which tends to exclude technologies.

6.7.2 Science and Technology

Science and engineering practitioners and colleges of science and engineering are also sources of technology management research, teaching, and practice. These have very different research traditions, concerns, and values from those typical of business schools, and here technology management has evolved from respective cognate disciplines. This training has one great advantage—it inherently encompasses technologi-

---

*While we use "risk" in a rather generic sense, observers like Steele point out that risk, properly defined, can be managed, while uncertainty presents a much more difficult challenge.

cal dimensions. But, while it also covers at least the basics of conventional management, it suffers from management's lack of an integrative approach to technology. Typical engineering school curricula also ignore or underemphasize key issues such as standards setting, institution building and management, interplay between public policy and technology, technology forecasting, or management of specialized personnel. In addition, graduate science and technology education encourages narrower and deeper specialties, not the broader perspectives required of technology managers; producing those is mostly incidental.

6.7.3 Business

Technology management needs have scarcely perturbed "professional management" in most U.S. firms. Assumptions persist that managers are interchangeable within and between firms, industries, and businesses, despite conglomerate and other disasters. Managers are frequently recruited from one industry into another with very different technology knowledge bases, but are still expected to perform effectively (with the aid of technical experts).

Time and again, even in high-tech firms, this has produced managers who never were, or are no longer, familiar with technologies they are managing. Generalist managers' responsibilities for technology-based companies include Robert Miller's attempt to turn around Wang Computers, John Scully's failure to leverage the Macintosh architecture and Apple Computer's transient market advantages, and Lou Gertsner's apparent failure to leverage IBM's technological strengths. In each case, the individual had enjoyed remarkable success in firms with different technological bases. Selection of these highly competent generalist managers obviously did not require their understanding technology evolution or leveraging.

However, American industry produces many excellent technology managers. One path takes engineers and scientists, perhaps after 5 to 10 years of specialist experience, and exposes them to a broader vision of management, primarily through the M.B.A.[55] The technologist is provided a conventional (atechnical) management education with little opportunity to integrate that with existing technology skills. Most come away with a better appreciation of the needs and challenges of general management and a suitable set of tools for addressing technology related issues. But overall, this conversion process remains ad hoc, even while producing outstanding technology-based general managers, like George Fisher at Motorola. Few firms or business schools offer a systematic approach for technologists, and there is no good model for converting generalist managers into technology managers.[56,57]

6.8 INSTITUTIONS AND TECHNOLOGY MANAGEMENT: COMPETITIVENESS

Competitiveness at any level, firm, industry, or national, cannot be understood outside its institutional context.

6.8.1 Definitions of Institutions

Definitions of institutions are almost as varied as those of management or competitiveness. However, institutions clearly denote some kind of semipermanent structure which has its own sets of rules, regulations, procedures, codes, cultures, knowledge, routines, technologies, beliefs, paradigms, etc. Institutions do not exist in vacuum, but are parts of larger structures: suites of institutions or, perhaps more accurately, suites of structural arrangements.[54] Firms and industries exist within, and are highly influenced by, complex, interacting, and sometimes competitive groups, suites, and sets of institutions like markets, insurers, government agencies, investor groups, standards groups, media, etc. Educational systems are rooted in different kinds, levels, and combinations of institutions. Governments (at various levels) include incredibly complex groups and suites of, again interacting and sometimes competitive, institutions of varying forms, levels, and types.

6.8.2 Management and Institutions

Management teaching underemphasizes structure and institutions, although there are some excellent studies on innovation and technological change,[41] important contributions like the *dominant-firm* concept[58] (technology, markets, and industry structure), and management frameworks like Porter's five forces (industry structure analysis). Studies of specific firms or industries[59,60] are ubiquitous, but the role of technological change is rarely explored within an institutional context.[61,62] Institutionally focused academic analyses have been largely supplanted by approaches which treat institutions as outcomes of other forces,[63] omitting their role as sources of pressures, influences, and norms. That compromises our understanding of strategic management, technology management, and competitiveness.

6.8.3 Technology and Institutions

Institutions also reflect changing technologies. Industrial age technology was supported by codevelopment of a relatively simple set of institutional arrangements: at the macro level were basic intellectual property instruments such as patents and copyrights; at the industry level were instruments such as the telecommunications act of 1937 and the Federal Communications Commission. Those institutions, seemingly complex at the time and frequently reviled as new-fangled interferences in commerce, had the virtue of clear definition. Businesses, products, and markets seemed distinct. One did not confuse the auto industry with railroads or the telephone company with gas suppliers. Regulation could be relatively focused and straightforward.

Such distinctions have blurred, less from regulatory pressures than from technology. For example, overlaps and spillovers in silicon chip data processing blur previously singular product, market segment, and business categories. Similar microchips can be found in automobiles, shipboard navigation gear, airborne radar, telephone switching networks, fax machines, home computers, children's games, VCRs, etc. The institutional infrastructure that has developed around this type of product is incredibly complex. Suites of institutional arrangements exist to set technical standards, to ensure operational compatibility, to promote noncompetitive research and development, to maintain related national competitiveness, to inhibit exporting advanced technologies, to promote exporting advanced-technology products, to ensure producer and user safety, to allocate satellite positions or frequency spectra, to revise intellectual property laws, etc. The component parts of these suites are reactive, proactive, competitive, cooperative, passive, active, hyperactive, obvious, invisible—all interacting with products and with each other.

A critical problem is the inability of many institutions to keep up with technological changes, let alone anticipate them. Another significant challenge is to managements' ability to understand, navigate, leverage, manipulate, or merely survive this complexity. For example, the intricacies and politics of industry standards setting are now intimately a part of the management of technology. Most managers lack the tools to understand such changes and how they affect their business, or how to leverage these processes to their advantage.

6.8.4   Institutions, Management, and Competitiveness

The United States still dominates in number, scope, and capabilities of technology-based firms, a dominance being extended in fields like semiconductor and information technologies. That is less an indication of the health of U.S. technology management, than a reflection of structural and institutional advantages which have been widely discussed.[4] By and large, American firms still benefit from first-mover advantages in most commercially important fields of technology, but that depends on ad hoc processes at many firms.

For example, until recently a large semiconductor manufacturer managed its microprocessor business in much the same fashion as its analog components business even though the microprocessor market is based on substantially different technologies and exhibits very different institutional characteristics. As a result, it missed an opportunity in microprocessors. The firm once dominated semiconductors, but allowed leadership in this key segment to slip to a company with an inferior architecture and suboptimal technologies. Similarly, the advantages of the Ohono system invented at Toyota were well known to the U.S. automobile industry by the mid-1980s, yet as of 1995 there are still many American plants which have yet to adopt essential underlying technologies.

U.S. Management and Institutions.   American institutions' tendency to promote a managerial elite mostly lacking systematic training in technology management was tolerable when there were few competitors in technology-based industries. However, with their expansion and globalization and the end of the cold war, technology competition has accelerated and will further intensify. Some suggest that future U.S. technology industries will be characterized by a few large firms which are internationally competitive over a broad range of products. Industry structure will resemble that of a "computer bus" in which those large "anchor" companies are highly dependent on alliances with a shifting set of smaller and midsized firms that can develop, adapt, and implement technologies quickly. American economic, social, and political systems and institutions inherently encourage such firms, and features of this trend are already apparent in "virtual companies," strategic alliances, and other new corporate formations.

However, there are important caveats. Such industry alignments require managers that are alert, adroit, and well-trained in identifying, assessing, utilizing, and leveraging technologies. In a world of rapid change and exploding technology diffusion paths, the evident American dominance of technology industries could rapidly dissipate. It is vital to fully and quickly embed technology management into appropriate institutional contexts for management training, education, and development.

Japanese Management and Institutional Framework.   Automobiles, consumer electronics, semiconductors, and other products indicate that Japan will remain a fierce technology-based competitor on the world scene, but there are limited materials available in English which provide insights into Japanese technology management. One recent study[43] of a group of major Japanese corporations concluded that those firms approach technology management from individualistic and very different perspectives that are products of their institutional arrangements. Technology management in eight large Japanese firms showed surprising characteristics. Japanese firms are known for pursuing aggressive business strategies, but they tend to pursue conservative technology strategies (although there is evidence that may be changing). Their society is seen as homogeneous and conformist, but their corporate cultures reflect considerable diversity.

This diversity is underpinned by the absence of elite business schools "filtering" and "homogenizing" corporate cultures, as described earlier. With stable personnel and industry structures each firm develops a culture which is robust, unique, and reflects its organizational structure, technology management approach, and overall strategy. Since managers are trained internally, they have an intimate appreciation of their firm's technological capabilities. The strong individualistic culture, incorporating that knowledge base, produces individualistic technology strategies.

Mitsubishi Electric's major customer base is the electric utility industry, and its strategies are focused on close collaboration with utilities to develop and supply equipment and technologies which meet their needs. These relationships led Mitsubishi Electric to develop systems technology and management capabilities, in contrast to its previous concentration on heavy electrical equipment. As required, it adds specific technological capabilities to its limited set of core technologies, either through internal development or more often in alliances with customers, in areas like software and systems architecture.

Sony defines itself by technical agility. It emphasizes long-range research and visioning of new products with large potential markets, obsoleting its present products despite any technical obstacles. It concentrates on a few product functionalities and masters or acquires whatever technological capabilities those demand, rather than fitting product plans to a core of existing technologies.

NEC's goal is leadership in communications and computer (C&C) technologies over the maximum number of markets that its technology-focused strategies can penetrate. It grows, not by diversifying into other types of businesses which its core technologies might support, but by extending C&C into other application areas, such as the home.

Institutions drive certain common development paths. Diffusion-oriented intellectual property regimes encourage widespread patent copying and filing around basic inventions. The concomitant lack of technology market development (a hallmark of American competence) and dominance of industry—as opposed to government-funded research—also contribute to the perpetuation of the system. Related competitiveness issues, such as overall technical and innovation agility, are less clear, but understanding these also requires an institutional perspective.

6.8.5   Summary: Institutional Influences in the United States and Japan

The Japanese model encourages major privately owned research laboratories in each conglomerate, whose size, concentration, corporate cultures, and technical capabilities are inherently suited to (1) systematic technology development and (2) internally leveraging technologies into new businesses. By contrast, the American model, while it neglects intrafirm technology management, encourages tremendously agile technology markets. A free-enterprise orientation, relatively open markets, and deeply rooted concepts of equality before the law promote opportunities for a wider spectrum of people and firms. Suites of American institutional arrangements that define and support institutions like stock markets extend opportunities far beyond limits found in countries such as Japan. There, culture and closely held values require rules written more to exclude certain groups or entities, rather than to conform to western concepts of equity.

In the United States, complex webs of important legal, social, cultural, and public policy infrastructures are critical to adoption and implementation of new technologies regardless of source. While many U.S. businesspeople lament an excessive number of lawyers, it is precisely the competence of the American legal establishment that allows rule structures to be crafted quickly. Without them, many pioneering American technologies such as the cellular phone could not have been brought effectively and efficiently to market.

It is also clear that it would be counterproductive for the United States to adopt Japanese models, even if that were possible. Rather, improvements in technology management through evolutionary extensions of traditional strategic management appear more attractive.

6.9   COMPETITIVENESS

6.9.1   Technology and Competitiveness

Technology's ability to redefine competitiveness at all levels has long been recognized by philosophers, scholars, and businesspeople from Adam Smith to John Kenneth Galbraith to Michael Porter to Bill Gates. At industry levels, technology-driven impacts can be obvious: solid-state devices replace vacuum tubes; passenger traffic shifts from railroads to aircraft. Successes and failures also can be obvious at the firm level, but specific linkages among management, technology, and competitiveness are less apparent.

We will use, as a general guide, concepts broadly rooted in a "resource-based" model of the firm. Terminologies are still vague, but in a simplified view, the competitiveness of a company can be measured by the economic rents derived from certain capabilities (bundles of combined know-how and resources) which it possesses. These capabilities give the firm the means to provide value to customers in some differentiated manner (by scarcity, nonreplicability, etc.) which creates competitive advantages that provide the rents.

Two postulates are easily accommodated by conventional strategic management:

1. A firm's *competitiveness* is defined largely by specific *competitive advantages.*

2. A primary purpose of *strategic management* is *creation of competitive advantages.*

6.9.2   The Firm as Portfolio

The firm's task is to (1) evolve specific competencies in management (e.g., accounting, human resources, marketing), specific competencies in technologies (e.g., electronic assembly, chemical engineering, microchip design), and other necessary resources (e.g., financial assets, plant and equipment, reliable suppliers) and (2) effectively and efficiently employ these to meet customer needs. Sources of management and technical competencies can be internal—there when the firm was established, developed within it, or a combination of the two—or introduced by acquiring relevant specializations from outside the firm. Usually firms perfect those competencies by

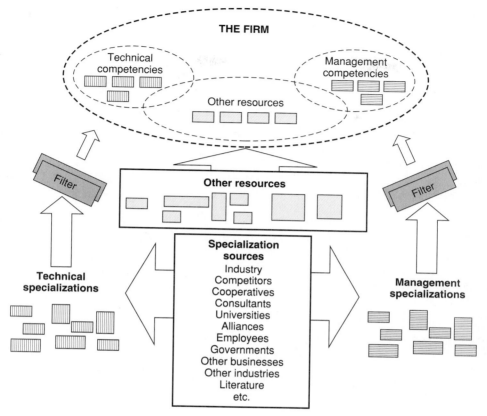

FIGURE 6.2   Specializations and resources are inducted into the firm.

some blend of acquisition and internal development. Similarly, other resources are acquired and/or developed or modified by the firm.*

Thus, a firm is a *portfolio* of management competencies, technical competencies, and other resources. Acquired management and technical specializations originate in, develop in, and are largely formed by the institutional frameworks of sources like universities, laboratories, competitors, other businesses, governments, technical and trade organizations, etc. They are inducted into the firm through filtering processes (which modify them) and are then further adapted (Fig. 6.2). Thus, the firm's competencies, technical or management, are products of complex and highly idiosyncratic processes peculiar to their sources, to the firm's specific needs, and to its unique institutional dynamics, culture, and structure.

---

*Endowments, in which the firm has a particular advantage not directly attributable to and/or under the control of current management (protected markets, valuable patents, etc.), represent a special type of resource.

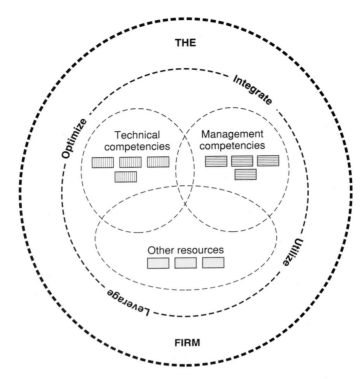

FIGURE 6.3   Specializations, competencies, and resources in the firm.

6.9.3 Management and the Portfolio

Management's task is to optimize, integrate, utilize, and leverage firm competencies and resources to competitively satisfy customer needs and demands (Fig. 6.3). The higher the level of individual competencies and the richer the resources, the more competitive the firm should be. However, observation suggests that management's optimization, integration, utilization, and leverage of these competencies and resources are more important. Those are reflected in the firm's overall competency (its metacompetency[2]) and competitiveness. While it is easier to visualize this in high-tech companies, it is equally true of "nontechnical" firms since both have portfolios of management competencies, technical competencies, and other resources.

The need for competent management of that portfolio becomes starkly apparent when one of its three elements is inadequate. Stories abound regarding incompetent management or lack of resources. However, the consequences of lacking updated technical competencies can be seen even in nontechnical firms. One example is the graphic arts business which has largely shifted from mechanically assisted hand-crafted graphics to computer-generated graphics. Many firms, failing to recognize a link between creativity and craftsmanship and what was then considered mere data-processing equipment, suffered serious competitive damage.

High-tech firms are not immune to such problems. American computer mainframe producers, for a variety of reasons, kept ignoring the personal computer. As a result, Unisys, NCR, DEC, Honeywell, and other major firms were never players in, have never made serious profits in, or have effectively disappeared from the PC market. Even IBM, which set the industry standard, lost its leadership position and is struggling despite the richest of technology portfolios.

6.10 STRATEGIC ARCHITECTURE: AN INTEGRATIVE APPROACH

A more recent strategic management approach from Hamel and Prahalad provides a step toward dynamic analyses. *Strategic architecture*[2] is a more holistic approach and better lends itself to inclusion of technological issues. It is not a detailed, rigid plan, but rather a broad, top-level, opportunity-oriented blueprint of the firm's future: what will the firm provide its customers (what "functionalities" will be included, what value or benefit added); what resources will it need (new skills, technologies, capabilities or *core competencies*); how does the interface between firm and customer (or the interface between firm and supplier) need to be changed?

The capstone of strategic architecture is *strategic intent,* the "animating dream" that provides intellectual and emotional bases for thrusting the firm forward and for completing its journey. It obviously encompasses what is generally called "vision," but goes beyond it by including strong motivational elements; many vision and mission statements tend to be generic, interchangeable, and bland. It defines where and what the firm wants to be and at least implicitly promotes a stretch from existing capabilities and strengths to meet new challenges. Thus, strategic architecture incorporates traditional strategic management elements, but forces a more strategic point of view. It also addresses what the firm must do at the present time to achieve the future position it has defined, thereby effectively mating short-term needs with longer-term opportunities.

6.10.1 Core Competencies

Central to strategic architecture is the firm's portfolio of *core competencies,* bundles of individual skills and capabilities which enable it to provide value or benefit to customers. These differentiate the firm from its competitors and underlie whatever market leadership it enjoys.[2] Strategic management must be the driver for optimizing, integrating, utilizing, and leveraging the firm's competencies and resources into its core competencies (Fig. 6.4).

The identity, self-image, market image, and public image of a firm are almost always based on specific products, services, or business units. Those are important, but the firm's portfolio of competencies, integrated sets of specific skills, technologies, and capabilities, are central or "core" to its meeting strategic goals and objectives. They are not conventional assets, mere infrastructure, or endowments. Not every distinctive competitive advantage is a core competency. They can become irrelevant with time or they may evolve into industrywide baseline capabilities. They do not wear out, but should become more effective with use. They should also be extendible or expandable into gateways to future competitive advantage.[2]

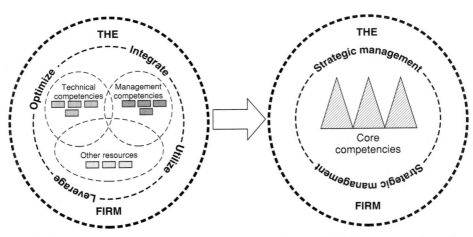

FIGURE 6.4   Optimization, integration, utilization, leverage: specializations, competencies, and resources become core competencies.

6.10.2   Competencies, Capabilities, and Competitiveness

Core competencies may seem clear in simple situations. Core competencies *A, B,* and *C* are used in different combinations to produce the firm's products (or core competency platforms) 1, 2, 3, and 4 (Fig. 6.5). These platforms create competitive advantage

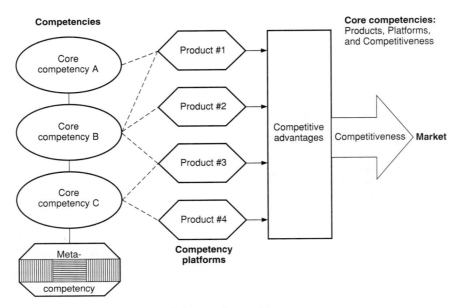

FIGURE 6.5   Core competencies, platforms, and competitiveness.

and firm competitiveness, but those also come from the firm's metacompetency and strategic architecture. This common type of core competency diagram may overemphasize products.

We use the term *core competencies** to specifically encompass that optimization, integration, utilization, and leveraging of technological competencies, management competencies, and other resources. Technologies, specializations, management skills, and other resources are thus effectively combined in core competencies.

They are by no means easy to identify and isolate.[2] Logically grouping organizational, financial, technical, and a host of other constituents into true core competencies is far from a trivial task, and results may be more instructive than precise. It is a process that requires the time and attention of a range of knowledgeable and competent people who have been given the freedom and the proper tools for the task. In that regard at least, core competencies, while a distinct concept, share some elements with value-chain analyses.

6.10.3 Core Competencies and Individual Firm Competitiveness

There exist a myriad of approaches, paths, or decision patterns, by which specializations, competencies, and other resources are converted into individualistic sets of core competencies. Those differences reflect situational and institutional factors at different times in different businesses. There is a wealth of information and material regarding transferring external technologies or developing or transferring internal technologies,[64,65] outsourcing key operations or core competencies,[16] and developing new management approaches.[66] All reflect efforts toward defining and establishing competitive advantages and metacompetencies, and the more recent particularly stress the importance of the integrative role of management. That idea is clearly central to this framework and particularly highlights what an impediment a management gap can become.

6.10.4 Extending Core Competency Analysis

For complex operations, it may be necessary to further break down single core competencies. The next stage we will call *specific competencies.* These are narrowly based sets of skills, processes, resources, and technologies that, although important, are unlikely to significantly influence the metacompetency of the firm. For even more complex operations, these can be further subdivided into highly specific *individual competencies.* This analytical fine structure can uncover elements which are more easily identified and isolated. Technological change impacts on them are frequently more readily analyzed and evaluated. Thus, such analysis starts with a top–down deconstruction of firm activities, examines environmental changes, determines individual impacts, and then reconstructs those to estimate overall firm impacts.

This technique is illustrated in Fig. 6.6. It begins by (1) identifying and isolating core competencies, then (2) breaking each down into specific competencies, as required, then (3) further breaking each of those into individual competencies (specific management tasks or needs, specialized technical skills, particular plant or equipment requirements, etc., as required), (4) identifying relevant change indicators from environmental scanning data, and (5) matching those to individual competencies (or specific or core competencies, depending on the degree of complexity).

---

*The term *capabilities* is often used, rather than *competencies,* when other resources are included.

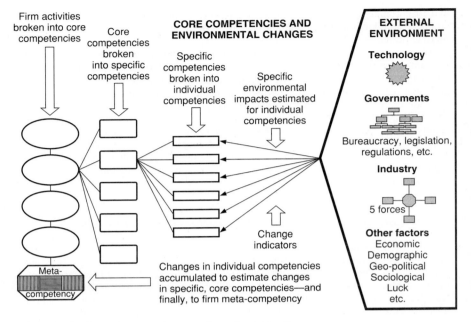

FIGURE 6.6 Core competency deconstruction and reconstruction.

Especially for technological changes, standard technology deconstruction and decomposition techniques[67] can be used to gauge probable trajectories into at least the near future. For example, the primary technologies underlying a given process (or instrument or production tool) can be determined, and technology roadmaps established for each (anticipated rates and directions of change, possible limits, etc.). For other types of individual elements (personnel or skill requirements, new plant or equipment needs, etc.), similar attention to relevant skill levels, training programs, personnel policies, staffing, financial plans, etc., is required.

In the most complex cases, anticipated effects of such changes on individual competencies are accumulated, and those results projected to find the likely impacts on that specific competency (less complex analyses might start with changes in specific competencies). Once this information is assembled for all specific competencies which constitute a core competency, the impacts of all such changes on that core competency can be estimated. If this can be done for each core competency, potential changes in the firm's metacompetency—and thus its competitiveness—can be deduced.

Obviously, such a process can consume appreciable time and resources. However, top management may find much of the detail work already being done by individual technical or management specialists working at the specific competency level. Creation of a supportive structure which permits systematic, continuous accumulation, evaluation, and use of such information is vital.

A Simple Example. The metacompetency of a microchip manufacturer is supplying high-complexity microprocessors. While other competencies are certainly required of

TABLE 6.4 Competency Analysis of Semiconductor Manufacturing

Activity level	Competitiveness significance	Typical activities	Activity selected for analysis	Activity change drivers
Metacompetency	Metacompetency (or small set of metacompetencies) defines overall firm competitiveness	Device design, marketing, and manufacturing	Manufacturing complex, submicrometer integrated circuit chips ("front end")	Markets, technologies, economies, governments, etc.
Core competency	Core competencies interact in complex, interlocking patterns to define firm metacompetency	Wafer preparation, Oxide deposition, Photolithography, Diffusion, Ion implantation, Contamination control	Photolithography	New technologies, government regulations of materials, equipment and operating costs, etc.
Specific competency	Specific competencies interact with others to form core competencies—insufficient by themselves to impact metacompetency	Photoresist selection, Photoresist preparation, Photoresist application, Alignment and exposure, Photoresist processing, Etching (wet and dry), etc.	Alignment/exposures	Increasing demand for smaller geometries, new illumination sources, larger wafers, new alignment techniques, etc.
Individual competency	Individual competencies are impacted by changes in firm and external environments—may or may not appreciably change specific competency	Photomask selection, Photomask handling, Stepper selection, Stepper maintenance, Exposure process, etc.	Photomask selection	Impacted by new technologies, supplier changes, substrate materials, mask materials, geometry requirements, wafer surface characteristics, etc.

Change indicators: Technology roadmaps, staffing and training plans, product announcements, etc.

6.25

such a firm, we will examine only its manufacturing metacompetency (Table 6.4). Manufacturing complex microchips with submicrometer feature sizes requires serial-processing silicon wafers through many discrete manufacturing operations. In many of these the firm may well possess a competency equivalent to that of its primary competitors, a standard industry-level competency. For this example, we assume the firm has a core competency in the critical photolithography operation, that set of skills, resources, and technologies which produce precisely defined and aligned photographically derived patterns on the chip surfaces. That would be a core operation since a substantial advantage here could translate directly into technical, operational, economic, or other competitive advantages and enhance the firm's metacompetency.

The core competency of photolithography is made up of specific competencies such as the application, processing, exposure, and subsequent removal of photoresist (a photosensitive coating), the alignment of different pattern layers, wet (chemical) or dry (plasma) etching steps, etc. (Table 6.4). Each of these specific competencies is identifiable, separable, and important (but not important enough by itself to impact competitive advantage). Each requires its own subset of individual competencies: component skills and technologies which are supported by particular personnel, facilities, equipment, materials, controls, etc. Most of these also may be industry standards, but a leading firm will likely have some which are distinctly different and competitively advantageous. It is also the selection, development, integration, synthesis, and management of these interacting and overlapping individual competencies (individual management skills, technologies, support elements, etc.) and resulting specific competencies that create core competencies.

Each of these individual competencies is subject to influences and change pressures from the external (and likely from the internal) environment. Specific chemicals or operations can be restricted by environmental regulations, new technologies are developed (e.g., x rays replacing ultraviolet exposure systems), managers or managerial policies mutate (perhaps as firm ownership changes), economic conditions change (impacting plant and equipment investment decisions), market demand shifts to different products, and so on. The probable impact of each such change on each individual competency can be estimated, the accumulated effects of all relevant changes on all individual competencies of a specific competency gauged, and resulting impacts on core and metacompetencies judged.

Extending the Example.   In actuality, the overall competitiveness or metacompetency of such a firm would require more than just manufacturing. Device design would be critical and would include (1) the skills of creative engineers, (2) highly advanced CAD (computer-aided design) and simulation hardware and software, (3) enlightened design management, and (4) up-to-date physical resources and other technologies, skills, and capabilities. Again one could find interacting and overlapping sets of individual competencies, specific competencies, and core competencies. Marketing requires yet another set of competencies, again interacting and overlapping with others, including those in manufacturing and design. Thus, a full-scale core competency analysis of such a complex firm would itself be complex—and time-consuming. But even partial analysis, or examination of only part of the value chain, can prove enlightening.

6.10.5   Summary: Using This Type of Analysis

As noted, such impacts are scarcely limited to high-tech or technology-based or manufacturing firms. Service, craft, or virtually any type of business is continually modifying its choices of technologies (broadly defined). Perhaps a new software release displaces a familiar word processor in an insurance office, or a more sophisticated copy machine is acquired. These affect people, previous technical choices, and firm operations, but are not enough by themselves to have much impact on a core competency (e.g., selling homeowners' insurance policies) or firm competitiveness. Over a long enough period of time, of course, failing to keep pace with such technical opportunities could have serious consequences.

New core competencies are sometimes required. In the United States, improved steel minimill technologies and practices allow these firms to produce higher-quality and more specialized products, taking market share from integrated steel producers. Those improvements required considerable reworking of technologies, equipment, and procedures—and retraining of personnel—to develop new core competencies. Many other firms have had to add data processing, global marketing, or communications competencies in order to survive, grow, and prosper, changes often reflected in new equipment, personnel, and facilities.

Finally, entire industries can be disrupted; as semiconductors replaced vacuum tubes, none of the principal U.S. vacuum-tube manufacturers survived to be significant semiconductor suppliers. Here, one sees the impact of major technological discontinuity. The invention of the transistor and integrated circuit within a 10-year span created new products requiring entirely new sets of core competencies and many new supporting specific competencies. Earlier core competencies—and factories and firms—were either made obsolescent or reworked. Schumpeter's "winds of creative destruction" are most apparent in such situations in which technology directly and suddenly redefines products. However, they are also blowing, albeit more softly, in the more continuous and less visible changes discussed above.

6.11 WHAT IS TECHNOLOGY MANAGEMENT?

What do we mean by *technology management?* If *management* is about getting other people to do what you want, *technology management* is about getting people and technologies working together to do what you want. Technology management is a collection of systematic methods for managing the process of applying knowledge to extend the range of human activity and produce defined products (goods or services). It is not about managing only technical specialists in technology-based businesses, but includes that conventional, but very limited definition in a holistic and integrative approach. Effective technology management synthesizes the best ideas from all sides: academic, practitioner, generalist, or technologist.

6.11.1 The Technology Manager

Managing technology is about systematically applying know-how—both technical and managerial—to meet needs by getting people to produce particular goods and services and thereby create competitive advantages for the firm. Thus, a technology manager must have three critical and interlocking skill sets.

First, the manager must be a change manager, in both the usual sense of possessing skills for implementing effective organizational changes and also in a technology-change sense. Second, as with any good change manager, the technological change manager must have a deep and practical understanding of the institutional structures being affected by the change and those affecting the change and must be familiar with the technologies involved, but also comprehend the institutional basis in which these technologies exist or will operate. Third, the technological change managee must understand the fundamentals of relevant technology change mechanisms and trajectories and effectively operate within relevant technological, political, economic, and management systems.

Thus, the profile of a good technology manager should be that of a person who combines solid competence in (1) general management (hopefully available at most business schools), (2) change management (taught, but generally underemphasized at business schools), and (3) overall technological literacy (well beyond that available from almost any business school). Table 6.5 represents a partial, certainly not a comprehensive, review of the skills required. No individual skill is really confined to the category selected, but typically spills over and onto many others in complex, situational patterns.

6.11.2   The Critical Nature of Context

Again we emphasize the institutional dimension. A competent technology manager must be institutionally informed and alert and understand the suites of institutional arrangements that influence the industry, the firm, and managerial actions. These must include, but not be limited to, those involved in strategic alliances; the import and export of goods, services, and technologies; significant sources of management and technical specializations—and modification mechanisms—within and outside the firm; regulatory and economic institutional structures; sociopolitical structures; etc. These include all the various elements and players one would expect: governments, trade groups, labor unions, educational institutions, technical and industry associations, competitors, etc. Managerial requirements also include specific skills in cross-cultural negotiations, knowledge of technology change and diffusion patterns and mechanisms, influences and establishment of industry and product standards, etc. Effective technology management cannot exist within hermetically sealed organizational units.

Obviously, the ideal technology manager is rare. Such a manager has to have excellent managerial skills, competent political skills, and a good technical network in addition to the normal requirements of an excellent general manager. The previous discussion regarding the institutional context in which American technology managers are educated and operate reveals at least some of the historical reasons why American institutions have not been successful in generating a large cadre of such individuals.

6.12   SUMMARY AND CONCLUSIONS

Technology management remains the stepchild of professional management, both academic and practitioner. Technology plays an ever-growing role in our increasingly complex business environment. Whether because of simple ignorance, chronic technophobia, or institutional rigidity, management education in the United States has, for the most part, neglected this primary business and economic driver. But it is scarcely unique; businesses large and small have largely followed a similar course.

We have taken a narrow approach: seeking linkages which connect strategy, technology, and competitiveness. In our critique of conventional management theory and practice and in defining what approaches appear beneficial, we emphasize institutional contexts. This area of study has been badly neglected in technology management,

TABLE 6.5 Proposed Skill List for Technology Managers

Change-related skills	Institution- and structure-related skills	Analysis and/or synthesis
Understanding and competence in managing technical, organizational, and other types of changes demanded by evolving competitive environments and firm capabilities	Familiarity with and ability to use basic intellectual property mechanisms and institutions	Imagination, understanding, and creativity in finding new ways to leverage the technology base of the firm
The capacity to understand and learn the capabilities and limitations of new technologies rapidly; a broad-based, not a narrow, technical expertise	A keen appreciation of both "history" and "structure," or the fit between existing technologies and systems and new or potential technologies and systems	Familiarity with tools such as relevance trees and technology deconstruction in discerning significant performance and cost drivers, basic underlying technologies, trends, limits, etc.
Understanding patterns of technical change—e.g., incremental vs. discontinuous—and strategic implications	Ability to utilize a systems perspective on technology—how spillovers occur, how a new technology must fit into an overall system (or be powerful enough to overturn it)	Familiarity with basic technology forecasting concepts and methods
Membership in a network of competent, trustworthy technical experts capable of providing both information and "calibration" on internal expertise	Familiarity with and an understanding of the organizations, functions, and significance of industry standards setting (including knowledge of how to use these to advantage)	Ability to identify, supervise the creation of, and understand the relevance of technology roadmaps
Sufficient knowledge and understanding to do realistic and reliable environmental scanning—or to know how and where to get the necessary information	Ability to comprehend and manage processes of system integration	Ability to analyze and objectively evaluate the performance of technical experts (or the ability to have that done effectively and competently by others)
	A clear understanding and competence in managing the interplay between public policy, standards-setting institutions, and business development	
	A similar understanding and competence in regard to other areas of government, at all levels and on a global basis, regarding assistance, regulation, and intervention	
	An understanding of the mechanisms of technology transfer (diffusion, technology push/pull, typical paths, etc.)	

compromising our understanding of its real scope and role. Also, we have not explored the vivid tapestry of what comprehensive technology management curricula and practice can encompass: forecasting techniques, technology-change patterns, technology diffusion and spillover mechanisms and trajectories, intellectual property issues and approaches, technology transfer scenarios, linkages with business process reengineering and quality management paradigms, and many others. This is an area of incredible richness for both academics and practitioners—and of incredible value for firms, industries, and nations.

Our conclusions are straightforward:

1. Traditional strategic management paradigms, models, frameworks, and techniques have limited usefulness for technology management. Indeed, their shortcomings in accommodating change and for dynamic analyses are ever more clearly recognized.

2. Treating technologies and technologists as something separate and apart from mainstream management is both risky and futile. The management gap must be closed.

3. In similar fashion, ignoring the institutional contexts of management—especially technology management—sharply limits the credibility of any analysis.

4. Newer strategic management thrusts, particularly strategic architecture and its associated elements, have capabilities for illuminating technological change impacts in management. These concepts may not yet provide a cohesive dynamic theory of strategic management, but studies of technological change trajectories and impacts which employ these frameworks may well point the way.

Finally, there remains the central problem of technology management: generalist managers' widespread ignorance regarding technology. Enhancing the value of an M.B.A. for technologists appears relatively straightforward, using some of the integrative materials discussed. Turning generalists into technology managers raises more difficult problems. Perhaps one clue lies in the advice of an accounting professor who told his M.B.A. class, "I don't really want to turn you into accountants; I just don't want you to be snowed by an accountant." We must find methods which provide generalist managers with accessible, practical means to acquire a working understanding of the nature and characteristics of technological change. As Lowell Steele put it, "...nontechnical managers dare not leave technology to the specialists."[8] To do so would present the greatest risk, by any definition of that term, of all.

6.13   REFERENCES

1. Richard A. D'Aveni and Robert Gunther, *Hypercompetition: Managing the Dynamics of Strategic Maneuvering,* Free Press, New York, 1994.

2. Gary Hamel and C. K. Prahalad, *Competing for the Future,* Harvard Business School Press, Boston, 1994.

3. David Angel, *Restructuring for Innovation: The Remaking of the U.S. Semiconductor Industry,* Guilford Press, New York, 1995.

4. John Kanz, "Origins of the U.S. Semiconductor Resurgence: Products, Perceptions, Paradigms," *Internatl. J. Technology Management* (1995).

5. Shaker A. Zahra, Sarah Nash, and Deborah Bickford, "Transforming Technological Pioneering into Competitive Advantage," *Acad. Management Executive,* **IX**(1): 17, Feb. 1995.

6. J. Fred Bucy, "Technology Transfer and East-West Trade: A Reapprisal," in *National Security and Technology Transfer: The Strategic Dimensions of East West Trade*, Gary K. Bertsch and John R. McIntyre, eds., Westview Press, Boulder, Colo., 1983, pp. 198–200.

7. Shoshana Zuboff, *In the Age of the Smart Machine*, Basic Books, New York, 1988.

8. Lowell W. Steele, *Managing Technology: The Strategic View*, McGraw-Hill, New York, 1989.

9. Michael Hammer and James Champy, *Re-Engineering the Corporation: A Manifesto for Business Revolution*, Harper Business, New York, 1993.

10. Taichi Sakaiya, *The Knowledge-Value Revolution*, Kodansha America, New York, 1991 (originally published in Japanese by PHP Kenkyujo, Kyoto, 1985).

11. Simon Ramo, as quoted in Devendra Sahal, *Patterns of Technological Innovation*, Addison-Wesley, Reading, Mass., 1981.

12. Vannevar Bush, *Endless Horizons*, Arno Press, New York, 1975 (reprint of original publication by Public Affairs Press, Washington, D.C., 1946).

13. Richard Swedberg, *Schumpeter: A Biography*, Princeton University Press, Princeton, N.J., 1991.

14. Eliot Marshall, "Nobel Prize for Economic Growth," *Science*, 754, Nov. 6, 1987.

15. Jon Elster, ed., *Karl Marx: A Reader*, Cambridge University Press, Cambridge, U.K., 1986 (excerpt from *Das Kapital*, vol. 1).

16. James Brian Quinn, *Intelligent Enterprise: A Knowledge and Service Based Paradigm for Industry*, Free Press, New York, 1992.

17. Don Tapscott and Art Caston, *Paradigm Shift*, McGraw-Hill, New York, 1993.

18. Jaclyn Fierman, "What Happened to the Jobs?" *Fortune*, 40, July 12, 1993.

19. Larry O. Natt Grant II, "An Affront to Human Dignity: Electronic Mail Monitoring in the Private Sector Workplace," *Harvard J. Law Technol.*, **8**(2): 345, 1995.

20. Lester Thurow, *Head to Head: The Coming Economic Battle Among Japan, Europe, and America*, Morrow, New York, 1992.

21. Clyde Prestowitz, *Trading Places: How We Allowed Japan to Take the Lead*, Basic Books, New York, 1988.

22. Paul Krugman, "Competitiveness: A Dangerous Obsession," *Foreign Affairs*, 28, March–April 1994.

23. d'Aveni, op. cit. (Ref. 1), Preface.

24. Ibid., p. 17.

25. Richard A Goodman and Michael W. Lawless, *Technology and Strategy: Conceptual Models and Diagnostics*, Oxford University Press, New York, 1994.

26. Thomas Peters and Robert Waterman, *In Search of Excellence*, Harper & Row, New York, 1982.

27. Robert Hayes and William Abernathy, "Managing Our Way to Economic Decline," *Harvard Business Review*, **58** (4): 67, July–Aug. 1980.

28. C. Chet Miller and Laura B. Cardinal, "Strategic Planning and Firm Performance: A Synthesis of More than Two Decades of Research," *Acad. Management J.*, **37**(6) 1649, Dec. 1994.

29. Henry Mintzberg, *The Rise and Fall of Strategic Planning*, Free Press, New York, 1994.

30. Michael Porter, "Towards a Dynamic Theory of Strategy," *Strategic Management J.*, **12**: 95, Winter 1991.

31. David Kearns and David Nadler, *Prophets in the Dark: How Xerox Reinvented Itself and Beat Back the Japanese*, Harper Business, New York, 1992.

32. Michael Porter, *Competitive Advantage: Creating and Sustaining Superior Performance*, Free Press, New York, 1985.

33. Michael Porter, *Competitive Strategy: Techniques for Analyzing Industries and Competitors,* Free Press, New York, 1980.
34. Jane C. Linder and H. Jeff Smith, "The Complex Case of Management Education," *Harvard Business Review,* 16, Sept.–Oct. 1992.
35. Ronald Stone and Richard Wines, "Management Education Reform: Opportunities for Metropolitan Business Schools," *Metropolitan Universities,* 20, Winter 1992.
36. Richard Pascale, *Managing on the Edge: How the Smartest Companies Use Conflict to Stay Ahead,* Simon & Schuster, New York, 1990.
37. "Re-Engineering Reviewed," *Economist,* 66, July 2, 1994.
38. Jong-Song Chiang, "The Research Strategy in Management of Technological Innovation," *Technol. Forecasting Social Change,* **37**(3) 32, Nov. 1993.
39. Robert Burgleman and Modesto Maidique, *Strategic Management of Technology and Innovation,* Irwin, Homewood, Ill., 1988.
40. David I. Cleland and Karen M. Bursic, *Strategic Technology Management: Systems for Products and Processes,* AMACOM, New York, 1992.
41. James Utterback, *Mastering the Dynamics of Innovation,* Harvard Business School Press, Boston, 1994.
42. Charles B. Wang, *Techno Vision: The Executive's Survival Guide to Understanding and Managing Information Technology,* McGraw-Hill, New York, 1994.
43. Lewis M. Branscomb and Fumio Kodama, *Japanese Innovation Strategy: Technical Support for Business Visions,* CSIA Occasional Paper no. 10, University Press of America, Lanham, Md., 1993.
44. Peter Drucker, "The Theory of the Business," *Harvard Business Review,* **72**(5): 95, Sept.–Oct. 1994.
45. John Alic, Lewis Branscomb, Harvey Brooks, Ashton Carter, and Gerald Epstein, *Beyond Spinoff: Military and Commercial Technologies in a Changing World,* Harvard University Press, Boston, 1992.
46. John Holusha, "The Risks for High Tech, When Non-Techies Take Over," *New York Times/National Ed.,* p. F7, Sept. 5, 1993.
47. Mary Good, "Communication of R&D to Management: A Two-Way Street," *Research Technol. Management,* **34**(5): 42, Sept.–Oct. 1993.
48. Paul S. Adler and Kasra Ferdows, "The Chief Technology Officer," *Calif. Management Review,* 55, Spring 1990.
49. Priscilla O'Clock and Marily Okleshen, "A Comparison of Ethical Perceptions of Business and Engineering Majors," *J. Business Ethics,* **12**(9): 677, Sept. 1993.
50. S. Ram and Hyung-Shik Jung, "'Forced' Adoption of Innovations in Organizations: Consequences and Implications," *J. Product Innovation Management,* **8**: 117, 1991.
51. Urs Daellenbach and Michael Waterhouse, "Management Systems for R&D and Technology: A Study of Oil and Gas Companies," University of Calgary, Faculty of Management, Working Paper WP94-47, Oct. 1994.
52. John Kanz and Michael Waterhouse, "Technology Management: An Escalating Challenge for Business and Academia," *Business and the Contemporary World* (special issue on competitiveness), 79, May 1995.
53. David Teece, "Contributions and Impediments of Economic Analysis to the Study of Management," in *Perspectives On Strategic Management,* James W. Frederickson, ed., Harper Business, New York, 1990.
54. Cal Clark and Danny Lam, "The Competitiveness Debate: Recognizing and Transforming the Institutional Context of National Economic Behavior," *Business and the Contemporary World* (special issue on competitiveness), 12, May 1995.
55. Alan Chapple, "Technology Management: New Kind of MBA," *Cost Engineering,* **35**(10): 25, Oct. 1993.

56. William A. Weimer, "Education for Technology Management," *Research Technol. Management,* **40**, May 1, 1991.

57. James Hall et al., "Training Engineers to be Managers: A Transition Tension Model," *IEEE Transact. Engineering Management,* **39**(4): 296, Nov. 1992.

58. William Abernathy and James Utterback, "Patterns of Industrial Innovation," *Technol. Review,* **80:** 40, June–July 1978.

59. Michael Detrouzos, Richard Lester, and Robert Solow, *Made in America: Regaining the Productive Edge,* MIT Press, Cambridge, Mass., 1989.

60. Michael Borrus, *Competing for Control: America's Stake in Microelectronics,* Ballinger, Cambridge, Mass., 1988.

61. Bela Gold, *Explorations in Managerial Economics: Productivity, Costs, Technology, and Growth,* Basic Books, New York, 1971.

62. John Kanz, "Strategy Evolution and Structural Consequences in the U.S. Semiconductor Industry," *Technovation,* **14**(4): 221, 1994.

63. James G. March and Johan P. Olsen, *Rediscovering Institutions: The Organizational Basis of Politics,* Free Press, New York, 1989 (see p. 10).

64. Stephen K. Markham et al., "Champions and Antagonists: Relationships with R&D Project Characteristics and Management," *J. Engineering Technol. Management,* **8:** 217, Dec. 1991.

65. Fumio Kodama, "Technology Fusion and the New R&D," *Harvard Business Review,* 70, July–Aug. 1992.

66. Innumerable citations can be given: Peter M. Senge, *Fifth Discipline: The Art and Practice of the Learning Organization,* Doubleday (Currency), New York, 1990; Rosabeth Moss Kanter, *When Giants Learn to Dance:The Post-Entrepreneurial Revolution in Strategy, Management, and Careers,* Simon and Schuster, New York, 1989; George Stalk, Jr., and Thomas M. Hout, *Competing Against Time: How Time-based Competition Is Reshaping Global Markets,* Free Press, New York, 1990.

67. J. O. Hilbrink, "Technology Decomposition Theory and Magnetic Disk Technology," *IEEE Transact. Engineering Management,Ó* **39**(4): 284, Nov. 1, 1990.

CHAPTER 7

THE CONTRIBUTION OF TECHNOLOGY TO COMPETITIVE ADVANTAGE

Theodore W. Schlie
Center for Innovation Management Studies
Lehigh University
Bethlehem, Pennsylvania

7.1 INTRODUCTION

An essential component of managing technology is recognizing the role that technology plays in the competitive success of a firm in a free-market economy, and acting to ensure that technology decisions and policies contribute to the firm's competitive advantage. This chapter provides a framework which the manager can use to analyze and understand the linkages between technology and competitive strategy and/or competitive advantage of a firm in any given situation.

7.2 DEFINING COMPETITIVE STRATEGY AND COMPETITIVE ADVANTAGE

Many scholars, including Chandler,[1] Mintzberg,[2-4] Andrews,[5] and Quinn,[6,7] have contributed to the concept of *strategy*. The traditional approach to strategy has emphasized setting goals and developing the means to achieve them by matching the resources of the firm (strengths and weaknesses) with opportunities (and associated risks) in the external environment (which includes, especially, customers and competitors), and deciding which industries, businesses, or product-market segments to compete in. For example[8]

> ...corporate strategy is the pattern of decisions in a company that (1) determines, shapes, and reveals its objectives, purposes, or goals; (2) produces the principal policies and plans for achieving these goals; and (3) defines the business the company intends to be in, the kind of economic and human organization it intends to be, and the nature of the economic and non-economic contribution it intends to make to its shareholders, employees, customers, and communities....

7.2.1 Corporate Strategy

Strategy is a term that has become popularized, however, with terms such as *marketing strategy, manufacturing strategy, competitive strategy, corporate strategy, business strategy,* and *technology strategy* being widely used, and it is necessary to sort through their different meanings by considering different levels of organization of the firm. At the highest, corporate level, there is the multiindustry-business firm, the conglomerate or the diversified firm, of which there are many examples, such as General Electric. At this level, corporate strategy addresses issues such as choosing which industries or businesses to be in, utilizing portfolio management techniques to achieve a balance among the industries or businesses chosen, and—in the case of a diversified firm—achieving synergy among the industries or businesses chosen.

7.2.2 Competitive Strategy

Moving down a level, there is the single industry-business firm such as Bethlehem Steel, or the division of the multiindustry-business firm such as GE Aircraft Engines. At this level, the issue of which business or industry to be in is moot—one is in the steel or the aircraft engine industry, and the only issue is how to compete successfully in it, or get out. This is the level at which we are concerned with *competitive strategy,* a concept that is perhaps most closely associated with Michael E. Porter,[9,10] who expresses it as follows:[11]

> Essentially, developing a competitive strategy is developing a broad formula for how a business is going to compete, what its goals should be, and what policies will be needed to carry out those goals.

7.2.3 Functional Policies

Moving down one more level, there are the functions of the firm—the familiar R&D, engineering, manufacturing and production, marketing, sales, service, personnel and human resources, purchasing, accounting, finance, planning, etc., functions. Although the term *strategy* is widely used at this level, in order to avoid confusion I will use the term *policies*—e.g., *manufacturing policies,* instead of *manufacturing strategy.* Policies must be formulated and implemented in any function of the firm, such as how to structure the organization, how to allocate resources, and how to reward people. The important point to note is that functional policies must all support, reinforce, or contribute to the competitive strategy of the firm in order for the firm to compete effectively.

7.3 CHARACTERISTICS OF COMPETITIVE STRATEGY

Competitive strategy is a top-management responsibility—it is difficult to imagine something more important to a firm than deciding how it will compete in its industry—but there must be organizationwide participation in both its formulation and implementation. Information must be communicated upward regarding the capabilities and competencies, the strengths and weaknesses, of the firm in order for a feasible competitive strategy to be formulated, and once a competitive strategy has been formulated, it must be implemented through policies and decisions made in all the functional activities of the firm. In practice it is always difficult to resolve the chicken-egg issue: which should come first—the strategy or the capabilities and/or competencies. In reality, it is an iterative process in which the chicken-egg issue is irrelevant.

Competitive strategy is a long-term phenomenon—a firm cannot have a "strategy of the month." There is a long-term need for stability *and* a short-term need for continual updating in response to evolutionary changes in the environment. But what is *long-term* can vary from industry to industry, depending on its internal dynamics. A long-term strategy in the electronics industry might be 2 to 3 years, but in the chemical industry it might be 8 to 10 years or more. When there is a discontinuous change in the environment—a world oil embargo, the ending of fixed exchange rates, a world war—of course, all strategies need to be examined for major changes or reformulation.

7.4 COMPETITIVENESS

7.4.1 Competitiveness Definition

In a free-market economy, I stipulate that the generic goal of industrial firms is *competitiveness,* which I define as follows: *Competitiveness is the ability to get customers to choose your product or service over competing alternatives on a sustainable basis.*

This definition of competitiveness has the virtue of being operational, as compared with other definitions which include factors of *free and fair market conditions, standard of living,* and *maintaining or expanding the real income of its citizens,*[12,13] and are too complex to be measured. Competitiveness as defined above is measured by market share trends over time, and can be described in terms such as increasing, decreasing, or stable. The scope of the market share is dependent on the scope of competition—for most manufactured products today, it is a global scope and a world market share that are relevant. Competitiveness, then, is not a single point in time number. To say that the U.S. commercial large transport aircraft industry accounted for a 65 percent share of the world market in year X may indicate that the United States dominates this industry, but it does not indicate how the U.S. industry is doing competitively. For this, one also has to know what the U.S. world market share was in years $X-1$, $X-2,...,X-n$, and in this particular case, although the United States may still dominate the industry, it has been losing competitiveness (world market share) to the European Airbus Consortium over the past 10 years or so. Why this is happening is a different matter, a matter that is the subject of *competitive analysis* or *competitive assessment.*[14]

7.4.2 Sustainability

There is an important caveat contained in the definition described above, however, which is that competitiveness must be built *on a sustainable basis.* It is possible, in the short run, for a firm to get customers to buy its products or services over competing alternatives on an unsustainable basis by, for example, mortgaging its assets and using the proceeds to subsidize and lower prices, thus attracting customers until the proceeds run out and the firm collapses. In this context of sustainability, a discussion of profits and the tradeoff with growth and market share is useful.

7.4.3 Profitability

A certain threshold of profits is necessary for any firm to be able to compete on a sustainable basis. In the absence of profits, firms cannot continue to exist—even the largest of firms. Eventually the patience of lending or investing institutions, of stockholders, even of governments, will run out and the firm will die. So even as our focus of competitiveness is on market share, we cannot ignore profits—they must be maintained at least at some threshold level in order to sustain the competitiveness. What this threshold level is may vary from industry to industry, and—more importantly—from nation to nation, and may be influenced by national policy. But I would argue that the goal of competitiveness is not to maximize profits, but to sustainably increase market share, and I would argue the converse point even more strongly—that it is very possible to be losing competitiveness even as a firm is achieving record levels of profit.

7.4.4 Customer Focus

Many observers of competition begin with a focus on firms and what firms are doing in R&D, manufacturing, or marketing to improve themselves. This is the wrong place to begin. Competitiveness is determined by customers, by the collective decisions they make in purchasing one product or service over competing alternatives. If we can assume that such purchasing decisions have some rational basis (which may not always be the case—consider the purchase of a pet rock), then customers will have *reasons* for making this decision which reflect what they value. These reasons I call *competitive advantages*.

7.5 COMPETITIVE ADVANTAGES

If we ask why customers choose the products or services that they do, a number of generic reasons come to mind:

The *price* is lower.

The *quality* is higher.

Availability is sooner, or more dependably just in time.

Customer service is better.

Attractiveness is greater.

Awareness is greater.

The *stability* of long-term relationships is important.

There are other social, psychological, and ideological reasons.

7.5.1 Price

A lower *price* is perhaps the most obvious of the reasons why customers choose one product or service over competing alternatives. One must be careful to point out that from the customer's perspective it is normally the price that is important, not the cost that it took to provide the product or service—i.e., lower costs. Usually, in fact, the customer doesn't even know what the costs were, and lower prices might have been achieved through a change in exchange rates (in international competition) or internal or government subsidies. The exception is cases in which industrial customers are entering into long-term partnering arrangements with suppliers, when a lower cost structure may, indeed, be known to customers and be the more relevant advantage.

7.5.2 Quality

Quality is another term that has become widely used, so much so that one has to ask what is meant specifically by a customer who claims to have purchased a particular product or service over competing alternatives because of its *higher quality*. In this framework, quality will refer to one of two different meanings:

- Higher *reliability* at a given level of performance, i.e., conformance to specifications
- Higher level of *performance*

Having made this assertion, one must recognize that there is an aesthetic dimension to quality as well. For example, people will often prefer natural leather or wood materials to plastics or other synthetic materials because of their higher quality, without regard necessarily to either reliability or performance.

Reliability. *Reliability* is the ability of a product or service to perform at a specified, promised level over a reasonable useful life under normal conditions of use. It is not, however, expected to perform at levels higher than what was promised. A product or service is defective when it does not meet the conditions listed above; and when a customer's reason for a purchase is higher reliability, this means that, of the number of products or services they have purchased from a supplier and used, very few—if any—have been defective.

There are, however, two very different ways to achieve high reliability: (1) to inspect it in by identifying and removing defects from the product-services stream before they reach the customer and (2) to build it in by improving production processes and getting them under control. There are costs that result from poor reliability—the costs from waste, scrap, rework, warranties, loss of customer goodwill, and product liability. Improving reliability also has costs—the investments needed to inspect or build it in. In the latter case, getting production processes under control may require process simplification and redesign or the implementation of an effective statistical process control (SPC) program, but if done well, it should produce significant savings in all the cost categories resulting from poor reliability. It is in this sense that "quality is free," as Philip Crosby puts it,[15] that the benefits from greater reliability through improved process control can—at least in the initial stages of improvement—far outweigh the costs. Although some complex, high-cost-of-failure products may still require testing and inspection, the result of improved process control should be higher yield rates, which means that fewer defects to the customer correspond to less waste, scrap, and rework. Therefore, higher reliability—at least up to the point where processes have been improved so much that diminishing returns have set in—is *not* a tradeoff with cost. The achievement of both lower costs and higher reliability is possible.

Performance. *Performance* level refers to a property, feature, or characteristic of a product or service which customers value, and therefore having more of it than competing alternatives do can be a reason for their purchase decision. Performance has multiple dimensions depending on the specific product or service and the customer or market segment. For industrial customers of fabricated parts, dimensional tolerances are a performance dimension that is often cited along with surface finish (corrosion resistance—durability), weight, etc. In the case of process industry customers, purity and uniformity might be relevant performance dimensions. The range of performance dimensions is much wider for final consumers. For a teenage hot rodder, the relevant automobile performance dimension might be acceleration, but for an older person it might be the degree of riding comfort; for a hacker the relevant computer performance dimension might be processing speed, but for the neophyte it might be the degree of user-friendliness. The diversity of what different customers value, and the range of product, service, and product-service combinations possible, make performance a very fertile form of competitive advantage.

7.5.3 Availability

Availability is a time-related competitive advantage. All other things being equal, many customers would prefer the competitive alternative that is available soonest—instantaneously, if possible. Firms with new products which have no competing alternatives available until competitors can copy or catch up to or leap-frog over them have a special *availability* advantage. For industrial customers operating in a just-in-time mode, availability translates into dependable delivery at precisely the scheduled time. Availability also encompasses time-based convenience in purchasing such as one-stop shopping and the advantages that accrue to suppliers who have broad product-service lines.

7.5.4 Customer Service

Customer service enhances the utility of a product or the social relationships that complement its sale and use. Traditional forms of customer service have included applications engineering, training of employees, and service-maintenance contracts. Financing services (time payments, leasing, trade-ins, etc.) that enable the customer to purchase the product are also included. Good customer service can also make up for a lot of customer ill will caused by product defects. And the social nature of person-to-person interactions that occur in connection with the sale and use of the product can also be very crucial in competition—consider the restaurant with great food but poor service and vice versa. In any of these cases, customers may be choosing among competitive alternatives based on the service that comes along with a product.

7.5.5 Attractiveness

Attractiveness applies principally to consumer products, although even industrial customers may be turned off by a product's unattractive appearance. Attractiveness obviously encompasses *style* and has an aesthetic component that transcends the annual style changes of, for example, the fashion apparel industry, although what is perceived as being *attractive* may have some cultural basis. An attractive product design may also be functional, of course, and for some consumers functionality itself may constitute attractiveness. Because of its highly individualistic and abstract nature, attractiveness operates something like the Supreme Court's view of obscenity—it's hard to define, but you know it when you see it.

7.5.6 Awareness

Awareness is a factor in all other competitive advantages as well as a possible reason in its own right why customers choose one product or service over competing alternatives. If customers simply know more about one product or service than competing alternatives, they may choose it because of the comfort level which that knowledge brings them compared to the relative uncertainty of the alternatives. If the knowledge is positive and is repeatedly reinforced through experience and marketing or advertising activities, brand-name loyalty may be created in customers who continue to choose it apart from any objective evaluation of the actual facts. On the other hand, even if one product or service has a lower price, greater reliability, higher performance, sooner availability, better customer service, or more attractive designs than competing alternatives *but* customers are not aware of these *facts,* they cannot influence customers' choices.

7.5.7 Stability

Stability of long-term relationships probably applies only in very specific situations where—again—that stability provides a comfort level to customers which they prefer over the relative uncertainty of short-term or temporary supplier-customer relationships. For example, in the case of strategic materials or—recently—petroleum, the stability of long-term supply contracts may be preferred by a customer over temporary supply arrangements that offer lower prices or other advantages. This stability may also apply to long-term social relationships between customers and suppliers in cases where customers value the relationship itself apart from the product or service the supplier is providing.

7.5.8 Other Advantages

Finally, there are always a number of reasons why customers buy one product or service over competing alternatives that are not easily classified or described—the *other* category. These other reasons are based more on sociopsychological or political-ideological factors than on rational decision making by customers. *Snob appeal,* for example, would appear to be in this other category, as would whatever is involved in the *impulsive purchase* of a pet rock. "Buy national" behavior applied as an expression of patriotism rules out an entire set of competing alternatives from foreign sources, and the ideology which prohibits Cuban cigars from being imported into the United States rules out what is reputed to be the high-performance competing alternative for U.S. cigar smokers. Although many more of these idiosyncratic reasons behind customers purchasing decisions undoubtedly exist, we will be focusing on the major competitive advantages of price, quality, and availability.

7.6 SHOPPING AROUND

In practice, customers *shop around* among competitive alternatives by setting parameters for some of these advantages and then making the final choice on the basis of the key or critical advantage. To illustrate this, consider how I might approach buying a new business suit. For many of the parameters I would set limits: in terms of *price,* I would not go above $350; in terms of *performance* (durability), I would not go below a certain threshold as indicated perhaps by the thickness of the material or the use of double stitching; in terms of *availability,* I might need it right away—off the rack—or as soon as I could get it—or if *customer service* tailoring were required, *within 24 hours.* But having met all these parameters, I would make my final choice on the basis of *attractiveness*—on how well it fit my body, on the color, on the styling.

Customers today expect high reliability and low prices—these are mutually reinforcing attributes that a supplier is expected to achieve just to be in the competitive ball game. But these are seldom enough. The winning competitor must have either the lowest price and highest reliability, or achieve one of the other competitive advantages that customers value.

7.7 GENERIC COMPETITIVE STRATEGIES

7.7.1 Low-Cost Leadership and Differentiation

In his treatment of competitive strategy, Porter defines and discusses three generic strategy types: *cost leadership, differentiation,* and *focus.*[16,17] *Low-cost leadership* means essentially what it says, achieving the lowest-cost position possible in each and every operation of the firm, not just manufacturing, through such means as vigorous cost reduction programs, strict cost and overhead controls, economies of scale, and learning-curve efficiencies. *Differentiation* refers to a uniqueness of a product or service as perceived by a customer, which leads the customer to prefer it over competing alternatives. A differentiating uniqueness is typically achieved through such means as design features, establishing a brand-name identity, or offering superior customer service.

The linkage between the earlier list of competitive advantages and Porter's generic strategies should be obvious—low-cost leadership corresponds to a potential low-price competitive advantage; differentiation corresponds to uniqueness in terms of higher quality, sooner availability, better customer service, etc., competitive advantages.

There are some important qualifications that Porter makes regarding low-cost leadership and differentiation strategies that deserve greater emphasis than they usually receive. The first is that the successful low-cost low-price competitor cannot ignore the differentiating advantages in its pursuit of low costs. If quality, availability, customer service, or other parameters of differentiated competition fall below a threshold of acceptability held by customers, then the low-cost low-price competitor may sink to a lower category of discount or low-quality competition. Therefore, Porter says that low-cost leaders need to achieve *parity* with competitors in differentiating advantages.[18] A similar qualification applies to the generic strategy of differentiation. The differentiating uniqueness must be something that is valued by customers, and that value must be great enough to persuade customers to pay the price premium that is normally associated with the additional costs required to create the differentiating uniqueness. Therefore, the successful differentiator does not ignore costs—in Porter's terms, it needs to achieve cost *parity* or *proximity* relative to competitors through cost reductions in activities that do not affect its differentiating uniqueness.[19]

As indicated above, a firm cannot normally pursue both low-cost low-price and differentiating strategies at the same time. Attempting to do so is one example of how, according to Porter, a firm winds up being *stuck in the middle.*[20] There can be exceptions, usually in cases of major process innovation, but the achievement of a differentiating uniqueness normally comes at the expense of higher costs. Firms that routinely claim that they are the lowest-cost, highest-quality, quickest-delivery, best-everything competitor are the rare exceptions, or are engaging in rhetorical hyperbole. Since differentiators also lower costs in activities unrelated to achieving their uniqueness, and low-cost low-price leaders also pursue improvements in quality, delivery, etc., in order to remain in the mainstream of competition, the only sure way to distinguish between them is to examine tradeoff decisions where potential differentiating uniqueness is at stake. If those decisions are made consistently in favor of lower costs, the obvious conclusion is that the firm is pursuing a low-cost low-price strategy, and vice versa.

7.7.2 Focus

Porter presented his cost leadership and differentiation generic strategies in two contexts—a broad-scope or industrywide competition, and a narrow-scope or focused competition. At the same time, he defines his third generic competitive strategy of *focus* in terms of cost focus and differentiation focus, where focus can be based on geography, market, or product-service segments, or on some combination of these.[21] Therefore, it is probably more productive to think of competitive advantages in terms of low cost–low price and differentiation, which can apply to firms competing industrywide (in all or most segments) or on a focused basis (one or two segments only). In more recent work, the competitive scope dimension has been expanded to include multiniche competition,[22] which is applicable in situations currently referred to as mass customization[23] or agile manufacturing[24] (see Fig. 7.1).

Generic Competitive Strategies

	Low Cost - Low Price	Differentiation
Industry Wide		
Focus		
Multiple Niche		

FIGURE 7.1   Strategy-scope framework.

7.8   THE VALUE CHAIN

7.8.1   Defining the Value Chain

Thus far I have dealt with competitive strategy by emphasizing customers and their ultimate role in determining competitiveness by collectively choosing to buy the products or services of one firm over competing alternatives. In this section the focus shifts to firms and what firms do to achieve competitive advantage in implementing their competitive strategies. To address this set of issues, another *tool* developed by Michael E. Porter is utilized: the value chain. In Porter's words, the value chain is "a systematic way of examining all the activities a firm performs and how they interact...for analyzing the sources of competitive advantage."[25]

Porter's value chain can be used to organize all the activities of a firm into categories of primary and support activities. *Primary activities* constitute the processes by which firms receive inputs (inbound logistics), convert those inputs into outputs (operations), get those outputs to customers (outbound logistics), persuade customers to buy the outputs (marketing and sales), and support customers in using the outputs (service). Support activities are processes which provide support to the primary activities and to each other in terms of purchasing inputs (procurement); developing new and improved ways of doing activities (technology development); dealing with personnel (human resource management); and general management, accounting, finance, and other activities which support the entire organization rather than individual activities (firm infrastructure).

Using Porter's value chain and focusing on categories of activities enables one to see the firm as a collection of activities rather than as an organization chart with administrative units such as the purchasing department, the R&D labs, the manufacturing division, and the personnel department. This is important because critical procurement activities can occur outside the purchasing department, technology development activities outside the R&D labs, and so on.

7.8.2   Value-Chain Activities

Activities are processes, things that are done in a firm. In order to be identified and labeled, they must be distinct—have a beginning and an end which distinguishes them from other activities or operations. In using the value-chain tool to analyze sources of competitive advantage in a firm, one would first identify its nontrivial activities and assign them to the most appropriate category. A second useful step would be to examine specific activities from an input/output perspective.

7.8.3   Input/Output Analysis

Input/output analysis can be used at various levels of an organization—including the organization as a whole—to get a better idea of what the entity is and does. The organizational unit of analysis—in this case, the specific activity—is represented as a black box, with inputs going in and outputs coming out (see Fig. 7.2). The amounts and types of inputs will vary from organization to organization and activity to activity, and in addition to the intended output there will always be other outputs, some portion of which will be waste. The activity can then be viewed as the transformation or conversion of inputs into outputs, and the efficiency or productivity of the activity can be measured by the ratio of intended output to inputs.

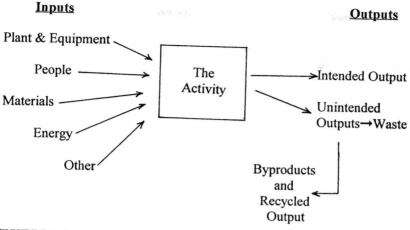

FIGURE 7.2   Input/output analysis.

7.9 *TECHNOLOGY*

7.9.1 Defining Technology

One more piece of information is needed to finish this input/output analysis: *technology*. In this context, I will define *technology* as *a way to do something, a way to perform a value-chain activity*. This process view of technology allows one to carefully distinguish between technology and products or hardware—technology is not a product, a tangible entity. But when a product is used to do something, it is that use which defines it as technology for the user. Thus a computer is a product, a thing. Computer users, however, may regard it as, for example, a data-processing technology, a text editing technology, a scheduling technology. Indeed, what makes the computer so powerful a tool is the ubiquitous and wide range of uses to which it can be put.

7.9.2 Alternative Technologies

For any given value-chain activity, then, there are alternative technologies—alternative ways of doing it. There are new and old, labor-intensive and capital-intensive, appropriate and inappropriate, and unknown technologies yet to be developed.

   For example, let us take the inbound logistics activity known as *inbound materials handling*—the movement of material goods from where the supplier gives them over to the firm to when they enter into operations. There are a wide range of alternative ways to do these activities—alternative technologies that might be used, including manual labor, manual labor supplemented by tools (hand carts), conveyor systems, forklift trucks, automated guided vehicles, robotic loaders/unloaders, stacking cranes, and pneumatic hoses. The nature of the inbound material would, of course, limit the feasible use of some of these technologies, but an input/output analysis of inbound materials handling utilizing forklift trucks vs. conveyor systems technologies would point out the differences in inputs required, intended outputs achieved, unintended outputs produced, and the efficiency of the conversion process. One could look similarly at the inbound materials storage or the inbound materials inspection activities.

7.9.3   Ubiquitousness of Technology

The power of using this definition of *technology* is that it applies to every value-chain activity, not just the activities in manufacturing or engineering that we usually associate with the term. Thus there are technologies involved in accounting and cost management activities (the traditional direct labor plus overhead; activity-based costing or management), in personnel selection activities (psychological and aptitude testing, personal interviews), in market research activities (focus groups, consumer surveys, test marketing), and even in R&D activities (computer-aided molecular design, electron microscopes, gene-splicing equipment). This view of technology allows us to include management technologies (ways to motivate personnel, allocate resources, and structure organizations) and even what might be called *social technologies* (ways to integrate people and work, support the whole person, and provide for the people who lose out in the course of doing business) in our analysis of technology's contribution to competitive strategy.

7.10   CREATING COMPETITIVE ADVANTAGE

Using the value-chain analysis, firms can then create competitive advantage in at least three ways:

1. By placing greater or lesser emphasis (allocation of resources, management time and attention) on specific activities than competitors do
2. By performing specific activities better (better management, more highly trained people, better-maintained equipment) or differently (using an alternative—presumably new or improved—*technology*) than competitors do
3. By managing *linkages* among activities better than competitors do

7.10.1   Technology Choice

In item 2 (above), the decision of which technology will be used to perform a specific value-chain activity corresponds to a major element in what Porter calls *technology strategy*.[26] The technology choice for the activity should obviously correspond to the competitive advantage being pursued by the firm, taking into account the qualifications discussed earlier—i.e., the low-cost technology should be chosen by the firm pursuing a low-cost low-price advantage; the high-performance technology for the high-performance advantage, and so on. This point is discussed further later in this chapter.

7.10.2   Managing Linkages

Defining Linkages.   The subject of *linkages* among value-chain activities requires further elaboration. According to Porter, value-chain activities are not independent, but form chains of interdependent, *linked,* activities, in which "Linkages are relationships between the way one value chain activity is performed and the cost or performance of another."[27]

Earlier it was indicated that inbound materials inspection was a specific activity that might be found in the inbound logistics category of the value chain. The way this activity is done—the technology used, the amount of emphasis placed on it, how well managed it is—could have major impacts on the cost or performance of downstream activities such as assembly, systems testing, after-sales service, and warranty claims administration.

Coordination and Optimization of Linkages.   For a firm to manage its *linkages* better than competitors do, it must first recognize that such interdependencies exist. This is often made difficult by organizational boundaries that separate inbound inspection activities from manufacturing, systems testing, or after-sales service activities. If the recognition is present, then better *coordination of linked activities* can be a source of competitive advantage.[28] For example, better coordination of the inbound inspection and systems testing activities could help to identify causes of systems failures and preferred components suppliers. Moving beyond coordination, investments can often be made in one activity to improve the cost or performance position of linked activities. For example, improving inbound inspection activities through investments in more sophisticated technology, the training of inspectors, or the assignment of better managers to this activity should avoid costs incurred by defective components in the downstream activities mentioned earlier. But even when the downstream benefits are recognized and quantifiable, it can be difficult to make such investments because of the perverse effects of organizational boundaries and reward systems; in many organizations, functional managers can be driven by internal reward systems to optimize performance at the subsystem level (suboptimization) to the detriment of the system as a whole. Not making the investment to improve inbound materials inspection may make the performance numbers of the purchasing department or the warehouse unit look better, for example, because all the benefits of the investment accrue to downstream organizational units and show up only in their performance numbers. Therefore, being better able to *optimize such tradeoff investments in linked activities across organizational boundaries* at the level of the total organization is another possible source of competitive advantage.[29]

Upstream and Downstream Linkages.   Furthermore, any organization with its own unique value chain has upstream suppliers (of inputs) and downstream distribution channels and customers (of outputs) with their own value chains and unique configurations of value-chain activities. The concept of *linkages* can then be extended to interrelationships between supplier or customer value-chain activities and the value-chain activities of the organization.[30] For example, through supplier certification activities of the organization, its own inbound materials handling and inbound materials inspection activities may substantially decline in importance or disappear altogether as suppliers' shipping and finished goods inspections activities are linked directly with the organization's assembly activities.

Information Technology and Linkages.   Linkages are perhaps more easily recognized and managed today because of the pervasive and ubiquitous influence of information technology—the marriage of sensor, computing, communications, and software technologies. Information systems are in place that collect more data than ever before, integrating across organizational boundaries and into supplier and customer organizations, with real-time analytical capabilities that can establish and quantify cause and effect linkages, thus making them observable and manageable.

7.11 LINKING TECHNOLOGY TO COMPETITIVE ADVANTAGE

On the basis of the preceding discussion, the choice of which way to perform a value-chain activity—which technology to use—should be governed by the competitive advantage(s) that the firm is pursuing in implementing its competitive strategy. In other words, if low cost–low price is the strategy, then low-cost technologies should be used, consistent with maintaining acceptable levels of quality, availability, attractiveness, and so forth. (Of course, technology choice interacts with other strategic variables—e.g., low unit costs are often achieved through economies of scale which in the past have depended on mass-production technologies and large customer markets demanding standardized products and services.) Similarly, if a differentiating uniqueness is the strategy, then technologies which maximize the specific competitive advantage in terms of higher performance, sooner delivery, better customer service, etc., should be used, consistent with the price premium customers are willing to pay for the uniqueness.

Technology choice is therefore a more complex decision for firms pursuing a differentiation strategy. The number of potential differentiating competitive advantages is very large, and much more attention must be paid to determining which value-chain activities are the major sources of competitive advantage and to examining technology alternatives in terms of the specific competitive advantage being sought. To say that one is pursuing the advantage of higher performance is not enough. One must specify the performance dimension that is predicted to be of value to a segment of customers, look at which activities create and deliver that performance value, and make technology choices for those activities accordingly. In trying to understand and illustrate these concepts, we often find ourselves limiting our thinking to manufacturing activities, manufacturing technologies, and manufactured products. But Porter's value-chain analysis and the definition of technology used in this chapter should open our thinking to a much broader range of concerns.

The technology development activity is an important case in point. Thus far we have considered a technology choice from among existing technological alternatives. Technology development activities, however, have the promise of creating new ways to do things—new technologies—that can contribute even more to higher performance, sooner delivery, better customer service, etc., than existing alternatives, or that can lower the cost penalty of achieving that higher performance, sooner delivery, etc. Obviously, decisions about which technologies to focus on in development activities should be guided by strategic considerations, which is also an element of *technology strategy*.[31] In rare but extremely significant cases, major, radical, and discontinuous technology developments can be the basis for totally new ways of competing in an industry and can transform industry structure!

Having said all of this, I will now illustrate the linkage between technology and competitive advantage more explicitly by focusing on advanced manufacturing technologies, pertaining primarily to discrete-parts fabrication and assembly activities, and on management technologies usually labeled as Japanese manufacturing management techniques such as TQC, JIT, and Kanban pull systems. These technologies will be related to the three competitive advantages of low price, higher quality, and sooner availability.

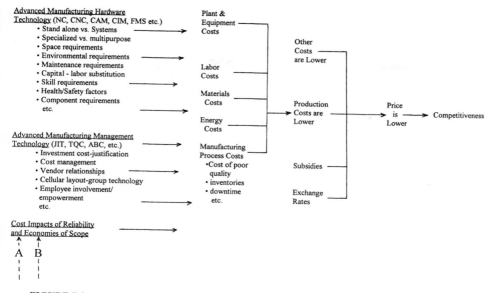

FIGURE 7.3   Competitive advantage of low price.

7.12   TECHNOLOGY AND LOW COST–LOW PRICE

Figure 7.3 suggests one way to examine the competitive advantage of *low price.* Working from right to left, the first thing to point out is that customers don't normally care about what a supplier's costs are—they only care about and often only see the price that is charged. (The important exception is industrial customers entering into close relationships with suppliers who want to know everything about their cost structures, quality systems, etc.) It is important to note, therefore, that there are other ways to achieve lower prices than lowering costs. Subsidies—either internal to the firm or provided by governments—can be used to lower prices, and in international competition, changes in exchange rates can lower or raise prices overnight without any change in the firm's costs. In Fig. 7.3, *production costs are emphasized,* but as indicated earlier, low costs must be pursued in each and every activity of the low-cost leadership firm. Wringing the last penny of cost out of production activities may be more than counterbalanced by "fat" in distribution (outbound logistics) activities, for example.

7.12.1   Production Costs and Advanced Manufacturing Hardware Technology

Working further to the left, *production costs* are divided into the familiar categories of *plant and equipment, labor, materials, energy,* and *manufacturing process* costs. When an examination is made of how advanced manufacturing hardware technologies

[AMHTs—e.g., CAM, CAD/CAM, CIM, FMS, NC, CNC (computer-aided manufacturing, computer-aided design/manufacturing, computer-integrated manufacturing, flexible manufacturing systems, numerical control, computer numerical control)] might affect these production costs compared with alternative, traditional manufacturing technologies, there is no clear and easy answer. The answer is the all-too-familiar "it depends!"

Plant and Equipment Costs.   On what do these cost impacts depend? A partial list might include whether the AMHT was implemented in systems or in a series of stand-alone pieces of equipment—e.g., replacing an entire machinery system vs. replacing a worn-out lathe with a new NC or CNC lathe. Especially in product flow layouts, the AMHT systems approach might cost less than traditional technology because of the potential cost savings due to systems redesign, while the stand-alone approach would probably cost more because the NC/CNC lathe costs more than the manually controlled lathe and there are no cost savings from systems redesign. Similarly, equipment costs might be affected by the choice of specialized vs. multipurpose AMHT—e.g., a CNC lathe vs. a CNC machining center. The lathe might fit into a system providing one of a number of required machining operations, while the machining center might do all the required operations itself, so the comparison of AMHT equipment costs with traditional manufacturing technology equipment costs has to be adjusted accordingly.

But this latter example illustrates other potential cost impacts as well. The multi-purpose CNC machining center should also take up much less space than the system with the CNC lathe alternative (and in growth-capacity expansion situations, avoiding the cost of building additional factory floor space can be quite considerable) and should avoid WIP inventory costs associated with product flow through a series of specialized CNC operations.

The impact of AMHT on plant and equipment costs might also be affected by how sensitive the AMHT equipment is to environmental factors such as temperature, humidity, and vibration. This sensitivity can imply plant infrastructure costs that might not normally be considered. In obvious cases like cleanroom AMHT, these infrastructure costs are known up front, but there is also the case where a sophisticated piece of CNC equipment installed and adjusted in the winter didn't work right in the summer because of the higher angle of the sun shining through the windows!

Labor Costs.   The traditional justification for using more advanced manufacturing equipment has been the substitution of capital for labor, and there is little doubt that utilizing AMHT usually lowers direct-labor costs. However, there are other cost implications to consider. Although the amount of direct labor might be reduced, what are the skill requirements of the labor that remains? If the AMHT requires less but more highly skilled direct labor, there may be additional costs associated with employee training or higher wages. In addition, what happens to indirect labor with the utilization of AMHT compared to traditional manufacturing technology? AMHT maintenance is again an "it depends" situation. AMHT equipment is normally more complex and sophisticated so that maintenance costs might be presumed to be greater, but on the other hand much of the complex electronics is modularized on printed-circuit boards which are simply replaced when a circuit goes bad, so maintenance costs might be less.

AMHT might have more positive cost impacts when the health and safety of workers is considered. Automating dangerous or environmentally hazardous operations utilizing AMHT (e.g., spray painting, welding) may require more expensive equipment but provide savings in lost labor hours, workmen's compensation premiums, and other employee health and safety costs. Similarly, automated operations that are difficult for humans to do because of their physical makeup or boring for humans to do because of their psychological makeup can both lower health and safety costs and prevent human error that results in scrap, rework, or other costs of poor quality.

Materials Costs.   Materials costs might or might not increase with AMHT and less direct labor. The human operator may lack the data input and processing speed of the computer, but is more flexible when it comes to dealing with unexpected problems. Using AMHT in assembly operations, for example, may require higher-tolerance and higher-cost components so that feeders don't jam and robots don't position parts off center, whereas human operators using traditional manufacturing technology could have easily handled these kinds of exceptional problems when they occurred.

   The point of all this discussion is that the contribution of advanced manufacturing hardware technology to the competitive advantage of low cost–low price is ambiguous and situation-specific. One would not want to precipitously rush into AMHT pursuing a low-cost low-price competitive strategy!

7.12.2   Production Costs and Advanced Manufacturing Management Technology

Japanese Management Practices.   When we look at *advanced manufacturing management technologies* (AMMTs), advanced ways to manage manufacturing operations, however, we see a more straightforward linkage with various production costs. There are a number of so-called Japanese manufacturing management practices that can affect production costs; for instance, JIT lowers WIP inventory and plant costs; TQC [*andon, poka-yoke* (Japanese terms meaning automobile hazard lights and foolproof mechanisms), etc.] lowers scrap, rework, and other costs of poor quality; Kanban scheduling systems lower finished-goods inventory and WIP inventory costs; the reported preference of Japanese managers for multiple copies of smaller, less expensive, more mobile, less sophisticated machinery rather than "supermachines" lowers equipment costs; total productive maintenance lowers labor, equipment, and process downtime costs; and cellular layout and group technology lowers labor and manufacturing process costs.[32,33]

Management Policies.   Many of these "technologies" might seem to be "policies," but they are policies that govern the way activities are done. For example, the way in which investments in new equipment are cost-justified—high-hurdle-rate discounted cash flow vs. strategic cost management techniques—can impact the decisions of the firm to purchase and implement any new technology;[34,35] the way in which overhead costs are defined and allocated—the traditional way of basing this on direct-labor vs. activity-based costing and management—can affect management's understanding of the real costs of the firm's operations;[36] vendor relationships that are based on cooperative partnering rather than arm's-length confrontation can lead to lower materials and manufacturing process costs;[37] negotiated partnering with labor unions (trading job security for work-rule flexibility) vs. confrontational bargaining over wages and benefits can lead to lower labor and manufacturing process costs;[38] and giving operators the power and the responsibility for quality, routine maintenance, and other activities associated with their principal work tasks rather than relying on specialized quality control or maintenance staffs can lower labor and manufacturing process costs.

Hardware and Management Technologies.   These kinds of illustrations can go on and on. Two general points to make are as follows:

1. Advanced manufacturing management technologies appear to have a more direct and a larger impact on lowering production costs than do advanced manufacturing hardware technologies—indeed, in numerous cases the cost and reliability benefits from implementing AMMT are regarded as the real key to manufacturing improvement.[39]

2. Even when AMHT is successfully implemented and has a positive impact on production costs and other competitive factors, it usually occurs after or in tandem with the successful implementation of AMMT—what one author has called *synchronous innovation.*[40]

Finally, in examining Fig. 7.3, we note the cost impacts of reliability and economies of scope. The sources of these cost impacts arise out of the quality and availability advantages and will be discussed below. However, the cost impacts are felt principally in lowering manufacturing process costs.

7.13 TECHNOLOGY AND HIGH QUALITY

Continuing this discussion of advanced manufacturing technology and the competitive advantage of low price, Fig. 7.4 sets out a similar way to consider advanced manufacturing technology and the competitive advantage of *higher quality*. As discussed earlier, a customer who expresses a preference for one particular product or service over competing alternatives because of its higher quality might be conveying one of two different meanings:

1. The *reliability* of the product or service is higher—a production-process-related phenomenon.

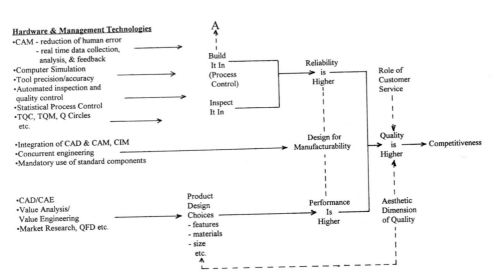

FIGURE 7.4   Competitive advantage of high quality.

2. The *performance* of the product or service is higher—a design-related phenomenon.

These two aspects of quality are interrelated. For products, this interrelationship is known as *design for manufacturability,* which refers to situations in which a design which is unnecessarily complex or sets performance levels unnecessarily high may cause process problems which decrease the reliability of the resulting product. A similar concept could relate to service design and the processes necessary to provide that service.

7.13.1 Reliability: Building It in

Working from right to left, there are two basic ways to improve reliability: to *build it in* or *inspect it in.* Building it in is usually the preferred way to improve reliability because, as discussed earlier, this involves improvement of the production process itself and therefore has the added competitive effect of lowering manufacturing process costs at the same time—see the earlier discussion of reliability. Thus the dashed arrow (A) in Fig. 7.4 shows up also in Fig. 7.3.

Process Control. Improving the production process is a matter of getting it *under control or stabilized* so that operations are *repeatable*; thus, variations in operations due to special causes arising out of specific circumstances not inherently part of the process are removed, and only the acceptable variation that is inherently part of the process remains. (If a process to produce a product or service at designed performance levels cannot be brought under control, then the process or the product or service must be redesigned—or both.) Getting a process under control results in increasing yield rates (or declining defect rates), and this is the source of the net reduction in production process costs.

Advanced Manufacturing Hardware Technology. What are the benefits of advanced manufacturing hardware and advanced manufacturing management technologies on building reliability in to production processes? The first benefit is that all computer automation in CAM which removes human beings from direct contact with the production process reduces *human error* as a source of special variation affecting the process. The greater the potential for human error in operating a process, the more important CAM is to improving reliability. Second, the operations of sensors, computers, analytical software, and output devices make it possible for vast amounts of process data to be collected and analyzed in real time. The results of the analyses can be fed back to automatically adjust process parameters or to alert human operators or computer monitors to the need for action—e.g., predictive maintenance. In either case, reliability is improved and process costs are reduced compared with the use of more traditional and conventional manufacturing hardware technology. Advances in materials science embodied in cutting tools also might improve process control and reliability.

Advanced Manufacturing Management Technology. Perhaps the best-known advanced manufacturing management technique affecting the production process is *statistical process control* (SPC). SPC alerts human operators to the fact that special sources of variation have crept into the process and that steps must be taken to bring the process back under control, thus preventing reliability losses. The variety of so-called Japanese *total quality control* techniques such as hazard lights (*andon*), production control boards, and the use of foolproof mechanisms (*poka-yoke*), should also work in the same direction. Although *quality circles* and *total quality management* (TQM) techniques both contain the word *quality,* their scope of meaning goes beyond the narrower definition of quality being used here, and they more aptly fit under the title of *continuous improvement.*

7.13.2   Reliability: Inspecting It in

Inspecting quality in to the process without building it in results in higher reliability as perceived by the customer because few or no defects get past the inspection system, but there are no reductions in production process costs because the process itself remains unchanged—warranty, service, loss of customer goodwill, and liability costs of poor quality are merely shifted to scrap and rework.

A firm can inspect quality in by providing more inspection—hiring more QC inspectors—or by providing better inspection, which can be the result of using advanced technology. Automated quality control and inspection equipment such as coordinate measuring machines or automated test equipment can improve the inspection activity by reducing human error that would result in mistakenly letting defects get through (or mistakenly weeding out nondefects), by doing inspections faster than humans can, and by doing inspections with greater accuracy and precision than humans can. Advanced training technology or improved inspection methods design might also be used to improve manual inspection, but in all these instances the investment made to achieve the higher reliability (as perceived by the customers) is a direct cost tradeoff since there are no cost reductions from bringing the process under control.

7.13.3   Performance: Designing It in

Improving product performance is largely a matter of product design and the choices that are made in the product design activity. These design choices include the tightness of the tolerances that are specified, the special features that are included, the choice of materials utilized, and the size of the product or service offerings.

Advanced Manufacturing Hardware Technology.   *Computer-aided design/engineering* (CAD/CAE) technologies would appear to have a significant impact on performance in terms of reducing the potential for human error and in the enhanced ability to design more sophisticated and complex (assuming that sophistication and complexity contribute to performance!) products. At the same time unit design costs should be significantly reduced because of the labor productivity gains that CAD/CAE offers. However, higher performance normally is achieved in tradeoff with higher costs. Hardened metal cutting tools or purer materials may also offer performance gains in precision and accuracy or product consistency.

Advanced Management Technology.   As was pointed out earlier, however, the higher performance provided by the product must be valued by the customer, and valued highly enough to cover any price premium associated with it which covers the higher costs needed to achieve that performance. Advanced management techniques in market research such as the use of *focus groups* or *quality functional deployment* help ensure that the customers' values are reflected in product design decisions. Traditional design techniques such as *value analysis* and *value engineering* are also applicable in this regard.

7.13.4 Integrating Reliability and Performance

The integration of CAD/CAE and CAM provides a hardware linkage between reliability and performance, the two dimensions of quality. This hardware linkage gives design and manufacturing engineers real-time electronic sharing of information that supports concurrent or simultaneous engineering. Each group can feed back cost and capability information to the other so that in the end, the product-process combination is optimal for the firm, rather than one or the other. This hardware integration can be supported by management policies to, for example, use standardized components whenever feasible or justify why not. These policies can be programmed into the CAD/CAE and the computer process simulation systems to prompt either group of engineers to consider the viewpoint of the other. Thus hardware and management technologies are themselves integrated into design for manufacturability.

7.13.5 Aesthetics and Customer Service

Two side issues complete this discussion of quality. In terms of performance, there is an aesthetic dimension that must be taken into account apart from objective facts. Many people, for example, prefer natural leather or wood products over plastic, relating this in their minds to *higher quality,* even though on any given performance dimension (hardness, durability, cleanability, etc.), the plastic product might objectively score higher. Market research techniques help identify these aesthetic preferences. Second, there is also a customer service dimension to the reliability issue. When customers unfortunately experience a product defect, the loss of goodwill can be prevented and customer loyalty even enhanced through appropriate customer service activities. Information technologies which speed up the service to fix or replace the defective product combined with management policies which remove the "hassles" from the experience for the customer can go a long way toward alleviating customer complaints, but this service again comes at additional cost only unless the process is first brought under control.

7.14 TECHNOLOGY AND SOONER AVAILABILITY

The third and final competitive advantage to be examined in the context of advanced manufacturing hardware and management technologies is that of *availability.* Figure 7.5 proposes a structure to examine this advantage.

7.14.1 New Products

Figure 7.5 shows two situations from which to examine the availability advantage: that of new products and that of expanded or extended product lines. Perhaps the ultimate availability advantage is to come to market first with a *new product* that is not available from any competitor. For that period of time when no competing alternative is available, a firm can charge what the market will bear, can create *first-mover advantages*[41] which will endure even after a competing alternative is available, can obsolete its current new product with an even better or cheaper one to stay ahead of the competition, and so forth.

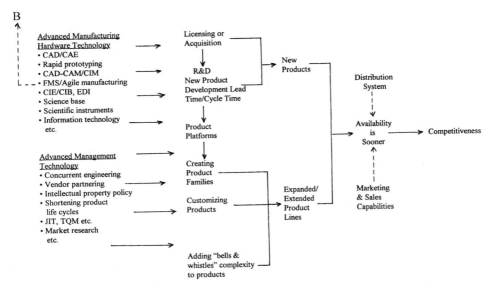

FIGURE 7.5   Competitive advantage of sooner availability.

Licensing or Acquisition.   There are a couple different ways to obtain new products. One way is to *license* or *acquire* them as products, or to license or acquire underlying technologies which feed into the second way—the firm's internal *new-product development process* which normally is managed by the R&D function. Licensing or acquiring technology—*external technology sourcing,* as some call it—have become more popular activities as trends toward firm downsizing and focusing on *core competencies*[42] have increased, and information technologies have improved firm capabilities to monitor and evaluate external technology development on a global scale.

New-Product Development Cycle Time.   The conventional way to achieve new products is still the internal development process, however, which begins with R&D activities but extends through manufacturing and marketing and sales—through commercialization. Because competition is becoming more intense and product life cycles in many industries have been shortening, there is increasing pressure to shorten *new-product development cycle times* or to achieve a new product lead time over competitors.[43] How might advanced manufacturing hardware and management technologies impact new product development cycle times so as to achieve the competitive advantage of *availability?*

The short answer to this question is, potentially, "a very great deal!" In fact, I would contend that the biggest contribution that these technologies can make to competitive advantage is in this area of availability.

Advanced Manufacturing Hardware Technology.   Earlier we mentioned the performance impacts of CAD/CAE technologies, but surely a bigger impact of CAD/CAE pertains to the speed with which product design and engineering analysis activities can be done. *Fast prototyping* is another technology that has the potential to speed up product design. New products must be manufactured before they can be sold, however, and CAM/FMS technologies can be used to change manufacturing systems over from existing to new products—as long as the new products are within the *envelope* of the systems—almost instantaneously and at almost no cost. CIM technology, which integrates the product design and manufacturing processes electronically, has the potential to make the transition from design to manufacturing go even more quickly and smoothly.

Other activities and new technologies also have the potential for expediting the new-product development process. The more fundamental research that goes on in industry as well as in universities and national laboratories can result in an iteration between science and technology that results in new and faster ways of doing R&D. Advances in microscopy and spectroscopy technologies have resulted in much more powerful scientific instruments which have advanced the scientific study of molecular and atomic structures related to materials and resulted in computer-aided molecular design technologies for developing new pharmaceuticals, for example.

Advanced Management Technology. New management technologies can also play an important role in speeding up new product development. Concurrent or *simultaneous engineering*—the parallel or joint development of new-product and accompanying new-process technology—not only improves design for manufacturability as discussed earlier but also can significantly speed up the development cycle. Management policies to deliberately shorten product life cycles by obsoleting current products with new developments can speed things up as well. Many of the process improvements that result from TQM-type programs in new product development also tend to speed up the process. All of these new product decisions can, of course, be affected by the firm's approach to intellectual property protection, and market research activities still play an important role in identifying the customer values that can be satisfied through new products.

7.14.2 Expanded or Extended Product Lines

The second situation in which the availability advantage can apply is with *expanded* or *extended product lines,* and there are three cases of how this can happen.

Product Platforms and Families. The first is with *families of products* created off of *product platforms.*[44] A new-product platform, developed in R&D/engineering design activities, is a basic product design from which different family member products are spun off to appeal to specific market segments. All family members share the base design and production process, but differ in performance levels, features, and price. The platform product family is a planned approach, although new members of the family can always be added in response to market segment demand. Examples of this would include Chrysler's LH platform and the Chrysler Concorde, Dodge Intrepid, and Eagle Vision product family, the Boeing 747 and all its derivatives, and the Sony Walkman.

Customized Products. In contrast with planned product families, the second case of *customized* products refers to design and production activities responding to individual customer orders. In many industries, customers have unique requirements which can be satisfied only with a customized product. The extent of the customization may range from minor modifications to product features to totally new designs of major components or subassemblies. These would not be considered new products, however, as the customization is based on current knowledge and skills. The customized order might be for a single product, or for thousands of units, but there is no commitment to that specific customized design beyond that order. Examples of customization occur in many machinery and equipment industries in which the equipment—and the major components and subassemblies—have to be customized to fit the specific size and functionality requirements of the customer.

Adding Bells and Whistles.   The final case of expanded or extended product lines is where complexity of the "bells and whistles" type is added to products. This is a special case of customization, but is very different in its motivation. Customization as described above is done to meet the unique functionality requirements of customers; adding bells and whistles is done to appeal to customers' personal tastes and preferences, or impulses. A popular example would be a T-shirt shop which allows customers to print pictures of their faces on shirts in real time when making their purchases. More complex cases might include accommodating customers' desires for coffee-cup holders, round vs. square corners, narrow lapels and cuffs vs. wide lapels and no cuffs, and so forth in various products.

Hardware and Management Technology.   The availability advantage for these expanded or extended product lines also depends on speeding up design-manufacturing cycle times as was indicated above in the discussion of new-product development. Therefore all the advanced manufacturing hardware and management technologies discussed above also have the potential for greatly enhancing the availability advantage in these situations.

Market Feedback and Information Technology.   Expanded or extended product line situations are more dependent on market information and the ability of the firm to quickly act on this information, however, and on market information on customers' needs and wants just to get started, but then, perhaps even more importantly, on market feedback to be able to quickly adjust to winners and losers. Information technology to interact directly with customers—such as *electronic data interchange*—and to respond very quickly to their signals can be very powerful in this regard. Wall-Mart is often cited as a pioneer in applying information technology for this purpose.[45]

7.14.3   Agility

The recently developed concept of *agility* is also inherently related to the competitive advantage of availability: "...to be agile is to be capable of operating profitability in a competitive environment of continually, and unpredictably, changing customer opportunities."[46]

Agility is flexibility combined with strength—the agile gymnast as compared with the flexible rubber band. As such, *agile technologies* would speed up the design-manufacture cycle much as flexible technologies would.

7.14.4   Economies of Scope

The agile-flexible CIM technologies used to create availability advantages with expanded or extended product lines also give rise to a cost impact called *economies of scope,* which is defined as cost savings that can result from multiproduct—i.e., flexible-agile—manufacturing systems.[47] What are some of these potential cost savings? Since flexible-agile manufacturing systems can switch from one product design to another (within the parameters or envelope of the system) with little or no cost or time penalty, the risk that an investment in a manufacturing plant might be rendered obsolete by a sudden change in market demand is reduced considerably, the ability to meet quick delivery requirements without finished goods inventories is increased, and the downside of the level scheduling tradeoff—the accumulation of finished-goods inventories to handle seasonal fluctuations in demand—can now be avoided by the judicious selection of a product mix whose individual demand patterns are complementary.[48] These cost impacts of economies of scope are shown in Fig. 7.3; see arrow B.

7.14.5 Distribution and Customer Education

The activities and technologies shown in Fig. 7.5 are not the only ones affecting the availability advantage, of course. Products normally go through a *distribution system* to reach a customer, and these activities and the technologies used in them can obviously affect how soon they reach the customer. Marketing and sales activities and technologies may also be critical in *educating customers* about the existence and use of an expanding choice of products available to them.

7.15   SUMMARY AND CONCLUSIONS

To summarize the framework presented in this chapter for use in analyzing and understanding the contribution of technology to competitive advantage of the firm, please refer to Fig. 7.6.

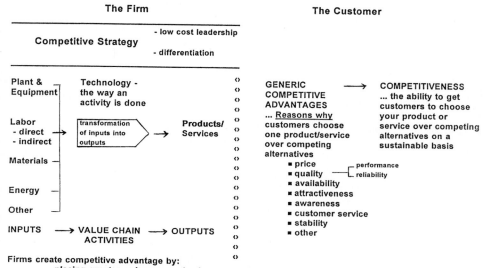

FIGURE 7.6   Technology and competitive advantage.

On the right side of Fig. 7.6 is the customer, representing the decision makers who collectively determine the winners and losers in competition when they choose to buy one product or service over competing alternatives for reasons which are called *competitive advantages*. On the opposite side of Fig. 7.6 is the firm which is competing in the marketplace in accordance with its *competitive strategy,* pursuing *low-cost leadership* or *differentiation.* It does so by carrying out *activities* which can be organized into a system called the *value chain.* Each activity—and the value chain in total—can be viewed as a black box into which inputs flow and from which outputs flow. Inside the black box, inputs are transformed into outputs, and *technology* is the way in which this happens—the way an activity is done. One of the most important ways to create competitive advantage lies in the choice of technologies used to perform those activities that are most important to the advantage being sought. Information technologies are also very important in managing linkages among activities to create competitive advantage. On the left side of Fig. 7.6, the sources of competitive advantage are addressed. If the firm is successful, customers on the right side of Fig. 7.6 will value its products and services more than competing alternatives and make their choices accordingly.

7.16   REFERENCES

1. Alfred Chandler, *Strategy and Structure: Chapters in the History of Industrial Enterprise,* MIT Press, Cambridge, Mass., 1962.

2. Henry Mintzberg, "Strategy Making in Three Modes," *Calif. Management Review,* **15**(2):44–53, Winter 1973.

3. Henry Mintzberg, "Generic Strategies: Toward a Comprehensive Framework," *Advances in Strategic Management,* JAI Press, Greenwich, Conn., 1988.

4. Henry Mintzberg and James A. Waters, "Of Strategies, Deliberate and Emergent," *Strategic Management J.,* **6**(3):257–272, July–Sept., 1985.

5. Kenneth R. Andrews, *The Concept of Corporate Strategy,* Irwin, Homewood, Ill., 1987.

6. J. B. Quinn, "Strategic Goals: Process and Politics," *Sloan Management Review,* **19**(1):21–37, Fall 1977.

7. J. B. Quinn, *Strategies for Change: Logical Incrementalism,* Irwin, Homewood, Ill., 1980, p. 93.

8. C. Roland Christensen, et al., *Business Policy: Text and Cases,* Irwin, Homewood, Ill., 1982, p. 93.

9. Michael E. Porter, *Competitive Strategy,* Free Press, New York, 1980.

10. Michael E. Porter, *Competitive Advantage,* Free Press, New York, 1985.

11. Porter, 1980, op. cit. (Ref. 9), p. xvi.

12. *Global Competition—The New Reality,* The Report of the President's Commission on Industrial Competitiveness, vol. I, Jan. 1985, p.6.

13. *Building a Competitive America,* First Annual Report to the President and Congress, Competitiveness Policy Council, March 1, 1992, p. 1.

14. David A. Garwin, "Competing on the Eight Dimensions of Quality," *Harvard Business Review,* **65**(6)101–109, Nov.–Dec. 1987.

15. Philip B. Crosby, *Quality is Free,* McGraw-Hill, New York, 1979.

16. Porter, 1980, op. cit., chap. 2.

17. Porter, 1985, op. cit. (Ref. 10), chap. 1.

18. Ibid., p. 13.

19. Ibid., p. 14.

20. Ibid., pp. 16–20.

21. Ibid., p. 15.

22. Theodore W. Schlie and Joel D. Goldhar, "Advanced Manufacturing and New Directions for Competitive Strategy," *J. Business Research,* **33**(2):103–114, 1995.

23. B. Joseph Pine II, *Mass Customization: The New Frontier in Business Competition,* Harvard Business School Press, Boston, 1993.

24. Steven L. Goldman, Roger N. Nagel, and Kenneth Preiss, *Agile Competitors and Virtual Organizations,* Van Nostrand Reinhold, New York, 1995.

25. Porter, 1985, op. cit., p. 33.

26. Ibid., chap. 5.

27. Ibid., p. 48.

28. Ibid., p. 48.

29. Ibid., p. 49.

30. Ibid., p. 50.

31. Ibid., chap. 5.

32. Kiyoshi Suzaki, *The New Manufacturing Challenge—Techniques for Continuous Improvement,* Free Press, New York, 1987.

33. Richard J. Schonberger, *Japanese Manufacturing Techniques,* Free Press, New York, 1982.

34. Robert S. Kaplan, "Must CIM Be Justified by Faith Alone?" *Harvard Business Review,* **64**(2):88–95, March–April 1986.

35. John K. Shank and Vijay Govindarajin, "Strategic Cost Analysis of Technological Investments," *Sloan Management Review,* **34**(1):39–51, Fall 1992.

36. Manash R. Ray and Theodore W. Schlie, "Activity-Based Management of Innovation and R&D Operations," *J. Cost Accounting,* **6**(4):16–22, Winter 1993.

37. Suzaki, op. cit. (Ref. 32), chap. 13.

38. Ann Majchrzak, *The Human Side of Factory Automation,* Jossey-Bass, San Francisco, 1988.

39. Schlie and Goldhar, op. cit. (Ref. 22).

40. John E. Ettlie, *Taking Charge of Manufacturing,* Jossey-Bass, San Francisco, 1988.

41. Porter, 1985, op. cit., pp. 186–191.

42. C. K. Prahalad and Gary Hamel, "The Core Competence of the Corporation," *Harvard Business Review,* **62**(3):79–91, May–June 1990.

43. Christopher Meyer, *Fast Cycle Time,* Free Press, New York, 1993.

44. Marc Meyer and James Utterback, "The Product Family and the Dynamics of Core Capabilities," *Sloan Management Review,* **34**(3):29–47, Spring 1993.

45. Thomas H. Davenport, *Process Innovation,* Harvard Business School Press, Boston, 1993, p. 255.

46. Goldman et al., op. cit. (Ref. 24), p. 3.

47. J. C. Panzer and R. D. Willig, "Economies of Scope," *American Economic Review,* **7**(2):268–272, 1977.

48. Schlie and Goldhar, op. cit., p. 111.

CHAPTER 8
TARGETED BASIC RESEARCH
INDUSTRY-UNIVERSITY
PARTNERSHIPS

Frederick Betz
National Science Foundation
Washington, D.C.

8.1 INDUSTRIAL NEED FOR SCIENCE

Since industry directly uses technology in its production of goods and services and not science, industry needs science when (1) new basic technologies need to be created from new science, and (2) technological progress in an existing technology cannot be made without a deeper understanding of the science underlying the technology.

8.1.1 Technology Bottlenecks

Industry knows how to incrementally develop new technology for its needs. Technologies are systems, and the morphology and logic of technology systems can be described. Industry can identify technology bottlenecks in current technology systems and organize research programs in industrial laboratories to address these bottlenecks. It is when industry needs new science that major problems in the R&D infrastructure arise and when industry-university research partnerships prove valuable.

8.1.2 Industry's Use of University Research

Science has become expensive and seldom contributes to near-term profitability in a direct sense. The average time from scientific discovery to technological innovation has varied historically, from as long as 70 years to as short as 10 years. In the twentieth century in the best cases, it has usually taken at least 10 years from laboratory to product. Also, science has become primarily the responsibility of modern universities. In universities, scientific research is integrated with graduate education, and projects are divided into "thesis" size and duration. Scientific research in universities is performed by doctoral candidates for their thesis requirements. Doctoral research is seldom at a scale, breadth, or timeliness suitable for industrial needs.

This is the problem—when industry needs science, (1) it can do it itself at great expense, high risk, and trouble, or (2) it can turn to universities to do the science, but performed in forms and timing unsuitable to industrial needs. Thus, while there is a natural basis for cooperation between industries and universities, the problem is to effect the cooperation.

8.1.3 University Science and Industrial Technology

Fundamental science has a long-term, generic perspective on what is important; and scientists are motivated by achievement of scientific fame. Technology has a shorter-term view, focused on solving a particular problem; and technologists are motivated by the satisfaction of solving a problem and being rewarded by commercial and financial success. Fundamental scientists and technologists form distinct research communities with distinctly different cultures and values.

Industry has adopted and evolved the culture of technology within industry because it directly uses technology, but industry has never wholly adopted and evolved the culture of science within industry because it only indirectly uses science.

In contrast, universities have adopted and evolved the culture of science because they directly use science in education. But universities have only partially adopted and evolved the culture of technology (and then only in their professional schools of engineering and medicine, and not in their schools of arts and science). For effective long-term corporate technology strategy, the resolution to the problem of how best to organize research in industry requires attention not only to managing corporate and divisional research labs but also to managing industrial involvement in university-based science.

8.2 NEXT-GENERATION TECHNOLOGIES

In the early studies on innovation, Don Marquis identified three types of innovation: radical, incremental, and systems (Marquis, 1969). Now, however, we distinguish four types:

1. Radical component innovations
2. Incremental innovations to existing technologies
3. Radical systems innovations
4. Next-generation technology innovations

A *radical innovation* provides a brand-new functional capability, which is a discontinuity in then current technological capabilities. An *incremental innovation* improves the existing functional capability of an existing technology through improving performance, safety, and quality and lowering cost. A *systems innovation* is a radical innovation providing new functional capability but based on reconfiguring existing technologies. Incremental innovations within a system can also sometimes create new technical generations of a system. Such an innovation is still a kind of system innovation, but not a radically new innovation. It is a systemic innovation which is called a *next-generation technology* (NGT).

Industry and university research cooperation can provide an effective means to improve management's capability of exploiting discontinuous technological change: NGTs.

8.2.1 Technology Strategy

Although all firms use technologies in products and production and services, not all gain a positive competitive advantage from technology. There are many factors in competition, and technology is only one factor. Yet some firms effectively use technology as a competitive advantage, and others do not. One important factor in the successful use of technology has been the role of general management in technology strategy. In particular, it has been management's ability to foster corporate core technical competencies.

An example of this is Joseph Morone's study of companies that have turned technology leadership into long-term competitive advantage for business success, from which he concluded (Morone, 1993, p. 17)

> The similarities among the cases of success (in using technology for a competitive advantage) begin with the way general management defines the domain in which their businesses compete. In every case, a strategic focus has been consistently articulated and applied for decades.

The central idea here is that a business can be developed around a long-term, consistent focus on a core technological competency. What it means to have a core corporate technical competency is to lead in both innovating new-technology products and improving manufacturing quality and lowering cost of these products. Not only can products be improved in future generations of technology, manufacturing processes can also be improved in future generations of technology.

For example, in Corning (Morone, 1993, p. 20)

> Creating entirely new businesses through radical innovation is a hallmark of Corning's history. The optical fiber business is the most dramatic recent example...(such) radical innovations were followed by generational innovations, and within each generation by incremental improvements. In optical fibers, Corning made its own manufacturing capability obsolete by developing and introducing seven distinct generations of processing technology over the course of 15 years.

Morone and others have argued that businesses that have successfully pursued technological innovation have done so through a combination both of incremental and generational innovation (Morone, 1993, p. 20):

> To be sure, all these businesses have vigorously pursued incremental improvements to their product lines. But they have done so *in combination* with an equally vigorous pursuit of radical and generational innovations.

Businesses using technology only with incremental innovations are seldom as successful as businesses using technology with both incremental and generational innovations. But managing both types of innovation processes is difficult. Abernathy and Clark first made this point in their study of innovation in the automobile industry (Abernathy and Clark, 1978, p. 168). Morone (1993, p. 21) also reemphasized the difficulty of managing both incremental and discontinuous technological change:

> There appears to be an inherent conflict between the managerial attributes required to manage the continuous improvement of already established product lines...and those required to manage the discontinuous leaps into new businesses and markets.

8.2.2   Impact of Technological Change on Firm Competencies

Technological change can alter the basic capabilities of a firm in either production or marketing, or both. Technological change can alter corporate operations *to preserve or to destroy* existing competencies in production or marketing.

Abernathy and Clark classified innovations by their impact on such existing competencies (Abernathy and Clark, 1985):

1. *Regular innovations*—preserve production competencies and market competencies.
2. *Niche-creation innovations*—preserve production competencies but disrupt market competencies.
3. *Revolutionary innovations*—obsolete (phase out) production competencies but preserve market competencies.
4. *Architectural innovations*—obsolete production competencies and disrupt market competencies.

For normal strategic processes at the executive level, a firm can deal with the regular and niche-creation innovations—since the technical skill base of the organization is not affected by either type of innovation. However, the event of either a revolutionary or an architectural innovation requires a strategic business reorientation. Discontinuities in technology always result in revolutionary or architectural innovations, because these discontinuities obsolete the current technical skills in a firm, thereby impeding the firm's future ability to produce and to serve markets.

The relationships between the nature of innovation and markets are

1. Incremental technological change allows a firm to grow and hold markets in an industrial structure.
2. Discontinuous technological change allows a firm to alter an existing industrial structure and to destroy competitors.

8.3   MANAGING INCREMENTAL AND DISCONTINUOUS INNOVATIONS

Research has usually been organized in a diversified firm in three general patterns:

1. A single, centralized corporate research laboratory
2. Completely decentralized division labs without a corporate research lab
3. Combination of a central lab and decentralized divisional labs

Alfred Rubenstein observed the organization of research in many firms over a long period of time and concluded that no organizational form alone could provide a "best" solution (Rubenstein, 1989, p. 41).

The purely decentralized divisional labs form facilitates and encourages research to focus on the current businesses of the corporation and requires special management attention to focus on the future of the corporation itself. The purely centralized form of a corporate research lab facilitates and encourages research to focus on the future of the corporation itself and requires special management attention to make its research relevant to any of the current businesses of its divisions. It is difficult, therefore, for the corporation-level research to be directly relevant to any of the businesses of the firm without being duplicative of the divisional labs. Since new businesses may emerge only infrequently from corporate research, in any given period of time it is usually difficult to measure the corporate research lab contributions to profitability.

In a combination pattern (3), divisional labs are responsible for the engineering product and process development, and the central research lab is responsible for the generic fundamental and applied science of the corporation. Most companies have tried this combination pattern, with at first the corporate labs being generously funded and later put through a harshly critical review by divisional managers.

Rubenstein (1989, p. 41) described a typical history of the mixed form:

> As the operating divisions begin to flex their decentralized muscles and start acting as though they were indeed independent enterprises…, they begin to become impatient with the level of quality or relevance of the work being done in the central R&D activity. For those division managers who see a real need for strong, direct technical inputs to their division's operation, the central lab seems unwieldy, distant, and not very responsive to their immediate and near-term future needs.

When a diversified firm adopts such a mixed form (having both a corporate research lab and divisional labs), corporate management can still expect the two forms of research organizations to snipe at each other. The divisional labs will always be impatient with the longer-term focus and more distant relevance of the research in the corporate labs. The corporate research lab will always be impatient with the short-sighted focus of the division labs and their low-tech, NIH (not-invented-here) syndrome.

The conclusion is that a mixed form may not be "best" for everybody's purposes, but it may be necessary for all purposes. The organizational conflict in centralized versus decentralized industrial research will always exist and cannot be eliminated by any management procedure but should be managed constructively.

8.4 TARGETED BASIC RESEARCH

Sometimes the ideas for new directions of research for a technology grow out of research in previous areas. Also, often the performance for this early exploratory research will have occurred in both universities and in some industrial corporate laboratories, as well as in some governmental mission laboratories. The support for such research now often comes from federal programs and industrial firms. The best source of information about the *longer-term* directions of new technologies can be found in research projects that are basic in nature but technologically oriented, called *targeted basic research,* in which government programs support exploratory research in broad areas. In the United States, some federal agencies that have significant programs of supporting basic research include the Department of Agriculture (USDA), the Department of Defense (DoD), the National Science Foundation (NSF), and the National Institutes of Health (NIH).

8.4.1 Planning Targeted Basic Research

When groups of researchers began to see that particular exploratory projects are opening up a new technological area, the federal government, in collaboration with industry and university researchers, can organize national workshops which identify the new opportunities and propose topics for future research. If government and industry subsequently fund research programs on these topics, projects funded within these programs can be monitored in order to identify both the directions and paces of technical advances.

Research for technology cannot be planned until after a basic technological invention occurs for the technology system. After the basic technology invention, research can be planned by focusing on the generic technology system, production processes, or underlying physical phenomena.

Thus technology-focused-and-targeted basic research can be planned for

1. Generic technology systems and subsystems for product systems
2. Generic technology systems and subsystems for production systems
3. Physical phenomena underlying the technology systems and subsystems for product systems and for production systems.

8.4.2   Basic Research for Improving Technology

In any of the technology systems, basic research can be focused on improving the system through inventing improvements in any aspect of the system:

1. Improved system boundary
2. Improved components
3. Improved connections
4. Improved materials
5. Improved power and energy
6. Improved system control

The physical phenomena underlying a system can be focused on any of the system aspects:

1. Phenomena involved in the system boundary
2. Phenomena underlying components
3. Phenomena underlying connections
4. Phenomena underlying materials
5. Phenomena underlying power and energy
6. Phenomena underlying system control

Since science focuses on phenomena, targeted basic research is scientific research motivated by any of the above.

8.4.3   Targeted Basic Research for Next-Generation Technology

For targeted basic-research planning, procedures should (Betz, 1993, p. 425)

1. Characterize a technology as a system

2. Identify technical bottlenecks in the way of improving the system
3. Imagine if and how new research instrumentation, instrumental techniques, algorithmic techniques, or theoretical modeling could be focused on the bottlenecks
4. Imagine how to improve the understanding and manipulation of the phenomena of the bottlenecks or to develop alternative phenomena to substitute new technical manipulations at the bottlenecks

Research planning for the long term should try to envision and research discontinuities as a next generation of the technology. The first step in research planning for an NGT system requires a vision.

A vision of NGT requires not only envisioning a major improvement in performance but also research imagination:

1. A description of the functionality, performance, and features that are very desirable but cannot be obtained in the current generation of technology
2. Imagination of how current applications could be better accomplished with the new technology and new applications that might be possible which cannot be accomplished with the current technology
3. Some new methodological research advances that might be used in a new generation of the technology

Illustration: Research Planning for Biotechnology Processing in 1990. As an illustration of planning targeted basic research for a next-generation technology, we look at the research planning in 1991 in an Engineering Research Center for Biotechnology Process Engineering at the University of Massachusetts directed by Professor Daniel Wang. In this center, university and industrial researchers had planned basic research for the technologies of producing biotechnology products grown in mammalian cell cultures (Betz, 1993, p. 445).

The technology system divides into a *bioreactor* (which grows the cell cultures) and a *recovery system* (which recovers the desired proteins produced by the cells).

Within the bioreactor, the technological performance variables for protein production are

- Number of cells
- Functioning of each cell
- Quantity of the protein produced by the cells
- Quality of the protein produced by the cells

The products of biotechnology cell production are particular proteins that have commercial use in therapeutic medicine. The *quantity of protein* is a measure of the productivity. The *quality of protein* depends on being properly constructed and properly folded. Proteins are long chains of organic molecules that fold back in particular configurations. The usefulness of the protein depends not only on having the right molecules link together but also on the ability to fold up into the right pattern. One of the technical bottlenecks in protein production was to get the proper folding.

For the design of bioreactors (for the production of proteins from mammalian cells), the biotech engineer needed to design reactors that

- Grow a high density of cells

- Formulate the proper medium for their growth
- Provide proper surfaces for mammalian cells to attach to
- Control the operation of the reactor
- Provide for recovery of the proteins that are grown in the cells

In 1991, bioreactor designs for growing mammalian cells were fiber-bed, fixed-bed, and ceramic matrix designs. Bioreactor control required sensors for monitoring cell growth and growth medium, expert systems for deciding on control strategy, and actuators for controlling the bioreactor temperature and material flows.

For the design of the recovery processes of the protein product from the bioreactors, the biotech engineer needed to control the physics and chemistry of protein structures, including

- Protein–protein interactions that avoided aggregation of proteins or mutations of proteins
- Protein–surface interactions that affect aggregation of proteins or absorption of proteins to separation materials
- Protein stability that affected the durability of the protein through the separation processes

In 1991, the recovery processes for separating the desired proteins from the output of the bioreactor included chromatographic procedures and filtration procedures and protein refolding procedures.

With this kind of system analysis of the bioreactor and recovery system and the important technological variables, the MIT Biotechnology Process Engineering Center had identified the research areas necessary to provide improved knowledge for the technology.

For the bioreactor portion of the system, they listed the following scientific phenomena as requiring better understanding:

1. Extracellular biological events
   - Nutrition and growth factors of the cells
   - Differentiation factors of the cells
   - Redox (reduction-oxidation) oxygen conditions in the cellular processes
   - Secreted products of the cells

2. Extracellular physical events
   - Transport phenomena of materials in the medium
   - Hydrodynamics of fluid–gas interactions of the medium
   - Cell–surface interactions
   - Cell–cell interactions

3. Intracellular events and states
   - Genetic expression of the proteins in the cells
   - Folding of the proteins and secretion from the cells
   - Glycosylation of the proteins
   - Cellular regulation of secretion
   - Metabolic flows in the cells and their energetics

We see this is a list of the biological and physical phenomena underlying the cellular activities in the bioreactor.

Similarly, MIT Biotechnology Process Engineering Center listed the scientific phenomena underlying the recovery process of the system that required better understanding:

1. Protein–protein interactions
   - Aggregation of proteins into clumps
   - Mutations of protein structure
2. Protein–surface interactions
   - Aggregation of proteins through denaturation
   - Adsorption of the proteins to surfaces
3. Protein stability
   - Surface interaction
   - Chemical reaction
   - Aggregation in the solvent
   - Stabilization

Accordingly, the Center organized their research into two areas:

- Engineering and scientific principles in therapeutic protein production
- Process engineering and science in therapeutic protein purification

In 1990, the research projects in the first research area of protein production included the following projects:

- Expression: transcription factors
- Protein trafficking and posttranslational modifications: glycosylation and folding
- Redox potential
- Pathway analysis
- Intercellular energetics
- Regulation of secretion
- Hydrodynamics: gas sparging
- High-density bioreactor designs
- Substrata morphology for cell attachment

Also in 1991, the research projects in the second research area of protein separation, included the following projects:

- Protein adsorption: chromatography and membrane
- Protein aggregation
- In vivo protein folding
- Protein stability in processing, storage, and delivery

In this example, we see that in research planning of the center (1) the technology system was described, the critical technology performance variables identified, and the underlying scientific phenomena identified, and then (2) research projects were formulated for improved understanding of the underlying scientific phenomena that could be used for inventions, design aids, and control procedures of the technology system.

8.4.4 Demonstrating a Next-Generation Technology

A vision of an NGT is a description of the functionality, performance, and features desirable that cannot be obtained in the current generation of technology and includes imagination of how current applications could be better accomplished with the new technology and new applications that might be feasible which cannot be accomplished with the current technology. A research strategy leading to a technical feasibility demonstration of NGT is a set of interdisciplinary research thrusts containing projects that are scheduled for integration into an experimental testbed embodying the functionality of an NGT system.

Because the form of an NGT system is complex, one should try an example of a system on an application in order to see which advanced research ideas will be useful, which won't, and what cannot yet be done. A research testbed for an NGT system allows the demonstration of advanced ideas to reduce risk by making more clear what is the most useful engineering pathway to the future generations of a technology system.

Once an NGT vision is created, the next step is to formulate a long-term research strategy:

1. The current state of the technological system should be described, delineating present limitations on functionality as performance limits and existing features and current costs, along with present applications of the technology.

2. A desirable significant change in performance and features is then planned which would constitute a dramatic leap forward, to alter existing markets and create new markets. The NGT goals are then expressed as planned performance, features, and cost.

3. The research issues are identified which could lead from the present state of technology toward the NGT vision. These issues are then grouped into a set of interdisciplinary research thrust areas, which together constitute a program of basic science and engineering research for the next generation of the technological system.

4. Initial research projects are proposed in these thrust areas to address the range of research issues and are then coordinated in time to result in prototype subsystems of the NGT system. A milestone chart is formulated which plans the integration of these prototype subsystems into an experimental testbed for the NGT system.

These steps lay out a targeted basic research program aimed at demonstrating an experimental prototype of the NGT system.

8.5 INDUSTRY-UNIVERSITY RESEARCH COOPERATION FOR NGT

Industrial researchers are very sophisticated about current technology and, in particular, about its problems and locating and identifying the roadblocks to technical progress. However, because of the applied and developmental demands on industrial research, they have limited time and resources to explore ways to leapfrog current technical limitations. On the other hand, academic researchers have time, resources, and students to explore fundamentally new approaches and alternatives that leapfrog technologies.

Together, industry and university researchers can see how to effectively bound a technological system in order to envision a next generation of technology. This boundary is an important judgment, combining judgments (1) on technical progress and research directions which together might produce a major advance and (2) over the domain of industrial organization that such an advance might produce a significant competitive advantage.

8.5.1 Industry-University Research Centers

To effect industry and university research cooperation on NGTs, a bridging institution is necessary—because industries and universities live in almost completely different universes. The industrial universe is a world of technology, short-term focus, profitability, and markets. In contrast, the university universe is a world of science, long-term view, philanthropy, and students.

An industry-university research center (IURC) for cooperative research bridges these two world views, creating a balance between (1) technologically pulled research and scientifically pushed research, (2) short-term and long-term research focus, (3) proprietary and nonproprietary research information, and (4) vocationally relevant education and professionally skilled education. These are the issues inherent in industry and university cooperation. Properly handled, these provide creative tension:

1. Linking technology and science in real-time operation
2. Creating progress in knowledge and developing the technological competitiveness of nations

Requirements for Industrial Use of University Science. For an industry to effectively support and use university research centers, several requirements must be met:

1. A diversified firm must have a corporate research lab as well as divisional labs.
2. The corporate research lab must be tightly linked into divisional labs.
3. The company should work with multidisciplinary strategically focused university research centers.
4. Such centers must be large enough to perform a critical mass of research useful to industry.
5. Such centers should link with both the corporate research lab and the divisional labs of a company.
6. Company personnel should participate in both the governance and research of the university center.
7. The corporate research lab and divisional labs should be performing joint applied and development projects parallel to the research projects of the university center.
8. The firm should be hiring university graduates from the university center.
9. The firm should join with several other firms and with government in financially supporting the university center.
10. The firm should participate in and support several university centers sufficient to provide it with a long-term competitive edge in strategic technologies.

University Requirements for IURCs There is a complementary set of requirements that the university must satisfy for an industry-university research center to operate:

1. The university needs to provide contiguous space for the center and arrange for the instrumentation needs of the center to be at a cutting-edge state of art in science and technology.

2. The university needs to have recruitment, tenure, and promotional policies that reward and properly balance faculty contributions to both scientific achievement and technological progress.

3. The center needs to be multidisciplinary and complementary to the departmental educational structure of the university.

4. The center needs to actively coordinate educational requirements of programs and students with research requirements of industry.

5. The center needs to formulate research projects that result in publication in the open literature.

6. The center needs to inform faculty and students how to cooperate with industry while still fulfilling the principal missions of the university in education and research.

7. The center needs to work with large firms as well as small firms, with the large firms providing the principal industrial support.

8. The center needs to be capable of strategic research planning.

9. The center needs to be capable of performing scientific research at the systems level of a technology.

10. The center needs a dual-structure organization providing capability for performing scientific research and for technology development capabilities for industry.

These requirements arise from much experience in starting and operating university research centers that receive substantial industrial support. They have been tested in at least 75 university centers in the United States from the mid-1970s into the 1990s.

8.5.2 Technology Transfer of NGT in Industry-University Centers

The following procedure facilitates the appropriate kinds of cooperation for technology transfer of NGT:

1. Corporate research should strategically plan next-generation technology, and this is best done within an industrial-university-government research consortium.

2. Corporate research should plan NGT products jointly with product development groups in the business divisions.

3. Marketing experiments should be set up and conducted jointly with corporate research and business divisions, with trial products using ideas tested in the consortium experimental prototype testbeds.

4. The CEO team should encourage long-term financial planning focused on NGT.

5. Personnel planning and personnel development are required to transition knowledge bases and skill mixes for NGT.

Consensus on vision and testing technical feasibility in a program of technology transfer between corporate research and strategic business units should pose and address the following questions:

1. What will be the boundaries of the next generation of a technology system?
2. What NGT ideas are technically demonstrable and should now be planned into product strategy and what NGT ideas must still be technically demonstrated and should not yet be planned into product strategy?
3. What is the pace of technical change, and when should the introduction of products based on NGT be planned?
4. What professional development and training should be planned for product development groups in order to prepare for NGT products?

8.6 SUMMARY

Next-generation technology research planning is an excellent way for industry to utilize university-based science. It improves management capability of exploiting technology discontinuities. For technology transfer between industry and university to be effective, it is necessary for both the strategic business units' research labs and corporate research to cooperate with university researchers in formulating and reaching a consensus on an NGT vision and research plan. Industry-university research cooperation can facilitate good cooperation between corporate and divisional research labs and help management with its long-term research needs.

8.7 REFERENCES

Abernathy, William J., and D. B. Clark, "Mapping the Winds of Creative Destruction," *Research Policy,* **14** (1):2–22, 1985.

Betz, Frederick, *Strategic Technology Management.* McGraw-Hill, New York, 1993.

Marquis, Donald G., "The Anatomy of Successful Innovations," *Innovation*, November, 1969.

Morone, Joseph, "Technology and Competitive Advantage–the Role of General Management," *Research-Technol. Management,* **36**:16–25, March–April, 1993.

Rubenstein, Albert H., *Managing Technology in the Decentralized Firm,* Wiley, New York, 1989.

CHAPTER 9
MANAGING TECHNOLOGY-BASED INNOVATION

Hans J. Thamhain
Bentley College
Waltham, Massachusetts

9.1 THE ROLE OF TECHNOLOGICAL INNOVATION

Technological innovation has become the strongest engine driving society since the 1980s. We are enjoying a continuous stream of new products and services, from consumer electronics to automobiles, aircrafts, telecommunications, and pharmaceuticals, to name but a few. Yet, technological innovation is not a new phenomenon which suddenly emerged as part of the space age. It has been around and shaped our life for thousands of years. The Egyptian pyramids, the Great Wall of China, the inauguration of a king, navigational methods, Noah's ark—all these undertakings involved a great deal of technological innovation. The basic organizational tools, techniques, and systems for managing these innovative efforts, such as planning, budgeting, directing, scheduling, motivating, and task integrating, were already known in these early times and formed the foundation for today's management systems. What has changed gradually is the ability of organizations to synthesize components, products, and processes on a worldwide basis. That is, organizations today have the capacity of combining materials, components, knowledge, skills, and processes to create new products, processes, and services with new characteristics and higher performances that did not exist before.* This continuously increasing base of technology, together with the dynamics of the free market, has created tremendous business opportunities for companies around the world. But it also has increased competitive pressures, risks, and the need for business strategies, leadership, and management styles that are radically different from the traditional methods.

Traditionally, the competitive advantage of a company was derived from efficien-

---

*This process is described by Dan E. Kash as *perpetual* or *synthetic innovation* (see Kash[15]).

cy, such as low cost of products and services. Today's companies gain their competitive advantage and economic benefits largely from innovation. Those who can leverage technology to achieve superior performance, new features, *and* lower cost will add the largest value to their products and compete most effectively in the market. A significant part of the success formula includes resource utilization, business process effectiveness, and speed. Today, managers recognize the important role of technological innovation for a company's business success.[7–9] It can create a competitive advantage for one firm while eroding the market position of another.[10,11,15] Yet, in spite of this reality, for most managers, innovation is a risky process! Many companies everywhere are concerned that their investments in R&D and other innovative activities produces adequate returns. Managers wrestle daily with complex issues of choice and balance:

How much to invest in plant and facilities and how much to invest in people and their development?

How much to focus on process versus product technology efforts?

What technology to develop in-house and which to buy?

How much cooperation and alliance should be promoted with suppliers, customers, and competitors, and when does it become too risky, ending up in costly litigations?

Despite this thicket of thorny issues, the realities of our highly dynamic global business environment point at innovation as one of the most critical determinants of overall business performance in their companies.*

Innovation might be the last competitive frontier. However, its power is being harnessed only when it creates value for the stakeholders of an organization, its customers, owners, and employees.

This statement also provides a starting point for defining *innovation* in contrast to *invention*. Innovation is a complex, multistage, multiperson process, composed of two parts: (1) the generation of an idea or invention and (2) the conversion or exploitation of this idea into a useful application, which is often called *commercialization*. A more detailed discussion is provided in Fig. 9.1.

The companies that will survive and prosper in the decades ahead will be those that can manage innovation and derive business benefits from it. And they must do this in spite of the complex organizational processes, rapidly changing technology, increasing risks, uncertainties, cost, demands for better market responsive, and relatively low barriers to entry into almost every business.

The new realities also confront businesses with a double challenge of keeping up with currently committed developments and starting new efforts to position the company favorably in the future. In addition, increasing complexities and continuously changing technologies, components, support functions, materials, and methods make it

---

*Between 1984 and 1995 the author interviewed 364 senior managers from 47 technology-based companies regarding the significance of innovation to their firm's business performance. As part of the interview, these managers indicated the specific significance of innovation on a four-point scale: (1) very significant, among the most crucial factors of survival and growth; (2) important; (3) somewhat important among many other factors; and (4) not too important. Of these managers, 88 percent indicated significance level 1 or 2, and 100 percent indicated levels 1, 2, or 3. No one supported a "not too important" position.

The invention process includes all aspects leading to the creation of a new concept which, at least in principle, is workable. The innovation process takes a new concept, or combines several new or old concepts into a new scheme (another invention), and then develops it into a commercially useful product, process, or service. While the lines between invention and innovation are often blurred in business practice, the distinction focuses on the exploitation of a new concept toward commercial application and value. For technology-based innovations, this commercialization process involves the transfer of technology from its idea generator to a paying customer. As an example, the Englishman Swan invented the incandescent lightbulb in 1900, a working lab model that produced light for a few minutes before burning out. Edison, an innovator, took this lightbulb idea and enveloped it, together with other schemes such as power distribution, into a commercial product for consumer markets. In more recent history, Shockley and others discovered the junction effect in semiconductors in 1947, inventing the transistor. This led to a broad range of innovative efforts and outcomes, creating new industries and products that radically changed our lifestyle forever.

Technologically innovative outcomes come in many forms: radical, breakthrough, or incremental. They can be directed top–down or originate at the operations level of a company, such as continuous improvement efforts. Innovation can involve new product and service developments as well as product modification, extensions, and combinations of systems and technologies.

In today's organizations, most technologically innovative activities are beyond the idea generation. They seldom produce radical breakthroughs, but rather incremental advances which need to be systematically transferred and integrated throughout the various organizational systems in order to produce desired technological change.

FIGURE 9.1 Technology-based invention and innovation: some definitions and examples.

virtually impossible for one company to have the resources to develop all the technologies needed to support all of their products and services. In this context innovation must be defined more broadly than just R&D. The need for innovation exists in all functions, at all levels throughout the organization. Especially in technology-based environments, innovation cannot be confined to selected organizations but must be encouraged and nurtured at all levels and with all people.

Indeed, technology continuously complicates the business process. It has become a significant and dominating factor which affects every company from small to large, and from service to manufacturing. New technologies, especially computers and communications, have radically changed the workplace. While making the business process more intricate and complex, it also leads to a higher mobility of resources, skills, and processes, and provides more opportunities for innovative performance throughout the organization. In fact, innovation impacts a company in all functions and business processes—from the conception of an idea to product research, development and engineering, transferring technology into manufacturing the market, product distribution, upgrading, and services. It is further interesting to note that as companies or industries mature, the focus of innovation seems to shift from the front part of the product life cycle—where the emphasis is on concept development—to manufacturing, marketing, and services. Part of the reason for this trend is related to the distribution of capital within the company or industry. Traditionally companies have targeted relatively large portions of their "creativity budgets" toward new-product concept development, with lesser resources going to manufacturing, marketing, and services. Thus most of a company's R&D effort would be directed toward new products. In today's environment, especially within mature businesses, companies gradually shift their spending in favor of manufacturing and services, leading to expanded innovative efforts in these areas.

This trend acknowledges that innovation can create value in two ways: (1) by directly *improving the value of a product or service* to its customers and (2) by *improving the work process* of creating, developing, producing, delivering, and servicing the product.

9.2 THE PROCESS OF TECHNOLOGY-BASED INNOVATION

As the crucial role of innovation for a company's business performance became clear, management research and practice first focused on the qualities and characteristics of the individual. Starting the 1950s, many studies looked into the traits that would help in identifying and developing innovative individuals. These early studies had already identified the important role of work-related knowledge and skills, the need for risk sharing, and a supportive work environment.[22,31] However, until the 1970s, innovation was rarely integrated strategically and operationally into the company's business process. Few managers took a top–down look when assessing the role of innovation for their company's overall performance, nor did they have a unified business policy that includes the management of *innovation* as an integrated resource similar to engineering, manufacturing, and marketing. In this modus operandi, which is often described as *first-generation management,* innovative performance depends on an organization's ability to

- Hire and assign qualified people
- Provide them with the necessary resources and most suitable tools and techniques
- Manage them toward desired results

Under such a system, the burden of developing, enhancing, or reengineering the innovation process falls largely on resource managers, with top management's role confined mostly to being a catalyst or facilitator. This was an established business practice which worked reasonably well until the 1970s. Even today many companies, largely as a result of revenue-producing pressures, still work in this first-generation management mode. However, with the increasing dynamics and complexities of the organizational environment, management recognizes the need for better understanding and orchestrating innovation as an integrated part of the total business process.

In recent years, a number of researchers, such as Johne et al.,[14] Cooper,[4] Crawford,[6] Quinn,[18] Van de Ven,[31] and Gupta et al.[12] have explored the enormous breadth and depth of subsystems and variables involved in the management of innovation. These subsystems involve broad-ranging organizational processes, such as strategic and product planning, entrepreneurship, top management, manufacturing, marketing, product and project management, and most importantly, the people who create and run these processes. Further, it is quite common among managers to emphasize the need for dealing with the nonlinear, intricate, often chaotic and random nature of innovation, which involves all facets of the organization, its members, and its environment.[8,15,17]

Beginning with the 1990s, many companies adopted innovation management practices that are distinctly more systematic and analytical and more specifically focused to business and market needs. They ensure that the operational processes are well linked with the market, technology, and administrative systems, and that innovation can be transferred to an application which adds value to the company. This so-called *second-generation management mode*[8] also requires a larger degree of power sharing and resource sharing, and self-directed control at the operating level of the company.

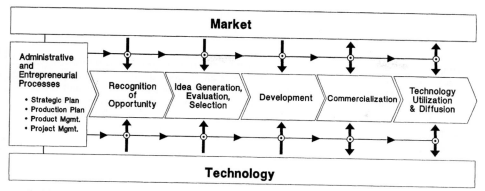

FIGURE 9.2   The innovative process and its interfaces with the market, technology, and administrative subsystems.

The process model of innovation that evolved with this mode, shown in Fig. 9.2, is based on the pioneering work of Edward B. Roberts.[20] This model shows that innovation is a multistage process which is strongly influenced by the prevailing market, technology, and administrative processes. Specifically, Fig. 9.2 presents the innovation process in five stages, described in the following paragraphs. The precise number and labeling of these stages may depend on the specific business and organizational settings. Managers by and large use the model to recognize and control the factors that influence the process and the linkages among the stages.

Stage 1: Recognition of Opportunity.   In most cases, the innovative process is prompted by an opportunity to fill a market need (market-pull) and/or exploit a technology (technology-push). These opportunities could be for new or improved products, processes, or services. The potential customer could be internal or external to the organization. Recognition of an opportunity will set administrative processes and/or entrepreneurial systems toward *idea generation* in motion.

Stage 2: Idea Generation, Evaluation, and Selection.   This stage is dominated by the search for ideas to capture the opportunity identified in stage 1. This might include formal RD&E processes or informal thinking. Results may vary from an orally communicated idea to formal concept papers, designs, prototypes, and feasibility studies. Further, the idea generation process may vary, depending on company culture and management philosophy. They range from incremental to breakthrough innovation, and from top–down direction to bottom–up innovative efforts, typical for continuous productivity improvements. Furthermore, idea generation typically produces *creative results,* therefore, by definition also involving invention. For most business situations, the new concepts that have been generated in this stage are being evaluated regarding feasibility, value, and desirability, and eventually either selected for commercialization, redirected, or terminated.

Stage 3: Product Development.   This stage involves transfer of the new concept to the market. It is a problem-solving stage which takes the advanced concepts and ideas generated in stage 2 and develops them into a working prototype or pilot production run. Strong cross-functional linkages must be established and maintained among all functions engaged in this stage 3 technology transfer which usually involves highly coordinated efforts among R&D, product development and engineering, prototyping, manufacturing, marketing, and a host of support functions such as finance, product assurance, field services, and subcontractors.

Stage 4: Full-Scale Development, Volume Production, and Commercialization. This stage takes a proven concept from stage 3 and transforms it into a final product according to predefined specifications, reliability, cost, production volume, and schedules. Well-established organizational linkages are crucial to transferring technology into the market and to leveraging an organization's production capabilities, as well as to integrating all company resources into the total innovation process, throughout its five stages.

Stage 5: Technology Utilization and Diffusion into the Marketplace. This stage involves the manufacturing, market promotion, distribution, and technical support of the new product or service. This stage usually requires the largest investment of resources, often far exceeding the combined cost of stages 1 through 4. It is also associated with a large risk factor, as demonstrated by the statistical realities that, on average, only one-third of products entering this state ever achieve a break-even return on their investment. Successful companies recognize the complexity and multifunctionality of the underlying process. They involve people from R&D, engineering, manufacturing, finance, marketing, sales, and field services during the earlier stages to ensure that products are conceptionalized and designed for manufacturability, quality, service, and overall value to the customer. Successful companies also understand that such complex business processes do not perform well by themselves, but must be managed carefully. This includes the continuous study of these processes, defining measurements, documentation, comparison, standardization, and control toward continuous improvement.

9.3 MEASURING INNOVATIVE PERFORMANCE

Innovative performance involves complex sets of interrelated variables, which fluctuate with the cultural and philosophical differences among departments and companies. Most managers agree, however, that certain metrics, such as (1) the number of innovative ideas, (2) the number of new product concepts implemented, (3) cost and performance improvements, and (4) patent disclosures, are important factors in contributing to a company's innovative performance. Yet, the individual measurement of quality and effectiveness of such innovative contribution is very fuzzy and often impossible to obtain with any degree of confidence. These measurements of innovative performance are even more difficult in nonengineering areas, such as manufacturing, marketing, and product assurance. In these areas innovation and creativity are often critical to meeting customer expectations or delivering results according to plan. However, traditional measures of innovative performance seldom apply. In many cases, outcomes are part of collaborative efforts among many departments and individuals without any useful metrics for measuring important contributions such as agility, change orientation, multifunctional cooperation, and customer satisfaction. Therefore, it is not surprising that managers in technology-based companies, for most functions, including R&D and engineering, use *overall judgment* as the only principal measure of innovative performance. However, in support of such an overall judgment, specific subsets of parameters, some of them quantifiable, can be developed, as shown in Fig. 9.3, and used to (1) articulate desired innovative behavior and characteristics to members of

Judgment measure	Performance			
	Low	Medium	High	Very high
1. Contributions toward functional or operational improvement	[ ]	[ ]	[ ]	[ ]
2. Innovativeness in meeting established budget and schedule objectives	[ ]	[ ]	[ ]	[ ]
3. Degree of innovations in meeting established technical objectives	[ ]	[ ]	[ ]	[ ]
4. Degree of individual collaboration and teamwork	[ ]	[ ]	[ ]	[ ]
5. Degree of technical, schedule, and/or budget improvement	[ ]	[ ]	[ ]	[ ]
6. Effectiveness and sophistication in dealing with change	[ ]	[ ]	[ ]	[ ]
7. Effectiveness and sophistication in dealing with risk and uncertainty	[ ]	[ ]	[ ]	[ ]
8. Effectiveness and sophistication of transferring technology	[ ]	[ ]	[ ]	[ ]
9. Effectiveness, sophistication in using computer support: CAD, CAE, CAM	[ ]	[ ]	[ ]	[ ]
10. Generating ideas that leverage core competencies	[ ]	[ ]	[ ]	[ ]
11. Helping others develop professional skills	[ ]	[ ]	[ ]	[ ]
12. Leadership in facilitating collaboration and teamwork	[ ]	[ ]	[ ]	[ ]
13. Quality and effectiveness in facilitating cross-functional communications	[ ]	[ ]	[ ]	[ ]
14. Quality and effectiveness of decision making	[ ]	[ ]	[ ]	[ ]
15. Significance and sophistication of product ideas, concepts, and improvements	[ ]	[ ]	[ ]	[ ]
16. Significance and sophistication of patent disclosures	[ ]	[ ]	[ ]	[ ]
17. Significance and sophistication of professional papers and presentations	[ ]	[ ]	[ ]	[ ]
18. Significance of new ideas contributing to core business strength	[ ]	[ ]	[ ]	[ ]
19. Sophistication of problem solving	[ ]	[ ]	[ ]	[ ]
20. Utilizing established business processes effectively: PM, CE, MRP	[ ]	[ ]	[ ]	[ ]
Overall judgment of innovative performance	[ ]	[ ]	[ ]	[ ]

FIGURE 9.3 Sample of typical judgment measures of innovative performance for self-evaluation or benchmarking.

the work team; (2) benchmark innovative performance, especially on the organization or team level; (3) engage in focus-group discussions toward organizational improvement of innovative performance; and finally (4) support managerial judgment of overall innovative performance and salary reviews.

In addition to the samples shown in Fig. 9.3, you can develop your own list of measures of innovative performance:

Step 1. Assess, or ask your management, how your department's performance will be measured, and what goals and objectives you are expected to meet.

Step 2. Form a focus group of seasoned professionals from your department and ask them to define the factors that are crucial for meeting these objectives and for functioning at high performance levels.

Step 3. Combine and integrate the measures of steps 1 and 2, and define a draft list of performance measures.

Step 4. Fine-tune and ratify the list of innovative performance measures with your focus group. Review and revise as necessary.

9.4 CHARACTERISTICS OF AN INNOVATIVE WORK ENVIRONMENT

One of the major distinctions of today's technological innovations from earlier periods lies in the organizational process. Innovation is no longer the product of individual geniuses. Rather, innovations are delivered by teams of people and support organizations interacting in a highly complex, intricate, and sometimes even erratic way. The process requires experiential learning, trial and error, and risk taking, as well as the cross-functional coordination and integration of technical knowledge, information, and components. It is often a fuzzy process that cannot be objectively described, or its specific results predicted. Yet research shows that certain characteristics of the work environments are conducive to generating technological innovative results which can be transferred to the ultimate user and support the mission objective of the sponsoring organization.

In spite of the complexities of the innovation process and the differences among companies, research shows that specific bridging approaches are helpful in stimulating and enhancing innovative performance in technology-based organizations. Figure 9.4 summarizes the conditions that seem to have the strongest influence toward innovative performance. The influences that drive these favorable conditions can be organized into three categories: *process-, people-,* and *organization*-oriented influences.

Business-process-oriented influences include the organizational structure and the technology transfer process which relies by and large on modern project management techniques. It provides proper planning of the activities which should benefit from innovation with joint participation of cross-functional support groups, joint reviews and performance appraisals, and the availability of the necessary resources, skills, and facilities. Other crucial components that affect the process are team structure, managerial power, and control and its sharing among the team members and organizational units, autonomy, and freedom, and most importantly technical direction and leadership.

People-oriented influences seem to have the strongest effect on the innovative performance of an organization. The most significant drivers are derived from the work

A work environment conducive to innovation must have the ability to

- Anticipate future trends and operate proactively
- Create project ownership and commitment to established plans
- Deal with risks, uncertainties, and conflict
- Develop solutions incrementally and concurrently
- Form effective cross-functional linkages for information transfer: data, work in process
- Integrate multidisciplinary work
- Make collective multifunctional decisions
- Measure project status; provide metrics for tracking, status reporting, and control
- Operate flexibly and be change-oriented
- Produce solutions that create economic value
- Provide checks and balances and early-warning systems within its business process
- Provide reasonable job stability
- Resolve conflict, mistrust, power struggle, and confusion
- Respond quickly to changing requirements and customer-user needs
- Self-develop the work team and its management system, tools, and techniques
- Self-direct projects according to plan
- Share power and resources
- Utilize resources effectively

FIGURE 9.4   Characteristics of an innovative work environment.

itself: the personal satisfaction with the professional challenges, results, accomplishments, and recognition of the work. Other important influences include effective communications among team members and support units across organizational lines, good team spirit, mutual trust, low interpersonal conflict, and personal pride and ownership. All these factors help in building a unified project team that can exploit the organizational strengths and competencies effectively, and produce integrated results that support the organization's mission objective.

Organization-oriented influences include many of the variables that are primarily within the domain of the senior manager's control, such as organizational stability as perceived by the people, availability of sufficient resources, management involvement and support, personal rewards, and the stability of organizational goals, objectives, and priorities. Since all these influences are derived as a result of personal perception, it is important for management to create the desired perception through proper communication. For example, a company merger might be perceived as an opportunity or threat, or as a stabilizer or destabilizer, depending on how it is communicated. The relationship between managers and their staff and the people in their organizations and the mutual trust, respect, and credibility all are critical factors toward effective communication.

To foster an innovative work environment, most companies use these bridging approaches in an integrated, often overlapping way. In essence, they create a work environment which has the characteristics shown in Fig. 9.4.

9.5   MANAGING PEOPLE AND PROCESS

Managers have always explored opportunities for improving business performance. Yet, in spite of all the research and conceptional insight, only in recent years have we begun to understand the processes which underlie innovation and its management. The need for strong integration and orchestration of cross-functional activities is particularly stressed by those researchers and practitioners who see the product innovation process as a sequence of interrelated multifunctional efforts which span the complete product life cycle: from recognition of an opportunity and creation of new knowledge and concepts, to product research, development and engineering, transferring technology into manufacturing and the market, product distribution, upgrading, and service.

Successful organizations and their managers pay attention to the human side. They are effective in fostering a work environment conducive to innovative work, in which people find the assignments challenging as well as leading to recognition and professional growth. Such a professionally stimulating environment also seems to lower communication barriers and conflict and enhances the desire of personnel to succeed. Further, it strengthens organizational awareness of environmental trends and the ability to effectively prepare for and respond to these challenges.

In addition, technologically innovative workgroups have good leadership. That is, management understands the factors crucial to success and makes the proper provisions. Management is action-oriented, provides the needed resources, properly plans and directs the implementation of their programs, and helps in identifying and solving problems in their early stages. In fact, many early warning signs of low technical team performance can be identified as summarized in Fig. 9.5. Effective team leaders monitor such feedback and focus their efforts on problem avoidance. The effective team leader also recognizes the potential interrelationships among drivers of and barriers to innovative performance. While assessing the exact impact of all potential problems may be difficult, the effective manager can keep an eye on situations that may cause problems, and proactively intercept and minimize them wherever possible.

In summary, the effective manager of an innovation-oriented work team is a social architect who understands the interaction of organizational and behavioral variables and can foster a climate of active participation and minimal dysfunctional conflict. These abilities require carefully developed skills in leadership, administrative techniques, organization, and technical expertise. In addition, the effective manager must possess the ability to work effectively with upper management to assure organizational visibility, resource availability, focus, and overall support for the innovation-oriented activities and programs throughout their life cycles.

9.6   RECOMMENDATIONS

A number of recommendations can help increase the manager's effectiveness in building high-performing innovative technology-based organizations.

Influence Factors.   Managers must understand the various influence factors that drive innovation and build a work environment conducive to innovative team performance. Specifically, they should pay attention to the factors in Figs. 9.4 and 9.5.

Goals and Objectives.   Management must communicate and update the organizational goals. The project objectives and their importance to the organizational mission must be clear to all personnel who get involved with innovation-oriented activities.

Complaints about insufficient resources
Disinterested, uninvolved management
Excessive conflict among team members
Excessive documentation
Excessive requests for directions
Fear of failure, potential penalty
Fear of evaluation
Lack of performance feedback
Little team involvement during project planning
Little work challenge (not stimulating professionally)
Low level of pride and project ownership
Low motivation, apathy, and team spirit
Mistrust, collusion, and protectionism
No agreement on project plans
Perception of inadequate rewards and incentives
Perception of excessive change
Perception of technical uncertainty and risks
Poor communication among team members
Poor communication with support groups
Poor recognition and visibility of accomplishments
Problems in attracting and holding team members
Professional skill obsolescence
Project perceived as unimportant
Strong resistance to change
Unclear mission and business objectives
Unclear requirements
Unclear role definition, role conflict, and power struggle
Unclear task or project goals and objectives

FIGURE 9.5   Early warning signs of problems with innovative team performance.

Senior management can help develop a "priority image" and communicate the basic project parameters and management guidelines. Moreover, establishing and communicating clear and stable top–down objectives helps build an image of high visibility, importance, priority, and interesting work. Such a pervasive process fosters a climate of active participation at all levels, helps attract and hold quality people, unifies the team, and minimizes dysfunctional conflict.

Planning.   Effective planning early in the life cycle of a project will have a favorable impact on the work environment and team effectiveness. Because project managers have to integrate various tasks across many functional lines, proper planning requires the participation of the entire project team, including support departments, subcontractors, and management. Phased project planning (PPP), stage-gate concepts (SGCs), and modern project management techniques provide the conceptional framework and tools for effective cross-functional planning and organizing the work toward innovative execution.

Process and Involvement.   Managers should encourage the involvement of personnel at all organizational levels. This involvement will lead to a better understanding of the task requirements, stimulate interest, help unify the team, and ultimately lead to commitment toward the project plan. The proper setup and management of technology transfer processes, such as concurrent engineering, SGP, CAD/CAE/CAM, and design-build, is often important for enhancing cross-functional linkages necessary for innovative performance.

Team Structure.   Management must define the basic team structure and operational process for each project early in its life cycle. The project plan, task matrix, project charter, and operating procedure are the principal management tools for defining organizational structure and business process.

Image Building.   Building a favorable image for an ongoing project, in terms of high priority, interesting work, importance to the organization, high visibility, and potential for professional rewards is crucial for attracting and holding high-quality people. It is a pervasive process that fosters a climate of active participation at all levels, and also helps unify the work team and minimizes dysfunctional conflict.

Interesting Work.   Whenever possible, managers should try to accommodate the professional interests and desires of their personnel. Innovative performance seems to increase with the individual's perception of professionally interesting and stimulating work. Making work more interesting leads to increased involvement, better communication, lower conflict, and higher commitment.

Senior Management Support.   It is critically important that senior management provide the proper environment for an innovative team to function effectively. At the onset of the development program, the project leader needs to notify management about the needed resources. The project leader should also obtain a commitment from management that these resources will be available. The relationship of resource managers and project managers to senior management and the ability to develop and sustain senior management support critically affect perceived credibility, visibility, and priority.

Clear Communication.   Poor communication is a major barrier to teamwork and innovative performance. Management can facilitate the free flow of information, both horizontally and vertically, by workspace design, regular meetings, reviews, and information sessions.

Team Commitment.   Managers should ensure team member commitment to the project plan and its specific objectives and results. If such commitments appear weak, managers should determine the reason for such lack of commitment of a team member and attempt to modify possible negative views. Because insecurity is often a major reason for low commitment, managers should try to determine why insecurity exists, then work to reduce the team members' fears. Conflict with other team members and lack of interest in the project may be other reasons for lack of commitment.

Management Commitment.   Managers and team leaders must continuously update and involve management in order to refuel their interest and commitment to the technical venture or project.

Leadership. Leadership positions should be carefully defined and staffed at the beginning of a new project. The credibility of project leaders among team members, with senior management, and with the program sponsor is crucial to the leader's ability to manage the multidisciplinary activities effectively across functional lines.

Team Building Sessions. Such meetings should be conducted by the team leader throughout the project life cycle. An especially intense effort might be needed during the team formation stage. The team should be brought together in a relaxed atmosphere to discuss such questions as

> How are we operating as a team?
>
> What are our strengths?
>
> Where can we improve?
>
> What steps are needed to initiate the desired change?
>
> What problems and issues are we likely to face in the future?
>
> Which of these can be avoided by taking appropriate action now?
>
> How can we "dangerproof" the team?

Focus-group concepts and benchmarking provide useful frameworks and toolsets for leading these team building sessions.

9.7 A FINAL NOTE

The successful management of technological innovation involves a complex set of variables that are related primarily to the task, the people, and the organizational structure and environment. The role of the technology-based manager is a difficult one. That person must be skilled enough to lead task specialists toward innovative, quality-oriented results that can be integrated according to established business plans.

The manager must understand the interaction of organizational and behavioral variables in order to foster a work environment conducive to the individual needs of the people and the team as a whole. This understanding will facilitate a climate of active participation, minimal dysfunctional conflict, and effective communication. Such a climate is goal-oriented and conducive to change and commitment.

No single set of broad guidelines exists that guarantees instant managerial success. However, by understanding the variables and the interrelationships that drive innovative performance in a technology-oriented environment, managers can develop a better, more meaningful insight into effective organizational performance, which may help in fine-tuning their leadership styles, actions, and resource allocations.

9.8 REFERENCES

1. A. Abbey and J. W. Dickson, "R&D Work Climate and Innovation in Semiconductors," *Acad. Management J.,* **26:** 362–368, 1983.
2. William J. Abernathy and Kid B. Clark, "Innovation-Mapping the Winds to Creative Destruction," *Research Policy,* **14**(1): 3–22, 1985.

3. H. Allan Conway and Norman W. McGuinness, "Idea Generation in Technology-Based Firms," *J. Product Innovation Management,* **3**(4): 276–291, 1986.

4. Robert G. Cooper, "Third Generation New Product Process," *J. Product Innovation Management,* **11**(1): 3–14, Jan. 1994.

5. Robert G. Cooper, "How New Product Strategies Impact on Performance," *J. Product Innovation Management,* **1**(1): 5–18, Jan. 1984.

6. C. M. Crawford, *New Products Management,* Irwin, Homewood, Ill., 1983.

7. George S. Day, Bela Gold, and Thomas D. Kuczmarki, "Significant Issues for the Future of Product Innovation," *J. Product Innovation Management,* **11**(1): 69–75, Jan. 1994.

8. Jean-Philippe Deschamps and P. Ranganath Nayak, "Lessons from the Juggernauts," *Prism,* Second Quarter, 5–24, 1993.

9. Jean-Philippe Deschamps, "Creating a Product Strategy," *Prism,* Second Quarter, 1993, 25–42 .

10. P. F. Drucker, *Innovation and Entrepreneurship: Practice and Principles,* Harper & Row, New York, 1985.

11. John H. Friar, "Competitive Advantage through Product Performance Innovation in a Competitive Market," *J. Product Innovation Management,* **12**(1): 33–42, Jan. 1995.

12. Ashok K. Gupta, S. P. Raj, and David L. Wileman, "Managing the R&D-Marketing Interface," *Research Management,* 38–43, March–April 1987.

13. Watts S. Humphrey, *Managing for Innovation,* Prentice-Hall, Englewood Cliffs, N.J., 1987.

14. F. Axel Johne and Patricia A. Snelson, "Success Factors in Product Innovation: A Selected Review of the Literature," *J. Product Innovation Management,* **5**(2): 114–128, 1988.

15. Don E. Kash, *Perpetual Innovation,* Basic Books/HarperCollins, New York, 1989.

16. Edward F. McDonough, "Faster New Product Development: Investigating the Effects of Technology and Characteristics of the Project Leader and Team," *J. Product Innovation Management,* **10**(3): 241–250, June 1993.

17. Thomas J. Peters and Robert H. Waterman, *In Search of Excellence,* Harper & Row, New York, 1982.

18. James Brian Quinn, "Managing Innovation: Controlled Chaos," *Harvard Business Review,* pp. 73–84, May–June 1985.

19. James Brian Quinn, "Technological Innovation, Entrepreneurship, and Strategy," *Sloan Management Review,* **20**(3): 19–30, Spring 1979.

20. Edward B. Roberts, "Managing Inventions and Innovation," *Technol. Management,* **31**(1): 11–29, Jan.–Feb. 1988.

21. Edward B. Roberts, *Generating Technological Innovation,* Oxford University Press, New York, 1987.

22. A. H. Rubenstein, A. K. Chakrabarti, R. D. O'Keefe, W. E. Souder, and H. C. Young, "Factors Influencing Innovation Success at the Project Level," *Research Management,* 15–20, May 1976.

23. William E. Souder, "Managing Relations between R&D and Marketing in New Product Development Projects," *J. Product Innovation Management,* **5**(1): 6–19, March 1988.

24. Hans J. Thamhain and Judith Kamm, "Top-Level Managers and Innovative R&D Performance," in *Handbook of Innovation Management,* Blackwell, Oxford, U.K., 1993.

25. Hans J. Thamhain, "Managing Technology: The People Factor," *Technical and Skill Training J.,* Aug.–Sept. 1990.

26. Hans J. Thamhain, "Managing Technologically Innovative Team Efforts toward New Product Success," *J. Product Innovation Management,* **7**(1), March 1990.

27. Hans J. Thamhain, "Managing Engineers Effectively," *IEEE Transact. Engineering Management,* **30**(4): 231–237, Nov. 1983.

28. Hans J. Thamhain, "A Manager's Guide to Effective Concurrent Engineering," *Project Management Network,* **8**(11): 6–10, Nov. 1994.

29. Hans J. Thamhain and David L. Wileman, "Building High-Performing Engineering Project Teams," *IEEE Transact. Engineering Management,* 130, Aug. 1987.

30. Michael L. Tushman and William L. Moore, *Readings in the Management of Innovation,* Ballinger/Harper & Row, Cambridge, Mass., 1988.

31. A. H. Van de Ven, "Central Problem in the Management of Innovation," *Management Science,* **32:** 590–607, May 1986.

32. Robert H. Waterman, *The Renewal Factor,* Bantam Books, New York, 1987.

CHAPTER 10

INNOVATION: MANAGING THE PROCESS

Marv Patterson
President, Innovation Resultants International
San Diego, California

10.1 INTRODUCTION

New-product innovation is the primary means by which a business enterprise creates new value for its customers. The performance of its new product innovation processes, relative to those of the competition, thus has a direct impact on a company's financial success. The purpose of this chapter is to describe key principles critical to the management of competitive new-product innovation activities in a business.

Management of new-product activities takes place at multiple levels in a company. The discussion here covers opportunities and issues that exist at these various levels. Insights based on industry experience and best practices are offered on how managers of new-product innovation, from executive through project manager, can contribute to greater business success for their company.

A common theme throughout this chapter is that the purpose of new-product innovation is to create business success by ultimately increasing both revenue and profits. A business context for new-product innovation is described for both the entire enterprise and for the single project. This, in effect, provides business perspectives on new-product activity from both 30,000 and ~5000 ft. In each case, the links between new-product processes and financial performance are outlined.

An attempt has been made to keep the discussion general in nature, not tied to any specific industry. It is slanted, however, more toward product manufacturing rather than process industries. The emphasis here is on useful management principles rather than specific practices.

10.1.1 Overview

The organization of this chapter is depicted graphically in Fig. 10.1. As shown in the diagram, this introduction is followed by a brief historical background that describes how the focus of new-product activities has evolved over time. This historical perspective sets the stage for the study of improved techniques for management of new-product efforts. Next, the business context for new-product innovation is established

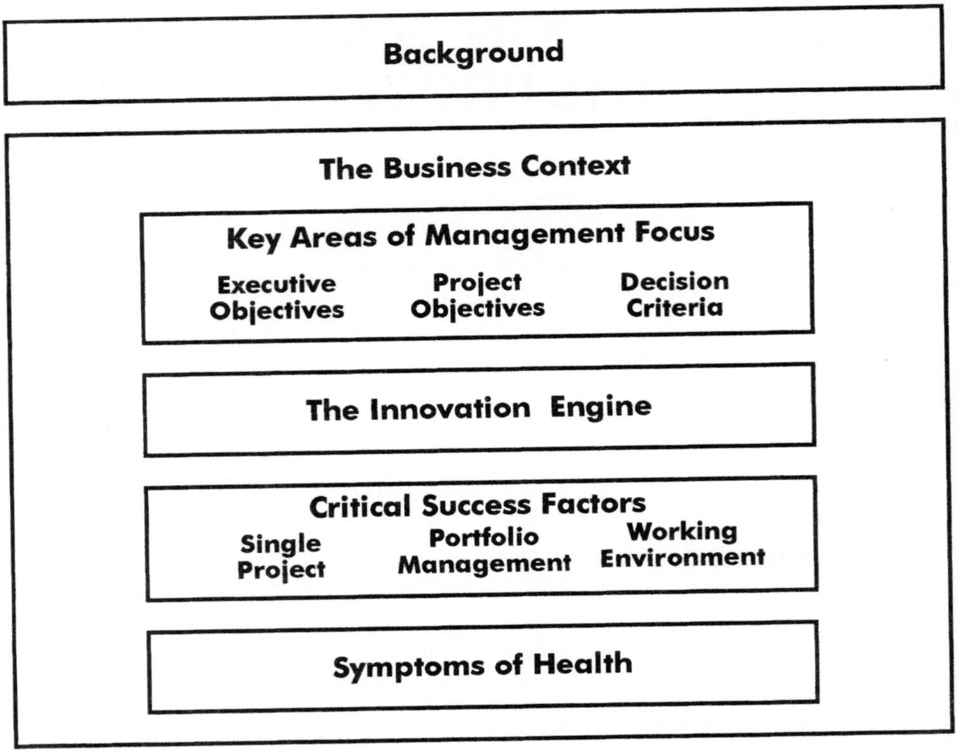

```
┌─────────────────────────────────────────────────────────┐
│                      Background                          │
│                                                          │
└─────────────────────────────────────────────────────────┘
┌─────────────────────────────────────────────────────────┐
│                 The Business Context                     │
│   ┌─────────────────────────────────────────────────┐   │
│   │          Key Areas of Management Focus           │   │
│   │   Executive         Project         Decision     │   │
│   │   Objectives       Objectives       Criteria     │   │
│   └─────────────────────────────────────────────────┘   │
│   ┌─────────────────────────────────────────────────┐   │
│   │             The Innovation  Engine               │   │
│   │                                                  │   │
│   └─────────────────────────────────────────────────┘   │
│   ┌─────────────────────────────────────────────────┐   │
│   │           Critical Success Factors               │   │
│   │     Single        Portfolio       Working        │   │
│   │     Project      Management     Environment      │   │
│   └─────────────────────────────────────────────────┘   │
│   ┌─────────────────────────────────────────────────┐   │
│   │             Symptoms of Health                   │   │
│   │                                                  │   │
│   └─────────────────────────────────────────────────┘   │
└─────────────────────────────────────────────────────────┘
```

FIGURE 10.1   Graphical map of chapter contents.

for both the business enterprise and a single project with special emphasis on the value of time in the product delivery cycle.

An understanding of how new-product innovation links to business success leads next to a discussion of key areas in which management attention can make a significant difference. Objectives for executive leaders of the enterprise are outlined, as are those for managers who work on individual projects. The criteria needed to make sound decisions during project execution are reviewed as well.

Next, an overview of a generalized version of the new-product innovation process is provided. This process is described as an enterprisewide, cross-functional information system. The sole purpose of this system is to create, as rapidly as possible, a high-quality information set that describes how to manufacture, sell, and support a new product. Information inputs and outputs are described for each subprocess in the system, as is the value that each subsystem adds to the eventual information set.

Critical success factors for both management of single projects and the collective management of all new-product activity are discussed next. These success factors evolve from the relationships that link new-product activity to business success. To balance the emphasis on process management, the importance of a suitable working environment is highlighted. The crucial roles that managers play in creating this environment are outlined as well. This chapter concludes with a discussion of the symptoms of health that characterize a competitive new-product innovation activity.

10.2 BACKGROUND

The way that technology companies compete has changed over time. Twenty or more years ago many companies focused on simply bringing new technology to bear on customer needs. The company that could offer the greatest functionality, the most capable implementation of a new technology, could count on success. This success was characterized by relatively long product lives and substantial profit margins. Research and engineering were the owners of the innovation process, and their primary concern was the technical feasibility of achieving competitive levels of performance in their product. As products became more complex, they became more expensive and reliability suffered. These were acceptable sacrifices, however, in return for more competitive technical specifications.

Eventually customer demands for more cost-effective and reliable products began to be heard. Cost of ownership and product downtime became important design goals, thus making design for reliability, manufacturability, and cost essential engineering disciplines. Technical performance was not any less important, but the successful company had to deliver its technology in a less expensive and more reliable package in order to be competitive. The ongoing cost of maintaining product functionality became a competitive consideration. As these changes occurred, the process of innovation expanded to include not only research and engineering but also the marketing, manufacturing, and customer assurance functions.

In more recent times the competitive edge has shifted to manufacturing companies that can bring new technology to bear on customer needs in the shortest time. Again, nothing has been given up. Functionality, cost, and reliability are still important, but they have become common fare among competitors. Speed of innovation is now the competitive differentiator.

As these changes have come about, the competitive arena has changed as well. Local markets that were once available only to local players have been invaded by foreign enterprises. These invaders often compete more effectively through the application of new methods and tools. With the fusion of local markets into global markets, the number of competitors increases at first and then declines to a few who dominate the market. The performance levels required for survival increase dramatically in a worldwide marketplace. As companies fail to keep up, they drop out of contention, leaving only those capable of adapting rapidly to ever-changing competitive ground rules.

Customer and user needs and expectations have changed as well. Initially product functionality was of primary concern. Then reliability and cost of ownership were added to the list. Now customers are also interested in ergonomics, product safety, and environmental impact considerations as well as compliance with applicable standards. The track record of a company for business integrity and worldwide customer support is often taken into account as well before purchase agreements are signed.

As customers and competition have left them behind, one company after another has had revenues flatten that previously grew without bound. As this happens, profits usually fall off dramatically. Efforts to improve quality and productivity help but are not sufficient to restore the company to its former financial performance levels. No amount of improvement in the same old way of operating can turn a company around when it has come up against new competitive paradigms in the marketplace. Instead, management must view the new-product innovation process from a new perspective and be willing to invest in fundamental changes to the way the company provides new value to its customers. More than any other aspect of the business, the new-product innovation capacities of a technology company must be competitive for the company to survive for long in today's rapidly changing, global markets.

10.3 THE BUSINESS CONTEXT

To successfully participate in leading their company to healthy growth in performance, managers must understand both the importance of new-product innovation to business success and their own role in supporting the new-product innovation process. The purpose of this section is to outline how new-product innovation affects business performance.

10.3.1 Innovation Drives Growth

New-product innovation is a fundamental business process whose purpose is to create and sustain growth in both revenues and profits. This process, depicted in Fig. 10.2 as the innovation engine, gathers, creates, and transforms information in a way that enables the operational part of the business to manufacture, sell, and support successful new products. The innovation engine gathers information on topics such as new technology, market trends, customer needs, and new manufacturing techniques. The processes inside the innovation engine transform this information, adding value in proprietary ways until an information set has been created that is so valuable that it enables operations to introduce new products that stimulate desired growth and shape the future of the business.

When the innovation engine is competitive, the loop depicted in Fig. 10.2 creates exponential financial growth. Information from the innovation engine enables the introduction of a steady stream of new products into the customer and user world. The value that these products provide stimulates an ever-increasing customer base to make

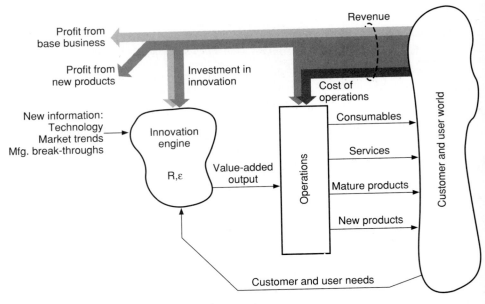

FIGURE 10.2   New-product innovation drives growth.

purchases that result in steadily increasing revenues for the company. Opposing this growth, however, are the declining sales of mature products that have become obsolete. Revenue increases due to new products must continually overcome the impact of product obsolescence for healthy financial growth to occur. A part of the incoming revenue stream is used to cover the cost of operations required to manufacture, sell, and support products. The remainder is available to the company as gross profit to be used for other purposes. Business leaders generally decide to spend some of this profit on further investments in new-product innovation, thus completing the loop.

A key point to be made here is that each year, top-level business managers decide whether to invest in new-product innovation and how much of the company's money to spend this way. They select this investment over other possibilities because they believe that the long-term return in stakeholder benefits will be greater than with any other investment that they could make with the same money. If they stop believing this, they will stop investing in new-product innovation. A key objective of managers of new-product innovation activities should be to make sure that high-level executives keep believing that their innovation engine is the best investment opportunity available.

There is an inherent time dynamic to the new-product/revenue loop introduced by both the longevity of existing products and the innovation cycle time. Products already in the market will continue to generate revenue, although at some steadily declining rate, regardless of whether the innovation engine continues to introduce new products. Likewise, from the time that a decision is made to invest in new-product innovation and initial information is gathered that begins to shape an eventual new product, a period ranging from several months to several years must pass before that product can generate revenue. These effects give the revenue loop the equivalent of inertia such that it neither increases nor declines instantaneously.

In a very healthy enterprise, gross profit created by newly introduced products will be so great that it more than pays for the cost of current investments in further innovation. In this case, the new-product stream actually generates extra cash for the business. In other cases, the innovation engine will require more funding than is available from new-product profits. Cash will then have to be taken from profits from the established base business or some other source to operate the innovation engine at desired levels. In a start-up company, for instance, no revenues or profits exist at all and the innovation engine must be funded by investment capital. This, in effect, "primes the pump" to get the revenue loop started.

10.3.2 Performance Measures for the Innovation Engine

Business measures that indicate a healthy innovation engine include rates of growth for both revenue and profits. Some companies measure the percentage of revenue due to new products—say, those products introduced within the past 2 years. Measuring the ratio of profit from new products to the total innovation investment is another interesting metric. A ratio greater than 1 means that the innovation activity is self-supporting and a net supplier of cash to the enterprise. An interesting variation of this metric is to compare current profit from new products to the size of the investment required to create today's products. This can be estimated by using current profits as the numerator but time-shifting the investment levels used in the denominator back by the duration of a typical product development cycle, say, 2 years. This version of the metric comes closer to displaying the true productivity of the innovation engine.

10.3.3 Single-Product Cash Flow

The revenue streams discussed above are, in general, created by a large collection of individual products, each one making its own contribution to the overall revenue picture.

To create a healthy enterprise, individual products must, on average, succeed in creating profit. A primary objective of the innovation engine then must be to ensure that the cash-flow cycle for each new product, depicted in Fig. 10.3, creates a suitable return on investment. This, as it turns out, is very dependent on achieving competitive innovation cycle times, as defined in the diagram. Although idealized, the cash-flow waveform in Fig. 10.3 is useful in conveying several concepts that are key to successful management of the innovation process.

Each new-product innovation activity in general addresses a new opportunity in the marketplace. As defined here, an opportunity exists when congruence occurs between a customer need and the ability of technology to create a solution for that need. Sometimes new technology is invented that can better satisfy existing needs, and sometimes new needs emerge that can be satisfied by existing technology. In either case, a new business opportunity exists and, at least philosophically, there is a specific time at which it came into existence. This point in time is the beginning of the cash-flow cycle illustrated in Fig. 10.3.

If someone could be there when a new opportunity emerges, they could start the clock on the innovation cycle time. The clock would be running while the enterprise takes time to perceive the new opportunity. Even more time would tick away before it begins to invest in related innovation activity. Once the enterprise decides to respond, initial investment levels are generally low while a small team decides on both the nature of the opportunity and the feasibility of the eventual product. When feasibility and product definition have been established, the investment level increases as the enterprise commits to getting the product to market. Eventually the new product is released for manufacturing. Sales and shipments begin to ramp up, initial customers try the product, and, hopefully, these customers are satisfied with the solution that it provides. The innovation cycle time clock continues to run through all this activity,

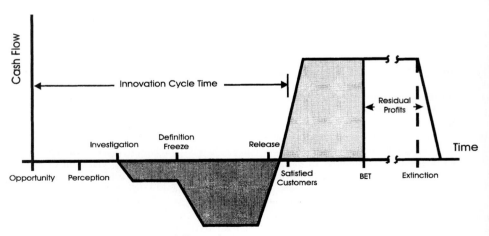

FIGURE 10.3 Single-product cash-flow cycle.

right up to the point when the first customers are satisfied that the new product solves their need.

While physically impossible to measure, this demanding definition of innovation cycle time includes some important elements of the innovation process that other time-to-market measures leave out. It thus provides an important philosophical backdrop to the discussion that follows.

An essential objective of every business is to create sufficient value for its stakeholders. These stakeholders include customers, employees, and shareholders. Reducing innovation cycle time provides major benefits to all three. Customers are more satisfied when solutions to their needs are provided earlier rather than later. They will give their loyalty more often to an enterprise that is first to effectively bring new technology to bear on their needs. Employees enjoy working for a winner, and the company that can more quickly recognize opportunity and get corresponding products to market will most often be the winner. Career and salary growth, job security, and the working environment are generally better at successful companies. Finally, as discussed below, return on investment is dramatically improved in most cases when innovation cycle time is reduced. This creates greater revenue and profit for a given level of investment and thus provides the value that shareholders are seeking.

Once the product is released and shipments begin, the positive cash-flow cycle begins. The amplitude of the waveform in this part of Fig. 10.3 is the gross profit generated by the new product—in other words, its contribution to that part of the revenue stream in Fig. 10.2 that remains after operating costs are paid. When gross profit accumulated by the product equals the total investment required to get the product to market, the break-even time (BET) for the product occurs. This is the time when the return on investment (ROI)—that is, the ratio of accumulated gross profit to total development cost—is just equal to one and the project has paid for itself. In Fig. 10.3, this is the point where the two shaded areas are equal. The area under the remainder of the positive cash-flow waveform, from BET to extinction, represents the residual profit created by the new product.

A key management goal is to direct innovation activity in a way that maximizes business success. At this point in the discussion it is clear that, if a manager wants to increase total project profitability, then priority should be given to reducing BET, moving extinction time as far out as possible, and taking steps to maximize gross profit.

Management control over extinction time exists only when the product is being defined and the product architecture is being established. Decisions made in this early project phase will cause extinction time to move around out in the future. For example, a choice to include the next-generation memory chips, under design by a vendor, in the product design will move extinction time out, but possibly at the expense of added risk. A choice to go with memory devices currently available on the market will reduce project risk but move extinction time closer in. The forces that will ultimately cause product extinction can only be estimated, but the defense that the product has against these forces is established very early in the development cycle by management and engineering decisions. Many of these decisions will affect extinction time, but only in approximately knowable ways. Once the product definition and design approach are frozen, however, extinction time also freezes as well, at some relatively obscure point in the future.

The fact that extinction time is frozen by earlier decisions makes reducing innovation cycle time a key factor improving total ROI. Every month that can be removed from the innovation cycle time adds a month at peak profit levels to the residual profit period. Conversely, for every month a project slips, a month of peak level gross profit is lost.

When people think about reducing time to market, they most often consider ways to make the development engineering team more productive. Taking a month out of the development time can be done but it is relatively difficult. Review of Fig. 10.3, however, reveals that a month removed from the idle time between the occurrence of an opportunity and the beginning of project investigation will also extend the residual profit period by a month, since extinction time is frozen. Removing time in the "fuzzy front end" is generally a lot more feasible and a lot less expensive than taking a similar amount of time out of the development engineering activity.[1]

10.3.4   The Value of Time

The value of project schedule time can be quantified in financial terms by estimating project financial parameters and applying the relationships illustrated in Fig. 10.3.[2] The financial impact of a month lost or gained varies, however, with the nature of the marketplace. Figure 10.4 illustrates graphically the financial impact of a 1-month slip in the project schedule for various kinds of products. The last two examples in the figure show how, in general, a slip in the project schedule squeezes the financial return against an unyielding product extinction time. In each case, a delay adds to the total development cost which will reduce the total ROI ratio. The major impact on total

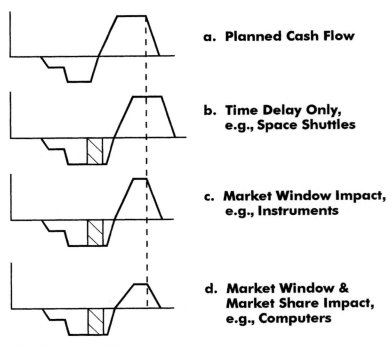

a. Planned Cash Flow

b. Time Delay Only, e.g., Space Shuttles

c. Market Window Impact, e.g., Instruments

d. Market Window & Market Share Impact, e.g., Computers

FIGURE 10.4   Financial impact of project slip.

ROI comes, however, from the lost return, most dramatically when both market window and market share are impacted by the slip.

The financial value of schedule time has been quantified for a number of business situations ranging from heavy-equipment development to the design of silicon chips. These estimates of the value of time vary from $500,000 per month to over $30 million per month. Time is often the most expensive commodity in new-product innovation, but this fact is often overlooked. Making the value of time clear can do a lot to shape decisions so that they more effectively lead to business success.

Reducing innovation cycle time provides a wide range of other business benefits besides improved financial returns. Improvements in productivity, understanding of the marketplace, customer loyalty, and organizational learning rate also result when time is removed from the innovation cycle.[3]

10.4   KEY AREAS OF MANAGEMENT FOCUS

New-product innovation not only creates financial return for the enterprise; it shapes the future of the corporation. To create a successful outcome, the objectives of everyone involved in new-product innovation must be well aligned and aimed in the right direction. The purpose of this section is to outline, in general terms, objectives and decision criteria that are important at both the executive and project levels of management.

10.4.1   Executive-Level Objectives

Top-level business executives are responsible for both achieving acceptable business performance today and steering their corporation to a successful future in a competitive world. Pressure from shareholders for short-term results often makes investment in those actions that will ensure a successful future a difficult choice. Appropriate executive guidance of the innovation engine can help achieve both the short-term and long-term needs of the corporation.

Figure 10.5 provides a view of ROI for the entire enterprise that offers a road map of how to use the innovation engine most effectively.[4] Investing in steps that will increase this ratio should help keep shareholders satisfied both now and in the future. Each term in the corporate ROI equation is heavily influenced by executive direction. The notes attached to each term indicate the directions that leading corporations are pursuing. In particular, effective shaping of the innovation engine can help achieve significant improvements in *revenue and cost of sales* as well as in *plant and equipment* (P&E). Effective cooperation between the innovation engine and operations will help achieve competitive performance in raw material and work in process (WIP) as well.

The following objectives will help engage the innovation engine with the task of moving the corporation in these directions.

Establishing Effective Strategic Plans.   To improve future revenue performance, executives must do their part in improving the selection of opportunities for investment. This means establishing strategic directions for new-product innovation activity that guide the corporation toward markets that will support future growth.[5] These plans should be managed so that they are effective in guiding the selection of individ-

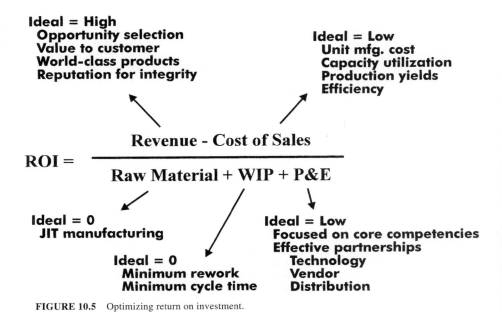

Ideal = High
   Opportunity selection
   Value to customer
   World-class products
   Reputation for integrity

Ideal = Low
   Unit mfg. cost
   Capacity utilization
   Production yields
   Efficiency

$$ROI = \frac{\text{Revenue - Cost of Sales}}{\text{Raw Material} + \text{WIP} + \text{P\&E}}$$

Ideal = 0
   JIT manufacturing

Ideal = 0
   Minimum rework
   Minimum cycle time

Ideal = Low
   Focused on core competencies
   Effective partnerships
   Technology
   Vendor
   Distribution

FIGURE 10.5   Optimizing return on investment.

ual project opportunities in a manner that moves the corporation incrementally toward its desired future.

Establishing High Standards for New-Product Innovation.   Executives must communicate their intentions for superior performance with regard to new-product activities. These intentions must establish the expectation that (1) new products will provide a high rate of return, excellent competitive performance, and superior value to the customer and (2) all dealings with customers will reflect the high integrity of the corporation.

Establishing Manufacturing Excellence as a Goal for the Innovation Engine.   To compete, a manufacturing enterprise must implement and maintain a world-class manufacturing infrastructure capable of manufacturing, distributing, and supporting the lowest-cost and highest-quality products worldwide. The goals of this infrastructure should include minimizing cost of sales, WIP, and inventories. This infrastructure will include not only the internal capacity of the company to manufacture products but also the network of vendors and partners that the company utilizes. The innovation engine must anticipate and support this manufacturing infrastructure so that newly introduced products fit naturally within it with minimal effort and disruption.

Focusing Plant and Equipment Investments on Strategic Core Competencies.
Leading corporations have learned that they cannot do it all. To be competitive, they must focus their energy and investment on only those capabilities that they are best at doing, those core competencies that differentiate them from their competitors and that will provide the necessary competitive edge for the future.[6] They must then find and rely on vendors and partners to fill in the rest of the capacity they need to provide products and services to their customers. The innovation engine, too, must focus its

investments on a selected set of strategically important core competencies and then find and rely on technology partners and vendors to provide whatever else is needed for new-product innovation. The desired effect of this strategy is to reduce overall investments in P&E to lower levels while keeping the intensity of investments in core competencies critical to the future at competitive levels.

10.4.2 Management Objectives within the Innovation Engine

Managers operating within the innovation engine must translate executive expectations into detailed project activity. Within the context set by the high-level strategies and objectives of the corporation, all managers of new-product innovation activity must strive to optimize the cash-flow waveform for their projects. The following objectives will help accomplish this.

Focusing Investment on the Best Opportunities. The financial success of a new-product innovation project is critically dependent on the size of the opportunity that it addresses. Given that an opportunity fits with corporate strategic plans, the best opportunities will have the following attributes:

A large population of potential customers

A substantial profit potential

Technical, market, and manufacturing feasibility

The potential for a sustainable competitive advantage

Defining the Most Competitive Product Offerings. Spending the time and energy initially to describe the best possible product response to a given opportunity will help a great deal to ensure the desired financial return. Extinction time will be extended out as far as possible, and market share and profit margins will be increased. In addition, a clear, exciting, and stable product definition is a major contributor to the productivity of the development team.

Making and Achieving Competitive Commitments. New-product requirements and schedules should be realistically aggressive and competitive with norms in the given marketplace. The size of project teams should allow efficient achievement of best time to market. The goals of management should be to provide the team with what they need to meet or beat these commitments and to remove any obstacles that get in their way.

Ensuring that Innovation Efforts Support the Manufacturing Infrastructure. The competitive manufacturing infrastructure described above is essential to business success. Innovation activities should compliment this infrastructure and help move it in desired strategic directions. The responsibility for ensuring that the output of innovation efforts fits within required manufacturing and regulatory constraints rests with managers of the innovation activity.

10.4.3 Decision Criteria

The new-product innovation works best when everyone involved clearly understands the criteria that will be used to make decisions as the project unfolds. If these criteria

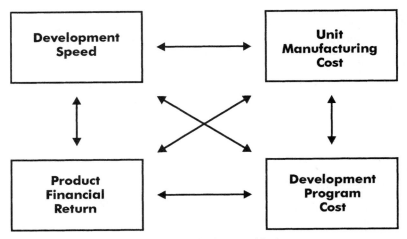

FIGURE 10.6   Tradeoffs between product development objectives.

are clear, consistent, and rational, then most decisions can be made quickly, at the lowest levels.

Figure 10.6 outlines six tradeoffs that are characteristic of new-product innovation projects. For example, development speed can often be improved, but only at the expense of either program cost overruns or perhaps leaving out product features that may hurt financial performance in the marketplace. Questions often arise during the course of a project about whether to slip the project schedule or take other steps to prevent delays. Alternatives might include spending money to off-load work to outside contractors or, perhaps, relaxing unit manufacturing cost goals. Effective decision criteria will provide quantitative financial guidelines for making these tradeoffs.

The six tradeoffs in Fig. 10.6 relate directly to the cash-flow waveform in Fig. 10.3. Increases in unit manufacturing cost will, in general, affect the amplitude of the positive cash-flow waveform by reducing either profit margin or sales volumes or both. Delays in product delivery will extend the development time and shrink the market window, the time between product release and extinction time. Product financial performance includes the full area under the positive cash-flow portion of the waveform but also includes total ROI, the ratio of this area to the area under the negative part of the waveform. Development program cost relates to both the amplitude of the negative cash-flow portion and to the total area under this part of the curve. The effects of any given tradeoff can be translated into financial terms by estimating its overall impact on the cash-flow waveform. Guidelines for managing these tradeoffs can be developed by estimating project parameters and then applying the relationships highlighted in Fig. 10.3.[2]

10.5   THE INNOVATION ENGINE

A lot has been said so far about the innovation engine, but little has been said about what goes on inside it. To meet the expectations described earlier, the system inside the innovation engine must look something like that depicted in Fig. 10.7.

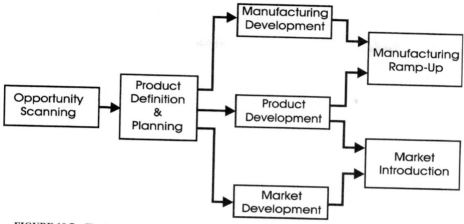

FIGURE 10.7   The innovation engine.

This diagram depicts the system that enables a business enterprise to continually redefine itself so that it can sell, manufacture, and support an ongoing stream of successful new products. Every element in this system is a cross-functional process, and many of the elements shown are executed concurrently. The innovation engine is a system in that every element shown in the diagram must both perform individually at competitive levels and work smoothly together for long-term business performance to remain competitive. If the system is not currently performing competitively, it must be diagnosed as a system and the weak elements corrected. Many businesses, for some reason, look to product development engineering whenever they have concerns with the new product process. That works only when every other part of the innovation is in good shape.

The innovation process begins with the search for new business opportunities and continues until (1) manufacturing is capable of shipping product at mature volumes and (2) sales and product support activities are also operating at mature levels. People from almost every function in the business are involved in some way or another with new-product innovation as described here and thus are a part of this system. The innovation engine must therefore be owned and managed by the top-level leadership team of the business unit.

The purpose of this section is to describe the innovation engine in general terms and to discuss key relationships that are essential to its successful operation.

10.5.1   Innovation as an Information Process

The inputs to the innovation engine are various forms of information about technology, market needs, competitive products, regulatory constraints, and so forth. The output of the innovation engine is product-specific information in a form that enables the mature manufacture, sales, and support of new products. In other words, the only value provided to the business by the innovation engine is related to (1) the information that it gathers and brings into the business, (2) the value that it adds to this information, and (3) the information that is created within the engine.[9] This system does not deliver new products; it delivers information about new products. The innovation engine is fueled with money invested by the business enterprise and also with the human resources, equipment, and facilities that the corporation provides.

The primary information flow in the process is with the arrows shown in the diagram. However, supporting information flows in every direction, between every subprocess, all the time. In addition, each block depends on a continuous flow of information coming in from sources external to the innovation system. The blocks in the diagram are major centers of activity needed to improve the value of incoming information in specific ways as it flows through the process. While there are blocks labeled "manufacturing development," "product development," and "market development," these are not intended to be the old idea of functional silos in which one functional department holds sway. Instead, these are concurrent information processes accomplished by cross-functional teams. The process of developing the product, for instance, must involve expertise from both the manufacturing and marketing functions to ensure that the right product is designed into a market competitive and manufacturable form. To develop the capacity to manufacture a new product requires close links to both development engineering and marketing expertise. In general, every subprocess in Fig. 10.7 involves a concurrent, cross-functional effort.

10.5.2   Elements of the Innovation System

This section describes the subprocesses of the innovation engine in terms of their information inputs and outputs and in terms of the value that each subprocess adds to the product information set. In a few cases, schematic diagrams of the subprocess are provided.

Describing the innovation engine at this level establishes the basis for both effective management of the engine and improvement of its performance level. Once the inputs and outputs of a given subprocess are understood—along with how it must relate to other subprocesses—its internal workings needed to support the extant innovation environment can be readily described. These internal elements are, in general, specific to the products of interest in each reader's situation and so will not be detailed here. The initial elements in the innovation engine are so critical and so often overlooked, however, that a schematic for each of these subprocesses will be provided for clarity.

Opportunity Scanning Process.   As with other elements of the innovation engine, opportunity scanning is best implemented as a cross-functional process. Figure 10.8 depicts one possible way of representing some of the key elements of the opportunity scanning process. Inputs to this process include nonspecific tidbits of information from the technical, market, and manufacturing arenas in which the business enterprise is involved.

Involvement in these communities of practice includes both learning and sharing of information. Each of these three arenas is capable of producing a business opportunity on its own, but, more likely, opportunities will involve integration and correlation of information from two or more of these sources. A technical breakthrough, for example, may not represent much of a business opportunity, but, when it is combined with an understanding of a broad base of customer needs, opportunities are more likely to emerge.

The desired output of this subprocess is a prioritized queue of potential business opportunities that is understood by both the business leadership team and key players

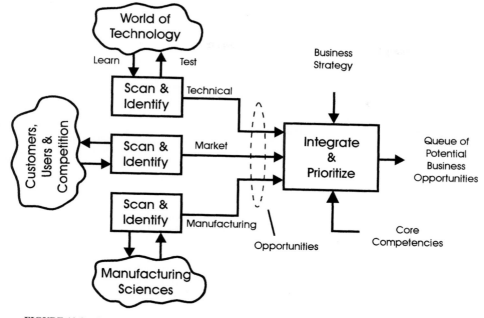

FIGURE 10.8   Opportunity scanning process.

throughout the innovation engine. This queue provides input to the project selection process considered later in the discussion of portfolio management. It also makes tangible possible alternatives that can realize the objectives outlined in strategic plans. When a decision is made to invest in the highest-priority opportunity, this subprocess provides initial direction and input that launches the investigation and definition of a potential new product.

As shown in Fig. 10.8, possible opportunities from various communities of practice must be integrated to best discern potential business opportunities. This integration step must consider the business strategy and both existing and desired core competencies in establishing both the viability and the priority of a potential business opportunity. Opportunities that align with and provide substantial progress toward strategic business objectives will move to the top of the priority list. Likewise, opportunities that facilitate development and maintenance of desired core competencies will be most important.

There are many possible implementations of the schematic in Fig. 10.8, some quite formal, some not so formal. One informal example that has worked very well is described here.

The opportunity scanning process that was used to get Hewlett-Packard Company into the large format drafting plotter market in the early 1980s involved primarily a midlevel R&D manager and a product marketing manager who worked together at an interested business division. These two individuals committed to spending about 1 week per month visiting potential drafting plotter customers and users. They learned to lead customers into discussing thought-provoking questions such as

What are the primary benefits that you provide your customers?

What is your vision for how you want to serve your customers in the future?

What obstacles keep you from being more competitive?

Opportunities emerged as the answers to these questions, provided by a wide range of prospective plotter customers, were considered in the light of available technologies.

Discussion of potential opportunities with the division's business leaders provided the sense of alignment with strategies and core competencies needed to establish the right priorities. Final integration and prioritization took place largely through informal discussions during travel. This process not only provided the desired queue of opportunities for the drafting plotter business; it developed the judgment needed to operate effectively in this new business arena.

Product Definition Process.   Once an opportunity has been selected for investment, the product definition subprocess describes the most competitive product offering possible that will address the given opportunity. As shown in Fig. 10.9, information inputs to the product definition process are (1) information from the opportunity scanning process that describes the selected business opportunity, (2) information about customer needs, and (3) information about applicable new technologies.

The output of this subprocess should be viewed as a business proposal that describes (1) the product that will be developed, (2) the schedule and resources required to bring this product to market, and (3) the business benefit expected if this product is introduced into the market. This proposal should be presented to the business management team to enable them to make a well-informed decision on whether to invest the necessary resources.

Quality function deployment (QFD) has proved useful as an information tool for

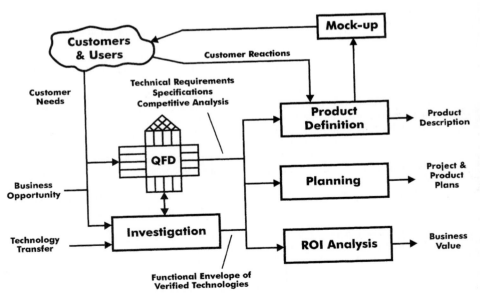

FIGURE 10.9   Product definition process.

transforming customer needs and technical capability into technical requirements and specifications. It also provides a method for analyzing proposed product performance levels and comparing them with those of competitive products. This information enables definition of product features, project and product planning, and evaluation of expected product financial performance.

Because this subprocess sets the direction for all work that follows, the quality of the information it provides is critical. The mockup loop in Fig. 10.9 is one means of ensuring high information quality in the product definition. Once a preliminary feature set has been defined for the product, a mockup of the actual product is created that can be shown to customers. This mockup might take the form of a cardboard model painted and made up to look like the actual product. Simulated displays, front-panel buttons, and knobs are in place to create the impression of a finished product. Sometimes the mockup takes the form of a computer screen simulation that acts like a real product with which the customer can interact. The idea is to allow a cross section of customers and users to interact with the mockup in a way that gets them mentally engaged with how the proposed product might work in their environment.

Their feedback and level of excitement is carefully noted and used to modify the product definition. This loop is repeated as quickly as possible until the mockup is successful in generating the desired level of customer enthusiasm. The final product definition and detailed customer responses are then used to create final plans and financial analyses.

Product Development. Once project plans and the product definition are in hand, they must be transformed into information that describes a fully integrated and verified product design. The details of this design must be documented as needed so that the information is useful to the operational end of the business. Design details, theory of operation, and product performance specifications must be captured in final form so that manufacturing, product marketing, and product support functions can access this information as needed.

Figure 10.10 provides an overview of some of the key elements of an effective product development subprocess. Detailed product requirements provide the information necessary to initiate a system design activity that determines how each product

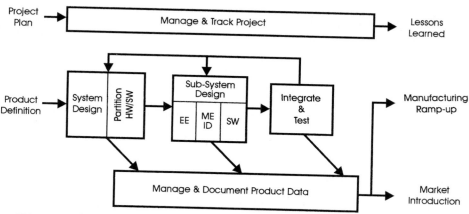

FIGURE 10.10   Product development process.

feature will be implemented. The basic architecture of the product is established here, and many decisions and tradeoffs are made that will largely determine product performance, cost, and reliability. These decisions will also determine the ultimate extinction time for the product in the marketplace.

Once the system architecture has been determined, product implementation is partitioned between hardware and software design activities. Specific design requirements are generated for the various engineering disciplines involved in the project. These design requirements provide the initial information needed to launch subsystem design work. Here, electrical, mechanical, and software engineers translate requirements into verified and documented designs. Prototypes of subsystems are implemented, tested, and modified until they meet design requirements. Once performance testing of a prototype is finished, the unit is sometimes strife-tested to determine failure modes under extreme operating conditions. Strife tests often yield valuable new information that enables quick closure on a robust design.

When all subsystems meet their design requirements they are integrated into a complete product prototype. This prototype is tested to (1) ensure that the subsystem elements work effectively as a system and (2) verify that the product meets performance specifications. As variances occur in the test results, information is fed back to earlier stages in the subprocess that initiates corrective action. Modified subsystem designs are incorporated into the prototype and tested. This loop continues until the product prototype meets product performance and reliability requirements.

The cost and time required to complete these corrective loops and achieve a successful product design are a direct function of the quality of the initial work done in both the system design and subsystem design stages. They also depend on the quality of the information available to engineers as they do their initial design work. A well-managed product development activity will include explicit steps to verify the quality of designs before they are prototyped and tested. Design quality assessment techniques might include

Breadboarding key elements of the design

Peer reviews of each engineer's design work

Computer simulation of key design elements

Each stage in the development activity described above contributes to the information set that describes the product. This information is the result of a significant investment of both time and dollars and should be handled carefully. Every product development operation has its horror stories of lost engineering investment because the wrong file was purged or, perhaps, design changes were made to the wrong version of a part. The information created in product development should be captured and managed by an effective product data management system that controls access to the information, manages revisions, and provides systematic backup and safe storage of the information.

Manufacturing Development. This subprocess develops information that describes the new capacity in manufacturing operations needed to produce the product under development. Launched by the product definition and project plans and utilizing an ongoing flow of information about the unfolding product design, manufacturing engineers develop fabrication and assembly tools, assembly lines, and automated fabrication equipment. Materials engineers work with vendors to establish trustworthy sources for parts and materials unique to the new product. Manufacturing test programs, fixtures, and tooling are created. Product unit manufacturing cost estimates are developed and tracked.

In leading businesses, this work involves continual dialogue between manufacturing engineers and product development engineers. The engineering teams work closely together as design and manufacturing approaches for various subsystems and parts are adjusted to satisfy product requirements while utilizing manufacturing resources most effectively. The tradeoffs and decisions made in these interactions largely establish the manufacturability, unit manufacturing cost, and reliability of the new product.

The output of manufacturing development is an information set that describes how to put in place all that is needed to manufacture the new product and provide a ubiquitous supply of spare parts. This information launches the actual fabrication of assembly lines, tools, and other physical facilities needed to build the new product. It also initiates necessary training of the manufacturing workforce.

In earlier times, the transfer of this information could be relatively informal as manufacturing operations were usually part of the business and were in the same location as the manufacturing development activity. People could meet face to face and easily accomplish needed information exchanges. Now, more often than not, manufacturing must be established on a global scale with fabrication centers located in various countries around the world. It involves both internal efforts and external vendors and partners. Transfer of manufacturing information must now be at least as formal as the documentation and transfer of product design information.

Marketing Development. Initiated by both the product definition and product plans, this activity prepares the business to introduce, sell, and support the new product. Support activities include efforts such as applications support, distribution of consumables, and repair services. Development of the marketing capacity required by the new product depends on an ongoing exchange of information with both the product development and manufacturing engineering efforts.

Information from the product development activity about product features and performance specifications is essential to market introduction, advertising, and sales training activities. Details of the product design, assembly, and theory of operation are needed to prepare manuals and train customer support and field repair people. Information from the manufacturing development effort about product cost and assembly-line capacity is needed to develop product pricing models and shipment schedules. In addition, this activity acquires all kinds of information from outside sources in such areas as distribution channels, competitive products, customer characteristics, and market trends.

These various forms of incoming information are transformed in the marketing development subprocess into various forms of value-added output information needed to support the product launch. Training for distributors, repair technicians, and sales personnel, and distribution plans for consumables, application notes, and advertising brochures are but a few examples of the information that flows from the marketing development activity. One of the most critical outputs from this subprocess is predicted initial sales volumes. This data is used in the manufacturing rampup activity to determine initial parts and materials procurement levels and to establish the unit volume of initial production runs. Forecasting errors can dramatically affect the shape of the cash-flow cycle for the product.

The Product Information Set. The combined information delivered by the manufacturing development, product development, and marketing development subprocesses represents most of the proprietary intellectual property sought by the business when it decided to invest in the project. If its quality is adequate, this information set will provide the business with the repository of knowledge, in easily accessible and usable form, that it needs to support expected levels of business performance throughout the life cycle of the product. This extremely valuable accumulation of intellectual property should be treated with great respect, recorded in final form, and archived in a safe place.

Manufacturing Rampup.   Using the product information set as its primary information source, this subprocess acquires the final knowledge needed to manufacture the product at mature volumes with confidence. If all that was done before this step were of perfect quality, no manufacturing rampup would be required. With flawless a priori knowledge, the product could be launched into full-volume production with total confidence. Realistically, though, new tools must be tried on a limited scale, new assembly processes must be tested, and people must receive practice with unfamiliar manufacturing roles. The quality of products produced in a production environment for the first time must be verified. As these things are accomplished, the manufacturing pace and volume can be steadily increased to mature levels. The output of this subprocess is the information gained from experience that allows the business to proceed confidently into the mature production phase.

Market Introduction.   This subprocess also uses the product information set as a key source of information. The intent of this activity is to create awareness and acceptance of the new product among its target customers. Advertising campaigns are launched, and initial feedback on their effectiveness is analyzed. New distribution channels are initiated, and the salesforce gains initial experience with the product in front of customers. Product support personnel obtain early experience with product maintenance and customer application issues. Like manufacturing rampup, the useful output of this activity is the information gained from experience that allows sales, distribution, and support efforts to achieve and maintain operations at mature volume.

10.6   CRITICAL SUCCESS FACTORS

This section addresses some specific topics of interest to managers responsible for the effective execution of new-product innovation endeavors. First, selected topics important to the execution of a single project are covered. Next, consideration is given to the management of the entire collection of new-product activities—the so-called project portfolio. Finally, the roles that managers play in creating an effective environment in which new-product innovation can flourish are discussed.

10.6.1   Single-Project Considerations

While no means complete, this section will highlight a few areas that are critical to the success of an individual project. These include project management skills, the engineering process, and project staffing.

Project Management Skills.   In most businesses, those individuals who have been put in charge of new-product innovation projects got that assignment because they distinguished themselves as technologists and engineers. They invested years of education and experience to reach a high level of competency in their chosen technical profession. That competency earned them an invitation to enter a new profession for which they have usually had little training at all—that of project manager. This is probably the single most critical job in a new-product innovation activity, and its incumbents are generally the least prepared to succeed at their assignments.

The project manager is responsible for defining the most competitive product, assembling and motivating a team to develop that product, breaking the work into specific tasks, and assigning those tasks to specific members of the team. Once the project is under way, the project manager must monitor all activity to ensure that it is on track. As contingencies or obstacles are encountered, this manager must develop alternative plans and approaches and find the resources necessary for implementation. This person must stay in touch with the engineers as they work through their assignments, help them pick effective approaches, act as mentor, and coach them as they encounter difficulties. Finally, the project manager is the primary point of contact for the project to target customers and other outside resources. As such, this individual must play the roles of ambassador, negotiator, liaison, and partner.

In view of these considerations, development of project management skills is perhaps one of the most important investments that a business can make. This requires providing more than a course in project management fundamentals and some project management software tools. An effective development program must also include interpersonal skills, technical leadership skills, marketing skills, and negotiation skills as well. Transforming an engineering professional into an effective project manager requires an investment that is comparable to that required to achieve an additional degree.

Perhaps the most important work that a project manager does is to see that crucial project success factors are in place. Critical success factors for well-run innovation projects include (1) a clear, stable product definition; (2) a consistent and pervasive understanding of both customer and user needs; and (3) a common understanding of priorities and decision criteria. All individuals involved with the innovation project must be led to internalize this information so that it guides their day-to-day decisions and behavior.

The Engineering Process. Engineers involved with new-product innovation generally follow similar processes in accomplishing their work. This general process includes

Understanding and internalizing requirements

Conceptualizing approaches to the solution

Designing the preferred approach

Prototyping and verifying the design

Designing, implementing, and testing corrections

Documenting the final solution

The level of productivity and success of an innovation project depends largely on how well each engineer accomplishes each of these tasks. In managing engineering efforts, the project manager needs to see that each engineer has the help needed to succeed at each of these steps.

In understanding and internalizing requirements, engineers must understand not only what their task requires but also the needs of internal customers who are depending on them to deliver effective results in usable form. They must understand how their work relates to the needs of the ultimate user of the product and also how the information that they produce will be used by others in the innovation process.

When conceptualizing possible approaches to their task, some engineers often leap to implementation of the first approach that comes to mind and that seems feasible. There are usually many feasible approaches to a problem, and some offer far greater benefits than others. Engineers should be encouraged to involve others in brainstorming possible approaches and to consult with others in choosing the best approach to their task. The objective is to assemble a rich set of alternatives in a reasonable amount of time before a selection of the approach to be implemented is made. Engaging others who have more experience in the selection process can help ensure the highest quality direction for the subsequent design.

Engineers should be provided with the most competitive methods and tools to use in their design work. The output formats created by these tools should be compatible with those expected by the recipients of this information elsewhere in the innovation process. Having engineers redoing their design work by hand to get it into a useful format is both wasteful and time-consuming. Besides, this kind of work is not rewarding at all and leaves engineers feeling frustrated.

In addition, engineers should be provided with easy access to the information that they need to create an effective solution to their design requirements. Requiring engineers to search for needed information when others could find it for them more efficiently is wasteful. The danger is that they will proceed without the information that they need and thus create a flawed design.

Once designs are complete, they need to be prototyped and verified. This verification should be accomplished in the quickest and most effective manner available. Prototyping is an expensive and time-consuming process and should be managed closely. Only those parts of the design that are uncertain should be included in a prototyping effort. Furthermore, the quickest and easiest means should be employed to verify a design.

Prototyping methods should be selected with the objective of obtaining information about design effectiveness as quickly and effectively as possible. The objective of prototyping is to verify that the design performs within expected limits and that it works well within the overall system. Sometimes computer simulation tools are good enough to verify a design. If this is true, then no further prototyping is needed. If physical prototypes are needed, then they should be as limited to those elements needed to accomplish design verification.

Testing of prototypes should be designed to yield information about design performance and robustness as quickly as possible. Performance tests should explore the limits of design performance and verify that they are within expected bounds. Wear and reliability tests determine whether the design meets specifications in these areas. Environmental tests ensure that the design will operate within specified temperature and humidity ranges as well as in the presence of static discharge, shock, vibration, or other conditions.

As flaws in the design are detected, corrections must be designed and implemented. These corrective loops are simply abbreviated instances of the preceding steps and should employ the same effective practices. The number of corrections required and the time each corrective loop imposes on the innovation cycle are key measures of the quality of the overall new-product innovation process.

Engineers should document their final designs in accordance with standards established by the business enterprise. These standards should be well conceived and pervasive throughout the organization. Changes to final documents and files should be managed by a product data configuration control system. Finally, the intellectual property that is documented as a final design should be archived for protection against fire, earthquake, or other natural calamities.

Project Staffing.   Proper staffing of a project can help a great deal to ensure success. The project needs both the right people in the start-up phase and the right level of resources during development phases.

The critical decisions made during product definition and planning require a small, highly effective cross-functional team. Often the first ones on a project are the most expendable engineers released early from the previous project. These people want to contribute and will do what makes sense to launch a new project. The quality of the decisions they make may fall short, however, and can lead to rework later when more qualified project leaders are in place. A better practice is to find other work for these people and hold off on product definition activities until the right team can be assembled. Once this start-up team has established an effective direction, other resources called for in the project plan can be added.

The level of staffing on the project during the development phase is critical to the financial success of the product. As shown in Fig. 10.11, the ROI and time to market are dramatically affected by staffing level. The data in this figure are drawn from a simulation of a product having cash-flow relationships similar to those shown in Fig. 10.4c. The simulation model was adjusted to exhibit time-to-market and ROI dependencies similar to those of projects in the author's career experience. The resulting curve reflects the interrelationships that are characteristic of real-life projects. This data, however, will vary widely in scale, depending on the nature of the business and on the details of the actual project. The curve in Fig. 10.11 should therefore only be used to understand relationships, not to predict performance on the reader's project.

The ROI graphed (on ordinate; vertical axis) in Fig. 10.11 is the total area under the positive part of the cash-flow curve generated by the product divided by the total cost required to get the product to market. ROI increases and time to market (TTM)

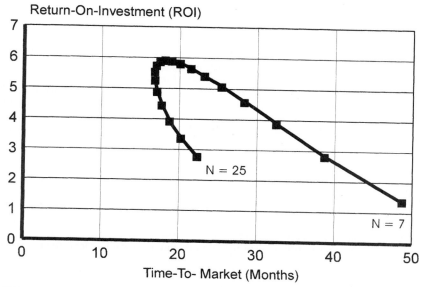

FIGURE 10.11   Effect of staffing level on project performance.

decreases as the project staff raises from a low of 7 to an optimum level of about 14. In this range of staffing levels the total project cost is roughly constant, while the extra people on the project simply get it to market quicker. Shorter TTM creates a wider market window and thus improves the ROI.

As more people are added, however, the beneficial effect on TTM weakens and then reverses. The workload for an individual improves to a point when people are added to a project. When too few people are assigned to a project, each individual has too much to do and has to cascade tasks in a time sequence. As people are added, individuals reach a point when their workload is optimum and they can get all of their tasks accomplished in the allotted time. They are not burdened with excessive coordination and can work several tasks concurrently. As even more people are added, each individual has fewer tasks assigned than they could finish in the time allowed. Any spare time they might have, however, is taken up in meetings, coordinating their efforts with an ever-increasing number of coworkers. Additional people fail to improve TTM and add cost to the project, therefore reducing ROI. As the project becomes even more overstaffed, TTM actually increases and further brings down ROI.

10.6.2 Portfolio Management Issues

The financial success created by the innovation engine depicted in Fig. 10.2 depends heavily how well the portfolio of projects that it contains are selected and supported. This section addresses several key issues that relate to the successful management of the project portfolio.

Project Selection.   Selection of projects for investment happens at several levels, as depicted in Fig. 10.12. As opportunities are scanned, articulated, and prioritized, the range of possible project investments is narrowed. When resources become available, the business leadership team decides to invest limited resources in the investigation of a specific opportunity. Later, when the investigation, product definition, and product planning are complete, the business team decides whether the resulting proposal merits further investment. Opportunities that fail to attract investigation resources are discarded, as are project proposals that fail to receive funding.

The objective of this selection process is to create and maintain a relatively small portfolio of projects that has the highest possible future business value. This occurs when resources are focused on the highest-priority opportunities and each project is staffed to optimize ROI. In addition to fully staffed projects, about 10 percent of the available workforce should be working at any given time on early investigations and various support activities. The number of projects in the portfolio at any point in time will thus be somewhat smaller than the ratio of the total number of people available for new-product innovation to the average number of people required to minimize TTM for a typical project.

There are several earmarks of an effective selection process that are worth discussing. First, the selection of investments should be explicit rather than enigmatic. Too often, projects are started in haphazard and sometimes mysterious ways. No one can explain how the portfolio came to look the way it does. The investments in the new-product innovation portfolio are among the most important in the business. The process of selecting these investments must receive due respect and attention.

The best available business judgment should be applied to portfolio decisions. In particular, the decision process should employ insightful business viewpoints as well as keen technical and manufacturing insight. The process should strive to utilize the best available information and, in fact, should overtly assess the quality of the infor-

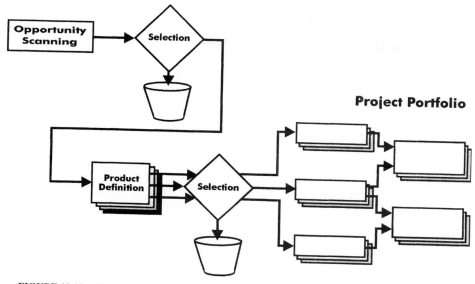

FIGURE 10.12   The quality of selection processes establishes portfolio value.

mation at hand for a particular decision. Finally, portfolio decisions are best made by a "benevolent dictatorship"—perhaps one person or a selected few key leaders. Committee decisions made by consensus take too long and tend to be of lower quality and less binding.

The Portfolio Information Environment.   All the projects in the portfolio have similar information needs. Figure 10.13 outlines typical information needs that are common across the project portfolio. These needs can often be best satisfied by common information support efforts. Providing this support improves the quality of results, saves time, and reduces wasted effort.

Each project, for example, has requirements to optimize designs for manufacturability, reliability, serviceability, and so forth. These "-ilities" are referred to in Fig. 10.13 as "design for X standards and guidelines." Providing a common source for these design standards improves productivity and consistency throughout the innovation engine. Without a central source for this information, individual engineers on each project must either dig out the information they need or move ahead without it. The result is wasted effort and inconsistent, poor-quality results.

Figure 10.13 outlines a range of information commonly needed by all projects. An important objective of the portfolio management activity should be to provide this information centrally, in an effective and systematic fashion.

Managing Bottlenecks.   As the reality of new-product innovation unfolds, critical projects will occasionally stall for one reason or another. Some unforeseen obstacle occurs that slows the progress on a critical element in the development activity. This creates a bottleneck that, unless resolved, will establish the pace for the rest of the project at an agonizingly slow rate. The project manager, by shifting the limited

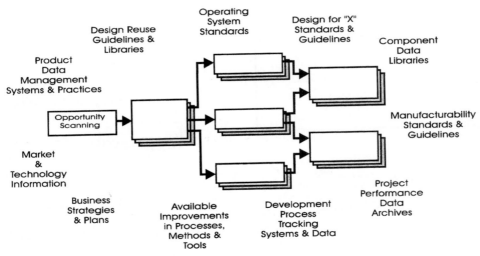

FIGURE 10.13 Project portfolio information needs.

resources available on the project, can sometimes resolve the bottleneck. If not, the problem must be escalated to those in charge of portfolio management.

Bottleneck resolution at the portfolio level involves first identifying the nature of the bottleneck and the resources needed to resolve it. Next, possible sources of help are identified: (1) on the project that is in trouble, (2) anywhere inside the innovation engine, and (3) external to the business enterprise. A key source of resources that can be used to resolve bottlenecks is the 10 percent of the innovation workforce mentioned above that is assigned to the investigation of new business opportunities and support efforts. Often these activities can be put on hold without serious disruption, and their technical staff can be used temporarily to get the stalled project back on track.

If resources cannot be freed up internally to resolve bottlenecks, then management must decide whether to hire external expertise. This is always expensive, but then so is the business impact of the time lost on the stalled project. The decision can be made on a purely financial basis if the cost of time for the errant project is known.

10.6.3 Fostering an Innovative Environment

This chapter has dealt with management of the new-product innovation process. Experience with a wide range of product innovation situations, however, teaches that focusing on the process alone does not ensure competitive performance. Equally as important is a truly empowered innovation workforce. An *empowered workforce* is defined here as one in which all individuals (1) are excited and motivated by the vision and goals of their project and (2) have both the freedom to decide what needs to be done and the power to take action. A new-product innovation team that is empowered in this sense is unstoppable—a source of competitive advantage every bit as important as the innovation process. The purpose of this section is to outline several key success factors essential to a supportive environment for innovation.

To be sure, the term *empowerment* has often been overused and poorly applied. Some firms believe that they can create an empowered workforce by simply declaring publicly their intent to do so and perhaps offering a 2-hour workshop on empowerment. One of the key enablers of an empowered workforce, however, is a supportive set of business values that is repeatedly and consistently exemplified by management behavior. If past management conduct has not been supportive of empowered action at the individual level, and nothing happens to change this, then declarations and workshops will not have any beneficial effect on the performance of individuals: "Keep on doing what you're doing, and you'll keep on getting what you're getting."

Leadership Roles. Empowerment of the innovation workforce depends almost entirely on the environment either created or, at least, tolerated by management leadership. Individuals must feel trusted, confident, and responsible before they will exhibit the decisiveness and propensity for action that empowerment promises. To be willing to take risks and be creative, they must be free from fear—fear of both reprisal and ridicule. Their self-esteem must be both high and safe. Ideally, they will feel challenged by their work assignment but also competent to meet that challenge. Finally, they must know that they can get help when they need it. All these factors are within the power of management to provide.

Leaders of new-product innovation from the project manager up through the executive level must balance their behavior across a range of roles to create an environment such as this. These roles include

Taskmaster

Obstacle remover

Cheerleader

Coach

Direction setter

Depending on their individual styles, different managers will be better at some of these roles than others. An empowered workforce, however, needs to feel the influence of each of these roles in their environment.

Taskmaster is a role that is usually well covered in most environments. The purpose of this role is create a sense of accountability for results. All employees need to feel some level of tension related to how well they are doing on the work that has been assigned them. When they deliver good performance, this tension ought to dissolve into satisfaction.

In an empowered environment managers actively pursue the role of *obstacle remover.* A manager is expected to get work done through the efforts of others. In an ideal situation, once assignments are made, a manager would have nothing to do but hand out accolades for results achieved. In the real world, however, people encounter obstacles beyond their control that slow or stall their forward progress. A key role of managers is thus to learn about these obstacles and bring all the power that they can muster to bear on eliminating them.

Some people who are assigned difficult or risky tasks need frequent encouragement—*cheerleading.* This moral support helps keep them motivated and confident that they will ultimately achieve their objectives. Managers who play this role well reinforce the environment in favor of empowerment and generally increase employee productivity.

A *coach* is generally someone who has prior experience at an activity and is able to increase the performance of others by watching their work and offering advice on the basis of their experience and knowledge. A coach does not do the work but acts as mentor for others who actually do the work. Individuals deeply involved in a task often lose sight of better ways to proceed. A manager who can observe and offer advice from a higher perspective can often help individuals greatly improve their performance and avoid unnecessary work or pitfalls.

A final and very important role that managers must play to create an empowered environment is that of *direction setter*. A good manager communicates a vision for the project outcome that excites and motivates the team. The manager then determines clear and appropriate objectives for the work to be done by the team, breaks those objectives into individual assignments, and communicates these assignments to each employee. Empowered employees must understand both the team's objectives and how their own objectives fit in to be self-directed in their work.

Part of a good direction setting process will be the communication of clear priorities and decision criteria for the work at hand. When these are in place, individuals can make most of the decisions needed on the way to the achievement of their objectives, without intervention from management.

The objective of empowerment is to create a workforce populated by highly focused and motivated employees who independently achieve their objectives and deliver expected results with little help or direction from management. To achieve this objective, managers must wear many hats and balance their behavior between the essential roles discussed above.

Recognition and Rewards.   The system of recognition and rewards in an organization has a great deal of influence on employee behavior. Generally there is both a formal system and a de facto system, and both are important. If an empowered workforce is sought, then the recognition and reward system should be adjusted so that individuals perceive empowered behavior as an essential factor in their personal success.

If management wants their employees, for instance, to pursue objectives independently and aggressively, even to the point of taking some risks, then they must consistently provide enviable recognition of those individuals and teams who are willing to make decisions and take appropriate risks. A common failing is to recognize those who achieve a successful outcome but admonish others who take reasonable risks that lead to failure. Punishment of failure sends strong signals to the workforce, telling them risk should be avoided at all costs.

One electronics firm, for example, recognizes managers who have had their projects canceled by publicly awarding them a plaque. The plaque is made up of a picture of the project manager surrounded by a toilet ring as a frame and is affectionately known as the "in the toilet" award. These awards are presented by the business unit manager whose words praise the manager and the project team for their willingness to take on risky projects even though those efforts sometimes fail to result in a marketable product. Managers who receive these awards proudly display them on their office walls.

The de facto recognition and reward system includes the common interchange that occurs when an individual goes to the boss for a decision. Some one-on-one time and the manager's personal involvement with the individual's problem are forms of de facto recognition. This recognition is gratifying to the employee, and the interchange can also provide a boost to the manager's self-esteem.

Empowerment, however, is often hindered by these interactions. They reinforce a pattern of dependence on management for decisions. When appropriate, a manager should instead offer a response such as, "This looks like a decision that you can better make than me. Why don't you figure out what needs to be done and do what makes sense?" Consistent encouragement for employees to "do what makes sense" is, at the core, what empowerment is all about. Managers, of course, need to wander around, stay in touch with what is happening, and provide assistance when an occasional individual gets off track.

10.7 CONCLUSION: SYMPTOMS OF HEALTH

Just like a healthy person, a successful new-product innovation system has characteristic symptoms of health. The purpose of this section is to describe some of the key indicators that new-product innovation is working well.

Revenue and Profit Growth. This chapter began by making the point that the purpose of new-product innovation is to create growth in both revenue and profits. Strong financial growth then is certainly one symptom of a healthy innovation engine. This is particularly true if the growth is due, at least in part, to (1) capturing market share from competitors and (2) generating revenue growth from new market opportunities.

Revenue and profit growth are, however, lagging indicators of new-product innovation performance. Once improvements have been made in the new-product innovation capability of an enterprise, a time lapse equal to several product development cycle times may be required before the impact on financial performance is significant. If revenues and profits have been declining over time, improvement of the innovation engine will precede the turnaround in financial performance by perhaps 1 to 3 years. Managers who fail to allow for this time lag have been known to kill perfectly good innovation improvement efforts because they have not delivered the expected improvement in the bottom line quickly enough.

Smooth New-Product Introduction. The nature of new-product introduction activities is another symptom of health for the innovation engine. In the ideal case, new products generate excitement in the marketplace and sales ramp up quickly. The product moves into production smoothly with few problems. An easy product launch such as this is, to a large degree, the result of excellent work and decisions accomplished in the initial innovation steps: opportunity scanning, product definition, and systems design. Conversely, a product introduction that is rife with problems reflects flaws in early planning and decision-making processes.

How high-level managers interact with the new-product processes is a good measure of the health of the innovation engine. If the time and effort that they invest is devoted mostly to strategic planning and direction setting, the engine is probably doing well. On the other hand, if they are instead heavily involved in the final introduction phases, dealing with one crisis after another, the engine undoubtedly needs work.

Low-Overhead Work Effort. In a healthy innovation environment most of the work people do is focused on creating value that will benefit the ultimate customer of the product under development. Overhead activities that add little or no value are kept to a minimum. These might include status reports and non-project-related meetings. The workload imposed by meetings is kept low by (1) efficient meeting processes and (2) limiting attendance to the essential few. Information services such as a technical library and materials engineering support gather information needed by the development team, thus allowing them to focus on getting the product out.

Competitive Commitments Made and Met.   Product definitions and plans are aggressive and realistic. Managers have good judgment with regard to both the capability of their development teams and the level of performance needed to compete. They commit to aggressive product requirements, plans, and schedules and then manage their activities to ensure that these commitments are met. Variances in schedules, budgets, unit manufacturing cost, and product performance are low and never threaten product success.

Discarded Opportunities That Create Anguish.   A healthy innovation engine will always identify more good business opportunities than it can address. Of those opportunities that it chooses to investigate, more will prove feasible than can be transformed into fully staffed projects. As a result, the discard buckets depicted in Fig. 10.12 will be filled with excellent potential opportunities and projects that will never come to fruition. While people anguish over these lost opportunities, those that do get into the project portfolio are clearly superior and provide the highest possible business value. The degree of agony felt for these abandoned opportunities is thus a good measure of the viability of the innovation engine.

Self-Directed and Motivated Team Members.   The environment created by management of the innovation engine fosters empowerment to the point where employees understand what needs to be done and have the self-confidence to take action on their own. Two qualitative measures reflect whether this environment is in place:

The grandparent metric. All employees can describe with pride to their grandparents what their organization does and why their own job is important.

The Monday morning metric. When employees come to work on Monday morning, they know, without being told, what will best move the project ahead and are eager to get started on these tasks.

10.8 REFERENCES

1. Preston G. Smith and Donald G. Reinertsen, *Developing Products in Half the Time,* Van Nostrand Reinhold, New York, 1991, chap. 3.

2. Ibid., chap. 2.

3. Marvin L. Patterson, *Accelerating Innovation: Improving the Process of Product Development,* Van Nostrand Reinhold, New York, 1993, chap. 1.

4. University of Michigan Executive Program, 1992.

5. Gary Hamel and C. K. Prahalad, *Competing for the Future,* Harvard Business School Press, Boston, 1994.

6. C. K. Prahalad and Gary Hamel, "The Core Competence of the Corporation," *Harvard Business Review,* pp. 79–91, May–June 1990.

7. Patterson, op. cit. (Ref. 3), chap. 5.

8. Philip R. Thomas, *Getting Competitive: Middle Managers and the Cycle Time Ethic,* McGraw-Hill, New York, 1991.

9. Patterson, op. cit., chap. 6.

METHODOLOGIES, TOOLS, AND TECHNIQUES

CHAPTER 11

TOOLS FOR ANALYZING ORGANIZATIONAL IMPACTS OF NEW TECHNOLOGY*

Ann Majchrzak, Ph.D.

Associate Professor
University of Southern California
Los Angeles, California

This chapter is a survey of some of the tools that are on the horizon for helping companies analyze the organizational impacts of new technology. The chapter proceeds by first explaining why a manager should worry about organizational impacts of new technology and why technology designers and managers rarely consider organizational issues in a technology design process. Given the importance of considering organizational issues, the chapter proceeds by discussing the capabilities companies need to have if organizational issues are to be included in technology design decisions. The current state of tools to facilitate these capabilities are described, and some new tools on the horizon are presented. The chapter closes with a discussion of how an organization can become ready to utilize these tools to achieve their benefits.

11.1 ORGANIZATIONAL IMPACTS OF NEW TECHNOLOGIES AS A REASON FOR CONCERN

11.1.1 The Increasing Pace of Technology Adoption

Technology, in the form of computer hardware and software, is being implemented as fast as new computer hardware and software are introduced into the marketplace. Managers with a focus on competing in tomorrow's twenty-first-century marketplace believe that technology will be a key to provide a competitive advantage.[1] For those enterprises in which project management and systems integration of supplier parts will become the core competencies in the twenty-first century, for example, technologies will be needed that facilitate integration such as project management software, soft-

---

*The author would like to thank the National Center for Manufacturing Sciences for their support in preparation of this chapter.

ware that coordinates distributed group work, open data-exchange standards, common databases, and an internet infrastructure with limited firewalls. For manufacturers that expect to be agile by the year 2000, technologies will be needed that promote rapid product development, rapid product rampups, and significant part mix flexibility. Such technologies as automated inventory systems, computer-aided engineering tools, flexible assembly systems, reconfigurable equipment, CAD/CAM links, and on-line programming capabilities will become more and more critical to the firm's core business. The question for these companies is, then, not whether to implement new technology but what to implement and when and how.

11.1.2 High Failure Rate of Technology Adoption

When these organizations proceed down this path of implementing new technology, the sad truth is that most will fail. The American Productivity and Inventory Control Society and the Organization for Industrial Research have estimated the failure rate of computer-automated technologies to be as high as 75 percent.[2] In a 1990 survey of 1200 U.K. manufacturing companies using computer-integrated technologies, A. T. Kearney found that as few as 11 percent of the installations had met their intended objectives such as inventory reduction for material requirements planning (MRP) systems or enhanced quality for CIM.[3] In another survey of 2000 U.S. firms that had implemented new office systems, at least 40 percent of these systems had failed to achieve the intended results.[4] As one example of failure, a large electronics manufacturer implemented automatic test equipment for printed-circuit boards in order to speed up the production process. As the equipment was brought in, production did not speed up; instead, production dropped off for 2 years while the process adjusted to the new equipment.

These failures will arise for a variety of reasons, ranging from data incompatibility to management resistance. However, several studies have documented that new technologies are most likely to yield productivity gains when they are coupled or integrated with changes to the organizational and human resource systems. In one such study, Kochan[5] examined six sites involved in the International Motor Vehicle Program[6] and placed them on a grid drawn between innovation in human resource management practices (e.g., few job classifications, flexible work organization, and extensive communication) and degree of technological automation (including information and manufacturing technologies). He found that the plants with the poorest productivity and quality performance were the plants who were lowest on the human resource management dimension, regardless of automation level.

In another study sponsored by the Swiss Federal Institute of Technology, 60 companies were studied. The most productive companies were found to be those that held a human-centered management style (in which efforts were made to integrate technology and organizational designs) rather than a technology-centered style.[7] In a study on U.S. competitiveness, the MIT Commission on Industrial Productivity concluded that future productivity growth is possible only with the effective integration of human resources with changing technologies.[8]

In another study which surveyed 2000 firms purchasing office automation, the respondents attributed less than 10 percent of the failures to technical problems, with the majority of the reasons given as human and organizational in nature.[4] Finally, in a 1990 Ernst & Young American Competitiveness Survey, a majority of the respondents reported that organizational and people issues were the key constraints on new technology. This result led the authors of the study to conclude that the success of technology is very dependent on a solid foundation of nontechnical characteristics concerning the business.[9]

These studies and others suggest, then, that technology implementation is more likely to be successful when the technology, organization, and people (TOP) issues have been designed to complement and integrate with each other and that such integrative planning is rarely done successfully.

11.2 REASONS FOR LACK OF SUCCESSFUL TECHNOLOGY-ORGANIZATION-PEOPLE (TOP) INTEGRATION

There have been several studies examining why TOP issues are often poorly integrated in industry. In one study,[10] the lack of TOP integrative planning was attributed to a culture in which automation designers do not understand the problems that shop-floor workers must solve, they use design approaches that focus too narrowly on the material transformation process rather than such processes as continuous improvement and problem solving, they do not consider the importance of the workers in designing systems, and they do not consider the characteristics of the organization in which the system will be used. In another study focused on failures to integrate JIT and CIM technologies,[11] the failures could often be attributed to managers making inappropriate assumptions about the organizational context in which the systems are implemented. For example, managers often incorrectly assume that they can implement JIT without modifying their organizational structure.

These studies and others suggest that high failure rates of technology change can be attributable to managers and technology designers lacking an understanding about the organizational and human resource changes that might be needed with new technology. In particular, several issues are often poorly understood. These include that

- Design of process and product technologies affects how productive people are on their jobs, how effectively they communicate with others, and how they feel about their work.

- Effective implementation of a new technology demands sufficient advanced planning so that job design, training, and organizational structural changes can occur in conjunction with the technology change (rather than after the implementation).

- Involvement of technology users in the design of the technology and the concomitant organizational changes is an essential ingredient to ensuring adequate user motivation and commitment.

- Firms that effectively integrate the design of new technology with the design of organizational and human resource systems are more successful than those that have incompatible systems.

What this means, then, is that organizations preparing for the twenty-first century will need to improve their knowledge about TOP integration if they are to successfully utilize those technologies that they believe are so critical to their future success. The remainder of this chapter describes tools that have been created to help organizations increase their knowledge of TOP integration.

11.3 HOW ORGANIZATIONS CAN INCREASE THEIR KNOWLEDGE OF TOP INTEGRATION

Research into the effective design of technology for organization-technology integration has shown that organizations more effectively integrate their organizational and technology designs when they have the following capabilities resident in house:

- There is *broadly diffused expertise* among workers, engineers, technicians, and managers about different TOP integration options and their impacts on each other.
- *Different disciplines* and stakeholders work together as a team to create a comprehensive picture of the as-is and to-be TOP aspects of the enterprise.
- There are *modeling techniques* that allow the team to explore and assess the consequences of different integration options so that intuitive predictions about effects of technological or organizational change can be tested against best-practice models.
- *Designs are documented* with the expectation that they will *evolve* over time in response to ongoing learning and adjustment so that as workers gain experience with new technical and organizational designs, they can improve their designs.

Ensuring that an organization has these capabilities resident in house is not easy. There are many reasons why these capabilities are so difficult to achieve. First, the amount of knowledge needed about relationships among TOP features is often too great to keep in one's head; as a result, the knowledge is often filtered such that only that knowledge which has been personally experienced or agrees with implicit tacit assumptions is remembered. Another difficulty is that the knowledge is often probabilistic and highly contingent on particular circumstances. For example, a worker empowered to intervene in a process *might* intervene contingent on organizational norms and rewards encouraging or discouraging intervention, as well as a personal commitment to achieving quality outcomes from the process. A final difficulty is that planning cycles are long; a design decision made today may not yield observable results for months or years later.[12] These factors together suggest that TOP integration is sufficiently difficult that tools that aid the design process can be of benefit, especially tools that aid an organization to have all the capabilities needed to perform TOP integration.

11.4 THE CURRENT STATE OF TOP INTEGRATION TOOLS

According to the National Institute of Standards and Technology (NIST), current design tools to aid a collaborative TOP integrated design process are quite primitive.[13] Frank Emspak, of NIST, states: "Tool producers, such as Autodesk, DEC, Intel, and National Instruments, recognize that current design tools are only at the beginning of their evolution" (Ref. 13, p. 2). These tools are primitive because they provide only limited support for each capability needed by a company to successfully do TOP integrated design.

Broadly Diffusing Expertise.   To broadly diffuse expertise to a variety of constituents in an organization about different TOP design options, the most typical tools used today are classroom-type training or "learning by trial and error." That is, the constituents may be trained in different design options by receiving formal training classes; visiting benchmarking sites and receiving presentations about what the site is doing; or trying out one design, finding that it doesn't work, and replacing it with another. This last approach is particularly used with structural changes when "organization chart boxes" and job responsibilities are moved around among managers on a fairly frequent basis in an attempt to see "what works." There are three ways in which these tools provide only limited aid to broadly diffusing TOP expertise. First, these tools rarely diffuse the knowledge about the design options to as large a group of people as necessary. For example, engineers and workers might attend the site visits, go for training, or try out different design options, but managers may not be sufficiently involved to understand which design options were considered and their relative effectiveness. Second, the design options considered with these aids are limited. For example, learning about design options through benchmarking tends to limit understanding of design options to just those tried in a particular setting, and learning about design options through trial and error often limits consideration of design options to those most quickly implemented, rather than those with the highest long-term and broadly conceived payback. Third, the learning format provided by these aids limits the depth of learning and the long-term impact of learning on actual behavior and attitudes of the design team. For example, learning about TOP integration in a classroom or at an off-site meeting will yield learnings that do not necessarily get reinforced once the participants are back on site with the daily pressures of the jobs; instead, aids that broadly diffuse expertise in vivo while TOP design is occurring on a daily basis are needed.

Different Disciplines Working Together.   For different disciplines to work together as a team to design TOP-integrated designs, the most typical tools and techniques used today are facilitated meetings. Often such aids as role plays, expert process facilitators, brainstorming, team building sessions, nominal group technique, and group problem-solving techniques are used at these meetings to help individuals share ideas in a group and come to some consensus about the design options and their relative impacts. These aids provide useful general guidance about how to resolve conflict and share information cross-functionally. However, they do not provide specific and explicit guidance to team members about how to overcome differences in education, decision-making style, experience, pay policies and status, languages, cultures, norms, technical assumptions, and tradeoff values to share equally in the TOP integrative design process. For example, these tools do not provide explicit guidance to a design team in which an engineer defines the term *fixture* differently from how maintenance technicians define it by identifying these differences in definitions early in the design process, determine which differences in definitions are critical to creating an effective design, and determine how to resolve them quickly. Nor do these aids help shop-floor workers without the technical background to articulate and vision about the technical possibilities for improving their jobs. As a result, the general guidance provided by these aids may be helpful but limited for ensuring that a host of different disciplines contribute equally to the TOP design process.

Modeling Capability.   The third capability, of having TOP-integrated modeling tools that allow for exploring and assessing the consequences of different integration options, is facilitated today by aids that promote procedures for conducting an analysis and specific techniques on how to conduct that analysis. The modeling capabilities for assessing organizational and people (human resource) impacts of alternative technology designs tend to be primitive in that there has been no effort to automate an analysis or to create a database of known impacts of different technologies. For example, sociotechnical analysis has a traditional nine-step process beginning with the environmental scan, proceeding to social and technical analyses, and concluding with the organizational and technology design phase.[14,15] The analysis techniques are all manual, including a variance control matrix and a specific instrument to complete for the social analysis. Modeling impacts of alternative designs is also done manually and is very labor-intensive if done with care and rigor. While these tools are helpful, they provide limited support to a modeling capability for two reasons. First, according to Mintzberg, design is a combination of art, science, and inspiration so that laying out specific steps and procedures actually inhibits the creative design process rather than encourages it.[16] Second, modeling of impacts done by hand is sufficiently time-consuming that it creates a barrier to encouraging alternative scenarios, thus further limiting the creative process to known TOP options.

Documentation.   The fourth capability, of documenting design choices and learnings to help future design evolution is facilitated today primarily by writing case reports or making presentations about the learning experience. Using case reports as the primary tool for documentation provides only limited support for documentation for design evolution because it does not ensure that similar information is contained in different cases or that the information is documented to the same depth. In addition, the rationale for design decisions and alternative design options that could have been made but were discarded are rarely included in these documents. Finally, as a manual method, updates to the document as experience continues to grow are cumbersome. As a result, an ongoing learning process is stilted.

In sum, the tools currently in widespread use to facilitate the integration of technology and organizational issues are sufficiently primitive that new technologies and tools have needed to be developed.

11.5   BRIEF DESCRIPTIONS OF STATE-OF-THE-ART TOOLS FOR TOP INTEGRATION

A variety of different organizations have realized this need to develop new tools. Many of these new tools are in fact in pilot use in a variety of locations. The rest of this chapter describes some of these new tools. While this list is not meant to be exhaustive, it does present descriptions of some of the state-of-the-art thinking in tool development for TOP integration.

11.5.1   Model Learning Workplace

The Model Learning Workplace[17] is a joint venture of Xerox and Digital, based on a concept similar to Virtual Reality Assessment Center at Steelcase. The Model Learning Workplace is currently under development. In this workplace, the resident engineering design team will be technologically and organizationally supported as a living laboratory or testbed for trying out new design ideas for cooperative work and assessing their probable organizational, technological, and people impacts. One of the tools to be used in the Model Learning Workplace is project management software that will provide a group memory for maintaining problem lists in a common database, for keeping track of schedules and dependencies, for managing input/output relations for all the engineering subsystems under design, and for capturing events such as results of laboratory experiments, outcomes of meetings, and records of decisions. In addition, the Model Learning Workplace will contain full multimedia to support the capturing, indexing, organizing, and accessing of audio and video data to support telecollaborations. With these tools, the impacts of engineering design decisions will become immediately apparent to the designer. This enhances the first needed capability of more broadly diffusing TOP integration because the impacts of technology designs on organizational and people issues will be experienced reasonably immediately by the design team itself. There are few specifics on the tool as yet, so how it will be used to overcome the barriers of multiple disciplines working together is yet to be seen. The modeling capability provided by the living laboratory concept is enormous provided that the design ideas are quickly implementable. Finally, the documentation capability is a real strength of this tool, provided the indexing is quite flexible as design ideas and purposes change creating the need to change indexing requirements.

11.5.2 GRIPS

The Swiss Federal Institute of Technology (ETH) in Zurich has been supporting a multiyear effort to develop research findings and tools for small and medium-sized Swiss firms to conduct diagnoses of their technology, organization, and people integration and to make recommendations for redesigns. The project is called GRIPS.[24] The GRIPS project first conducted research to support the assertion that effective organizations focus not only on technical aspects, but also on the design of work organization and the use of skills and qualifications. This approach to the implementation of new technologies has been termed *work-oriented* and is characterized by the organizational features of decentralization at the enterprise level, functional integration at the organizational unit level, collective regulation at the group level, and complete and challenging tasks at the individual level.

Once this assertion was examined and found to be supported by an empirical study of 90 companies,[7] instruments were developed for the sociotechnical analysis of organizations. The instruments help users, such as engineering managers, conduct a diagnosis of their own organizations for TOP integration and offer suggestions for redesign. For example, in one organization, the instruments helped the organization determine that it needed to redesign its structure to integrate the functions of sales, assembly planning, engineering design, process planning, purchasing, operational planning, production control, technical planning, tool preparation, manufacturing, quality control, and maintenance into four different production islands: shipping, design, planning, and manufacturing. While these instruments have not yet been translated into English, translation and U.S. distribution are likely to happen. The GRIPS project has provided valuable tools because the instruments can be completed by a variety of people in the organization and not just engineers; the training provided on criteria of effective sociotechnical systems (including, e.g., the independence of organizational units, multiskilled employees, technology-organization convergence, and self-regulation) helps diffuse the knowledge base about effective integration options, and the instruments provide a documentation of design choices. Figure 11.1 presents a sample output of the GRIPS work system analysis instrument for three different work systems. A high value on the sociotechnical criteria should be interpreted as a sociotechnically optimized design. Thus, the work system on the top of the figure should be interpreted as doing much better than the work system represented at the bottom. The modeling capability is still quite primitive in GRIPS, however, as this was not an intention of the GRIPS instruments.

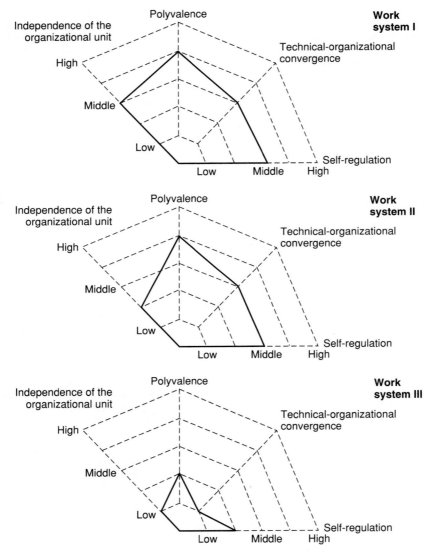

FIGURE 11.1   Sample results of a GRIPS analysis.

11.5.3   HITOP

This tool, known as *high integration of technology, organization, and people* (HITOP),[18,19] is a methodology to conduct an analysis of a TOP system, like the GRIPS instruments. There is no software involved. The HITOP analysis process was developed specifically for the circumstances in which a technology is being designed and the organizational and people implications of that technology need to be understood. A HITOP analysis involves the design team completing a series of checklists and forms that describe their organization and current technology plans, and then helps the design team identify the implications of those plans on organizational and people issues. The HITOP analysis contains no knowledge base, nor does it recommend specific design options; these are expected to be brought to the analysis by the design team itself. To complete an analysis, the design team proceeds through seven sets of analyses:

1. Organizational readiness. How ready is the organization to make changes recommended by a HITOP analysis? For example, a HITOP checklist question queries whether the design team can tie the reason for doing a HITOP analysis to a clear business purpose of the organization. If this as well as other questions are answered "no," then the design team is encouraged to either change the organization's readiness or wait before proceeding with the analysis.

2. Critical technical features. These are features of the planned technology that are most likely to impact the integration of the technology with organization and people. The features comprising the HITOP analysis include degree of information integration with other systems; degree of mechanical integration with other systems; equipment reliability; flexibility of the technology to perform different processes for different inputs; capability of the technology to monitor, diagnose, and correct its own errors (i.e., the inverse of the need for human monitors); and the fault tolerance of the technology. This next step in the HITOP process, then, is to rate the design team's technology plans on each of these features.

3. Essential role requirements. For the four primary functions in a manufacturing workforce (core, support, supervision, and management), HITOP identifies eight role requirements, including degree and type of interdependence, information exchange, decision authority, and involvement and complexity of strategic goal setting. The analysis team rates each primary function by each of these role requirements.

4. Job designs. The HITOP analysis requires the team to develop a set of job design values, with statements such as "workers should have control over resources for those areas over which they are responsible." The values are combined with the role requirements to determine how work should be divided into specific job classifications.

5. Skill requirements. These include selection and training. The minimal skill requirements (categorized by perceptual, conceptual, manual dexterity, problem solving, technical, and human relations) for each role requirement are determined in this step and a determination of which skills will be trained versus selected.

6. Reward systems. Forms are provided to help the design team make three decisions about rewards: basis for pay (e.g., merit, hours, performance), basis for nonfinancially recognizing and rewarding performance, and future career paths.

7. Organization design. Forms are provided to help the design team work through five organizational design changes typically seen with the implementation of new technology: changes in reporting lines, procedural formality, unit grouping, cross-unit coordination mechanisms, and organizational culture.

HITOP has been used by a variety of U.S. companies (including, for example, DEC and HP) as well as translated into foreign languages for use in Italy, Germany, Switzerland, and Spain. A sample worksheet is presented in Fig. 11.2. HITOP provides a specific set of generic procedures and contains no knowledge base for modeling what-if scenarios; as such, then, HITOP does not represent a significant gain in modeling capability. However, by providing a common language of discourse and by including variables that necessitate the involvement of multiple disciplines, and by provid-

(Sample) **Partially Completed Matrix 1** To-be job requirements and design			
Essential task requirements	Cross-CTF role requirements	Solo or team	Number of individuals to perform responsibilities
Task interdependence	Low except high with programmers	**Core work** Team with maintenance	3 people per team for 5 CNCs
Information exchange	Low except med with maintenance		
Manual work	Low except in emergencies		
Decision authority	High		
Complex problem solving	Low		
Task interdependence	High with operators	**Technical support (maintenance)** Team with operators	
Information exchange	High with operators		
Manual work	Low		
Decision authority	High		
Complex problem solving	Medium		
People motivation and management		**Supervisory-managerial**	
Proactive opportunity seeking			
Goal setting			
Possible O/P interrelationships			
Reporting lines, skill, training, pay			
Implications for organizational readiness			
No history of teamwork			

FIGURE 11.2   Sample HITOP worksheet (*from Ref. 19*).

ing worksheets to document the results of design decisions, HITOP extends existing tools in important ways.

11.5.4 COSAT

Developed by the Industrial Technology Institute with funds from the National Institute of Standards and Technology, the *Cross-Organizational STEP Adoption Tool* (COSAT)[20,21] is a computer software tool designed to assist managers to make the organizational and human resource changes necessary to gain the maximum benefit from technologies which implement the new Standard for the Exchange of Product Model Data (STEP) as part of concurrent engineering. COSAT is a hypermedia computer system that provides on-line guidance to concurrent engineering (CE) by walking the user through cases on best practice in CE, presenting a relatively traditional planning process for organizational change (assess as is, design new system, and implement), and presenting design principles for optimizing work and information flow and organizational structure. COSAT is useful because it guides a team through the assessment and design process with structured activities at each step in the process. COSAT provides matrices, worksheets, and guides for working through the planning and change process. COSAT has been used at Sandia National Labs, which credits COSAT with facilitating the development of an enterprise model. COSAT is useful because it helps diffuse knowledge about TOP integration. In addition, it provides the documentation and memory capability because the user's as-is and new systems can be described on the screen. It provides suggestions for cross-functional collaborative integration work although the suggestions are rather general. It is not intended as a modeling tool so its modeling capabilities are limited. A sample screen print of a COSAT run is shown in Fig. 11.3.

11.5.5 ACTION or TOP-Integrator

To provide organizations with the capability to model the impacts of different organizational, technology, and strategy choices, ACTION[12,22] was developed. ACTION is an interactive software system, with an accompanying methodology for use, which embodies a complex knowledge base about effective relationships among different technical, organizational, and strategic features. For example, ACTION specifies the best practice skills, information, technology characteristics, performance measures, rewards, norms, and empowerment needs for 141 activities, 18 process variance control strategies, and 7 business strategies (including new-product development flexibility, minimizing throughput time, and maximizing process quality). Complete explanations are provided for each needed feature. The user interacts with the software by choosing among business and variance control strategies. Given these choices, a profile of the ideal organization is invoked from the knowledge base in terms of activities, information, skills, technologies, etc. needed. Users can then either review the ideal profile and determine whether their strategies are too ambitious, or begin a process of describing their current organizations relative to this ideal. As data describing the existing organization are entered, the gaps between the existing organization and the ideal profile are visually displayed, as shown in Fig. 11.4.

After the organization is comprehensively described, all the gaps are listed and alternative priorities of the gaps, based on different prioritization criteria, are offered to the user. For example, the gaps can be prioritized by the number of different strategies they negatively impact, or the likelihood that the negative impact will be seen across all situations, or the significance of the impact on any single strategy relative to other gaps. This last prioritization is calculated based on a probability model in which a high probability of accomplishing a particular business strategy (such as minimizing throughput time) is calculated on the basis of the number of gaps meeting particular criteria of successful organizational designs: minimized coordination needs, maximizing unit capabilities, and motivating workers through appropriate performance metrics and rewards. Prioritizing gaps in this way may determine, for example, that among three gaps, if two are related to minimizing coordination needs, only one of those two may need to be resolved to achieve a high probability.

Cossat, Inc. cross-functional communication matrix: As-is and to-be

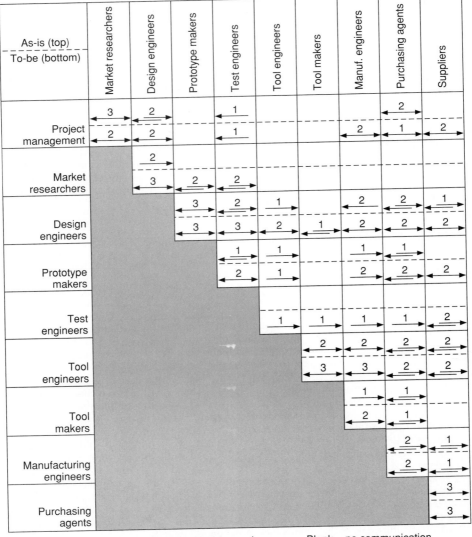

FIGURE 11.3   Example COSAT screen (*from Ref. 20*).

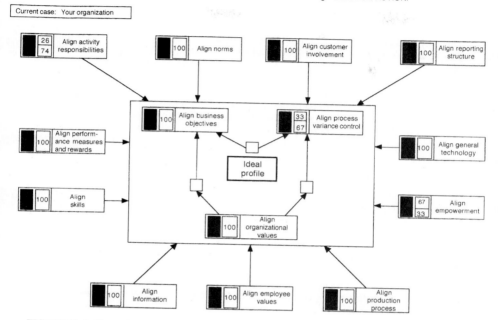

FIGURE 11.4   Example ACTION screen.

ACTION has been used in a variety of organizations, including Texas Instruments, Hewlett-Packard, General Motors, and Hughes. The use of the tool has varied from an aid to strategic planning (such as choosing between business strategies), strategic visioning (such as in deciding whether strategic priorities should be placed on automation or skills or organizational design), organizational diagnosis, and organizational design.

ACTION supports the four capabilities for an organization doing TOP integration. Knowledge about TOP integrated options is diffused through ACTION's user-friendly explanations of the knowledge base. Multiple disciplines and stakeholders are encouraged to become involved in the TOP design process because of the breadth of variables that ACTION includes. Modeling of alternative TOP designs is promoted in ACTION by a computer-accessible knowledge base of best practices, an ease of changing inputs to describe different alternative ideas, and not having a specific order in which data or design steps must be inputted, processed, and outputted. Finally, documentation for design evolution is facilitated by having a case directory in which design decisions as inputs to ACTION are easily retrieved and compared to later design ideas.

11.6 DETERMINING WHETHER THESE TOOLS WILL WORK

The basic assumption underlying all these tools is that collaborative and TOP-integrative designs are more likely to occur when there are tools to help with the design process. Since most of these tools are relatively new and not diffused widely, there is little widespread evidence of direct cost savings from their use. However, there are selected case examples in which significant benefits have been achieved.

As one example of a company benefiting from these new tools, a Hewlett-Packard plant preparing to introduce a new product used HITOP with the new-product introduction teams. The teams included production-line workers who would manufacture the new product, maintenance technicians, and process engineers. The tool was used to help the team systematically work through the various technology, organization, and people issues to make sure that nothing fell through the cracks during planning for the new product rampup. With the use of the tool in conjunction with the new-product introduction teams, the plant experienced few problems and engineering change orders when the new product line was "turned on" compared to previous new-product rampups. As a result, quality and volume goals were achieved within the first 4 weeks of manufacturing release of the product rather than the 5 to 12 months typically needed to achieve quality and volume goals. This represented a fivefold increase in rampup time. In addition, the quality at introduction was better than the quality achieved by mature products already existing in the factory, and the quality has remained consistently better than any other product in the factory.[23]

As another example of a company benefiting from these tools, a manufacturer of turbine generators had designed a computerized labor reporting tracking system as the first stage of a new computer-integrated manufacturing (CIM) system. Since this was the first step in a major modernization program, the organization was concerned that the tracking system would encourage employee enthusiasm about the modernization and would be properly implemented and used. One of the tools described above was used to analyze the tracking system for its TOP fit. As a result of the analysis, several suggestions for increasing the fit were made. The company has moved forward to implement many of these recommendations for the tracking system and the organization with an estimated cost savings of $600,000 on a $3.1 million modernization program.

Finally, as an example of the anticipated benefits from tools such as these, Boeing reported that 10,000 design changes were made as the Boeing 777 went into production, at an average cost of $10,000 for each change. Boeing asserts that any design process or tool that could reduce the number of design changes by even 20 percent would easily be worthwhile.[13]

Thus, limited experience with these tools suggests that they may be worthwhile. However, such experience will not come easily. Since these tools represent a new technology in and of their own right, they must be adequately integrated with their own sociotechnical environment if they are to be effectively utilized. An environment most conducive to effectively using these tools appears to be one in which

- Managers are enthusiastic about innovation in all aspects of the way work is done (i.e., they expect continuous process improvements not just on the shop floor but in other functions as well).
- There has been some history in the organization of involving shop-floor workers in process design decisions.

- There is some sense of urgency among managers and engineers to design processes and technology correctly at the outset so as to avoid engineering change orders, crisis management, and the need to redefine objectives and specifications after implementation.

- Engineers and managers have experienced pain in past efforts at implementing new technology (i.e., they have their share of "failures" to wince about).

- Engineers, managers, and shop-floor workers recognize that the integration of technology, organization, and people is sufficiently difficult that tools have some utility above and beyond what trust, personal intuition, and goodwill can generate.

- Engineers and managers are willing to accept that the results of a TOP-integrative analysis may suggest changes to their own "pet" technologies and not just changes to the organization and people on the shop floor.

How many environments exhibit these characteristics and thus are ready to use these new tools? Probably not many. However, in a recent (Fall 1994) series of meetings hosted by the National Institute of Standards and Technology to assess the interest of industry in developing tools such as these, the following major companies and unions participated: Bath Iron Works, International Association of Machinists, Texas Instruments, General Motors, Hewlett-Packard, Hughes, Communication Workers of America, U.S. West Communications, Digital Equipment Corporation, Xerox, New Balance Athletic Shoe Inc., Mill Cabinet, Heath-Tecna Aerospace, Textron, National Instruments, and Jet Propulsion Laboratories. So clearly there is interest. With such interest, the sophistication and utility of these technologies are just years away. It may be time to watch them unfold.

11.7   QUESTIONS TO ASK YOURSELF TO PREPARE FOR THESE TOOLS

You may be able to determine whether your staff is ready for or receptive to these new tools by addressing the following issues:

1. Is your management and engineering staff aware of the technology failure rates internal to your organization?

2. Have there been any efforts in your organization to document learnings from your failures to confirm that the failures have been attributable to lack of planning of how the technology would be integrated with the people and organization? Have these learnings been disseminated to managers and engineers?

3. Have there been any efforts to examine in your organization how technology design, especially as it relates to organizational and people issues, is conducted today? Has there been any effort to benchmark how your organization does technology design with how other companies do it?

4. Have there been any efforts to examine how tools could be used to help facilitate the TOP integrative design process? Has a list of criteria for a worthwhile tool been generated?

5. Have there been any efforts to search for available tools that have been or could be used to help facilitate a TOP integrative design process?

6. Have there been any efforts, albeit small steps, to involve shop-floor workers in continuous process improvement efforts in which they work collaboratively with engineers?

The more questions you answer "no," the harder it will be for your organization to utilize these tools when they are ready for broad diffusion. So why wait?

11.8 REFERENCES

1. S. L. Goldman, R. N. Nagel, and K. Preiss, *Agile Competitors and Virtual Organizations: Strategies for Enriching the Customer,* Van Nostrand Reinhold, New York, 1995.

2. M. Works, "Cost Justification and New Technology Addressing Management's No! to the Funding of CIM," in *A Program Guide for CIM Implementation,* 2d ed., L. Bertain and L. Hales, eds., SME, Dearborn, Mich., 1987.

3. J. Bessant, P. Levy, C. Ley, S. Smith, and D. Tranfield, "Organization Design for Factory 2000," *Internatl. J. Human Factors in Manufacturing,* 2(2): 95–125, 1992.

4. R. J. Long, "Human Issues in New Office Technology," in *Computers in the Human Context: Information Technology, Productivity, and People,* T. Forester, ed., MIT Press, Cambridge, Mass., 1989.

5. T. A. Kochan, "On the Human Side of Technology," *ICL Technical J.,* 391–400, Nov. 1988.

6. J. P. Womack, D. T. Jones, and D. Roos, *The Machine That Changed the World,* Rawson Associates, New York, 1990.

7. L. Leder, O. Pardo, and E. Ulich, "Interrelationships between Strategies of Use and Development of Human Resources and the Design of Computer-Aided Integrated Manufacturing Systems," in *Advances in Agile Manufacturing,* P. T. Kidd and W. Karwowski, eds., IOS Press, Amsterdam, 1994.

8. M. Dertouzos, R. Lester, R. Salon, and the MIT Commission on Industrial Productivity, *Made in America: Regaining the Productive Edge,* MIT Press, Cambridge, Mass., 1989.

9. T. R. Ozan and W. A. Smith, *The American Competitiveness Study: Characteristics of Success,* Ernst & Young, New York, 1990.

10. C. Kukla, E. A. Clemens, R. S. Morse, and D. Cash, "Designing Effective Systems: A Tool Approach," in *Usability: Turning Technologies into Tools,* P. S. Adler and T. A. Winograd, eds., Oxford University Press, New York, 1992.

11. P. R. Duimering, F. Safayeni, and L. Purdy, "Integrated Manufacturing: Redesign the Organization Before Implementing Flexible Technology," *Sloan Management Review,* 47–56, Summer 1993.

12. L. Gasser and A. Majchrzak, "ACTION Integrates Manufacturing Strategy, Design and Planning," in *Advances in Agile Manufacturing,* P. T. Kidd and W. Karwowski, eds., IOS Press, Amsterdam, 1994.

13. F. Emspak, "Focused Program Recommendation: Integrating Workers in Design Engineering," paper presented to the National Institute of Standards and Technology, Gaithersburg, Md., Oct. 24, 1994.

14. J. C. Taylor and D. F. Felton, *Performance by Design: Sociotechnical Systems in North America,* Prentice-Hall, Englewood Cliffs, N.J., 1993.

15. E. Trist and H. Murray, eds., *The Social Engagement of Social Science,* University of Pennsylvania Press, Philadelphia, 1993.

16. H. Mintzberg, *The Rise and Fall of Strategic Planning,* Free Press, New York, 1994.

17. C. Kukla and G. Zak, "Integrative Design Engineering and the Workforce: An Industry Initiative," invited presentation at the Boston Regional Meeting of the National Institute of Standards and Technology, Oct. 1994.

18. M. Fleischer and A. Majchrzak, "HITOP: A Method for Developing Organization and People Requirements for Manufacturing Systems," paper presented at the Conference on Design Productivity, Oahu, Hawaii, Feb. 6–8, 1991.

19. A. Majchrzak, M. Fleischer, and D. Roitman, *A Reference Manual for Performing the HITOP Analysis,* Industrial Technology Institute, Ann Arbor, Mich., 1991.

20. M. Fleischer, *Cross-Organizational Assessment for STEP Adoption,* report submitted to National Institute of Standards and Technology, Industrial Technology Institute, Ann Arbor, Mich., 1993.

21. M. Fleischer, "Reengineering the Product Development Process: Structured Methods," paper presented at CALS Expo '94 Conference, Long Beach, Calif., 1994.

22. A. Majchrzak and L. Finley, "A Practical Theory and Tool for Specifying Sociotechnical Requirements to Achieve Organizational Effectiveness," in *Symbiotic Approaches: Work and Technology,* J. Benders, J. de Haan, and D. Bennett, eds., Avebury Publishing, Gull, U.K., 1995.

23. A. Majchrzak and S. Winby, *What Do Agile Manufacturing, Reengineering, and Concurrent Engineering Have in Common: It's the Integration that Will Make Them Succeed,* technical report, National Center for Manufacturing Sciences, Ann Arbor, Mich., 1994.

24. O. Strohm, C. Kirsch, L. Leder, O. Pardo, P. Troxler, and E. Ulich, "Work-Oriented versus Technically-Oriented Manufacturing Systems: Methods and Results of a Case Study, in *Advances in Agile Manufacturing,* P. Kidd and W. Karwowski, eds., IOS Press, Amsterdam, 1994.

CHAPTER 12

FORECASTING AND PLANNING TECHNOLOGY

Frederick Betz
National Science Foundation
Washington, D.C.

12.1 APPROACHES TO FORECASTING

Three approaches have been proposed for anticipating or forecasting technological change: (1) extrapolation of past and current trends, (2) structural analysis of underlying factors determining these trends, and (3) exploration of possible changes in the structural factors.

One begins with the first approach, extrapolating trends. Then one can use the second approach to probe deeper into these trends by identifying the underlying structures which influence the trends. Having identified structural factors, one can then attempt to determine what could alter these structures. At this point, one begins technology planning.

12.1.1 Logic Tree of Forecasting Approaches

Periodically there have appeared reviews of the literature of the topic of technology forecasting, such as in Landford (1972) or Godet (1983). In 1988, Jack Worlton offered a logic tree as a guide to that literature, which is sketched in Fig. 12.1 (Worlton, 1988, p. 312).

By the term *exploratory,* Worlton meant: "starting from the present and advancing step by step toward the future" (Worlton, 1988, p. 312). Methods which have been proposed to do this have all mostly been extrapolation of current trends into the future. The two most important methods have been (1) the Delphi method for what he called "subjective" and "group" and (2) the technology S curve for what he called "objective" and "quantitative."

The logic line which Worlton called "subjective" and "individual" has not provided a useful technique, because there is no sound method for calibrating which individuals at any time in history are good predictors of the future. Moreover, even after a basic innovation, some experts may be too pessimistic about progress because they are very familiar with the problems and difficulties of new technology. Thus one does require more objective and systematic procedures for technology forecasting, than simply the subjective judgments of individuals (no matter how brilliant and successful).

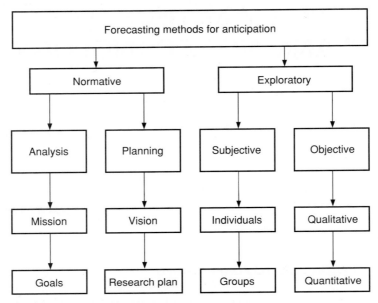

FIGURE 12.1   Worlton logic tree for forecasting approaches.

By the term *normative,* Worlton meant "inventing some future and identifying the actions needed to bring that future into existence." The approaches "normative" and "analysis" look at the underlying structures of current trends, not merely extrapolating trends. These approaches ask what alterations in structures can create new technology? The principal method here is a morphological analysis of a technology as a system, which we will review later in this chapter. Finally, on the logic branch of "normative" and "planning," this approach involves actual formulation of technology strategy and research programs to implement such strategy.

In summary, from Worlton's logic tree, we can see that technology forecasting is a companion piece to technology planning. One should try both to anticipate the future and try to make happen a future that one desires. This is the essence of all action, and particularly the essence of technology forecasting and planning.

12.2   *TECHNOLOGY PERFORMANCE PARAMETERS*

A technology is describable by a functional transformation which transforms purposely defined input states to purposely defined output states. The measure of performance of this transformation is the efficiency of the transformation. The rate of technical change in a technology can be expressed as changes in rate of improvement in a generic *performance parameter* for the technology. Generic performance parameters indicate that technical improvement will be useful to a wide range of applications.

All technologies are systems whose operations are controlled by many different variables and specifications. The appropriate performance measure for a technology is one which directly relates technical progress to customer utility. The choice of the technology performance with which to characterize the utility of technical progress is therefore a critical choice in forecasting technological change.

12.2.1 Rates of Change of Technology Performance Parameters

One can plot the rate of change of a performance parameter over time, and one will likely see a kind of S-shaped curve, such as in Fig. 12.2. This common pattern that has historically occurred has been called the "technology S curve" and has been used as a basis for extrapolative forecasts of technology change (Twiss, 1968; Martino, 1983; Foster, 1982). This curve has three periods:

- An early period of new invention
- A middle period of technology improvement
- A late period of technology maturity

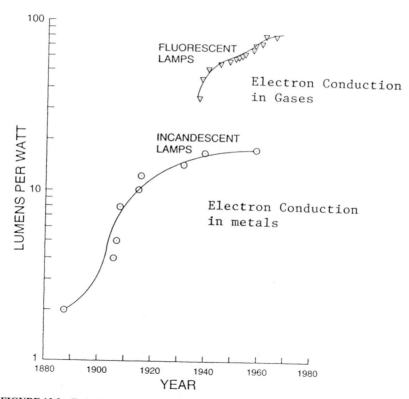

FIGURE 12.2   Technology S curve for progress in lamp technology.

12.2.2 Types of Rates of Technology Progress

There are two kinds of rates of technical progress: (1) *incremental progress* along an S curve as improvements are made in a technology utilizing one physical process and (2) *discontinuous progress,* jumping from an old S curve to a new S curve as improvements in a technology use a new and different physical process.

12.2.3 Scientific Bases of Technology S Curves

Underlying any technology is a basic physical phenomenon. The range of technical performance of the technology is determined by the nature of the underlying phenomena. Technical progress in a technology based on one physical process will show the continuity of form along an S-curve shape. However, technical progress that occurs when a different physical process is used will jump from the original S curve to a new and different S curve. Every technology S curve is expressive of only one underlying physical phenomenon (Fig. 12.2).

12.2.4 Technique for Fitting Technology S Curves

The following technique may be used to fit technology S curves to technical data on the rate of progress of a technology in the following manner:

1. Identify a key technical performance parameter.
2. Collect existing historical data on technical performance since the date of innovation of the technology and plot them on a time graph.
3. Identify intrinsic factors in the underlying physical processes that will ultimately limit technical progress for the technology.
4. Estimate the magnitude of the natural limit on the performance parameter, and plot this asymptote on the graph of the historical data.
5. Estimate the time differential of two inflection points between the historical data and the asymptotic natural limit (first inflection from exponential to linear rate of progress and second inflection from linear to asymptotic region).
6. The reliability of expert forecasting of the exact times of inflection will likely be more unreliable than their anticipation of the research issues required to be addressed for inflection points to be reached.

12.2.5 The Technology S Curve as an Analogy

The technology S curve requires that the new ideas for improving a new technology be proportional to an existing set of ideas in the beginning of the new technology for an exponential growth to occur. Since the creation of new ideas is not necessarily proportional to the quantity of preexisting ideas, one must take the technology S curve formula not as a model of technological progress but as a historical *analogy.*

It is true that for a brand-new technology, when ideas are new, they stimulate new thinking and thus new ideas (but not in a quantitative relation). Thus this analogy states that there will be an explosion of new thinking, which can be roughly described as exponential—as if it were quantitatively related—an analogy. When the ideas of a radically new technology are so exciting and intriguing, many new people start to think also about how to improve it. The mathematical form of the technology S curve is not a model of the process of technological change but an analogous pattern frequently seen in the histories of technical progress.

Since the technology S curve is not a model, it cannot explain the rate of change of a technology's progress. Such explanation must be derived from understanding the morphology of the technology, and how progress altered that morphology. Accordingly, the technology S curve is merely a *descriptive analogy*. But as an analogy it can be used for an intelligent "guess" at the rate of change, that is, anticipation and forecasting, but cannot be used to "predict" how that change may occur.

12.2.6 Transition from Technology Forecasting to Planning

Any forecast is an attempt to anticipate the course of future events. But this attempt to anticipate can proceed with different levels of knowledge about the nature of the events being forecast and can result in deepening levels of forecasts: (1) extrapolation, (2) generic patterns, (3) structural factors, and (4) planning agenda (Betz, 1993, p. 311).

A forecaster who has almost no knowledge about the events except historical data on past occurrence can do little more than *extrapolate* the direction of future events from past events. Extrapolation forecasting consists of fitting a trend line to historical data.

A forecaster who has some knowledge about the general pattern of a class of events but little knowledge about the specific exemplar of that class at hand can use the *generic pattern to fit the extrapolation* of the specific exemplar case. The technology S curve of this chapter is just such a generic pattern for the class of innovations. Fitting a generic pattern to an extrapolation involves more knowledge than mere extrapolation because one knows beforehand the form of the curve to be extrapolated.

In addition to knowing the generic pattern of an event, knowing something about the kinds of factors that influence the directions and pace of the events provides the basis for even better anticipation. Extrapolations from past data always assume that the *structure* of the future events is similar to the structure of the past events. Changes in structural factors will render extrapolation meaningless and create the most fundamental errors of forecasting.

The deepest level of forecasting requires understanding not only the generic pattern of the class of events to be anticipated but also the structure of the events and then proceeding to intervene in the future by *planning* to bring about a desired future event. A research agenda provides an anticipatory document required to bring about a technological future.

12.3 USE OF EXPERTS IN TECHNOLOGY FORECASTING

Within R&D infrastructures, there exist experts (researchers) in science and engineering who create knowledge and invent new and improved technologies. One technique some have used to forecast technology has been to gather these experts and ask them to predict the future. One of the first efforts at this was a study commissioned in the 1930s by the National Research Council and carried out by the National Resources Committee, chaired by a sociologist William Osburn (Osburn, 1937). Ayres (1989, p. 52) commented on this report: "this ambitious study used technological trend curves almost exclusively to illuminate the past rather than to forecast the future."

After World War II, the U.S. military began supporting formal methods of forecasting technology. Theo von Karman chaired a group to produce a technology forecast for the U.S. Air Force. The Rand Corporation (in Santa Monica, Calif.) was then created to continue and extend this work: "It (Rand) carried out many such forecasts and pioneered in developing technological forecasting methodologies such as the 'Delphi' method" (Ayres, 1989, p. 52).

12.3.1 Delphi Technique for Using Experts

Experts can be brought face to face to produce a report, such as the above-cited von Karman report. However, N. C. Dulkey and Q. Helmer, in developing what they called the "Delphi method," also saw a usefulness sometimes in querying experts by questionnaire and recirculating questions until a consensus among experts was obtained (Dulkey and Helmer, 1963). Joseph Martino (1983, p. 16) has succinctly summarized the Delphi technique: "Delphi has three characteristics that distinguish it from conventional face-to-face group interaction: (1) anonymity, (2) iteration with controlled feedback, and (3) statistical group response."

During a Delphi technique, the members of the group do not meet and therefore may not know who else is included in the group. Group interaction is carried out by means of questionnaires which are circulated to members of the group. Each group member is informed only of the current status of the collective opinion. Reiterations are continued until some consensus is reached.

When used for forecasting, such Delphi groups have attempted to predict what technologies will occur and when. However, technology-predicting forecasts by experts have often turned out to have been notoriously wrong, particularly about dates. Gene Rowe, George Wright, and Fergus Bolger reviewed the studies that have been performed on the efficiency of the Delphi methods versus any method of consulting experts. They concluded that the technique is probably no more useful than any technique for gaining group consensus among experts: "We conclude that inadequacies in the nature of feedback typically supplied in applications of Delphi tend to ensure that any small gains in the resolution of 'process loss' are offset by the removal of any opportunity for group 'process gain'" (Rowe et al., 1991, p. 235). By "process loss" they meant factors in group decision activities that included such factors as, in some circumstances, the group's collective judgment may be inferior to the judgment of the group's best member, premature closing of group judgment may result in poor results, or individual jockeying for power may skew results. Any refinement of techniques for forecasting which does not focus upon eliciting information about the underlying structures of forecasted events is not likely to improve the forecast.

In 1981 Martino, Haines, and Keller studied about 2000 technological forecasts, and found that about 700 of them used methods of extrapolation, leading indicator, causal models, and stochastic forms (Martino, 1981). All forms (Delphi consensus, extrapolation, leading indicator, etc.) for technology forecasting depend on expert opinion. Therefore it is important to understand what opinions experts are really experts about.

12.3.2 Capability of Experts in Forecasting

Are experts expert in predicting the future? Probably not! What research experts really know are (1) a technological need, (2) scientific theory underlying the technology, and (3) imaginative research approaches to attempt to improve a technology.

Therefore, the proper use of experts in technological anticipation and creation is to:

1. Identify future technological needs

2. Identify technology areas that require improved understanding in order to advance technology
3. Propose research directions to study these areas
4. Use detailed information about the research directions to forecast the direction and pace of technological progress
5. Use the research directions to structure research programs

12.4 PLANNING TECHNOLOGICAL PROGRESS

Merely forecasting a rate of technological change is not sufficient for technology strategy; one must also forecast the direction of change. Wherein will the changes occur in the technology? The way to do this is to describe the technology as a system. The concept of a "technology system" provides general principles by means of which a technology can be varied and improved.

12.4.1 Technologies as Transformational Systems

All technology systems are functionally defined open systems, accepting inputs, and transforming these into outputs. The boundary of a technology system is the points of the physical structure which receive inputs and which export outputs into the environment. The *kind* of transformation taking inputs into outputs defines the *functional* capability of the technology system.

All functional open systems, such as a technology system, must be described with two levels mapped to each other:

Logic schematic—a logical scheme of the functional transformation

Physical morphology—a constructed physical structure whose processes map in a one-to-one manner with the logic scheme

Logic Schematic of a Technology System.   The *logic schematic* is a step-by-step laying out in order of the discrete transformations required to take a generic input to a generic output of the technology's functional transformation. A logic scheme for a technology system is represented as a topological graph of the parallel and sequential steps of unit transformations.

Physical Morphology of a Technology System.   The *physical morphology* of a technology system is the assembly and connection of the physical structures and processes whose physical operation can be interpreted in a one-to-one manner with the logical steps of the schematic. The morphology of a technological device may be also analyzed as an open system defined by a boundary and containing components, connections between the components, and a control subsystem.

12.4.2 Progress in a Technology System

The general way of identifying potential or actual opportunities for technological advances lies in progress in either the logic schematic or the morphology. Technology advance may occur by

1. Extension of the logic schematic
2. Alternate physical morphologies for a given schematic
3. Improvement of performance of a given morphology for a given schematic by improving parts of the system

Technical progress may occur by adding features and corresponding functional subsystems to the logic. Technical progress may occur in alternate physical processes or structures or configurations for mapping to the logic. Technical progress may occur in improving parts of the system—its subsystems, components, connections, materials, energy sources, motivative devices, or control devices.

Thus, viewed as a system, technical progress can occur from changes in any aspect of the system:

1. Critical system elements
2. Components of the system
3. Connections of components within the system
4. Control subsystems of the system
5. Material bases within the system
6. Power bases of the system
7. System boundary

Technical progress in a system may occur from further progress in the components of the system. Technical progress in a system may also occur in the connections of the system. Technical progress in a system may also occur in the control systems of a technology. The material bases in the framework, devices, connections, or control of a system are also sources of technical change and progress. Finally, energy and power sources may provide sources of advance in a technology system.

12.5 MORPHOLOGICAL ANALYSIS OF A TECHNOLOGY SYSTEM

F. Zwicky introduced the concept of morphological analysis as a systematic way of exploring alternatives in the physical forms of a technology system (Zwicky, 1948). He proposed that one can systematically explore sources for technical advance in a system, by logically constructing all possible combinations of envisioned alternatives in any of these aspects of the system. First, the function of the technology must be clearly stated. A general logic scheme of the functional transformation must then be designed. Then all the physical processes mappable into the logical steps must be listed.

Morphological alternatives can then be searched by all conceivable variations of the logical architecture, and all possible physical substitutions in the logical steps. Then an abstraction must be made of the physical form of the technology system in terms of components, connections, materials, control, and power. One then lists (1) all conceivable alternative forms of each of these in terms of the physical processes and (2) all possible combinations of these alternative constructs—each offering an alternate morphology for the system.

12.5.1 Illustration: Morphological Analysis of a Jet Engine

Zwicky used the example of alternative structures for jet engines, identifying each major abstracted element (Jantsch, 1967, p. 176):

1. Intrinsic or extrinsic chemically active mass
2. Internal or external thrust generation
3. Intrinsic, extrinsic, or zero-thrust augmentation
4. Positive or negative jets
5. Nature of the conversion of the chemical energy into mechanical energy
6. Vacuum, air, water, and earth
7. Translatory, rotatory, oscillatory, or no motion
8. Gaseous, liquid, or solid state of propellant
9. Continuous or intermittent operation
10. Self-igniting or non-self-igniting propellants

Next Zwicky took all possible combinations of these features and came up with 36,864 alternative combinations. Now, many of these alternatives are not technically interesting, and these Zwicky eliminated. Actually, when Zwicky performed this evaluation in 1943, he did use many fewer parameters and reduced his alternatives to be examined to only 576 combinations. Zwicky's procedure in its full combinatorial form is cumbersome.

12.5.2 Research Planning as Based on Morphological Analysis

All technological research is actually focused as a limited form of morphological analysis, considering alternatives for features in a technology system. Industrial research laboratories use a modified version in organizing their research areas. They describe their principal product lines and production processes as technology systems and then organize programs of research to improve different aspects of these systems. Some form of morphological analysis provides the basis for all technology planning. What constitutes the boundary of a technology system is an important judgment, as it focuses on the system.

12.5.3 Procedure for Morphological Analysis

To summarize a procedure for conducting a morphological analysis of a technology system, one must start with the problems and technical limitations of a current technology system. One can then redefine the desired functional and performance capabilities of the technology (using current and potential embodiments of the technology in product systems and in applications systems). One must next abstract the principal parameters of the desired functional transformation. The different logical combinations of these parameters (that are mutually physically consistent) and ordered by hierarchy of physical influence provide a logic tree of alternative morphologies for the technology system.

To examine these alternatives:

1. Begin with an existing boundary and architecture which provides the functional transformation defining the technology.

2. Then use the applications system forecast to classify the types of applications and markets currently being performed by the technology system and also to envision some new applications and markets which could be performed with improved performance, added features, and/or lowered cost of the technology system.

3. Identify the important performance parameters and their desirable ranges, desirable features, and target cost to improve existing applications and to create new applications.

4. These criteria then provide constraints and desired ranges under which alternative morphologies of a technology system can be bound.

5. Next examine critical elements of components and connections whose performance would limit the attainment of the desirable performances, features, and cost.

6. The performance of critical elements which do not permit desired system performance can then be identified as technical bottlenecks.

7. Examine alternative base materials and base power sources for alternative morphologies, again selecting only those which would fulfill desirable system performance.

8. Examine alternative architectural configurations and alternative types of control that might attain the desired higher performances.

9. Then combine and recombine various permutations of critical elements, base materials and power, and alternative architectures and control systems to provide a systematic way of enlarging the realm of possible technology developments.

12.5.4 Procedure for a Functional Context Analysis

Complementary to a morphological analysis, a functional context analysis of a technology system also should be made from the application systems perspective:

1. Identify the major generic applications of product, production, and service systems that embody the technology system.

2. From the applications perspective, identify desirable improvements in performance, features, and safety that would be recognized by the users (final customers) of the product, production, and service systems.

3. Determine the prices which would expand applications or open up new applications.

4. Determine the factors that facilitate brand-name loyalty when product, production, and service systems are replaced in an application.

12.6 SUMMARY

Progress in technical performance parameters for a technology system can be accomplished through improvement in the logic or morphology of the technology. Research planned to improve the points within the system that limit the overall system performance—technical bottlenecks—brings about technical progress. Anticipating these bottlenecks allows one to know where progress must be made if the overall system is to be improved. And research planned to address these bottlenecks provides the structural basis for incremental improvement up the "technology S curve" for the system or discontinuous jumps to a new curve.

12.7 REFERENCES

Ayres, Robert U., "The Future of Technological Forecasting," *Technol. Forecasting Social Science Change,* **36**(1–2): 49–60, Aug. 1989.

Betz, Frederick, *Strategic Technology Management,* McGraw-Hill, New York, 1993.

Dulkey, N. C., and Q. Helmer, "An Experimental Application of the Delphi Method to the Use of Experts," *Management Science,* **9**: 458–467, 1963.

Foster, Richard N., "A Call for Vision in Managing Technology," *Business Week,* pp. 24–33, May 24, 1982.

Godet, Michel, "Reducing the Blunders in Forecasting," *Futures,* **15**: 181–192, June 3, 1983.

Jantsch, Erich, *Technological Forecasting in Perspective,* Organization for Economic Cooperation and Development, Paris, 1967.

Landford, H. W., "A Penetration of the Technological Forecasting Jungle," *Technol. Forecasting Social Science Change,* **4**: 207–225, 1972.

Martino, Joseph P., *Technological Forecasting for Decision Making,* 2d ed., North-Holland, New York, 1983.

Martino, Joseph P., P. A. Haines, J. L. Keller, et al., "A Survey of Forecasting Methods," research report (UDRI-TR-81-36P), University of Dayton Research Institute, Dayton, Ohio, 1981.

Osburn, William F., *Technological Trends and National Policy,* research report, National Research Council, Washington, D.C., 1937.

Rowe, Gene, George Wright, and Fergus Bolger, "Delphi: A Reevaluation of Research and Theory," *Technol. Forecasting Social Change,* **39**: 235–251, 1991.

Twiss, Brian, *Managing Technological Innovation,* Longman Group, Harlow, Essex, England,1980.

Worlton, Jack, "Some Patterns of Technological Change in High Performance Computers," *Proceedings Supercomputing '88,* IEEE Computer Society: Computer Society Press, Los Alamitos, Calif., 1988, pp. 312–319.

Zwicky, F., "The Morphological Analysis and Construction," *Studies and Essays,* Courant Anniversary Volume, Interscience (Wiley), New York, 1948.

CHAPTER 13

KNOWLEDGE MAPPING:

A TOOL FOR MANAGEMENT OF TECHNOLOGY

Karol I. Pelc

Michigan Technological University
School of Business and Engineering Administration
Houghton, Michigan

13.1 INTRODUCTION

One commonly accepted assumption is that *knowledge* constitutes a precondition for good management. This is also valid in the domain of technology management.

The purpose of this chapter is to analyze knowledge mapping as a tool for management of technology. *Mapping* is considered here as a process that allows one to identify the elements of knowledge and their configuration, dynamics, mutual interdependencies, and interactions. Different forms and principles of knowledge mapping are reviewed and discussed with regard to their applications in management of technology.

To manage technology means to be able to establish and influence a number of activities of different types, including R&D, product engineering, process engineering, manufacturing, systems engineering, and services, and to make decisions concerning technology. Those decisions may have different character—strategic, operational, tactical, local, global, short or long term, and so on. To be able to deal with that complexity, we should have a clear view on what kind of *knowledge base* is needed for specific tasks and decisions and how it should be structured, updated, and accessed. Instead of viewing the expertise in management of technology as a set of knowledge modules and blocks, we shall try to represent it rather as a *territory of knowledge.* This territory may be characterized by some distinct elements, and their clusters, either situated at a distance from each other or overlapping, interconnected, and interacting. In this chapter we distinguish two classes of territories and corresponding knowledge maps: maps of technology and maps of technology management. In *maps of technology* we represent the knowledge domain of a specific technology; in *maps of technology management,* we represent knowledge of models, methods, and tools required and used for managing technology. Both technological knowledge—in almost all fields of technology—and knowledge of management of technology are growing rapidly. Mapping of knowledge becomes necessary for several practical purposes and, at the same time, is a thought-provoking theoretical concept.

We have to "navigate" across the expanding universe of knowledge and use maps to better understand current "position," to detect opportunities and paths as they appear, and to be able to reach our destinations.

Growth of technology management knowledge is driven by practical needs. It is a result of accumulation of managerial experience, methodological studies, and experimentation. There are more and more publications, conferences, academic institutions, and professional groups in this domain. Knowledge of technology management develops according to a general pattern of knowledge growth, which can be viewed as a process of continuous *accumulation*. According to A. Reisman, this process includes *consolidation* of knowledge within each scientific discipline and across disciplines, to create still new opportunities for further *expansion of knowledge*.[1,2] We can also view the process of knowledge growth as a result of two subprocesses. The first subprocess, *divergent growth,* is driven by the human need to understand the world, i.e., natural need of cognition, reflected by our curiosity. At the same time, this tendency is restricted by boundaries of individual intellectual capabilities and hence a limited scope of individual expertise. This leads to divergence of scientific orientations, with splitting of disciplines into a growing number of subdisciplines, specialties, etc. It is a *disciplinary branching effect* in the knowledge growth process. The second subprocess, *convergent growth,* is driven by the need to solve practical problems dictated by the reality of the economic, social, and technological environment. Those problems are usually trespassing boundaries of disciplines and specialties. They require one to pull knowledge from different scientific disciplines and professional specialties together, and they lead to growth of interdisciplinary, multispecialty knowledge. It is an *interdisciplinary crystallization effect* in the knowledge growth. This view on growth of knowledge is illustrated in Fig. 13.1. It may constitute a framework for mapping of knowledge.

Systematic representation of technology management knowledge has been a subject of earlier publications.[3–5] Those publications indicate some extent of overlap between the areas of "engineering management" and "technology management." The first is centered around management of engineering projects, engineering organizations, and engineering activities, while the other emphasizes issues of technology, strategy, evaluation of technology, and processes related to technology development, implementation, and transfer. For practical reasons, both areas or regions of knowledge should be *integrated* because many managerial solutions have broad impact or may depend on a broader knowledge context. The knowledge mapping approach allows us to integrate those different components of the knowledge base, needed for management of technology, without losing their specificity.

13.2 APPLICATION OF KNOWLEDGE MAPPING FOR TECHNOLOGY MANAGEMENT

Knowledge mapping may have several applications in the practice of technology management. Four of them, as illustrated in Fig. 13.2, seem to be most evident and important:

- Definition of research programs
- Decisions on technology related activities
- Design of knowledge-base structures
- Programming of education and training

Definition of research programs is usually driven by the need to fill the gaps in current knowledge and technology. Identification of those gaps may be easier when a map is

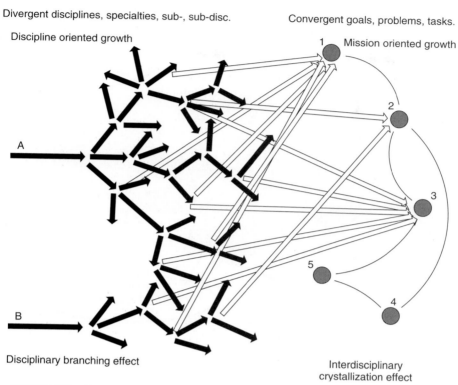

Divergent disciplines, specialties, sub-, sub-disc.   Convergent goals, problems, tasks.

Discipline oriented growth   Mission oriented growth

Disciplinary branching effect   Interdisciplinary crystallization effect

FIGURE 13.1   Knowledge growth as combination of disciplinary branching and interdisciplinary crystallization. (*Reprinted from Pelc.*[5])

made to represent the current state of a respective area of scientific, technological, or managerial knowledge. For example, setting a research agenda for a scientific institution or a corporate research lab requires a review of the state of the art and an evaluation of current intellectual potential of the institution or the research lab. This review, combined with discussion of individual interests of members, may constitute a basis for mapping of existing knowledge resources and revealing knowledge gaps which need to be eliminated through research. Examples of such applications have been described by several authors.[6–8] Courtial and Remy used knowledge mapping to evaluate thematic linkages among research projects and interfaces between research projects and more general economic and application criteria set for a large research organization, which consists of many specialized labs. The same study allowed the authors to determine relative "centrality" of projects in the broader context of French agricultural research.[6] Healey et al. presented results of an experiment in applying the knowledge mapping method for research planning at a national level. In this case the subject of evaluation was the state of various scientific disciplines, and identification of potential areas for useful investment and subsidies by a national foundation in the United Kingdom.[7] Ijichi et al. conducted analysis of R&D network dynamics in a selected field of technology (videocassette recording).[8] They built maps of R&D team relations within a corporation to identify thematic clusters, professional linkages among authors, and time sequences of knowledge contributions (patents, publications).[8] This kind of mapping may serve

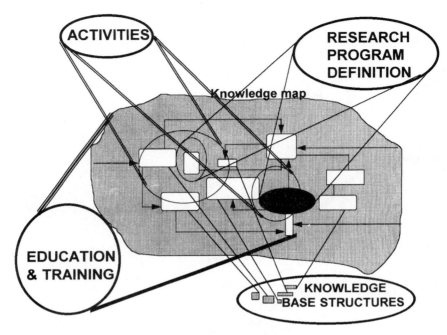

FIGURE 13.2   Applications of knowledge mapping in technology management.

for evaluation of roles played by individual professionals within R&D teams, assessment of the R&D group capacities, and contributions of R&D teams and groups to the company's technology. As a result, the mapping of R&D may provide input to procedures of allocation of R&D resources in the future.

Decisions on technology related activities include selection of R&D projects (e.g., in a corporate R&D lab), setting goals and allocating resources for development of new products and processes, acquiring and implementing new technologies, organizing cooperative engineering projects, and establishing new labs and testing facilities. Such decisions typically require a combination of technical and managerial expertise. Knowledge and data have to be drawn from different sources simultaneously. They must be processed, evaluated, and applied. Compilation of knowledge components may be easier when their relative positions in the knowledge base are well defined. We should be able to determine the most common patterns of knowledge needed in managing technology. In many cases, when coordination of technical and managerial activities is critical, it might be facilitated by representing both the problems and potential solutions on a common map. Mapping could also be used for creating an inventory of cases from the past experience. Coordination of activities may be efficient and effective if at least some knowledge and beliefs are shared by the parties involved. This is sometimes difficult to achieve. There are some attempts to use mapping of knowledge and beliefs as a tool for analysis of decisions being made in management of technology. Swan and Newell analyzed factors affecting the adoption of technological innovation by mapping knowledge and beliefs of individual managers.[9] In that case mapping served as a tool for analysis of decisions made by individual managers. Mapping is also useful for coordination of projects and activities realized in a distributed system, for example, when engineers and researchers, supported by an interconnected AI (artificial intelligence) system, work on the same project in different locations, and cooperate by providing current information inputs from many parallel tasks. In such a system, decisions are the result of a coordinated flow of information without centralized control. Zhang et al. developed a software program for coordination of distributed cooperative "agents." In the distributed environment, "each agent must reason about the knowledge, actions, and plans of other agents, yet the system can act only collectively. Therefore, knowledge representation and reasoning techniques at a local agent must achieve the properties of cognitive integrity, transparency, and simplicity."[10] A common platform of reference for all agents is based on a knowledge map. In recent years this approach became technically feasible and is considered to be essential for advances in such new fields as distributed artificial intelligence (DAI).[11–13]

Design of knowledge-base structures also assumes existence of a knowledge map as a frame of reference. New information technology allows one to package data, information, and knowledge with an increasing density. The structure of the knowledge base should be determined in such a way as to allow for easy and fast access to specific knowledge as well as to make it possible to update and expand. For example, the structure should be *flexible* enough and *open* to be able to accommodate and filter large amounts of new publications (including computer-based reports) that appear every day. Knowledge mapping is needed to provide a background for establishing the knowledge base structure in a given domain of knowledge.

Education and training may be programmed efficiently, if we know which components of knowledge are "centrally located" and what kind of interconnections exist among different parts of the knowledge domain. The programming of education corresponds to tracing some preferred or recommended paths across the map of a knowledge domain. The education program should be viewed as a dynamic process in which respective elements of knowledge are added, consolidated, or eliminated, in accordance with changes taking place in knowledge contents of a given scientific discipline or technical specialty. A knowledge map may also serve for making comparisons between different educational programs and studying their evolution and verifying mutual correspondence and compatibility among them. We can find, and represent graphically on the map, which parts of knowledge "territory" are "visited" in a course or curriculum, and how that corresponds to their central or peripheral location in the context of the knowledge domain under consideration.

Building of knowledge maps may be based on different principles and methods. The rest of this chapter is devoted to explanations of those methods.

13.3 METHODS OF KNOWLEDGE MAPPING

13.3.1 General Remarks about Knowledge Representation

It is very difficult to define knowledge as an object of mapping. Bertrand Russell, one of the most distinguished philosophers of the twentieth century, characterized *human knowledge* as an "ambiguous notion." He wrote:[14]

> Let us remember that the question "What do we mean by 'knowledge'?" is not one to which there is a definite and unambiguous answer, any more than to the question "What [do] we mean by 'baldness'?"

The formal theory of knowledge constitutes a special branch of philosophy called *epistemology.* The term *knowledge* usually reflects cognition and understanding, which is based on human experience and observations of facts or causal relations between them. We use that term in many contexts, such as personal knowledge or educational knowledge. Polanyi described several aspects and forms of *personal knowledge.*[15] He emphasized the importance of knowledge mapping as a tool in processing of knowledge at the personal level: "When we find our bearings by aid of a map we gain an understanding of the region represented by the map, and from this conception we can derive an indefinite number of itineraries."[15] Howard described the value of knowledge mapping in the context of management education.[16]

Knowledge representation became one of central issues in knowledge engineering, which emerged as a new type of engineering, concerned with advancement of AI and expert systems (in their multiple applications). The latter involve procedures for acquisition and processing of professional knowledge (of experts). In recent years, we are experiencing conceptual transformations taking place in the computer-based systems: from those operating with numerical *data* (data processing), through those capable of also operating with textual and graphical *information* (information systems), to *knowledge-based systems* which operate with both explicit and implicit forms of knowledge.[17,18] Literature on knowledge engineering does not concentrate on definition of knowledge. Bibel et al., who described a number of formal schemes for knowledge representation, admitted that "the notion of knowledge is ill defined and thus has a different meaning for different people," but they also assume existence of the *"cognitive form of knowledge."*[19] For practical purposes, regarding methods presented later in this chapter, we should be interested in representation of connections and interactions between different elements of knowledge. One such connection corresponds to *knowledge association,* which is sometimes essential for problem solving, in which knowledge is drawn from many, sometimes remote, contexts. Niwa suggests that the phenomenon of knowledge association plays an important role in managerial decision making and, as such, should be a subject of modeling.[20] Knowledge association might be represented as an incidental leap between remote elements of the knowledge map (making remote connections). Knowledge maps, which show both the elements of given domain(s) of knowledge and connections among those elements, may be designed to represent different contexts of knowledge. The definition of elements and their location and interconnections (impacts) will be different in each map and will depend on the methodological principle used for mapping. Four different methods of knowledge mapping will be discussed here:

- Chronological mapping
- Co-word-based mapping
- Cognitive mapping
- Conceptual mapping

13.3.2 Chronological Mapping

To represent history and the current state of knowledge in a specific discipline of science or technology, it is practical to track back the chronological sequences of such events as discoveries and inventions and to represent them in a form which would correspond to their temporal and logical interdependencies. As a result, we create a representation of different contributions of knowledge which leads to the state-of-the-art technology. Each individual scientific or technological fact is linked with its followers in accordance with the *chronology* of their occurrence. Graphical representation, or a map, consists of *nodes,* which represent events and contributions, and *arrows or branches,* which represent influences and linkages. In the case of technology mapping, the graph will be directed usually from earlier fundamental scientific observations and theories to more recent applications, and combinations of the latter. For example, one could easily draw a map showing sequential events leading to the contemporary radio-communication technology. It would include, as nodes, such discoveries as electro-magnetic induction by Faraday, electromagnetic wave theory by J. C. Maxwell, dipole oscillation experiments by H. Hertz, and invention (patent) of wireless telegraphy by G. Marconi. This is an elementary or root sequence leading to the radio. In most cases, however, it is necessary to analyze multiple, simultaneous, and sequential developments in several fields of scientific research and many technological innovations, to arrive to the current situation. Martino used this approach to describe the laser technology as resultant from the knowledge growth in such fields as molecular beams, magnetic resonance, microwave spectroscopy, and masers.[21] This example is shown in Fig. 13.3. Flows of information and impacts of subsequent contributions are depicted in the form of a "knowledge tree" for those many fields. It also includes many "side branches" which might not be essential to creation of the laser but were developed simultaneously in respective disciplines of science.

This mapping method is applicable not only for historical studies of technology or for tracing the sources of knowledge that we currently apply but may also serve as a *tool for technology strategy considerations.* In this case it should combine analysis of historical development with that of generating strategic options for a company. Goodman and Lawless described the technology mapping as an effective means for analysis of particular technologies and potential for their development.[22] The authors emphasize that it is necessary to ensure multidimensional analysis of technology. This requires that the product-technology matrix be developed, which includes evaluation of criticality (e.g., on a qualitative scale of three levels of criticality) of respective technologies for each group of products. This analytical information may be combined with chronological mapping of those technologies which are essential for the company.[22] An example of such a chronological map for videotape-recording technology is given in Fig. 13.4, which shows that parallel advances, in such fields as magnetic materials, theory of magnetic phenomena, magnetic recording technology, electronics, and frequency modulation techniques before 1950, contributed to acceleration in videorecording technology after that year. During the next 6 years two paths of development were followed.

The chronological mapping of technology requires a large amount of technical information, combined with opinions of experts. The following procedure may be used in building the chronological map of a technology:

- An initial overview of basic sources of knowledge and main branches contributing to current (or future) technology under consideration. This first step requires discussion with specialists in the field and their active involvement.

- Collection of detailed data on patents, publications, citations, and cross-references in the professional literature. These data are accessible through the scientific information system (citation indexes, databases, source documents, etc.).

- Identification of major events (nodes) which initiated new branching or constituted connections to other source specialties. To evaluate the importance of those events (milestones, or breakthroughs), judgment of individual experts or the group expertise is needed (via questionnaires, interviews, etc.). It is essential for definition of the main branches and nodes in the map of technology.

- Location of each node (event) at the appropriate branch and along the time scale, to reflect both the causal relations and the chronology.

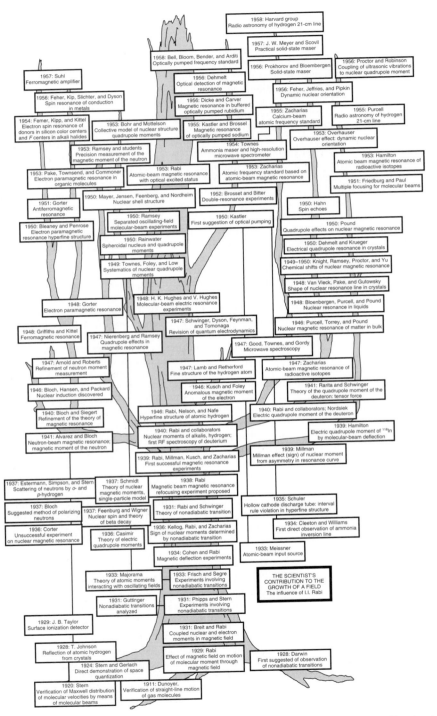

FIGURE 13.3 Chronological map of knowledge growth leading to masers and lasers. (*Reprinted from Martino.*[21] *Adapted from "Basic Research in the Navy," Report of Naval Research Advisory Committee, 1959.*)

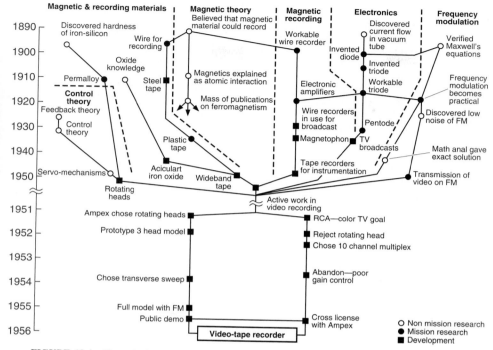

FIGURE 13.4 Chronological map of knowledge growth leading to videotape recording technology. (*Reprinted from Goodman and Lawless.*[22] *Adapted from Batelle, 1973.*)

- Verification of the technology map by a group of experts representing all major source disciplines or specialties and the state-of-the-art technology.

13.3.3 Co-Word-Based Mapping

Quantification of knowledge growth is based on numbers of publications and patents that appeared in a given period of time. With all reservations regarding the unequal importance of individual publications and contents of new knowledge they represent, in this way we can monitor the dynamics of research activity in various disciplines, observe shifts of research interest (and related investments) across a whole spectrum of subject areas, and assess changes in intensity of results which are put in the public domain (patents and publications). The same data can be used for knowledge maps. Bibliometric calculations of frequency of words appearing in the documents (or their titles and/or abstracts) allow us to determine intensities of information in each subject area or technical specialty. These intensities may be used next as indicators of importance of respective areas to be represented in the map. If those calculations also include relative density of publications and/or patents in which some words cooccur, then we can discover clustering effects and determine "proximity" of corresponding elements of knowledge. A method of concurrent word mapping (also called *co-word-based mapping*) and its application for mapping of technology has been described by Engelsman and van Raan.[23] The authors presented their experience in applying the word cooccurrence principle to study the diffusion of knowledge among different fields of technology (knowledge diffusion has been defined, in this study, as the use of field-specific knowledge in fields other than the original field, i.e., "knowledge-sup-plying" and "knowledge-absorbing," fields of technology have been distinguished). Empirical data and calculations were based on large collections of patents. The set of patents (for a given period of time) was analyzed with regard to frequency of individual words and frequency of cooccurrence of those words. Maps of technology, created in this way, reflect both the importance of each subject area or knowledge element (represented by the frequency of words) and relative distance or proximity between those elements or subjects (represented by the frequency of cooccurrence of words). The closer the elements of knowledge (subjects), the higher frequency of word cooccurrence. Depending on the patent collection available, it is possible to obtain different maps over time and to study evolution of the map for a specific field of technology. It is also possible to make comparisons between maps representing situations of a given technology in different countries.

More general principles and a review of bibliometric methods, which may be used for knowledge mapping in science and technology, have been presented in the book by Callon et al.[24] A good example of knowledge mapping for a discipline has been described by Peters and van Raan, who applied concurrent word-based method for mapping of chemical engineering as a knowledge domain.[25] They used the following three indexes to evaluate dimensions of knowledge map elements on the basis of word cooccurrence: the Jackard index, the inclusion index, and the proximity index.

The *Jackard index*[25] was defined as

$$J(ij) = \frac{c_{ij}}{c_i + c_j - c_{ij}}$$

where c_{ij} = number of cooccurrences of words i and j
   c_i = frequency of word i
   c_j = frequency of word j

The *inclusion index*[25] was defined as

$$I(ij) = \frac{c_{ij}}{c_i}$$

assuming $c_i < c_j$. The *proximity index*[25] was defined as

$$P(ij) = \frac{N\, c_{ij}}{c_i\, c_j}$$

where N is the number of publications. These authors compared maps obtained with each index. An example of such a map is given in Fig. 13.5. The size of characters used for each word in this map (based on a 3-year search in *Chemical Abstracts*) indicates the relative frequency of that word. Distances between elements and their linkages are based on the values of the *inclusion index* for pairs of words.

This method may be used effectively for mapping of fields of technology (or scientific disciplines) when we have access to a large database of patents or publications, ready for computerized access, so that both the search of words and estimation of their frequencies and cooccurrences may be done automatically (using an appropriate computer software). The following procedure may be used for such mapping:

- Selecting the basic source of data, e.g., a set of publications or patents which will be analyzed for word frequency and word cooccurrence. This selection should be based on expertise of technologists and scientists who are actively involved in current research in the domain under consideration.

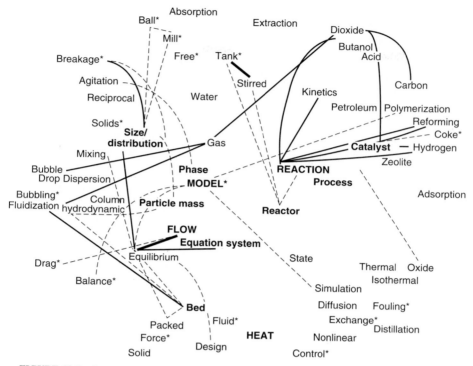

FIGURE 13.5   Co-word-based map of chemical engineering knowledge. (*Reprinted from Peters and van Raan.*[25])

- Initial review and estimation of the frequency of words appearing in the *titles* of source documents and then narrowing the list of words to those with the highest frequencies (with assistance of specialists).

- Computerized search of words in the set of *abstracts* and computation of selected index values for word cooccurrences.

- Constructing the graph of the knowledge map representing the most frequent words as nodes and linkages among them as lines. Those lines should differ in thickness or type to represent different values of the word cooccurrence index.

- Validation of the map by a group of experts representing the field of technology; discussion and interpretation of the map by comparison with intuitive expectations of the experts.

13.3.4   Cognitive Mapping

Cognitive mapping has been studied for a long time as a method of personal knowledge representation. Later it developed into a framework for *systems thinking* and system dynamics studies. It is a way of graphical expression of one's understanding of causal relationships between elements or factors influencing situations in a given environment. It is also considered to be useful in management. According to Langfield-

Smith and Wirth, "a causal cognitive map is a directed network representation of an individual's beliefs concerning a particular domain at a point in time. The nodes and the arcs joining them indicate causal beliefs."[26] In the 1960s, this tool was introduced by Forrester for industrial system analysis.[27] This concept has also been used to describe the *system dynamics* method for managerial applications.[28] Roberts proposed the same approach to be used in the R&D project analysis and management.[29] Senge used the cognitive mapping method to explain a set of system archetypes, which are essential to develop a system view of any organization.[30] Recently, cognitive mapping has been used for representing management problem situations and typical causal patterns of managerial problems in a company, including those related to management of technology. A practical advantage of cognitive mapping is that it allows representation of feedback mechanisms (both positive and negative) and causal loops at any level of complexity. Hence it may serve as a knowledge representation tool to show causal relations in both technical and managerial problems and for both types together. An example of a cognitive map for a technology management problem situation is given in Fig. 13.6. Three causal loops are presented in this graph. Nodes of the cognitive map are connected by arrows representing influences and reactions as perceived by a manager, who should be able to assess direction, relative strength, and possible delays in each causal relationship.

This method of mapping allows us to create models of problem situations in technology management at any level of decision making. A typical procedure to develop a cognitive map includes

- Listing all major elements (nodes; e.g., factors, variables, objects, actors) that play a role in the problem situation under analysis
- Creating a cross-impact table which shows direction and strength (and delay, if any) of the individual causal relation between any two elements from the list

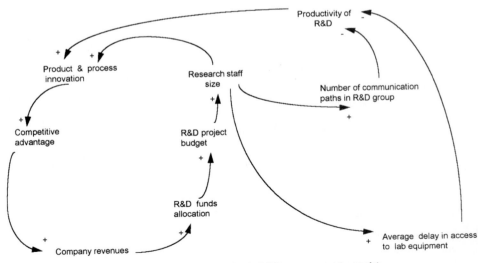

FIGURE 13.6   Cognitive map of problem situation in R&D management (example).

- Drawing a diagram presenting all elements (nodes) connected by arrows which are oriented in accordance with the table of causal relations
- Inserting numerical values representing strengths of each impact or linkage

In case of less complex problem situations (few elements), we can start with an intuitive graphical representation of impacts and feedbacks between a limited number of elements (without creating a list and a table) and gradually refine the graph, by adding either more elements or introducing additional connections (arrows). In this way we can conduct an exercise in mental modeling of the problem situation, identify the critical elements, or find those causal relations which are most important. In this sense the cognitive mapping may be viewed as a tool serving for an immediate or approximate analysis of almost any problem in technology management.

13.3.5 Conceptual Mapping

Methodological Background of Conceptual Mapping. *Conceptual mapping* may be used for representing a *whole domain* of knowledge (not just a problem situation, as in the case of cognitive mapping) in order to identify the gaps and areas of special interest. The object of mapping in this case may be either a scientific discipline or a technology, or an interdisciplinary domain of knowledge. We shall now concentrate on mapping of the knowledge domain of technology management, as a representative *interdisciplinary domain of knowledge.* To build the map, we have to first identify elements of the knowledge domain and connection between them. There are three basic categories of knowledge elements in any domain of knowledge (i.e., scientific discipline, field of technological expertise, etc.):

- A set of terms and concepts, a dictionary, etc.
- A set of statements, descriptive and prescriptive information and data, reports on observations, experiments, facts, events, etc.
- A set of methodological tools, models, and theories

These categories are functionally related. Terms and concepts are used to formulate statements. Those statements, empirical observations, rules of behavior, etc., are either a result of application of some models or constitute a basis for new models and theories, which, in turn, are used to develop new statements or to create new concepts. All those processes and interactions are reflected in publications, research reports, documents, databases, etc.

Mapping of knowledge that represents those elements has been used earlier. In a morphological analysis of knowledge in astronomy, Zwicky used categories of celestial objects and methods of astronomical measurements as dimensions which define mapping of this discipline.[31] A table of mathematics, developed by Z. Janiszewski, identifies major theories as basic elements of the discipline and represents their interconnections in terms of intensity and direction—specifically, the extent to which an advance in a given mathematical theory may contribute to solution of problems in another subdomain of mathematics, or which theory represents a source of knowledge for other.[32] A similar mapping principle has been used by Dunford and Schwartz in their "interdependence table" recommended as a learning guide in mathematics.[33] Both astronomy and mathematics are well-defined and traditional disciplines of science. Knowledge mapping of an interdisciplinary domain, such as knowledge of technology management, involves a few additional aspects. One of them lies in the very nature of interdisciplinary knowledge. Such a domain of knowledge grows under the influence of several *source disciplines* which provide initial concepts, terminology, and methods. To represent this phenomenon (evolving over time) on the knowledge map, we have to determine the source disciplines and affinities of elements of the interdisciplinary knowledge domain with each source. It is practical to view the source disciplines as an *environment* for the knowledge territory under consideration. This mapping principle is illustrated in Fig. 13.7.

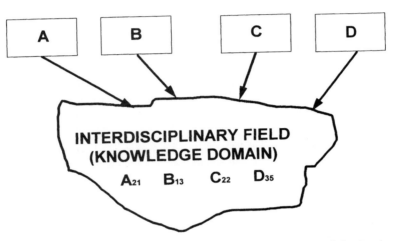

FIGURE 13.7 Source disciplines and elements of an interdisciplinary knowledge domain. Descriptors A_{21}, B_{13}, C_{22}, and D_{35} pertain to examples of knowledge elements with affinities to respective disciplines (numbers are arbitrary) for illustration only.

In the case of technology management knowledge, we are interested in developing a map which would be flexible enough to accommodate changes (additions, consolidations, expansions, etc.) in a whole spectrum of knowledge elements either adopted from other disciplines or created independently. In connection with the dynamics, we should be able to evaluate and represent the *level of knowledge* in each particular element of the domain. Distinction between knowledge levels has been emphasized by several authors and considered to be a difficult task of knowledge representation. Lenat and Feigenbaum analyzed the thresholds of knowledge in the context of artificial intelligence (AI) and used this concept to evaluate potential for learning and discovery in AI systems.[34] Mehrez et al. proposed a sequence of five levels of knowledge for classification of contributions to management science and distinguished the following levels of knowledge:[35]

Level 1: reality—empirical data on reality, perceptions, descriptions

Level 2: reality to model—conditions of similarity, approximation, assumptions of modeling

Level 3: models—representation of reality by the model

Level 4: model to statement—verification technique, algorithms, and rules of reasoning

Level 5: statements—theory, inference, explanations, and judgment

It is possible to adopt this sequence for purposes of knowledge mapping in technology management.

Building a Conceptual Map of Technology Management Knowledge. The conceptual mapping approach, proposed by the author and discussed in an earlier publication,[5] assumes the following properties and rules of graphical notation in a knowledge map:

- Each subdiscipline or specialty, considered as a knowledge element of a given domain, should be shown on the map as a single box (frame).
- The relative size of knowledge content in an element, e.g., as measured by the number of publications, patents, active authors, etc., should be represented by size (or thickness of frame) of the element in the map.
- The level of knowledge, as characterized by type of the most advanced knowledge available (e.g., classified according to the five levels described earlier), should be represented by darkness (or color) of each element.
- The proximity of respective knowledge elements, assessed by experts or measured by the bibliometric indices of proximity,[29] should decide on relative location of elements.
- The location of elements on the map should also reflect their "origin" from and affinity with external disciplines (knowledge sources).
- Connections between knowledge elements should reflect direction and intensity of impacts, or knowledge flows. They should be shown as arrows and lines of adequate thickness. Assessment of those linkages may be based on data on citations, word cooccurrence, and/or opinions of experts in the respective subdisciplines or specialties.

To build a map of knowledge in the domain of technology management, one has to identify all elements and connections (as described above). The following steps are suggested:

- Create a list of source disciplines for technology management knowledge. This step involves expertise and extensive analytical work (review of literature, citations, etc.).
- Create a list of knowledge elements to be included in the map (knowledge subdomains, major areas of expertise, problem areas, etc.), and group them according to affinities with respective source disciplines. An example of such a list for technology management knowledge is given in Table 13.1 (which includes 7 source disciplines and 33 elements). In this step, an expert group is needed to assure completeness of the list for each subset of elements.
- Assess the current "size" of each element. It may be expressed in qualitative terms (e.g., qualitative judgment of the amount of information on scale: small, medium, or large).
- Assess the level of knowledge (on a scale of 1 to 5, as described earlier).
- Create a cross-impact matrix and evaluate strength of connections between ele-

Table 13.1   Initial List of Knowledge Elements for Mapping of Technology Management

Affinity with economics	Technology forecasting methods
Engineering economy	Technology policy analysis
Technological change theory	Technology strategy models
Technology diffusion models	Technometrics
Technology implementation	Affinity with computer science
Technological innovation models	Artificial-intelligence systems
Technology investment analysis	Engineering decision support systems
Technology transfer methods	Engineering information systems
Affinity with management science and operations research	Affinity with ethics and law
Concurrent engineering management	Technology property law
Manufacturing management	Technological risk analysis
Project selection methods	Affinity with psychology
Project planning and scheduling	Cognitive process theory
Project control	Design theory
Product and process development	Engineering communcation
R&D project management	Affinity with sociology
Affinity with systems science	Engineering culture analysis
Environmental impact analysis	Engineering education models
Technology assessment methods	Technology marketing methods
Technology dynamics analysis	Technological organization models

Source: Reprinted from Pelc.[5]

ments (importance of impact). A scale of 1 to 3 with qualitative descriptors (e.g., weak, medium, strong) may be used.

- Identify clusters of elements according to their "proximity." assessed either intuitively or on the basis of bibliometric measures. Clusters of elements in the map may also emerge because of their methodological or application similarity.
- Construct a map which depicts all source disciplines as neighbors to the knowledge domain and all elements located according to rules described above. An example of such a map for the domain of technology management is given in Fig. 13.8.
- Validate and verify the map by a group of experts, including practitioners in technology management.

This mapping technique requires a multidisciplinary effort. In almost every step, experts representing individual disciplines of science and practitioners of technology management are involved. In many cases the bibliometric studies and syntheses are also needed.

13.4   CONCLUSION

Mapping of knowledge in technology management is a process that allows for definition of the scope and contents of this new discipline. The same process can be used for

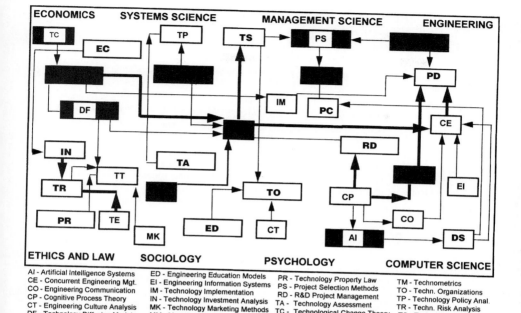

FIGURE 13.8 Conceptual mapping of technology management knowledge (example). (*Reprinted from Pelc.*[5])

monitoring and evaluation of tendencies in expansion of knowledge due to new practical problems that arise. Review of potential applications and methods presented here suggests that knowledge mapping is potentially powerful but still not a mature methodology. Four methods presented in this chapter need not be used in separation. Some experience exists in combining and modifying some of them for a specific purpose. For example, the Japanese study of R&D network dynamics,[8] which we refer to in the first section of this chapter, used a combination of word cooccurrence mapping (modified to analyze coauthors of patents and publications) with a chronological mapping technique (modified to represent a relatively narrow technological field and its development during a short period of time).

Further experimentation and research on knowledge mapping should provide more empirical material to verify individual methods and to determine the range of most effective application of each of them. Potential applications of knowledge maps for tracking the past trajectories and selecting future directions of research, structuring of databases needed for technology management, analyzing the skills and knowledge profiles of research teams, and designing new educational programs seem to be challenging enough to stimulate interest in this new field.

13.5 REFERENCES

1. A. Reisman, *Management Science Knowledge: Its Creation, Generalization and Consolidation*, Quorum Books, Westport, Conn., 1992.
2. A. Reisman, "Creativity in MS/OR: Expanding Knowledge by Consolidating Knowledge," *Interfaces*, **24**(3): 91–99, May–June, 1994.
3. D. F. Kocaoglu, "Research and Educational Characteristics for the Engineering/Technology Management Discipline," *IEEE Transact. Engineering Management*, **37**(3): 172–176, 1990.
4. K. I. Pelc, "Knowledge System of Engineering and Technology Management," in *Technology Management: The New International Language. Proceedings of PICMET '91*, D. F. Kocaoglu and K. Niwa, eds., IEEE, New York, 1991, pp. 550–553.
5. K. I. Pelc, "Knowledge Mapping in Technology Management," in *Management of Technology IV: The Creation of Wealth*, Tarek M. Khalil and Bulent A. Bayraktar, eds., Institute of Industrial Engineers, Norcross, Ga., 1994, pp. 880–889.
6. J. P. Courtial and J. C. Remy, "Towards the 'Cognitive Management' of a Research Institute," *Research Policy*, **17**: 225–233, 1988.
7. P. Healey, H. Rothman, and P. K. Hoch, "An Experiment in Science Mapping for Research Planning," *Research Policy*, **15**: 233–251, 1986.
8. T. Ijichi, T. Yoda, and R. Hirasawa, "Mapping R&D Network Dynamics: Analysis of the Development of Co-Author and Co-Inventor Relations," paper presented at the Annual Meeting of the Academy of Management, Technology and Innovation Division, Dallas, Aug. 1994.
9. J. A. Swan and S. Newell, "Managers' Beliefs about Factors Affecting the Adoption of Technological Innovation: A Study Using Cognitive Maps," *J. Managerial Psychol.*, **9**(2): 3–11, 1994.
10. W. R. Zhang et al., "A Cognitive-Map-Based Approach to the Coordination of Distributed Cooperative Agents," *IEEE Transact. Systems, Man, Cybernetics*, **22**(1): 103–113, 1992.
11. L. Gasser, "Social Conceptions of Knowledge and Actions: DAI Foundations and Open Systems Semantics," *Artificial Intelligence*, No. 47, 107–138, 1991.
12. R. G. Smith and R. Davis, "Frameworks for Cooperation in Distributed Problem Solving," *IEEE Transact. Systems, Man, Cybernetics*, **SMC—11**(1): 61–70, 1981.
13. S. D. Sheetz, D. P. Tegarden, K. A. Kozar, and I. Zigurs, "A Group Support Systems Approach to Cognitive Mapping," *J. Management Information Systems*, **11**(1): 31–57, 1994.
14. B. Russell, *Human Knowledge: Its Scope and Limits*, Simon & Schuster, New York, 1948.
15. M. Polanyi, *Personal Knowledge*, University of Chicago Press, Chicago, 1962.
16. R. A. Howard, "Knowledge Maps," *Management Science*, **35**(8): 903–922, 1989.
17. D. C. Berry, "The Problem of Implicit Knowledge," *Expert Systems*, **4**(3): 144–151, 1987.
18. I. M. Neale, "Modelling Expertise for KBS Development," *J. Operations Research Society*, **41**(5): 447–458, 1990.
19. W. Bibel, J. Schneeberger, and E. Elver, "Representation of Knowledge," in *Knowledge Engineering*, vol. 1, H. Adeli, ed., McGraw-Hill, New York, 1990, pp. 1–27.
20. K. Niwa, *Knowledge-Based Risk Management in Engineering: A Case Study in Human-Computer Cooperative Systems*, Wiley, New York, 1989.
21. J. P. Martino, *Technological Forecasting for Decision Making*, Elsevier, New York, 1972.
22. R. A. Goodman and M. W. Lawless, *Technology and Strategy: Conceptual Models and Diagnostics*, Oxford University Press, New York, 1994.
23. E. C. Engelsman and A. F. J. van Raan, *Mapping of Technology*, Ministerie van Economische Zaken, The Hague, 1991.

24. M. Callon, J. Law, and A. Rip, eds., *Mapping the Dynamics of Science and Technology*, Macmillan, London, 1986.

25. H. P. F. Peters and A. F. J. van Raan, "Co-word-based Science Maps of Chemical Engineering. Part I: Representations by Direct Multidimensional Scaling," *Research Policy*, **22:** 23–45, 1993.

26. K. Langfield-Smith and A. Wirth, "Measuring Differences between Cognitive Maps," *J. Operational Research Society*, **43**(12): 1135–1150, 1992.

27. J. W. Forrester, *Collected Papers of Jay W. Forrester*, Wright-Allen Press, Cambridge, Mass., 1975.

28. E. B. Roberts, ed., *Managerial Applications of Systems Dynamics*, MIT Press, Cambridge, Mass., 1984.

29. E. B. Roberts, *The Dynamics of Research and Development*, Harper & Row, New York, 1964.

30. P. M. Senge, *The Fifth Discipline*, Doubleday, New York, 1990.

31. F. Zwicky, *Morphological Astronomy*, Springer-Verlag, Berlin, 1957.

32. Z. Janiszewski, "Conseiller pour Autodidactes—Introduction Generale," in *Oeuvres Choisies*, PWN, Warsaw, 1962, pp. 212–257.

33. N. Dunford and J. T. Schwartz, *Linear Operators, Part I: General Theory*, Interscience (Wiley), New York, 1964.

34. D. B. Lenat and E. A. Feigenbaum, "On the Thresholds of Knowledge," *Artificial Intelligence*, No. 47, 185–250, 1991.

35. A. Mehrez, G. Rabinowitz, and A. Reisman, "A Conceptual Scheme of Knowledge Systems for MS/OR," *OMEGA Internatl. J. Management Sci.*, **16**(5): 421–428, 1988.

CHAPTER 14

THE PROCESS OF DEVELOPING AN R&D STRATEGY

Robert Szakonyi, Ph.D.
Director, Center on Technology Management
IIT Center
Chicago, Illinois

14.1 INTRODUCTION

The selection of R&D projects is the most important of all R&D management activities. Unless an R&D organization is working on the right R&D projects, all its efforts could come to naught. Consequently, most R&D organizations spend a great deal of time trying to select the right projects.

What is seldom seen in these efforts to pick the right R&D projects, however, is the importance of having an overall R&D strategy before selecting R&D projects. An R&D strategy is important because it helps an R&D organization select projects in terms of a broader perspective rather than just piecemeal. When an R&D organization has an overall strategy—and thus has some strategic goals—it will more likely select projects that match its technical strengths, the capabilities of its company, and the demands of the marketplace.

In order to develop an R&D strategy, certain preconditions must be in place. These preconditions will be described first. Then some examples of R&D strategic goals will be presented. Finally, the limited progress that most R&D organizations have made in developing an R&D strategy will be examined.

14.2 PRECONDITIONS FOR EFFECTIVE DEVELOPMENT OF AN R&D STRATEGY

In order to develop an R&D strategy, an R&D organization first must make sure that certain conditions are in place. There are seven preconditions (prerequisites) for developing an R&D strategy:

1. A perception that an R&D strategy can solve a problem

2. Creation of a planning staff within a large R&D organization—or the establishment of a strong commitment by the R&D line managers in a small R&D organization to devote enough effort to doing R&D strategic planning

3. A way of linking R&D strategic planning to R&D operations

4. A way of getting strategic marketing done

5. The active support from senior business managers

6. One or more previous efforts to develop an R&D strategy

7. A series of concrete efforts that produce tangible results on their own, but also allow R&D people to get better at R&D strategic planning

14.2.1   Perception That an R&D Strategy Can Solve a Problem

Although the idea of developing an R&D strategy may be accepted, usually the development of an R&D strategy will not come about *unless* an R&D strategy is perceived as solving a problem. That is, the effort that goes into developing an R&D strategy—and the conflicts that may occur about how R&D resources should be allocated—are so great that unless an R&D organization needs to have an R&D strategy in order to solve a problem, it rarely will make the effort or face the conflicts.

On the whole, R&D organizations require an R&D strategy when they *develop* technology for future products. If, on the other hand, an R&D organization focuses almost solely on utilizing existing technology to develop new products, then it will probably not need an R&D strategy.

Usually, R&D managers will recognize that an R&D strategy will help solve a problem when they have doubts that R&D resources are being used in the right way. For example, the head of an R&D organization in a chemical company championed the need to develop an R&D strategy because she was anxious about her R&D organization's ability to deliver the new businesses that senior business managers expected from the R&D organization. When this R&D organization's company had sales of $100 million, this R&D manager was always confident that her R&D organization could develop one new business worth $40 million each year. After the company reached $1 billion of sales, however, this R&D organization was expected to develop four new businesses, each worth $40 million every year. Although this R&D manager was confident that her R&D organization could develop one or two of these new businesses each year, she was not confident that it could develop four of them per year. In addition, at this time the R&D organization was also under pressure to be more productive in its use of R&D resources. For these reasons, therefore, this R&D manager perceived developing an R&D strategy as a way of solving her problems related to what the R&D organization was expected to produce.

The R&D managers of a food-processing company perceived the need to develop an R&D strategy for other reasons. They were forced to consolidate within one laboratory all the R&D on coffee that previously was carried out in several laboratories. To accomplish this task, they recognized that they had to have an R&D strategy to help establish priorities and coordinate all the R&D being done on coffee. Later these R&D managers perceived the need to develop an R&D strategy related to R&D on beverages when they were forced to consolidate the R&D being done on beverages in a few laboratories.

In sum, although developing an R&D strategy is a nice idea, this idea usually will not be put into practice unless R&D managers perceive that an R&D strategy will solve a problem.

14.2.2 Creation of a Planning Staff within a Large R&D Organization

Creating a planning staff or establishing a strong commitment by the R&D managers in a small R&D organization to devote enough effort to R&D strategic planning is essential. Although line managers within a large R&D organization by themselves can develop an R&D strategy, in practice, if there is no planning staff to facilitate the development of an R&D strategy, there will seldom be an R&D strategy.

For example, in a natural-resources company R&D managers have recognized the need for an R&D strategy for quite a while. Because no one was appointed to facilitate the development of an R&D strategy, however, this R&D organization still has not developed one.

In contrast, in a pulp-and-paper company there has been an R&D planning group for 8 years. Although this company now has an R&D strategy, it was not easy to develop one. The R&D planners ran across two types of problems: analytical problems and organizational problems.

The analytical problems surfaced when the R&D planners first attempted to facilitate the development of an R&D strategy. They found that there was no accepted methodology for developing an R&D strategy. Although R&D managers in some companies, consultants, and academicians all had their opinions on what an R&D strategy should be and on how it should be formulated, almost none of them had a complete picture of the process. In addition, the opinions of these various R&D managers, consultants, and academicians often were in conflict with each other or were irreconcilable. Thus, the R&D planners in this company had to develop their own methodology.

In addition, these R&D planners also found that members of the R&D staff were not interested in developing an R&D strategy. Because of this, these R&D planners had to spend 70 percent of their time during the first few years first persuading the R&D staff to do R&D strategic planning and then educating them with regard to how an R&D strategy can be developed.

The experiences of these R&D planners are typical. Someone usually has to develop the methodology to be used in doing R&D strategic planning. Someone also has to be the day-to-day champion of R&D strategic planning, or it will not get done. Theoretically, R&D managers in a large R&D organization can handle these responsibilities. In practice, R&D line managers in a large R&D organization normally have so many other responsibilities that they neglect R&D strategic planning. Thus, an R&D planning group usually proves to be necessary to getting R&D strategic planning done.

In a small R&D organization, on the other hand, R&D line managers are the only ones who can develop an R&D strategy because staff positions rarely exist. Thus, to get R&D strategic planning done, the R&D line managers in a small R&D organization must add the responsibilities of an R&D planning group to their normal responsibilities. These R&D managers usually will not be able to devote much time to developing a planning methodology. On the other hand, because they have responsibility for managing the R&D groups, if the R&D managers in a small R&D organization do develop an R&D strategy, they should have less difficulty in getting this strategy accepted and implemented.

The key to planning in either situation is that the *doers,* not the planners, must do the planning. The role of the "planners" is to facilitate the planning process, which is an important, although seldom appreciated, role.

14.2.3 Linking R&D Strategic Planning to R&D Operations

To get an R&D staff to develop an R&D strategy, R&D planners (or R&D line managers) have to find ways to relate R&D strategic planning to R&D operations. This connection between R&D strategic planning and R&D operations has two aspects.

1. Members of the R&D staff have to be able to see that their interests can be served through developing an R&D strategy. Thus, R&D planners must find a mechanism that involves R&D strategic planning and at the same time serves the interests of the R&D staff. For example, at a candy company R&D planners established a cross-disciplinary forum involving a variety of R&D people. One of the purposes of this forum was to get these R&D people to talk with each other. Although they all carried out R&D on chocolate, they seldom talked with each other. By chance, this cross-disciplinary forum turned out to be a useful mechanism not only for improving communication, but also for getting R&D strategic planning accepted. Once involved in this interdisciplinary forum, these R&D people became interested in coordinating their technical activities. They lacked, however, a common language to describe their technical activities and an analytical framework according to which they could evaluate their various technical activities. With the help of the R&D planners, they learned how to use the tools of R&D strategic planning, which helped them in both areas. Thus, through this mechanism R&D people not only found a way to coordinate their technical work but also, in the process, learned how to do R&D strategic planning.

2. The R&D projects that are actually selected and carried out have to be linked to the R&D strategy to be meaningful. In other words, an R&D strategy is not worth much if it does not affect which R&D projects are selected and carried out.

An R&D organization in a household products company addressed this problem by viewing the development of an R&D strategy as involving two phases. During the first phase the senior R&D managers defined the overall direction of the R&D strategy. During the second phase middle-level R&D managers defined the specifics of the R&D strategy through the projects they selected and carried out.

A rule of thumb for judging whether R&D strategic planning is linked to R&D operations involves looking for a connection between any plan and the budget. Unless a plan is linked to the use of R&D resources, a plan is only an analytical exercise.

14.2.4 Getting Strategic Marketing Done

Although it is important for an R&D organization to link its strategic planning to its operations, this is not enough. For an R&D organization's strategic plans to pay off, those strategic plans must be clearly linked to customers' future needs. Therefore, besides doing R&D strategic planning, an R&D organization must also get strategic marketing done in its company. Two of the hard questions that must be addressed in a strategic marketing are

1. Who will the company's customers be in the future?

2. What will those customers in the future need?

In answering these questions well, one must understand economic, regulatory, demographic, and social trends and opportunities.

Three approaches that R&D organizations have taken to get strategic marketing done are described below.

1. In a chemical company an R&D planner also serves informally as the strategic marketing planner. This R&D planner gained these responsibilities—at least informally— because (a) no one in marketing was doing the job and (b) this R&D planner had the capability of dealing with marketing questions. This R&D planner spends 40 percent of his time dealing with marketing questions, such as what the trends may be like in 3 to 5 years and how the public may think about environmental issues in the future.

2. In a dairy-products company, a marketing person who is knowledgeable about technology was assigned to the R&D organization to do strategic marketing as part of investigating opportunities for new-business development. This marketing person leads a group made of four technical people at the company's headquarters and six other technical people who work in various regions of the country. An example of one of this group's goals is to help the company go from selling food products that are commodities to selling specialty products in other areas, such as selling fats and oils to the cosmetic industry. Good strategic marketing is one of the keys to helping this company make such a transition.

3. At an appliance company market research people are invited to serve on R&D project teams. The advantage of this approach is that many individual market research and R&D people communicate with each other. The disadvantage is that because each project team focuses on only a specific area, strategic marketing is done in a piecemeal fashion. The challenge in making this approach work is getting the market research and R&D people to actually develop the close collaboration that teams are supposed to foster. One way in which this company tries to make this happen is through conducting very critical reviews of project plans. Not only senior R&D managers but also senior business managers ask difficult questions related to the market implications of the R&D being carried out.

14.2.5 Active Support of the Senior Business Managers

To get its R&D strategy integrated with business plans, an R&D organization must have the active support of senior business managers. To gain the active support of senior business managers, an R&D organization must explain the value of R&D in business terms—for example, with regard to how R&D will help the company (a) satisfy customers' needs, (b) cut costs, (c) expand into new markets, or (d) minimize distribution problems.

Those R&D organizations that have been able to gain the active support of senior business managers for their R&D strategy were helped by organizational factors.

For example, in a chemical company the R&D organization was able to gain the active support of senior business managers because the chief executive officer is one of the major proponents of R&D in the company. As opposed to most of the other senior business managers in this company, this chief executive (CEO) previously managed the division of the company that sells to industrial customers. Because the customers of this division are more knowledgeable about what R&D can contribute (e.g., many of these industrial customers want their suppliers to do more R&D), this CEO realizes how valuable R&D can be and thus has not only supported R&D strategic planning but also has actually encouraged the R&D organization to do it.

At an aerospace company, the R&D organization was able to gain the support of senior business managers because it deals with strategic business executives rather than the managers of the individual businesses. These strategic executives gained their position through a reorganization. The management positions that they were given were created to make sure that the strategic issues that normally lie outside the interests of the managers of the individual businesses are addressed effectively. Therefore, one of the strategic business executives' main responsibilities is to approve the individual business managers' business plans, particularly with regard to how these business managers plan to utilize new technology to grow their businesses. Having senior business managers with this kind of assignment has made it much easier for the R&D organization to gain the support of senior business managers.

At an electronics company, a key factor that allowed the R&D organization to first gain and then maintain the support of senior business managers was continuity in leadership among senior business managers. Continuity in leadership was important because it took a few years for the senior business managers to appreciate the benefits of having an R&D strategy. By learning year after year what the R&D organization was trying to accomplish with an R&D strategy, these senior business managers were able to appreciate the benefits. The R&D organization, in turn, could build on the progress that it had made with these senior business managers in previous years.

14.2.6   One or More Previous Efforts to Develop an R&D Strategy

An R&D organization's first efforts at developing an R&D strategy normally will not yield the results than the R&D organization had desired. Unavoidably, there is a learning process through which an R&D organization has to go. This learning process may take at least 3 to 5 years. Whatever the period of time, the key to learning is having a commitment to accomplishing something better, such as being able to allocate R&D resources more effectively. Without this commitment, false starts will make the process of developing an R&D strategy seem hopeless. With this commitment, an R&D organization becomes open to learning from these false starts.

For example, in looking back at his R&D organization's first efforts at developing an R&D strategy, the head of R&D in a chemical company called these first efforts "naive." To use his words, during the first year of R&D strategic planning his R&D organization just categorized the obvious. In the second year, his R&D organization began understanding R&D strategic planning better and consequently gained new insights. For example, it recognized for the first time that new technology for improving the company's manufacturing processes was far more important than it had previously thought. After 5 years of doing R&D strategic planning, the R&D organization in this company has more-or-less institutionalized the process of R&D strategic planning. One of this R&D manager's greatest challenges now is maintaining the vitality of the process so that the R&D people continue to learn new things by doing R&D strategic planning.

14.2.7   Concrete Efforts that Produce Tangible Results on Their Own

One way in which an R&D organization can maintain the vitality of the R&D strategic planning process is through carrying out a series of concrete efforts that produce tangible results on their own, but also allow R&D people to improve their R&D strategic planning expertise. The experiences of an R&D organization of a candy company illustrate what a series of such efforts might be like.

The first effort consisted of the establishment of a cross-disciplinary forum involving R&D people who looked at different technical aspects of chocolate, as mentioned above. This effort was successful because it allowed the R&D people to coordinate their activities. It also introduced them to the tools of R&D strategic planning.

A second effort involved a benchmarking study in which the R&D people evaluated the strengths and weaknesses of 36 of their company's technologies in relation to the strengths and weaknesses of six competitors' technologies. The R&D people gained two things through this effort: (1) they gained a much better understanding of how their company stood in relation to competitors and (2) they learned how to analyze their technologies. For example, they learned how to think more precisely about how their technical work could improve the performance of the company's products, such as enhance the flavor of cocoa beans.

A third effort that the R&D organization is considering involves using the techniques of portfolio management to evaluate the potential and payoffs of various technologies. Since the effort is not under way yet, the exact benefits that would result are not yet known.

As illustrated, the key to taking each of these steps is picking a problem that the R&D organization is facing that also calls for a more systematic analysis of the use of R&D resources. The value of carrying out a series of concrete steps is that together they serve as stepping stones to improve R&D strategic planning. Moreover, because they produce tangible results along the way, they also help elicit support for the R&D strategic planning process.

14.3 EXAMPLES OF SOME STRATEGIC GOALS OF R&D ORGANIZATIONS

Some strategic goals of R&D organizations are described below. For example, the R&D organization of a household cleaning products company identified infection control as one of its major strategic goals because it realized that 80 percent of its products involved infection control in some way. The value for this R&D organization of focusing on infection control is that by emphasizing the prevention of infection, rather than just the removal of dirt and stains, this R&D organization's company can better differentiate its products from the competitors'.

In order to make this strategy viable, the R&D organization had to demonstrate to the marketing department and to senior business managers the value of emphasizing infection control. As part of its mission, this R&D organization educates the general public about infection control through a variety of seminars. It also sponsors many clinical studies that investigate the conditions underlying infection control.

One of the factors that led this R&D organization to focus on infection control was having a head of R&D who is a biochemist rather than chemist. For over 15 years, the heads of R&D in this company had been chemists. Partly for this reason they had focused more on typical areas of new-product development in a consumer product industry. Having a head of R&D who viewed the issues related to new-product development differently laid the basis for developing new strategic goals in R&D.

The R&D organization of a consumer product company which produces toothpaste has identified one of its strategic goals as cavity prevention. Having such a strategic goal means much more than just improving toothpaste in order to cut down on tooth decay. It also means researching how cavities could be prevented with or without the use of fluoride, in case government regulations ever force the company to change what it puts in toothpaste. Researching how to prevent cavities also involves considering how less well-known, but still important, kinds of tooth cavities can be prevented. For example, one challenge that this R&D organization is addressing now has to do with preventing tooth decay in the roots of teeth, which normal brushing of teeth cannot prevent.

R&D organizations of food-processing companies also have identified strategic goals. For example, an R&D organization in a soup company is focusing on new ways of preserving foods. Because many of the company's products are preserved through chemical means (which result in an excess of fats and oils), this R&D organization has set a goal of helping the company produce food that is preserved through other ways.

One of the strategic goals of an R&D organization of a breakfast cereal company is the improvement of packaging. Rather than be bound by the traditional packaging solution involving a box and a liner, this R&D organization is looking for ways to package breakfast foods in a simpler way.

Finally, an R&D organization that supports company businesses related to the retail sale of cooked food has the strategic goal of remaking the company's businesses by developing a system of precooked food for the retail outlets. A new system of precooked food would improve the quality and variety of the company's products. In addition, it would allow the staff in the retail stores to not be so tied to the manufacturing aspects of cooking food and instead to focus on serving the consumer.

14.4 LIMITED PROGRESS OF MOST R&D ORGANIZATIONS IN STRATEGY DEVELOPMENT

Few R&D organizations have an R&D strategy—or have established the preconditions required for developing an R&D strategy. For example, few R&D organizations have perceived an R&D strategy as a way to solve a problem. Many large R&D organizations do not have an R&D planning group, and the R&D managers of most small R&D organizations do not do R&D strategic planning. Also, few R&D organizations have found a way to get strategic marketing done. Finally, in most companies, efforts to do R&D strategic planning do not have the active support of senior business managers.

In addition, if one looks closely at the strategic goals of those R&D organizations that have them, one also finds that many of these goals were arrived at intuitively, not analytically. That is, many of the strategic goals that R&D organizations have are intuitively obvious. For example, most of the strategic goals that food-processing R&D organizations have formulated are similar and quite predictable: (1) to ward out fundamental threats to the company's businesses, (2) to improve food safety, and (3) to develop salt-free and low-fat foods.

Few R&D organizations have conducted benchmarking studies aimed at comparing their technologies to companies' technologies. Few R&D organizations carry out technology forecasting studies aimed at understanding (1) what technological advances may be occurring in the next 5 to 10 years and (2) what those technological advances may mean.

Even those R&D organizations that have made significant progress in developing an R&D strategy admit that they still have much to do to improve the R&D strategic-planning process. For example, the R&D organization of an electronics company has an R&D strategy, but this R&D strategy is not accepted by the managers of the company's businesses. The R&D organization of a chemical company also has an R&D strategy, but it sets priorities only *within* product categories, not *across* product categories. Because much of the value of an R&D strategy lies in setting priorities across product categories, this R&D organization still has much work to do in getting R&D strategic planning accepted. The R&D organization of an appliance company also has developed an R&D strategy, but it has not found a way to integrate its R&D strategy with the marketing and manufacturing strategies in the company. In addition, parts of the R&D strategy in this company can be altered at the whim of anyone who at a later date becomes involved in the planning process—without necessarily informing anyone else about the changes in the plans.

In sum, some R&D organizations have made significant progress in developing an R&D strategy. Most R&D organizations, however, have barely started developing an R&D strategy. The most immediate challenge facing most R&D organizations, therefore, has to do with first establishing the preconditions required for developing an R&D strategy. After doing this, they will then be able to meaningfully deal with issues as to what the R&D priorities should be and how R&D resources should be allocated.

CHAPTER 15

DECISION SUPPORT SYSTEMS IN R&D PROJECT MANAGEMENT*

A. W. Pearson

Director of R&D Research
University of Manchester
Manchester, England

M. J. W. Stratford

Visiting Fellow
Manchester Business School
Manchester, England

A. Wadee

Eltec

A. Wilkinson

Administrative Director
Manchester Business School
Manchester, England

15.1 INTRODUCTION

The historical background of *decision support systems* (DSSs) in R&D is extensive in the literature, but limited in practical usage in industrial situations. This chapter reviews the literature and sees hope in the growing use of information technology to cope with the real complexity of the problem. It also reviews some specific research on a new DSS, the development of which has thrown fresh light on possible reasons for the past lack of success. As a result, conclusions are drawn which should be of benefit in the development of further systems in the future.

*The research on which this chapter is based was supported through a CASE award provided by the U.K. Science and Engineering Research Council. The industrial sponsor was the pharmaceutical division of ICI plc.

15.2 BACKGROUND IN THE LITERATURE

15.2.1 Early Work

During the last 25 years, management scientists have expended substantial effort in developing, testing, and implementing a variety of quantitative techniques for R&D management in general, and particularly for project management involving the screening, evaluation, selection, budgeting, scheduling, and control of R&D activities.

In the early 1960s methods such as *program evaluation and review techniques* (PERTs) and *critical-path methodologies* (CPMs) had found extensive use in R&D project scheduling.[1,2] Further developments led researchers to add stochastic flexibility in network modeling, resulting in *graphical evaluation and review techniques* (GERTs),[3] and their extensions; and, in addition, to Monte Carlo and *general-purpose systems simulation* (GPSS) techniques.

Other techniques in common use at the time were aimed at rank-ordering the available candidate projects so as to select the best ones. Profile models[4] rated candidate projects subjectively on a series of criteria, displaying the results visually for easy comparison. In checklist and scoring models,[5-7] a total value can be developed for each project, with the various criteria being weighted to modify the total rating of a project. Economic models[8] rank projects on criteria such as rate of return and present worth and this has led to the use of *constrained optimization models,* which seek to optimize some objective function subject to specified resource constraints.[9]

The late 1960s saw the appearance of risk-analysis methods which typically are based on a simulation analysis of input data in distribution form and provide output distributions of such factors as rate of return and market share.[10] Unfortunately, with this series of methods, as the need of data is increased, the complexity of treatment is also increased, thereby reducing the ease of use.

15.2.2 More Recent Work: Objective Functions Representing Multiple Objectives

One direction in which research on R&D decision making moved was toward the construction of objective functions to represent multiple objectives or attributes for use under either certain or uncertain conditions. As mentioned earlier, checklists or product profile charts addressed the issue of multiple objectives or attributes directly by rating individual projects in terms of a number of attributes, and then comparing the projects by inspection of the sets of attribute rates.[11] Scoring models attempted to improve this procedure by establishing weights to be assigned to each objective so that the project ratings on different attributes could be combined into a single measure of project merit.[12]

The literature on decision making has set out conditions under which projects could be evaluated on the basis of multiple objectives or attributes and the resultants combined into a single measure of project merit that reflects the decision maker's preference on values. If the objective function included the decision maker's feeling toward risks, the objective function was referred to as a *multiattribute utility function* (MAUF). However, if the decision maker's preferences were modeled under certain conditions, the objective function was referred to as a *multiattribute value function.*[13]

Conditions were derived under which the MAUF might be used in R&D project management.[14,15] This was done by assuming that decision makers were rational and sought to minimize their expected utility by properly allocating resources among research activities.

The contribution of *multiattribute utility theory* (MAUT) to R&D was said to be twofold. The theory recognized the problems of projects in which different scales were appropriate for measuring the attributes of the decision maker, and that the decision was not indifferent to the uncertainty surrounding the value of the R&D projects. Quite often, decision makers were reluctant to express their feelings on a common scale such as dollars. MAUT defined the conditions under which R&D project values could be ascertained on separate scales and the results combined into a single measure of project merit. In addition, the methods of MAUT were designed to capture the decision maker's feelings about taking risks. One would expect decision makers to be averse to taking risks unless the expected returns in the risky situation were significantly higher.

The most serious shortcoming of this approach was its data requirements. Decision makers had to provide a considerable amount of information concerning how they felt about each attribute and the relative importance of each attribute compared with the others. Another difficulty with this approach was that it assumed that every decision maker was qualified to evaluate the program or project on each attribute being considered. Typically the evaluation of a research proposal might cover a broad range of technical issues and economic questions as well as personnel, managerial, and finance considerations. It might then be totally inappropriate to assume that each decision maker was qualified to address all the questions as assumed in the multiattribute decision approach.

The behavioral issues raised by the assumptions made in the MAUT were taken up,[11] and it was argued that a fundamental shortcoming in most project selection models was that they, indeed, ignored the behavior of people in organizational settings. Project selection decisions, they claimed, were often performed in organizational and group environments that were influenced by human emotions, human desires, and departmental loyalties.

Challenged by the need to take these aspects into account, some project selection model builders have proposed two approaches: *behavioral decision aids* (BDAs) and *decentralized hierarchical modeling* (DHM). In the BDA approach, structured process aids were overlaid on carefully selected combinations of classic and/or portfolio models to "upgrade" their organizational settings.[16]

In the DHM model, which originated in mathematical programming theory, the involved parties had a dialogue through computer terminals, iterating until a consensus was arrived at. Typically, the information review process was recycled several times until all parties came to an agreement.[17]

Like the BDA, the DHM approach focused on a nontraditional philosophy of project selection: the use of operations research to catalyze new depths of interdepartmental and interpersonal interaction.[11]

More recently a knowledge-based decision support tool for project assessment has been described.[18] This has been used very successfully for the decision to go to full-scale development in a project. It makes use of the *analytical hierarchy process* (AHP) in place of MAUA (multiattribute utility analysis) as discussed below for the evaluation of projects with some simple rules built in. Proposals were made for obtaining feedback from case histories on projects, both successful and failed, and translating it into rules for KBDSSs (knowledge-based decision support systems).

15.2.3 The Use of Evaluation or Selection Procedures in Industry

Despite all these techniques, there have been few empirical studies of their use in an industrial setting. The R&D evaluation and control procedures used in British companies in the late 1960s were investigated using a mail survey, supplemented by some personal visits.[19] The following conclusions were based on 112 responses:

1. Most organizations used at least one standard financial method for project evaluation.
2. About one-third used mathematical models and/or weighted checklists or project ranking indexes as part of their project assessment procedure.
3. Only 26 percent did not use any scheduling techniques.
4. PERT and CPM were used by about 60 percent of respondents, with heaviest usage in the fine chemicals, dyestuffs, and pharmaceutical industries.

The study further concluded that although many of the "textbook" methods were used in most large organizations, the extent of use was limited. Also, several organizations expressed dissatisfaction with the available resource allocation and scheduling procedures and discontinued their use.

The growth in available quantitative methods, and advances in computer technology for data acquisition, storage, retrieval, and analysis, led to a more recent study on management science practices in R&D management.[20] A nonrandom sample of 40 respondents from 29 major U.S. industrial firms from the *Fortune 500* group was selected to represent various industrial sectors. Nearly all the respondents used a few of the standard measures of financial analysis. PERT and CPM found selective use almost exclusively for development and engineering projects. R&D departments which performed new product-process and/or embryonic-exploratory research exclusively seemed to use less formalized techniques, and sought rough screening and scheduling methods. Many respondents were dissatisfied with their project monitoring, scheduling, and control, but several expressed interest in using user-friendly interactive systems for resource allocation, and multiproject tracking and control, while others expressed interest in improved DSS techniques and systems, preferably involving project termination criteria and multiproject control methods.

The study concluded that several changes may well occur over the next few years, increasing the familiarity and usage of quantitative project management techniques by R&D budget heads and high-level staff. This will lead management scientists to develop techniques to assist the decision-making process rather than attempt maximization.

Such a change was conjectured during the 1970s:[21] "The trend in application appears to be away from decision models to decision information systems." Thus the optimization phase of the involvement of management sciences in R&D project management has developed understanding but has had limited success in real assistance to management. There is, however, a basis for moving into the decision-aid (knowledge-based) approach.

From the empirical studies discussed above, it can be concluded that apart from the use of some financial analysis techniques and scheduling techniques (e.g., CPM), the various project selection methods discussed earlier (scoring methods, linear programming, etc.), although providing valuable insights and uses in particular circumstances, have not been subject to a general acceptance by R&D managers. Indeed, there is no great evidence of some of these models being used in industry.

In the study referred to earlier,[19] it was concluded that in many large organizations the use of textbook methods was limited. The reservations were that, while many models were adequate for part of the selection process, they were of limited further use. The reason may well be that little attention had been focused on what exactly a project was and what it involved. The concept that a project is a single entity proceeding from its inception to its completion, and the assumption that the same criteria for evaluation and termination processes apply for the whole of the project, must now be questioned.

It would therefore be of interest if DSSs were designed to provide guidance, with clearer insights into the evaluation process for the whole of a project's life cycle, and also supportive evidence, or otherwise, on the suggestions that the individual phases of a single overall project should be judged by different criteria.[22]

15.3 SPECIFIC TECHNIQUES USED LATER IN THIS CHAPTER

15.3.1 Introduction

In the development of a practically useful decision support system, three pieces of research and/or practice were considered to be of potential value. These were multiattribute utility analysis (MAUA),[23,24] *discriminant analysis,*[25,26] and *distance from target,*[27] which are covered more fully in the following sections.

15.3.2 Multiattribute Utility Analysis

The basic question raised by MAUA is how best to address relative preferences for decisions that can be represented by some aggregated value of attributes. Several methods for assessing such relative preferences for multidimensional values have been suggested in the literature.[28,29] MAUA is currently regarded as being one of the best and most widely used methods for evaluating a number of choice options in the face of many evaluation attributes. The attributes can be arranged in a hierarchical fashion.

Creating a hierarchy of evaluation criteria is advantageous because it enables the user to disaggregate highly complex and generic attributes into their measurable components. The user systematically judges the value of each alternative on a 0 to 100 utility scale for each bottom-level attribute: 0 and 100 are the so-called anchor points defining respectively the worst and the best situations for each attribute. Word models are set up at, say, 25-point intervals on each 0 to 100 scale to give a clearer definition of the scale. The user then judges the relative contribution of each attribute to the whole by determining the weights at each level in the hierarchy.

Working through this systematic procedure permits the user to make a small number of relatively simple judgments to determine the relative value of the alternatives. The necessity for the user to make unaided the highly complex and often unreliable, overall judgment of preference between alternatives is avoided. The score at any level in the hierarchy is simply the weighted sum of scores of attributes at that level. Repeating this calculation at each level in the hierarchy gives the total score for the option.

The use of a hierarchy of criteria and scoring on utility scales with the aid of word models is seen as a method giving sharper definition with reduced ambiguity to a process involving considerable subjectivity. Numeric data should be used in defining 0 to 100 utility scales wherever possible.

An important point that needs to be highlighted is that for the procedures outlined above to be used with confidence, assumptions such as utility independence and preferential independence, as mentioned earlier, should be verified, as well as the measure of sensitivity related to the final output figures.

Considering the technical aspects of *project portfolio management* (PPM), a basic

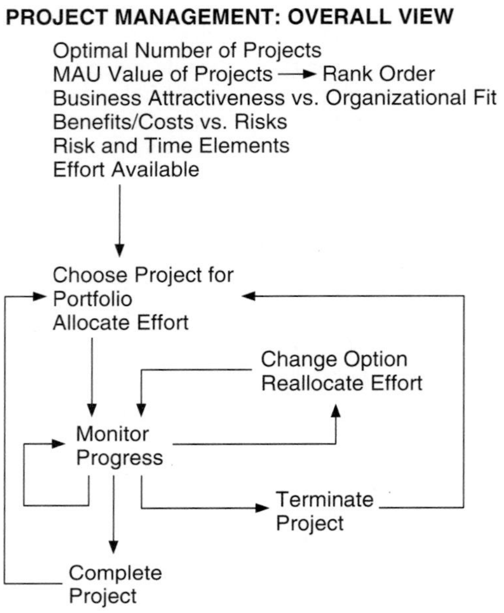

PROJECT MANAGEMENT: OVERALL VIEW

Optimal Number of Projects
MAU Value of Projects ⟶ Rank Order
Business Attractiveness vs. Organizational Fit
Benefits/Costs vs. Risks
Risk and Time Elements
Effort Available

Choose Project for
Portfolio
Allocate Effort

Change Option
Reallocate Effort

Monitor
Progress

Terminate
Project

Complete
Project

FIGURE 15.1   Overall view of project management.

model was developed which uses MAUA and is shown in Fig. 15.1. This model gives guidance on how to modify an existing portfolio when projects are either successfully completed or terminated, thereby releasing resources. In the evaluation of projects to which the released resources have to be allocated, the first six factors in Fig. 15.1 are taken into account.[23]

This information is used to advise management on project selection and the appropriate allocation of resources. Monitoring and control of progress on each project can be achieved by tracking the change in MAU value, and expected time to completion, over time, and taking the appropriate management action. If progress is unsatisfactory, a change in research approach and/or effort would first be made and monitoring continued. If progress continues to be unsatisfactory, project termination may be necessary with release of resources and selection of new work. A rule-based expert system (ES) based on MAU values was envisaged for project termination.[23]

Lists of strengths and weaknesses for projects are seen as providing additional useful information for the selection process. A "strength" for a project is defined as an attribute with a value greater than, say, 90 on a 0 to 100 MAU scale and a "weakness" as value less than 10 on the same scale. Successful applications of MAUA in the area of PPM are discussed.[23,24]

15.3.3   Discriminant Analysis

Looking at the factors highlighted in two papers[25,26] as being determinant to project termination, it is clear that certain factors were considered important by both. More distinct information on project termination per se, in the form of a conceptual model and a suggested scheme for applying it, was given in the later paper.[26] It proposed that the decision to terminate an R&D project was distinct from the selection decision both temporally and conceptually—temporally distinct because a number of different factors had to be considered by the decision maker in order to determine whether to terminate the project.

Balachandra and Raelin[26] also stated that an important feature for inclusion in the termination model was an ability to evaluate the changes in the attractiveness of a project since its inception. To assist in this process, a set of 10 quantitative and 13 qualitative factors was developed. Eight of the quantitative factors measured the change in some characteristic of the project by computing the ratio of the new or current value to the old value estimated at the time of project inception or at its last review. The qualitative factors, by and large, measured the changes that had taken place in the environment, including market and organizational dimensions, and these were rated on an ordinal scale.

By questionnaire and personal interview, an extensive list of managerial factors perceived by R&D managers to be relevant to the project termination decision was obtained. A wide range of companies were approached, and each gave data on an equal number of successful and unsuccessful projects. The projects were then divided into two halves, each containing successful and unsuccessful projects. One half was subjected to a discriminant analysis to determine the really important factors and apply to each a discriminant coefficient to arrive at a single equation which would distinguish between successful and unsuccessful projects. They determined such an equation with a satisfactorily high level of discrimination, and this was validated by applying it to the other half of the projects with a 92 percent correctness.

When the DSS reported in Sec. 15.5 (below) was being developed, the work on discriminant analysis[26,27] was available and gave clear guidance on the key parameters for project evaluation. Having become conscious, however, of the dynamic nature of a project, the investigators decided to build into the associated questionnaire the ability to check the discriminant factors. The main purpose in doing so was to see whether they varied with the stage of a project, or with the technology or industry. If evidence of variation became available, then the DSS could be refined to improve the quality of advice to managers.

15.3.4 Distance from Target

The final factor considered was the distance from the target.[28] For this to be used, it is necessary to identify the key parameters which are likely to determine success in a specific project.

This can be done by asking the team to state the "definitive design" in the marketplace which will have to be equaled or bettered if success is to be achieved, and then identify the key determinants. For each such determinant, three values can be sought: the target value, the current value, and the minimum value for the particular factor which would be deemed acceptable if all other factors were at their target level. For each factor this would give two distances: d, for the distance of the current value from target; and m, for the distance of the minimum acceptable level from target. The overall distance from target is then taken to be the square root of the sum of the squares of (d/m) for all the individual factors.

Again, the projects can be rank-ordered by their distances from target, with the better ones having the lower values. At the same time, however, the manager is forced to identify which factors are causing the most difficulty, and clearly these should be the ones for serious management concern and attention.[29]

This approach had the great virtue of focusing attention on industrial parameters and how far short they were of preestablished objectives. This technique was also adapted to identifying "killer" parameters within a project (see Sec. 15.5.4).[30]

15.4 EXPERT SYSTEMS

Research on artificial intelligence led in the 1970s to the concept of expert systems. A useful definition of such a system has been provided.[31]

> An ES is an intelligent computer program that uses knowledge and inference procedures to solve problems that are of sufficient difficulty to require significant human expertise for their solution.
>
> The knowledge of an ES consists of facts and heuristics. The facts constitute a body of information that is widely shared, publicly available, and generally agreed upon by experts in the field. The heuristics are rules of good judgment that characterise expert-level decision making in the field.

Recent ESs are in marked contrast with earlier algorithm-based programs which lacked the present human self-awareness of the techniques now being employed and could not reason about or explain their mechanisms.[32] Now the focus is on knowledge in the ordinary meaning of the term, i.e., facts about the task domain, and heuristics or rules of thumb that guide the use of knowledge to solve problems in the domain. Most current knowledge-based systems are divided not into code and data, but into a corpus of knowledge and a comparatively simple mechanism for applying the knowledge in such a way as to solve the problems.

15.4.1 Building an Expert System

The traditional way of building an ES is to interview the expert(s) and then represent their knowledge base and logic in a program. A prototype is then built on the basis of the results of the interview. This is tested against actual cases and modified after consultation with the expert, until a satisfactory performance is achieved. Such an approach was considered, for example, to build a system for selecting a statistical tool.[33]

However, this "rapid prototyping" approach is inherently very time-consuming, and satisfactory performance is rare in practice.[34] Knowledge acquisition, and knowledge representation or processing, have both been regarded independently as the determining factor in the performance of an ES. In fact, both are essential.

Acquisition should not be guided initially by the representation scheme (as in the case of the traditional approach), but should reflect the expert's way of problem solving. A representation scheme can then be chosen on the basis of this initial knowledge. This approach has been termed *slow prototyping*,[34] based on work on the development of a methodology for knowledge acquisition.[35]

15.4.2 Choosing an ES Inference Engine

Two tools are currently available for constructing ESs: shells and high-level programming-language environments. Both suffer from a number of shortcomings which limit their use. *Shells* are usually obtained by abstraction of the knowledge base specific to the problem of an existing expert system and thus normally consist of an inference

engine and an empty knowledge base and some usually primitive debugging and explanation facilities. Claims for the breadth of utility of shells are often exaggerated.[36]
The commonest complaints about shells[37] are

1. The inference engine which was successful in one application need not be successful in other applications.
2. The knowledge representation scheme often makes the expression awkward, if not impossible.

In other words, shells have proved too rigid, and it is difficult for *one* scheme to cover a large number of domains and tasks.[37] In *high-level programming environments,* such as KEE,[38] ART,[39] or LOOPS,[40] the "knowledge engineer" is provided with programming tools in order to construct an inference engine with a minimum of difficulty. While systems such as KEE are useful as tools for program development, they are difficult to use. They provide the knowledge engineer with a bewildering array of possibilities and little guidance for their use. There is therefore a need for close understanding of the problem by the programmer.[41] There are no easy solutions, and a wide awareness of the shells and languages available is required so that a combination might yield the optimum result in generating a new system.[36]

These problems were very apparent in the DSS development work described below, which began by experimenting with shells, and later some of the languages. It was difficult to see how the complexity of questionnaire structure and analysis was to be achieved easily and yet meet the great desire to be readily available to a wide range of managers with varying computer skills.

One of the "experts" accepting a primary role as a systems designer suddenly realized that it was possible to write the whole DSS structure, as then conceived, in Lotus 1-2-3, albeit with some quite complex macros built into the system. This was a tremendous advantage, giving broad acceptance, since many in the R&D field are familiar with the Lotus software or one of the closely related packages. We could not find any evidence for Lotus having been used before for handling a questionnaire with immediate analysis of responses, but it worked extremely well.

15.5 THE DEVELOPMENT OF A NEW DSS

15.5.1 Introduction

The work which is discussed here had its origins in a working group of R&D managers who wished to explore the use of expert systems in their own organizations. Considerable knowledge about such systems was available within this group, who then focused their attention on the expertise needed to enable them to apply these systems effectively within the R&D function.

In this section we will discuss the progress of this work leading to the development of a system which has now been tested in a wide variety of situations and is being used as part of the planning process within a major technological organization.

15.5.2 Gathering the "Expertise"

Having chosen "project selection" as the first subject, the panel of R&D managers, the "experts," set out to share the approaches followed by their individual organizations.

While differences were expected, no one was prepared for the almost total lack of harmony between the individual approaches. Against this, it had to be assumed that none of the approaches was wrong since each organization could point to past and present success. After a number of meetings, the following conclusions were drawn:

1. *There had to be some rationale for the differences.* The length of a typical research project for a particular company, and the capital intensity of the project, seemed to be the key determinants dictating the differences of approach. The two extremes present within the experts were a pharmaceutical company often thinking of a 12-year project life, and a food products company with numerous projects of 6- to 12-month duration. As a consequence, the decision-making processes also vary in that, for the pharmaceutical company, many of the decisions in the early stages of the project are substantially within R&D, whereas multifunctional decisions dominated the food company's approach to managing a project from its inception.

2. *Any system would have to reflect these differences if it were to be of broad acceptability.* Rationalizing the process after the event, the approach was to frame the structure of the system, and the questions and statements within it, using terms that were clearly understood by managers using the system, which were imprecise but capable of being scaled in a qualitative way. For example, an axis labeled "time to complete project" in the eventual model, with a range from high to low, can be interpreted intelligently but with a totally different absolute scale by the managers within the two industries quoted in the previous paragraph.

3. *A clearer definition of objectives was desirable.* The initial concept was to create an expert system to help in the selection of projects. This clearly raises questions as to what "help" means. Who is helping whom? Was the system to give a definite answer, or be advisory to a manager? If the latter, what was the nature of the advice? Was the manager being advised only within R&D, or was the system to be open to use by any function, and, indeed, might it be a means of communication between functions? It was decided that the expertise to be captured was that of the senior managers in R&D with substantial experience of the process, and the system was to be advisory to more junior managers or project leaders who might be employed primarily in any function. It was also seen as desirable that the system should provide the means of comparison of one project against others so that priorities for budget allocations could be established systematically. It was at this point that attention turned from an ES toward the development of a decision support system.

The initial lack of agreement by the experts on the selection process meant that the first need was to agree on the parameters to be used in assessing a project, and to develop a structure within which to consider these parameters. It was further assumed from the beginning that for a successful project, there would be a range of acceptable values for each parameter. Furthermore, the parameters and the ranges of acceptability might vary with the nature of the project and its corporate environment. Such variations in treatment were rarely envisaged in the work reviewed in Sec. 15.2.

15.5.3   Stage: 1: Classification of Projects

Previous work had not really classified projects by their nature, so this became a serious point of discussion in considering individual parameters. Eventually a simple three-dimensional model based on the axes in Fig. 15.2 was developed. When tested against a number of projects, this framework was seen to be very helpful. Given, how-

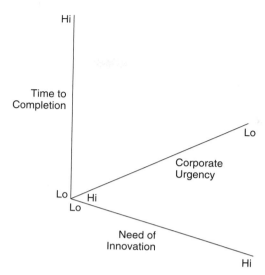

FIGURE 15.2 Coordinate axes for three-dimensional DSS model.

ever, that there could be a continuous scale on each axis of variable units, the experts initially considered the model reduced to a cube with a 2 × 2 × 2 matrix as in Fig. 15.3. This led to desk research in which the model was used to classify projects differing in nature, and Fig. 15.4 gives the kind of analysis which proved to be possible.[22] This encouraged the experts to agree that this model should be pursued.

As the work continued, however, the cube came to be seen less as a means of classifying projects, and more as a tool for assessing progress in the management of a single project. This was triggered by the realization that a project was a dynamic activity within the cube space, starting in the top, right corner of Fig. 15.3 and finishing, if successful, in the bottom left corner. Optimal progress would be along the diagonal, and the extent of deviation from the diagonal would give some measure of failure in managing the project.

Whereas the initial hypothesis was that projects clustering within the cube space, i.e., projects classified as similar, would have similar criteria for success, this changed so that projects clustering together would need similar managerial treatment to move them back toward the optimal track.

A more dynamic perspective of a project became apparent, and in considering individual parameters, it was realized that the significance of any parameter would almost certainly change with each stage within the life cycle of a project. This might seem obvious, but it is a factor substantially overlooked in the earlier work reviewed above (Sec. 15.2). In almost all cases, a project had been taken simply as the complete span from idea inception to completion in an operational process or a commercial product.

This led to a further conclusion, namely, that the structure of any system would have to reflect stages in a project. Provisionally, a list of six stages was developed for exploratory work, and these were:

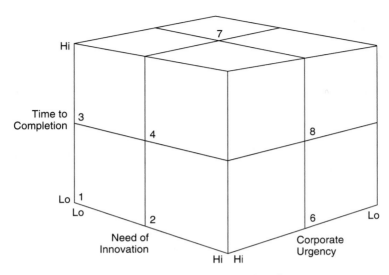

FIGURE 15.3   Three-dimensional DSS model in 2×2×2 matrix.

Idea definition

Exploration and evaluation of the idea

Product or process definition

Pilot plant, or product prototype

Scaleup to full production

Full installation and operation

These stages were not sacrosanct or exclusive, but simply suggestions for evaluation. Subsequent work, however, with a pilot system and several stages in developing the present operational system, have not caused any change from these original six-stage definitions.

15.5.4   Stage 2: Managerial Decisions

The "experts" were R&D directors or senior managers, and they all agreed that a major purpose of R&D was to ensure the continued success of the organization through development of the existing business, and in pursuing innovation into new areas of activity. This had already been recognized in Stage 1, in that two of the axes of the cube relate to a corporate perception of a project. "Corporate urgency" reflects the strength of interest by the organization for the project to be successful so that the product or process, which is the subject of the work, can be utilized in the furtherance of corporate objectives at a time desired within the corporate plan. Furthermore, this time element is a key determinant for the axis "time to completion."

Thus the DSS must be designed to encourage project managers or their supervisors to interact with their colleagues in other functions to monitor a project's continuing relevance within the overall corporate context. The competitive environment must also

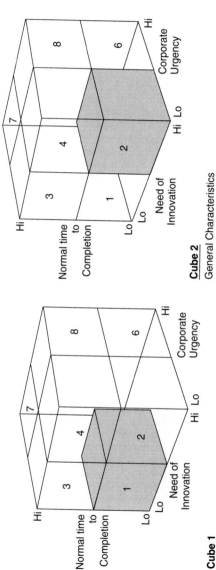

Cube 1

General Characteristics
 – Normally short-term projects
 – Not very innovative ones
 – Need to be implemented quickly

Examples
 – Product modifications
 – Consumer products
 – Industrial blends for effects, or for
   adding to other products

(a)

Cube 2

General Characteristics
 – Normally short-term projects
 – Highly innovative
 – Short time to market

Examples
 – Fashion goods
   Hoola Hoops, Rubik Cubes,
   Souvenirs linked to a date
 – Quick responses to environmental/
   health hazards
 – Emergencies

(b)

FIGURE 15.4 Three-dimensional DSS model analyzed in eight cubic representations to classify projects varying
in nature: Cubes 1 and 2.

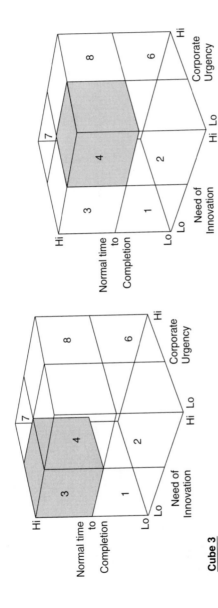

Cube 3

General Characteristics
- Long-term projects
- With relatively low innovation
- Short time to market

Examples
- Company is in long-term game but
  the competition has sprung surprise
    - by a radical product innovation or modification
    - by a radical process development
    - or both
- Any project on the back burner
    - Licensing/trading/agency
    - Environmental clean-up projects

(c)

Cube 4

General Characteristics
- Normally long-term projects
- Of high innovation
- But wanted urgently

Examples
- Similar to Cube 3 but the need for high innovation
  probably minimizes the opportunity to take something
  off the back burner
- Therefore
    - Licensing
    - Trading/agency

(d)

FIGURE 15.4 (*Continued*) Three-dimensional DSS model analyzed in eight cubic representations to classify projects varying in nature: Cubes 3 and 4.

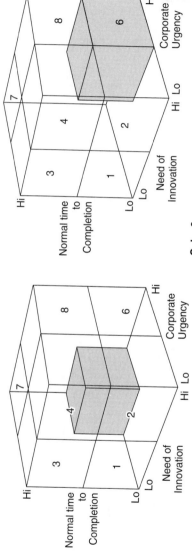

(e)

Cube 5

General Characteristics

- Shortish time to completion
- Of low innovation
- With no great urgency

Examples

- Such projects could be potboilers taken up to utilize temporarily surplus facilities
- Exploration of hazy ideas for later, fuller work, probably "related" to existing business

Cube 6

General Characteristics

- Projects of shortish duration
- Of high innovation
- But no great urgency

Examples

- Exploratory meeting on products and processes meeting longer-term strategies probably "unrelated" to existing business activities
- Radically new products and processes
- Attempts to get into a new business activity

(f)

FIGURE 15.4 (*Continued*) Three-dimensional DSS model analyzed in eight cubic representations to classify projects varying in nature: Cubes 5 and 6.

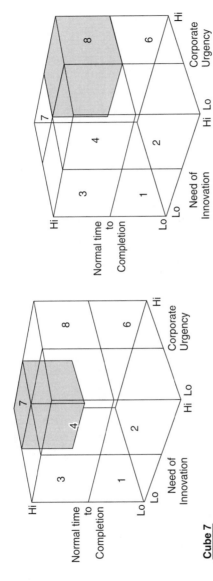

Cube 7

General Characteristics
– Projects of long duration
– Of low innovation
– Plenty of time to completion

Examples
– Normal longer-term process
  development
– Large scale-ups of existing know-how
– Academic work in pursuit of knowledge

(g)

Cube 8

General Characteristics
– Long-term projects
– Of high innovation
– With plenty of time to complete

Examples
– Projects with a high "R" content
– Often with strong elements of unrelatedness

(h)

FIGURE 15.4 (*Continued*) Three-dimensional DSS model analyzed in eight cubic representations to classify projects varying in nature: Cubes 7 and 8.

be taken into account. Is the development within the project such that it will leapfrog over what is known of a competitor's current technical position, or any hints of their direction of development, and so give real advantage? Awareness of competitive developments external to the organization by those already in the business area, or by encroachment from other organizations, possibly based on applying different or new technology, is a major responsibility for the R&D function. Again, the DSS should cover such aspects in its questioning and analysis.

Most of the methods already described for the evaluation of projects competing for resources tend to result in a single score for each project so that ranking by the score is possible. It was a clear desire of the experts that they should not be blindly dependent on such scores without being able to look back over their derivation and reconsider the scaling of two or more projects to adjust their ranking. In effect, the DSS was to be advisory to managers and not a closed decision-making system. This was seen as essential since no DSS of broad application can reflect all the special circumstances surrounding individual projects. There must be scope for managers to inject their understanding of such extra parameters into the comparison process.

The ultimate sanction in the competitive evaluation of R&D projects is the decision to kill a project. This was not an initial objective of this work, but its importance became more evident as the work proceeded. As investment in R&D to achieve a particular result becomes increasingly expensive, there is clear economic advantage if projects can be closed down with some degree of confidence earlier rather than later, and the resources switched to other projects needing acceleration to give corporate advantage, or to the early exploration stage of longer-term, more risky projects.

Comparative work can give indications of projects which should be terminated. If a project loses out in several successive portfolio evaluations, the corporate urgency cannot be high, and the decision to terminate should be made. More difficult circumstances necessitating a termination decision were noted. In all methods of comparison, it is possible for a project to go on achieving a high score, and so appear to be a candidate for continuing work. Within a good score, however, there may be a single factor or parameter which is so low that commercial development would be totally unwise. This was called a "killer" factor, and the identification of these factors had to be a key target for an effective DSS.

The DSS was now developed using the cube structure and the three techniques for generating advice to the manager. Each of the four components gave different aspects of guidance, and the overall process is well illustrated by Fig. 15.5.

Thus, along the path "monitor progress/adjust plan" in Fig. 15.5, the techniques that can be used are

1. The cube model which provides a reference map of the location and state of the project relative to the corporate objectives. If there is deviation from the optimum path, the model will make managerial suggestions to get back.

2. The multiattribute utility model, which tracks progress by following changes in the project attribute scores and expected time to completion. If this is unsatisfactory, then management action would be required (e.g., change of research option and/or effort, and continued monitoring to check for improvement, as described in Sec. 15.3.2).

3. The calculation of distance from target (DFT), measuring how far a project is away from meeting defined targets. A change of research program may be necessary if the distance is not decreasing with time.

Along the "terminate path" axis in Fig. 15.5, the techniques that are available to ascertain the continuance of the project are

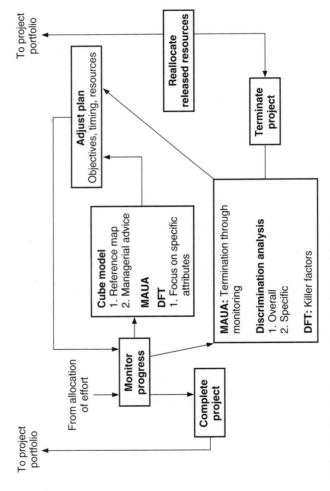

FIGURE 15.5 Project management: monitoring and termination.

1. Use of an expert system, made up of rules which are based on the attribute values of projects which have been successfully completed, or failed. For a project with a given set of attribute scores, the expert system would advise on whether work should continue. Such a system has yet to be developed.

2. Discriminant analysis which determines the discriminating factors and provides a cutoff value below which termination ought to be considered.

3. Use of DFT calculations to determine the minimum values of single attribute scores or combinations of low attribute scores which would terminate a project. These are the so-called killer factors.

15.6 SUMMARY AND DISCUSSION

In the field of R&D project evaluation and selection, there have been many attempts to develop techniques for more effective management. The quality and uncertainty of the results have, in themselves, deterred many from using the systems. Furthermore, the considerable development work, and learning time, needed to use the systems did not seem justified. Practical use in R&D laboratories has been insignificant compared with the research and literature in the field.

The authors, who are very familiar with this background, still proposed to a group of experts in industry that the approach was worth pursuing further. The proposal was clear and simple in its initial concept, but as work developed every aspect needed challenging. The end result has been the development of a satisfactory DSS which is now being applied in industry. In addition, an insight has been gained about the process, giving a better understanding of the reasons for past failures, and guidance for the future.

15.6.1 What was the initial concept?

The initial proposal was apparently simple and clear in its objective: *to develop an ES for the selection of projects.*

As work progressed, however, every significant word was challenged and fresh questions raised, which are discussed in the following sections:

• What is a project?

• For whom is the selection being made?

• Does management really want to select projects?

• What system is required?

• Why did the methodology succeed against a background of failure?

15.6.2 What is a project?

The historical answer, and the one envisaged in the initial concept in the work described above, was that "a project is the whole span of activity, predominantly within R&D, from the initial idea generation to a successful product or process contributing to the overall benefit of the organization." Also historically, a project was perceived as a single entity with little inherent structure.

In fact, a *project* is a dynamic activity, with a recognizable life cycle of its own. This activity often has its origins in the R&D function, but has its reason for existence, and its purpose, embedded in the dynamic life of the whole organization, contributing to its future goals. Furthermore, the wider organization is also operating in an even wider, and possibly harsh, competitive environment of suppliers, rival producers, and customers all pursuing their own corporate objectives.

It is necessary to recognize the life cycle of a project with its own structure because the managerial considerations of the project change significantly as the project progresses. For convenience, the life cycle of a project was split into six stages, but experience and practice has not indicated any need for change (see Sec. 15.5.3).

This separation was necessary because the parameters taken into consideration, and their weightings, were varied with the stage.

15.6.3   For whom is the selection being made?

The R&D function would undoubtedly have been the answer initially in the work described in Sec. 15.5, and in much of the historical research described in Sec. 15.2.

With the definition of a project set out in Sec. 15.6.2, however, the R&D function should be in close communication with other functions in the organization so that every single project is defined and planned to meet the corporate goals in the wider competitive environment. The only projects which can be handled exclusively within R&D are exploratory ones designed to determine technical possibilities as a precursor to a full project definition within the wider context.

Thus, in developing any managerial system in the project area, the system must recognize, and involve, the interests of the other functions. Desirably, it should be capable of use by other functions, and might be a means of encouraging dialogue between functions.

It should also be accepted that many decisions, even at all stages of a project, should be multifunctional ones.

15.6.4   Does management really want to select projects?

The selection of projects, as a term, is redolent of bygone days when research budgets were relatively generous and not subject to cost-benefit analysis, and when competition was usually seen as national rather than global. In present times, the corporate objective to which an R&D project is addressed, and the competitive advantage to be achieved, should both be crystal clear. Furthermore, there are now always more projects for consideration, which can be met by resources in cash, workforce, or facilities.

In these circumstances, project selection has to give way to the evaluation of projects competing for resources. Furthermore, the competitive position of a project has to be considered at each stage in the project's life cycle. This is essential in recognition of the understandable psychological resistance to killing a project which, in simple technical terms, is going well. In cost-benefit terms, however, it is better to kill a project as soon as it begins to lose out in competition with other projects, in order to transfer the resources that it is—and will continue—consuming, to projects of greater corporate advantage.

Thus, a new system had to be aimed at determining competitive advantage between projects, in corporate as well as technical terms, in language and content acceptable to all functions, and giving warning, as early as possible, for killing a project.

15.6.5   What system is actually required?

In common with most earlier work in this field, the system in the research described in Sec. 15.5.3 was initially predetermined. Probably because of their novelty, the objective was to develop a number of ESs relevant in the management of R&D. Furthermore, the ES had to be in a widely used, or easily mastered, program so as to be available to a wide range of managers of widely variable computer skills.

The point of decision was soon reached, however, in which the existing, simple ES programs, while being relatively easy to master, were seen to be too limited in structure to cope with the variety of questions to be asked, or for the mathematical analysis to be included. To have changed to a lower-level language, and developed an independent program, would have seriously reduced the accessibility to a range of managers.

Furthermore, ESs appear to have their best use where a fairly clear and definable answer is required, or where the user is to be directed to a solution to a problem, or to a source of data. In project evaluation, the variety of parameters, and their differences in units of assessment, lead to a fuzzier decision environment, and one in which managers desire to have access to adjust the parameters and weighting in response to their perceptions of competitive factors which might be difficult to quantify.

Thus the control of the system was vested totally in the problem owners, and the system designers developed their outline response first on paper. The designers then found that all the operations and processes could be programmed in Lotus 1-2-3, which would be broadly acceptable to potential users.

15.6.6   Why did the methodology succeed against a background of past failure?

A knowledge of the literature would suggest that in nearly all cases of the studies reviewed, the system or process was predetermined. Furthermore, the studies were usually conducted by the system developer or owner, looking to industrial managers to provide the data, and approve, or disapprove, the results.

The lack of understanding of the true nature of a project as set out in Sec. 15.6.2 was hardly evident in any of the historical work.

In the research described in Sec. 15.5, the experts not only made their experience available but also owned the problem and defined the nature of the results they wanted and how they should be presented. It was for the system designer to meet the requirements as best as possible, regardless of system or program. In practice the division was not as clear-cut as stated here in that the problem owners were quite well versed in the systems, while the systems designers, although based in a business school, had considerable experience in R&D in industry.

The clarity of both problem definition and the nature of the advice to be given, was given the highest priority. The system simply had to achieve the ends by the most appropriate means. It was immaterial what techniques or programs were used, so long as the desire for easy access was met. Such a problem-led approach seems to have been the key determinant in achieving success.

15.7   CONCLUSION

Although it is dangerous to generalize from one extensive piece of work, we believe that the principles that have been spelled out above appear to be of broad significance and should contribute to the more effective development of DSSs in the future.

15.8 REFERENCES

1. H. Eisner, "A Generalised Network Approach to the Planning and Scheduling of a Research Program," *Operational Research,* **10**(1): 115–125, 1962.
2. E. W. Davis, "Resource Allocation in Project Network Models—a Survey," *J. Industrial Engineering,* **17**(4): 177–188, 1966.
3. A. Pritsker and W. Happ, "GERT: Graphical Evaluation and Review Technique—Part I, Fundamentals," *J. Industrial Engineering,* **17**(5): 267–274, 1966.
4. W. Souder, "A System for Using R&D Project Evaluations Methods," *Research Management,* **21**(5): 21–37, 1978.
5. D. Augood, "A New Approach to R&D Evaluation," *IEEE Transact. Engineering Management,* **22**(1): 2–10, 1975.
6. I. Ansoff, *Evaluation of Applied Research in a Business Firm,* Harvard Business School Press, Boston, 1962.
7. J. Moore and N. Baker, "Computational Analysis of Scoring Models for R&D Project Selection," *Management Science,* **16**(12): B212–B232, 1969.
8. B. Dean and S. Sangupta, "Research Budgeting and Project Selection," *IRE Transact. Engineering Management,* **9**: 158–169, 1962.
9. R. Freeman, "A Stochastic Model for Determining the Size and Allocation of the Research Budget," *IRE Transact. Engineering Management,* **7**: 2–7, 1960.
10. E. Pessemier, *New Product Decisions: An Analytical Approach,* McGraw-Hill, New York, 1966.
11. W. Souder and T. Mandkovic, "R&D Project Selection Models," *Research Management,* **29**(4): 36–42, 1986.
12. B. Jackson, "Decision Methods for Selecting a Portfolio of R&D Projects," *Research Management,* **26**(5): 21–26, 1983.
13. R. Keeney and H. Raiffa, *Decisions with Multiple Objectives,* Wiley, New York, 1976.
14. K. Golaby et al., "Selecting a Portfolio of Solar Energy Projects Using Multi-attribute Preference Theory," *Management Science,* **27**(2): 174–189, 1981.
15. D. Keefer, "Allocation Planning for R&D with Uncertainty and Multiple Objectives," *IEEE Transact. Engineering Management,* **25**(1): 8–14, 1978.
16. W. Souder, *Management Decision Methods for Managers of Engineering and Research,* Van Nostrand Reinhold, New York, 1980, pp. 137–190.
17. N. Baker et al., "A Budget Allocation Model for Large Hierarchical R&D Organisations," *Management Science,* **23**(1): 59–70, 1976.
18. M. J. Liberatore and A. C. Stylianou, "The Development Manager's Advisory System: A Knowledge-Based DSS Tool for Project Assessment," *Decision Sciences,* **24**(5): 953–976, Sept.–Oct. 1993.
19. J. Allen, "A Survey into the R&D Evaluation and Control Procedures Currently Used in Industry," *J. Industrial Economics,* **18**: 161–181, 1970.
20. M. Liberatore and G. Titus, "The Practise of Management Science in R&D Project Management," *Management Science,* **29**(8): 962–974, Aug. 1983.
21. R. Baker and J. Freeland, "Recent Advances in R&D Benefit Measurement and Project Selection Methods," *Management Science,* **21**(10): 1164–1175, June 1985.
22. A. Wilkinson, "Developing an Expert System on Project Evaluation: Part I, Structuring the Expertise," *R&D Management,* **21**(1): 19–29, Jan. 1991.
23. M. J. W. Stratford, "R&D Project Portfolio Management," *Internal ICI Report,* Aug. 1991.
24. G. Islei, G. Lockett, B. Cox, S. Gisbourne, and M. J. W. Stratford, "Modelling Strategic Decision Making and Performance Measurement at ICI Pharmaceuticals," *Interfaces,* **21**(6):

4–22, Nov.–Dec. 1991 (awarded the TIMS/ORSA Prize, 1991, for outstanding DSS application and achievement).

25. C. Buell, "When to Terminate a Research and Development Project," *Research Management*, **10:** 275–284, July 1967.

26. R. Balachandra and J. Raelin, "How to Decide When to Abandon a Project," *Research Management*, **23:** 24–29, July 1980.

27. R. Balachandra and J. Raelin, "When to Kill That R&D Project," *Research Management*, **27**(4): 30–33, 1984.

28. A. G. Baker, "Distance from Target," paper presented to the R&D Management Conference, Ghent, Sept. 1989.

29. W. Edwards, "Use of Multi-attribute Utility Measurement for Social Decision Making," in *Conflicting Objectives in Decisions*, D. Bell, R. Keeney, and H. Raiffa, eds., Wiley, Chichester, U.K., 1977, pp. 247–275.

30. V. Belton, "Use of a Simple Multi-criteria Model to Assist in Selection from a Shortlist," *JORS*, **36**(4): 265–274, 1985; also V. Belton, *Multi-Criteria Decision Analysis—Practically the Only Way to Choose*, Operational Research Tutorial Papers 1990, Operational Research Society Publication, pp. 53–101.

31. E. Feigenbaum, *Knowledge Engineering for the 80's*, Computer Science Department, Stanford University, Stanford, Calif., 1982.

32. J. Quinlan, "Learning Efficient Classification Procedures and Their Application to Chess End Games," 1983.

33. D. Hand, "Statistical Expert Systems: Design," *The Statistician*, **33:** 351–369, 1984.

34. L. Hackong and F. Hickman, "Expert Systems Techniques: An Application in Statistics," *Proceedings of the Fifth Conference of the British Computer Society*, Specialist Group on Expert Systems, University of Warwick, Dec. 1985.

35. J. Breuker and B. Wielinga, *Structured Knowledge Acquisition for Expert Systems*, University of Amsterdam, 1985.

36. H. Reichgelt and F. Van Harmelen, "Relevant Criteria for Choosing an Inference Engine in Expert Systems," *Proceedings of the Fifth Conference of the British Computer Society*, Specialist Group on Expert Systems, University of Warwick, Dec. 1985.

37. P. Alvey, "Problems of Designing a Medical Expert System," *Proceedings of Expert Systems*, 1983, pp. 20–42.

38. J. Kunz et al., "Applications Development Using a Hybrid AI Development System," *AI Magazine*, **5**(3): 1984.

39. P. Harmon and D. King, *Artificial Intelligence in Business: Expert Systems*, Wiley, New York, 1985, Chaps. 4, 7–10.

40. M. Stefik and D. Bobrow, "Knowledge Programming in LOOPS: Report on an Experimental Course," *AI Magazine*, **4**(3): 1983.

41. B. Chandrasekaran, "Towards a Taxonomy of Problem Solving Types," *Artificial Intelligence*, **4:** 9–17, 1983.

CHAPTER 16

ENTERPRISE ENGINEERING IN THE SYSTEMS AGE

John E. Juhasz
The Micron Group
Cleveland, Ohio

Baldwin-Wallace College
Berea, Ohio

16.1 INTRODUCTION

The winds of change in the business world are, once again, seemingly everywhere. One observer of management trends in corporate America postulated recently that a significant indicator of change is the rate at which new management fads come into vogue, thereby obsoleting previously cherished change mantras. This dubious indicator, in which the observer noted that American management clearly leads the world, had seen some alarming rates of change in recent years. While it isn't clear when or if the old paradigms had "died out" (like old soldiers, they just seem to "fade away"), they are simply being swept aside by the emergence of the new, presumably superior "paradigm shifts."

As a recent example, even before the ink had dried on the first report cards of the ubiquitous TQM movement, the swift ascendance of reengineering [along with its many variations such as reinventing, restructuring, or the less ominous business process reengineering (BPR)] as the new, improved management mantra eclipsed almost all previous trends. To achieve the fastest possible change, it was argued, continuous improvement was *out,* and radical change was *in,* whereby it was encouraged (or even required) that the "old" process be obliterated in order to pave an unencumbered way for the "new" process to be defined by the reengineering team. Exactly *how* that new process would emerge was left to the reengineering team with a blank sheet of paper, to be guided by the omnipresent reengineering guru. The only commonly accepted reengineering principle seemed to be in its definition suggested by Hammer and Champy: "the fundamental rethinking and radical redesign of business processes to achieve dramatic improvements in critical performance...such as cost, quality, service, speed, etc."[1]

In spite of the "dramatic improvement" claims and the apparent financial commitments from top management, the sun seems also to be setting on the reengineering paradigm, at least in the way it is currently practiced. Adding to previous dismal results, no less than two out of three efforts have been projected to result in failure in the next few years, according to various expert opinions. As the frequency of reported failures increases, so does the inevitable postmortem speculation on the causes of failure and the suggested quick fixes. The usual change debates rage on incessantly over who should lead (the change initiative) and who must follow, the role of technology whether one should really start with the proverbial blank sheet, focused only on the desired future state (vision), or how much effort to expend on the current state definition, or whether the attempted change should be radical or incremental, and what constitutes the dividing line between the two. Even Hammer and Champy have openly disagreed on how to define this divergence and how to mitigate the risks of each path. Paradoxically, as the autopsies of failed projects proceed, reports of spectacular successes also continue to arrive, perhaps with less frequency or fanfare but with very impressive gains. A logical question arises: Why such huge discrepancies in the results?

Surely a part of the answer must be attributable to differences such as strength of leadership, sound management, a credible vision, measurable goals, and the usual host of other success factors. Conversely, Senge identifies various organizational handicaps that he calls "learning disabilities" which create barriers to effective change management and teamwork.[2] There may plausibly be a variety of such disabilities submerged deeply in the corporate psyche or in individual belief systems, which must be skillfully excavated and neutralized for progress to take place. There is also perhaps a more fundamental answer which lies below the surface, one best expressed by Jay Forrester, Senge's former mentor at MIT:[3] "I believe the [discrepancy] is in the failure of management to understand that corporations and institutions are indeed Systems."

Forrester maintains that our management methods and concepts have not kept pace with the explosive changes driven by technology, thus rendering much conventional wisdom obsolete. A natural but potentially lethal consequence of this myopia, especially in turbulent times, Forrester concludes, is the lack of understanding of system behavior, the interdependencies of its parts, and its relationship to its dynamic environment.

Ignoring any of these factors could derail even a nominal reengineering initiative, no matter how well intentioned, to say nothing of large-scale, radical redesign. It seems logical to conclude that the masters of successful corporate change management have managed to grasp the essentials of system behavior and have learned to direct the disciplined practice of systems methods to their favorable adaptations. Forrester further argues for the adoption of a "system design" approach to the enterprise, including the creation of models and simulation processes, as a means of dealing with change and complexity. He concludes by admonishing the business schools to undertake the teaching of "enterprise design" in addition to the standard fare of operations management. This chapter, focused on system engineering of the enterprise through structured enterprise models, is intended to introduce concepts in response to this challenge.

16.2 SYSTEMS: THE MODERN ERA

16.2.1 The Systems Age

The term *age* is generally intended to reflect a period in which there is a prevailing, commonly held view of the world and the nature of things. While there is yet no commonly used term for this era, most of our modern philosophers and futurists agree that the world has become a fundamentally different place in the last few decades. Ackoff writes about the "second industrial revolution," another label for the modern period he also calls the "systems age." Describing major phases of human evolution, Toffler writes about the "third wave,"[4] replacing the previous industrial period, which he called the "second wave." Drucker refers to our time as the "postcapitalist society." There is much more harmony than discord in these and other writings, giving us pause to ponder the emergence of this "new age."

Russ Ackoff coined the term "systems age" to differentiate the present period following World War II from that preceding, which he identifies as the "machine age."[5] Ackoff characterized the 300 or so years of the machine age, also known as the period of the industrial revolution, as one in which the world was perceived as a clocklike mechanism, composed of hierarchically arranged sets of parts, and all parts operating through a sequence of cause-and-effect relationships. The principal mode of analytical thought was based on *reductionism,* in which the analyst would seek knowledge by taking apart (decomposing) a system in order to focus on the simpler part, then further decomposing the part into subparts, etc., thereby hoping to achieve understanding of the system through a focus on the parts, based on cause-and-effect observations. Just as the ancient Greeks explained their world from the view of four basic elements—fire, earth, air, and water—reductionist thinkers sought greater understanding of the universe through its smallest particles. This mode of thinking led to the relentless search in science to discover the basic "elements" of nature, the lowest level of particle which was presumed indivisible. The elementary particle focus of research and much dramatic discovery throughout the machine age was the atom in physics, the molecule in chemistry, the living cell in biology, etc. All of these collectively, however, failed to provide much insight to the nature and behavior of systems.

In the industrial context, reductionist principles were focused on decomposing the factory work process to elementary "task" levels. The industrial designer would seek to automate these via a suitable machine, or if this were not feasible, would assign it to a human operator. Thus the work of individuals was largely "dehumanized" by treating the human as a machine, an interchangeable element of the larger factory system. Workflow design in some modern facilities is in many ways still based on labor concepts derived from the machine age, with all the attendant human problems. These are some of the residual consequences which have fueled the current passionate motivation for reengineering.

The seeds of systems thinking were sown long before the transition to the systems age, in the form of scientific dilemmas and discrepant behavior not explained by reductionism or cause-effect relationships. Most such dilemmas were largely ignored until a major global event (World War II) gave impetus to a period of high intensity and chaos as each side sought maximal competitive advantage. The stress and turmoil of this era gave rise to numerous critical technologies such as computers and the semiconducting transistor, as well as several new paradigms of management thought. Operations research, born of the urgency to gain advantage in wartime intelligence activities and logistics planning, emerged as a credible, interdisciplinary field of study. The modern teaching of management science is essentially derived from this discipline. Of greater significance, however, was the increasing awareness of the new concept of a purposeful system, which did not conform to machine age principles, but resolved many prior philosophical dilemmas. While in the machine age even humans were viewed as machines, the concept of systems accounts for the behavior of machines and the free will of humans and as interdependent parts of a purposeful entity. The search for indivisible parts was superseded by the new awareness of the indivisible whole, whose significance was much greater than the sum of its parts.

A simple definition of *system* is suggested by Ackoff as a set of parts or collection of elements which constitute an indivisible whole satisfying three conditions:

1. The performance of the whole is affected by every one of its parts.
2. The contribution of each part is related to or dependent on some other part.
3. Combinations of parts into subgroups will cause the subgroups to have the same properties as parts (i.e., definitions 1 and 2).

The emerging awareness of systems and the relevant theories of their purposeful behavior gradually replaced the previous notions of reductionism with a concept of *expansionism*. The cognizance of the whole system as more significant than the sum of its parts motivated shifting our preoccupation from the indivisible element to the indivisible whole. This fundamental shift in thinking taught that in order to understand systems, instead of decomposition by analysis and focusing on the parts, one should understand first how the systems interact with their larger wholes of which they are a part.

Our modern systems pioneers resolved that the only rational way to comprehend the whole system is to first grasp the system's "purpose," which addresses the reason for its existence and explains its dynamic interaction within its environment, the larger whole. The central principle of systems thinking—that the whole is unique, indivisible, and much greater than the sum of the parts—has profound consequences in our design of the modern enterprise. Ackoff postulates the main theorem of systems behavior and its corollary in the following description:[6]

> If you take a system apart to identify its components, and then operate those components in such a way that every component behaves as well as it possibly can, then there is one thing of which you can be sure. The system as a whole will not behave as well as it can. The corollary is—if you have a system that is behaving as well as it can, then none of its parts will be [behaving optimally].

The dominant philosophical implication for enterprise change management thus becomes self-evident. It we seek to "optimize" the system, then all its parts must be designed to conform to primary requirements, which, in turn, must be derived from goals linked to the system purpose and applicable constraints. We must also learn to accept that some of its parts, and possibly even all parts, will not be and may never be operating in their "optimal" manner. Contrast this philosophy with the actual practice of our two most (recently) cherished change movements. In TQM, the emphasis was on continuous improvement of each part, presumably toward some ultimate concept of infinite perfection, but the dynamic interaction of the parts was largely ignored, and the derivation of each part's requirements from the purposes of the whole was rarely connected or traceable. In BPR, the focus on the end-to-end process seems to offer a more enlightened approach, until we attempt to deal with complexity by making the difficult process more "manageable," i.e., breaking it down into its simpler subprocesses and into their lower-level processes. By allowing each to be "reengineered" separately, many companies have been hoping for a beneficial improvement of the whole. Some have proudly flaunted their "commitment" to reengineering by pointing to the sheer number of simultaneous BPR projects under way. The failure in BPR to examine process interdependencies and requirements relationships to the whole leaves us in the same unsatisfactory (at best, suboptimized) overall state. Gravely lacking in these endeavors is a systems view, which requires the entire enterprise to be considered a purposeful process interacting with the set of external processes that make up the environment, from which a consistent, rigorous set of requirements could be derived to guide all internal subprocesses.

16.2.2   Convergence of Systems Approach

Since the beginning of the systems age, two powerful streams of technological evolution have been moving quietly but inexorably toward convergence. Common methods of system analysis are being applied increasingly in the disciplines of systems engineering and systems management. This revelation may not be surprising, since management science (MS) and engineering are both firmly rooted in mathematics and natural law. Many of the techniques developed in one field have found useful application in the other, and have frequently crossed over to support new approaches to problem solving. Significant examples in MS include the increasing use of decision support systems with comprehensive data and model bases, and expert systems employing various artificial intelligence schemes, all derived from methods "engineered" in computer science. Conversely, utility theory and probabilistic decision models have found their way into engineering design evaluation and risk assessment.

Although generally increasing in sophistication, many such crossover applications have been mostly limited to special areas or cases of quantitative analysis with limited significance in a business context. The ultimate fusion of the two evolutionary streams may depend on yet another vehicle which is emerging on the horizon of systems thinking. This vehicle, powered by the engine of modern computing technology and fueled by rigorous systems methods, seeks to combine the capabilities of tools from each field with comprehensive, structured analysis to fulfill a broad set of business needs and expectations. It is defined here as the *structured enterprise model*.

The concept of enterprise modeling is not new, as it, too, has been evolving for some time. It has been advocated by many and is periodically manifest in limited form in various strategic plans, manufacturing process flows, or information systems models. Ackoff identifies several strategic planning models which are based on a solid foundation of system principles, and which portend the emergence of such system models. In its fully developed form as advocated herein, the structured enterprise model is focused on an extensive systems view of the target enterprise, defined as any entity purposefully engaged in a goal-seeking process. The definition extends from the smallest identifiable functional unit to an entire company, global enterprise, regional economy, or industry.

A sophisticated model-based methodology is likely to have profound implications as a strategy paradigm for the business enterprise. A top-level system model, with clearly established purpose and identified interactions with its environment, would serve as a beginning framework to ensure goal alignment of the lower-level subsystems with the whole. Measures of performance or goal attainment would then be the key drivers of requirements to the successively lower subsystems or processes. Linkages of the dynamic interfaces with the environment model would permit numerous "what if" observations of behavior to be conducted via simulation. The enterprise model provides a foundation for disciplined application of systems methods coupled with management science, to yield a powerful decision support capability for the continuous management of change and complexity.

16.2.3   The Systems Method

The road toward building any complex system is usually a rocky, twisted path, full of the potholes of tradeoff analysis and decision making. The fundamental challenge for both the systems engineer and the systems-oriented manager is to anticipate and solve problems in their respective domains of expertise and "design responsibility." Both seek to apply the tools and skills of their craft in a way which provides an optimized result within the constraints of the problem boundary. Their most basic role, performed consciously or otherwise in a continuous, cyclical pattern, is represented by a rudimentary decision process. It is to evaluate competing alternatives and select the best choice, while anticipating the probable occurrence and likely effects of certain uncontrollable conditions on each alternative.

A common simplified definition of the systems method which applies to both engineering and management disciplines is suggested as:

> The directed application of resources and knowledge to a specified problem via rigorous methods to achieve a desired outcome.

Directing the effort toward a desired outcome encompasses the leadership responsibility of defining purpose and establishing compatible goals. Specifying the problem and rigorous methods, i.e., the enterprise modeling process, is the principal focus of this writing, along with qualitative approaches to assessing need and forming a vision to drive goals and requirements. The two key input ingredients, resources and knowledge, require further consideration.

Quantitative methods of management science present a variety of decision models involving the use of resources, specifically their "optimal" allocation among competing demands. The many variations of mathematical programming provide a rich set of alternative methods for arriving at the preferred solutions under given resource constraints. Modern awareness of enterprise resources is no longer confined to the traditional land, labor, and capital, but also extends to the use of space, time, energy, raw materials, the physical environment, computing power and storage capacity, human skills and intellectual potential, and organizational learning capacity, to name just a few. It also includes that ubiquitous, modern strategic resource called *information.* These resource definitions are now commonly accepted and recognized as "decision variables" within the domain of control of the enterprise decision maker. Since resource consumption usually relates directly to a firm's critical financial resources, these variables need to be identified and accounted for in terms of how they contribute to the systems' achievement of purpose. In all applications of the systems method the cognizance and quantification of resource constraints remain fundamental prerequisites to formulating meaningful requirements.

The definition and acquisition of knowledge, a presumably inexhaustible resource and the other critical ingredient for systems methods, poses greater challenges. Its constraints and limitations remain unknown, and its application is not always well addressed by quantitative methods. A paradox of modern science is the correlation of increasing knowledge to a widening perception of the knowledge gap, i.e., the awareness of how much more there is yet to know. Hawking[7] explains that to the early part of this century, science had assumed that the universe was essentially static and completely deterministic, consistent with the cause-effect beliefs of the machine age. In the early 1800s the French scientist Laplace had postulated the existence of a set of scientific laws that would allow us to predict everything that would happen in the universe, if one could only measure completely its state at any one point. Some nineteenth-century scientists were so persuaded and impressed with their knowledge accumulation that they arrogantly predicted the precise timeframe, circa 1920, at which point science would have discovered all laws governing the behavior of the universe.

The doctrine of scientific determinism remained unchallenged until two major discoveries early this century. The astronomer Edwin Hubble discovered that the universe was far from static, and in fact that the galaxies, which he loosely estimated at about

100 billion, instead of being stationary, were actually moving away from us and each other. This revelation of an expanding universe clearly emphasized the constancy of change, that nothing can be assumed to remain static over long periods of time. A second profound discovery was when Werner Heisenberg showed the impossibility of accurately measuring related variables such as position and velocity of an object, that in fact the greater the attempt to precisely measure one, the more uncertain would be the measure of the other. The now famous *uncertainty principle* formulated by Heisenberg completely refuted Laplace's model based on the principle of determinism and led to the "new" science of quantum mechanics and its close associate—chaos. While many of its discoveries are both fascinating and terrifying, it has irrevocably introduced uncertainty (or risk) as a fundamental, inescapable property of the world. Not only for science but for all human organizational entities, Margaret Wheatley[8] expresses the new reality that the world "...has become a strange and puzzling place where I cannot rely on what I knew and I don't yet feel secured by the new sources of confidence. It makes things much more interesting, expecting there to be new ways of working without being able to discern them clearly.... I've become aware of how difficult it is *not* to be certain."

For system builders in pursuit of their "ideal" designs for long life at minimum risk, clearly the perpetual challenge is to verify the scope, the sources, and the authenticity of knowledge to be applied in the system process, and to expect (and value) occasional chaos along the way. The constancy of change and the uncertainty of measurement are principles of particular interest in the system context. Management science teaches that risk can be reduced through acquisition of relevant information, provided the decision maker knows what is relevant and has the knowledge to apply it to the problem. Today's system developers are challenged to recognize and plan for change and quantify elements of risk in their knowledge as well as their systems due to subtle but omnipresent uncertainties in both external conditions and internal behavior. The modern systems enterprise must be designed to be ceaselessly adaptive in response to perpetual change, and continuously seeking relevant information and improving knowledge (learning) to minimize risk.

16.2.4 The Systems Boundary

We are frequently reminded by chaos theorists and quantum philosophers that in the complex place we live called the "universe" all things are somehow connected to each other through a web of relationships. The mythical butterfly wings flapping in China can, indeed, be estimated to affect the tornado in Texas, as chaos theory has shown that the world is far more sensitive than we had long contemplated in our mechanistic newtonian thought. For the enterprise system builder, consideration of Heisenberg's principle is essential for risk assessment, but some degree of simplification is equally essential in order to overcome paralysis and facilitate any form of progress. The inevitable tradeoff penalty for simplifying assumptions is the acceptance of risk, that some unforeseen nominal, or even catastrophic, event might seriously jeopardize our system.

A logical distribution for simplification and risk assessment, and a practical point of beginning the system understanding, is the *system boundary*. To preclude our being overwhelmed by the vastness and uncertainties of the universe, systems thinking provides a judicious principle for partitioning from the infinite to a limited, problem-focused view, i.e., the control domain (system) boundary. This sober principle rationally suggests that for a given problem we can and should focus our system analysis on decisions and solutions which fall within our domain of influence and control.

While definition of this domain also has some associated uncertainties, these are presumed resolvable to a rational, recognizable boundary. Outside this boundary is the eternal and universal vastness called the "environment," which we can neither hope to fully understand nor control, but within which we can identify likely patterns of interactions. As long as we recognize these potential influences from the outside environment, we can and should account for their consequences on the behavior of our purposeful system.

As for those "unforeseen" chaotic events, our system must be willing to tolerate a certain level of acceptable risk, even catastrophic failure, which we as decision makers can assign a value based on some postulated probability. As living, purposeful systems we humans subconsciously accept catastrophic risk on a daily basis when we venture into traffic, play sports, step into a bathtub, eat a burger, choose to live near seismic fault lines, or fly the sometimes unfriendly skies, because we assume (or hope for) a relatively low probability of a harmful consequence. We accept these risks because we know that the alternative, spending life in a "protective cocoon," poses consequences more certain (and certainly worse) than death.

Our control domain can thus be defined by a virtual boundary, shown in Fig. 16.1, which separates our system from the rest of the universe, from external influences and conditions which we cannot control, to internal behavior within this boundary over which we must, by design, establish a reasonably well-ordered control. It is important to recognize that an organization, the collective set of enterprise resources including humans, consists of subsystems which may be in pursuit of their own purposes, perhaps at times inconsistent with the overall system purpose. We also recognize that all "open" systems, a definition which distinctly applies to any modern enterprise, exchange matter, energy, economic value, and information with the external environment across this boundary. Our well-designed system therefore must remain responsive to interactions between our system and the external environment, guide the interactions of the purposeful subsystems contained within the system, and maintain perpetual awareness of the consequences of such interactions.

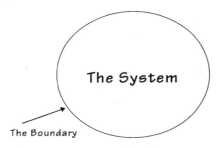

FIGURE 16.1   The system boundary.

Enlightened by the principle of definable boundaries, we can now unburden our systems thinking from the infinite to the finite, from the eternal span of time to the relative present. Our parochial interest here is in today's synthetic (human-made) systems, in particular the enterprise "design" and its intended purpose. The time span of concern is the proposed or assumed life cycle of the system, and the external environment to be defined is that which is reasonably "known" or expected to influence our system behavior.

Two additional theories are essential to provide a firm foundation for systems-based enterprise development and change management. The first is the principle of successive refinement, which explores the cyclical processes and mechanics of system evolution. The second is the broad set of general systems principles which have been observed over time to be consistent explanatory factors influencing system behavior.

16.2.5 System Evolution

In the evolution of living things, as in all synthetic designs—in particular enterprise systems—the underlying "system purpose," whether clearly evident or deeply submerged, plays a central role. Maslow's famous "hierarchy of needs" defines a hierarchically layered set of needs-driven purposes and their relationships. While societies, corporations, and humans have many complex expressions of needs and associated purposes, the most basic and instinctive of these is survival, a common denominator for all living species. (Since organizational systems are created mainly as economic entities, survival should be defined in economic terms, but the "living" metaphor is just as useful.) The most primitive of system purposes is to first and foremost satisfy the perpetual survival need, only after which other needs come into consideration.

The dawning of human consciousness emerged at first slowly but resolutely in an environment of chaos and uncertainty. Humans' first encounters with the laws of system behavior, also perhaps their first experience differentiating them from animals, most likely occurred about 1.5 million years ago, according to best estimates of paleontologists. On walking erect and attaining some level of self-awareness, humans soon became cognizant of the hostile environment which perpetually threatened their survival. Given their instinctive purpose to survive and some awareness of newfound excess capacity in the form of two nimble hands, early humans must have truly rejoiced in their discovery of the first tools, possibly a sharp rock or well-balanced club, provided them by some fortuitous circumstance of nature. They quickly "learned" that with these primitive implements they were able to gain leverage over their environment and some much-needed competitive advantage over the other lifeforms. The early successes (and conceivably many failures) further provided "feedback" which shaped their beliefs and motivated the design refinements and improvements which followed, from which whole generations of tools evolved. This early experience (depicted in Fig. 16.2) established the first series of closed-loop, reinforcing evolutionary cycles of humans and, simultaneously, that of their tools. Each individual cycle, born of a stressful condition of need which was linked to purpose (and accompanied by some chaos), motivated a vision of improvement that, following the inevitable work and toil, yielded an eventual "design solution" which attempted to satisfy that need. The experience of evaluating and interacting with this solution generated feedback essential to "learning" which completed the cycle of successive refinement.

In his 1984 work "Technostress," author Craig Brod advances a bold, somewhat radical notion of this evolutionary cycle. In warnings against the misuses of modern technology, Brod argues that as humans have created tools, their refinements in turn have "re-created" humans, or in other words, humans not only evolved along with their tools, but evolved *because* of them.

Tools have always set in motion great changes within human societies. Tools create us as much as we create them. The spear, for example, did much more than extend the hunter's reach; it changed the hunter's gait and the use of his arms. It encouraged better hand-eye coordination; it led to social organizations for tracking, killing, and retrieving large prey. It widened the gap between the unskilled and the skilled hunter and made the pooling of information more important as hunting excursions became more complex. There were other, less obvious effects: changes in the diets of hunting societies led to the sharing of food and the formation of social relationships. The value of craftsmanship increased. People began to plan ahead, storing weapons for reuse. All of these tool-related demands, in turn, spurred greater development of the brain. Brain complexity led to new tools, and new tools made yet more complex brains advantageous to the survival of the species.

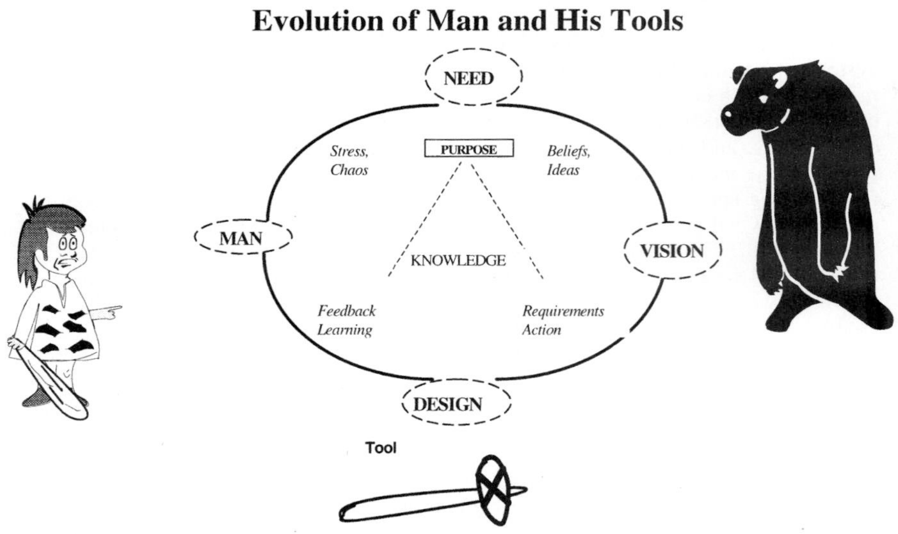

Evolution of Man and His Tools

FIGURE 16.2   The virtuous cycle of successive refinement.

Brod effectively traces an evolution scenario in which tools played a critical interactive role leading to successive cycles of human refinement and tool development. Senge also describes such reinforcing interactions in dynamic systems, where feedback cycles create a cause-effect-like regenerative system loop. But he identifies dual consequences of such feedback as potentially positive or negative, leading either to beneficial growth (a "virtuous" cycle) or to undesirable decline (a "vicious" cycle). Both types of system behavior are regulated by a third cycle which introduces self-correcting, goal-seeking mechanisms to create stability. Such balancing forces interact with the regenerative systems by imposing limitations, constraints, and delays to cause inevitable slowing or reversal in the regenerative cycles.

The evolutionary spiral described by Brod may itself reveal inherent contradictions or limiting tendencies when considered with historical evidence of human behavior. We discern that the tools did not themselves "re-create" humans, but it was the mental, physical, and emotional interactions with the tools, the inspiration from success (or the desperation from failure) which drove humans to continually seek better solutions to their relentless needs. We also know that the cycle slowed or halted whenever a society reached certain comfort levels of satisfaction with its progress, presumably whenever its perceived needs were met. The cycle was probably reversed into "vicious" decline when the focus of human concentration shifted from need to greed, when their irrational beliefs obscured their vision, or when critical effort was lacking or inadequate to yield a meaningful result. A perhaps more subtle but equally malevolent decline resulted when feedback and learning was obstructed as they fell into smug complacency or narcissistic arrogance over their magnificent tools and technologies, their celebrated, presumed flawless theories, or their ideal government laws and social solutions.

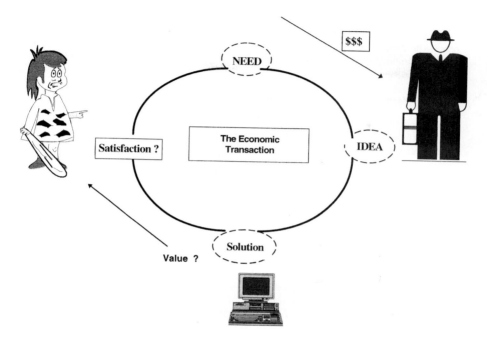

FIGURE 16.3   The economic transaction.

A further concept of system evolution is of importance to appreciate driving forces of enterprise development. The maturing of humans the toolmakers and humans the hunter/gatherers in a trusting, social context led to the inevitable specialization of skills, which foreshadowed the arrival of another evolutionary milestone not chronicled by archeologists, but equally significant to subsequent human development. Soon after the first appearance of tools, "economic humans" inexorably emerged from the bartering interaction between the toolmaking artisan and the tool-using hunter. As each recognized and appreciated the other's superior skills, they negotiated and engaged in mutually beneficial economic transactions in which both of their needs could be satisfied through a bartered exchange of value. At this critical point, humans realized that they could achieve their own purposes by fulfilling the needs and purposes of others. From this mutual value perspective the customer-provider relationship was born and growth via purposeful, economic transactions became the norm of humankind's long evolutionary march (see Fig. 16.3). As many more needs were identified and specialized skills evolved over the millennia, bartering was replaced by more convenient media of exchange through commonly accepted units of value. Economic development accelerated greatly as a result and specialization of skills further proliferated, as it continues to do today.

The business enterprise, the human's ultimate tool for the creation and allocation of economic wealth, is a modern manifestation of an economic system engaged in the fundamental pursuit of its purpose. It integrates specialized skills through providing value in the form of goods and services which are intended to satisfy the needs of its customers, as it seeks continued survival and satisfaction of its own higher needs. The customer-provider relationship born in prehistoric times remains today as the principal process to guide and evaluate economic transactions. A mutually satisfactory exchange of value remains the key criterion of success, resulting in mutual growth of both parties.

16.2.6 General System Principles

Throughout the human's collective knowledge acquisition of systems behavior, several key principles have become self-evident over time and accepted today as fact. We have discussed the constancy of change, universal uncertainty, and evolution by successive refinement and economic exchange as keys to understanding system behavior, and therefore critical considerations of enterprise modeling. It is worth restating for emphasis that systems are purposeful s , perpetually seeking to achieve a goal, even though it may not always be obvious. Further, although the purposes of subsystems should be naturally derived from the larger system purpose, we recognize the potential "optimization" conflict between the overall system and its parts.

In addition to these concepts it is beneficial to review some general systems principles which have been compiled over time by numerous system observers, along with their implications for enterprise designers. Although no formal "proofs" can be presented here (and some may not be possible), these principles are offered to facilitate presentation and illumination of system behavior, and thereby provoke further "systems thinking" interaction and assimilation of cognitive awareness.

An old and well-worn adage of communication is that a picture is worth a thousand words. If pictures and graphic symbols truly enhance communications, then a corollary should be that the communication of potent ideas is clearly enhanced by powerful symbols. The pyramid, itself an ancient graphic symbol and forceful conveyor of concepts over the eons, is again found useful in illustrating key system principles. Maslow used it to portray the hierarchy of human needs. Anthony represented multiple levels of organizational management via the symbolic pyramid. Martin[12] used it effectively throughout his extensive work on information engineering to relate data and functions. The pyramid icon symbolizes many of the core system principles and ideas which form the foundation philosophy of structured system models. These principles are listed below, along with expanded descriptions of each, and some of the implications for system design (*in italics*):

1. Systems are hierarchical and relational. All systems "belong" to or are contained in some higher-order system, of which they are a subsystem interacting with other subsystems. The set of the interacting subsystems constitute a unique system property of "wholeness" which is manifest only within the complete system. The interdependencies of the interacting parts constitute subsystem relations which impact the performance of the whole. All systems can be functionally and architecturally decomposed into smaller processes, subsystems, and parts, thus constituting function and structure hierarchies which resemble a pyramid.[13]

Implicit in this principle is the recognition that the whole is more than the sum of its parts. All observations of dynamic behavior must therefore be with consideration of the whole system, and subsystem improvement must be defined and verified within the purpose and context of the whole.

The system "context diagram"

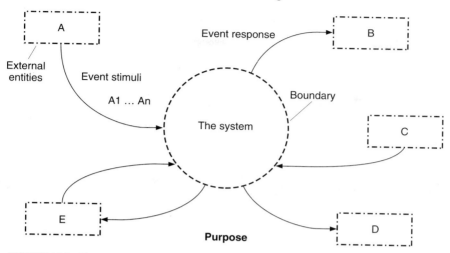

FIGURE 16.4   The system context diagram.

2. Systems interact with their environment. An *open system* is by definition[14] one which exchanges matter, energy, or information with various "entities" in its outside environment as shown in Fig. 16.4, the view of the system in the context of its environment. Virtually all living systems, and certainly all economic enterprises, are considered "open" to their surroundings.

> *System processes must be viewed and accounted for in terms of their response to "events" which, in turn, are the results of external scenarios, unique conditions or states of the system environment. The manifestation of an event is recognized via a "stimulus" that is perceived by the system as it crosses the boundary. The logical system reaction to the perceived stimulus is a "response" to one or more entities in the environment.*

3. Systems are multifaceted. Depending on one's vantage point of observation and interest, a system will appear differently, presenting diverse views of its behavior to the observer. Four such views, including operations (event-response), functions, architecture, and risk are shown in Fig. 16. b5, from the perspective of the top of the pyramid. Although additional views may be postulated for some systems, this graphic shows a minimal set of essential views for any system.

> *In the modeling and analysis process, each facet of the "system pyramid" must be fully developed to reflect a complete, integrated view of internal and external behavior, and integrated with the other facets. Verification of model and system integrity is facilitated through examining the matrix of intersections of each view and accounting for all required flows of stimuli, information, and responses.*

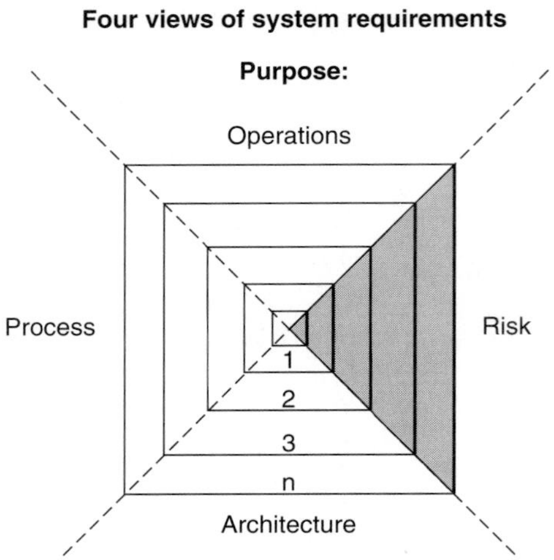

Four views of system requirements

Purpose:

Operations

Process   Risk

1
2
3
n

Architecture

FIGURE 16.5   Multiple system views.

4. Systems law of growth. All systems "grow" through a phased life cycle, including birth, growth, maturation, and eventual decline or transformation.[15] Evolution of the system progresses by successive refinement through a process of cyclical iteration, guided by feedback.

> *As system designers our perpetual mission is to design robust, adaptive systems which can survive over their "design life." The desired lifespan of a system must be specified, so that credible scenarios and potential environmental conditions anticipated during its life cycle can be considered and accounted for in the design. The lifespan of each external entity with which the system interacts is also finite, which ensures an ongoing change in scenarios and events to be anticipated.*

5. Systems perform transformations. Systems create a transformation of the inputs which cross the boundary into outputs by a specific process or transfer function. Given a precise mathematical expression of inputs and desired outputs, the rigorous definition of a system "transfer function" can be expressed as the ratio of the output variables to the input variables.

> *Although such mathematical formulation of system behavior is not always feasible or necessary, the qualitative relationships between inputs and outputs, i.e., cause-effect relationships, must be understood in order to define a process.*

6. System law of adaptability. This law states that the more narrowly specialized a design is toward serving a specific mission or objective, the less adaptable it will be over a broad range of conditions or influences during its life cycle.[14]

> *Depending on the life-cycle objectives for the target system, adaptability to changing environments demands more flexible, less rigid design to accommodate potential change.*

This may further compromise the "optimization" of design, since dynamic changes quickly displace today's optimal.

7. System law of maintenance. The larger, more complex the system, the greater the proportion of its resources which must be allocated to its own maintenance. The diminishing portion which remains is left for application to productive endeavors.[14]

This principle teaches the avoidance of overly complex system design and its associated life-cycle maintenance burden through intelligent partitioning into appropriate subsystems, applying the laws of hierarchy, but retaining cognizance of the purposeful whole.

8. System law of entropy. All systems are potential victims of the laws of chaos and entropy, creating a tendency toward increasing disorder and degeneration leading to potential failure.[11]

The active prevention of such tendencies must be done via a continuous process of assessment and abatement to minimize risk arising from performance degradation or premature failure.

While these principles may not all be readily self-evident for every manner of system, their validity can be substantiated from numerous historical examples. The systems builder's task is to apply the positive aspects of these principles in the enterprise design, and maintain vigilant awareness of the negatives for risk avoidance.

16.3 THE ENTERPRISE

16.3.1 Classic Systems Development

The evolution of the modern economic systems called *enterprises* somewhat parallels development of other synthetic systems, complete with the many pitfalls and miscues typical of complex systems programs which are frequently and prominently derided in the media. Figure 16.6 shows the sequential order of systems development steps in what is considered the classic waterfall life cycle, so named because of the apparent "flowdown" of progress and information to successive steps of development. Such "sequential" ordering of life-cycle phases has been found to be lacking in significant ways, owing largely to the misconception that user needs, external conditions, and available technologies are given and remain static throughout the life of a system. Experience has revealed some of the well-documented, notable shortcomings of the waterfall life cycle as

1. The traditional lack of patience with disciplined mission definition and early requirements development. Eagerness to "press on" with the project causes frequent confusion of purpose and shortcuts in requirements, to the later detriment of the system.
2. The tendency to require a "design freeze" at each stage before proceeding to the next stage, creating premature inflexibility, and assuring the freezing in of errors.
3. The cumulative cost of recovery from errors. Later phases impose a much greater

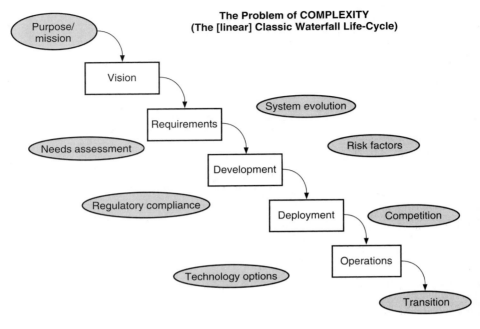

FIGURE 16.6 The waterfall life cycle of development.

recovery cost penalty than if the error is discovered in the early (e.g., require-ments) phase.

4. The "systems" approach is generally confined to the engineering development activity, while other "specialists" concentrate on the other phases. This often leads to the attempted optimization of the parts, and the consequent suboptimization of the overall design and inadequate focus on risk assessment.

5. The prospects for alternative evaluation or implementation diminish rapidly beyond the requirements phase, due to the substantial cost penalty of "backing up."

In spite of such shortcomings, the classic linear phase method prevails as a common paradigm in development of systems as well as enterprise. Part of the difficulty is the introduction of change in culture, the same problem that has impeded implementation of management science techniques for decades. As Machiavelli predicted with remark-able insight, "Nothing is more difficult nor more dangerous to undertake than to attempt to introduce a new order of things."

In recent times concepts of "concurrent engineering" and "rapid prototyping" have emerged in various government and commercial endeavors as alternative development paradigms. These methods overcome many of the linear-phase shortcomings by con-centrating engineering disciplines simultaneously on development and early risk assessment. Prototypes of high-risk designs for evaluation of alternatives before the "freeze" points is a principal feature. Several professional societies and universities have embraced these themes as they reemphasize the "systems approach." The recent-ly revitalized "reinventing" movement in the federal government affirms boldly (albeit prematurely) the notion of a systems approach. Concurrent development and "virtual prototypes" are natural by-products of a model-based system development.

16.3.2 The Enterprise Challenge

Driven by demands for increased productivity and competitiveness, modern tools and methods are making positive impact on structured development of complex systems. NASA, for example, has published standards to be levied on prime contracts for the evolution of its complex systems which are based on solid foundations of systems principles, successive refinement, and concurrent engineering.[9] (It is too soon to determine whether these are actually intended for internal practice.) Some of the highly systems-cognizant companies seem to be on an aggressive track in the right direction.

However, the typical modern economic enterprise, the grand instrument of value and wealth creation and the most important design task of the manager as entrepreneur, is still too often allowed to evolve by chance. While the TQM and BPR movements have brought some clarity and sanity to managing change, there is still widespread confusion among different internal camps, with only infrequent reference to a unified strategic plan or vision. Adrift in rough seas without a compass and buffeted by waves and storms, it is small wonder that so many businesses "fail" to reach their destination. Even greater difficulties exist at various societal institutions such as health care and education, and at the many bureaucratic levels of government, where the elusive "vision thing" remains well out of focus, and conflicting political agendas of different factions obscure the very notion of a common purpose.

Enterprise development poses further complications to the system builder confronting the integration of technology with the human element. People are unquestionably the critical factor as operating subsystems in any enterprise, but reliable predictability of human behavior remains the mystifying domain of the behavioral sciences or even metaphysics and astrology. Managers often seek to cultivate an appropriate "culture" in which human skills are properly blended and carefully balanced with essential technologies and motivation and productivity is presumably optimized. Such cultural solutions tend to become "sacred cows" as they seem to thrive successfully for a time, as long as internal commitments and external conditions remain static. However, as inevitable changes challenge the organization and its bottom-line, new technologies, procedures, or "paradigms" may be introduced which upset the delicate balance of order. Today's "cults" tend to quickly become yesterday's fads in such times, while a new order is hastily sought under management by crisis.

Prominent reengineering experts have advocated radical, revolutionary approaches to redesign of the enterprise, starting with the renowned "clean sheet" approach. The difficulty with such ideas is too often not in the theory but in the implementation. Severe problems may indeed require drastic solutions, much like a critically ill patient requiring radical, life-saving surgery. Along with Hammer and Champy,[1] Deming[10] pulls no punches nor does he offer comfort in his imperative to "demolish" the system or to "obliterate" the process which has failed, but the "how to rebuild" part is left to the entrepreneur. In more moderate cases where the enterprise problems appear presently tolerable, the natural inclination of management is to defer the pain of analysis and redesign until some "later, more convenient" occasion. Such opportunities of course never arrive as the degeneration continues until only the radical option is left. Thus we tend as an enterprise, as whole industries, and even as a society to fall into the "boiled frog" syndrome explained by Senge, in which the frog placed in the pot of tepid water is still comfortable but fails to notice the very gradual rise in temperature, until it is too late and he is thoroughly boiled.

Revolution is surely the swiftest means for implementing change, but it is also the most destructive. An evolutionary process of adaptation, guided by specific goals linked to purpose together with predicted future scenarios of the changing environment, is certainly preferable to and more compatible with the "natural" process of system evolution by successive refinement. The tools of modern technology, coupled with disciplined systems methods together with management science techniques and a clear, consistent purpose, promise such a better alternative.

16.3.3 The Enterprise Model

Slater[11] exemplifies in his lucid work an approach to manufacturing which emphasizes a new process design method in manufacturing, reflected in the title of his book as the *integrated process management* (IPM). The essence of this paradigm, although focused on manufacturing, is the *model* of the *processes* through which all end-product results are achieved. Consistent with Deming's principles on process-driven quality, Slater argues that control of the key process variables will always assure that the desired result is achieved, and therefore contends that an intense, model-based focus on the process is warranted. He postulates the potential applicability of the IPM concept to a broad scope of processes, from making steel to hamburgers to health-care claim processes. In the past few years, numerous initiatives and methodologies have been advanced in a variety of industry and professional society forums. Improved tools and techniques have spurred the development of sophisticated models for development of enterprisewide computing architectures known as "client/server" systems. Increasingly these concepts, aimed at providing management with significant improvements in reengineering decision support, advocate methods which follow a systems approach in manufacturing as well as service industries. A logical extension of Slater's IPM and these emerging methods, captured in a comprehensive computer database and capable of representing dynamic behavior through simulation, is the structured enterprise model.

Structured enterprise modeling is a modern paradigm based on creation of a comprehensive behavioral "system model" of the enterprise. It is built on principles similar to IPM, with the concepts of "system processes" extended into all (practical) segments of enterprise activity. It further captures the nature and flow of data in the organization as it is processed and "consumed" in decision making. It also establishes an opportunity for communication throughout the enterprise, an essential requisite for sound teamwork, with unparalleled clarity and precision.

A fundamental characteristic of this new paradigm is the emphasis on use of graphics to describe system behavior, as opposed to lengthy narrative text. A major problem of "conventional" communication, usually text or speech, is often language itself. Words tend to get in the way of intelligibility, because of the diverse meanings and ambiguous sentence structure. An inevitable consequence of narrative information is that it frequently requires "interpretation" and "clarification" to resolve the ambiguity, often with no more success after many "clarification meetings" than before. Deming used a simple process flow graphic over and over to illustrate convincingly the major aspects of a process and their relation to customers. He clearly understood the power of graphic communication and used it very effectively throughout his distinguished career.

The more complex the system, the more unproductive is the reliance on communication by narrative language, particularly in international enterprises where language impediments and cultural barriers abound. Instead of narrative documents to describe simple or complex systems, ponder the common use of drawings, such as electronic circuit schematics or architectural design blueprints. As long as the team members understand the basic symbols depicted on the drawings and the fundamental concepts which govern the interaction and relationships of these symbols, communication is quickly established and ambiguity is avoided, allowing a convergence of focus to be drawn to the design itself. An electronics engineer in New Mexico, using a schematic representation of a circuit, has less difficulty communicating with other engineers in New Delhi, Bogotá, or Budapest, or all the above simultaneously, provided the symbols are commonly understood and the requisite competency for comprehending the nature of circuits exists. In our global, interconnected enterprises, business communication will be enhanced similarly by use of "symbolic" language supported by minimal textual description. The alternative, a narrative description of a building, a circuit, or a business process, becomes unthinkable and surely unworkable as a means of communication without the supporting drawings.

In structured systems modeling and development, an enormous communications advantage results from minimizing or eliminating laborious narrative documents. A structured system model can be represented entirely by symbols and interpreted like a schematic or generic blueprint, with limited textual descriptions where needed. Assimilation becomes a parallel rather than a serial process, a vital aid to comprehension. The power of such graphic models of systems for detailed, unambiguous communication across the enterprise or around the world, representing the processes, data, functions, and logical architecture plus a host of relationships, may prove to be one of its greatest benefits.

16.3.4 The Vision Thing

Clarity of purpose and its derivative, the vision of what is to be accomplished, is the crucial responsibility of enterprise leadership. It continues to remain a major stumbling block in systems evolution. Vision is one of those elusive metaphysical concepts which tends to perplex many of our leaders. This problem remains in spite of the near universal admonition of philosophers, motivational psychologists, and systems thinkers on the importance of goals and sharing of vision. The elusive "vision thing," a pejorative expression attributed largely to President Bush in reaction to frequent queries about administration policy in the "new world order" (another Bush expression), still plagues politicians, corporate leaders, and system builders alike. Many pundits attribute the lack of vision as having been a significant factor affecting the 1992 election outcome.

The initiation of an enterprise modeling effort, intended to mirror the commencement of an enterprise itself, must begin with a clear vision, translatable into a definition of measurable, verifiable goals to be achieved. By previous definition, the systems method requires that we first "specify the problem" and establish the "desired outcome." All subsequent requirements and design solutions which constitute the emerging system pyramid must be linked and traceable to the quantitative metrics of that desired goal. But exactly how that tangible goal, the capstone of the pyramid, comes into existence in a quantifiable way is beyond the scope of structured logical analysis. Rigorous methodology may never answer the fundamental question of *what* we choose to do or *why*; it can help only with the *how*.

The challenge of vision formation remains a critically important prerequisite to systems development as well as to successful enterprise change management. It can be neither bought off the shelf in the vision department of your local supermarket, nor captured in a quick-trip seminar. It typically resides only in the deep recesses of the enterprise leader's imagination, where—commingling with purpose and beliefs—it manifests itself in some perceptible form. In his discourse on personal mastery, Senge links the process of vision creation to a skill which is submerged in the subconscious. He argues that the integration of rational thought with intuition, sifted through the subconscious mind, will ultimately reveal a visionary purpose, which he stresses must become a "shared" vision throughout the "learning" organization. Slater refers to a similar process as "imagineering," letting the imagination soar with lofty ideas and then engineering them back to earth.

None of these definitions is logically complete or inherently satisfactory. Yet for almost any goal-seeking enterprise, it is virtually impossible to overstate the power of shared vision, derived from purpose, and supported by a credible base of beliefs. By whatever label it may be defined, Peters and Waterman[16] clearly indicate in their work that the successful companies have mastered its essence, both in creation and sharing across the enterprise. A loose compilation of the process is attempted in Fig. 16.7,

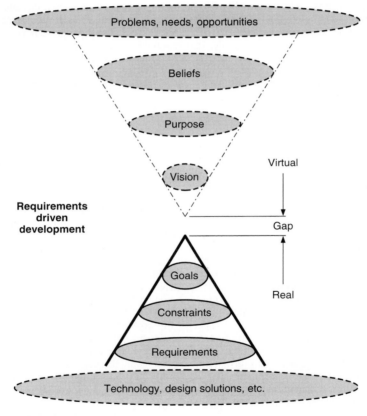

FIGURE 16.7   Flowdown of vision.

which shows a vision flowdown from a virtual "desired" domain, linking vision to a concept of purpose and need through a filter called "belief," that amorphous and vast organizational porridge of assumptions, values, guiding principles, and suspected limitations, often referred to as the "corporate culture." While much has been written on the power and importance of vision, precious little is known about the essence of organizational culture and its foundation of beliefs. The important postulates are that the vision must be shared with the enterprise team, revisited, and reexamined periodically in light of purpose and beliefs, and the enterprise goals which exist in the tangible, visible, and measurable domain must be quantified and linked to the mystical vision.

16.3.5 Requirements Flowdown

Given the meaningful expression of enterprise goals linked to a vision and purpose, firmly grounded in basic beliefs and core values, and armed with the key principles governing system behavior, the "engineering" focus must shift to the capture and formulation of the system requirements. In his seminal work on general systems theory, Ludwig von Bertalanffy argues that any dynamic system can be represented (at least in theory) by a set of simultaneous differential equations from which all system states and behavioral relationships can be derived.[15] While this approach may be overly troublesome or impossible in a complex enterprise, there is a useful value in at least attempting to establish some quantitative framework in preparation for enterprise modeling. Some quantitative or subjective measures of goal formulation will be necessary to evaluate progress and to "see when we get there." The communication and ensuing interactions on a team wrestling with this formulation can uncover obscured relationships and promote keen insight which unleashes creative energies in systematic problem solving. A great deal of progress can be achieved through effective communication even if the "system equations" are never completely defined or "solved" in the classic sense.

The enterprise goals or mission statement ought first to be expressed in the classic *management science goal programming* format, maintaining a sharp focus on the need which defined the original purpose. At this early stage, any reference to "solution" or "technology" must become subordinated to understanding the need, from which requirements will be developed to drive the design. It is important to link the statement of goals or objectives as directly as possible to the enterprise purpose, and thus avoid the pitfalls of specifying a premature solution in advance of deriving requirements. Top-level goal statements should be devoid of "design" orientation, such as striving "to achieve a level of 90 percent automation" or "acquire laser technology." These constitute only some alternative means, perhaps among potentially thousands, to satisfy requirements which must be understood before exploring the means. Premature focus on the solution, or the "right technology," poses the risk of obscuring the true need which is fundamental to our definition of purpose. It also tends to "lock out" other, potentially superior alternatives, since we have already "locked in," by stated objective, an approach which reflects our "preferred" solution.

Where objectives can be conceived only in qualitative form (e.g., employee morale, customer satisfaction, company prestige), the concepts of utility theory should be employed to convert these goals into some numerical form, even if no more elegant than the usual "scale of 1 to 10" convention. The simple point to all this is that there must be an acceptable *basis for evaluation* of the system design or changes which is understood to rate the level of satisfaction achieved. As the frequently quoted proverb teaches, "If you can't measure where you are on the road to your goals, how will you know when you get there?"

Quantitative methods suggest a rich set of alternative approaches to the meaningful formulation of these concepts, which are well supported by computer tools. In broad, general form, this enterprise goal formulation minimally includes the following information:

1. *A clear statement of purpose and goals—what* is to be achieved, and in what order/priority? *Why* is this important? How will it be measured?

2. *An itemization of the decision variables*—which elements can be "controlled"? What is the range of their values?

3. *An assessment of the constraints*—which elements, e.g., resources, are "limiting" factors? What are the limits?

4. *A formulation of the objective (goal) function*—what are the (mathematical) relationships between the objectives and the variables?

5. *A formulation of the constraint expressions*—what are the (mathematical) relationships between the variables and the constraints?

If all the suggested formulations were feasible in precise mathematical form, one could readily apply a variety of optimization algorithms to arrive at the "best" overall decision. In any complex system and particularly in a people-based organization, this is usually impractical or impossible, and therefore rarely attempted. A fundamental tenet of systems thinking is that humans are after all *not* machines, and should not be so considered. This argument nevertheless misses the point that organizations are purposeful systems, and the human elements of that system are willing participants (or ought to be) in the pursuit of the systems goals, even if that requires "suboptimization" of the human (individual) purpose. The dilemma of achieving balance between system and subsystem goals remains one of the major challenges of the systems age, and frequently confounds reengineering efforts.

From this top-level set of goal formulations, quantitative or otherwise, the process of requirements extraction proceeds to successively lower levels. The basic question which is central to this flowdown is: "Given that I must achieve (objective A), how can I best do that?" Tradeoff analysis and decision making is an essential aspect of resolving the barrage of issues which inevitably surface during this process. Although the enterprise modeling methodology is not specifically software or computer-dependent, a practical project implementation clearly requires support of a suitable computing platform and graphically structured knowledge base. A computerized database captures the refined requirements and vital model data in a central repository, including all historical decisions and knowledge which become the foundation of the model.

16.4   POWER TOOLS OF THE SYSTEMS ERA

16.4.1   Workflow Maps

The previously described principle of purposeful, successive refinement in system evolution, coupled with the premise of mutual economic growth based on satisfactory transactions between "partners," supports a new, graphical view of enterprise behavior. The combined concept yields an illuminating view of the essential interactions between people, both external and internal to the organization. In some of its variant forms, the concept has become a cornerstone in the accelerating trend to transform companies into "internal market"–based units, centered around teams of "intrapreneurs."[17] Halal identifies this evolution in corporate America as basically attributable to the same types of forces which recently toppled the rigidly centralized hierarchies of eastern Europe. Enabled by the information revolution, these corporate transitions hold forth the promise of vastly improved productivity, at the likely expense of increased complexity.

The tools and methods of the enterprise modeling paradigm provide means of managing change and complexity, regardless of the form or size of the enterprise unit. Since people are still the principal decision makers in any enterprise, it is rational to

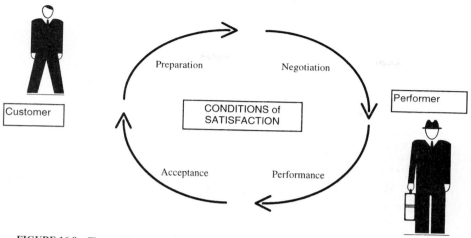

FIGURE 16.8   The workflow process map.

explore how commitments to achieve results and decisions between alternatives take place among people within the organization. Figure 16.8 illustrates an efficient yet powerful method of process logic and flow definition, capturing organizational interactions based on the concept of "workflows" and successive refinement. With the rigorous focus on both customer and performer roles in all transactions, and with computer-aided mapping logic and consistency checking, these techniques are considered superior to standard flowcharting and block-diagram representation of process logic. Unlike flowcharts, which were originally intended to specify flow of data or logical sequence in computing systems, workflow maps emphasize transactions and commitments between people (humans), the essence of the previously defined economic cycle of growth. Thus, completed workflow maps develop into effective "blueprints" of enterprise activity, compact graphic representations of organizational commitments and responsibilities, providing the added side benefit of precise, nonambiguous communications throughout the enterprise.

Basic Workflow Process Map.   The essential concept of workflows is depicted in Fig. 16.8 showing the basic element of a workflow map, called a *workflow cycle.* The definition of each workflow cycle begins by identifying the customer and the performer in their respective roles. A key criterion of this partnership emphasis is that activities are completed only by customer acceptance of work which is compliant with the stipulated "conditions of satisfaction" (i.e., requirements), evidenced by the information item which indicates acceptance. Comprehensive discussion of workflow mapping is provided by White and Fisher.[19]

All work proceeds through the cycle by passing through four sequential phases, identified as preparation, negotiation, performance, and acceptance. Where secondary or subsidiary workflows are used to further define process logic to a lower level, the link arrows identify the path of sequential activity. The power of this representation is in recognizing that all flows begin with the customer (entry to preparation) and end with the customer (completion of acceptance). This perspective ensures that the original focus (*why* are we doing something) is not likely to be obscured, regardless of the complexity of the maps. An example of a top-level map is shown for illustration depicting the generic (simplified) operations process.

16.4.2 The EMSA Methodology

The workflow maps described above (see also Figs. 16.9 to 16.11), which are well supported by available software tools, provide an ideal starting point for the collection of data governing enterprise purpose, process activities, and conditions of satisfaction. The mapping process itself provides an effective means to foster team communication through the unambiguous use of the workflow symbol. The identification of customer and performer roles and the rigorous focus on customer satisfaction is an essential step of enlightenment and knowledge capture.

The complete *enterprise modeling by structured analysis* (EMSA) paradigm (see also the context diagram in Fig. 16.12), the rules from which the structured enterprise model will evolve, are graphically depicted in the simplified diagram of Fig. 16.9. The

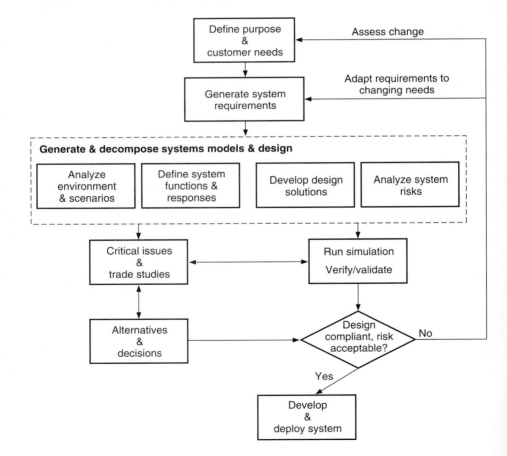

FIGURE 16.9 The EMSA paradigm.

System development process

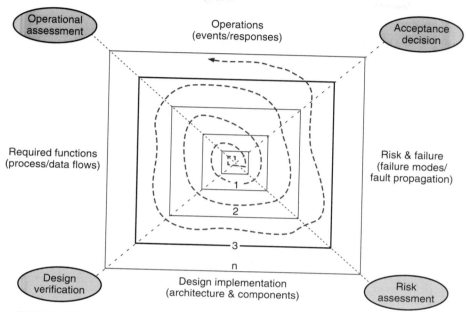

FIGURE 16.10 Spiral development.

diagram can be read like a process flowchart, with time flowing from top to bottom. The process begins with data captured from interviews, workflow maps, the formulated goal statements, and other available enterprise information. A modified system context diagram as shown in Fig. 16.12 with focus on the enterprise system and its environment is an essential supplement to the process. With these tools the task is to examine each of the facets of the enterprise pyramid in circular order (see Fig. 16.10) in a first pass around the top of the pyramid. It should be noted that all four tasks of facet analysis can in reality proceed concurrently. Since written text does not provide us the flexibility of concurrent description (graphic diagrams overcome this obvious limitation), they will have to be described in sequential narrative phases. Examples related to a generic enterprise "system" for each phase are shown (*italics*).

Operations View. The first facet of the pyramid is the operations view, which reflects the desired behavior of the system in response to external conditions. These conditions, derived from the stated mission requirements, will be defined as operational scenarios at the higher levels of hierarchy, and will be the source of events and stimuli to which the system must respond. A fully developed context diagram (Fig. 16.12) becomes an indispensable aid in this process. Porter[18] defines scenarios as "an internally consistent view of what the future might be." He identifies scenario building for both enterprise and industry as a valuable means of exploring uncertainty and its consequences, and describes it as a key strategy planning tool. Quantitative decision theory refers to such conditions as states of nature, and seeks to assign probability values to their occurrence as part of a decision process. The critical point at this level is

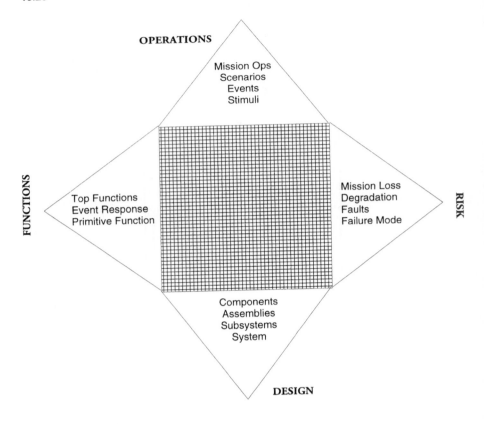

FIGURE 16.11   Inside the pyramid.

to consider thoroughly the entire life cycle of the enterprise, and anticipate all credible scenarios which may be imposed.

Scenarios can be considered the end result of correlated cause-effect relationships, and can be thus analyzed or statistically inferred by suitable methods. They will be refined at lower levels of hierarchy into more detailed events, which, in turn, produce discrete, observable stimuli crossing the system boundary. At each level of iteration, consistency checking must be performed to ensure that all stimuli are hierarchically "connected" to top-level scenarios, and traceable to the "internal" system functions.

> *Typical enterprise operations scenarios include various economic indicators such as the climate for improved business conditions, attributable to lower interest rates (the cause), and producing opportunity for increased orders (the events). Numerous other events may emerge which are traceable to this singular scenario. The discrete stimulus resulting from a "customer order" event becomes the specific purchase order.*

Functions View.   This facet is the view in which the system's internal processes are hierarchically and horizontally defined, in accordance with the stated requirements

Purpose:

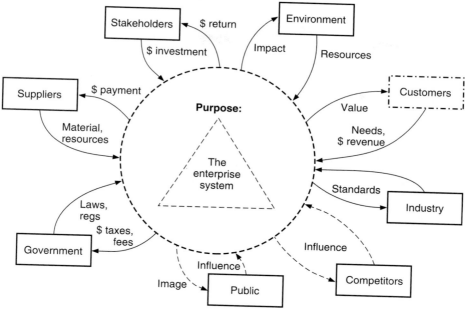

FIGURE 16.12   The enterprise context diagram.

and the expected operational scenarios. For each scenario, a top-level function must be defined which produces a satisfactory system response, compliant with all predefined conditions of satisfaction. Each set of such required functions must be internally "connected" to reflect the entire process, the sequence of activity, and the flow of data in producing the desired response. The complete functional decomposition will show a hierarchy of required functions, from top-level scenario responses to low-level "primitive" functions, internally connected in appropriate order.

> *The enterprise functional response to the triggering event (the purchase order) includes all internally "connected" activities or functions which the enterprise must undertake to deliver the expected output result (product or service) in order to support the mission (earnings requirement). The overall response may include more specific "outputs" beyond the actual product or service. There may be various forms and documents expected or required as a condition of the order, such as acknowledgments, shipping information, product details, and project status, all traceable to the single event "trigger."*

Architecture View.   The architecture (or implementation) view shows the deployment of all resources, including pieces, parts, facilities, equipment, humans, and software elements, which constitute the system implementation. The hierarchy of design indicates the logical partitioning of these elements into subsystems. For consistency verification at each level, for each function which has been identified, an allocation of the function to an architectural element must be established. The specific interface

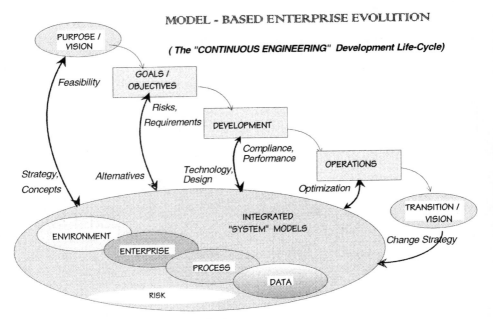

FIGURE 16.13   Model-based life-cycle development.

between architectural elements, whereby exchange of material, data, or energy occurs, is also identified within this facet. The set of all such allocations can be identified in a matrix form, which provides an ideal means of verification planning.

> *The resources needed to implement the required functional response of product delivery include all tools and equipment, the people, the facilities, and associated elements used in producing the item as well as related essential outputs (such as support and billing). For consistency, all allocation of resources must be traceable to specific functions identified in the functional hierarchy. The set of architectural elements will have vertical relationships which are depicted in the architectural hierarchy.*

Risks and Failure View.   The fourth facet is depicted as the dark side of the pyramid, in which the myriad uncertainties must be pondered and reconciled. Consideration must be made for possible faults, both in equipment and in humans, along with their propagating consequences, as well as the effects of increasing entropy which may lead to degeneration and premature failure.

For each architectural element, a set of failure modes or degradation states can be postulated along with the associated loss of function. This will allow the fault propagation paths to be identified, and their effects on the overall mission predicted. The purpose of the fault assessment process is to completely understand the nature and magnitude of risk which may arise with a given design under assumed scenarios. Such risk assessment may include the consequences of total or partial failure, loss of profit or opportunity, the potential for safety hazards, injury to the environment, product liability situations, labor strikes, and other undesirable disasters. According to Porter, the behavior of the competition under certain scenarios may also pose suggested risk factors to be contemplated. The risks discovered in this facet become revised requirements for operations planning.

Enterprise risk factors to be considered which may impact the functional response of product delivery include failure modes of elements (equipment and people), which may cause schedule slip, product defects, cost overruns, etc. Propagation of such faults and their end result, and the potential loss of order or customer, can be estimated to a range of probabilities. At a strategic level, risk assessment will focus on factors which may be detrimental to the achievement of goals, e.g., liability issues of significant exposure, loss of market share, adverse regulation, and disruption of resources.

Review and Acceptance. Following the first and each subsequent level of pyramid definition, a procedure of analysis is performed to increase confidence in the emerging design. This validation procedure consists of integrity checks which ensure that each operational condition has been accounted for, that a functional response exists for each such condition, that an architectural element exists for each required function, and that failure mode risk assessment has been done. Projected performance analysis and life-cycle cost estimations are further key components of this review. Enterprise management must contemplate the emerging design and the consequences of potential failure modes. If the probabilities of lost customers (due to poor products or performance) is at an unacceptable level, a new requirement must be stipulated to improve internal processes. These iterations continue until the risk level becomes "acceptable," as shown in Fig. 16.9.

The completion of each layer is dependent on management acceptance of the previous level results in both compliance with requirements and calculation of risk, and most importantly in compatibility with the shared vision. Before proceeding to the next level, management may elect to consider alternative solutions, or a modification to requirements. In either case, the loop is repeated at the same level, taking into consideration the desired changes and recording the different results to support the management decision.

The use of the computer-model database greatly facilitates such "what if" types of tradeoff analysis, providing a rich set of alternatives for the decision process. All the tools of scientific management, both quantitative methods and qualitative reasoning, can be brought to the table and focused intelligently on the critical issue. A wealth of data results from dynamic simulation of model behavior under a broad set of scenarios. The uncertainty of decision making will be significantly reduced, permitting numerous risk-abatement solutions to be evaluated. Only after the preferred alternative is chosen and the risks accepted should the modeling proceed to the next level. However, the decision process can be replayed at the higher level any time as external conditions change, providing an excellent vehicle for continuous change evaluation.

16.4.3 Inside the Pyramid

Explorers of the ancient pyramids have often related tales of the vast treasures of the Pharaohs stored inside the pyramids, their "eternal palace." The enterprise pyramid, the "store of knowledge" for the comprehensive enterprise model, presents similar riches to the enterprise decision makers. The "successive refinement" of the increasingly detailed model continues to add gems of critical insight into system behavior. The downward spiral of model evolution to lower levels of the pyramid concentrates the focus of the system designers as they continually resolve critical issues for the emerging design. The stored knowledge which accumulates in increasing detail from contributions of all model developers becomes a literal enterprise encyclopedia of knowledge, providing a ready source of information on each facet of operation and design. Complete operating procedures will emerge in the process, along with policy manuals, functional unit requirements descriptions, and all manner of equipment and facilities specifications, mostly in graphical form, but easily convertible to the "standard" narrative document. Most important, and consistent with worldwide quality standards for maintaining central configuration management of processes, this data will be maintained in a central repository, eliminating the frequently "tailored" versions of documentation manually kept in many organizations.

A clearer perspective of some of the contents of this knowledge base can be seen in Fig. 16.11, the view from inside the pyramid. Each of the four facets shows its own view arranged in a hierarchical order, with verified consistency and connectivity from top to bottom. The base of the pyramid is a multilayered virtual matrix, each plane of which shows the associations and relationships between paired, adjacent views. This includes operations-function, function-architecture, architecture-failure modes, and failure risk-decision views, as they are conceptually "mapped" to each other. Since the pyramid "depth" is one of some arbitrary initial judgment and continuously evolving to increasing depth, the multilayered mapping relationships can be generated at any horizontal layer, with increasing detail at each level. These matrix maps provide an ideal strategy and resource planning tool at the higher layers, and for the conduct of operational decision verification activity and ongoing change evaluation.

16.4.4 Simulation

A further "treasure," perhaps the most precious of all to be mined from the model pyramid, is the vast wealth of information to be derived from the "dynamic execution" of the model, i.e., the capabilities of simulating system behavior under nonstatic conditions. Performance studies under numerous dynamic scenarios can be explored to estimate performance in goal achievement, resource consumption, and inherent "structural" problems of a candidate organizational design. Various alternative solutions can be investigated and compared with "baseline" performance, generally intended to reflect the current state of the enterprise.

The potential of a graphic EMSA model with supporting workflow maps and simulation results for promoting enterprisewide communication is of noteworthy importance. Rather than becoming a faceless sacred cow, as the model grows, it becomes a visible, tangible entity, unambiguously displaying enterprise behavior through symbolic notation and increasingly precise simulation results. It begins to assume an identity of its own, with ownership shared by all model team members and contributors. The model and its content become the appropriate target of constructive critique and discrepancy resolution, leading to increasing accuracy and refinement in reflecting true enterprise behavior, and to the conceptualization of alternatives. In the process, a broad cognizance of processes, interfaces, and interrelationships naturally develops across the enterprise.

16.4.5 Life-Cycle Decision Support

Of momentous significance is the capability of the model to provide a suitable decision framework, a structure within which the "culture" of the organization can be

MODEL - BASED CHANGE MANAGEMENT

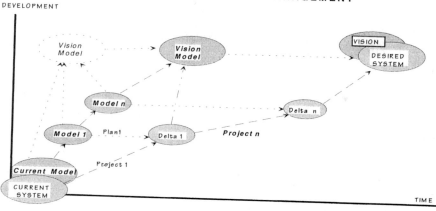

FIGURE 16.14 Model-based change management.

properly defined and practiced over its life cycle as shown in Fig. 16.13. As the organization proceeds through its life stages, many decisions will be based on model data or simulation studies under numerous scenarios. As it nears its apparent end of life, the model may be a vital source of vision or ideas for transition to a new form embodied in the "vision model," in essence, the guiding of the change management process through incremental or radical change with a model-based roadmap for project mangement as shown in Fig. 16.14.

Many will argue that human behavior is too unpredictable to yield to such structured analysis, or that there is something unethical or "dehumanizing" in attempting to do so. Unquestionably the human element of the enterprise will always bear the burden of decision making and, together with the appropriate culture, must continue to be the source of the mortar which binds the stones of the pyramid in place and allows for purposeful, continuing progress. The antimodeling argument ignores the fact that such models have long been successfully applied to the production process in which human functions and machine processes are all integrally represented, or that information system models routinely identify and allocate functions to computers as well as human operators. It also misses the mark on the ethics argument since decision makers have a perpetual need for improved decision support in pursuit of enterprise goals, and models reflecting predicted enterprise behavior present a most valuable decision support means.

Given a fully visible and simulation-ready enterprise model, with the shared vision and requirements flown down to the lowest process, each team member will know exactly how to approach the "continuous improvement" objectives of TQM or the radical change imperative of a major reengineering process seeking rapid transition. The effective mitigation of risk through prior model-based evaluation of decisions is an added significant benefit. People as decision makers will be empowered through the model to effectively manage change and minimize risk through continuous enterprise engineering, which benefits all members of the enterprise.

16.5 APPLICATIONS

Applications of the EMSA method based on computer models are continually evolving, but early experiences indicate fertile and prolific utilization potential. Models have been in development for NASA's original Space Station Freedom (SSF), which was an initial driver for the development of the EMSA methodology. SSF models include the core systems on orbit, developed individually as subsystems and later aggregated into a composite whole. The combined models are expected to eventually support integrated simulation for mission operations analysis and for design verification.

In spite of the complexity and sophistication of systems such as a space station, it is a virtual certainty that it pales in comparison to the complexity of a normal-sized business enterprise. A logical question is, therefore, "If sophisticated tools are available to facilitate complex system design, why can they not support the design of complex enterprises?" This question is being resoundingly answered in the affirmative through numerous, emerging software offerings. Commercial applications of EMSA have been in development with the use of such tools, focused on a variety of manufacturing and service sector models. Some salient examples are:

1. Preparing an economic behavior analysis of an enterprise "cluster" (a small group of interacting companies bound by a common purpose). The objective of the modeling is an attempt to evaluate and predict the likely successful outcome of prospective investment strategies, and possibly predict return on investment. Several detailed manufacturing process flows have emerged from the modeling exercise which have shown positive benefit, as has the clarification of interfaces and interactions between members of the cluster.

2. Health-care facilities models are under development, beginning with the enterprise model of a family medical practice, and gradually evolving to the larger representation of health-delivery systems. The eventual model is intended to include the entire network of community resources of integrated health delivery. Quality concepts of improved patient care are the intended goals, within cost containment constraints.

16.6 THE NEXT MILLENNIUM

16.6.1 Challenges of the Systems Age

In his concluding remarks on the systems age (also known as the "second industrial revolution"), Ackoff[6] poses four fundamental problems, and their implications for management, which are the consequences of how we view the world in this new age. Unfortunately, he argues that today's management is ill-prepared to address these challenges due to the continuing pervasiveness of machine age thinking. He cites these crucial challenges as developing the means for dealing with problems in terms of:

1. *Interdependency*—the process of dealing with problems as interactive sets rather than as individual, isolated conditions
2. *Dynamic adaptation*—the process of developing systems which are capable of adaptation, i.e., maintaining or improving effectiveness under changing conditions
3. *Humanization*—serving the purposes of the individual members and subsystems while simultaneously serving the needs of the overall enterprise

4. *Environment*—serving the needs of the environment by minimizing negative impacts while serving the purposes of the enterprise

These are weighty, ponderous challenges which directly confront the enterprise leaders as they prepare for the close of this tumultuous millennium and await the next, hopefully more orderly era. While projections of victory may be premature, and the degree of relevance still arguable, clearly a well-developed, validated model of an enterprise system and its environment addresses each of these challenges to some meaningful extent. While much work remains and the road ahead is long and difficult, the path has been illuminated, and the vehicles of technology and systems methods are ready to transport us. The structured enterprise model opens the way for continuous engineering of the enterprise as a means of addressing the current challenges and for adaptation to perpetual change.

16.6.2 Mirrors of Tomorrow

By now we should be inclined to question whether all this is really possible. A few years ago this was a question worthy of pondering and debate. Could an entire large-scale enterprise—say, General Motors—or a metropolitan regional economy, or the entire U.S. government, be effectively stored on a hard disk and depicted on a monitor in a computer behavior model? Could such a model provide sufficient depth and adequate sophistication to explore behavior under different scenarios, and permit tradeoff analysis for decision making? With today's state of advanced technology, the question is no longer "if," but only "how soon" and "how much." An equally critical issue is whether we can continue to evolve and compete in our complex world without such tools and methods. Although the first attempts at adaptation are not always successful, no enterprise can today afford to be the last to adapt. With the shifting of alliances in our globally competitive industries, it is perpetually imperative to stay one step ahead of the competition.

Today's technology does have its limitation, and the size of the model database grows rapidly as evolution to the greater details at lower levels takes place. However, it is a virtual certainty that any size of enterprise would benefit from even a few work-flow maps or structured levels of modeling. The law of hierarchy is also in our favor in permitting distributed development of smaller models at a subsystem level, with one additional top-level model reserved for "system integration." This has been the approach used on Space Station Freedom, which, with its distributed core systems and ground operations, has often been referred to as "the most complex undertaking of humankind."

And what about tomorrow? What opportunities will the relentless rush of technology offer to system builders at the dawn of the third millennium? In a fascinating new work titled *Mirror Worlds*,[20] subtitled "the day that software put the universe into a shoebox," Gelertner offers a futuristic view of very complex software models, showing comprehensive behavior in "virtual reality" of entire cities. He describes a vast network of sensors and instruments generating "oceans of data" on numerous activities, feeding on demand into computers running multimedia presentations and real-time models reflecting a variety of behavioral views of activity. The observer of this model could tour from his desktop any number of places in this "software city" and witness reality almost as it occurs. What is truly astonishing is his estimation that such systems and models could be constructed with today's software and hardware technology, and could be commercially feasible in a few years. Projections of future computing power continue to promise an astounding rate of increase, along with correspondingly more powerful software than today's awe-inspiring programs. Technology for effective and efficient decision support is available in abundance, and still getting better. What is needed is empowered teamwork and leadership willing to forge resolute commitment, guided by clear purpose and the courage of shared vision.

16.6.3 The Human Factor

Earlier in this work we have attempted to capture a philosophy of systems development in what is now called the "systems age," beginning with a historical perspective of its evolution and culminating in its destined manifestation, the structured enterprise model. Only the most salient features and benefits of the EMSA method have been described, as the full scope of the process and its potential wealth of payoffs is beyond the bounds of this chapter. Applications of the method are the essential means of providing feedback for improvement of the process, but these are still in the early phases.

It must also be recognized that there are numerous variations of emerging systems methodology supported by ever more sophisticated tools, prompting further successive refinement. Workflow analysis techniques continue to proliferate, as increasing demand for groupware tools in support of team-oriented enterprises continues. The arguments over feasibility and ethics of attempting to model "human behavior" continue to grow along with use of these tools. Behavioral scientists perspectives must be integrated into the methodology of modeling, and the full impact of human reactions to modeling and enterprise integration needs further assessment. In spite of the sophistication of the tools, the practical modeling of the human element of the enterprise, the identification of the failure modes and associated risks and ethics considerations will remain a major challenge. Lest we, too, fall into the trap of paradigm paralysis, we must continue to view the entire methodology process as evolutionary and perpetually seek its refinement.

16.7 CONCLUSION

In our complex, dynamic world struggling with its myriad of problems, the systems method is considered to offer a significant hope and a ray of light as our civilization gropes and stumbles in darkness toward the next millennium. Of all the needs and imperatives which confront our families, industries, and nation and the world community, none is more compelling than the need for economic health and continued growth. A strong, vibrant national and world economy is the only means to address global environmental issues and social issues, and make cooperative progress toward the realization of the greater purpose of humankind. It is the persistent underlying, beneficial cause in the desired virtuous cycle of our continued evolution.

The process of enterprise modeling described in this chapter achieves essential leverage in an evolutionary enterprise cycle through disciplined systems methods. It works by focusing energies and technology solutions on the essential goals and requirements, driven by a vision linked to purpose and belief. It promotes optimization by exposing waste, both in useless activities and in confusing, confounding communications which threaten to slowly boil us like Senge's frog. It imposes structure and order which can retard chaos and entropy and through which human motivation and productive culture can flourish, achieving economic benefit for all.

In the dark days of World War II, Franklin D. Roosevelt stated that "we have nothing to fear but fear itself." Although there are many in our society who still fear advancing technology, it should be indisputable that our civilization has nothing to lose from experimenting with better methods. Indeed, it should be the responsibility of leadership, in industry, academia, and government, to promote the most enlightened methods and best possible solutions which technology can offer to our many needs and challenges. Ultimate failure comes not from unsuccessful experiments, but as in Senge's frog, from ignorance to changing conditions and the eventual inability to respond, leading to the catastrophic failure of nonadaptation. The admonition for continuous improvement of the enterprise, a vital cornerstone of the TQM movement, must be extended from a slogan or religion to a disciplined process, consistent with the enterprise vision. Tough realism and scrutiny must be imposed on radical change initiatives to focus them sharply on the mission objectives. The means to bring both objectives to harmonious reality exist in the structured enterprise model.

16.8 SUMMARY

The deployment of modern technologies, no matter how sophisticated, will not of itself provide management with the essential means to secure the enterprise future. In spite of the undeniable, proven value provided by advanced technologies, they are principally only potential solutions to the underlying problems of enterprise evolution. Unfortunately, rather than focus attention on the core problem and its derivative requirements, management often expends great energy in seeking the silver bullet of the "right technology," as if tomorrow's semiconductor revelation or the newest computing system will magically transform the company to its coveted nirvana state. The modern "reengineering" movement has regrettably reinforced this misconception by frequently holding forth the promise of information technology (IT) and other advanced technology solutions as the swiftest way to achieve radical change. Thus we have created many fine solutions looking for the "right" problem.

To gain more valuable insight for strategic management, technology solutions must be viewed in the context of the broader perspectives of managing the adaptation of the enterprise to continuous change and increasing complexity. These factors, characteristics of the "environment" which is usually beyond management control, are the primary drivers of requirements to which technology solutions must be carefully matched and integrated with existing enterprise resources, human considerations, and various cultural, legal, and economic constraints. Such insight comes only from consideration of the whole enterprise "system," dynamically interacting with its environment while engaged in pursuit of some specified purpose. "Systems thinking" has been a hallmark of the "systems age" to which we have been gradually transitioning since World War II. "Machine age" thinking is the leftover residue of the prior age which created many of the conditions we are currently struggling with in our attempted, often ill-fated reengineering initiatives.

While systems thinking has been a dominant theme in business schools and an essential prerequisite for enlightened management, we find it may no longer suffice for dealing with relentless change and complexity. With these twin burdens of the systems age come also new needs and methods beyond mere "thinking" in systems. In the past decades, the availability of new technologies has fueled the demand for increasingly sophisticated defense, aerospace, and commercial "systems." This demand, in turn, has spawned an impressive array of high-tech "systems tools" for dealing with the inherent complexities of such systems, even including complex enterprises. With the application of rigorous methods and some adaptation of these tools, it is now feasible to practice "continuous enterprise engineering," supported by sophisticated computer models, to achieve the desired business process design or enterprisewide adaptation to change.

This chapter has traced our historical transition to the systems age and illuminated the difficulties of transforming our managerial and social systems for continuous adaptation. A perspective on system principles and evolution illustrated these points. It also provided a focus on a rigorous methodology for change management through continuous enterprise engineering, based on the discipline of systems thinking, application of systems methods, and the deployment of modern systems technologies.

16.9 REFERENCES

1. Michael Hammer and James Champy, *Reengineering the Corporation*, HarperCollins, New York, 1993.
2. Peter M. Senge, *The Fifth Discipline*, Doubleday, New York, 1990.
3. Jay Forrester, "The Next Great Frontier: Designing Managerial and Social Systems," *The Systems Thinker*, 4:1Feb. 1993.
4. Alvin Toffler, *The Third Wave*, Bantam Books, New York, 1980.
5. Russ Ackoff, *Creating the Corporate Future*, Wiley, New York, 1981.
6. Russ Ackoff, "The Second Industrial Revolution," speech to AT&T, 1989.
7. Steven Hawking, *A Brief History of Time*, Bantam Books, New York, 1988.
8. Margaret Wheatley, *Leadership and the New Science*, Barret Kohler Publishers, San Francisco, 1992.
9. *NASA System Engineering Standards Handbook*, 1992.
10. Mary Walton, *The Deming Management Method*, Putnam, New York, 1986.
11. Roger Slater, *Integrated Process Management*, McGraw-Hill, New York, 1991.
12. James Martin, *Information Engineering*, Prentice-Hall, Englewood Cliffs, N.J., 1989.
13. Derek Hatley and Imtiaz Pirbhai, *Strategies for Real-Time System Specification*, Dorset House, London, 1988.
14. Edward Yourdon, *Modern Structured Analysis*, Yourdon Press/Prentice-Hall, Englewood Cliffs, N. J., 1988.
15. Ludwig von Bertalanffy, *General Systems Theory*, Penguin University Books, Harmondsworth, Middlesex, England, 1973.
16. Tom Peters and Robert Waterman, *In Search of Excellence*, Warner Books, New York, 1982.
17. William Halal, *From Hierarchy to Enterprise: Internal Markets are the New Foundation of Management*, vol. 8, Academy of Management Executive, Pace University, Briarcliff Manor, N. Y., 1994.
18. Michael Porter, *Competitive Advantage*, Free Press, New York, 1985.
19. Thomas White and Layna Fisher, eds., *The Workflow Paradigm*, Future Strategies, Alameda, Calif., 1994.
20. David Gelertner, *Mirror Worlds*, Oxford University Press, New York, 1991.

CHAPTER 17

MANAGING THE "TECHNOLOGY GRADIENT" FOR GLOBAL COMPETITIVENESS

David J. Sumanth
Professor and Founding Director
UM Productivity Research Group
Department of Industrial Engineering
University of Miami
Coral Gables, Florida

John J. Sumanth
Research Assistant
UM Productivity Research Group
University of Miami
Coral Gables, Florida

17.1 INTRODUCTION

In today's rapidly changing technological environment, there is a great need for proper implementation and supervision of new technologies. Increasingly, we are seeing major restructuring within corporations in which personnel positions are either removed or reconfigured to suit the incoming foreign technology(ies). Therefore, the issues of technology transfer and the methodologies to achieve successful technology transfers need to be considered.

17.1.1 Technology Transfer Issues

Okko and Gunasekaran (1994) define technology transfer as "the spread of technology from one culture, country, or region to another." However, other definitions of technology transfer abound, depending on the perspective of an organization. Johnsrud (1994, p. 341) states that "industry perspectives generally portray technology transfer as an internal or *intraorganizational* technology management problem," whereas "government and academic organizations more often view technology transfer as an *interorganizational* activity." This difference in opinion can pose barriers to interorganizational technology transfer, and as a result, reduce the competitiveness of the industry. Too often, the modern industrial giants of our day have been guilty of "overisolating R&D, engineering, manufacturing, and marketing through organizational compartmentalization" (Stewart, 1989). Wood and EerNisse (1992) also presented a number of reasons for different bottlenecks in the technology transfer; one of the more important ones was the cultural gap between industry and the federal laboratories. The seemingly uncatchable "technology train" is causing companies to reevaluate which technologies they decide to keep and which they discard. A prime example of this is the health-care field. Medical breakthroughs and technological advances are almost daily occurrences, some of them costing millions of dollars. As Clemmer (1991) points out, these discoveries force us to answer tough questions such as "Which technologies do we choose? Where do we draw the line?"

17.1.2 Technology Transfer Methodologies

Because of these important technology transfer issues facing enterprises, several methodologies and solutions have been proposed by various authors in an attempt to achieve efficient and cost-effective means of technology transfer. A traditional approach, advocated by many federal laboratories, is the *"technology–push"* strategy. This method "pushes" patented discoveries and inventions out of the lab and into the marketplace (Johnsrud, 1994, p. 342). In contrast to technology push are *"market–pull"* approaches to technology transfer. "Market pull approaches emphasize that technology will only be transferred as a result of companies seeking specific technology from laboratories that can solve particular problems or meet strategic product development opportunities" (Johnsrud, 1994, p. 344). Another methodology of technology transfer was based on the *integration* of the concepts of manufacturing strategy and international technology transfer. Suite (1992) contends that joint ventures between local and foreign enterprises should be sought out in an effort to speed up technology transfer. He further argues that factors such as consulting entities, construction entities, and training organizations define the role of joint ventures in technology transfer. "Hence, joint venture provides a dynamic mechanism for developing countries to improve technological progress" (Okko and Gunasekaran, 1994).

In the rapidly changing business panorama, characterized by dramatically increased rates of interconnectivity between enterprises all over the world, and the overwhelming rate of advances in technologies, at least five tough questions need to be addressed:

1. *When* should a transnational enterprise withhold its proprietary technologies from its alliances?

2. *What* corporate *policies* are appropriate when the technology transferor has more advanced technologies than the transferee?

3. *Which* technologies must a company invest in, and *why?*

4. *Where* should a company draw a line in the growth or decline of a particular technology in use?

5. *How* should *national policies* with regard to technology transfer affect an enterprise's transfer policies?

The extent to which a transnational enterprise withholds its proprietary technologies from its alliances may well determine the rate of market penetration, the ability to provide barriers for competitive entry, and the ability to generate future alliances. The issue is *when* such an enterprise should withhold its proprietary technologies.

When a company has transferred its latest technologies to another enterprise, the transferee initially might have a greater dependence for know-how on the technology provider or transferor (e.g., licenser), but as time goes on, the transferee may advance the imported technologies to a point where the licenser may lose the competitive edge. The issue here is *what* type of corporate policies should be put into place to prevent such a scenario.

One of the greatest challenges is in determining *what* types of technologies a company must invest in, and what the justification is. The implications of a decision to invest in technologies of a particular kind can be far-reaching—economically, ecologically, politically, and culturally.

When a particular technology is being used, we know that it is only a matter of time before it reaches that portion of the "S curve" in which only marginal revenues are realized. When technology discontinuities occur while going from one technology to another, decisions have to be made as to the timing of new technologies to substitute the existing ones, with minimal negative impact of such discontinuities.

Depending on the nature of technology transfer by a multinational company, the national policy can affect the management of such technologies. The question then becomes one of assessing what impact such national policies would have on technology transfers.

Clearly, the answer to these five issues and questions are neither easily understood nor well documented in the open literature. We believe that our concept of "technology gradient" (TG) can provide a relevant and meaningful approach to addressing these questions either partially or fully.

17.2 DEFINITION OF THE TECHNOLOGY GRADIENT

Just as a thermal gradient exists between two bodies of different temperatures, there exists, in our opinion, a "technology gradient" between a *technology transferer* and a *technology recipient*. It represents the rate of change in a technology's advantages over those of the known ones. At least three possible scenarios are likely to occur for a technology gradient to exist, as shown in Fig. 17. 1.

Referring to this figure, the first situation occurs when a *technology imitator* tries to modify or adapt the technology concepts and methodologies from the original *technology introducer.* This is particularly true in the initial stages of the life cycle of a technology. For example, when IBM introduced its first personal computer in 1981, only a few dozen small companies began to emulate IBM's open architecture of PCs, and develop their own clone versions. As time went on, by the latter part of the 1990s, more than a hundred manufacturers of PCs have emerged in the world. Today, PCs are assembled even in small shacks and boats in some Far Eastern countries! The technology gradient substantially diminished between a technology introducer (IBM in this case) and the manufacturers of the clones. The second situation for technology gradient occurs when a technology transferee has a technical collaboration agreement with the technology transferer. Here, the gradient is more pronounced in the earlier stages of the agreement than in the later stages. Consider an American company transferring a new product technology to a company in a developing country. This could be the usual 16-year, technical collaboration agreement. In the first 5 years, the technology transferee is greatly dependent on the transferer, in fully understanding the technology in question. During this time, the technology gradient is somewhat predominant in a

Situation 1: Technology Imitator tries to modify or adapt the technology concepts/methodologies from the original Technology Introducer in the initial stages of the life cycle.

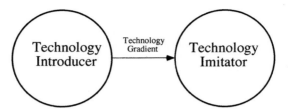

Situation 2: The Technology Transferee has a collaboration agreement with the Technology Transferer.

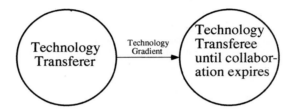

Situation 3: Technology Innovator/(Improvisor) goes beyond the capabilities of the original Technology Introducer in the later stages of the life cycle.

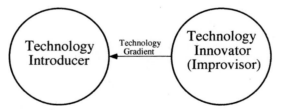

FIGURE 17.1   The "technology gradient" concept in some basic situations.

positive sense, from the American company's standpoint. However, in the last 2 or 3 years of the agreement, this gradient could have substantially decreased as the transferee gained considerable experience and know-how while working with the American company.

In a third situation, a *technology innovator* goes beyond what the original *technology introducer* has achieved; the gradient even flows in the reverse direction, with a positive advantage to the innovator. A familiar example is the television technology. Up until the late 1960s, many American companies, including General Electric, Zenith, RCA, and Magnavox, were major TV producers, but Japanese companies such as Sony, Hitachi, Mitsubishi, Panasonic, JVC, Toshiba, and Sharp have been innovating TV technology, both in product and process, so much so that they now have outsurpassed the American companies. As of this writing, only Zenith manufactures TV sets in the United States, but a major part of even this company has been purchased by a foreign TV parts supplier!

It must be pointed out that the three situations discussed and depicted in Fig. 17.1 are only *some* of the *basic* scenarios, not all the possible ones.

17.3 TECHNOLOGY PROVIDERS AND RECIPIENTS

Figure 17.2, the technology gradient chart, shows four *basic* types of technology providers and recipients, namely

- Technology leader
- Technology yielder
- Technology gainer
- Technology loser

17.3.1 Technology Leader

In the context of the technology gradient, we define a *technology leader* company as one which consistently shows an upward technology gradient trend in the proactive zone (Fig. 17.3).

Conceptually, the lower control limits of the "technology gradient threshold" separate the proactive zone from the reactive zone, based on the "technology advancement score."

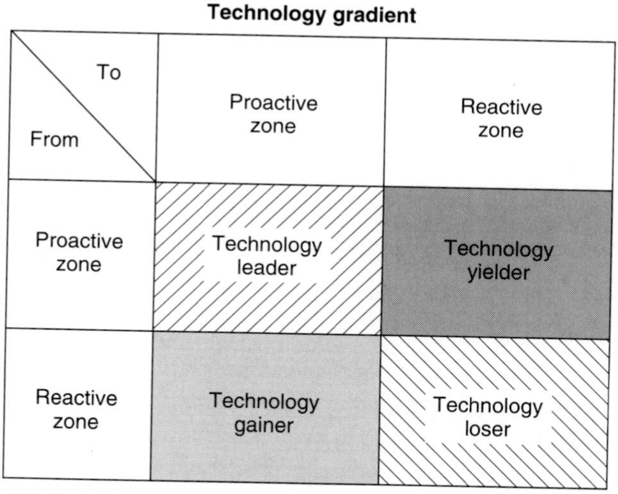

FIGURE 17.2   The four basic types of technology providers and recipients, based on the technology gradient.

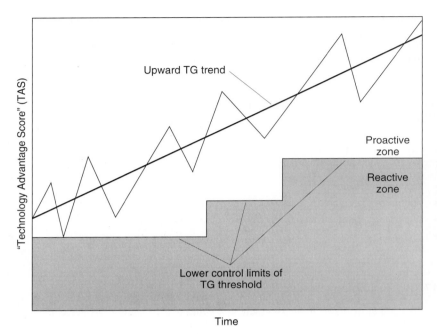

FIGURE 17.3 Example of a technology leader.

17.3.2 Technology Yielder

A *technology yielder* company starts out in the proactive zone, but its technology gradient rapidly moves it into the reactive zone (Fig. 17.4).

17.3.3 Technology Gainer

A *technology gainer* company starts out in the reactive zone, but maintains an upward TG trend into the proactive zone eventually (Fig. 17.5).

17.3.4 Technology Loser

A *technology loser* company starts and ends in the reactive zone, wherein the TG has a downward slope (Fig. 17.6).

17.4 MEASUREMENT OF THE TECHNOLOGY GRADIENT (TG)

We shall now present a numerical approach to measuring TG, using data from two hypothetical companies, company A and company B. For these two companies, the

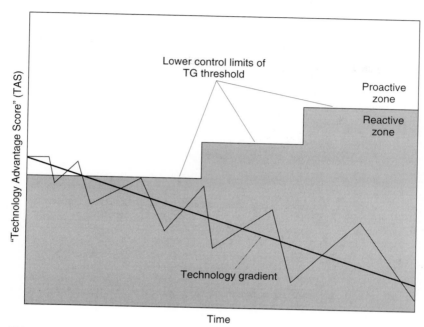

FIGURE 17.4 Example of a technology yielder.

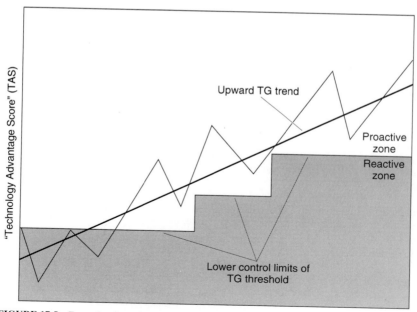

FIGURE 17.5 Example of a technology gainer.

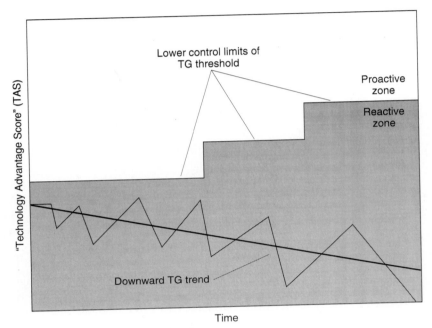

FIGURE 17.6   Example of a technology loser.

basic data in terms of commonly monitored technology indicators are shown in Tables 17.1, and 17.2, respectively.

Referring to these two tables, indicators numbered 1, 2, and 4 are usually available from the R&D department or function; indicators 3 and 6, from the comptroller or the accounting–finance function; and indicator 5, from the marketing function. Additional indicators, as relevant to the technology under consideration, can be included in the tables.

However, for the purpose of showing the computational scheme for TG, let us just work with the six indicators. For this illustration, we are assuming that a particular technology was introduced in year 1 for company A, and in year 2 for company B. A rank value on a 1 to 5 scale is assigned to various realistic ranges of each indicator, as shown in Table 17.3. A team of experts within a *technology group* (representing the areas affected mainly by the technology) confer and come to consensus about these rank values.

Now, each indicator is assigned a weight, reflecting the relative importance of the indicator, as perceived by the technology group. The sum of the weights must be 1.0. Using the rank values and weights, we compute, as shown in Table 17.4, the *technology advantage score* (TAS), and the *technology advantage index* (TAI) from the following formulas:

$$TAS = \Sigma(rank \times weight)$$

$$TAI = \frac{TAS \text{ for a particular time period}}{TAS \text{ for the base period}}$$

TABLE 17.1 Raw Data for Company A

Indicator	Description	Year 1*	Year 2	Year 3	Year 4	Year 5
1	R&D as percent of sales	3.0	4.7	5.1	5.3	5.4
2	Number of patent applications submitted	2	6	8	9	7
3	Net income per R&D dollar invested, $/$	101.5	150.7	169.6	171.2	172.8
4	Number of new products developed	2	5	6	6	5
5	Market share, %	13.4	17.7	18.1	18.3	18.5
6	Net profit margin, % of sales	1.7	5.7	7.1	7.5	7.7
	Other relevant, "user-definable" factors (e.g., sales growth, buyer concentration, earnings on net worth as required)					

*Year of introduction of the technology.

TABLE 17.2 Raw Data for Company B

Indicator	Description	Year 2*	Year 3	Year 4	Year 5	Year 6
1	R&D as percent of sales	1.0	1.6	3.1	4.2	5.7
2	Number of patent applications submitted	1	3	7	9	11
3	Net income per R&D dollar invested, $/$	25.2	41.2	93.2	171.1	200.3
4	Number of new products developed	1	2	4	6	9
5	Market share, %	3.1	6.3	12.4	15.8	19.7
6	Net profit margin, % of sales	5.6	8.3	9.7	10.6	12.4
	Other relevant, "user-definable" factors (e.g., sales growth, buyer concentration, earnings on net worth as required)					

*Year of introduction of the technology.

For example, TAS for year 2 for company A is equal to

$$(4 \times 0.1) + (2 \times 0.2) + (4 \times 0.1) + (3 \times 0.2) + (3 \times 0.3) + (2 \times 0.1) = 2.9$$

Assuming year 1 to be the *base period* (period of reference), the TAI for year 2 = 2.9/2.1 = 1.38. Then, the technology gradient (TG) is calculated as the average percentage change in the technology advantage index. Thus, for example,

$$\text{For year 2:} \quad TG = \frac{1.38 - 1.00}{2} \times 100 = 19.0\%$$

$$\text{For year 3:} \quad TG = \frac{1.71 - 1.00}{3} \times 100 = 23.8\%$$

Similar computations are shown for company B in Table 17.5.

By graphically plotting the TAS for each year, we obtain Fig. 17.7. The graphical plots for the TGs are shown in Fig. 17.8.

From Fig. 17.8, while comparing the TGs for the two companies, we can see that

- Company B has a much larger TG than company A, beginning in year 3.

TABLE 17.3 Rank Values as Determined by the Technology Supplier–Technology User

Indicator	Description	Unit of measure	Range	Rank value
1	R&D as percent of sales	%	0–1.0	1
			1.1–3.0	2
			3.1–4.0	3
			4.1–5.0	4
			≥5.1	5
2	Number of patent applications submitted	#	0–2	1
			3–6	2
			7–9	3
			10–11	4
			≥12	5
3	Net income per R&D dollar invested, $/$	$/$	25.0–75.0	1
			75.1–100.0	2
			100.1–160.0	3
			160.1–210	4
			≥210.1	5
4	Number of new products developed	#	0–1	1
			2–4	2
			5–8	3
			9–10	4
			≥11	5
5	Market share, %	%	0–5.0	1
			5.1–10.0	2
			10.1–15.0	3
			15.1–30.0	4
			≥30.1	5
6	Net profit margin, % of sales	%	0–5	1
			5.1–7.0	2
			7.1–9.0	3
			9.1–12.0	4
			≥12.1	5

- Company A's TG improved moderately from year 2 to year 3, but then declined drastically, all the way through year 5.
- Company B had a dramatic positive growth in TG up to year 5, but then leveled off in year 6.

17.5 MANAGING NEW AND EXISTING TECHNOLOGIES BY INCORPORATING THE TGs

In Fig. 17.9, a conceptual framework is presented to manage any technology—new or existing—by applying the technology gradient concept, described above. To fully understand this conceptual framework, the reader is advised to refer to Chap. 3 in this handbook.

TABLE 17.4 Computation of Technology Advantage Index (TAI) for Company A

Indicator	Weight	Rank value				
		Year 1	Year 2	Year 3	Year 4	Year 5
1	0.1	2	4	5	5	5
2	0.2	1	2	3	3	3
3	0.1	3	4	4	5	5
4	0.2	2	3	3	3	3
5	0.3	3	3	4	4	4
6	0.1	1	2	3	3	3
Technology advantage score (TAS) = Σ(rank×weight)		2.1	2.9	3.6	3.3	3.3
Technology advantage index (TAI)		1.00	1.38	1.71	1.57	1.57
Technology gradient (TG), %/year		0.0	+19.0	+23.8	+14.3	+11.4

TABLE 17.5 Computation of Technology Advantage Index (TAI) for Company B

Indicator	Weight	Rank value				
		Year 2	Year 3	Year 4	Year 5	Year 6
1	0.1	1	2	3	4	5
2	0.2	1	2	3	3	4
3	0.1	1	1	2	4	4
4	0.2	1	2	2	3	4
5	0.3	1	2	3	4	4
6	0.1	2	3	4	4	5
Technology advantage score (TAS) = Σ(rank×weight)		1.1	2.0	2.8	3.6	4.2
Technology advantage index (TAI)		1.00	1.82	2.54	3.27	3.81
Technology gradient (TG), %/year		0.0	+41.0	+51.3	+56.8	+56.2

Referring to the lower right portion of Fig. 17.9, we see that technology planning is accomplished primarily through the "technology cycle." Then, this technology planning is combined with the business planning, thereby resulting in a powerful, realistic, and objective integrated business strategy. Many companies are only beginning to develop such integrated business strategies in recent years. Sumanth, in his earlier work (1984, 1985, 1990, 1994), demonstrated the power of profit planning and performance measurement through his *total productivity model* (TPM). This model, apart from many useful capabilities, can provide "what if" scenarios to predict profits at different levels of total productivity (total tangible output divided by total tangible input in constant dollars with respect to a base or reference period). The business planning exercise is strengthened considerably by interfacing it with this type of profit planning based on the TPM. The upper portion of Fig. 17.9 depicts six possible technology gradients, depending on the various conceivable technology levels a company might be operating at, at any particular time. Within the context of this statute, a company can

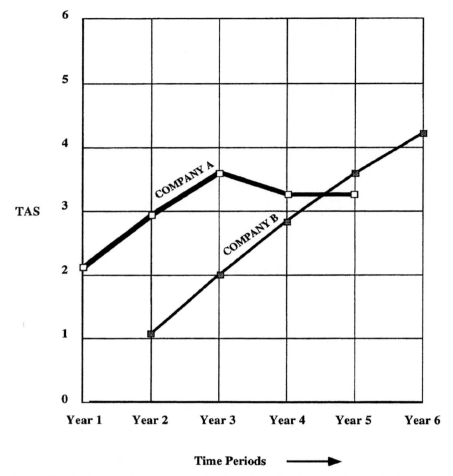

FIGURE 17.7   A graphical representation of the technology advantage score for years 1 to 6.

develop an integrated business strategy by applying the 10-step systematic procedure outlined in Fig. 17.10. The framework becomes a pragmatic tool when this procedure is followed; Fig. 17.10 is self-explanatory.

17.6 BENEFITS OF THE TG APPROACH

The TG approach has some unique benefits while managing new and/or existing technologies, particularly for multinational enterprises:

1. Enterprises of any kind—and in particular, companies willing to have transnational alliances—can manage the nature of their business relationships much better by having a clear understanding of their roles in managing their respective technologies,

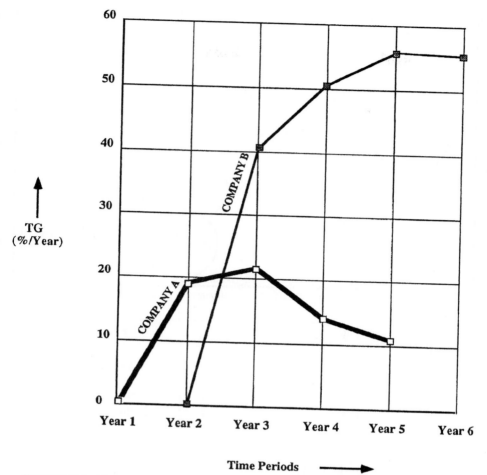

FIGURE 17.8   The technology gradients for a particular technology competed by company A and company B.

because they can predetermine their TGs and position themselves in the market for a mutual advantage.

2. While introducing new technologies, a company can determine a TG profile for the short-, intermediate-, and long-term periods. This will enable better strategic planning, as the company will know some of the major challenges as well as opportunities confronting such technologies in the global markets.

3. Decisions to phase out existing technologies can be made more objectively than at present, by using the TG computations as part of the economic and technical feasibility studies.

4. Measures taken to advance the present technologies can be based on the possibility of success from the TG standpoint. This would avoid expensive capital investments. For example, if a company has a TG much better than its competition for one

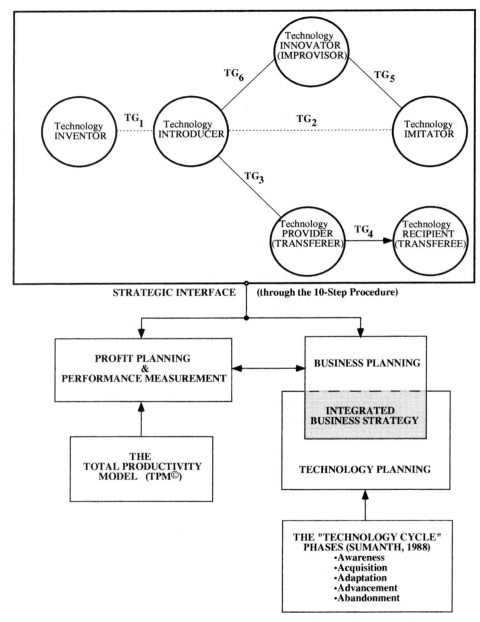

FIGURE 17.9   A conceptual framework for managing the technology gradients through a strategic interface.

Notation: new technologies (type N); existing technologies (type E)

Step 1 Awareness possibility

- Determine the expected TG for each of the type N technologies that are commercially feasible for adoption. Do the same for the type E technologies that are in the "advanced" phase of the technology cycle.
- Rank-order these technologies by the TG potential.

Step 2 Acquisition capability

- Conduct technical and economic feasibility studies for all the type N and type E technologies rank-ordered in step 1 above.
- Select the type N technologies that will provide the greatest and fastest market share, customer retention, and return on investment.
- Purchase the selected type N technologies and install them.

Step 3 Adaptation potential

- For each of the type N technologies, determine the potential for adaptation to your particular sociotechnical environment. If the potential is high, set aside such technologies for adaptation.

Step 4 Advancement potential

- Out of the present type E and type N technologies, determine the possibility for advancing these technologies to result in the highest possible TG value.

Step 5 Abandonment possibility

- Make a systematic assessment of the TG for the type E technologies and rank-order them from the lowest to the highest TG.
- Eliminate the type E technologies with the lowest TG, based on a minimum desired TG threshold.
- Generate technology transfer strategies to transfer the remaining type E technologies to those countries or operations *if* the end user is interested.

Step 6 Business planning

- Formally introduce the "TG analysis" into the business planning strategies to maximize and/or leverage the technology advantage as a strategic business weapon. This is almost essential in today's global enterprises.

Step 7 Profit planning

- Apply Sumanth's "total productivity model" (TPM) or his CTPM, to determine the impact of various TGs on the profits of your enterprise for the next 5 to 10 years.

Step 8 Reward system

- Apply a total productivity-based reward system to motivate employees and suppliers, customers, and stockholders, community and society.

Step 9 Contingency planning

- Using the TG concept, prepare contingency plans for the most unforeseen and the most unexpected appearances of technologies that never were envisaged.
- For every product line or service unit, do the impact analysis in the event of "surprise technologies."

Step 10 Consistency and benchmarking

- Benchmark your product technologies and process technologies using the TG approach. Do so regularly on an international basis.
- International benchmarking is a necessary condition for global competitiveness. Have this function undertaken as a routine one.

FIGURE 17.10 Managing new and existing technologies while incorporating the TG in an established enterprise: the 10-step procedure.

of its core technologies, then it does not have to spend millions for new technologies at the present time, but when such a technology becomes a "technology loser," the company may have to acquire a new one.

These unique benefits are summarized in Fig. 17.11.

---

1. Provides a framework to determine the timing to withhold proprietary technologies, particularly in transnational alliances
2. Helps in formulating corporate policies with regard to the nature and extent of technology transfer from the transferer to the transferee
3. Assists in the determination of the investments needed to acquire new technologies
4. Provides a basis for a company to determine the upper limits for growth and advancement in existing technologies
5. Makes available a mechanism for a national technology policy group to understand the implications of a company's technology transfer strategy in global settings

---

FIGURE 17.11   Unique benefits of the TG approach.

17.7 CONCLUSIONS

Thirteen years ago, John Naisbitt (1982) predicted what he called the 10 *megatrends,* the first three of which were:

- Industrial society—to information society
- Forced technology—to high-tech/high-touch
- National economy—to world economy

These three trends are a well-accepted reality today. It will be interesting also to note that technology in general, and information technology in particular, have been major drivers to shape a global economy. Five years ago, Naisbitt and Aburdene (1990) proposed "Megatrends 2000: 10 new directions for the 1990's," the first of which reemphasized the impact of the "booming global economy of the 1990's." As we continue to move rapidly toward integrated information systems, technology transfers, and transnational technology alliances, the need is growing dramatically to manage both existing and new technologies for maximum economic leverage. In this chapter, we have attempted to present the concept of technology gradient (TG) as a novel tool to achieve this maximum economic leverage. As we approach the new century, global enterprises will have the challenge to manage *multiple* technology gradients, *simultaneously,* at a rate we have never seen before.

In the spirit of intellectual humility, we suggest at least three possible areas for further research and investigation, with respect to TGs:

1. Since the "technology cycle" will continue to rotate much faster than ever before, it would be beneficial for companies to model the behavior of TGs with the relevant variables.

2. We expect a situation in which global enterprises will have to manage "TG networks" (TGNs)—not just a few individual TGs. Dealing with just one TG at a time will be totally unrealistic, and even suboptimal. Thus, it would be appropriate to develop methodologies that can *optimize* such TGNs.

3. In determining the TGs, with the methodology we presented in this chapter, it would also become necessary to include many more indicators, many of which are unimaginable at this time, because of the complexity of the technology issues. Refinements to our approach will be a constant need, not a luxury.

17.8 SUMMARY

Today's enterprises not only have to be competitive in domestic markets, but increasingly so in the global market arena. With more countries emerging as free-enterprise economies, there has been a dramatic increase in the rates of technology transfer from the developed to the developing economies. At the same time, the economic capital of the developing economies has become scarce, thus making it necessary for a careful analysis of technology investments *before* they are made. In this chapter, we propose the concept of technology gradient as a tool for companies to address some of the technology transfer issues facing them, particularly in determining the candidate(s), timing, level, extent, and cultural setting, for technology transfer among transnational enterprises. This concept is illustrated by a numerical example and its benefits are enumerated. A 10-step methodology to apply the technology gradient is proposed to manage new and existing technologies, and the scope for further research and refinement of this methodology is suggested.

17.9 REFERENCES

Clemmer, J., *Firing on All Cylinders,* Macmillan of Canada, Toronto, 1991.

Johnsrud, C., "Industry, University, and Government Perspectives on Technology Transfer: Market Pull, Technology Push, and Organizational Heterarchies," *Management of Technology IV, Proceedings of the Fourth International Conference on Management of Technology,* Miami, Feb. 27–March 4, 1994.

Naisbitt, J., *Megatrends,* Warner Books, New York, 1982.

Naisbitt, J., and P. Aburdene, *Megatrends 2000,* William Morrow, New York, 1990.

Okko, P., and A. Gunasekaran, "An Investigation on the Technology Transfer and Diffusion Processes," *Management of Technology IV, Proceedings of the Fourth International Conference on Management of Technology,* Miami, Feb. 27–March 4, 1994.

Stewart, H. B., *Recollecting the Future,* Dow Jones-Irwin, Homewood, Ill., 1989.

Suite, W. H. E., "Joint Ventures and Contract Award to Overseas Contractors—as Technology Transfer Mechanisms," *Management of Technology III, Proceedings of the Third International Conference on Management of Technology,* Miami, Feb. 17–21, 1992.

Sumanth, D. J., *Productivity Engineering and Management,* McGraw-Hill, New York, 1984, 1985, 1990, 1994.

Wood, O. L., and E. P. EerNisse, "Technology Transfer to the Private Sector from a Federal Laboratory," *Management of Technology III, Proceedings of the Third International Conference on Management of Technology,* Miami, Feb. 17–21, 1992.

EDUCATION AND LEARNING

CHAPTER 18

THE LEARNING ORGANIZATION

Thomas G. Whiston

Senior Fellow, Director of Studies
(Technology and Innovation Management)
Science Policy Research Unit (SPRU)
University of Sussex
Falmer, Brighton, United Kingdom

18.1 INTRODUCTION

If knowledge is power, learning is the key to power. Without learning there is no progress, no chance of wisdom, no relaxation which affords creative contemplation and further "learning," further advance—in short, learning permits the obtaining of mental spare capacity. And such a "spare capacity," if it is utilized well, is the cornerstone of further improvement, the arbiter of smooth performance and quality—spare capacity, as we shall see, is not wasteful (although it can be if not used well); rather, it is essential to robust performance. Indeed, learning is about the development of spare capacity, and spare capacity begets, or permits, further learning.

Now since learning is the "key to power," it is not surprising that in recent years an extensive, but by no means exhaustive, literature has appeared on the subject of so-called organizational learning. We say "extensive," but not "exhaustive," because in many ways the extant literature is highly unsatisfactory. Thus, while numerous examples of insightful journal papers and, indeed, whole texts devoted to the subject exist, to this writer's knowledge, no organic, holistic or satisfactorily "theoretical overview" has yet emerged. Thus the extant literature is "bitty," anecdotal, and, to say the least, generally uses such words as "learning," "skill," and "skill acquisition" in the most perfunctory and unquestioning manner. This difficulty is accentuated by the fact that at present the topic has become one of the "in" subjects...resulting in a plethora of popular books which offer to provide the magic recipe for improved company performance.

This is a great pity and a potential danger of enormous magnitude. For we could posit that no subject is more important than company or organizational learning (just as with human or individual learning), since if our subject matter is more fully understood in a theoretical, or holistic, sense, then much economic and social gain is to be obtained. The danger is, however, that the fashion of today will (or could) degenerate and recede into yet another lost cause. As indicated earlier, this would be a tragic loss.

Why is there not a strong theoretical framework, a robust basis, presently available from which further analytical and empirically researched studies can emanate? To a large degree the answer to that question relates to the background and disciplines of the researchers, analysts, or writers who have presently been addressing the topic. They comprise economists, technologists, "systems" people, sociologists, policy analysts, and rarely, too rarely, cognitive psychologists or information theorists. It is to the latter disciplines that we must turn if we are to posit a theoretical understanding of learning—it matters not whether this is seen as individual, organizational, or corporate learning for the issues are, at bottom, the same.

Thus there is a wealth of information, theoretical understanding of learning, skill acquisition, unlearning, forgetting, memory encoding, and performance improvement deriving from those latter disciplines which can suggest the intellectual framework that is so badly needed if we are to develop our subject matter in a satisfactory, noneclectic manner. Such a framework, as we shall note, will not be perfect—and it will not be a simplistic extrapolation of an understanding of individual human performance to a more macro socioorganizational learning level. Modification and attenuation will be required. Nevertheless, there will be *direction* and *coherence* to the thoughts which shape and direct subsequent examination of that macro-organizational structure, less arbitrariness—and the chance of fuller understanding.

In the sections which follow, in relation to the broad critique which we have just posited, we will therefore locate our discussion into three broad areas. First, in Sec. 18.2, we will consider the nature of *individual* learning, skill acquisition, information processing, and information usage. We shall also make several points relating to "forgetting," learning deficiencies, motivation, stress, and vigilance curves. Having laid down our framework, we will then examine (in Sec. 18.3) what this points to (and demands) in relation to the broader cousin of individual learning: macro-organizational learning (and nonlearning). This will then be followed, in Sec. 18.4, by examples from the extant literature which at least provide us with some insight and data which we can begin to fit into the framework, or alternatively permit us to modify the framework—for much is, at present, unknown.

Finally we detail future research and analytic tasks to be undertaken. For it is almost a virgin landscape, thereby offering rich pickings for dedicated researchers and much practical gain to companies, public organizations, and the marketplace in general, if applied to good purpose.

18.2 TOWARD A THEORETICAL FRAMEWORK OF LEARNING AND SKILL ACQUISITION: FACTORS TO CONSIDER

It is neither palpably possible, nor wise to attempt to summarize the present state of knowledge regarding our theoretical understanding of human learning and skill acquisition in the small space that we can allot here. We therefore merely *outline* one main, but nevertheless useful, paradigm and signal some of the more important concerns and ideas. How can we approach this? First, let us recognize that learning and skills are intimately concerned with *information flow,* decision procedures (both explicit and implicit, viz., algorithms and heuristics—or tacit knowledge). Learning is dependent on sensing, encoding, and organizing a flow of *incoming* information, perceiving patterns, and decoding on appropriate responses, actions, and judgments, which is in essence *outflowing* information.

Before we amplify on this, let us consider a "simple" (but nevertheless complex) example of the learning involved in a quasi-complex task: namely, that of learning to drive an automobile. At first our potential learner is overwhelmed with competing "inputs" (vision; road tracking; engine sound; gear movement requirements; coordination of clutch, brake, and accelerator; responsiveness; etc.). The task, simple as it may appear (the same applies to flying an aircraft), is almost daunting. To the basic task add (say) light signals, heavy flow of traffic, external distractions, and so on. Under such conditions there is little chance of the learner driver holding a *parallel* conversation, listening to the radio, daydreaming, or planning ahead—as the competent driver is later able to do. In short, there is no *spare capacity*. But the potential driver "learns" and matches the right responses to the right *selection* of relevant inputs. A flow and pattern to the skilled driver emerges. A beautiful orchestrated pattern of input/output information flow develops and, most importantly, spare capacity develops. The driver can, ultimately, drive the car almost without thinking about it. (This can be a little unnerving—and many of us may have sometimes wondered who, or what, was driving the automobile.) Spare capacity is developed such that it is easy to undertake the primary task (driving the car) while also, in parallel, performing numerous secondary tasks (listening to the radio, engaging in conversation, daydreaming, planning ahead, or whatever). It is also possible, should the driver wish, to use that spare capacity to *direct* the spare attentional power to the *primary* task and thereby lead to a significant improvement in performance. It is this later characteristic which distinguishes good from excellent, the mediocre *amateur* from the highly skilled, continually learning, *professional.*

But we should not be so complacent. For if road conditions get tough—bad weather, poor vision, very heavy traffic, a worrisome day—accidents and errors abound, indicating a limit to performance, an erosion or restriction of the spare capacity—in effect, the informational channel capacity of the driver becomes overloaded. (The more ergonomically designed the automobile—indicator panels, seating, and gearshift arrangements—the less is the threat.)

Now all the above is descriptive but it can be expressed in a highly detailed mathematical, procedural, and decision-oriented form.[1] It is to this more theoretical framework that we now turn. Thus, the main requirements in relation to learning and skill acquisition (especially for perceptual-motor skills, but we can extend our repertoire to more creative identical skills and learning) are to understand the following:

1. The translation of inputs and outputs of information flow into a controlled pattern (albeit permitting flexibility of response and appropriateness of action).
2. The overcoming of a "limited channel capacity" to permit spare (mental) capacity.
3. The use of the spare capacity to permit continual improvement in performance. Thus we are *not* speaking of a robotic automaton (which was the downfall of "stimulus-response" behavioristic accounts of learning and also incidentally of "Taylorism" in trying to understand skilled performance in the workplace). The *continual* improvement in performance is *learning,* not just competence. Thus, without continual learning, mistakes emerge and forgetting, declining performance, or whatever ensues.
4. The ability to cope with stress, boredom, and fatigue repetitivity and the need to recognize that such characteristics are always potentially present.
5. The need to recognize that many aspects of complex skills are essentially *heuristic* in nature (which is a deeper and more complex thing than so-called tacit knowledge) and that despite our best efforts—indeed, that of tens of thousands of cognitive psychologists over a period of decades—we are still only at the brink of understanding the true nature of those "heuristics." [Thus AI (artificial intelligence), expert systems, complexity theory, and cybernetics are at only an embryo stage in their development.] Nevertheless, *if* we can reduce a heuristic to a set of

TABLE 18.1   Attentional Load and Skill Stage

Stage	Attentional characteristics
1. Early descriptive	*Almost total attentional load directed to external events.* Considerable weighting on self-awareness, and critical attention to one's own performance.
2. Intermediate association	*Intermittent attention to external cues.* This could be followed by "internal attention" to temporal and phasing relationships. Partial self-criticism and accompanying self-awareness may still occur.
3: Final autonomous	*Minimal attention to external errors,* although this is a function of task requirements. At this stage, there is minimal self-criticism and analysis, unless the subject is so motivated.

Source: Whiston.[1]

quantifiable or understandable algorithms, we gain much in our understanding. In many ways this is a "pattern recognition" problem, and much can be gained by a judicious application of information theory.[2]

How much do we understand regarding these factors? Can we integrate that understanding into a coherent model on which to base (and test) further exploration and comprehension? Such an integration has been attempted by several analysts,[3] in essence, with information theory and communication theory[4] as a language tool. Let us return to the car-driving example which we outlined briefly above. The skill development stages which the driver moves through can be expressed in simplified form as shown in Table 18.1.

In stage 1 (Table 18.1), the early descriptive stage, the individual cannot take on "extra" tasks, and is overwhelmed by the learning task. By stage 2, the intermediate associative stage, things are more relaxed. At the final stage (Table 18.1, stage 3) a task can *seem* effortless. Attention can be paid to "external secondary features." This *permits* but does *not* guarantee further learning or improvement. The aim is to direct the "spare attentional capacity" to further skill improvement, not to dissipate the spare energy.

18.3   THE NEEDS OF LEARNING ORGANIZATIONS

Having laid down the foundations of a theoretical framework for "learning" (and other terms of reference are possible), how can we connect this with a "learning *organization*"? Can we utilize the theoretical framework plus the supporting constructs as a means of furthering our understanding of the learning organization? And with respect to the supporting constructs, does this prescribe what exactly it is that we *should* be paying attention to if we are to be more prescriptive and analytic with regard to how any learning system functions? Do such concepts as information flow; limited channel capacity; spare capacity; heuristics and algorithms; serial-to-parallel processing; stress and vigilance curves; rigidity, boredom, and reduced performance; information overload (or underload); and primary and secondary tasks help in our need to describe the conditions of continual learning, enhanced performance? In particular, do such concepts or constructs possess meaningful interpretation and counterparts at the macro-organizational level? As we shall note in this section, there are, indeed, numerous parallels, gainful interpretation—not just in a metaphorical sense—but in a much more meaningful sense which then permits a robust, scientific understanding of the manner and needs of organizational learning.

However, with regard to the examples which we highlight here, we should be cautious not to overclaim with regard to our full understanding of organizational learning. Thus no company, no social organization, "learns" to the same degree or with the same proficiency as that capable and exhibited by an *individual.* The marvelously smooth, patterned behavior, the wondrous mental models which reflect an individual's innate learning capacity, far exceeds the most "organized corporation." As to the reasons for this, we reserve comment until the concluding section of this chapter.

Now what exactly is it that an organization needs "to learn"? The answer to that question is: "Everything!" Thus any company or organization needs to learn, continue to improve on, its ability to (for example) derive its corporate plans and strategies; derive and formulate its product portfolio; obtain temporal matching and synergy between the output from its R&D portfolio and the subsequent product portfolio; and improve manufacturing and production procedures, maintenance and service schedules, and market analysis and marketing expertise. None of this learning needs to stand alone. Indeed, their *interdependence* (e.g., design for manufacture, exigencies of "lean production") is something again which has to be "learned" and continually improved on. To all this, an era in which "payback windows" are shorter than ever before, in which new generic technologies [IT (information technology), biotechnology, new materials] proffer discontinuities in knowledge, or the relevance of primarily hard-earned "tacit" knowledge, introduces new challenges to learning.[5]

Under such circumstances it is essential that *information absorption,* company intelligence gathering and information sensing, and diffusion of information and knowledge across all company functions be as efficient and effective as possible. This requires (as discussed in Chap. 30) new multidisciplinary and interdisciplinary skills, new organizational structures, continuous retraining, and personnel updating. In particular, the information flow implied requires that the learning organization is flexible, adaptive, but not overloaded with information to the point of inducing functional stresses and subsequent losses or breakdown in performance. Hence the importance of such concepts (as referred to earlier) of "channel capacity," limited channel capacity, and the organization of information (and structure) to induce spare capacity, which can then be used to further improve performance.

Indeed, within a multidisciplinary, integrative cross-functional environment (see also Chap. 30) and where interdependence of numerous subfunctions demands integrative cross-communicative skills and information exchange, then this reinforces the notion of skilled performance as smooth integrated *patterned* performance, which is the essence of any complex skill.[6]

Thus the learning organization has to be viewed *holistically,* as do the learning pattern and skill of an individual. We will note below how such aspects as channel capacity, spare capacity, and vigilance and stress reduction can be given meaning at the company organizational level, but first it is useful if we comment on the different properties, structures, and demands of various organizational settings, in particular, the disparities of large and small organizations.[7] For just as with individuals, there is not one specific pattern of learning (although a generalizable model is possible), so, too, with commercial organizations, there are significant differences between large and small companies. This, in turn, implies differential analysis in terms of skill and learning requirements and differences as to how they can best improve on internal information flow.

Consider Table 18.2, which illustrates some of the main characteristics, strengths, and weaknesses of large and small enterprises.

TABLE 18.2 Advantages and Disadvantages of Small and Large Firms in Innovation*

Activity or function	Small firms	Large firms
Marketing	Ability to react quickly to keep abreast of fast changing market requirements. (Market startup abroad can be prohibitively costly.)	Comprehensive distribution and servicing facilities. High degree of market power with existing products.
Management	Lack of bureaucracy. Dynamic, entrepreneurial managers react quickly to take advantage of new opportunities and are willing to accept risk.	Professional managers able to control complex organizations and establish corporate strategies. (Can suffer an excess of bureaucracy. Often controlled by accountants who can be risk-averse. Managers can become mere "administrators" who lack dynamism with respect to new long-term opportunities.)
Internal communication	Efficient and informal internal communication networks. Affords a fast response to internal problem solving; provides ability to reorganize rapidly to adapt to change in the external environment.	(Internal communications often cumbersome; this can lead to slow reaction to external threats and opportunities.)
Qualified technical staff	(Often lack suitably qualified technical specialists. Often unable to support a formal R&D effort on an appreciable scale.)	Ability to attract highly skilled technical specialists. Can support the establishment of a large R&D laboratory.
External communication	(Often lack time or resources to identify and use important external sources of scientific and technological expertise.)	Able to "plug in" to external sources of scientific and technological expertise. Can afford library and information services. Can subcontract R&D to specialist centers of expertise. Can buy crucial technical information and technology.
Finance	(Can experience great difficulty in attracting capital, especially risk capital. Innovation can represent a disproportionately large financial risk. Inability to spread risk over a portfolio of projects.)	Ability to borrow on capital market. Ability to spread risk over a portfolio of projects. Better able to fund diversification into new technologies and new markets.

18.8

Economies of scale and the systems approach	(In some areas scale economies form substantial entry barrier to small firms. Inability to offer integrated product lines or systems.)	Ability to gain scale economies in R&D, production, and marketing. Ability to offer a range of complementary products. Ability to bid for large turnkey projects. Ability to finance expansion of production base; ability to fund growth via diversification and acquisition.
Growth	(Can experience difficulty in acquiring external capital necessary for rapid growth. Entrepreneurial managers sometimes unable to cope with increasingly complex organizations.)	
Patents	(Can experience problems in coping with the patent system. Cannot afford time or costs involved in patent litigation.)	Ability to employ patent specialists. Can afford to litigate to defend patents against infringement.
Government regulations	(Often cannot cope with complex regulations. Unit costs of compliance for small firms often high.)	Ability to fund legal services to cope with complex regulatory requirements. Can spread regulatory costs. Able to fund R&D necessary for compliance.

*The statements in parentheses represent areas of potential *disadvantage*.

Source: Rothwell and Zegveld.[7]

What we should note, from the purview of our subject matter here, is that in many cases *small enterprises* are organic in nature; they are (or can be) dynamic and entrepreneurial, innovative in nature. They are *capable* of integrated behavior with the involvement of a large proportion of their personnel. From a learning standpoint this is excellent. Internal information flow is important in learning. But also there are severe difficulties to be faced by small enterprises. Staff can be overloaded, under pressure. They often cannot be released for retraining and updating. If staff are retrained, there is a danger of losing them through "poaching" of their enhanced skills and increased market value. Often a small enterprise does not have a significant R&D arm or function. Thus the sensing, the decoding, the absorption, the utilization of the large bank of information in the external environment is no easy task. Within such a context we begin to see the limitations of information channel capacity, the lack of spare capacity, the danger of stress, and information overload.

However, if we turn to large organizations, we can observe (at the risk of some oversimplification) a somewhat different pattern of strengths and weaknesses which can affect or dominate their learning capability. Large companies have in many cases taken on formidable hierarchical structures which are almost militaristic in nature—numerous levels of reporting layers, centralization of power and authority "at the top" (the text "On a clear day you can see General Motors" was well titled!),[8] centralization of R&D, multidivisional structures, specialized functions, bureaucratic regimes, and so on. In various ways each or all of the aspects just listed can compromise company performance. There are, of course, strengths as well as weaknesses to this essentially Fordist-type structure. Thus, under conditions of slow change, mass production, mature life-cycle circumstances, the economies of scale and low unit costs have shown undoubted benefits. Big might not have been beautiful, but it did produce the goods. General Motors, IBM, Du Pont, or whatever prospered for many, many years. However, as the marketplace becomes ever more dynamic, as the rate of technical change increases, as new generic technologies emerge, as IT systems permit and encourage new and more flexible work patterns, as the strength and potential of total workforce involvement in micro decision making becomes ever more obvious (as presently exemplified in Japanese "lean-production procedures," quality circles, JIT procedures, or whatever),[9] and as multifunctional skills and new design principles become ever more apparent for market success—then we begin to observe the need of entirely new anatomies for large organizations.[10] With those new anatomies we can observe new learning needs, new learning capability, and new learning potential. We can observe the need to overcome internal informational bottlenecks and to enhance channel capacity, in which differing functions can better communicate with each other. This latter requirement implies the ability to *encode* and *decode* information across functions in the most effective way. Since different functions (accountants, designers, engineers, marketing people, scientists, R&D) often—indeed, usually—speak very different languages, this requires major shifts in large company structure and mode of operation if learning, *mutual* learning, is to be effective. It demands more organic integrative structures (viz., the "strengths" of the small enterprise anatomy) and interdisciplinary and multidisciplinary training (see Chap. 30) and communication channels to "gate" these new multidisciplinary skills.

The strength of large organizations (in comparison to small enterprises) is that they possess large R&D functions. Thus they can "sense," decode, and act on new potential scientific and technical information. However, the (often) physical, geographic, and intellectual isolation or separation of the R&D department means that both the remittance and the subsequent dissemination of R&D is by no means optimal. It is not just "even here" that we see the need of new integrative forms, it is *particularly* here that the need is obvious—but not always acted on. Thus a strong potential strength of large organizations (viz., the R&D function) with respect to corporate learning can, in fact, be, if not an encumbrance, at least far, far less than its true potential.

We therefore see differing strengths and weaknesses for large and small companies: differing learning challenges. Can we place these needs into a more organized format and in so doing relate the new needs, the comparative strengths and weaknesses to the theoretical framework which we outlined earlier? It is to that need for coherence that we now turn.

The terms and concepts which we emphasized earlier with respect to learning and skill acquisition included the following characteristics:

- *Channel capacity.* A term developed to understand how much information can be transmitted along a particular channel or corridor (often measured in bits per second or some such unit).

- *Limited channel capacity.* The upper limit to transmission rate. What is important is the coding system that is used in determining bandwidth and the influence of extraneous noise.

- *Selective attention.* The ability to focus on important features of the external environment. There is a subtle problem here since continual attention to numerous information sources arriving in parallel, in real time, requires selective filters. The sorting (filtration process) ensures that the system is not overloaded. Selective attention has to be learned and continually improved on.

- *Informational overload (or underload).* Trying to deal with *too fast* an information flow causes breakdown or stress: certainly loss in quality in performance. A situation with continual informational *underload* equally can produce boredom and fatigue. Neither is compatible with good performance or enhances learning capability.

- *Heuristics and algorithms.* Most individual skills are poorly understood. The intuitive, implicit, or tacit way in which the brain (or an individual) performs reflects "heuristics"—it works—but we don't know how. Continual study reveals some of the underlying algorithms—patterns, decision processes which we can codify.

- *Vigilance curves (and performance decrement).* Continual attention to a repetitive task usually results in a very large decrement in performance (it can be as high as 40 percent). Task diversification (if meaningful) helps considerably.

- *Spare capacity.* This is essential to learning. If an individual (or system) is "glued," "tied" to incoming signals and related information load, then no learning ensues—the action and responses are machinelike and robotic, however impressive the performance. However (as noted earlier), if the individual learns to encode information, selectively filter, and impose realistic patterns on the environment—then "spare capacity" develops. That spare capacity *can* facilitate learning—but only if it is directed or motivated to do so. Spare capacity can reduce strain, stress, and fatigue—it can provide relief—but it is also essential to effective learning. The more that is usefully learned, the greater the spare capacity. The greater the space capacity, the greater the learning potential. (If you want something done quickly, why give it to a busy person to do?)

- *Serial-to-parallel processing.* In processing information, sequential decision making is far less efficient than ability to process information in parallel. But the latter requires that the more complex gestalt patterns be learned and understood. Ultimately such learning, in effect, increases the channel capacity of a system and permits, through the implicit spare capacity induced, a further ability to learn.

- *Patterned performance.* The distinguishing feature between skilled and highly skilled (professional) performance is the smooth pattern of behavior exhibited by the latter individual. The patterned performance permits attention to be paid to the "fine tuning" of performance as unpredictable events occur (or to correct internal faulty actions). In one sense patterned performance reflects a "giant internalized and much tested standard" (the concert pianist, the professional car racing driver). But such pattern is *not* rigid, predetermined. On the contrary, the ability to see patterns (see serial-to-parallel processing above)—to impose and construct patterned performance—permits and is the essence of spare-capacity and continual learning. The danger occurs when a pattern becomes "obsolescent."

How do these terms (and remember from our discussion earlier that in describing *individual* performance they form an interrelated coherent whole) relate to macro-organizational or company learning? As we indicated previously, we should not over-claim; the circumstances of macro-organizational performance (and hence learning needs) may not be *exactly* the same as those that describe individual learning; nevertheless, as we now indicate, there are numerous closely related issues (and hence challenges to understanding).

Thus, if we consider the terms just listed, and apply them to *company or organizational circumstances*, we can note the following learning and skill acquisition terms applied at the organizational level:

Channel Capacity. All companies depend intimately on smooth information flow between individuals, groups, departments, and functions. Numerous restrictions and blocks to information flow evolve over time, most especially in large bureaucratic hierarchically organized systems. Numerous reporting layers may facilitate the organization (but not necessarily the *effectiveness*) of vertical-communication command systems. However, at the same time nothing like enough attention is paid to horizontal communication (see Fig. 18.1).

This imbalance in the creation of communication channels reduces the overall company or organizational channel capacity, the subsequent flow of information, and hence the ability to learn. Some companies have introduced cell-like structures at lower levels of the company hierarchical structures to try to compensate for this difficulty as illustrated in Fig. 18.2.

But this is not enough. What is required in order to achieve significant improvement in total organizational learning capability is a more radical interlinkage of functions, with the development of numerous new multidirectional channels of information as illustrated in Fig. 18.3.

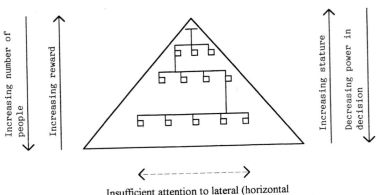

Insufficient attention to lateral (horizontal communication channels).

FIGURE 18.1   Hierarchy and decision making.

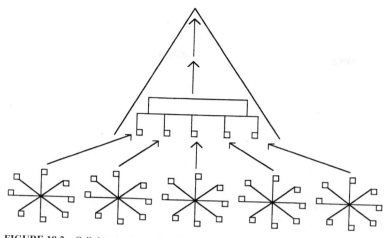

FIGURE 18.2 Cellular structures placed into hierarchical structures.

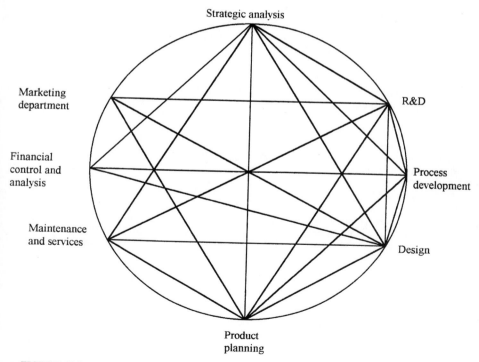

FIGURE 18.3 Further developments of multidirectional communication and involvement. (*Source: Whiston.*[1])

Limited Channel Capacity.   As the number and complexity of communication channels are increased, there exist numerous barriers to optimum use of the channels. This "extraneous noise" can severely hinder total company learning. The reasons or problems are various: overloaded individuals or subgroups (see information load below); divisions, departments, or functions who are not well integrated into the total organizational entity and thereby hinder communication and learning potential. (This can apply especially to "isolated" green-field-site R&D departments.) The need, the remedy, is threefold: (1) design of totally integrated, organic structures;[10] (2) encouragement of interdisciplinary linkages and focus of analysis; and (3) encouragement of interdisciplinary and multidisciplinary training (see Chap. 30).

Selective Attention.   *Selective attention* implies the need to focus, filter, and select, from a multitude of incoming (and internally generated) information flow. Organization of that data is the basis of subsequent analysis, formulation of corporate strategies, improvement of product design and product portfolios, and correction and improvement of past and ongoing procedures. How can this "filtration," this selective attention, be better organized at the company level to improve self-learning? Some companies employ technological gatekeepers; others recognize the importance of their R&D department in this role—but R&D departments who are actively linked and aware of all organizational needs. Ultimately whatever information is filtered, selectively attended to, must be reintegrated for use; hence the importance of integrated cross-linkage of departments and multidisciplinary communication skills.

Information Overload (or Underload).   All levels of participants in any organizational system can be subject to information overload (or underload). In either case learning and performance is compromised. To some degree it is more likely that more senior managerial positions are subjected to information overload; and shop-floor operatives, undertaking highly repetitive work, to information underload. This is illustrated in Fig. 18.4.

The nonlinearity of the information load indicated in Fig. 18.4 can be disastrous for company performance and learning capability. The implication is to introduce numerous measures which in essence tend to reverse the classic organizational pyramid.[10] Organic structures, matrix management, reduced number of reporting layers (flattened pyramids), lean-production techniques, quality circles, socioergonomic design of the workplace, "managers on the shop floor" (viz., much favored in Japan), total employee involvement, and so on are all features of such a trend to improve total organizational learning capacity.[10]

Heuristics to Algorithms.   Much company knowledge is tacit in form.[5] There are many advantages to this. Knowledge is necessarily embedded in individuals, and ultimately it is individuals who learn, and organizations and companies who benefit. The difficulty is that (1) much noncodified knowledge is lost on employee transfer, or closure of departments, etc.; and (2) the internal (tacit) nature of knowledge can hinder interdepartmental (or person-to-person) communication and mutual learning. The need, therefore, is to codify, to make knowledge "transparent" wherever possible. However, the potential danger and threat then arises in terms of external competition due to intercompany learning.

Vigilance Curves.   A large proportion of the literature pertaining to *vigilance curves* (and performance decrement) focuses on lower-level perceptual-motor skills—plant operatives, machinists, shop-floor employees, etc.—and recognizes the high-quality

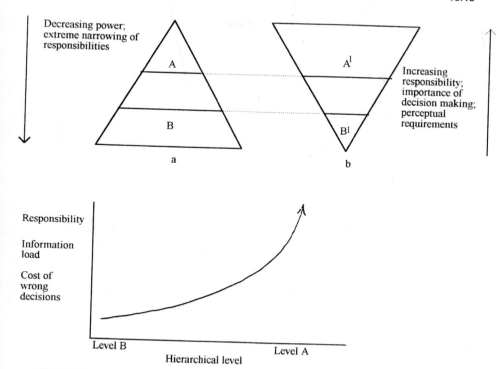

FIGURE 18.4   The nonlinearity of power and decision load.

loss which can accompany repetitive behavior. As a consequence, job rotation, skill enhancement programs, multiskilling, etc., are recommended.

However, what is not sufficiently recognized is that much higher skill levels—viz., managerial levels in all forms—can be subject to a similar problem. Indeed, there is a deep paradox. As learning ensues, as experience extends for the intelligent employee, limited scope of action can induce a "sameness" to the job which is equivalent to the unwanted characteristics of a vigilance curve. The subsequent loss to company learning capability can be large. Such a problem is reinforced in those organizations which have become inherently conservative: where taking risks may endanger a career and where mistakes are not forgotten and the safest promotional route is conservatism and conformity, not candor.

The challenge is to extend people, to encourage creativity, and to recognize that failure is not forbidden—or punished.

Spare Capacity.   *Spare capacity* is *not* the same as carrying extra staff; it is not "fat"; it is not massive financial reserves. Spare capacity is the outcome, the consequence of ongoing effective company learning. It is the integrative accumulation of internal knowledge (explicit and tacit), effective company sensing devices, *flexible* corporate strategies, and foresight and continuing analysis of the challenges and opportunities to be faced. The evolution of lean-production systems, of the expansion of concurrent

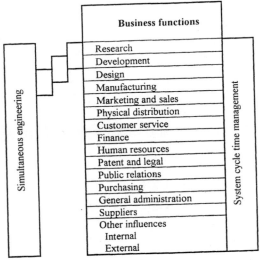

FIGURE 18.5 Differences in scope between simultaneous engineering approaches and system cycle time management. (*Source: Gaynor.[12]*)

engineering principles toward a much wider systems involvement (see Gaynor[11]) reflect the seeking of systems, company organization, that has cross-linked all its multifarious skills into a total company capability which can handle the new challenges and market dynamics (which are never constant). Spare capacity results from the application of all of the principles we have detailed under other headings, and thereby encourages and permits further learning.

Serial-to-Parallel Processing (and Pattern Recognition).   This demands the perception of pattern, the ability to handle complex information in a totality. It is the hallmark of skilled performance and the means of overcoming the limited channel capacity and communication problems (and hence learning problems) outlined above.

As indicated earlier, serial-to-parallel processing demands cross-department linkage, multidisciplinary training, and more "organic" structures. This can take many forms and has been the subject of extensive review.[10] An interesting feature is illustrated in the studies of Gaynor[12] with respect to the need to link all departments or functions in an organization (see Fig. 18.5).

General Implications.   If we now collate the above, we can more clearly see the *learning challenges* which face all companies (whether large or small) but we can also organize or group our subject matter, into classes or categories. Thus, in general terms, in order to optimize learning, the challenge is to

- Maximize channel capacity.
- Overcome limited channel-capacity problems.
- Direct selective attention to the most important areas, but simultaneously develop sensory filter mechanisms in order to prioritize all the parallel information which impacts on any organization (or system).

- Reduce information overload and possibly increase or improve on information underload. (The first induces stress, then fatigue.)
- Translate heuristics (tacit knowledge) into more overt algorithms and thereby offset the dangers of lost skills, company "forgetting," and opportunity costs due to staff transfer and employees leaving. In addition, the translation of implicit knowledge into explicit knowledge (heuristics to algorithms) yields a "transparency" in which criticism, improvement, diversification, wider dissemination of knowledge across all company functions becomes possible.
- Overcome the problems of repetitive, boring occupations which lead to qualitative loss in employee and systems performance. Although this problem applies particularly to less skilled employees (who often in mass-production systems undertake very repetitive work—and where, incidentally, most of the study of vigilance curves have been undertaken), related problems can also apply to more senior positions. Indeed, as an individual (or function) gets completely "on top of a job" and has thoroughly mastered the skill, boredom can set in. The reverse side of "spare capacity" implies the need of continuous challenges, continual learning, and so on.
- Seek mechanisms which induce or encourage spare mental capacity. This, above all, permits further learning. Its encouragement is a sine qua non. When an individual learns a skill, it happens "naturally." The equivalent challenge in an organizational setting is to create organizational circumstances, a culture, a cross-disciplinary climate, a training and retraining program, a linkage and involvement of R&D functions, such that all functions can better handle increasing information inputs, disassemble it, and understand and reapply such information as is needed in order to relate to the *overall corporate strategy*. This is *not* to say that all company functions can, or ever will, understand all the entire bank of skills of their colleagues; this is neither feasible or desirable. But what is required is a basal level of common knowledge far in excess of that which is usually the case. Thus different functions then know when to pass on information; what is relevant to them, when they should "add knowledge," add "value" to incoming, outgoing, or internally generated knowledge. They thus demonstrate the spare capacity to rise above the immediate exigencies of disparate information. They relate where necessary, they process and absorb (learn), but they do not overattend in situations where this is counterproductive.

At a more meta- or macro-company level, if all subfunctions are interlinked and performing well, the organizational *totality* is such that *the organization* as a whole is not overwhelmed when market conditions change rapidly, when competitors make significant advances or threats, or when new technological paradigms emerge.

Spare capacity *does not* mean extra staff; idle time; wasted effort. It means, as with skilled individual performances, that the organization has a smooth, controlled, partially autonomized level of skill, that is not phased out when under threat. Thus the "organization's skills" are patterned where they need to be, but on top of that pattern, because of that pattern, the organization is able to handle crisis, fast incoming information, and new stresses—and in so doing adds this to their repertoire of natural skills. Thus the learning curve becomes endless, always moving forward, as it has to.

Much of the above is enhanced if the transition of serial-to-parallel processing of information can be achieved. In technical terms as described earlier, this greatly encourages the limitations of a "limited channel capacity" and hence leads to the desired spare capacity. *Serial-to-parallel processing* implies the ability to recognize and act on more *complex patterns* in their totality. Interfunctional linkages, matrix overlap of functions, and greater academic-industrial linkage programs in which disparate skills and insights are brought together all assist. Parallel processing requires parallel judgments and hence integrative structures (see Figs. 18.3 and 18.5).

FIGURE 18.6 Requirements and examples of the means of improving the conditions of learning in large organizations.

All the observations discussed above are *generalizable*, common requirements of effective learning. In Figs. 18.6 and 18.7 we disaggregate these into the needs of large and small companies or organizations.

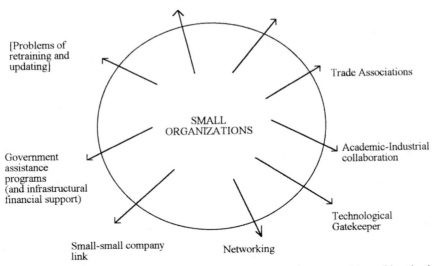

FIGURE 18.7 Requirements and examples of the means of improving the conditions of learning in small organizations.

18.4 CONTEMPORARY ANALYSIS

Numerous texts, studies, and papers are now available on the learning organization—so many, in fact, that the choice is bewildering.[10-32] The choice dilemma is that much greater because different authors and analysts (and perhaps even pundits) have their own favored point of view. To review in detail that enormous literature would not be possible in this short chapter. However, Table 18.3 does indicate some key authors together with the area of discourse that they tend to favor.

A large part of that extant literature is eclectic but often stimulatory and useful. What *is* required, however, as we have argued earlier, is a more integrative coherent

TABLE 18.3 Analytic Perspectives Regarding Organizational Learning

Dodgson[13-15]	In a series of articles, Dodgson has provided extensive reviews of contemporary issues and empirical detail of organizational learning. Several of the articles emphasize the relevance of academic-industrial collaboration, technical change, innovation studies, and organization theory—which demands a multidisciplinary approach if complete understanding is to be achieved.
Gaynor[11,12]	In these two texts Gaynor addresses the importance of a total-systems and functionally integrated organization in providing the basis for competitive development. Organizational learning and improvement is shown to be greatly influenced by such an approach.
Goold and Campbell[16]	In this text the authors examine the relationship "between corporate, divisional and business units and show how the centre can choose a style that adds value to the business in its portfolio." Building on Chandler's classic account of the role of the Centre (in "strategy and structure"), they provide a detailed analysis of the management process in diversified corporations.
Handy[17]	In this book, Handy goes beyond the organization or company and argues that the accelerated rate of change "is radical, random, discontinuous" and "doesn't make patterns any more." But this can be turned to an organization's advantage if they adapt and innovate: this requires individual working not only within the organization, but outside—being self-motivated and multiskilled. "New types of 'destructured organizations' will (therefore) arise." Adequate response demands wider educational change, with governmental and market changes.
Hayes et al.[18]	These writers (and the school they reflect) consider the new manufacturing and organizational principles available to companies. Their text provides an informative and proactive overview.
Morgan[19]	Morgan has placed the role of the individual as a key component in corporate learning.
Morgan[20]	This text provides a stimulatory examination of organizational life in terms of "metaphor." It suggests how this can be used as a means of diagnosing organizational problems. Comparison is made with the brain (as a self-organizing system), and the limitations of the brain metaphor. It also considers "organizations as information-processing brains...and as holographic systems."

TABLE 18.3 (*Continued*) Analytic Perspectives Regarding Organizational Learning

Patel and Pavitt[5]	These researchers consider the role of technological change, R&D, and innovatory development. In "technological competencies in the world's largest firms" (in which more than 400 of the world's largest firms are examined despite their diversified, but stable, multitechnological competencies), they indicate the nature and importance of the scope for managerial choice. They argue "the importance in technology strategy of integration (or `fusion') of different fields of technological competence." (See also Whiston.[10])
Peters and Waterman[32]	These writers placed considerable emphasis on "business success through organizational learning and strategic management"—and in reviewing seemingly "best practice."
Senge[21]	This author argues the importance of "systems thinking" and team learning. He distinguishes partnership as a form of "entitlement" from participation as a basis of learning. Shared vision and team learning are highlighted, as is the development of "mental models." Planning is thus a potential learning platform at both personal and corporate levels.
Thomson (ed.)[22]	In this volume numerous contributors examine how "leading firms are organized to learn and commercialize learning, which brings continuing technological advantages....Communication among institutions in leading countries fosters learning and innovation." It is argued that "Institutional change is needed to achieve and maintain leadership; institutions that supported one phase of technical change may block later phases."
Thurbin[23]	This author recognizes the importance of "employee learning on a continuous basis." Reference is made to the experience and lessons of the Rover Group in the United Kingdom. The importance of "*trigger* mechanisms" to stimulate the setting up of the learning process, *belief* that transformation can occur and that risks should be taken, *action* in part by trial and error within existing organizational strategies, and *benefits* and performance criteria that reinforce learning are considered as basic tenets.
Whiston[10]	This author has explored in some detail the importance of functional, managerial, and technological integration in providing a robust basis for company learning. On the basis of wide national surveys and empirical study, the barriers to achieving integration and improved learning are identified. National and corporate policy recommendations are then examined.

theme which permits the analyst, the company executive—and developing theory—to embed the "heterogeneous availability" into a consistent philosophy.

We can note in Table 18.3 authors or schools of thought who (variously) emphasize (1) the individual; (2) the organizational system; (3) deep or "meta" strategic and systems thinking; (4) wider environmental, societal, or educational features; (5) pragmatism and "best examples"; (6) theoretical overviews; and (7) technological overviews. None of these is mistaken or irrelevant, but all can be more fully understood and integrated if information flow, decision procedures or mechanisms, and strategic analysis viewed as an intelligence function, are utilized as coherent combinatory concepts. The organization, like the brain, must sense "data," process, not be too overloaded, be free to "play" with the data, learn by mistakes, be motivated to learn, and learn that it can learn. The greater the synergy between individuals within the organization and the organization itself, the greater the synergy between the company and the wider environment (whether this involves customers, industrial networking, government agencies, the education establishment, or whatever), the greater the potential for, and smoothness of, learning.

The conditions of such synergy are highly dependent on communication and information flow at all levels. Synergy, therefore, takes on the characteristics of integration, organicness, employee-employer identification, conditions to minimize stress and fatigue, cooperative mutuality of purpose, ethos, and identification.

It is not difficult to see how the various "classes of authorship" indicated in Table 18.3 variously characterize that synergy. No individual school of thought is uniquely correct; nor is any uniquely wrong. But the coherence emerges if we consider motivation, information flow, and removal of communicative bottlenecks and barriers (as discussed in earlier sections) as being of paramount importance.

Numerous Japanese and Scandinavian companies recognized this quite a while ago; hence the transformation from Taylorism or so-called scientific management toward a wider socioergonomic goal. Quality circles, team organization, lean production, and JIT are natural stepping-stones—but they are only stepping stones to a more organic organizational learning environment. As recognized by Handy,[17] the requirements for effective innovative learning do *not* stop within the organization. Indeed, there is no limit to the systems boundaries that must be considered and changed. Education, wider societal networking, governmental policies, and international and global marketplace changes ultimately compound back on and within and without the "learning organization."

At the time of this writing, new technologies and information networks and the enormous potential for parallel highly responsive and reactive information data networks and communication network links (whether this is, say, of the internet variety or "private" multinational corporate databases) only serve to challenge the sensory nerve endings, the communication-handling capabilities of learning organizations.

The danger of systems overload has never been greater. For here we now have a great social irony. More parallel information sources (as with the human brain) offer to the learning organization the potential for even more learning, but also the danger of systems overload and rigidity due to informational overload—a form of organizational fatigue. Only if the learning organization is relaxed, integrative, and internally communicative (which implies numerous smooth functioning interfunctional links and multi- or interdisciplinary trained staff) can that new source of information be properly utilized and learning accelerated (or even achieved).

A critical part of achieving such smoothness and facility in information handling, as well as encouraging a "creative learning intelligence" can be the research, the design, and the development functions or departments. But here again we see the need of coherence and systems linkage: both internally and externally.

The structure and pattern of linkage between research, design, and development functions is of critical importance in relation to overall organizational learning; but as we have just noted, it has both internal and external features (just as is so for individual learning). Also, much depends on the size and resources of an organization. The problems of "learning" for small companies, small manufacturing enterprises (SMEs), and large corporations are not exactly the same. The small organization (and usually the SME) tends to be organic and intimate. If these two enterprises *do* have a research arm, it is usually well linked to other departmental functions (this often is *not* the case with all large enterprises). We then begin to observe different challenges and needs to encourage total organizational learning.

Thus in the case of large organizations there is a need to link and cross-link the research, design, development functions with all the other departments in as rich a manner as is possible (see Figs. 18.3 and 18.5). In the case of smaller enterprises, the need—if there is a paucity of research capability—is to encourage external networking and linkage especially through academic-industrial collaboration, trade networks, and the like. But here there is encountered a new form of learning difficulty—the large enterprise with its large R&D departments has the means, the intelligence capability, if you will, to decode the enormous "intelligence," data, which science, academia is continually generating (the challenge then is to sensibly *filter* such information and to diffuse it throughout the whole enterprise—hence the need of integration and interdisciplinarity). However, for small enterprises such a decoding or internal assessment facility is often lacking. A research department, or even a so-called technological gatekeeper, may well be absent. Appropriate "external" academic-industrial collaboration can then be of critical importance to subsequent learning. (In Chap. 30 we give several examples of such contemporary collaborative linkage.)

In essence, therefore, the smaller organization has to learn to tap into a wider intelligence, while for the larger organization the need is to *share* internally its innate intelligence. Both perspectives point, at base, to *cooperative needs,* but of different forms. Learning is dependent as much on cooperation as on competition (and learning from explanation or mistakes). In one sense that is the common message of most, probably all, of the various views expressed in Table 18.3. Cooperation is also the means of overcoming the different problems of small and large enterprises indicated in Table 18.2. Thus the small enterprise can improve its learning capability through external linkage: if it is of an appropriate form; the larger enterprise via improved internal functional linkage.[24]

The linkage can be facilitated by organizational change,[25] improved use of technologies,[26] improved training of individuals, and greater recognition by *all* employees at all levels of the importance of learning. Thus learning becomes in the latter respect an open, intentional, purposive endeavor—not merely a by-product of how a company functions. Numerous companies have made and are continuing to make impressive strides in placing emphasis on company learning: much beyond the conventional human resource development programs of yesterday.

It is perhaps invidious to single out particular companies; nevertheless, we should note here that such enterprises as Rank Xerox; Shell, 3M; Canon Inc., Japan; most Japanese auto manufacturers; Volvo and Saab; Lucas and Rover; Du Pont; Motorola; and IBM—to name only a few—place increasing emphasis on the improvement of organizational learning capacity.

An extensive empirical literature attests to the various gains so made. There is "no single recipe"—which is the point made in this section. It is an ethos. However, each company, each enterprise has its own failings and disabilities which have to be overcome if continuous learning is to develop and be encouraged. Perhaps this is well exemplified in the case of the major U.S. automobile manufacturers: GM, Ford, and Chrysler. For over a decade the U.S. automobile industry has been assailed by the ever-ferocious competition of the Japanese and European auto manufacturers. Detailed accounts of the difficulties they faced—in part relating to the historical trajectories of their earlier development—are to be found in a variety of penetrating texts.[28,29] In addition, the changing conditions of the U.S. marketplace, and the ever-increasing regulation-bound sector, the environmental challenges to product development, implied that these gargantuan companies would either have to "learn" or die. Thirty or so years ago such a scenario, if not undreamed of, was not sufficiently widely recognized. Ever-increasing reliance on capital-intensive technologies, centralization, strong divisional structures, alienated workforces, Taylorist principles, and productivity gain at the expense of radical product innovation sowed the seeds of learning disability. Much reversal of company philosophy, much change in organizational structure, new internal communication patterns, new patterns and forms of linkage with suppliers, new forms of assessment of market need, and so on have, in total, if not reversed the earlier learning disabilities, at least now provided a more robust platform for organizational learning.

At one time the automobile industry was seen as a "mature" industry in terms of its place in the development cycle. All that changed in terms of both product and process development. The auto industry is now, and continues to be, an immature—viz., learning, innovative—industry. In face of the challenges of the twenty-first century,[30] it has to learn even more and go on learning. There is a metamessage here. No industry, no sector should consider itself "mature," inviolable, secure. That way lies not maturity, but obsolescence and death. Only continual open-ended learning ensures survival. Many of the rapidly expanding corporations of Southeast Asia have recognized that challenge and now continue to oust competitors.[31] To learn is to live.

18.5 FINAL COMMENTS

The area of discourse briefly outlined in this chapter suggests that the following be kept uppermost in mind:

1. There is a need for both *explicit* and *implicit* attention to be constantly paid to the overall needs of organizational learning. The former demands explicit, transparent corporate strategies; the latter demands structures, tie-lines—across departments, across individuals, and across the whole organizational entity. The explicit and implicit modes are self-reinforcing and touch *all* levels of a company's employees. It is *not* exclusive to senior or middle management. Japanese and Scandinavian countries have demonstrated, in numerous companies, the importance of total employee involvement in respect of "company learning."

2. To some extent, beyond (*a*) the *generalizable* principles of encouraging a good corporate "memory," of developing internal and external "selective attentive mechanisms" which improve the coupling of the company to both the external marketplace, to the science and technology base, to internal company process knowledge and product development (or more particularly its innovatory potential); there also (*b*) exist *particular* requirements dependent on company circumstance (viz., science-based, supply industries, scale-intensive sectors, traditional industries).

3. The organizational fluidity, structure, and circumstances of *small and large* companies differ as does, usually, their degree of *organicness*. This signals particular learning abilities (and deficiencies) which should be attenuated or remedied.

4. However, new organizational systems geared to the improvement of corporate flexibility (or, specifically in response to the potential of new technologies) have in some sense *altered the competitive boundaries* between small and large companies. Mass-production systems are moving back to more batchlike procedures. Attendant with those post-Fordist changes are new learning challenges for small and large companies alike.[10] Indeed, the new technical era within which all companies find themselves (in both product and service sectors), with all the attendant networking and new global market challenges, suggests that *heightened attention* be paid to organizational learning; to new modes of information intake (and hence improved external linkages to customers, to academia—both viewed as informational and sensory learning sources).

5. All the above accentuates the importance of interdisciplinary skills, interfunctional collaboration, and strategic coherence which involves all "departmental" functions, far in excess of (say) contemporary concurrent-engineering principles. It indicates the need of managerial, organizational, and strategic *integration* across all facets of a company's activity. This, in turn, implies new training needs of a more interdisciplinary nature—which again, in turn, can foster improved learning capability. (See Chap. 30.)

What is implied in research terms? What is now required is that research studies attempt to collate and direct further study within a coherent framework rather than amassing more and more eclectic, disparate studies. One framework has been outlined in this chapter which can serve to better organize our presently limited understanding. It is by no means *the* only framework. Nevertheless, the learning approach outlined earlier has stood the test of time in relation to individual learning; it is the result of the endeavor of tens of thousands of researchers and has much merit.

Extrapolation from individual to organizational learning is, however, fraught with difficulties. These new emergent properties can be observed as systems complexity develops; different structures and different sectors of activity no doubt demand differing optimal forms of operation. But this still demands that a guiding theoretical framework be developed. At present we are only at the threshold of that fuller understanding.

18.6 REFERENCES

1. T. G. Whiston, "The Role of 'Spare Capacity' in the Development of Perceptual Motor Skills," in *Readings in Human Performance,* H. T. Whiting, ed., Lepus Books, London, 1975, chap. 2.

2. P. R. Meudell and T. G. Whiston, "An Informational Analysis of a Visual Search Task," *Perception and Psychophysics,* **7**(4): 212–214, 1970.

3. P. H. Lindsay and D. A. Norman, *Human Information Processing,* Academic Press, New York, 1972.

4. C. E. Shannon and W. Weaver, *The Mathematical Theory of Communication,* University of Illinois Press, Urbana, Ill., 1949; N. Weiner, *Cybernetics,* Wiley, New York, 1948.

5. P. Patel and K. Pavitt, *Technological Competencies in the World's Largest Firms: Characteristics, Constraints and Scope for Managerial Choice,* STEEP Discussion paper no. 13, Science Policy Research Unit, University of Sussex, May 1994.

6. A. T. Welford, *Fundamentals of Skill,* Methuen, London, 1968.

7. R. Rothwell and W. Zegveld, *Industrialization and Technology,* Longman, Harlow, 1985.

8. J. De Lorean, *On a Clear Day You Can See General Motors,* Wright Enterprises, Grosse Pointe, Mich., 1979.

9. R. F. Conti, *Taylorism, New Technology and Just-in-Time Systems in Japanese Manufacturing,* Judge Institute of Management Studies, Cambridge, U.K., 1992; C. Edquist and S. Jacobsson, *Flexible Automation: The Global Diffusion of New Technology in the Engineering Industry,* Blackwell, Oxford, 1988; P. de Woot, *High Technology Europe: Strategic Issues for Global Competitiveness,* Blackwell, Oxford, 1990.

10. T. G. Whiston, *Managerial and Organizational Integration,* Springer-Verlag, London, 1992.

11. G. H. Gaynor, *Achieving the Competitive Edge Through Integrated Technology Management*, McGraw-Hill, New York, 1991.

12. G. H. Gaynor, *Exploiting Cycle Time in Technology Management*, McGraw-Hill, New York, 1993.

13. M. Dodgson, *The Management of Technological Learning*, De Gruyter, Berlin, 1991.

14. M. Dodgson, "Learning, Trust and Technological Collaboration," *Human Relations*, **46**(1): 77–95, 1993.

15. M. Dodgson, "Technological Learning, Technology Strategy and Competitive Pressures," *Br. J. Management*, **2**:133–149, 1991.

16. M. Goold and A. Campbell, *Strategies and Styles: The Role of the Centre in Managing Diversified Corporations*, Blackwell, Oxford, 1993.

17. C. Handy, *The Age of Unreason*, Business Books Ltd., London, 1989.

18. R. Hayes, S. Wheelwright, and K. Clark, *Dynamic Manufacturing: Creating the Learning Organization*, Free Press, New York, 1988.

19. G. Morgan, *Creative Organization Theory*, Sage, Beverly Hills, Calif., 1989.

20. G. Morgan, *Images of Organization*, Sage, London, 1986.

21. P. M. Senge, *The Fifth Discipline: The Art and Practice of the Learning Organization*, Doubleday/Currency, New York, 1990.

22. R. Thomson, ed., *Learning and Technological Change*, St. Martin's Press, London, 1993.

23. P. J. Thurbin, *Implementing the Learning Organization*, Financial Times (Pitman Publishing), London, 1994.

24. T. G. Whiston, *Managerial and Organisational Integration. Technovation*, Part 1, **9**(7): 577–606, 1989; Part 2, **10**(1): 47–58, 1990; Part 3, **10**(2): 95–118, 1990; Part 4, **10**(3): 143–161, 1990.

25. A. Pettigrew, ed., *The Management of Strategic Change*, Blackwell, Oxford, 1987.

26. G. Dosi, C. Freeman, R. Nelson, G. Silverberg, and L. Soete, eds., *Technical Change and Economic Theory*, Pinter, London, 1988.

27. M. Gibbons, "The Industrial-Academic Research Agenda," in *Research and Higher Education: The United Kingdom and the United States*, T. Whiston and R. Geiger, eds., Open University Press, Buckingham, 1992, chap. 6; T. G. Whiston, *The Management of Technical Change, Training for New and High Technologies*, 2 vols., a report to the UK Manpower Services Commission on the Programme for Directors and Senior Managers of Austin-Rover held at Warwick University 1985/86, Feb. 1986, Science Policy Research Unit, University of Sussex, Brighton; T. G. Whiston, New Technologies and UK Educational Policy Response, *The House of Lords Select Committee on Science and Technology: Education and Training for New Technologies*, HMSO, London, Dec. 1984, vol. III, pp. 472–478.

28. A. Altshuler, M. Anderson, D. Jones, D. Roos, and J. Womack, *Future of the Automobile*, George Allen and Unwin, London, 1986.

29. J. Womack, D. Jones, and D. Roos, *The Machine That Changed the World*, Rawson, New York, 1990.

30. T. G. Whiston, *Global Perspective 2010: Tasks for Science and Technology*, Commission of the European Communities, Brussels, 1992.

31. M. Hobday, "Technological Learning in Singapore: A Test Case of Leapfrogging," *J. Development Studies*, **30**(3): 831–858, April 1994.

32. T. J. Peters and R. H. Waterman, Jr., *In Search of Excellence*, Harper & Row, New York, 1982.

CHAPTER 19

KNOWLEDGE IMPERATIVE AND LEARNING PROCESSES IN TECHNOLOGY MANAGEMENT

Anil B. Jambekar and Karol I. Pelc
School of Business and Engineering Administration
Michigan Technological University
Houghton, Michigan

19.1 INTRODUCTION

In recent years, a new paradigm has been adopted for management of technology: *knowledge-based management* relying on evolutionary process guided by *knowledge value revolution.*[1] The knowledge is considered to be an essential asset of any organization requiring attentive leadership efforts for its management. An underlying belief is that to gain competitive advantage, a company must include building and developing knowledge resources to leverage improved long-term financial performance. This imposes new requirements and creates new opportunities for individuals, teams, and organizations for managing technology. Development, adoption, and improvement of a new technology involves several types of knowledge and skills, such as technological knowledge, economics, and organizational dynamics, as well as technical-engineering and systems integration skills. The capability for technological innovation is limited by the capability for organizational change or vice versa. Both involve mental transformations and capability of learning. A technology-based organization is required to establish an internal learning environment and to involve itself in learning together with such partners as customers, suppliers, educators, and legislators. Organizational and technological learning are closely related and hence must be understood in order to manage them.

If *organizational learning* is the same as acquiring new skills and adopting new behaviors to improve performance, then one must understand how individuals and groups learn and then generalize the process to the organization as a whole. But is there a theory of learning? For many organizations, their very survival depends on the ability of their employees to continuously improve, to learn and reflect on new information about changing customer needs, and to learn how to creatively apply new technology and solve problems. So it should be obvious that managers and leaders ought to know the process of changing themselves and others and have knowledge about how people and organizations absorb, adapt, and integrate technological changes. In this chapter we shall review two topics:

- Role of knowledge as imperative for management of technology
- Models of learning processes in a technology-based organization

19.2   KNOWLEDGE IMPERATIVE

Importance of knowledge as a source of competitive advantage has been discussed by several authors, who also tried to determine parameters and consequences of knowledge-based organization.[1-4] Charles Savage formulated the following guiding ideas for managers of the *knowledge era:*[2]

- Transforming both raw materials and *raw ideas*
- Leveraging both capital and *knowledge assets*
- *Cooperating and collaborating* within and between companies
- *Team and reteam* capabilities

Mitroff et al. proposed a new structure of company in which the *knowledge-learning center* constitutes one of the most important functional units and has responsibility for finding out "what the company needs to know to produce/deliver world class products/services."[4] According to Davis and Botkin, "in knowledge economies, the rapid pace of technological change means...that people have to increase their *learning power* to sustain their *earning power.*"[3] The very survival of an organization in today's competitive world can be tied to the knowledge imperative. Implementation of such concepts into practice may require some radical changes in almost all spheres of business operations and also in the educational system. After all, an educational institution creates its first critical imprints on the knowledge workforce.

19.2.1   Types of Knowledge for Management of Technology

Creating, adopting, and implementing of new technical solutions (in products, processes, and systems) represent a set of difficult tasks for managers. This is due to some level of unpredictability and unrepeatability of the processes and procedures involved. Almost by definition, all those tasks, as well as the coordination between business and technology strategies and between overall technology strategy and product-process-service development projects, are difficult to routinize. Managers are expected to understand, communicate, and decide on mutually dependent and mutually reinforcing technological and organizational solutions to succeed. That creates a need to develop a *holistic view* on technology and organization, and to *integrate knowledge* on all facets of a company. Professional knowledge, in both engineering and management, has been traditionally developed, packaged, delivered, and acquired in the form of discipline-based courses, the building blocks of the modular structure of academic curriculum. An unfortunate consequence is that integration is left up to the learners, the majority of whom treat technological problems independently of organizational problems and vice versa, because this is how they learned.

Currently, there is a practical need for managing technology in a company:

- To develop interdisciplinary communication and interfunctional cooperation
- To integrate technical and organizational systems
- To establish technology-based operations in all domains of the company's activity (not only in design or manufacturing)

To achieve these goals, new technology must be viewed from a business process perspective and not from a functional perspective. A business process perspective requires adequate human capabilities to adopt new technologies and derive all possible benefits from them.

A holistic approach to management of technology requires a complete rethinking of educational and training processes. The spectrum of knowledge that is used in practice of technology management is expanding beyond formal divisions of academic disciplines. More than ever before, a combination of *experiential and theoretical knowledge* is needed.

A combination of *technical and organizational knowledge* is required for managing even such, relatively narrow, tasks as implementation of new manufacturing technology.[5] However, for total integration of human and technical systems to achieve, it is not enough to add together the two types of knowledge, as represented by conventional engineering and management education. The knowledge base for management of technology should include *methodology of systems integration*. The latter may lead to transferable insights, which is a form of organizational learning. This methodology, however, is still not a mature scientific discipline. It is rather a domain of expertise, conceptually rooted in systems theory. It requires careful evaluation and further development through experimentation. It includes some elements of empirical knowledge about human behavior (psychological and sociological background) and theoretical models representing typical patterns of interaction between people, organizations, and technical systems. Because of this broad scope, which makes it difficult to derive knowledge from a single well-structured theory, the systems methodology is being advanced by learning from real situations, actions, projects, and programs. They need to be evaluated from multiple perspectives. Linstone proposed a framework for *multiple perspective analysis* of ill-structured systems.[6,7] This approach assumes three basic perspectives superimposed in analysis of dynamic characteristics of complex systems: technical perspective, organizational perspective, and personal perspective. The object of system analysis is viewed as a whole and is evaluated simultaneously from these three perspectives. It is just one example of possible methods for integrated system analysis.

The importance of technology development projects as a source of knowledge for technology management has been emphasized by Clark and Wheelwright.[8] They specified five types of knowledge that may be acquired (or verified):

- Procedure (new sequence of operations or rules that project members follow)
- Tools and methods (acquiring new skills related with tools and methods)
- Process (new sequence of major phases of development project)
- Structure (new structure and/or location of organization)
- Principles (new concepts and values applicable for decision making in future projects)

A similar conclusion has been derived from the study conducted by the Manufacturing Vision Group, as described by Bowen and Clark,[9] on the basis of analysis of 20 development projects in five companies. The authors suggest that the development projects may serve as "engines of renewal." "By wisely selecting the projects it undertakes, a

TABLE 19.1 Types of Knowledge for Management of Technology (MOT)

Knowledge type	Sources	Application for technology management
Technical knowledge	Engineering science, engineering practice	Definition and development of products, processes, and technical systems
Organizational knowledge	Management science, economics, management practice	Managing technical operations, projects, and organizations
Systems integration methodology	Systems theory, development projects, organizational learning processes	Integration of technical and human systems (holistic view), developing and managing the learning organizations

manufacturer can use them to develop new skills, knowledge, and systems."[9] This means that the development project should be evaluated not merely for its direct technical outcome (e.g., design to specifications) but also for its contribution to the empirical knowledge base for management of technology in the company.

Classification of knowledge types for management of technology is summarized in Table 19.1. If the fundamental input to knowledge assets is people, the source is existing educational programs.

19.2.2 Education and Training for Management of Technology

In response to the increasing demand for interdisciplinary education of engineers and managers, several universities developed academic programs in management of technology (MOT). Names of those programs may be formally different, to indicate special core areas of expertise or to satisfy more specific needs, but their substance is quite similar. Recent review of the MOT educational programs, presented in 1994 by a team of authors from the Stevens Institute of Technology, includes characteristics of programs existing in 32 universities in the United States.[10]

According to Kocaoglu, the total number of "engineering and technology management" programs in the United States was 94 in 1990.[11] The latter number appears to include some programs that represent rather narrow scope (industrial engineering, construction management, etc.) and might not be counted in the same category.

Configuration of courses and detailed program requirements differ substantially. However, there is a common set of major areas of study which includes

- Introductory courses in functional areas of business such as accounting, finance, marketing, manufacturing, human resources, and organizational behavior
- Selected topics of economics (macro and micro economics)
- Course(s) in information systems and/or decision support systems
- Course(s) in technology and innovation management (including technology strategy, technological planning, R&D management)
- Courses reflecting special focus of a given program (quality management, operations management, international business management, etc.)

Most of those programs are offered at the graduate level. Many of them are intended primarily for experienced managers or engineers who have interest in or responsibility for managing technical projects or operations. It appears that it is still too soon to evaluate those university programs and their impact on quality of technology management in respective companies. Although the menu of courses offered is quite extensive, it is difficult to prove a consistent pattern of MOT programs or to see any dominant paradigm in their design. Rather, design is modular and hence traditional. In many programs students are expected to have a previous educational background in engineering or sciences and some amount of managerial experience. MOT programs are then focused at strengthening their organizational and managerial knowledge with special orientation to applications in technical operations and projects. However, the methodology of systems integration is seldom a subject of special courses. In several cases it is presented only as part of strategic management projects or in general introduction to technology management. Both universities and their industrial customers are still searching for an optimal mix between theoretical and empirical knowledge that should be provided in those educational programs. This subject is frequently discussed in journals and at conferences of professional societies related with management of technology.[12–14] Also[15]

> The following professional organizations are actively involved in discussion on education in MOT: International Association for Management of Technology, American Society for Engineering Management, IEEE Society of Engineering Management, Academy of Management, Institute for Organization and Management Sciences (former TIMS/ORSA), American Society for Engineering Education. In 1994 this problem was also a subject of the First National Conference on Business and Engineering Education sponsored jointly by: National Consortium for Technology in Business, American Society for Engineering Education, and American Assembly of Collegiate Schools of Business.

In addition to the degree granting MOT programs, several schools and institutes offer shorter and/or more intensive programs for executives. Their contents are highly dependent on and adjusted to specific expectations of participants. Some of those initiatives are even intended as ad hoc training to fill the current gap in the executives' knowledge rather than to lead toward transformation of their mindset.

Formal education and training in management of technology can be considered to be a necessary ingredient to activate the process of learning to build knowledge assets influencing the productivity and technological competitiveness of a company.[16,17] The process expects a fundamental transformation requiring every employee to balance continual and deliberate learning with task performance. The rest of the chapter elaborates on what it takes to build a learning organization.

19.3 LEARNING PROCESSES IN AN ORGANIZATION

19.3.1 Toward Definition of Organizational Learning

Today's manager-leaders absolutely must understand some basic facts about learning if they are to have any expectations of steering themselves and their organizations into the future and escaping information indigestion. Individual learning is not the same as organizational learning, although the former is a prerequisite for the latter. Ultimately, organizational learning has to do with improving performance over the long haul. If an individual is learning, we expect that person to perform better. Similarly, if a team is learning, we expect the team to perform for the benefit of the entire organization. However, using organizationwide performance indicators to gauge effectiveness of learning at an individual or a team level can be a trap, because in short run, an individual or a team can do all the right things but the results will not show up immediately, because of either intrinsic delays or confounding forces outside their control that are depressing the hoped for performance. If management keeps demanding performance, an individual or a team can take shortcuts to improve appropriate performance indicators, but run a risk of creating problems in the distant future or creating unexpected side effects for other parts of the organization.

So what is the definition of *organizational learning?* A contemporary view is that a learning organization is able to adapt quickly to a changing marketplace. However, the view is limiting. Learning is more than the ability to adapt. It is really about improving ability to create and to continually replace dysfunctional operating norms with more productive and competitive ones. Creating a learning organization is a long-term process of fundamental change.[18]

Every earning crisis leads a managerial team to brainstorm a list of possible actions. All the actions are hard measurable undertakings that can most dramatically enhance earnings. The list includes such things as investing in a new technology, eliminating an unprofitable business line, expanding into new business, lowering cost through layoffs and limiting expenses, and increasing productivity through adapting new management technology. The interesting point about such a list of options is that the same items appear over and over again every time the earning performance threatens the organization. All along an assumption has been made that employees are continually learning through experience. New programs such as *total quality management* (TQM) involved people who are given more responsibility and accountability with the expectation that the organization's earning would be enhanced. Sure, the new management program or a new technology would enhance learning at an individual level and even at a team level. But *individual learning* or even *team learning* is not the same thing as organizational learning, although they are necessary conditions. An organizational learning, if occurred, should reduce the risk of a performance crisis. Implications for management are that organizational climate must be created to promote individual and group learning as path breakers for building a *learning organization.*

19.3.2 Individual Learning

When organizations want their employees to learn new skills, they send them to training programs. The problem is that training or enrollment in a local university course, by itself, does not guarantee that an individual will learn new skills. As cited by Senge, Dewey postulated that all learning involves a cycling through following four stages:[19]

1. *Discover*—the discovery of new insights
2. *Invent*—the creation of new options for action
3. *Produce*—following through new actions
4. *Observe*—seeing consequences, which leads to new discovery, continuing the cycle

This is how children learned to walk, to talk, to ride bicycles, etc. Interrupting the cycle interrupts the learning. The children are supported throughout their learning process.[19] In effect, learning is moving back and forth between the world of thought and the world of action. It is an interactive process of linking the two to expand our capability. A fundamental message here is that deep learning takes place only when an individual cycles through all four stages.

A closer look at Deming's fourteen points[20] shows that the core of the quality movement lies in assumptions about people, organizations, and management that have a unifying theme: to make continual learning a way of organizational life, especially improving the performance as a total system. This is possible only by dissolving the traditional, authoritarian, command, and control hierarchy in which top management thinks and the local employees act, to integrate thinking and acting at all levels. The famous *plan-do-study-act* (PDSA) cycle, the never-ending cycle of experimentation that structures all quality improvement efforts, is at the heart of TQM. It recognizes that learning requires practicing new skills and obtaining feedback on our performance so that we can adjust our behavior. The concept of *intrinsic motivation* lies at the core of Deming's management philosophy. By contrast, extrinsic motivation is the operating norm of many organizations. "People do what they are rewarded for" is actually contradictory to the soul of the quality movement. This implies, even though rewards or punishment are not irrelevant, that they are not a substitute for intrinsic motivation to learn. An organization's commitment to quality or new technology that is not based on intrinsic motivation is equivalent to planting seeds for a periodic motivational crisis leading to less-than-potential or less-than-hoped-for performance. Deep-seated curiosity and the desire to experiment generate a spiral of continual improvement that cannot be matched by external rewards alone. From the intrinsic perspective, there is nothing arcane or academic about learning and continuous improvement. If left to their own devices, people will look for better ways to do things using the knowledge they have and by learning. What they need is appropriate tools, adequate information, and a supportive, trusting, and caring environment. Total quality management offers a variety of simple to complex tools,[21,22] the subsets of which can be taught to people at all levels of an organization. Positive influence of such a process is that the tools and the concepts provide a communication language fundamental for enhancing organizational learning.

19.3.3 Group or Team Learning

Many organizations are very actively promoting teamwork in an attempt to foster joint interrogation into organizational problems. Groups provide a forum to exchange ideas. Groups are formed for many reasons. They may be formed as training clusters, as problem solving teams such as quality circles, or cross-functional teams to improve process or product development. Regardless of the nature of the purpose, teamwork can help individuals appreciate and learn how other parts of the organization interact, how select actions have far reaching consequences than individual group members would have realized on their own, and why certain approaches are appropriate or inappropriate for the organization as a whole. When different viewpoints are represented in a business context, teams generate higher-quality information, reinforce learning, and deliver a heightened action orientation.[23]

Although group learning is a requisite for today's organizations, it is not synonymous with organizational learning. At the local level, groups solve problems that are important locally. The groups do not confront the larger organizational reality, and their solution can potentially, although unintentionally, create problems for other parts of the organization, or they may be solving problems inadvertently created by some other unit of the organization.

Cross-functional group problem solving addresses wider issues spanning several units of an organization, but they are still the problem-solving teams to resolve an issue with defined boundaries and scope. The solution generated, if it has to be phased in over a longer time period, can create side effects not in the interest of the larger organization, because of factors outside the original scope of the group influence, it can influence the reality or the group's desire to succeed, it can unintentionally create ill effects on the system as a whole. Organizational learning cannot be assured just by having numerous cross-functional teams working on quality, technology introduction, and product or process improvement. It just means you are creating a culture of teamwork, but not necessarily continuously learning culture.[23,24]

A barrier to group learning can be the defensive routines many follow. Good team members, acting rationally within the organizational context in which they live, create and maintain defensive routines that prevent organizations from learning.[25-27] *Defensive routines* are habits of interacting that serve us to protect us or others from threat or embarrassment, but also prevent us from learning. According to Argyris, once they are started, these routines seem to take on a life of their own. They not only prevent learning, but also prevent people from identifying and changing causes of the embarrassment or threat. Organizations take such routines for granted. For cross-functional teams to foster organizational learning, it is essential to reduce defensive routines that prevent learning about the importance of mutual interdependency for organizational success.

From a group learning perspective, an environment must be created for members to uncover and challenge deep-rooted assumptions and norms. The organizational operating norms, practices, procedures, policies, behaviors, resource allocations, and staffing make up the organizational map which must be improved through the result of group learning. Detection and correction of dysfunctional behavior patterns and embedding the results for the group inquiry into the organizational maps alter organizational memory for the better. An organization learns when individuals within it have changed their shared mental models about how the organization should behave. Ray Stata, CEO of Analog Devices, Inc.,[28] states that

> First, organizational learning occurs through shared insights, knowledge, and mental models. Thus, organizations can only learn as fast as the slowest link learners. Change is blocked unless all of the major decision-makers learn together, come to share beliefs and goals, and are committed to take the actions necessary for change. Second, learning also builds on past knowledge and experience—that is, on memory. Organizational memory depends on institutional mechanisms (e.g., policies, strategies and explicit models) used to retain knowledge.

This implies the creation of an organizational environment or culture that interweaves learning and work. It is a process by which individuals are able to confront inconsistencies in action which have become predicaments. According to Argyris, organization learning takes place when the manager's espoused theories get closer and closer to that person's theories in action as evaluated by others.

To promote organizational learning, from a perspective of group learning, individuals require skills in recognizing their own deep-seated values, beliefs, and assumptions—mental models used to frame and react to situations and ability to balance inquiry and advocacy. This invites individuals to loosen their grip on their models of reality and consider alternative maps that can yield superior and effective organizationwide results. Being aware of one's own thinking process—the ladder of inference, and balancing advocacy with inquiry are among few *action science tools* promoted by Argyris et al.[24,27] and are absolutely essential skills required to promote organizational learning through group interactions.

19.3.4 Theory-Building Process—Learning

Individuals and groups are continually learning in organizations, and this means that they are also building a theory of reality. The word *theory* is too often considered as an academic or esoteric word with no practical significance; in fact, it is of extreme practical importance for a firm aspiring to be a learning organization. When individuals and teams learn, in essence, they build theories in the form of general principles drawn from a body of facts and observations. The principles in effect represent distillations of their knowledge and understanding of the world. The theories provide means and foundation that bring coherence to our observations.

Dewey's individual learning process represented by the discover-invent-produce-observe cycle and Deming's PDSA cycle of continuous improvement, mentioned before, can be thought of as natural rhythms of individual and group learning behavior. There are other representations of learning cycles. For example, Kim presents Kofman's assess-design-implement-observe cycle as an individual learning process.[29] Kolb[30] introduces a Lewinian learning model which also has four stages. Two common threads in all representations are that a learning process moves back and forth between *thinking* and *doing* as well as between *abstract* and *concrete*. Table 19.2 offers a comparative view of all learning cycles. Figure 19.1 shows a visual representation of individual and group learning process. Deming's PDSA cycle is used in Fig. 19.1 because it is much more widely known to today's managers than any other depiction.

If organizational culture and environment can be created such that individuals and groups can work through all stages of the cycle, learning can certainly be enhanced. However, the breakdowns can occur as a result of dynamic complexity, time delays, unknown structures of the real world and delays, information errors, distortions, and biases in the feedback. The lessons learned before can become obsolete with the fast-paced changes in technologies and the marketplace. This suggests that the lifespan of new theories, even if they are accurate, can be shortened by changes in environment and technology. This does not reduce the importance of creating organizational culture

TABLE 19.2   A Comparative View of Learning Cycles

Dewey*	Deming[†]	Kolb[‡]	Kofman[§]	Learning process characterization	
Invent	Plan	Formation of abstract concepts and generalization	Design	Abstract	Doing
Produce	Do	Testing implication of concepts in new situations	Implement	Concrete	Doing
Observe	Study	Concrete experience	Observe	Concrete	Thinking
Discover	Act	Observations and reflections	Assess	Abstract	Thinking

*Senge;[19] Senge attributes credit to Dewey.
[†]Deming.[20]
[‡]Kolb.[30]
[§]Kim;[29] Kim credits F. Kofman of MIT.

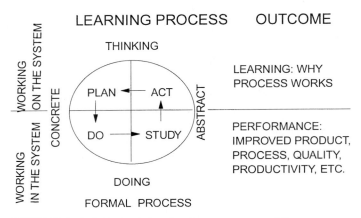

FIGURE 19.1   Individual and group learning process representation.

and environment to foster individual and group learning, but it adds additional challenges for employees to be able to design and redesign organizational structures in the face of new realities, while being part of the structure themselves. Thus organizational learning is enhancing capability to continually change processes and structures in the face of new realities and not the end goal.

19.3.5 Systems Thinking: The Foundation of the Learning Organizations

Systems thinking offers theory and tools for understanding complex systems, which have become increasingly threatening at the same time we are part of them.[31,32] The learning cycle from systems view is explained using feedback as a basic framework. We make decisions, these decisions change the world, we obtain some information feedback about what is happening, and on the basis of this new information, we make new decisions. The feedback processes, as shown in Fig. 19.2, are at the heart of the individual and group learning cycles discussed above. The outer loop shows the basic organizational learning loop, in which processes of dialogue, discussion, debate, and reflection eventually create changes in strategy, structures, and decisions, which through implementation process bring about changes in the real world.[33] The changes are interpreted through information feedback, which are analyzed in the organizational context to make changes in strategy, structures, and decisions. Figure 19.3 shows the relationship between the elements of learning feedback diagram and plan-do-study-act learning cycle.

   In typical organizations, there are hundreds of process cycles operating, all interrelated, all running on different time clocks, and many of them feeding back each other information, which get utilized to change the decisions to alter the reality. The PDSA view tends to be static when one tracks what is happening in a single process. The PDSA cycle works on a human-determined time clock without recognizing the real-world dynamics. Thus the PDSA cycle tends to work very well in the situations in which quick feedback is possible, such as on the factory floor or for the processes which are stable over extended periods. Systems thinking always looks at dynamic interrelationships. As dynamic complexity increases, the breakdowns in the learning are more likely. Kim[34] describes dynamic complexity for a particular project as a func-

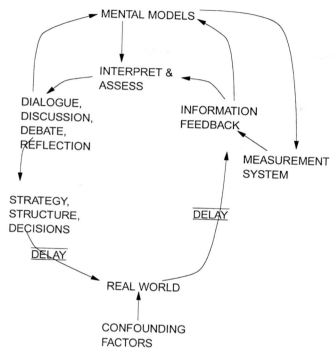

FIGURE 19.2   Feedback view of learning. (*Source: Morecraft.*[33])

tion of time lags of process and organizational complexity as shown in Fig. 19.4. By researching extensive literature in the field of systems dynamics, Senge has offered several principles useful for understanding behaviors of complex systems.[18]

- Today's problems come from yesterday's "solutions."
- Behavior grows better before it grows worse.
- The cure can be worse than the disease.
- The harder you push, the harder the system pushes back.
- The easy way out usually leads back in.
- Faster is slower.
- Cause and effect are not closely related in time and space.
- Small changes can produce big results—but areas of highest leverage are often the least obvious.
- Dividing an elephant in half does not produce two small elephants.
- You can have your cake and eat it, too—but not at once.
- There is no blame.

These principles hold true because cause and effect in complex systems generally are

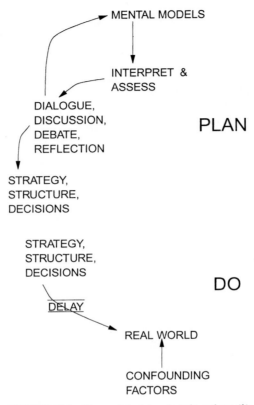

FIGURE 19.3 Plan and do (*a*) and study and act (*b*) stages of learning.

not close together in time or space, but in the PDSA process mode of operation, individuals or groups often assume and act as if they are. The methodology of systems thinking, with system dynamics as a core, offers many tools which vary from use of simple paper and pencil to draw causal loop diagrams or sketch behaviors over time to some sophisticated computer-based models and management flight simulators.[34]

Learning in complex systems is not as direct as the PDSA cycle suggests. Referring to Figs. 19.2 and 19.3, we can note that there are time delays between different parts of the feedback cycle: (1) between our decisions and their repercussion on the real world, (2) between their effects and our perception of information feedback, and (3) between the time we recognize the change and the time we decide how to intervene. Because the process or system is part of the larger system, which is constantly evolving according to its own dynamics, the other influencing factors change even as we are designing new interventions. Furthermore, quality of information feedback is a function of the measurement systems we have in place. The contents and information we select, measure, and pay attention to are imprisoned by mental models, and changes in mental models are influenced by what has been chosen to measure. Misperception of feedback can also account for another set of barriers. Inability to

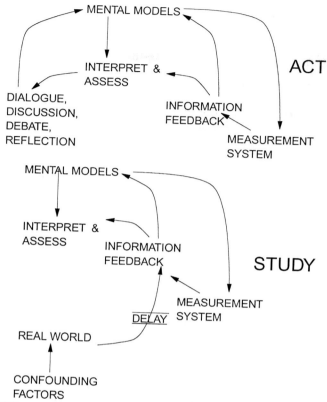

FIGURE 19.3 (*Continued*) Plan and do (*a*) and study and act (*b*) stages of learning.

deal with complexity results in the counterintuitive behavior of social systems.[35] People just do not have the mental capacity, because of cognitive limitations, to understand the dynamic complexity of the system. But people are good at identifying underlying structures. Thus computer-based simulations of correctly represented and widely agreed-on structures offer tools to overcome our inability to deduce long-term dynamics. Microworlds offer simulation based learning environment to understand recurring generic structures.[36] The recurring generic structures are also known as archetypes.[18] Understanding, identifying, and applying while engaged in the PDSA process can also deepen individual and group learning.

19.3.6 Challenge for Manager-Leaders

In order to sustain learning, manager-leaders need to create structures that enable the individual and group learning processes to become ingrained in the organization as part of its infrastructure. The learning infrastructure can be utilized to create widespread change. Figure 19.5 extends the learning process shown in Fig. 19.1 to include

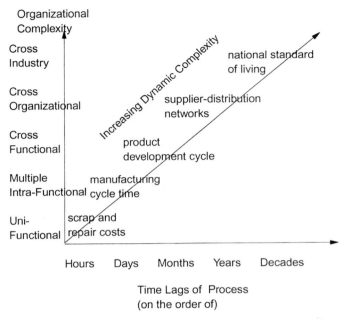

FIGURE 19.4   Complexity and time lags of organization. (*Source: Kim.*[34])

the systems thinking "LEARN" process. PDSA can be used to apply many of the quality management and action science tools, whereas LEARN can be utilized to improve systemic learning, supported by systems thinking tools.[37]

Implicit in the learning process model described are new roles for managers and employees. Employees have job security as long as they add value to the organization, and become responsible for finding ways to add value through continuous learning. The pace, nature, and scope of changes such as reengineering, downsizing, and new competitive maxims suggest that we do not know all we would like to know about the emerging model of technology-intensive corporate organizations. Block[38] defines *stewardship* as "willingness to be accountable for the well-being of the larger organization by operating in service, rather than in control, of those around us." He offers a vision of a future organization based on a belief that all employees are mature adults and can be held responsible for themselves and their actions. The new vision of the corporate model has an environment in which people can fully participate and contribute to the goals of the larger organization.

The organization learning process presented in this chapter is in alignment with the vision of the emerging view of an organization. The governance structure of most organizations will have to change from patriarchy to the concepts based on stewardship. It would require organizations to relinquish much of the control they have held over their employees and give them genuine authority to work in teams. *Stewardship* implies a belief that with better information and goodwill, people can make responsible decisions about what controls they require, and who they want to implement them. Obviously, having better information and goodwill may not be enough to make intelligent decisions, if people are not aware of the larger context in which decisions are being made, and if they do not have knowledge of appropriate tools.

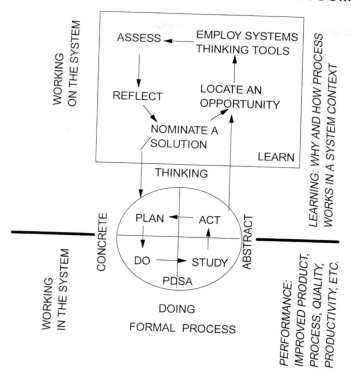

FIGURE 19.5   A framework for local and systemwide learning.

Stewardship comes not with a leadership position held in an organization, but leadership challenge of a moment, that requires individuals to make the choices and then live by them.[39] Recognition by all the requirements of leadership lie with everybody, and hierarchy becomes less of a system of power and control and more of a system of coordination of different types of work and responsibilities.

19.4 REFERENCES

1. T. Sakaiya, *The Knowledge-Value Revolution,* Kodansha International, New York, 1992.
2. C. M. Savage,"The Dawn of the Knowledge Era," *OR/MS Today,* pp. 18–23, Dec. 1994.
3. S. Davis and J. Botkin, "The Coming of Knowledge-Based Business," *Harvard Business Review,* pp. 165–170, Sept.–Oct. 1994.
4. I. I. Mitroff, R. O. Mason, and C. M. Pearson," Radical Surgery: What Will Tomorrow's Organizations Look Like?" *Acad. Management Executive,* **8**(2):11–21, 1994.
5. W. B. Chew and D. Leonard-Burton, "Beating Murphy's Law," *Sloan Management Review,* pp. 5–16, Spring 1991.

6. H. Linstone, "Multiple Perspectives: Concept, Applications, and User Guidelines," *Systems Practice,* **2**(3):307–331, 1989.

7. H. Linstone, *Multiple Perspectives for Decision Making,* North-Holland, New York, 1984.

8. K. B. Clark and S. C. Wheelwright, *Managing New Product and Process Development,* Free Press, New York, pp. 733–759, 1993.

9. H. K. Bowen and K. B. Clark, "Development Projects: The Engine of Renewal," *Harvard Business Review,* pp. 110–120, Sept.–Oct. 1994.

10. R. Jain, S. Jasti, T. M. Flaherty, and D. N. Merino, *Survey of Management of Technology (MOT) Educational Programs,* (discussion draft/working paper), Stevens Institute of Technology, Hoboken, N.J., 1994.

11. D. Kocaoglu, "Research and Educational Characteristics of the Engineering Management Discipline," *IEEE Transact. Engineering Management,* **37**(3): 172–176, 1990.

12. E. Howard, "The Management of Technology (MOT): Need for New Courses and Programs in U.S. Universities," *Proceedings of the 9th Annual Conference of ASEM,* Knoxville, Tenn., Oct. 1988, pp. 17–26.

13. T. M. Khalil and E. Berman, "Educational Programs in Management of Technology," *Proceedings of PICMET '91: Technology Management—the International Language,* Portland, Oreg., Oct. 27–31, 1991, p. 115.

14. Y. Shulman, "Graduate Programs in MOT—a Manual," *Proceedings of PICMET '91: Technology Management—the International Language,* Portland, Oreg., Oct. 27–31, 1991, p. 115.

15. *Engineering/Business Partnerships: An Agenda for Action,* Summary Report of the First National Conference on Business and Engineering Education, Auburn University, April 5–7, 1994.

16. P. S. Adler, "Shared Learning," *Management Science,* **36**(8): 938–957, 1990.

17. P. S. Adler and A. Shenhar, "Adapting Your Technological Base: The Organizational Challenge," *Sloan Management Review,* pp. 25–37, Fall 1990.

18. Peter Senge, *The Fifth Discipline: The Art and Practice of Learning Organization,* Currency/Doubleday, New York, 1990.

19. Peter Senge, "Building Learning Organizations," *J. Quality Participation,* pp. 30–38, March 1992.

20. W. E. Deming, *Out of the Crisis,* Center for Advanced Engineering Study, MIT, Cambridge, Mass., 1986.

21. Shoji Shiba, Alan Graham, and David Walden, *A New American TQM,* Productivity Press, Cambridge, Mass., 1993.

22. Don Clausing, *Total Quality Development,* ASME Press, New York, 1994.

23. Michele L. Bechtell, *Untangling Organizational Gridlock,* ASQC Press, Milwaukee, Wis., 1993.

24. Chris Argyris, *Overcoming Organizational Defenses,* Allyn & Bacon, Needham, Mass., 1990.

25. Chris Argyris, *Reasoning, Learning, and Action,* Jossey-Bass, San Francisco, 1982.

26. Chris Argyris and Donald Schon, *Theory of Action,* Jossey-Bass, San Francisco, 1974.

27. Chris Argyris, Robert Putnam, and Diana McLain, *Action Science: Concepts, Methods, and Skills for Research and Intervention,* Jossey-Bass, San Francisco, 1985.

28. Ray Stata, "Organizational Learning: The Key To Management Innovation," *Sloan Management Review,* Spring 1989.

29. Daniel H. Kim, "The Link between Individual and Organizational Learning," *Sloan Management Review,* pp. 37–50, Fall 1993.

30. D. A. Kolb, *Experiential Learning: Experience as the Source of Learning and Development,* Prentice-Hall, Englewood Cliffs, N.J., 1985.

31. Daniel H. Kim and Peter M. Senge, "Putting Systems Thinking in Practice," *System Dynamics Review,* **10:**277–290, 1994.

32. John D. Sterman, " Learning in and about Complex Systems," *System Dynamics Review,* **10:**291–330, 1994.

33. J. D. F. Morecraft, "System Dynamics and Microworlds for Policymakers," *Eur. J. Operational Research,* **35:**301–320, 1988.

34. Daniel H. Kim, " Toward Learning Organizations: Integrating Total Quality Control and Systems Thinking," *Pegasus Communications Reprint Series,* Cambridge, Mass., 1992.

35. Jay Forrester, "Counterintuitive Behavior of Social Systems," *Technology Review,* pp. 52–68, Jan. 1971.

36. Peter M. Senge and Colleen Lannon, "Managerial Microworlds," *Technology Review,* pp. 63–68, July 1990.

37. Anil B. Jambekar, "Frameworks for Integration of Systems Thinking with the Quality Management Practices," *International System Dynamics Conference—Change Management,* 1994, pp. 17–25, Systems Dynamic Society, Sterling, Scotland.

38. Peter Block, *Stewardship,* Berret-Koehler, Pergus Communications, San Francisco, Calif., 1993.

39. Daniel H. Kim and Colleen-Annim-Kim, "Stewardship: A New Employment Covenant," *The Systems Thinker,* **5**(6):1–3, August 1994.

CHAPTER 20
TECHNICAL LITERACY AND THE KNOWLEDGE IMPERATIVE

Kevin B. Lowe
Department of Management and International Business
College of Business Administration
Florida International University
University Park, Miami, Florida

Terri A. Scandura
Department of Management
School of Business Administration
University of Miami
Coral Gables, Florida

Mary Ann Von Glinow
Department of Management and International Business
College of Business Administration
Florida International University
University Park, Miami, Florida

20.1 INTRODUCTION

The pace of business change, characterized by globalization, higher consumer expectations, greater competitive pressures, and shorter process cycle times, has forced many organizations to fundamentally redesign work structures to meet rapidly changing marketplace requirements. Pressures for workforce productivity gains are intensifying. Organizations and industries now look beyond obvious efficiency gains from automation and personnel reduction to more systematic and breakthrough ways of being low-cost producers of high-quality products and services. The useful life of information is shrinking (McLagan, 1989), and organizations that are able to work in less time gain competitive advantage. Hierarchies are dissolving and being replaced by flatter, more flexible organizations that strive to generate new ideas and then transfer or generalize those ideas to action faster than their competitors.

In this dynamic environment, organizations increasingly view physical advantages in production technologies as a fleeting source of competitive advantage (Stein and Sperazi, 1991).

In search of a new winning formula, organizations increasingly focus on developing human asset competencies; such competencies have surprisingly become an increasing portion of value-added and, as many organizations are now discovering, the most difficult asset to duplicate (Casio, 1989). Some estimates suggest that even in manufacturing, which has traditionally placed strong emphasis on production and process technology, perhaps three-fourths of value added derives from knowledge (Losee, 1994). Jeffery Pfeffer of Stanford University writes in *Competitive Advantage through People* (1994, p. 6):

> Traditional sources of success—product and process technology, protected/regulated markets, access to financial resources, and economies of scales—still provide some competitive leverage, but to a lesser degree now than in the past, leaving organizational cultures and capabilities, derived from how people are managed as comparatively more vital.

This *primary* emphasis on managing human assets in the firm is a relatively new phenomenon (Ulrich, 1986). Organizations have always strived to manage scarce resources and accordingly have paid relatively less attention to those resources which are readily available. Traditionally scarce resources have been capital; abundant resources included skilled workers who have been highly capable of sustained performance. The easy availability of competent employees resulted in an emphasis on capital over competence. However, today's demographic trends such as an aging workforce, reduced growth in the rate of new workforce entrants, and declining educational system quality have made competence increasingly a scarce resource. In this scarce market for competence, firms are aggressively competing worldwide to attract, motivate, and retain competent employees.

As we approach the next millennium, business strategies will be more dependent on the quality and versatility of the competent human resource. Moreover, organizational resources will be expended toward identifying and assessing scarce human capital in much the same manner as obtaining physical assets and technologies have traditionally been sought as a means to achieve competitive advantage. One author has suggested that convergence of these forces in the 1990s has ushered in the era of "people power" as the key competitive force (Doyle, 1990). Another states "Strategies of the nineties will not be delivered if the organization's people aren't capable and committed. Organizations that apply only money and technology to problems, without bringing the people along will not survive" (McLagan, 1989).

Organizational changes in response to environmental pressures require corresponding changes in employee roles and the skill sets required to be effective in those roles (Katz and Kahn, 1979). As organizations have downsized their workforces in the 1980s and now the 1990s, many jobs have all but disappeared, leaving tasks that still need to be accomplished but fewer workers to perform the work (Hammonds et al., 1994). The tremendous effort now spent on organizational design initiatives is partially a reaction to headcount reductions. These redesign efforts frequently concentrate on self-directed work teams, process reengineering, and culture change to achieve a balance or "fit" between multiple organization tasks and competent workers who can perform these tasks (Fisher, 1993; Manganelli and Klein, 1994). Further, rigid jobs are displaced by team accountability and flexible, multiskilled job designs. These are more participative team-based designs which require a greater level of technical, business, and interpersonal skills than was required in a more rigid hierarchical structure with narrowly defined tasks. Now, organizational members are asked to do "knowledge work" which requires judgment, flexibility, and personal commitment to the job rather than compliance and submission to organizational procedures (McLagan, 1989). Peter Drucker has characterized the current era as an information society where knowledge workers constitute the primary source of competitive advantage (Drucker, 1985). To be effective in this fluid environment, employees must understand the organization's business objectives in order to take direction from the work itself in meeting those objectives rather than relying on policy manuals or being told what to do by their supervisor. The supervisor may have been replaced and the manuals may describe jobs which no longer exist or procedures which are obsolete in the newly configured workplace.

The incentive to develop and acquire competent employees with the requisite abilities is clear: Those high-performance organizations that are able to leverage the performance of individuals in their "jobs," as team members, and as organizational resources, will be the recipients of economic gain (Stein and Sperazi, 1991; U.S. Department of Education, 1992a). What is less clear is how and where American organizations will develop and acquire the type of talent required to be competitive. When productivity and quality are discussed, often the issue of the skill level of the workforce emerges as a potential barrier to improving organizational capacity (Charp, 1995).

It is essential that educational systems from kindergarten through university recognize increasing demands on workers to keep pace with a rapidly changing technological environment. Fewer workers leave the education system with the skills needed to do the job. This has been referred to as the "national crisis" in public education. As a result, highly technical organizations are undertaking a great deal of on-the-job training and education. What knowledge gaps are these companies increasingly having to fill? Primarily, the underpinnings of basic learning and thinking: research, data synthesis, and analytical skills (McKendree, 1991). Companies who for decades invested in technical training only for relatively specialized jobs can no longer ignore the need to teach the workforce how to investigate, analyze, and ultimately anticipate options and challenges that the competitive landscape will present in the years ahead (*Business Week*, 1994).

20.2 THE ROLE OF THE WORKER: A SHIFTING PARADIGM

The movement toward an emphasis on workplace literacy in the United States is in response to the shifts identified earlier from traditional production organizations to high-performance organizations. Traditional production organizations are based on nineteenth- and twentieth-century theories of management and productivity (Roth, 1993). Such organizations emphasize large-scale manufacturing to generate sufficient inventories, and focal attention is given to cutting costs and reducing per unit costs by increasing the number of units produced (Hodgetts et al., 1994). Product cycles are long in duration and new products are infrequently introduced giving the ultimate consumer limited product choice (Stein and Sperazi, 1991). Traditional industries rely on hierarchies in which multiple levels of management control workers in much the same manner as other tools in the production process. In traditional organizations, jobs are broken down into simple tasks, and the role of the worker is to repeat those tasks with machinelike efficiency (Bridges, 1994). Worker reliability and willingness to comply is valued. System improvements are the prerogative of an "elite" cadre of managers charged with reviewing processes (Walton, 1985). Effective in its time, the traditional organizational approach will be insufficient to meet competition in global markets with twenty-first century standards (Sasseen et al., 1994).

In traditional organizations, workforce learning was not viewed as a meaningful activity in relation to the production process (Stein and Sperazi, 1991). Workers in traditional organizations are expected to engage in only first-order learning, which involves improving the organization's capabilities to achieve known objectives and is often associated with routine and behavioral learning (Ulrich et al., 1993). Traditional workforce programs are problem-centered with outcomes measured in terms of short-term goal attainment. Training is viewed as a technique to prepare workers for action, a form of remedial activity designed to fill gaps in the abilities of workers to perform specific job skills. The "real" activity is viewed as job training which follows workforce education. There is a presumed inconsistency and conflict between education and production, and workers are not given release time for participation in learning (U.S. Department of Education, 1992b).

High-performance organizations view the production process differently. They are constantly reinventing themselves by emphasizing frequent product development even when it requires cannibalization of an existing market leader position. Customized products are built to order, inventories are small, and development time is short. Second order learning is expected of employees, requiring consistent reevaluation of the nature of objectives and the values and beliefs underlying them (Argyris and Schon, 1978). Second order learning, or double-loop learning, consists of "learning how to learn," something that even "smart managers" have difficulty with (Argyris, 1991). Every member of the workforce is responsible for product and process improvement, efficiency gains, and customer satisfaction. The emphasis in the production process is on continuous improvement, increased productivity, and growth. In high-performance organizations, managers function as coaches in "participatory processes" and workers are viewed as resources. Training prepares workers not only for jobs as currently defined but also for the job as it is expected to evolve and for future jobs. Workers are measured on working efficiently and smoothly in self-managed teams and for their ability to creatively solve problems (Fisher, 1993). Since improvements in the process and products are largely a result of worker inputs, perceived threat of job loss is minimized by the worker's role in developing process improvements. If the American economy is to transition successfully into the twenty-first century, then workplace education must respond and adapt programs to the educational needs of high-performance organizations (Stein and Sperazi, 1991).

20.3 CHALLENGES TO THE NEW PARADIGM

Organizations are moving into an era of system solutions where the concern is to resolve issues or make real changes, not just to implement programs (McLagan, 1989). Problem solving and change usually require multiple and diverse actions (training, policy change, job redesign). The number of new jobs in the United States is projected to increase to over 25 million by the year 2000, mostly in management, administrative support, sales, and service (Hudson Institute, 1987). These new jobs will require higher levels of formal education and technical literacy than are presently found, standards formerly expected only of managers and other high-level workers. Basic skill levels that formerly were adequate for assembly line production are inadequate in a workplace with just-in-time inventory processes, elaborate quality-control systems, flexible production, team-based work, and participative management practices (Hudson Institute, 1987). Charp (1995, p. 4) stresses the need for education beyond the "3R's" (reading, writing, and arithmetic) for organizations to maintain competitive advantage in an "information age": "Worker skills must mean more [than] job skills." She describes systems of continuous learning in place in Japan and Korea, which have increased organizational capabilities for learning that enable adaptation to change and innovation. These educational systems involve investment in training and development with a firm foundation in basic literacy skills at the K–12 (kindergarten through twelfth grade) level. A nation whose core workforce is only basic-skill-literate will not be competitive in the global marketplace.

20.4 BASIC SKILLS

Basic skills are traditional skills such as the ability to read signs, the ability to add three numbers to determine the amount of a bank deposit, and language fluency sufficient to articulate questions and understand their answers (Adams, 1993; Barton and Kirsch, 1990). Basic skills are a prerequisite, a necessary but not *sufficient* condition, to the skills necessary to be effective in twenty-first-century organizations. The following demonstrates one example of how basic skills provide access to higher-level skill learning (U.S. Department of Education, 1992):

> My employee was involved in the ESL (English as a second language) class wrote a first-line supervisor. I have seen a direct improvement in his confidence level. He will now come to talk to me, instead of having someone else come to ask questions for him. He talks much more freely. His initiative is greater and he looks more motivated. This same employee completed a 40-hour Robot Operating Training Course with two other employees who spoke English as their first language. He was able to participate equally in the training due to his increased English skills.

The quality of the American workforce, for both existing and anticipated entrants, falls woefully short of even this most basic requirement. One of every five current American workers reads at or below the eighth-grade level, and one of every eight lacks a reading competency above the first-grade level (Mikulecky, 1990). In international comparisons of student achievement in industrialized nations on 19 different academic tests, American students never finished first or second, but they scored last seven times. Of all high school graduates, 13 percent are currently illiterate, and among selected minorities, illiteracy is as high as 40 percent. Only 70 percent of U.S. students complete high school, as compared with 98 percent in Japan. The typical high-school graduate in Japan is better trained in the basic sciences and language than half the college graduates in the United States. And the average Russian high-school graduate has taken 5 years of physics, 5 years of algebra, 4 years of chemistry, 4 years of biology, and 2 years of calculus. In contrast, the typical American student has not taken physics or chemistry, and only 6 percent have taken calculus. And, finally, more than half of U.S. high-school graduates lack the sophisticated information processing, communications, teamwork, and analytical thinking skills that most of the coming decade's jobs will require. The U.S. Department of Education reports that by the year 2000 an estimated 17.4 million limited-English-proficient adults will be living in the United States. Immigrants will make up 29 percent of the new entrants into the labor force between now and the year 2000—twice their current share (U.S. Department of Labor, 1991).

Compare this with a cross section of *current* jobs where reading level requirements were found to be between the eighth- and twelfth-grade levels. Of the job-related material in these same jobs, 15 percent required even higher reading levels (Mikulecky, 1990). Clearly the problem of deficient basic skills is not new. What may be new is the inability of organizations to mask basic skill deficiencies within narrowly defined jobs and routinized job tasks. The competitive pressures of the global marketplace will not absorb the burdensome costs and lack of flexibility characteristic of the mechanistic organization operating in a more insular market (Hammonds et al., 1994).

The number of 18- to 24-year olds entering the workforce is shrinking, and thus reform in the schools, although relevant to longer-term competitiveness, will not eliminate current workforce deficiencies in basic skills (Hudson Institute, 1987). New entrants to the workforce are increasingly female and nonwhite adults (McLagan, 1989). Over three-fourths of those who will be working in the year 2000 are already out of school and most are already on the job (Hudson Institute, 1987). The literacy gap in basic skills continues to widen, and increasing portions of the population are being classified as "functionally illiterate" or unable to speak English (McLagan, 1989). The most lenient literacy standard in use today for employable adults is fourth- to sixth-grade skills, which include the ability to read simple text and street signs. Under this definition of literacy, estimates of the number of functionally illiterate Americans range from 16 to 27 million, with the upper figure representing approximately 20 percent of the U.S. adult population (Adams, 1993). The next-highest literacy standard requires eighth-grade reading skills which include the ability to read a driver's license manual, read a digest or newspaper article, and compute change from a purchase. Under this definition, nearly 45 million adult Americans or roughly one-third of the adult population are illiterates (Chall et al., 1987). These literacy rates can be contrasted with Japan, a major international competitor and trading partner, which reports a 98 percent literacy rate. American employers are thus being forced to reach out to less-qualified workers to develop entry-level workforces and the skills gap between current workforce competencies and future workforce competencies is continuing to widen (Mikulecky, 1990). The New York Telephone Company reported that it tested 57,000 job applicants in 1987 and found that 96.3% lacked basic skills in math, reading, and reasoning (Bradsher, 1990). Chemical Bank in New York reports that it must interview 40 applicants to find one that can be successfully *trained* as a teller (Bank of America, 1990). Filling the jobs that will be created by the year 2000 and in the decades beyond will require organizations to develop in the *existing* workforce the basic and advanced literacies needed (Hudson Institute, 1987).

Organizations must take an active role in moving its members along the continuum from basic skills to technical literacy to knowledge workers. Learning requires mastery of basic literacy and technical literacy at the workplace. Without basic literacy, technical literacy will be difficult to attain. It is clear that many organizations are investing in developing the basic literacy of their workforce to improve their competitive advantage. For example, Laabs (1993) describes Ruiz Foods' commitment to a *Comprehensive Competency Program,* which the company bought from the Ford Foundation (cofunded by the California State Employment Training Program). The company's 1200 employees have the opportunity to learn basic English, computer operations, and math skills. In addition, employees are encouraged to attend team-building seminars. The founder of this family-owned company believes that the investment in basic literacy of the workers is paying off for the company, which has more than tripled its sales during the past 5 years.

20.5 TECHNICAL LITERACY

An individual with basic skills possesses many of the building blocks required to upgrade to a standard we define below as *technical literacy.* Yet experts in technological education caution that the process of obtaining technical literacy may be fundamentally different than obtaining math and reading skills (U.S. Department of Labor, 1991). The deductive and problem-solving skills of an automobile mechanic may be more relevant to achieving technical literacy than reading level, although clearly a minimum performance standard in reading is relevant to utilizing repair manuals. Thus, even if we could magically eliminate the basic skills deficiencies of the workforce, we would still be faced with the development of a number of skills required for effective organizational functioning in the twenty-first century.

Definitions provided for *technical literacy* are often narrowly focused on the ability to use technological tools, especially computers (Filipczak, 1994; Stokes, 1993). A familiarity with computers may not be required of all jobs, but it is fair to say that the percentage of jobs requiring a human-computer interface is likely to increase rather than decrease. Given the exponential growth in computing power and a corresponding increase in the power of the computer to provide increasingly large amounts of data in the same time frame (Gross and Coy, 1995), a critical skill that will be needed is knowing how to produce *useful* data and how to analyze and interpret it. Too often in the name of technical literacy we have focused on *how* to use the computer for generation of the data with ever increasing speed and/or lower cost while paying less attention to educating employees on the *why* of data generation, whether they are truly useful, or whether alternative outputs could be more effectively utilized. Those who view the use of technology as a tool for doing the same tasks faster lose out on potential new ways of operating that redefine tasks rather than speed up the processing of existing tasks. Employees who understand the *why* of the process are equipped to reinvent the process to achieve the objectives rather than to simply repeat the process at increasing speed.

One narrow approach to measuring the technical literacy of a population would be the frequency or extent to which products of technology are used in daily lives and the assimilation of these technologies (Filipczak, 1994). According to research by Dell Computer Corporation, the Austin, Texas-based computer manufacturer, 55 percent of all Americans are technophobic to some degree, meaning that they resist the use of technology in their daily lives. Of the adults surveyed, 25 percent had never used a computer, set the timer on their VCR, or programmed stations on their car radio (Filipczak, 1994). These figures suggest a pronounced aversion of many individuals to alter routine processes through the use of technology to arguably improve their quality of life.

Walter Waetjen, president emeritus of Cleveland State University and chairman of the Technology Education Advisory Council, provides three dimensions to technological literacy in his definition:

1. You need enough knowledge to understand technological advances as they are reported in the media. You may not know how a modem works but do know that it is a tool that can get you on the Internet.

2. You should be able to solve basic technological problems. If nothing appears on the monitor or the printer will not work, do you know enough to check the cord connections?

3. You should know how to use basic low-tech tools to accomplish tasks (Filipczak, 1994). Ignorance of basic tools is a barrier to the development of higher level technical skills.

Another definition of technical literacy or "technoliteracy" as suggested by Filipczak (1994) that comes closer to our own is that to be literate in technology means that you understand what the technology does (a computer can replace both your typewriter and calculator) and you've overcome your fear of the machine in question.

Progressive levels of technoliteracy are then achieved as one begins using technology to solve problems, but the essence of technical literacy is the ability to develop skills requisite to understanding and using the technology to do work. O'Connell (1994) speaks of the management of "technology resources" as an umbrella term for the hardware, software, communications, and employee knowledge and skills needed to solve business problems. All resources, whether natural, human, financial, or electronic, need care and tending if they are to survive and be productive. One such nurturing process is the organization's emphasis and communication of the importance of technical literacy (Kouloupoulos, 1993).

We define *technical literacy* as a multidimensional toolkit of skills rather than a unidimensional familiarity with some particular aspect of a specialized technology or a broad familiarity with instruments of technology in general. A radar "technician" who can interpret blips on a computer screen but does not have an understanding of how the blips are generated and who could not rotate to an assignment using a different radar technology without extensive training is not technically literate. This employee is instead required to utilize a narrow skill set in a rigidly defined job where rote learning is more important for successful performance than the ability to engage in continuous learning and improvement. This person performs the job as specified within the boundaries of existing parameters but lacks the *why* perspective, which results in insufficient depth to address change and generate ideas for adaption of the job to business objectives (Reitzfeld, 1989). *Technical literacy* as we define it has three primary components:

1. The ability to understand how a technology can be utilized (i.e., modem as communications device for information access) and to obtain the required information to make an informed business decision (Sasseen et al, 1994; *Business Week,* 1994). This dimension most closely corresponds to what the typical layperson would label technical literacy. Managing technology calls for at least a modicum of technical knowledge, but the critical literacy is in how to apply technology to business problems (O'Connell, 1994).

2. A proficiency in the basic language of business and a familiarity with industry forces to ease categorization of information and facilitate communications (i.e., foundation in management, economics, and statistical process control). To be technically literate in the language of business is increasingly important as organizations engage in business process reengineering (Manganelli and Klein, 1994). Business process reengineering, with its emphasis on processes, requires more cross-functional business literacy than in the traditional hierarchical organization. Business literacy includes both basic generalized business literacy and a literacy that is more firm specific. The general business aspect of business literacy training can be effectively outsourced through the general business curriculum of local universities and community colleges or by bringing faculty on site (Philippi, 1993; U.S. Department of Labor, 1991). Stokes (1993) discusses a more localized, firm-specific, business literacy. He suggests that employees can be responsive to organizational objectives only when they know key industry players, understand developing industry trends, recognize jargon that is specific to the industry, and are attuned to what the enterprise sees as its current and future niche. He also includes in business literacy an understanding of the nature of the organization's political alliances (how things really get done), which increases capability in relating to internal clients and customers. To develop these firm-specific competencies and learning systems, the organization must provide specialized training on industry parameters and environmental forces that affect organizational objectives. Such training will likely have to be developed in house as relatively few outsiders will possess the expertise and knowledge of organizational objectives that would allow the outsourcing of training. Once employees understand firm and industry forces that impact the organization's business objectives, a need for developing informational systems and networks will arise to accommodate employee demands for the acquisition and processing of informational inputs from multiple levels. The demand for such inputs will present a great challenge for information systems groups who will be challenged to develop informational delivery systems that meet these needs.

3. A set of analytical skills which include problem identification, inference making, data reduction and synthesis, problem solving, and information presentation (i.e., solving organizational problems and conveying learning points to organizational members).

We regard individuals who possess these three technical literacies of technology awareness, business literacy, and analytical skills as possessing the core competencies to engage in "knowledge work." Knowledge workers are, in turn, those individuals who are capable of raising organizational capabilities to the level required of what has become known as the learning organization.

20.6   TECHNICAL SPECIALISTS MAY NOT BE TECHNICALLY LITERATE

The rate of technological change is escalating rapidly. As the workplace changes and employees are asked to continually improve processes, continually do more with less and continually learn, technological literacy is becoming a new learning imperative. Knowledge workers are keenly interested in understanding *why* their function exists in the organization. An understanding of the *why* component is especially important in an era of rapid change and cross-functional interaction. If individual jobs and processes are changing rapidly, then it becomes increasingly important to understand *why* certain things need to be accomplished. Workers who understand the why of the business objective are more likely to be able to evolve and adapt their roles to accomplish that objective. Technical specialists who are up to date on the latest developments in a narrow specialty area are the mainstay of today's organizations, but much of this technical acumen is lost if the individual has only a vague notion of how to apply these skills to meet the organization's business objectives. This person has a tool (technology), expertise using the tool (technical proficiency), but is unsure what the organization needs fixed (lacks business literacy and analytical skills).

Organizations must not only foster the acquisition of technical specialties but must also continually communicate business objectives to allow the employee to have a line of sight between a specialty area (role) and accomplishment of the organization's business objectives. At that point the employee will be able to be more responsive to accomplishment of business objectives by anticipating what needs to be "fixed" rather than being informed by someone with the "big picture" what needs to be fixed. Additionally, understanding internal customer needs is important in the communications process. The knowledge imperative for organizations is to develop in their employees the ability to "learn to learn" or to engage in what Argyris (1978) refers to as double-loop learning, and thereby continually adjust their "jobs" to meet organizational objectives. Workers who have "learned to learn" can, in turn, communicate to organizational members in a way that can be generalized to other organizational situations within relevant boundaries. Employees who understand the *why* of their jobs are more likely to know the *why* aspect of situational success or failure. By sharing this learning across the organization and embedding this information in the organization's collective thought processes, we have moved beyond basic skills and utilized our technical literacies to engage in knowledge work that fosters real organizational learning.

20.7 IMPROVING ORGANIZATIONAL LEARNING CAPABILITIES

The investment in basic and technical literacy is a component of a competitive strategy based on the ability of all workers to contribute to the production of better-quality products and improved service. This strategy emphasizes the key role of investment in human resources to create what have been termed the "workers of the future" (knowledge workers) (Harrigan and Dalmia, 1991). Harrigan and Dalmia (1991) describe these "knowledge workers" as employees that are problem solvers and innovators. They argue that these workers are the key to competitive advantage in a changing global economy. An individual's ability to learn and to convey learning to other organizational members is crucial to the organization's overall ability to compete. The ability to learn allows an individual to upgrade or acquire new skills in addition to organizing and taking from the experiences of others. These workers need not have high organizational rank or status. Ideally, innovation can be initiated at any level of the organization to improve processes and products (Harari, 1994). Organizations that create environments in which any employee can offer suggestions for improvements have been described as "Learning Organizations" (Senge, 1990a).

20.8 LITERACY AND THE KNOWLEDGE IMPERATIVE IN LEARNING ORGANIZATIONS

Ulrich et al. (1993) describe learning organizations as those that have the following characteristics:

- "Employees are continually challenged to help shape their organization's future."
- "The capacity of an organization to gain insight from its own experience, the experience of others, and to modify the way it functions according to such insight."
- "The process of improving actions through better knowledge and understanding."
- "An organization that is continually expanding its capacity to create its future."
- "An organization that continually improves by readily creating and refining the capabilities needed for success."

Learning organizations enable their employees to embrace their intrinsic motivation to learn. According to Senge (1990a, p. 4), "Learning organizations are possible because, deep down, we are all learners." Senge and others advocate the implementation of work-team systems to tap into employees' ability to learn and apply systems thinking. However, without the requisite basic literacy skills, work-team systems are doomed to fail. Prior to implementation of work teams as part of learning systems, an inventory of the requisite basic skills of the workers (reading, writing, basic arithmetic) is a necessary first step.

Given basic literacy, the next level of learning is an understanding of systems thinking (Senge, 1990a). This level requires a somewhat more abstract level of thinking and the ability to sort out cause-effect relationships. McKendree (1991) notes that organizations increasingly need to fill knowledge gaps in research, data synthesis, and analytical skills. The implementation of most total-quality programs requires employees to have at least this level of technical competency, since such programs are typically grounded in examination of processes in the search for ways to continuously improve them (Sashkin and Kiser, 1993). For example, in a health-care delivery system, a team was formed to examine ways to reduce the amount of time that patients spent waiting for appointments. The team determined that the root cause (80 percent of the situations with waiting periods that were found to be over the average waiting time) was retrieval of medical records and developed a plan to improve access of medical records in situations when patients were waiting for more than 15 minutes. Here, the team members needed basic literacy skills to read and understand the problem, basic math skills to analyze the data available, and basic written and oral communication skills to communicate to the team and prepare necessary reports on recommendations to management. Beyond this, the employee team members needed abstract reasoning skills to determine cause and effect relationships and to generate possible solutions to the problem. It seems clear that the implementation of new work systems based on learning organization principles necessitates basic literacy skills plus the capacity for abstract reasoning.

Many researchers present assumptions about learning organizations including the need for workforce competence (Senge, 1990a; McGill and Slocum, 1993; Ulrich et al., 1993). With knowledge becoming an increasing requirement for international competitiveness, basic and technical literacy have become imperative. Processes must be continuously improved which requires learning. Products must be improved or invented which requires learning. This requires organizations to continuously invest in the capability of their employees to acquire new skills and learn new ways of doing things and new ways of thinking. Ulrich et al. (1993, p. 63) define "learning capability" as the capacity of individuals to "generate and generalize" ideas with impact. "Generating" refers to the creative process through which new ideas are created. "Generalizing" refers to the ability for the idea to be shared with others and across boundaries of time and hierarchy, as well as outside the organization and geography. These abilities of generating and generalizing require abstract reasoning abilities, in addition to basic literacy.

The implications of basic literacy to the learning continuum are depicted in Fig. 20.1. Each level of this continuum assumes that competency at the previous level has

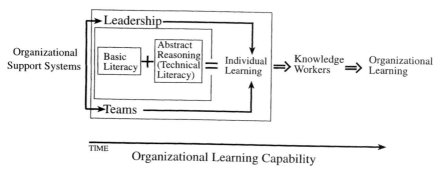

FIGURE 20.1   Technical literacy and organizational learning.

been attained by the worker. Abstract reasoning assumes that the basic tools are in place in order to work with concepts and data (basic literacy). Once abstract reasoning skills are developed, learning can occur because workers will be able to generate and generalize ideas (Ulrich et al., 1993). This enables some of these workers to become knowledge workers, who will be the innovators and problems solvers, which will create competitive advantage for the organization (Harrigan and Dalmia, 1991). In Fig. 20.1, fewer workers exist per level at any point in time in the development of the organization's learning capacity. However, the premise is that workers are learning and moving throughout the process, with the goal being the creation of as many knowledge workers as possible. At some point, a critical mass of knowledge workers, those with requisite knowledge, skills, and abilities (KSAs) will develop. Once this has occurred, a learning organization can emerge (Senge, 1990a). Literacy and knowledge literacy are an imperative for today's organizations and necessary for development of learning organizations.

20.9 ORGANIZATIONAL IMPLICATIONS OF THE KNOWLEDGE IMPERATIVE

The creation of knowledge workers and ultimately learning organizations will require investment in human resources at unprecedented levels (Doyle, 1990). Individual and organizational learning are not isomorphic, but the latter is dependent on the former. Individual learning occurs as *people* acquire tacit knowledge through education, experience, or experimentation. Organizational learning occurs as the *systems* and *culture* of the organization retain learning and ideas are transferred across organizational boundaries of space, time, and hierarchy (Ulrich et al., 1993). This is an important distinction as learning without organizational change can occur when individuals generate new ideas but the ideas are not generalized to organizational systems, or when units of the organization experiment with an idea, but fail to share their learning. In contrast, a learning organization adopts mechanisms to capture individual learning and develop in employees those ways of knowing that lend themselves to policy capturing. Individuals who are able to arrive at the "correct" answer but cannot elucidate the decision process offer the learning organization less than an individual who has been trained in systematic ways of knowing and therefore can assist in transferring knowledge throughout the organization. The organization must actively develop this policy capturing mechanism much the same way that "expert" systems capture aspects of individual learning. Such mechanisms are necessary to achieve learning organization status.

So what must organizations view as a prerequisite to achieving the status of a learning organization? First, the creation of knowledge workers requires basic literacy plus abstract reasoning abilities. To develop these capabilities, organizations will need to increase their investment in training and development. Assessment of knowledge gaps will become a necessary part of the training and development system. McKendree (1991, p. 101) comments that

> ...American companies are, by necessity, filling critical knowledge gaps in their new hires. Fewer workers are coming out of the education system with the skills they need to do their jobs. As a result, highly technical and complex organizations must undertake a great deal of training and education on their own.

Once gaps in knowledge have been identified, training must be provided in basic skills, plus abstract reasoning ability. These skills are necessary if the organization wishes to fully empower its workers (Taylor and Ramsey, 1991; Dobbs, 1993). *Empowerment* is allowing employees the authority to make decisions about their work without having to ask a supervisor for permission (Taylor and Ramsey, 1991). Too often, organizational decision makers rush to implement work teams, with disastrous results because the workers do not have the necessary basic skills to be fully empowered. Beyond basic skills, abstract reasoning skills are necessary to engage in creative problem solving. Brainstorming is an exercise in futility if the workers do not know how to critically evaluate the ideas generated at the appropriate time in the process or if good ideas have been lost for lack of facilitation skills. Critical thinking skills must be learned and practiced if the processes of brainstorming are to result in some useful product.

Second, learning organizations also employ teams of workers to solve problems and generate new products or services. Recent theories of knowledge suggest that learning organizations create "activity systems" within which workers engage in collaborative work, and share information and ideas (Blacker, 1993). If the team-based organizational form is being implemented, team skills are a prerequisite for all employees in addition to literacy skills (Brachule and Wright, 1993; Katzenback and Smith, 1993; Scholtes, 1993; Schonk, 1992; Sunstrom et al., 1990). Team and interpersonal skills are necessary for generalizing ideas and sharing information effectively within teams, in cross-functional teams, and throughout the organization (Katzenback and Smith, 1993; Schonk, 1992). Given the literacy and knowledge imperative, organizations will require new forms of leadership to implement systems that support and encourage learning. Technical literacy and the knowledge imperative have implications for training and development and leadership (Brachule and Wright, 1993; Katzenback and Smith, 1993; Senge, 1990b). Creation of a learning organization requires that employees have the requisite skills and the opportunity to apply them through participative leadership practices. Command-and-control autocratic leadership is not consistent with the learning organization, and all levels of management will need to unlearn old behaviors and learn new ways of leading (Conger, 1993). They will need to create learning environments in which workers take responsibility for the future of the organization (Fulmer, 1994). The leader's role will shift from telling workers what to do to that of teacher, coach, and role model (Manz and Sims, 1987; Senge, 1990a).

Leaders are responsible for shaping the culture of the organization (Deal and Kennedy, 1982). Leadership at the top is essential for establishing the vision of an organization that is committed to learning, and supporting the efforts of training and development in terms of necessary resources. Leaders in the middle level of management have a critical role in the learning organization, since it is their job to create high-quality work relationships with their workers that support innovative behavior (Graen and Scandura, 1987). Workers must be encouraged to learn, experiment, and keep on learning through courses and application of concepts to the work environment. Middle-level managers must learn to accept intelligent failures and learn a tolerance for employee mistakes. This is a very different type of leadership, in which the manager's job is to support, coach, and teach, rather than to command and control the workers. Senge (1990b) defined the roles of leaders in learning organizations as designer of vision, steward of employee empowerment, and teacher. The role of the leader as a role model is emphasized in the latter role of teacher. Leaders must be willing to "walk the talk" or employees will become disillusioned with the learning process, and perhaps dismiss it as another management fad. In sum, the role of top management is creating vision and support for the learning organization. Generating this idea and then generalizing it through translation to middle-level management is the next necessary step in the process. Next, it is middle management's role to translate this vision to the workers by being role models who engage in learning practices and serve as role models and teachers. As the organization's commitment to knowledge increases at these three levels, so does the organization's learning capability. Ulrich et al. (1993) provide key principles for extending learning across organizational boundaries.

1. Generate a large number of learning opportunities.

2. Generalize the learning beyond the individual.

3. Build in the desire and opportunity to learn from others.

4. Study failures and satisfactory episodes, not just successes.

20.10 RECOMMENDATIONS

20.10.1 Training and Development

Training and development within learning organizations must be continuous and not ad hoc "fixing up" of skill gaps in the workforce. To attain this objective, assessment of the employee basic and technical literacy skills defined in this chapter must be ongoing. As skill levels of employees improve, new training objectives should be set. Learning must be made a visible and central element of the strategic intent. Employees should be encouraged to challenge the status quo and constantly look for ways to improve their own basic and technical skills, as well as the products, processes, and services the organization provides.

Knowledge acquisition should be encouraged as a core competency, a valued commodity which is recognized and rewarded. Individual, team, and organizational competencies for learning should be a central objective in the organization's staffing and training initiatives. Similarly, the performance management system should encourage individual, team, and process learning by providing explicit feedback on accomplishments through appraisal and the provision of overt and tangible rewards. Continuous assessment and rewarding of individual capabilities that serve to increase organizational capability will encourage the development of a "learning mindset" that may become embedded in the organization's culture as a cherished value (Schein, 1993).

20.10.2 Basic and Technical Literacy as Gateways to Higher-Level Learning

Basic skills and technical literacy as gateways to higher-level learning should be emphasized. In developing our training sessions we must keep in mind that technologies are a tool, and that we need to learn problem-solving skills using that tool. Thus the computer (or any technical competence) must be made a part of training. Training employees on the tool and on needed skills separately is less effective than combining the learning of the skill required using the computer as a tool in problem solving. Research on adult learning (McKendree, 1991) suggests that children tend to learn first and then do, while adults tend to do first and then learn. If we want employees to integrate technology into their work processes, then training must integrate technology and problem-solving skills within a context relevant to the worker.

Computer skills training needs to incorporate the teaching of business skills. "Technical empowerment" is the idea that an employee needs to understand what the numbers on the spreadsheet mean instead of merely learning to manipulate them.

You've got to teach business skills along with technological competencies if you want to get the optimum productivity out of both employee and machine. The teaching of accounting in one course and the use of spreadsheets in another will not help employees blend the skills together, and that blend is the key to technical empowerment.

Training which focuses on problem solving with the computer will avoid confusion, eliminating the need to explain the "bells and whistles" available on many current software packages that may serve to intimidate the user rather than help to acquire basic skills.

20.10.3   Team and Interpersonal Skills

In addition to basic and technical literacy skills, training in team and interpersonal skills is necessary to develop a learning organization. Team-building skills such as knowledge of basic team processes, conflict resolution, and skills in running meetings should be taught. Interpersonal communication skills are a necessary underpinning of such training, in particular how to give and receive feedback and constructive criticism. Senge (1990a) describes team learning in contrast to team building as the necessary team-level process skill for the learning organization. In team learning, employees in teams become colearners and help one another to learn new skills, solve problems, and implement innovative ideas. In a team learning environment employees in teams learn how to learn, adjust, and self-regulate their own learning behaviors.

Implementing the learning organization under the literacy-knowledge imperative has implications for leadership. New roles for new managerial styles will be necessary. Senge (1990b) describes the new role of leaders in the learning organization as vision, steward, and teacher. Participative styles of leadership will increase with a corresponding decrease in the micro management of processes, as employee teams are empowered to make decisions regarding the work they do. The manager's job is to create and sustain activity systems, remove barriers to innovation, and assess and teach. To this end, a new focus on relationship development will be necessary. Mutual trust will become the hallmark of relationships in the learning organization. Employees with the basic and technical skills needed to do the job can be entrusted to do the job, with support and coaching from the leader. Interpersonal trust is needed for close interpersonal relationships and comprises faith of one person in another, dependability on one another, and predictability in the relationship (Rempel et al., 1985). Trust enables social systems to operate, since high levels of trust allow high reliability to be established through faith, dependability, and predictability. Underlying the development of trust is the development of basic and technical literacy of workers and leadership learning to trust them.

20.11   SUMMARY

This chapter reviewed the literature on the implications of basic literacy for work organizations and developed a definition of technical literacy. Next a model of the role of basic and technical literacy in the development of learning organizations is presented (Senge, 1990 a and b; Ulrich et al., 1993). Basic literacy is necessary for technical literacy and, in turn, technical literacy is necessary for learning.

Support systems for the development of basic and technical literacy in the learning organization were reviewed, including training and development, team learning, and leadership. Throughout the chapter, the role of learning as a strategic posture for competitive advantage is emphasized. Organizations that provide training, support systems, and leadership that enables them to learn how to learn and continue to learn will hold the key to competitive advantage in today's global and rapidly changing marketplace.

20.12 REFERENCES

Adams, J. S., "Research Methodology as a Social Issue: The Case of Illiterates in the Workforce and in Samples Used in Organizational Research," Academy of Management conference presentation, 1993.

Argyris, Chris, 1978.

Argyris, Chris, 1991.

Argyris, Chris, and Donald Schon, *Theory of Action,* Jossey-Bass, San Francisco, 1974.

Bank of America, "Human Resource Planning," *Perspectives,* **11:** 40, Winter 1990.

Barton, P. E., and I. S. Kirsch, *Workplace Competencies: The Need to Improve Literacy and Employment Readiness,* Office of Educational Research and Development, U.S. Department of Education, 1990.

Blacker, F., "Knowledge and the Theory of Organizations: Organizations as Activity Systems and the Reframing of Management," *J. Management Studies, 30:* 863, 1993.

Brachule, P. E., and D. W. Wright, "Training Work Teams," *Training and Development,* pp. 65–68, March 1993.

Bradsher, K., "U.S. Lag in Phone Trade Seen," *New York Times,* p. C3, Aug. 17, 1990.

Bridges, W., "The End of the Job," *Fortune,* pp. 62–74, Sept. 19, 1994.

Business Week, "The Rules of the Game in the New World of Work," *Business Week,* pp. 94–102, Oct. 17, 1994.

Carnevale, A., *America and the New Economy,* American Society for Training and Development; U.S. Department of Labor, Employment and Training Administration, Washington, D.C., 1991.

Casio, W., "Gaining and Sustaining Competitive Advantage: Challenges for Human Resource Management," *Research in Personnel and Human Resources Management,* Suppl. 1, JAI Press, Greenwich, Conn., 1989.

Chall, J. S., E. Heron, and T. Hilferty, "Adult Literacy: New and Enduring Problems," *Phi Delta Kappan,* **69**(3): 190–196, 1987.

Charp, S., "Workplace Literacy" (editorial), *J. Technol. Horizons Education," p. 4, 1995.*

Conger, J. A., "Personal Growth Training: Snake Oil or Pathway to Leadership?" *Organizational Dynamics,* pp. 19–30, Summer 1993.

Deal, T. M., and A. Kennedy, *Corporate Cultures,* Addison-Wesley, Reading, Mass., 1982.

Dobbs, J. H., "The Empowerment Environment," *Training and Development J.,* pp. 55–57, Feb. 1993.

Doyle, F., "People Power: The Global Human Resource Challenge for the '90s," *Columbia J. World Business,* pp. 36–45, Spring–Summer 1990.

Drucker, P. F., *Management: Tasks, Responsibilities, and Practices,* Harper & Row, New York, 1985.

Filipczak, B., "Technoliteracy, Technophobia, and Programming Your VCR," *Training,* Jan. 1994.

Fisher, K., *Leading Self-directed Work Teams: A Guide to Developing New Leadership Skills,* McGraw-Hill, New York, 1993.

Fulmer, R. M., "A Model for Changing the Way Organizations Learn," *Planning Review,* pp. 20–23, May–June 1994.

Graen, G. B., and T. A. Scandura, "Toward a Psychology of Dyadic Organizing," in *Research in Organizational Behavior,* vol. 9, L. L. Cummings and B. Staw, eds., JAI Press, Greenwich, Conn., 1987, pp. 175–208.

Gross, N., and P. Coy, "The Technology Paradox: How Companies Can Thrive as Prices Dive," *Business Week,* pp. 76–84, March 6, 1995.

Hammonds, K. H., K. Kelly, and K. Thurston, "The New World of Work," *Business Week,* Oct. 17, 1994.

Harari, O., "When Intelligence Rules, The Manager's Job Changes," *Management Review,* pp. 33–35, July 1994.

Harrigan, K. R., and G. Dalmia, "Knowledge Workers: The Last Bastion of Competitive Advantage," *Planning Review,* pp. 5–9, Sept.–Dec. 1991.

Hodgetts, R. M., F. Luthans, and S. M. Lee, "New Paradigm Organizations: From Total Quality to Learning to World-Class," *Organizational Dynamics,* **22**(3): 5–19, 1994.

Hudson Institute, *Workforce 2000: Work and Workers for the 21st Century,* Washington, D.C., 1987.

Katz, D., and R. L. Kahn, "The Taking of Organizational Roles," *The Social Psychology of Organizations,* Wiley, New York, 1978, chap. 7.

Katzenback, J. R., and D. K. Smith, "The Discipline of Teams," *Harvard Business Review,* pp. 111–120, March–April, 1993.

Koulopoulos, T. M., "Coping with Transformation in the Age of Empowerment," *Modern Office Technol.,* pp. 15ff., April 1993.

Laabs, J., "Business Growth Driven by Staff Development," *Personnel J.,* pp. 120–135, April 1993.

Losee, S., "Your Company's Most Valuable Asset: Intellectual Capital," *Fortune,* **130**(7): 68–74, Oct. 3, 1994.

Manganelli, F., and R. Klein, *The Reengineering Handbook: A Step-by-Step Guide to Business Transformation,* AMACOM, New York, 1994.

Manz, C. C., and H. P. Sims, "Leader Workers to Lead Themselves: The External Leadership of Self-Managing Work Teams," *Administrative Science Quarterly,* **32:** 106–128, 1987.

McGill, M. E., and J. W. Slocum, "Management Practices in Learning Organizations," *Organizational Dynamics,* **21**(2): 4–18, 1992.

McGill, M. E., and J. W. Slocum, "Unlearning the Organization," *Organizational Dynamics,* **22**(2): 67–79, 1993.

McKendree, W. G., "Sounding the Education Alarm," *Best's Review,* pp. 41ff., Dec. 1991.

McLagan, P. A., "Models for HRD Practice," *Training Development J.,* pp. 49–59, Sept. 1989.

Mikulecky, L., "Basic Skill Impediments to Communication between Management and Hourly Employees," *Management Communications Quarterly,* p. 3, May 1990.

O'Connell, S. E., "Five Principles for Managing Technology Resources," *HR Magazine,* pp. 35ff., Jan. 1994.

Pfeffer, J. L., *Competitive Advantage through People: Unleashing the Power of the Workforce,* Harvard Business School Press, Boston, 1994.

Philippi, J. W., *Literacy at Work: The Workbook for Program Developers,* Prentice-Hall, Englewood Cliffs, N. J., 1991.

Pucik, V., "The International Management of Human Resources," in *Strategic Human Resource Management,* C. J. Fombrun, N. M. Tichy, and M. A. Devanna, eds., Wiley, New York, 1984, pp. 403–419.

Reitzfeld, M., "Training is Necessary, Not a Reward," *J. Systems Management,* p. 25, Feb. 1989.

Rempel, J. K., J. G. Holmes, and M. P. Zanna, "Trust in Close Relationships," *J. Personality Social Psychol.,* **49:** 95–112, 1985.

Roth, W., *The Evolution of Management Theory,* Roth & Assoc., Orefield, Pa., 1993.

Sashkin, M., and K. J. Kiser, *Putting Total Quality Management to Work,* Berrett-Koehler, San Francisco, 1993.

Sasseen, J. A., R. Neff, S. Hattangadi, and S. Sansoni, "The Winds of Change Blow Everywhere," *Business Week,* pp. 93–94, Oct. 17, 1994.

Schein, E. H., "On Dialogue, Culture, and Organizational Learning," *Organizational Dynamics,* pp. 40–51, Fall 1993.

Scholtes, P. R., *The Team Handbook,* Joiner Assoc., Madison, Wis., 1993.

Schonk, J. H., *Team Based Organizations: Developing a Successful Team Environment,* Business One Irwin, Homewood, Ill., 1992.

Senge, P. M., *The Fifth Discipline,* Doubleday, New York, 1990a.

Senge, P. M., "The Leader's New Work: Building Learning Organizations, *Sloan Management Review,* pp. 7–23, Fall 1990b.

Stein, S., and L. Sperazi, *Workplace Education and the Transformation of the Workplace,* Centre Research, Boston, 1991.

Stokes, S. L., "Blueprint for Business Literacy," *Information Systems Management,* Spring 1993.

Sunstrom, E., K. P. de Meuse, and D. Futrell, "Work Teams: Applications and Effectiveness," *American Psychologist,* **15**: 120–133, 1990.

Taylor, D. L., and R. K. Ramsey, "Empowering Employees to 'Just Do It,'" *Training Development J.,* pp. 36–42, June 1993.

Ulrich, D. J., "Human Resource Planning as a Competitive Edge," *Human Resource Planning,* **9**: 41–50, 1986.

Ulrich, D., M. A. Von Glinow, and T. Jick, "High-Impact Learning: Building and Diffusing Learning Capability," *Organizational Dynamics,* pp. 52–66, 1993..

U.S. Department of Education, *Workplace Literacy: Reshaping the American Workforce,* Office of Vocational and Adult Education, Washington, D.C., 1992a.

U.S. Department of Education, *Voices from the Field: Proceedings of the September 1991 National Workplace Literacy Program Project Directors Conference,* Washington, D.C., 1992b.

U.S. Department of Labor, *What Work Requires of Schools,* The Secretary's Commission on Achieving Necessary Skills, Washington, D.C., 1991.

Von Glinow, M. A., "Diagnosing Best Practice in Human Resource Management," in *Research in Personnel and Human Resources Management,* Suppl. 3, B. Shaw, P. Kirkbride, G. R. Ferris, and K. M. Rowland, eds., JAI Press, Greenwich, Conn., 1994.

Walton, R. E., "From Control to Commitment in the Workplace," *Harvard Business Review,* pp. 77–84, March 1985.

CHAPTER 21
DEVELOPING TECHNOLOGY MANAGERS

Jeffrey C. Shuman and Hans J. Thamhain
Bentley College
Waltham, Massachusetts

21.1 THE NEW BUSINESS ENVIRONMENT

Technology has never been more important. Throughout history, technology has been a major driver of economic growth. It is also a driving force which makes borders between nations, between companies, and between compartments of our lives—transparent. A recent study of CEOs makes the point that "The strategic management of technology remains an unfulfilled goal and a pressing need for many multinational enterprises."[1] The CEOs stated that their businesses needed technology management education, particularly in the strategic incorporation of technology into business for shortening product development life cycles, capitalizing on innovation, and adopting or exiting technologies faster.

Clearly, part of the responsibility for technology management education rests with the nation's colleges and universities. For example, in the United States, approximately 75,000 M.B.A. and 250,000 undergraduate business and management degrees are awarded annually. In the widely heralded book *Made in America: Regaining the Productive Edge,* it was argued that[2]

> For too long business schools have taken the position that a good manager could manage anything, regardless of its technological base. It is now clear that this view is wrong. While it is not necessary for every manager to have a science or engineering degree, every manager does need to understand how technology relates to the strategic positioning of the firm, how to evaluate alternative technologies and investment choices, and how to shepherd scientific and technical concepts through the innovation and production processes to the marketplace.

Increasingly, both corporate executives and academicians have raised questions about, and debated how, business schools can add value to the business community:[3]

> The worldwide corporation in the 1990s is markedly different from its predecessors in the 60s, 70s, or even the 80s. Companies are now confronted by rapid globalization of

markets and competition, the increasing importance of speed and flexibility as key sources of competitive advantage, and the growing proliferation of partnership relations with suppliers, customers, and competitors. As a result these companies must respond with radically different management approaches to succeed.

However, relatively little attention has been paid to one of the most difficult corporate challenges of the 1990s: How to develop a new breed of senior managers who have knowledge, sensitivities, and skills necessary to lead corporations through the difficult times ahead. This is also a key concern for many business schools. Increasingly under fire for having lost their relevance to the practitioner community, business schools are grappling with the challenge of how they can fulfill their mandate of helping companies develop their next generation of leaders.

Technology management is, by its nature, global. Global markets, global financial institutions, global sourcing, and a global factory floor require successful business schools and ventures to change. As widely noted in the press, traditional business management paradigms have proven inadequate for the new technological age and are giving way to an enhanced set of principles which enable managers and organizations to meet these challenges. All institutions are rapidly becoming technology-intensive, and the need to integrate technology across traditional functions of the firm has become paramount, whether the firm produces technological products or utilizes process technologies to reduce costs or increase service.

This transition into the new technology-based business paradigm creates challenges to technology management education:[4]

> We live in an increasingly technological world. The U.S. is no longer the dominant technological innovator in the world, and the competitive position of many of America's companies has faltered. Many have lamented about U.S. managers' inabilities to harness the full potential of product and process technology. The ability to manage technology strategically will be an increasingly important competitive dimension during the remainder of this century, as more and more managers are being called upon to manage technical people and processes. Getting the product out of the lab and into the marketplace before the international competitors is vital, and managers are essential in this process.
>
> Organizations are moving away from vertically dominated structures where functions control resources, to horizontally dominated organizations in which processes own the resources. These core processes integrate the create-make-market processes to ensure that firms are competitive in reaching today's customers with new, high quality products, at reasonable prices, in the shortest possible time.

Today's technology-oriented businesses are more complex and multifaceted than conventional or traditional forms of organizations. They require effective planning, organizing, and integration of complicated, multidisciplinary activities across functional lines in an environment of rapidly changing technology, markets, regulations, and socioeconomic factors. Their management must share resources and power, and establish communication channels that work both vertically and horizontally, and intra- and interorganizationally, to integrate the many functions involved in modern technology-based business operations.

Managers who evolve with these technology organizations must confront untried problems to handle their complex tasks. In contrast to managers from more established organizations, these new managers have to learn how to move across various organizational lines, gaining services from personnel not reporting directly to them. They must build multidisciplinary teams into cohesive groups and deal with a variety of networks, such as line departments, staff groups, team members, clients, and senior management; each has different interests, expectations, and charters. They also have to cope with constant and rapid change of technology, markets, regulations, and socioeconomic factors. To get results, managers in technology-based organizations must relate socially as well as technically. They must understand the culture and value system of the organization in which they work. The days of the manager who gets by with only technical expertise or pure administrative skills are gone.

21.2 SKILL REQUIREMENTS

As we are undergoing a transition with our concepts for organization and management practices, many of the traditional principles of management and management development became obsolete. The kind of workplace managers were traditionally trained for was structured along stovepipe functions with rigid chain-of-command reporting relations, centralized decision making, and a reasonable predictable business environment. But as organizations are changing into flatter and more flexible structures with higher degrees of powersharing, distributed decision making, and a more self-directed, team-oriented workforce, traditional training methods no longer produce the management skills needed to function effectively in today's dynamic business environment. Specifically, managers in today's technology-based organizations must be skilled in a broad range of disciplines in order to deal with their complex challenges. They must understand the following:

- How to integrate technology into the overall strategic objectives of the firm
- How to get into and out of technologies faster and more efficiently
- How to assess and evaluate technology more effectively
- How to best accomplish technology transfer
- How to reduce new product development time
- How to manage large, complex, and interdisciplinary or interorganizational projects and systems
- How to manage the organization's internal use of technology
- How to leverage the effectiveness of technical professionals

There is no magic formula that guarantees success in managing technology-based organizations. But research consistently shows that high-performing managers have specific skills in three principal categories as shown in Table 21.1.

- Leadership and interpersonal skills
- Technical skills
- Administrative skills

21.2.1 Interpersonal Skills and Leadership

Effective leadership involves a whole spectrum of skills and abilities: clear direction and guidance; ability to plan and elicit commitments; communication skills; assistance in problem solving; dealing effectively with managers and support personnel across

TABLE 21.1   Skill Inventory of the Technical Manager

Leadership skills
    Ability to manage in unstructured work environment
    Action-orientation, self-starter
    Aiding group decision making
    Assisting in problem solving
    Building multidisciplinary teams
    Building priority image
    Clarity of management direction
    Communication (written and oral)
    Creating personnel involvement at all levels
    Credibility
    Defining clear objectives
    Eliciting commitment
    Gaining upper management support and commitment
    Managing conflict
    Motivating people
    Understanding the organization
    Understanding professional needs
    Visibility

Technical skills
    Ability to manage the technology
    Aiding problem solving
    Communicating with technical personnel
    Facilitating tradeoffs
    Fostering innovative environment
    Integrating technical, business, and human objectives
    System perspective
    Technical credibility
    Understanding engineering tools and support methods
    Understanding technology and trends
    Understanding market and product applications
    Unifying the technical team

Administrative skills
    Planning and organizing multifunctional programs
    Attracting and holding quality people
    Estimating and negotiating resources
    Working with other organizations
    Measuring work status, progress, and performance
    Scheduling multidisciplinary activities
    Understanding policies and operating procedures
    Delegating effectively
    Communicating effectively (orally and in writing)
    Minimizing changes

---

functional lines often with little or no formal authority; information-processing skills; ability to collect and filter relevant data valid for decision making in a dynamic environment; and ability to integrate individual demands, requirements, and limitations into decisions that benefit the overall project. It further involves the manager's ability to resolve intergroup conflicts and to build multifunctional teams.

21.2.2 Technical Skills

Most of today's work is technically complex. Managers rarely have all the technical expertise to direct the multidisciplinary activities at hand. Nor is it necessary or desirable that they do so. It is essential, however, that managers understand the technologies and their trends, the markets, and the business environment, so that they can participate effectively in the search for integrated solutions and technological innovations. Without this understanding the consequences of local decisions on the total program, the potential growth ramifications, and relationships to other business opportunities cannot be foreseen by the manager. Furthermore, technical expertise is necessary to communicate effectively with the work team, and to assess risks and make tradeoffs between cost, schedule, and technical issues.

21.2.3 Administrative Skills

Administrative skills are essential. Managers must be experienced in planning, staffing, budgeting, scheduling, performance evaluations, and control techniques. While it is important that managers understand the company's operating procedures and the available tools, it is often necessary for managers to free themselves from the administrative details.

The manager's effectiveness often depends heavily on personal experience, credibility, and understanding of the interaction of organizational and behavioral elements.

The significance of the *skill inventory* shown in Table 21.1 is in three areas. First, the spectrum of skill requirements remains relatively stable over time in spite of rapid changes in the field of management. Second, Table 21.1 could be used as a tool for assessing actual skill requirements and proficiencies. For example, a list could be developed, for individuals or teams, to assess for each skill component:

- Criticality of this skill to effective job performance
- Existing level of proficiency
- Potential for improvement
- Needed support systems and help
- Suggested training and development activities
- Periodic reevaluation of proficiency

Third, Table 21.1 can be useful in developing training programs by focusing on specific skill requirements and the development of appropriate training methods.

Learnability of These Skills. Formal studies investigating the learnability of project management skills reveal some good news. On average, 94 percent of the skills needed to perform effectively in technology leadership positions are learnable,[5] mostly on the job. In fact, as summarized in Fig. 21.1, 85 percent[6] of all management skills are derived from experience. Three-quarters are developed strictly by experiential learning, while one-quarter comes from more specific work-related methods such as observations, formal on-the-job training, upper-management coaching, and job rotation. Second to experiential learning, skills can be developed by reading the professional literature, such as books, magazines, journals, and research papers, as well as audio- and videotapes on related subjects. A third source for technical management skill development exists through professional activities such as seminars, professional

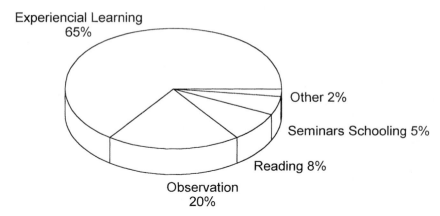

FIGURE 21.1   How do project managers develop their skills?

meetings, and special workshops. The fourth distinct category for managerial skill development consists of formal schooling. In addition, managers identify special sources, such as mentoring, job changes, and special organizational development activities, for building technology management skills.

These observations verify that skills can indeed be developed. The findings are further important as a basis for comparing the effectiveness of various training methods.

21.3   TESTING APTITUDE FOR TECHNOLOGY MANAGERS

Technology-oriented businesses have long recognized the importance of developing new managerial talent.[7] But despite enormous advances in evaluating, educating, and training technical professionals for managerial responsibilities, new managers often find themselves ill-prepared for their new assignments.

Even if they wanted managerial duties in the first place, many first-time supervisors are surprised at the scope and content of their new jobs. Often they suddenly feel unsure about their career development. Many studies have defined the type of skills and training an effective manager needs. Yet the formal education system that creates *technical* professionals, in most cases, does little to prepare them for advancement into management.

Since 1950, the number of technical workers has increased nearly 300%—triple the growth rate for the work force as a whole—to about 20 million. With one out of every four new jobs going to a technical worker, the Bureau of Labor Statistics forecasts that this army of technocompetence—already the largest broad occupational category in the United States—will represent a fifth of total employment within a decade.[8]

21.3.1   Aptitude Testing

There is strong interest among managers in tools to help assess technical management potential. The *aptitude assessment instrument,* shown in Fig. 21.2, can help technical professionals and managers to determine the potential of engineers for advancement into leadership positions, and for effective performance as managers. The instruments can help in career guidance, selection, and transitioning, as well as in the development process itself.

Further, the test questions provide some "predictor of success" by comparing the candidate's aptitude scores with those of the broader population of other test takers,[9] which is shown in graphic form in Fig. 21.3.

For managers, these instruments provide a set of criteria for initial assessment of an individual's potential for technical management, for defining career objectives and development plans, and for comparing candidates and attitude changes over time. *For the technical individual* the tests stimulate thoughts regarding the type of work, skill requirements, and changes in work habits, norms, and values required for effective role performance. These tests help in evaluating an individual's career wants and needs, as well as the development actions required for successful transition. Finally, these tests are also helpful to the *management researcher,* who can use them for comparative studies and refinement of selection criteria, as well as for statistical analysis of the aptitude scores regarding the probability of management success.

Suggestions for Using the Aptitude Tests. If used properly, aptitude tests can effectively facilitate career development. The following suggestions might help:

Realize the criteria for becoming a manager. The instruments should serve as guidelines rather than absolute measures for promotion. Personal judgment should remain an important factor in the final decision.

Don't use the numbers as a scientific measure. Realize that the scores, whether they come from you or others, are subjective and should serve as a basis for comparison and critical thinking, not as a mechanical selection tool or substitute for managerial judgment.

Use the instruments for career development. When technical people use these instruments to assess their own managerial aptitudes, they can get a better understanding of personal strengths, weaknesses, and desire, and of areas in which they need to increase their efforts. In other words, the test instruments may help in formulating career plans and specific action items.

Use the instruments for management development. In a conceptual manner, supervisors can use both the criteria and the instruments themselves to identify subordinates with management interest and potential. Remember to obtain data from multiple sources. In addition to your own evaluations of a candidate, solicit evaluations from other sources. Your peers, project leaders, customers, and support personnel who know the candidate may be good possibilities. You can solicit their input during informal discussions or by using formal questionnaires or letters.

Focus on managerial strength. If the aptitude scores indicate a strength in a specific area, the management candidate should seek opportunities for additional responsibilities in that area. That will build skills in areas of most likely success and provide favorable learning experiences. Management should encourage and support such learning experiences.

Use incremental skill building. Participation in such managerial activities as proposal development, task management, feasibility studies, and technology assessments helps a candidate sharpen skills in cross-functional communication, planning, and organizing. The candidate also can develop new skills, and test his or her desire for a career in management.

Score 1....10	Test Statement
	Use a 10-point scale to indicate your agreement or disagreement with each of the statements (1 = strong disagreement; 10 = strong agreement)
	Personal Desire To Be A Manager
....	1. Managing people is professionally more interesting and stimulating to me than solving technical problems.
....	2. I am interested and willing to assume new and greater responsibilities.
....	3. I am willing to invest considerable time and effort into developing managerial skills.
....	4. I have an MBA (or am working intensely on it).
....	5. I am prepared to update my management knowledge and skills via continuing education.
....	6. I have discussed the specific responsibilities, challenges, and skills required with manager.
....	7. I have defined my specific career goals and mapped out a plan for achieving them.
....	8. I would be willing to change my professional area of specialty for an advanced managerial opportunity.
....	9. Managerial and business challenges are more interesting and stimulating to me than technical challenges.
....	10. Achieving a managerial promotion within the next few years is a top priority, very important to the satisfaction of my professional needs.
	People Skills
....	1. I feel at ease communicating with people from other technical and administrative departments.
....	2. I can effectively solve conflict over technical and personal issues, and don't mind getting involved.
....	3. I can work with all levels of the organization.
....	4. I am a good liaison person to other departments and outside organizations.
....	5. I enjoy socializing with people.
....	6. I can persuade people to do things which initially they don't want to do.
....	7. I can get commitment from people, even though they don't report directly to me.
....	8. People enjoy working with me and follow my suggestions.
....	9. I am being frequently asked by my colleagues for my opinion and for presenting ideas to upper management.
....	10. I think, the majority of people in my department would select me as team leader.
	Technical Knowledge
....	1. I understand the technological trends in both my area of responsibility and the business environment of my company.
....	2. I understand the product applications, markets, and economic conditions of my business area.
....	3. I can effectively communicate with my technical colleagues from other disciplines.
....	4. I can unify a technical team toward project objectives and can facilitate group decision making.
....	5. I have a systems perspective in my area of technical work.
....	6. I have technical credibility with my colleagues.
....	7. I can use the latest design techniques and engineering tools.
....	8. I recognize work with potential for technological breakthrough early in its development.
....	9. I can measure work/project status and technical performance of other people on my team.
....	10. I can integrate the technical work of my team members.

Administrative Skills

.... 1. I don't mind administrative duties.

.... 2. I am familiar with techniques for planning, scheduling, budgeting, organizing, and personnel administration, and can perform them well.

.... 3. I can estimate and negotiate resources effectively.

.... 4. I can measure and report work status and performance.

.... 5. I find policies and procedures useful as guidelines for my activities.

.... 6. I have no problem delegating work even though I could do it myself even quicker.

.... 7. I don't mind writing reports and preparing for meetings, and I do it well.

.... 8. I can hand change requirements and work interruptions effectively.

.... 9. I am good in organizing social events.

.... 10. I can work effectively with administrative support groups throughout the company.

Business Acumen

.... 1. I would be good at directing the activities of my department toward the overall company objectives.

.... 2. I am productive.

.... 3. I enjoy long-range planning and find the time to do it.

.... 4. I am willing to take risks to explore opportunities.

.... 5. I feel comfortable working in dynamic environments associated with uncertainty and change.

.... 6. I would enjoy running my own company.

.... 7. I consider myself more of an entrepreneur than an innovator.

.... 8. In social functions, I tend to get involved more in business discussions rather than technical discussions.

.... 9. I enjoy being evaluated in part, on my contributions to my company's business environments.

.... 10. I have been more right than wrong in predicting the business environment

.... Total Test Score (Divided by 5) = Normalized Score [   ]

FIGURE 21.2 Aptitude test questionnaire.

21.9

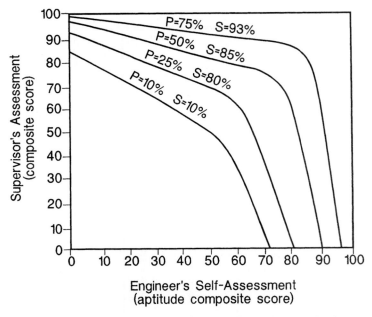

FIGURE 21.3   Probability of getting promoted *P* and probability of performing successfully *S* after promotion, based on aptitude assessment by superior and self-assessment by individual in line for promotion.

Recognize that self-assessment scores change with experience. The aptitude scores,
especially for self-assessment, improve with actual management experience. That
is, individuals who already made the transition into management evaluate themselves more realistically, and often even more favorably, than do individual contributors who have not yet experienced managerial responsibilities. Further, the
more favorable the management experience, the more favorable is the change in
overall aptitude score.[10]

Encourage discussions with technical managers. People who seek careers in management should discuss their ambitions with managers and get insight into related
tasks, responsibilities, and skills. Attending such functions as management conferences and business luncheons also can be beneficial in measuring management
abilities and providing a realistic background against which to take formal aptitude
tests.

Taken together, career decisions and selections for managerial training are complex.
To succeed in a challenging business environment, technology-based organizations
must be able to select and prepare future managers effectively.

Any decision regarding career paths, promotions, or training methods should be
based on a careful evaluation of the total situation. This should include considerable
personal judgment rather than mechanical scoring. The aptitude test may help to provide insight into a person's values and attitudes toward technical management. In the
final analysis, however, supervisors must make an integrated judgment that considers

all the factors against the person's professional background and the specific manageri-al challenges at hand.

The instruments for determining managerial aptitudes can be helpful in supporting professional development decisions, but we lack a method for accurately determining management potential. The challenge is for managers and engineers to develop enough understanding of the dynamics of their organizations and their managerial demands and to relate them to their candidates' personal strengths and desires, indi-vidual background, and the needs of the organization.

21.4 PREPARING FOR TECHNICAL MANAGEMENT

"Managers can be developed." This is the strong message of managers in technology-oriented companies.[11]

The action list in Table 21.2 describes how to prepare for a career in technical man-agement. It shows that personal preparation for a managerial position, supervisor's assistance, and organizational support have a significant impact on the ability to become a candidate for a technical management position, and ultimately succeed in the transition to technology-based management.

21.4.1 Criteria for Success

One of the key criteria for success, stressed by newly promoted managers and man-agement veterans alike, is a person's desire to become a manager. This desire seems to have a positive effect on many of the factors needed to transition into management and up the managerial ladder.[12] Yet, despite this favorable correlation, personal desire alone is insufficient to gain a promotion. In the final analysis, personal competence and organizational needs are the deciding factors. People who receive promotions usu-ally meet five key requirements:

1. They are competent in their current assignments. All individual contributors must master the duties and responsibilities of their current positions, must have the respect of their colleagues, and receive favorable recommendations from their super-visors.

2. They have the capacity to take on greater responsibility. The person must demonstrate the ability to handle larger assignments that have new and more challeng-ing responsibilities. Good time management, the willingness to take on extra assign-ments, and the expressed desire to advance toward a management assignment are usu-ally good indicators of a person's readiness for advancement.

3. They have prepared for the new assignments. A new management assignment requires new skills and knowledge. Candidates who have prepared themselves through courses, seminars, on-the-job training, professional activities, and special assignments will have the edge. Managers perceive such initiatives as evidence that a candidate is committed to the new career path, is willing to develop new skills, and wants to go the extra distance.

4. They are good matches with organizational needs. The candidate's ambi-tions, desire, and capabilities have to match both the current and long-range needs of

TABLE 21.2 Roles, Responsibilities, and Actions in Preparing for a Technical Management Career

Role	Actions in preparing for a career in technical management	Actions in transitioning into a technical management position
Individual preparation	Define specific objectives and plans Gain experiential learning Take on administrative assignments Practice team motivation and leadership Participate in task forces Seek out multifunctional assignments Participate actively in professional organizations Publish in professional journals and speak at conferences Maintain technical expertise Take courses and seminars Read management literature Complete an M.B.A. Talk to managers	Behave like a manager Develop more of a business perspective Build motivation and leadership skills Build credibility Practice delegation Develop a managerial style Observe other managers Seek advice and counsel Build peer support Build trust and respect with team and subordinates
Supervisor's assistance	Help the employee in assessing career ambitions Facilitate assignments that may provide management learning experiences Encourage leadership and assistance to management Facilitate dual-career ladders Encourage and support management training Use project management as a training ground Use temporary assignments and titles for incremental skill development (such positions include task manager, lead engineer, proposal manager, and team leader) Recognize the value of managerial skills for engineering assignments	Facilitate the transition with personal support Provide a charter of job responsibilities Help to establish communication channels Help to establish interaction with other departments Establish key performance objectives Build peer support Build respect and credibility for new manager
Organizational support	Establish policy guidelines for managerial development Develop managerial staffing plans Provide resources Establish dual-career ladders	Provide management development and training Provide networks for new managers Provide resources for assisting the transition

the firm. Obviously, a new job opening can create an immediate opportunity for advancement, but many companies make detailed, long-range plans for their managerial staffing needs. They also encourage managers to identify and develop future managers. In such environments, people with management ambitions must be recognized early as qualified candidates. Companies then may help them develop management skills with training and special job assignments. When the need for a new manager arises, the company often selects from the pool of prequalified candidates.

5. They have an aptitude for management. Taking an integrated look at the qualifications for transitioning into management, we see that the first four criteria are based on behavior in a *known* environment and on the assumption that the candidate will adapt to the new management situation. If these criteria are met, higher-level managers will have a lot of confidence in the candidate. However, there is yet no guarantee that the candidate will actually perform well in the new assignment. It is more difficult to evaluate the candidate's readiness for the new challenges in leadership, power, personnel administration, change, and conflict management. Therefore, technical contributors who want to be considered for leadership positions must *show* the actions and behavior at work which reflect strong aptitudes for management.

21.5 DIRECTIONS FOR THE FUTURE

As the issues of managing technology have become very real, companies have been responding to the challenges by intensifying their search for processes and programs potentially useful in developing technology managers. This becomes a challenge by itself since the type of knowledge and skills needed to function effectively in today's intense technology-based global business environment are not readily available through traditional training and development programs. Business leaders must go beyond the obvious and simple methods of developing managerial talent. They must recognize the highly multidisciplinary and cross-functional nature of skill requirements. They also must have business policies which integrate continuing professional education and skill development into the business process, and the individual performance appraisal and award system. A vital precondition for these multidisciplinary skill developments is interdisciplinary cooperation. Different disciplines often share the same business process, power, and resource basis, which often causes predictable and natural tension, anxieties, and organizational conflict. New ways and strategies have therefore to be found to bridge this gap.

An important role of senior management is to provide its career-oriented technical people with the knowhow to alter their destiny—to make career leaps, to break into new areas with higher and different skills and responsibilities, and to offer their employees wider horizons and far more opportunity than any generation of technology managers has encountered before. For the human race to survive, and for that survival to be worthwhile, we must master the skills and develop the knowledge required to manage technology.

21.6 REFERENCES

1. Jill E. Hobbs, "The First Global Forum on the Business Implications of Technology: Executive Survey," *Futurescope,* Decision Resources, Inc., Aug. 1992.

2. Michael Dertouzos, Richard Lester, Robert Solow, and the MIT Commission on Industrial Productivity, *Made in America: Regaining the Productive Edge,* HarperPerennial, New York, 1990.

3. Sumantra Ghosal, Breck Arnzen, and Sharon Brownfield, "A Learning Alliance Between Business and Business Schools: Executive Education as a Platform for Partnership," *California Management Review,* pp. 50–57, Fall 1992.

4. *The Resource Guide for Management of Innovation and Technology,* a joint publication of the American Assembly of Collegiate Schools of Business, the National Consortium for Technology in Business, and the Thomas Walter Center for Technology Management at Auburn University, Feb. 1993.

5. The managers claim that most of these skills are learnable was verified earlier via a field study by Thamhain and Wilemon which identified seven skill categories: (1) technical expertise, (2) organizational skills, (3) administrative skills, (4) leadership, (5) team building, (6) interpersonal skills, and (7) conflict resolution skills. On the average, 94 percent of these skills are learnable, as perceived by project managers. The learning proficiency varies depending on the particular category. For example, conflict resolution skills seems to be more difficult to learn than administrative skills. However, all categories taken together, 94 percent of project management skills seem to be learnable.

6. The percentile was determined by asking 355 technical managers to indicate for the skill inventory shown in Table 21.1: (1) what training methods helped to develop each skill and (2) how effective was the method in developing these skills relative to other methods employed. The data were evaluated to determine the contribution made by each of the five methods to leadership, technical, and administrative skill development. The distribution of sources was somewhat similar for all three skills. Figure 21.1 presents an aggregated view of the sources for managerial skill development. For detailed reporting of the study, see H. J. Thamhain, "Developing the Skills You Need," *Research/Technology Management,* **35**(2): 42–47, March–April 1992.

7. See H. Thamhain, "Managing Technology: The People Factor," *Technical and Skills Training,* Aug.–Sept. 1990. In a survey 85 percent of engineering managers considered the development of new engineering management talent crucial to the survival and growth of their businesses.

8. Louis S. Richman, "The New Worker Elite," *Fortune,* pp. 56–66, Aug. 22, 1994.

9. The statistical basis for Fig. 21.2 is a sample of 1500 technical professionals who were tested for managerial aptitude, using both supervisor's assessment and self-assessment. These technical professionals were then tracked regarding promotional actions and performance.

10. Consider also the confidence level and bias of the scoring; aptitude tests are more realistic after a candidate has gained some management-related experience.

11. This conclusion was reached during a study of 300 managers in 63 technology-based companies. These managers stated almost unanimously in one form or another that technical management skills don't just happen by chance, but are systematically developed through formal and informal methods.

12. This association was verified statistically using Kendall's correlation: the association between personal desire to become a manager and actual promotion was $\tau = .35$; between personal desire and subsequent managerial performance, it was $\tau = .30$. Both statistics are significant at a confidence level of 92 percent or better.

21.7 BIBLIOGRAPHY

Adams, John R., and Nicki S. Kirchof, "A Training Technique for Developing Project Managers," *Project Management Quarterly,* **24**(1), March 1993.

Alban-Metcalfe, B. M., "How Relevant Is Leadership Research to the Study of Managerial Effectiveness? A Discussion and Suggested Framework for Skill Training," *Personnel Review,* **12**(3), 1983.

Argyris, C., "Skilled Incompetence," *Harvard Business Review,* **64**(5): 74–79, 1986.

Badawy, M. K., *Developing Managerial Skills in Engineers and Scientists,* Van Nostrand, New York, 1982.

Beddowes, Peter L., "Re-inventing Management Development," *J. Management Development,* **13**(7): 40–46, 1994.

Bolster, C. F., "Negotiating: A Critical Skill for Technical Managers," *Research Management,* **27**(6): 18–20, 1984.

Brush, D. H., "Technical Knowledge or Managerial Skills?" *Personnel J.,* **58**,(11): 771–804, 1979.

Butler, F. C., "The Concept of Competence: An Operational Definition," *Educ. Technol.,* **18**(1): 7–18, 1978.

Carroll, S. J., and D. J. Gillen, "Are the Classical Management Functions Useful in Describing Managerial Work?" *Acad. Management Review,* **12**(1): 38–51, 1987.

Fottler, M. D., "Is Management Really Generic?" *Acad. Management Review,* **6**(1): 1–12, 1981.

Hague, D., "The Development of Managers' Knowledge and Skills," *Internatl. J. Technol. Management,* **2**(5–6): 699–710, 1987.

Harper, Kirke, "Trends in Management Education," *Public Manager* (Bur.), **23**(1): 19–21, 1994.

Katz, R. L., "Skills of an Effective Administrator," *Harvard Business Review,* **33**(1): 33–42, 1955.

Katz, R. L., "Skills of an Effective Administrator—Retrospective Commentary," *Harvard Business Review,* **52**(5): 94–95, 1974.

Kotter, J. P., "What Effective General Managers Really Do," *Harvard Business Review,* **60**(6): 156–167, 1982.

Lorange, Peter, "Back to School: Executive Education in the U.S.," *Chief Executive,* (92): 36–39, March 1994.

Morgan, P. G., "Central Skills of Successful General Managers," *Practicing Manager,* **8**(1): 13–15, 1987.

Peter, L. J., and R. Hull, *The Peter Principle,* Morrow, New York, 1969.

Rafail, Israel Dror, and Albert H. Rubenstein, "Top Management Roles in R&D Projects," *R&D Management* (UK), Jan. 1984.

Shenhar, Aaron J., and Hans J. Thamhain, "A New Mixture of Management Skills: Meeting the High-Technology Managerial Challenges," *Human Systems Management,* **13**(1): 27–40, 1994.

Smith, J. R., and P. A. Golden, "A Holistic Approach to Management Development," *J. Management Development,* **5**(5): 46–56, 1986.

Solem, O., "Professional Engineers' Need for Managerial Skills and Expertise," *Engineering Management Internatl.,* **3**(1): 37–44, 1984.

Stewart, R., "A Model for Understanding Managerial Jobs and Behavior," *Acad. Management Review,* **7**(1): 7–14, 1982.

Thamhain, Hans J., "Managing Engineers Effectively," *IEEE Transact. Engineering Management,* Aug. 1983.

Thamhain, Hans J., "Developing Engineering/Program Management Skills," *Project Management J.,* **22**(3), Sept. 1991.

Thornberry, Neal R., "Training the Engineer as Project Manager," *Training and Development J.,* Oct. 1987.

Waters, J. A., "Managerial Skill Development," *Acad. Management Review,* **5**(3): 449–453, 1980.

THE NEW-PRODUCT PROCESS

CHAPTER 22
MANAGING TECHNOLOGY IN THE SUBSTITUTION CONTEXT*

Clayton G. Smith
Formerly Associate Professor of Management,
Oklahoma City University
Meinders School of Business
Oklahoma City, Oklahoma

Major product innovations that create new industries are typically seen as offering substantial opportunities for growth and prosperity. But unless an innovation is tied to the creation of entirely new markets, it may also pose a threat of substitution to companies competing in a more established industry. When the new product first appears, its long-term potential is usually unclear; some potential substitutes never gain widespread acceptance, or find application only in relatively small market segments. Others, however, will develop in ways that can have devastating consequences, especially for firms whose largest or core business is threatened. Today, many companies (for example, Kodak and other manufacturers of conventional cameras) are contending with technological threats (i.e., electronic cameras); the executives of such firms must decide how to respond to an innovation that may have the capacity to destroy their existing business.[12,18]

This chapter considers the management of technology in the substitution context. To a significant degree, the discussion represents a synthesis of the author's previous research on the subject.[42,43] The specific focus is on major product innovations that are associated with the emergence of new industries, and that pose a substitution threat to firms in a more established industry. Innovations that create entirely new markets and substitution threats that are posed by other established industries (e.g., aluminum vs. steel) are outside the scope of this chapter.

Because of the importance of the substitution aspect of technology management to managers, a significant stream of research has developed regarding technological substitution (in which one technology displaces another in performing one or more functions in existing markets) and strategies for responding to substitution threats. In the

*The author would like to thank the editor, Gus Gaynor, for his valuable comments and suggestions, and also Daniel P. Hensley for his outstanding research assistance.

next two sections, established conceptions concerning substitution are considered and an emerging view of the phenomenon is explored. The discussion in these sections revolves around two basic strategies that are commonly used in responding to technological threats: (1) improving the traditional product and (2) entering the new industry (partially as a hedge against the risk of substitution).

In part, the emerging view of substitution indicates that while the range of functions for which an existing product is used may be narrowed, its sales can be relatively unaffected within its principal markets even as the new product gains widespread acceptance. In essence, the long-term relationship between the two products can be complementary, rather than competitive. Each product may have advantages in performing particular functions that customers value. Thus, it may not be necessary to stem the adoption of the new product to ensure that the traditional product will be able to survive and prosper.

There is also a growing recognition that it is important to look beyond the impact of a substitution threat on an incumbent firm's product and consider how the new product affects the value of the company's competitive capabilities. These capabilities—for developing and improving the traditional product, for making the product, and for marketing it—constitute the foundation of the firm's competitive position within the given markets. The extent to which these capabilities prove to be of value for the new product is a key issue for the firm's continued viability in those markets where the existing product is displaced.

Finally, there is an increasing awareness that the way in which a young industry develops can affect long-term substitution outcomes. Early in an industry's development, firms may pursue fundamentally different strategies, and sponsor alternative product designs that are based on different product concepts and technological approaches. How the new product comes to be defined and marketed can affect whether the traditional product is displaced, and the extent to which the technical and marketing capabilities for the old product prove to be valuable for competition in the new field.

With this emerging view of substitution as background, important technology management issues concerning the two response strategies mentioned above (improving the existing product and participating in the young industry) are considered. The specific topics that are explored concern

- The circumstances under which the traditional product will or will not be displaced within its primary markets
- The potential challenges that are associated with a strategy of participation in the young industry
- How an incumbent firm can improve its chances for competitive success in the new field, and reduce the risk that the existing product will be displaced
- How the traditional product may be improved so as to bring about a complementary relationship with the new

The discussion suggests that prior conclusions about the prospects of firms that are confronted with technological threats have been unduly pessimistic. It also suggests that the factors that affect the outcomes of the two most commonly used response strategies are not well understood. In the concluding section, the implications of the discussion are considered, and questions are posed for managers who find themselves having to make technology management decisions in the substitution context.

22.1 ESTABLISHED CONCEPTIONS

Substitution in the MOT context is the process by which one technology displaces another in performing a function or functions for one or more customer groups. Research concerning the substitution process has historically focused on the reasons underlying a new product's growing acceptance, and why the threat to the traditional product often increases over time. Past studies of how firms respond to substitution threats have also suggested that two very commonly used strategies—improving the existing product, and entering the young industry with their own versions of the new—often prove to be unsuccessful. Such research has viewed substitution mainly in product terms (i.e., the new product supplanting the traditional one in given customer groups), and has focused on situations in which the new and existing technologies were fundamentally different (e.g., transistors vs. vacuum tubes). While significant contributions have been made, the conceptions of technological threats that have arisen are limiting.

22.1.1 The Process of Substitution

Standard views of the substitution process are largely based on two parallel streams of research. The first, concerning innovation and technological threats, considers the role of continuing improvements in the new product following its introduction.[12,18,30,38,39,49] This research indicates that early versions of a new product are crude and expensive, and enjoy performance advantages in comparison to the existing product only in specialized niches. Over time, however, ongoing product and process advances bring about improvements in costs, quality, and functional performance. As a result, the new product normally penetrates additional customer groups. Such improvements may also extend the range of functions in which the new product offers price-performance advantages within a given group, thereby increasing the value that it provides relative to the existing product. For example, the first transistors were expensive and unreliable relative to vacuum tubes; they found early use only where their small size and low power consumption were inherent advantages (e.g., hearing aids). However, follow-on innovations steadily improved the transistor's cost and performance capabilities, and enabled it to invade other markets for vacuum tubes, such as portable and automobile radios and computers. Further, these ongoing advances eventually caused tubes to be supplanted in many applications where they had earlier been used together with transistors.[44]

The second stream of research, concerning the diffusion of innovations, highlights the role in the substitution process of changes in the propensity of buyers to adopt the new product as the young industry develops.[23,35-37] This research indicates that while early adoption is typically confined to adventurous, high-income buyers, and those who place a high value on the product's initial performance advantages, the proclivity of other customers to embrace the new product increases over time. Price-performance improvements, the development of an industry infrastructure and industry standards, knowledge of others who have had favorable experiences, and a growing familiarity with the product (due to word of mouth, etc.) all reduce the perceived risks, and make the purchase of the product a more attractive proposition. And as the diffusion process continues, social or competitive pressures to adopt the new product usually build.

22.1.2 Response Strategies

Prior research has suggested general strategies for responding to substitution threats. These include (1) finding new markets for the existing product that are unaffected by the substitute; (2) focusing marketing efforts on traditional customer groups that are likely to be later adopters of the new product; (3) attempting to maintain sales through actions such as increased promotion, greater customer service, longer warranties, or price cutting; (4) expanding work on the improvement of the existing product, in an attempt to stem the adoption of the new; and (5) entering the young industry. The latter two approaches appear to be used quite frequently, often in conjunction with each other.[12,18,35,49]

22.1.3 Improving the Traditional Product

The emergence of a substitution threat frequently induces vigorous and imaginative responses by incumbent firms that result in significant advances in what had been a stable technology.[12,18,38,39,49] While fundamental issues concerning the traditional technology may have long been considered settled, competition from the new product causes incumbents to reexamine their product's basic design characteristics, core components, and subassemblies, and the processes for its manufacture. Over time, this reassessment typically leads to considerable improvements in the product's functional performance, quality, and cost. As Rosenberg has noted (Ref. 38, p. 205), the emergence of a competing technology often is a more effective catalyst for improvements in an established technology than the more diffuse pressures of intraindustry competition.[38]

Nevertheless, while acknowledging that such advances are occasionally so substantial as to drive back the new technology, this research suggests that attempts to stem the new product's adoption will usually be unsuccessful. In general, this outcome is seen as being probable under the assumption that an existing product is normally much closer to the inherent limits of its potential in relation to the new.[18,49] While this may commonly be true in an overall sense, it should be noted that only limited attention has been given to the role in the substitution process of improvements in the existing product. The main focus of substitution research to date has been on continuing improvements that occur within the generic new technology. Rosenberg (Ref. 38, p. 203) noted that it is very common for researchers to fix their attention on the story of the new product as soon as its feasibility has been established, and to terminate all interest in the old.[38] This remains true today.

It will be argued below that efforts to improve the traditional product can be worthwhile, where the intent is to ensure that it is not displaced in its primary markets as the new product gains acceptance.

22.1.4 Entering the Young Industry

Many firms also respond to technological threats by entering the young industry, usually through internal development.[12,40] Their well-established brand names, customer relationships, and channels of distribution would suggest an ability to convert the threat into an opportunity. However, several studies have indicated that they often do not pursue the new technology aggressively, and frequently experience significant competitive difficulties in the new field. For example, Cooper and Schendel found that the new product's first commercial introduction was made by a company from outside the threatened industry in four of the seven cases examined.[12] Other studies have found that newcomers often play a lead role in improving the new product and in gaining market acceptance for it. Incumbent firms, in contrast, appeared to delay mounting a vigorous effort until after the potential of the new product had become apparent.[18,27,40,46]

Such research also suggests that an incumbent firm's historic experience can color its perceptions of how to compete in the new field. For instance, Tilton's study of the transistor industry's early years found that entrants from the vacuum-tube industry emphasized process innovation in their business strategies, an approach that had long been the norm in the tube business. In contrast, young semiconductor firms focused on product innovation, and sometimes obsoleted the transistors that the vacuum-tube companies were trying to produce in volume.[46] And finally, several studies have found that even leading incumbent firms are often unable to maintain a viable competitive position in the new field and are ultimately forced to withdraw.[12,18,27,46] Of course, there are exceptions, and the competitive position of some firms does appear to improve over time. But the general indication from this research is that success in the old technology by no means assures success in the new.

These studies have mainly considered situations where the new and traditional technologies were fundamentally different (e.g., transistors vs. vacuum tubes and electronic vs. electromechanical calculators). In this context, the patterns observed are partially attributed to concerns about the cannibalization of the existing product, and the obsolescence of investments in plant and equipment. They are also attributed to political resistance from important individuals and groups whose skills and influence are tied to the threatened business. However, the discussion below will argue that more basic issues are involved. Further, an incumbent firm will be better able to develop a viable competitive position in the young industry under some circumstances than others. The discussion will consider actions that can be taken in the new field to improve the chances of competitive success and, somewhat paradoxically, to reduce the risk that the traditional product will be displaced within its primary markets.

22.2 EMERGING VIEW OF SUBSTITUTION

The limitations of established conceptions of substitution are now becoming apparent. Increasingly, the substitution process is being explicitly considered in functional terms (i.e., the specific functions for which the new product proves to be superior). It is being viewed more in terms of the technical and marketing capabilities associated with the existing product as well (i.e., how the competitive value of these capabilities is affected where the product is displaced). Also recognized are the high levels of uncertainty that surround young industries, and the rivalry among firms that are seeking to structure the "framework of competition" in ways that favor their interests. The dynamics of this rivalry, and the rules of competition that are formed as the new industry develops, can have a significant impact on long-term substitution patterns. Overall, this emerging view of technological threats (summarized in Table 22.1) provides significant insights for decisions regarding efforts to improve the traditional product, and attempts to establish a position in the new field.

22.2.1 Functional Perspective

It has long been recognized that the *identification* of substitution threats involves focusing on products that perform one or more of the functions that the traditional

TABLE 22.1 Emerging View of Substitution

Developing perspective	Summary comments
Functional perspective	Vital for analyzing substitution dynamics and long-term substitution outcomes. Suggests that while the range of functions for which existing product is used may be narrowed, its sales can be largely unaffected within principal markets even as new product gains widespread acceptance. Thus, it may not be necessary to stem adoption of new product in order for existing product to be able to survive and prosper.
Resource-based perspective	Important to look beyond the impact on traditional product and view "threat of substitution" in terms of firm's existing R&D, manufacturing, and marketing capabilities. Such capabilities are the foundation of a firm's position in the given markets. Extent to which they prove to be valuable for the new product is a key issue for continued viability where existing product is displaced.
Competitive perspective	How an industry evolves can affect substitution patterns. Strategies of various firms may reflect different competitive capabilities and/or economic interests. Competing designs may vary in the range of functions performed, the potential effectiveness in performing given functions, the technical capabilities on which they are based, and the marketing requirements that they give rise to.
	Thus, how the new product comes to be defined and marketed can affect whether the existing product is displaced and the extent to which existing technical and marketing capabilities prove to be valuable in the new field.

product performs, rather than products that have the same form.[1,25,35] As Porter notes (Ref. 35, p. 275), a new product may perform a narrower range of functions than the existing product. It may perform a wider range as well.[35] (In addition to performing the functions of a typewriter, for example, a personal computer also performs other functions, including calculating and small-quantity copying.) And among the traditional functions that the new product performs, its price-performance capabilities relative to the existing product may be greater on some dimensions than others.

But only recently has it come to be fully recognized that a functional perspective is vital for analyzing substitution dynamics and long-term substitution outcomes.[42,43] While the range of functions for which an existing product is used may be narrowed, its sales can be largely unaffected within its principal markets, even as the new product gains widespread acceptance in those same markets. In essence, the long-term relationship between the new and traditional products can be complementary, with each having price-performance advantages in performing particular functions that customers value. Thus, it may not be necessary to stem the adoption of the new product in order for the existing product to be able to survive and prosper. Nowhere is the product (vs. functional) focus of prior research more apparent than in the notion of "switching costs," which are commonly viewed as one-time costs of changing from the old product to the new. These include the costs of learning to use the new product and of purchasing ancillary products such as software; they are typically seen as potential impediments to substitution.[35] From a functional perspective, however, they are viewed as "adoption costs," since buyers may come to use both products to satisfy the given set of needs. Such costs are considered as part of the price-performance comparison in evaluating the relative effectiveness of the new and traditional products on given functional dimensions.

22.2.2 Resource-Based Perspective

A growing body of research has indicated that while some innovations render a company's existing R&D and manufacturing capabilities obsolete, others permit the firm to utilize and build on them.[4,5,42,43,47] The latter can be true even for major product innovations (i.e., innovations that are associated with the creation of new industries). To the degree that a new product is rooted in the R&D and manufacturing capabilities for an existing product, Smith defines the underlying technology as being "related" to its traditional counterpart. (*Note:* Where the new and existing technologies are closely related, it is usually because the new has largely arisen from experience with the old. Examples include integrated circuits vs. transistors, and color vs. black and white television.[44,50])

Smith also explored the extent to which a firm from the established industry can benefit from its existing marketing resources (i.e., brand names, distribution channels, sales organizations, and service networks) and skills in the new field.[42] These marketing capabilities often provide a sizable advantage as an industry matures and product offerings become more standardized across competitors. However, this research suggests that the new technology and the strategies of newcomers can erode the value of such capabilities for marketing the new product to the firm's existing customers.

In sum, the emerging view of substitution indicates that it is important to look beyond the impact on the traditional product and also view the "threat of substitution" in terms of the firm's established competitive capabilities. These capabilities—for developing and improving the traditional product, for making the product, and for marketing the product—are the foundation of its competitive position within the given markets.[16] The extent to which they prove to be valuable for the new product is a key issue for a firm's continued viability in those markets in which the existing product is displaced.

22.2.3 Competitive Perspective

Early in an industry's development, firms often pursue different strategies and sponsor alternative product designs.[30] In some instances, the various strategies mainly reflect different assumptions about how to compete, and the design alternatives revolve around a single basic product technology. However, research concerning innovation and young industries indicates that there are also cases in which the diversity among strategies and product designs is more fundamental.[2,8,17,22] Here, the strategies pursued may reflect not only different assumptions about how to compete but also differing competitive capabilities and economic interests. And the various designs may differ in the range of functions performed, the potential effectiveness in performing given functions, the technical capabilities on which they are based, and the marketing requirements that they give rise to.

Indeed, a major reason for such diversity is the presence within the young industry of firms with differing backgrounds and competences, each trying to shape the framework of competition in ways that favor their interests.[9,17,22] The competition among firms is over how the industry will be structured, and which version of the new prod-

Dominant Designs vs. Compatibility Standards

This chapter draws a distinction between dominant designs and compatibility standards, which determine whether two products (e.g., personal computers) can share the same complementary products,* such as applications software.[21] In essence, compatibility standards should be viewed as part of the larger concept of the dominant design.

Where there are rival compatibility standards, the one that proves to be the industry standard can affect the competitive positions of rival firms. This impact stems from "installed base effects," in which the attractiveness to customers of a firm's (or group of firms') standard over others depends on the relative size of its installed base. A larger installed base normally gives rise to a greater variety and quality of, and lower prices for, complementary products for the given standard. A standard for which there is a large user base also provides a more effective means of social coordination. And over time, a standard's impact on the competitive positions of firms may be tied to the existence of buyer switching costs as well.[11,22]

Installed-base effects arise where a complementary product is essential for obtaining full-value use of the given product, and the complement is design-specific.[11] Where these conditions are both met, the emergence of a proprietary, brand-specific industry standard (vs. an "open" standard) can have particularly significant competitive consequences.

But beyond the issues that surround compatibility standards, dominant designs can also determine the extent to which an incumbent's capabilities are valuable for competition in the new field, and whether the existing product will be displaced. Today, for example, there are six competing technological approaches for high-definition television. Each is capable of meeting the proposed broadcast transmission standards of the U.S. Federal Communications Commission. However, the approaches differ substantially in the display technologies employed (e.g., advanced CRT technology, vacuum microelectronics, and active-matrix LCDs) and in the technical capabilities required for each.[7]

In the emerging microwave-oven field, the principal design alternatives were the countertop oven and the free-standing double oven. The countertop design (which proved to be the dominant design) viewed the new product as a supplement to a traditional range—one that would perform the functions of defrosting and heating, but not the function of cooking.† In contrast, the double oven combined a microwave oven and a conventional electric range in a single unit, a view of the product that implied the replacement of a household's existing range.[40]

Compatibility standards can have important effects on a firm's ability to develop and maintain a strong position in a new field. However, if dominant designs are equated with compatibility standards, there is a danger of overlooking important dimensions of design competition (e.g., competition among firms to define the product in ways that favor their capabilities). And there is a corresponding risk of overlooking the full significance of the design that wins the race for dominance.

*The term *complementary products* should be distinguished from the earlier usage of the term *complementary*, which focuses on the relationship between the new and traditional products.

†From the consumer's perspective, microwave energy is poorly suited for cooking meats and other foods, since it lacks the ability to brown food items.

FIGURE 22.1 Distinction between dominant designs and compatibility standards.

uct will be the "dominant design."[3] See Fig. 22.1. Where a fundamental diversity among strategies and/or product designs exists, the dynamics of this competition, and how the contest is ultimately decided, can have a significant impact on long-term substitution patterns. How the new product comes to be defined and marketed can affect whether the traditional product is displaced. It can also affect the extent to which the technical and marketing capabilities for the old product prove to be valuable for competition in the new field.[42,43]

22.3 TECHNOLOGY MANAGEMENT ISSUES

With this emerging view of substitution as background, four important technology management issues concerning the response strategies of improving the existing product and participating in the young industry are now considered. The specific topics explored concern (1) the circumstances under which the traditional product will or will not be displaced within its primary markets, (2) the potential challenges that are associated with a strategy of participation in the young industry, (3) how incumbent firms can improve their chances for competitive success in the new field and reduce the risk that the existing product will be displaced, and (4) how the traditional product may be improved so as to bring about a complementary relationship with the new.

22.3.1 Circumstances under Which Traditional Product Will or Will Not Be Displaced

Although an existing product may be close to the inherent limits of its potential in an overall sense, there are still circumstances in which it will not be displaced within its principal markets even as the new product gains widespread acceptance. While a threatened technology may be "mature," there can be significant room for price-performance improvements on important functional dimensions as incumbent firms reconsider the set of benefits that their product offers to the customer. For example, CT (computerized tomographic) scanners have proved to be very effective for imaging soft tissues, and for brain studies in particular. Nevertheless, the older technology of nuclear medicine has remained superior for organ function studies because of improvements that allowed it to better exploit its dynamic imaging capability.[15]

Moreover, while given versions of the new product may have clear and early advantages on some dimensions, they can also be fundamentally constrained in their ability to perform other traditional functions that customers value. As a result, the magnitude of improvements required of the existing product to maintain superiority on those latter dimensions may not be great. And finally, given versions of the new product may perform only some of the functions that are valued by the traditional product's mainstream customers. Under all three scenarios, the long-term relationship between the two products will ordinarily be complementary. Within the existing product's principal markets, the new product will supplant the traditional in performing the particular functions in which it has the advantage. But, as illustrated in Fig. 22.2, the traditional product itself will not be displaced.[42,43]

In general, the existing product will be supplanted only if the new product performs all the traditional functions that customers value more effectively (and it performs additional functions as well). (*Note:* In cases where the sales of the traditional product decline substantially, the time period from the new product's first commercial introduction to the point when its sales exceed those of the traditional appears to be about 5 to 14 years.[12]) The range of traditional functions in which the price-performance of the new product proves to be superior will depend on (1) the success of incumbents in bringing about improvements in the existing product on given functional dimensions, (2) the traditional functions that the new product performs, and (3) the ultimate effectiveness of the new product in performing those functions. (As discussed

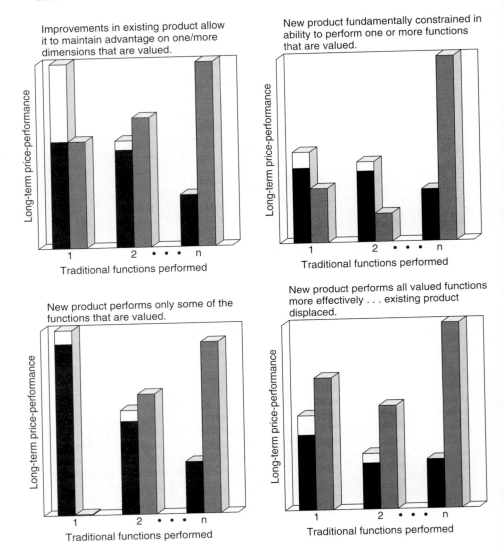

FIGURE 22.2 An existing product will be supplanted in its principal markets only where the new product performs all the traditional functions that customers value more effectively (and where it performs additional functions as well).

in the next section, where the diversity among alternative versions of the new product is fundamental, the latter factors may, in turn, be tied to the dominant design that emerges.[43] Stated differently, the range of traditional functions that "the" new product performs and its price-performance capabilities on those dimensions may depend on the design alternative that wins the race for dominance.)

22.3.2 Challenges of a Strategy of Participation

Young industries are characterized by high levels of uncertainty. Early on, no "right" strategy has been clearly identified, and firms—including entrants from outside the threatened industry—often utilize different marketing approaches and production technologies. Only time will determine the relative validity of the assumptions on which they are based. Competing product designs often vie for acceptance as well. [The rivalry among electronic watches with LED, LCD (light-emitting diode, liquid-crystal display), and analog displays in the 1970s is one example.] All the designs typically have serious shortcomings and technical obstacles that have to be overcome, the possible means for doing so are usually unclear, and the potential for improving their ability to meet perceived customer requirements—which are also evolving—is largely unknown. As a result, it is very difficult to predict which alternative will gain widespread acceptance, and emerge as the dominant design.[3,8,34]

Resource-Based Challenges.   From both functional and resource-based perspectives, these uncertainties sometimes have only limited significance. During the early 1980s, for example, PC designs had different operating systems, command systems (that used the mouse, touch-screen, or keystroke approaches), and microprocessors. However, the various designs performed the same basic functions, and there were substantial overlaps in the technical capabilities required for each. Further, the major competitors also followed parallel business strategies.[19] In such cases, there are risks that include the possible obsolescence of product inventory, raw materials, and components. But the uncertainties described above will not have a major bearing on whether the traditional product is displaced. Moreover, the value of existing technical and marketing capabilities in the new field will largely be tied to the basic technology that underlies the different versions of the new product.

A key issue here is the degree to which the new technology is related to its traditional counterpart. Where the two technologies are related to a significant degree, incumbent firms will be able to utilize and build upon their R&D and manufacturing capabilities as they attempt to surmount technical obstacles and establish a competitive position.[4,5,42,47] For example, this was true for transistor firms that became involved in the newer field of integrated circuits.[44] Under these circumstances, the incumbents' understanding of the new technology and their abilities for continued improvements should be at least as good as those of entrants from outside the threatened industry. Indeed, where the capabilities that are critical for the performance, quality, and cost of the new product are specific to the established industry, it may be difficult for new entrants to establish a beachhead.

Conversely, where the new and traditional technologies are fundamentally different, incumbent firms that enter the new field will face the challenge of developing the necessary technical resources and skills for the new product. Particularly if the new technology evolves at a rapid pace, they may find it difficult to offer a product that is state-of-the-art. This may place incumbents at a significant disadvantage, during the early stages of industry development, when product performance is usually the most important basis of competition. As the industry continues to evolve, knowledge about the new technology will tend to become more widespread, and the rate of technological change will eventually slow. Nevertheless, these early conditions will make it relatively more difficult for incumbent firms to develop a strong competitive position in the young industry.[30,34,42,48]

A further concern is the extent to which incumbent firms can benefit from their existing marketing resources and skills in the new field. Where one or more of the marketing assets noted in Table 22.2 were important for success in the established,

TABLE 22.2 Incumbent Firm Marketing Assets

Resources	Skills*
Brand names	Distribution methods
Distribution channels	Selling methods
Sales organizations	Marketing practices (pricing, promotion, and product policy)
Service networks	

*The delineation of marketing skills is based on Biggadike's[6] research.

threatened industry, such firms will often enjoy a significant advantage in the long term. As the rate of technical change slows and the products of competitors become more standardized, the value of having an established brand name, channels of distribution, etc., and strong marketing skills tends to increase correspondingly.[6,34,48] In general, this will help to strengthen the competitive positions of incumbent firms as the young industry continues to evolve.

But potentially, the new technology may render incumbent marketing assets less suitable or necessary, and erode the advantage that might have been realized. (Prior work has considered how technological innovations can create new markets in which incumbent marketing capabilities are of relatively little value.[4,12,49] In contrast, the focus here is on how a new technology can erode the value of such assets for marketing the new product to incumbent firms' traditional customers.) In some cases, the technology will create new marketing requirements that existing capabilities cannot adequately satisfy. This may occur, for instance, where a new product alters the dealer functions that have to be performed (e.g., personal selling vs. display) such that new channels are required to serve the given markets. A new technology may also relax existing marketing requirements and diminish the importance of incumbent capabilities. For example, while the distribution networks of vacuum-tube producers had been important as a means of quickly delivering replacement tubes, they became less valuable in transistors as the new product became increasingly reliable.[13,42]

Smith (Ref. 42, p. 26) argued that such patterns are due to differences in the characteristics of the new and threatened products, including differences in price, the range of functions performed, reliability, ease of use, and the frequency of required maintenance.[42] The value of incumbent marketing capabilities in the new field may become clear only over time, since the new product's characteristics often change as continuing advances are made in the basic technology. But because of their fundamentally different nature versus the existing technology, unrelated new technologies may have the greatest capacity to erode the value of incumbent marketing assets. As discussed below, the value of such capabilities may also be undermined by the actions of new entrants. Regardless of the reason, however, it will clearly be more difficult to build a viable position in the young industry where this occurs.

Competitive Challenges. As noted earlier, there are also cases in which the diversity among strategies and/or product designs is more fundamental. Under these circumstances, long-term substitution patterns cannot be viewed as a function of the characteristics of "the" new technology, since important issues are likely to be overlooked. The dominant design that emerges, and how the industry comes to be structured, can affect whether the existing product is displaced, and which firms' competitive capabilities prove to be valuable in the new field. Incumbent firms can be expected to pursue strategies and product designs that preserve their interests to the extent possible. But new entrants will have little interest in the status quo. Their actions are likely to be oriented toward capitalizing on their competitive capabilities, offsetting their weaknesses, and nullifying the advantages of the incumbents.[2,8,17,20,22,43,51]

In electromechanical calculators, for example, direct-sales organizations and service networks had been important requirements for success, and the major producers of such machines had concentrated on the business segment in their marketing efforts. As the electronic calculator field began to emerge in the 1960s, entrants from the threatened industry understandably sought to use their established sales and service networks for the new product as well. But the scientific and programmable models that they offered were aimed at statisticians, scientists, and engineers who had previously relied on large computers for their computational needs. As the division president of Monroe said later of his company's early participation, "Our effort in electronics, I think logically, was not to create competition for [our] electromechanical machines. It was not to take away the established base, but to seek new business over that base" (Ref. 27, p. 60).

However, new entrants targeted the mainstream business segment with the electronic calculators that they developed, and sold their models through office equipment dealers. Further, by exploiting the potential for dramatic reductions in the cost of electronic components, they drove market prices down to levels that made it difficult to cover the costs of direct-selling efforts. Several incumbent firms were forced to develop a network of office equipment dealers for the electronic business calculators that they belatedly began to manufacture. Finally, as the reliability of electronic calculators (which had few moving parts) improved, the need for a strong service network waned. Combined with the eventual displacement of electromechanical calculators by their electronic counterparts, the sales and service networks of the incumbents ultimately became irrelevant for competition in their traditional markets.[14,27]

Here, the marketing strategies of new entrants capitalized on the new technology's potential to create superior value for mainstream customers (in part, via lower prices and greater reliability), and to undermine incumbent marketing capabilities. In a similar fashion, the R&D and manufacturing strategies of newcomers can also undermine incumbent marketing capabilities. For example, American and Baldwin Locomotive (leading producers of steam locomotives) found that their new competitor in the diesel-electric field, General Motors, ignored their time-honored methods of custom manufacturing and built standardized locomotives for inventory instead. GM's success in gaining customer acceptance for this approach eroded the value of the incumbents' skills concerning selling methods and marketing practices.[13] In these instances, the different versions of the new product were based on a single basic technology, one that was unrelated to the traditional technology.

The emerging electronic-watch field illustrates the significance of competing designs that are based on different product technologies and concepts. During the early stages of industry development, semiconductor firms such as Texas Instruments concentrated on the all-electronic LED and LCD product designs, while traditional watch firms such as K. Hattori (Seiko) focused the majority of their efforts on the quartz analog. Most LED and LCD designs reflected a concept of the electronic watch as a "wrist instrument" that could perform a wide range of timekeeping functions. Instead of the traditional "hands" of the analog display, the technological approaches underlying these designs utilized a semiconductor (LED or LCD) display, which was well suited for the broad functionality of the wrist instrument concept. In addition, the approaches underlying the LED and LCD designs were largely unrelated to mechanical watch technology.

By comparison, the quartz analog design reflected a concept of the electronic watch as a timekeeping device that was also a piece of jewelry. While the technological approach underlying this design replaced the traditional mainspring and escape mechanism with electronic circuitry and a quartz crystal oscillator, it retained the gear train and hands of the mechanical analog watch. As a result, portions of the technical resources and skills for the old product remained relevant for this version of the new. The product concept underlying the quartz analog design was also more consonant with the brand images and jewelry store distribution channels of mainstream incumbent firms, such as Seiko. Ultimately, it was the quartz analog that emerged as the dominant design; the new entrants were largely unsuccessful in their efforts to alter the set of benefits that customers valued.[24]

Similar competitive struggles are being played out today in high-definition television and electronic cameras. In such cases, the strategies of competing firms and the dominant design that emerges can affect the extent to which incumbent technical and marketing capabilities prove to be of value for competition in the new field. Further, the range of traditional functions that the dominant design performs, and the price-performance capabilities that it proves to have on given functional dimensions, can affect whether the traditional product is displaced. In summary, the managers of incumbent firms need to be sensitive to the resource-based and competitive challenges that have been described. They need to carefully appraise the potential impacts of the new technology, alternative product designs, and the strategies of rival firms. And the challenges considered also have implications for actions that can be taken as part of a strategy of participation in the young industry. It is to this issue that the discussion now turns.

22.3.3 Improving Chances for Success in New Field and Reducing Risk of Product Substitution

How the new product comes to be defined and marketed depends on more than the latent potential of particular product designs and strategies. While a given design alternative may have the capacity to be the "best" means of meeting emerging customer preferences, the realization of this promise turns on the development efforts of the firms that pursue the design. The relative success of these firms, and of other firms, in unlocking the potential of the alternatives that they sponsor has a substantial impact on the outcomes of design competition. Industry participants also play an important role in shaping the formation of customer preferences in the first place. Market feedback affects the direction of product development activities. But at the same time, the designs that are available and the educational/promotional efforts of firms that seek to define the rules of competition in given ways influence the development of concepts that guide customer choice.[5,8,9,22]

Beyond the industry's direct participants, other organizations can also influence how the new field develops. Depending on the nature of the product, such organizations may include suppliers of complementary products (e.g., distributors of VCR cassette tapes), industry associations, and government regulatory bodies. Suppliers of complementary products will influence an industry's evolution where their decisions about which version(s) of the new product to support have a significant impact on the commercial viability of the competing product designs. And where the new product (such as high-definition television) is part of a larger system, decisions by industry associations and regulatory bodies regarding compatibility standards can have a sizable impact, if the decisions constrain the set of designs that are capable of meeting the standard.[21,45] Clearly, industry competitors and other organizations play important roles in determining the way in which technological and market uncertainties are resolved.

Influence Stratagems.   In many instances, it will be possible for an incumbent firm to pursue effective strategies and product designs that are consonant with its economic interests. This may take the form of strategies that seek to reinforce the value of the firm's established marketing capabilities in the new field. For instance, an existing service network might be used to maintain or repair other products of customers (beyond the product in question) for an attractive package price. An incumbent firm may also sponsor product designs that capitalize on its technical and marketing capabilities to the extent that this is feasible. (Sometimes, it will be possible to develop versions of the product that exploit existing technical and marketing capabilities. In other cases, there will be viable design possibilities that exploit existing capabilities for the traditional product but also require the development of new resources and skills. But there will also be situations where any feasible design will obsolete at least some of the traditional technical and marketing capabilities.)

Finally, versions of the new product may be developed with the primary objective of bringing about a complementary relationship with the existing product. (*Note:* It should be understood that the firm may have to develop new technical capabilities in order to bring such a relationship with the existing product about.) Such a version of the product might, for example, perform only some of the traditional functions that customers value. The basic objective would be to offer price-performance capabilities on a subset of the traditional functional dimensions that exceed those of both the existing product and rival versions of the new. Overall, these basic stratagems may enable an incumbent firm to enhance its chances for success in the new field (by preserving the value of established capabilities) and/or reduce the risk that the traditional product will be displaced. While none of the approaches will be viable under all conditions, each warrants consideration in most circumstances. Even where existing designs revolve around a single basic technology, and firms have been pursuing similar strategies, this does not necessarily mean that the limits of what is feasible have been defined; it may simply be a reflection of what has come forth, so far, from a larger set of possibilities.

Companies that have strong technical and marketing capabilities that are valuable for their design and/or strategy, and ample financial resources, should have a greater individual ability to influence how the new product comes to be defined and marketed. Relative to other entrants, such firms should have a greater capacity to unlock the potential of their design, to stimulate the process of its diffusion, and engage in aggressive educational and promotional efforts that help to shape the rules of competition. Further, the possession of strong competitive and financial capabilities may favorably affect market perceptions about the likelihood that the firm will be successful in the new field. This may influence consumer preferences by making buyers feel more secure in their purchase decision. Such perceptions may also affect the decisions of suppliers of complementary products, concerning the version of the product that they will support, and the decisions of industry associations and regulatory bodies, regarding compatibility standards.[9,40,45]

Time of Entry.   Research suggests that the timing of a firm's entry is important in the success of its efforts to influence how the young industry develops, and to establishing a strong position within it.[26,28,35,41] First-mover advantages can provide a firm that enters an emerging industry at an early date with greater leverage for shaping the way in which the new product comes to be defined and marketed. An early entrant will often gain an edge over firms that enter later from having more time to improve the performance, quality, and cost of its design. Order of entry can also affect the formation of customer preferences. To the extent that an early entrant is able to achieve significant consumer trial, it will be in a better position to define the attributes that are valued most within the product category. These advantages can, in turn, strengthen the firm's ability to influence the decisions of suppliers of complementary products, industry associations, and regulatory bodies. Early entry may also reduce the risk of being preempted by other entrants that have different economic interests and capabilities. Indeed, it may be possible to challenge such firms before they are able to develop a viable competitive position.[29,42]

From an organizational viewpoint, a decision to enter early may also be important. Where a firm's product design and/or strategy requires new competitive capabilities, early entry can facilitate their timely development, and their integration into the firm's existing knowledge base. This issue will be especially important if the technology underlying any feasible product design is unrelated to the traditional technology. Here, more time will be needed to develop the necessary technical capabilities. And because unrelated technologies are more likely to erode the value of existing marketing assets, significant amounts of time may also be needed to deal with this eventuality.

In addition, the various activities that are associated with participating in a young industry have important impacts on organizational learning. Early entry can lead to greater levels of cumulative learning over time, to a greater ability to assimilate new information that is related to what has already been learned and, as a result, to improvements in the quality of business strategy decisions.[10] While these organizational issues are somewhat intangible, they can still have a major bearing on the success of a firm's competitive efforts within the new field.

Interorganizational Actions.   As part of the influence stratagems described above, a company may undertake actions that harness the resources and skills of others. To the extent that a firm's product design requires new technical capabilities, it will often be valuable to augment internal R&D activities with "alliances" with entrants from outside the threatened industry. (Hamel and Prahalad refer to such arrangements as "competing through collaboration."[20]) This can facilitate the development of the necessary resources and skills, and thereby allow the firm to offer a more competitive version of the product. Such linkages with new entrants, which have been common between pharmaceutical companies and emerging biotechnology firms, for example, may take the form of contracts, licenses, joint ventures, minority equity investments, etc. And especially where the technology underlying any feasible product design is unrelated to its traditional counterpart, acquisition may also be a promising vehicle for gaining rapid access to needed technical capabilities.[31,33,42]

Collaborative efforts with other incumbent firms may also be undertaken in an effort to form an alliance around a basic product design, and to put the strength of the firms' combined resources behind it. Such companies could, for example, establish R&D partnerships or cross-licensing arrangements to foster more rapid advances in the performance, quality, and cost of the given design vis-à-vis competing alternatives.[9,32] In addition, a firm might participate (e.g., through traditional industry trade groups) in cooperative educational and promotional activities that are aimed at shaping the customer preferences that develop around the new product. In general, interorganizational actions—which influence buyer preferences and strengthen the competitiveness of the given version of the new product—will directly improve the firm's ability to shape the way in which the product comes to be defined and marketed. They may indirectly influence the decisions of suppliers of complementary products, industry associations, and government regulatory bodies as well.

22.3.4 Improving Traditional Product so that It Is Complementary with the New One

Technology-based efforts to bring about a complementary relationship with the new product involve an exploration of the functions on which the existing product may be able to maintain price-performance superiority, and a corresponding redirection of technical development activities. In some instances, such a relationship with the new product may be brought about as a result of previously unrecognized opportunities for product or process improvements in the existing technology. (The earlier discussion concerning nuclear medicine relative to the newer imaging technology of CT scanners provides one example in this regard.) In other cases, such a relationship may be created by exploiting limitations discerned in the ability of given versions of the new product to perform traditional functions that are valued by mainstream customer groups.

Frequently, as the firm learns what improvements are possible in the existing product, this strategy will evolve as it becomes more evident whether the shortcomings of given versions of the new product can be overcome and, in some cases, as the dominant design begins to emerge. The effectiveness of this effort to ensure that the traditional product is complementary with the new may be enhanced by sharing technical knowledge gained with other incumbent firms (e.g., through liberal licensing policies). It can also be enhanced by enlisting the aid of component and subassembly suppliers in the existing product's defense. In general, this strategy is most likely to succeed when the different versions of the new product cannot perform, or are fundamentally constrained in their ability to perform, some of the traditional functions that customers value.[43]

22.4 SUMMARY AND IMPLICATIONS

This chapter has presented an emerging view of technological substitution and has explored important issues concerning two strategies—improving the existing product, and participating in the young industry—that are typically used in responding to substitution threats. The discussion suggested the circumstances under which the traditional product will or will not be displaced within its primary markets, as the new product gains widespread acceptance within those markets. The basic ways in which the existing product may be improved to bring about a complementary relationship with the new product were also considered.

In addition, the challenges that are associated with participating in young industries were examined. Finally, actions that can be taken in the new field to (1) improve the chances of competitive success and (2) reduce the risk that the traditional product will be displaced were explored. (Questions for consideration by incumbent firm managers are posed in Table 22.3.) Overall, prior conclusions about the prospects of companies that are confronted with technological threats may have been unduly pessimistic. The factors that affect the outcomes of these two commonly used response strategies are only beginning to be understood.

The emerging view of substitution that has been presented has several implications. First, the focus of prior research has largely been on the role in the substitution process of continuing advances within the generic new technology. Future work could make important contributions by adopting a parallel focus on improvements that take

TABLE 22.3 Questions for Incumbent Firm Managers

---

1. Is the new product likely to be fundamentally constrained or unable to perform some of the traditional functions that mainstream customers value?
2. What actions can we take with respect to the traditional product to help bring about a complementary relationship with the new product?
3. To what degree will we be able to utilize and build on our existing R&D and manufacturing capabilities in the young industry?
4. What new technical resources and skills will need to be developed and integrated into the firm's knowledge base? How can this best be done? What existing technical capabilities may ultimately be rendered obsolete?
5. Is the new technology likely to create new marketing requirements that our existing marketing assets cannot adequately satisfy? In what ways might the new technology relax existing marketing requirements and diminish the value of these assets?
6. What are the characteristics of competing designs for the new product? What impacts could the widespread acceptance of a given design have on the traditional product?
7. How could the given design affect the competitive value of our existing technical and marketing capabilities in the young industry?
8. In what ways could the strategies that new entrants are pursuing erode the value of our existing marketing capabilities?
9. What strategic actions could we take to reinforce the value of our established marketing capabilities?
10. Are there viable designs that we could sponsor that would capitalize on our existing technical and marketing capabilities? Are there viable alternatives that would reduce the risk that the traditional product will be displaced?
11. What other actions can we take to influence the young industry's development in ways that favor our economic interests?

---

place in the traditional product in response to the substitution threat. How often, for example, and under what circumstances, do such improvements prevent the existing product's displacement? What are the commonalties among efforts that prove to be successful? And where the long-term effect is only to postpone the time when the traditional product is clearly obsolete, what is the typical length of the delay?

Efforts to improve the existing product are often viewed as a myopic response that stems from sunk costs and internal political difficulties.[18,49] But if such improvements often bring about a complementary relationship with the new product under given conditions—or postpone the existing product's displacement for extended periods—knowledge of this would be valuable to both researchers and practitioners. Moreover, investigations that focus on improvements in the traditional product could lead to a wider recognition that the substitution process is driven not only by advances that occur within the new technology but also by improvements in the existing technology.

Second, although research concerning innovation and young industries has recognized the diversity among strategies and product designs that often exists within emerging fields,[2,8,9,17,22] this has been less true of substitution research. While it is generally realized that firms often pursue different strategies and product designs, the implicit assumption has been that such diversity does not have a major bearing on long-term substitution outcomes. In some cases, this premise is true, either because the range of feasible designs and strategies is relatively narrow or, perhaps more likely, because they are the only ones that come forth from a broader set of possibilities. Nevertheless, it is important to understand that both classes of young industries exist—those that encompass fundamental diversity, and those that do not—and to distinguish between the two in substitution research.

Where the diversity among alternative strategies and/or product designs is funda-mental, assumptions that long-term substitution patterns are a function of the charac-teristics of *the* new technology are likely to be inappropriate. Researchers and practi-tioners that proceed from this premise are likely to overlook important dimensions of the substitution process. More generally, an explicit recognition by researchers of the diversity that can exist among different product designs and strategies should lead to a better understanding of the dynamics of technological substitution, and how compa-nies can influence long-term substitution outcomes through their strategic actions. For incumbent firm managers, such a recognition could lead to more effective responses to substitution threats, in part by causing them to consider a broader range of strategic options than what might have otherwise been the case.

Finally, in future research concerning strategic responses to technological threats, more direct attention should be given to strategies for influencing the way in which the new product comes to be defined and marketed. Basic stratagems for doing so (along with related issues concerning the time of entry and interorganizational actions) were outlined above. However, much needs to be learned regarding how incumbent firms can shape the framework of competition that emerges in ways that favor their economic interests, the factors that strengthen or weaken their ability to do so, and the difficulties that can be encountered in the process. The purpose of this chapter has been to explore central issues concerning the management of technology in the substi-tution context. To the extent that it helps stimulate the thinking of researchers and practitioners in a meaningful way, a contribution will have been made.

22.5 REFERENCES

1. D. F. Abell, *Defining the Business: The Starting Point of Strategic Planning,* Prentice-Hall, Englewood Cliffs, N.J., 1980.

2. W. J. Abernathy, *The Productivity Dilemma: Roadblock to Innovation in the Automobile Industry,* Johns Hopkins University, Baltimore, 1978.

3. W. J. Abernathy and J. M. Utterback, "Patterns of Industrial Innovation," *Technol. Review,* **80**(7): 41–47, 1978.

4. W. J. Abernathy and K. B. Clark, "Innovation: Mapping the Winds of Creative Destruction," *Research Policy,* **14**:3–22, 1985.

5. P. Anderson and M. L. Tushman, "Technological Discontinuities and Dominant Designs: A Cyclical Model of Technological Change," *Administrative Science Quarterly,* **35**:604–633, 1990.

6. R. E. Biggadike, *Corporate Diversification: Entry, Strategy, and Performance,* Harvard University Press, Cambridge, Mass., 1979.

7. K. B. Benson and D. G. Fink, *HDTV: Advanced Television for the 1990s,* McGraw-Hill, New York, 1991.

8. K. B. Clark, "The Interaction of Design Hierarchies and Market Concepts in Technological Evolution," *Research Policy,* **14**:235–251, 1985.

9. R. S. Cowan, "How the Refrigerator Got its Hum," in *The Social Shaping of Technology,* D. MacKenzie and J. Wajcman, eds., Open University Press, Philadelphia, 1985.

10. W. M. Cohen and D. A. Levinthal, "Absorptive Capacity: A New Perspective on Learning and Innovation," *Administrative Science Quarterly,* **35**:128–152, 1990.

11. K. R. Conner, *Strategic Implications of High-Technology Competition in a Network Externality Environment,* unpublished working paper, University of Pennsylvania, Philadelphia, 1991.

12. A. C. Cooper and D. Schendel, "Strategic Responses to Technological Threats," *Business Horizons,* **19:** 61–69, 1976.

13. A. C. Cooper and C. G. Smith, "How Established Firms Respond to Threatening Technologies," *Acad. of Management Executive,* **6**(2): 55–70, 1992.

14. Creative Strategies International, *Electronic Calculators,* Creative Strategies, Inc., San Jose, Calif., 1978.

15. Creative Strategies International, *Medical Diagnostic Imaging,* Creative Strategies, Inc., San Jose, Calif., 1978.

16. I. Dierickx and K. Cool, "Asset Stock Accumulation and Sustainability of Competitive Advantage," *Management Science,* **35**(12): 1504–1511, 1989.

17. G. Dosi, "Technological Paradigms and Technological Trajectories," *Research Policy,* **11**:147–162, 1982.

18. R. N. Foster, *Innovation: The Attacker's Advantage,* Summit Books, New York, 1986.

19. P. Freiberger and M. Swaine, *Fire in the Valley: The Making of the Personal Computer,* Osborne/McGraw-Hill, Berkeley, Calif., 1984.

20. G. Hamel and C. K. Prahalad, "Strategic Intent," *Harvard Business Review,* **67**(3): 63–76, 1989.

21. S. Hariharan and C. K. Prahalad, "Technological Compatibility Choices in High-Tech Products: Implications for Corporate Strategy," in *Proceedings of Managing the High Technology Firm,* L. R. Gomez-Mejia and M. W. Lawless, eds., The Graduate School of Business, University of Colorado, Boulder, 1988.

22. S. Hariharan and C. K. Prahalad, "Strategic Windows in the Structuring of Industries: Compatibility Standards and Industry Evolution," in *Building the Strategically-Responsive Organization,* H. Thomas, D. O'Neal, R. White, and D. Hurst, eds., Wiley, Chichester, U.K., 1994, pp. 289–308.

23. J. A. Howard and W. L. Moore, "Changes in Consumer Behavior over the Product Life Cycle," in *Readings in the Management of Innovation,* M. L. Tushman and W. L. Moore, eds., Ballinger, Cambridge, Mass., 1982.

24. D. S. Landes, *Revolution in Time,* Harvard University Press, Cambridge, Mass., 1983.

25. T. Levitt, "Marketing Myopia," *Harvard Business Review,* **38**(4): 26–37, 1960.

26. M. B. Lieberman and D. B. Montgomery, "First-Mover Advantages," *Strategic Management J.,* **9** (Special Issue): 41–58, 1988.

27. B. A. Majumdar, *Innovations, Product Developments and Technology Transfers: An Empirical Study of Dynamic Competitive Advantage: The Case of Electronic Calculators,* unpublished Ph.D. dissertation, Case Western Reserve University, Cleveland, Ohio, 1977.

28. R. G. McGrath, I. C. MacMillan, and M. L. Tushman, "The Role of Executive Team Actions in Shaping Dominant Designs: Towards the Strategic Shaping of Technological Progress," *Strategic Management J.,* **13** (Special Issue): 137–161, Winter 1992.

29. W. Mitchell, "Whether and When? Probability and Timing of Incumbents' Entry into Emerging Industrial Subfields," *Administrative Science Quarterly,* **34**:208–230, 1989.

30. W. L. Moore and M. L. Tushman, "Managing Innovation over the Product Life Cycle," in *Readings in the Management of Innovation,* M. L. Tushman and W. L. Moore, eds., Ballinger, Cambridge, Mass., 1982.

31. C. L. Nicholls-Nixon and C. Y. Woo, "Technological Responsiveness: Toward an Explanatory Model," paper presented at the annual meeting of the Academy of Management, Atlanta, 1993.

32. F.-J. Olleros, "Emerging Industries and the Burnout of Pioneers," *J. Product Innovation Management,* **1**:5–18, 1986.

33. G. P. Pisano, "The R&D Boundaries of the Firm: An Empirical Analysis," *Administrative Science Quarterly,* **35**:153–176, 1990.

34. M. E. Porter, *Competitive Strategy,* Free Press, New York, 1980.

35. M. E. Porter, *Competitive Advantage,* Free Press, New York, 1985.

36. T. S. Robertson, *Innovation and the Consumer,* Holt, Rinehart, Winston, New York, 1976.

37. E. M. Rogers, *Diffusion of Innovations,* Free Press, New York, 1962.

38. N. Rosenberg, *Perspectives on Technology,* Cambridge University Press, Cambridge, U.K., 1976.

39. D. Sahal, *Patterns of Technological Innovation,* Addison-Wesley, Reading, Mass., 1981.

40. C. G. Smith, *Established Companies Diversifying into Young Industries: A Comparison of Firms with Different Levels of Performance,* unpublished Ph.D. dissertation, Purdue University, West Lafayette, Ind., 1985.

41. C. G. Smith and A. C. Cooper, "Established Companies Diversifying Into Young Industries: A Comparison of Firms with Different Levels of Performance," *Strategic Management J.,* **9:**111–121, 1988.

42. C. G. Smith, "Responding to Substitution Threats: A Framework for Assessment," *J. Engineering Technol. Management,* **7**(1): 17–36, 1990.

43. C. G. Smith, "Understanding Technological Substitution: Generic Types, Substitution Dynamics, and Influence Strategies," *J. Engineering Technol. Management,* **9**(3–4): 279–302, 1992.

44. W. R. Soukup, *Strategic Response to Technological Threat in the Electronics Components Industry,* unpublished Ph.D. dissertation, Purdue University, West Lafayette, Ind., 1979.

45. D. J. Teece, "Profiting from Technological Innovation: Implications for Integration, Collaboration, Licensing and Public Policy," in *The Competitive Challenge: Strategies for Industrial Innovation and Renewal,* D. J. Teece, ed., Ballinger, Cambridge, Mass., 1987.

46. J. E. Tilton, *International Diffusion of Technology: The Case of Semiconductors,* Brookings Institute, Washington, D.C., 1971.

47. M. L. Tushman and P. Anderson, "Technological Discontinuities and Organizational Environments," *Administrative Science Quarterly,* **31:**439–465, 1986.

48. J. M. Utterback and W. J. Abernathy, "A Dynamic Model of Process and Product Innovation," *OMEGA,* **3**(6): 639–656, 1975.

49. J. M. Utterback and L. Kim, "Invasion of a Stable Business by Radical Innovation," in *The Management of Productivity and Technology in Manufacturing,* P. R. Kleindorfer, ed., Plenum Press, New York, 1985.

50. G. E. Willard, *A Comparison of Survivors and Non-survivors under Conditions of Large-scale Withdrawal in the U.S. Color Television Set Industry,* unpublished Ph.D. dissertation, Purdue University, West Lafayette, Ind., 1982.

51. G. S. Yip, "Gateways to Entry," *Harvard Business Review,* **60**(5): 85–92, 1982.

CHAPTER 23

A FRAMEWORK FOR PRODUCT MODEL AND FAMILY COMPETITION

Vic Uzumeri
Department of Management
Auburn University
Auburn, Alabama

Susan Walsh Sanderson
School of Management
Rensselaer Polytechnic Institute
Troy, New York

Our framework for competition among product models and families builds on two key aspects of management theory: (1) recognition of the importance of variety and (2) reliance on analogies to the biological life cycle. The closely related concepts of product differentiation and market segmentation, for example, recognize that product variety drives key manufacturing decisions:[1,2] in design, related to developing innovative products distinguished by features and aesthetics that better serve existing and attract new customers; in production, related to choices and applications of technologies that ensure quality manufacture and timely availability of these potentially diverse products; and in sales, related to matching products and product lines to the varied needs of individual customers.

Analogies to the biological life cycle recognize that these products and product lines are subjected to forces that occasion change that corresponds to the birth, growth, maturity, and death of living organisms.[3,4] Our framework hence supports the exploration of product competition in terms of variety and change and the *interactions* that characterize competition among products with different life cycles.

In the following sections we define units of analysis employed in the framework and analyze the key descriptive measures—product variety and rate of design change at two levels of analysis—for models and for product families. Empirical examples illustrate the application of these measures in a number of product categories. Finally, we identify the forces that drive product variety and rate of change and integrate these into a life-cycle model of product competition.

23.1 UNITS OF ANALYSIS

Models and *product families,* long and widely accepted in practice as basic units of analysis, have recently begun to receive attention in academic circles. Researchers have variously advanced examples of "design families" and "design variants" among automobiles, jet engines, and aircraft and hovercraft;[5] demonstrated effective management of incredible varieties of product models that target customer needs in distinct international markets;[6] suggested as a structural rationale a "design hierarchy" that includes a "core concept" and solutions to various "subproblems";[7] and developed an integrative typology for design projects that includes "platform" as well as enhancement, hybrid, and derivative projects.[8]

Model distinctions and family relationships are fairly easily established for sophisticated technical products, particularly assembled systems. Few would quarrel, for example, with Boeing's designation of the 747-200, 747-300, 747-400, and 747-SP as models that constitute a family.[9] Models across passenger-car product families tend to be consistent, usually including two- and four-door sedans, two-door coupes, three- and five-door hatchbacks, and perhaps a wagon ranging over the full spectrum of size and price.

But as one moves away from sophisticated assemblies, reliance on technical criteria increasingly gives way to subjective perceptions of subtler characteristics. At some point, the concept of family structure begins to have less to do with technical differences than with market preferences and industry conventions. For products such as handtools, furniture, and household linens, a given design can be assigned to a product family as much for the way it is made or used as for the technology it contains.

Boundaries are particularly fuzzy for product families that are based on recipes. House paint, for example, is sold in a handful of grades, each tinted at the factory to produce a few basic colors and a few dozen pigments that can be blended by retailers to produce an array of hues and tints. What constitutes a model—each basic color or every mixture a consumer takes home? Is each grade of paint a model or a product family?

Some contend that such subjectivity imperils any attempt at classification or categorization,[10] but others argue that the threat is overdrawn, that knowledgeable observers typically make consistent interpretations.[11] For most commercially manufactured products, we believe that the latter tends to be the case. Consensus among industry participants about appropriate model and product-family structures is often evident in the organization of suppliers' catalogs and industry buyers' guides.

We define a *model* to be a product design that differs sufficiently from other designs that the manufacturer assigns it a distinctive commercial designation and a *product family* to be a set of models that a given manufacturer makes and considers to be related (see Fig. 23.1). These definitions are subject to two limitations: (1) they rely on data provided by manufacturers that may be tempted to fudge these definitions to gain marketing advantage and (2) absent independent, corroborating technical information, manufacturer-provided definitions of models and product families cannot be used to investigate the *nature* of product-family and model differences and similarities.

Our objective is to develop a framework that describes broad patterns of innovation and competition; thus it is more important that units of analysis be easily applied than that they be fully understood. Lack of understanding can be shored up by supplementary industry investigations, but measures that are difficult to apply may make any meaningful investigation a logistic impossibility.

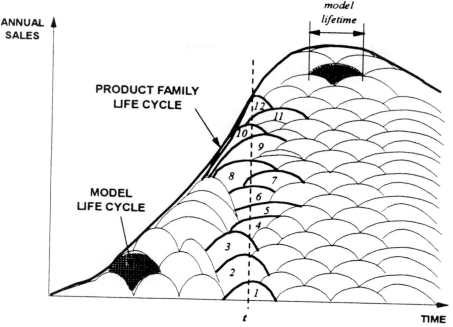

FIGURE 23.1 The product-family life cycle.

23.2 LEVELS OF ANALYSIS

Having established as our units of analysis product models and families, we analyze each in terms of the two key descriptors identified earlier: variety and rate of change. We examine first rate of design change for product models as a driver of model variety; the latter is the *de facto definition of product family*. In the abstract, the notion of variety makes no reference to volume of units sold. A model, whether it sells one unit or a million, counts equally in the estimation of variety. As it is a *number* measured at a particular time, model variety is highly likely to change over time and cannot be estimated unless the boundary of the product family is known, which reinforces the need to attend to industry criteria for family membership.

23.2.1 Model Variety and Rate of Change

A *model change*, simply construed, is an observed difference in a product examined at different times. Model change can be of two types: (1) *additive* change, in which an existing model is supplemented by one or more derivative models with a concomitant increase in model variety, and (2) *replacement* change, which implies discontinuation of the model that serves as the basis for a derivative and hence no increase in model variety.[12]

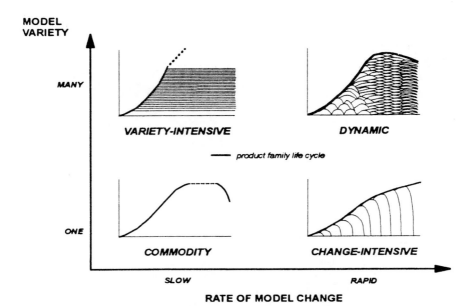

FIGURE 23.2   Patterns of model evolution.

Associated with every product model is a lifetime that is known ex post facto and, to the extent that a manufacturer can control its product planning, may be predictable ex ante facto. The reciprocal of a model's lifetime is an estimate of the rate at which it is replaced; a model with a 2-year lifetime, for example, will be replaced at a rate of 0.5 model per year.

Estimates of model variety and rate of model change contain a great deal of information that can be used to classify patterns of competitive activity. Juxtaposing model variety and rate of model change, as in Fig. 23.2, creates four distinctive patterns that reflect a firm's *competitive reaction* to external forces, particularly the introduction of new technologies and the actions of competitors. These patterns can be viewed as the "signatures" or "fingerprints" of the competitive forces acting on the firm.

The pattern of model evolution for *commodities,* eggs, carbon black, soda ash, chemical feedstocks, and other products that do not, or cannot, exhibit significant design variety is depicted in the lower left quadrant of Fig. 23.2. Design variants for these products are not offered, either, because no one can envision an alternative or because customers are unwilling to accept a change or substitute. That most early published descriptions of product life cycle resemble this pattern reflects contemporaneous mass production of, and slow rates of model replacement for, most products.

Products such as handtools, lightbulbs, and door hardware exhibit a *variety-intensive* pattern of model evolution (upper left quadrant in Fig. 23.2) characterized by additive model change. Stanley Works' provides an example. The company's handtools are offered in thousands of models and tend to change very slowly, a consequence, as one Stanley manager put it, of their market lives being "measured in centuries."[13] Packaging changes are frequent, new models much less so; fewer than 3 percent of the more than 1500 products listed in the 1990 *Stanley Hand Tool Catalog* were identified as new and most of those models reflected only cosmetic changes. Yet even slow rates of change can yield extraordinary model variety in a firm that remains in business for 150 years.

Products such as software that tend to be characterized by replacement change exhibit a *change-intensive* pattern of model evolution (lower right quadrant of Fig. 23.2). Model variety is rare because firms put all their effort into maintaining the pace of competitive redesign. This gives rise to serial model changes that amount to a race; a firm must be developing a follow-on model even as it introduces the current one. Microsoft's DOS, for example, has since 1981 been marketed in versions 1.1, 2.0, 2.1, 2.2, 3.0, 3.1, 4.0, 4.1, 5.0, and 6.0, each of which effectively replaced its predecessor as far as new sales were concerned.

Dynamic random access memory (DRAM) chips provide a more complex example. For more than a decade, DRAMs exhibited a persistent pattern of serial product and process innovation, with major new models appearing approximately every 3 years. To keep up this pace, DRAM manufacturers have had to introduce new models within a short time of the market leader, with the result that rates of innovation have necessarily been quite similar. No firm can pull very far ahead of the others in the industry, and any that fall behind must exit.

That DRAM generations until very recently exhibited little model variety is a function of the nature of DRAM manufacturing. Manufacturers rely on learning-curve effects to achieve the finer tolerances, higher yields, and increased profitability needed to reach the next density level.[14] Diverting factory production from high-volume learning-intensive products to marginally different designs that contribute little knowledge useful for process improvement can impair a firm's ability to achieve that next plateau.

DRAM design variety increased dramatically in 1989;[15] the reasons for the shift were primarily technological.[16] Previously, each generation of DRAMs had outdone its predecessor across all relevant dimensions of performance. This became harder with the much denser chips that began to emerge, attended by tradeoffs between, for example, information density and access time and voltage. The consequence was a proliferation of chip designs.

The DRAM example illustrates an important issue of interpretation. Most industry observers consider the radically increased density of DRAMs to represent not a model change, but the emergence of an entirely new product family. Semiconductor engineers, on the other hand, consider the technological changes in DRAMs to be fundamentally incremental (albeit expensive).

These contradicting viewpoints underscore the fact that change is harder to measure than variety. What determines whether a model change is so significant as to result in a new product family? Any answer to this question must be qualified by the facts of the industrial environment. That all the major chip makers have managed to keep pace with the successive generations of chip density we believe tips the balance in favor of viewing DRAM innovation as an example of change-intensive model competition. Reasonable people could credibly differ with our interpretation.

A complex and volatile mix of additive and replacement change yields a *dynamic* pattern of model evolution (upper right quadrant of Fig. 23.2). Design changes can increase variety, reduce variety, or leave it unchanged. In fact, model changes are both the source and the enemy of model variety. A growing number of globally competitive manufactured products—computers, machine tools, automobiles, specialized semiconductors, and medical equipment, among others—exhibit this pattern.

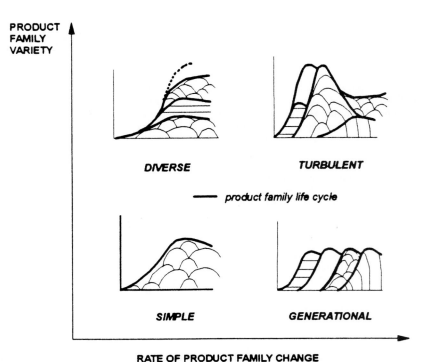

FIGURE 23.3   Patterns of product-family evolution.

23.2.2   Product-Family Variety and Rate of Change

Product-family as well as model life cycles can be plotted in terms of variety and rate of change (see Fig. 23.3).

For commodity products, model and product-family life cycles are essentially the same, but for most other products patterns of product-family evolution are independent of, and can be quite different from, patterns of model evolution.

Opportunities to build multiple product families simultaneously give rise to a *diverse* pattern of product-family evolution (upper left quadrant of Fig. 23.3).

Product families that exhibit this pattern can coexist and prosper, linked by a common technology or marketing insight that encourages observers to view them as a group. Stanley Works again provides a helpful example. Its handtool product family represents only a fraction of Stanley Works' overall product variety. During the 1980s, the company added thousands of new models as it expanded beyond the maturing carpentry handtool market, adding product families for industrial handtools, auto mechanic's tools, hydraulic and pneumatic power tools, glue guns, power fastening devices and fasteners, toolboxes, and a number of other hardware-related products (see Table 23.1).

Product families that replace one another in rapid, serial fashion exhibit a *generational* pattern of evolution (lower right quadrant of Fig. 23.3). Of all the patterns in model and product-family competition, this is the most confusing; the difficulty is in distinguishing generational replacement from change-intensive model competition.

TABLE 23.1 Evolution of Stanley Handtool Product Lines

Operating unit	Year(s) acquired	No. of SKUs*	New models	1989 sales†	Representative products
Stanley tools	1843	1,500+	50–75/year	$560M	Woodworking and construction handtools
Fastening systems: Bostitch, Hartco, Sutton-Landis, Halstead, Spenax, Parker Tools	1986–1989	4,000+	10–20/year	$320M	Fasteners and fastening systems
Mechanic's tools: MAC Tools, Proto Tools, Peugeot Tools, National Hand Tool and Beach Industries	1980–1987	11,000+	300–500/year	$400M	Auto mechanic's tools, industrial handtools, and toolboxes

*SKU = Stockkeeping unit.
†estimated, Shearson Lehman Hutton, March 8, 1989.

The logic advanced thus far suggests three criteria for generational product competition: (1) powerful and persistent market demand for continuous improvement (without which major change will not occur), (2) more than one technological way to satisfy the market need (given only one technological approach, change-intensive model evolution will dominate), and (3) strong market resistance to the simultaneous existence of more than one product family (without such resistance, diverse or turbulent product-family competition will result).

Products that satisfy all three criteria are difficult to find. The DRAM industry illustrates the potential for confusion, DRAMs almost meeting the criteria, except that the requisite critical improvements in processing technology are considered by many engineers to be incremental rather than radical, a violation of the second criterion. The shift from records to audiotape cassettes to compact discs and, finally, to digital tape is perhaps the best example of generational product-family change, but even here, product-family generations overlap. Are the different media in fact serving distinct niches, with analog audiotape the inexpensive solution, compact discs the choice for durability, and digital audiotape the new standard for fidelity and convenience?

Total turmoil in product-family competition yields a *turbulent* pattern of product-family evolution (upper right quadrant of Fig. 23.3). Both generational and turbulent product-family evolution present major challenges to manufacturers, inasmuch as each new product family represents a threat to or opportunity for every competitor. Each must manage its own complex pattern of product-family evolution, while contending with competition from the similarly volatile product families of its competitors. The generational and turbulent characterizations of product-family evolution agree with anecdotal reports of intensified product competition among globally marketed products. According to some observers, competitive pressures are forcing the product-family life cycle to contract and thereby gravitate to the diverse, generational, and turbulent quadrants shown in Fig. 23.3.[17]

An example can be found in the office electronics marketplace, which has evidenced an increase in product-family variety over the past 20 years.[18] Equally striking is the decline in relative sales revenue for some product families, notably portable clocks and calculators. The proliferation of relevant product families, many based on fundamentally different technologies, is capable of generating complex competitive dynamics.[19] New families can complement or replace existing families, be introduced individually or in groups, die quickly, or remain viable for years. Given market uncertainties, firms are seldom able to predict the effect of a particular product-family change. Persistent incremental changes at the model level, although they may not determine the long-range outcomes of product-family competition, may play a major role in the short-term battle for market share and revenue that enables a firm to remain a participant in product-family competition.

23.3   THE FORCES THAT DRIVE VARIETY AND RATE OF CHANGE

The experiences of Stanley Works and the DRAM industry—patterns of model proliferation that differ dramatically and markets that range from stagnant to those able to support enormous variety and a high rate of change—suggest that firms move around the variety-change frameworks in response to powerful forces that we do not yet fully understand. We distinguished four that seem to impinge on the interaction between rate of change and variety, (1) environmental forces associated with technological change, (2) environmental forces that originate in the marketplace, (3) internal cost constraints occasioned by the effects of the interaction of variety and rate of change on supplier organizations, and (4) forces related to product standardization and manufacturing and design flexibility.

23.3.1   Forces Associated with Technological Change

The notion that technology plays a major, perhaps even dominant, role in driving product change finds strong support in research. Anderson and Tushman's technology life cycle model, which is a respected summary of its impact, is shown in Fig. 23.4. The model, in which technology is presumed to pass through a series of life stages in strict order, offers powerful behavioral descriptions of the competitive activities that attend each stage of the cycle.

The cycle begins with a "technological discontinuity" that renders existing technologies and competencies obsolete. A discontinuity can be triggered by scientific advances or by profound technological insights. The discontinuity ruptures existing product development patterns and spawns a period of technological *ferment* during which competing firms explore different ways to package, deliver, and apply the new technology. In studies of three different products, Tushman and Anderson found that periods of ferment accounted for much of the measurable technical progress that occurred over the technology life cycle.[21]

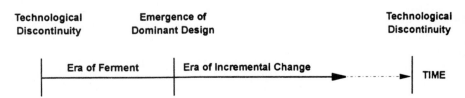

FIGURE 23.4   Anderson and Tushman's technology life-cycle model.[20]

As competing technological approaches clash, they are progressively eliminated until one or two survive to dominate the market. When the resulting *dominant design* emerges, it has a profound impact on the pattern of subsequent product competition. Competitive energies are directed away from the search for technological break-throughs and toward incremental refinement of the dominant design. This shift in technological activity produces an *era of incremental change,* with designers, both supported and constrained by the implicit standards of the dominant design, encouraged to make smaller, more frequent model changes that build on that design and on previous models.[22] Inasmuch as this era of incremental change may persist for an indefinite period, since it takes a new technological discontinuity to end it, Gomory suggests that as much as 85 percent of products are in an incremental stage at any given time.[23]

Technology may promote model variety, although this has been less thoroughly studied. A greater choice of technologies expands the design options available to product designers. Gasoline engines, for example, have been manufactured from steel, aluminum, ceramics, and high-temperature plastics; as each material finds a niche, it becomes a plausible foundation for a product family.

23.3.2 Market Forces

That the market pressures that drive model variety and rate of change are confusing is due to (1) the conceptual interdependence between model variety and rate of change and (2) the hidden nature of customer needs, which makes this interdependence difficult to unravel. Variety and rate of change both rely on the notion of difference; differences perceived at a particular time constitute variety, those that emerge over time, change. Confusion arises because the variants in a given variety must have originated as changes. This subtlety is not a serious concern when one examines model, product-family, or even technological changes, all of which are tied directly to tangible product designs.

In the marketplace, model and product-family variety and rate of change reflect the varying and changing nature of customer needs as well as customers' ability and willingness to buy new products or replace old models with new ones. Global markets bring new customers with differing technical, social, and economic agendas and growing diversity might reasonably be expected to generate pressures to increase model and product-family variety, as a number of authors are suggesting.[24,25] A slowdown of new offerings in the 1990s will be attributable to firms' efforts to cope with the dampening effects of a worldwide recession on customer demand for goods and services.

23.3.3 Cost Constraints and Other Internal Forces

Models today can become obsolete in as little as a few years or even a few months.[26] To cope, manufacturers have developed such techniques as "just-in-time manufacturing," "quick-response marketing," and "simultaneous engineering."[27] As product lives are compressed, planning horizons must contract; planned model variety must take account of the likelihood that many new models will be rendered obsolete shortly after being introduced. Timing of market entry has consequently become critical.

As this is the case, firms must allocate carefully these scarce resources—designers, budgets, and technology options—across the portfolio of planned new designs. We believe that this process forces firms into a tradeoff between model variety and rate of change. A firm pursuing a variety-intensive strategy will commit its design resources

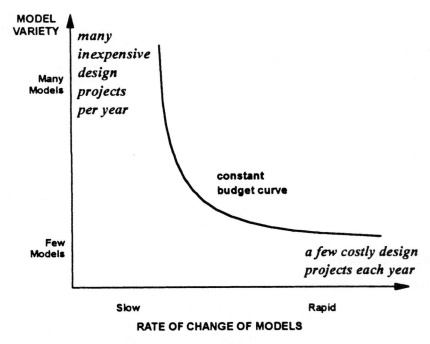

FIGURE 23.5   The tradeoff between model variety and rate of change.

to generating and maintaining variety; a firm pursuing a change-intensive strategy, to reinventing its models and product families. Figure 23.5 depicts the fundamental tradeoff implicit in the choice of strategy.

Competition for design resources suggests the existence of production frontiers within firms. These frontiers, the sets of all possible combinations of design projects that fully consume the firms' budgets, can be plotted in Fig. 23.5. Presumably, firms choose to operate at the point on the curve that maximizes their return on investment.

The precise shape and location of the production frontier remain a topic for future research. We depict the frontier as convex to the origin because (1) attempts to mix design projects of dissimilar size and differing objectives may occasion confusion and dysfunction and (2) on the basis of the limited empirical data we have gathered, competing firms seem to exhibit indirect evidence of such convexity; personal stereo and portable computer manufacturers are cases in point.

These production frontiers, although they may not obtain for individual firms, suggest that the boundary for the industry as a whole has moved away from the origin over the course of the study period, which is to say that variety and rate of change have intensified, an observation corroborated by anecdotal reports of the period.[28] In some industries, two competitors facing the tradeoff between model or product family (variety) and rate of change may choose sharply diverging paths, one electing to pursue a variety-intensive, the other a change-intensive, strategy and still compete effectively.

Abernathy and Clark's transilience map indirectly supports these speculations. Their model categorizes design changes in terms of their impact on a firm's existing technologies and markets.[29] Changes that rupture links with existing technologies or markets are viewed as more demanding than changes that build on those links. Changes that rupture links with *both* technologies and markets are deemed most onerous and risky. We previously generalized that change tends to be more strongly associated with technological forces, variety to be driven by market forces. A shift to a new technology pressures a firm to change. Abandonment of an old market for a new one leads a firm to discard its existing and generate new variety. Together, these forces might reasonably be expected to generate even greater pressures. One might expect that firms would elect to juggle variety and rate of change only if forced by changes in both technology and market.

Wheelwright and Clark's recent work on organizing for innovation suggests ways to create the flexibility needed to support the juggling of model variety and rate of change. Their framework, which would be particularly valuable for firms pursuing strategies that land them in the dynamic or turbulent quadrants shown in Figs. 23.1 and 23.2, suggests that design projects be divided into several categories, with non-product-specific research and advanced development projects laying the groundwork for the development of profitable products; radical breakthrough projects creating new types of products and supporting manufacturing processes (i.e., new product families); next-generation or platform projects generating major product design and manufacturing changes that provide a base for the refinement of a product or process family; enhancement, hybrid, or derivative projects that build on existing designs to create model variants; and alliance or partnered projects that enable a firm to tap into the technology and design resources of other firms and other product families.[30]

The variety-change framework can help firms identify the most suitable mix of project types. A high rate of model or product-family change will call for research and advanced development projects, and extensive model or product-family variety for platform and derivative projects. To compete in the middle of the variety-change framework would require an appropriate mix of project types.

We suspect that model and product-family variety and rate of change have organizational implications beyond the product design process. Firms operating in the various quadrants shown in Figs. 23.1 and 23.2 may be best served by varying manufacturing and marketing operations and differing organizational structures.

23.3.4 Standardization and Flexibility

Standardization and flexibility provide the means by which manufacturers and markets faced with the challenges of variety and rate of change can cope. Standardization enables manufacturers and customers to resist latent competitive pressures for variety and change; flexibility enables them to respond to those pressures without incurring unacceptable damage. Standardization is arguably the more powerful of the two mechanisms. It represents the only means by which industry participants can consciously influence the evolution of model and product-family variety and rate of change. To be fully effective, standardization requires that manufacturers and markets collaborate. Standardization employed by a manufacturer within its own product families will yield limited benefits; the full benefits of standardization accrue only when groups of manufacturers and customers agree to cooperate.

Firms may explicitly demand that a standard be set and that suppliers conform to it, as in the case of General Motors' MAP factory communications initiative.[31] But more often, customers create de facto standards, as they did in the home videorecorder industry.[32] If standard models work well or the costs associated with the failure of an untried approach are high, the market may be able to resist a departure from the standard for long periods (a case in point is the QWERTY keyboard).[33] When change eventually does occur, industry participants are likely to be surprised.

Standardization makes sense when market forces are exerting pressure for greater variety on manufacturers operating under constant design budget constraints. In its absence, market pressures will drive competing firms to increase their design budgets in order to provide greater model variety or change models more frequently than their competitors. There is evidence that Japanese automakers have already been pushed to the point of confusion and dysfunction, with new models priced out of reach of consumers.[34] Standardization offers relief from such pressure, if only temporarily.

Flexibility, on the other hand, enables a firm to push its production frontier further from the origin. At the same time, it is not monolithic. Slack has proposed a typology that divides manufacturing flexibilities into "range" flexibility, which enable firms to switch among existing model designs more quickly and cheaply, and "response" flexibilities that enable firms to introduce model changes more easily.[35] This distinction maps directly to the variety-change framework and suggests a fruitful avenue for future strategic research.

Wheelwright and Clark's framework reflects a strong regard for the power of both standardization and flexibility. In many ways, standardization lies at the heart of their prescriptions. Models based on "platforms" can be built from standard components and new models created easily by rearranging the components.[36] By maximizing such internal standardization, a supplier can respond with maximum flexibility.

23.4 THE PRODUCT COMPETITION LIFE CYCLE

Firms compete by offering product families that comprise models that vary and change across multiple levels. Figure 23.6 integrates the forces that operate on and

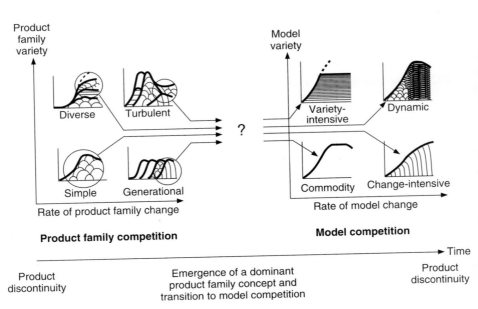

FIGURE 23.6   The product competition life cycle.

patterns of variety evident at each of these levels into a *product competition life-cycle model* that organizes the patterns of variety and rate of change depicted in Figs. 23.1 and 23.2.

The product competition life cycle is initiated by a discontinuity similar to a technological discontinuity, but not necessarily precipitated by technological change. Any idea sufficiently novel or important to give rise to a new product family can occasion the requisite discontinuity. When such an idea becomes industry knowledge, competing firms search for an integrating design concept to support the evolution of the resulting product family. A particularly volatile technology or market may necessitate the construction and abandonment of multiple design concepts before the right one is found. This period of *product-family competition* is analogous to the period of ferment in the technology life-cycle model proposed by Tushman and Anderson.

Innovations are not coordinated across suppliers during periods of product-family competition because each firm is intent on following its own development and marketing trajectory. Moreover, new products may meet resistance from manufacturers and customers that delays acceptance of one or more of the product families.[37] Rivalries can ensue at the model level, between families of similar models, and between families of dissimilar substitute products. When he traced the early history of the auto industry, Clark found considerable evidence of this type of product-family competition.[38] Inasmuch as young product families typically contain few models, this is also a period during which the distinction between product families and models is likely to be weakest. In fact, given that the earliest model may be identical to the early product family, the hallmark of this era is competition among product *concepts*.

Transition from product-family competition occurs when one or more product families achieves long-term stability. Often, a single product family will emerge as dominant. At other times, multiple product families may establish niches that permanently fragment the market and secure long-term futures for all.

As a product family achieves stability, industry forces align to support its position. Suppliers begin to offer materials, production equipment, and services tailored to the dominant design. If the product family captures market share, these suppliers enjoy increased sales volumes and production economies. Resulting cost reductions offer further encouragement to competitors to adopt the new dominant design. Growing availability and variety of attractively priced and configured models reinforce the dominance of the product family. In the absence of strong patent restrictions, most competing firms will adopt the dominant design and develop similar product families.[39]

Once the product family makes this transition, competition shifts to model rivalry which, although likely to generate less technical uncertainty, may be as intense as if it did. Because competing designs cannot deviate too much from the dominant product-family standard, firms must exploit small design differences to appeal to different sets of customers. Inasmuch as the same factors that make models easy to design make them easy to copy, firms compete by offering superior *patterns* or portfolios of model innovations.

Figure 23.6 depicts the transition from product-family competition to model competition with a question mark because any form of product-family competition can engender any pattern of model competition. This creates the potential for discontinuity in the middle of a product family's evolutionary development. It also makes it difficult to construct a single evolutionary model that explains more than a subset of all product evolutions.

The potential for uncertainty is demonstrated by the recent history of portable computers. Variety and rate of change in portable computer designs have been driven largely by the interaction of changes in component technologies and the influence of de facto user standards. Consumers generally have compared the performance of portable computers to that of their desktop counterparts. They have wanted desktop computer features in a package that is small, light, and rugged and can run for a long time on batteries.

From 1983 to 1986, manufacturers were hard-pressed to put any useful subset of desktop features in a transportable package, let alone one that ran on batteries. At the same time, major strides were being made in desktop computing standards, notably in microprocessor and display technologies. In 1986, component suppliers, operating at arm's length from computer assemblers, began to offer a stream of new-product models that enabled portable computers to inch closer to desktop performance. Meanwhile, the desktop computing standard stabilized on the Intel 80386 processor and VGA (video graphics array) graphics format.

As consumers decided which subset of desktop performance and features they were willing to accept, the market split into small niches. Because only they could afford to develop a new model as each new component emerged, large manufacturers such as Toshiba, NEC, Compaq, and Zenith dominated the industry from 1986 to 1990.

Then, in late 1990 and early 1991, the portable computer market underwent a massive change. A flood of small distributors entered the market, selling nearly identical "notebook" units sourced in Japan, Taiwan, and Korea. Firms with high variety suffered badly. Whereas their models were designed to supply compromise subsets of desired desktop features and levels of portability, the new notebook computers demanded little or no compromise. Widespread acceptance of notebook computers attracted dozens of new firms and triggered a dramatic drop in price. Toshiba, NEC, Compaq, and Zenith responded by cutting list prices by over 40 percent on all models, yet comparable units from smaller vendors continued to sell more than 40 percent below the new prices. Industry reports pegged Toshiba's losses at close to $50 million in the first half of 1991.[40]

By late 1991, nearly all manufacturers offered virtually identical notebook computers that weighed 6 to 7 lb, contained an Intel 80386SX/SL processor, and had a VGA display and 40- to 80-Mbyte hard disk. Such was the dominance of this single design concept that 46 Taiwanese manufacturers pooled funds to commission the design of a single model, dubbed "Project Teammate," that was shared among all contributing manufacturers.[41] This, more than any other event, illustrates the degree to which the portable computer had become a commodity.

The major manufacturers had begun to recapture market share by early 1993. These larger firms were subsequently able to inject new-product characteristics into the portable computer design, notably color displays, faster processors, longer battery life, smaller (<4-lb) subnotebooks, and pen-driven interfaces. The variety afforded manufacturers a renewed basis for appealing to radically different sets of customer needs while remaining firmly rooted in the notebook computer product family. Moreover, this newfound variety enabled Compaq, Toshiba, NEC, Zenith, IBM, and other large firms to reestablish control of the market's direction.

The portable computer example illustrates the speed and violence that can attend a transition in the product competition life cycle. The product and the industry went from turbulent product-family competition (1983–1990) to commodity model competition (1990–1992) to dynamic model competition (in 1993). These shifts, which occurred in the course of less than 3 years, were precipitated by external factors and blindsided some of the world's most savvy electronics manufacturers.[42] The story of portable computers stands as a cautionary tale to those who ignore the potential for rapid, fundamental shifts in product-family and model structures.

Intensified global competition is quickening the pace of migration along the framework depicted in Fig. 23.6. The interest that has been shown in Wheelwright and Clark's product development structure and the focus on management of core competencies in manufacturing firms suggest that firms are serious about wanting to reduce the uncertainty and costs associated with these transitions.[43] Also in evidence is strong industry interest in various forms of flexibility, cooperation, and standardization that might limit the violence of transitions.

Table 23.2 identifies 16 possible transition paths. All are potentially possible, if not equally probable. Technological and market constraints will limit some of the combinations. If only one product family and a small number of models are technically feasible, for example, a shift to dynamic model competition is unlikely. A product category may exhibit turbulent competition among competing product families, yet settle into commodity model competition a short time later. A simple, dominant product family may emerge quickly and painlessly, only to attract dynamic model competition that savages all the firms involved.

Firms' responses to these transitions will be limited by the cost constraints and organizational limitations of the manufacturers involved. A firm adept at managing competition among diverse product families may be ill-equipped to handle the sudden

TABLE 23.2 Transitions in Product Model and Family Competition

	Product-family competition	Model competition	Nature of transition
1	Simple	Commodity	Firm continues as before
2	Simple	Variety-intensive	Firm must learn to satisfy different customer demands that, at least, are not changing very quickly
3	Simple	Change-intensive	Firm must learn to change products quickly, but there are only a few to worry about
4	Simple	Dynamic	Firm must make extensive adjustments, very likely in the face of significant pressures for standardization from both customers and suppliers
5	Diverse	Commodity	Firm must switch focus from satisfying customer cravings for variety to reducing manufacturing cost
6	Diverse	Variety-intensive	Firm must carry over the same skill sets, but apply them to smaller design differences
7	Diverse	Change-intensive	Firm must totally change its behavior
8	Diverse	Dynamic	Firm must learn to innovate
9	Generational	Commodity	Firm must shift from managing rapid technological change to a ruthless focus on cost reduction
10	Generational	Variety-intensive	Firm must totally change its behavior
11	Generational	Change-intensive	Firm continues as before; changes become smaller and more frequent
12	Generational	Dynamic	Firm must learn to spread innovation expertise over more products
13	Turbulent	Commodity	Firm throws away all its flexibility in favor of ruthless cost reduction
14	Turbulent	Variety-intensive	Firm deemphasizes its innovation capability
15	Turbulent	Change-intensive	Firm focuses its innovation skills on fewer designs
16	Turbulent	Dynamic	Firm continues as before, but deals with smaller changes and differences

emergence of a single design approach that renders other models obsolete. Its accumulated skill at producing and marketing multiple models would be of little competitive value, and it would take time for the firm to build the production and marketing skills it would need to compete in a commodity product. This was precisely the experience of Toshiba and Compaq in 1991.

23.5 USING THE FRAMEWORK

The framework developed in this chapter adopts a multilevel description of product competition to help researchers and practitioners characterize patterns of product competition. This description utilizes model and product-family variety and rate of change to conceptually integrate the concepts of product innovation and choice, technological change, and diversity in customer perceptions. By explicitly representing these factors at two levels of analysis, the model is able to describe dynamic behavior not accounted for in previous frameworks and models.

We believe the framework makes two significant contributions to research on innovation and competition: (1) a sensitive and quantitatively applicable tool capable of *describing* patterns of variety and change reflected in product competition and (2) the framework's explicit identification of the *interactions* and *tradeoffs* between variety and rate of change, which offer new challenges for theoretical and empirical research into product innovation and competition.

The descriptive aspects of the framework can be used differently. Products for which variety and rate of change cannot be quantified can be subjectively classified by inspecting patterns of model evolution. If reliable industry definitions of models and product families are at hand, the framework can be used to quantitatively measure and map patterns of variety and rate of change. Variety-change patterns do not, of themselves, indicate competitive success or strategic causality; they do discriminate among the behaviors of competitive firms making the same product, even though the firms' approaches may appear quite similar by other criteria.

The examples presented in this chapter contribute to innovation theory by demonstrating that patterns of variety and rate of change, measured at the model and product-family levels, provide a means to differentiate among competing firms' product strategies and behaviors. We have seen that Toshiba pursued a variety-intensive similar strategy for its portable computer product family and outperformed other firms in the industry during the study period.

By eliciting differences in competitive behavior, the framework sheds light on a number of new and unresolved competitive issues. But although it makes a substantial contribution to the body of knowledge on innovation, it nevertheless leaves many questions unanswered. Perhaps the most important future research opportunity involves further identification and mapping of the technological and market forces that control the *dynamics of* and *interactions among* competing models and product families. These forces effectively staunched variety in DRAMs and continue to forestall changes in handtools. Given that failure to recognize and control these forces accounts for some of the world's most capable computer manufacturers being blindsided by the 1990 commoditization of portable computers, efforts to understand and manage them should reside high on the research agendas of management and innovation researchers.

The competitive dynamics of model competition represents a second important research opportunity. This form of product competition is of tremendous day-to-day commercial importance. Because it is more regular and structured, it may lend itself to greater generalization and theory building. Our framework and supporting examples argue that absence of major technological change does not imply that product competition is less challenging or less intense. If anything, product competition may involve greater competitive challenges as firms try to achieve a competitive advantage with smaller and smaller product differences.

A third research opportunity lies in the relationship among product variety, product change, and the internal structure and operation of firms. We have presented preliminary empirical data suggesting that extensive product variety and rapid product change are mutually exclusive. If confirmed by further research, it would follow that firms might gain a sustainable competitive advantage by specializing in one dimension or the other. If, as we suspect, specialization of this sort is difficult to reverse, firms may face critical strategic choices that have not previously been identified or articulated.

We believe that the product competition life cycle and variety-change framework can help researchers investigate these as well as many other issues. We suspect, for example, that the patterns of model and product-family competition described here may require very different organizational and managerial approaches. At the same time, a great deal remains to be done to perfect the framework and model, we are acutely aware that both constructs need to be defined with greater precision in order to be made more widely applicable. Such refinement can come only from a broader empirical testing that exposes latent contradictions and ambiguities.

With respect to the latter, we are particularly concerned that whereas our framework has only two, reality embodies dozens of levels of analysis. The model and product-family units of analysis gloss over many important competitive distinctions, including components, subassemblies, "platform" designs, brands, divisional product lines, business units, industry collaborations, and multinational cartels. Aspects of product competition play out in all these levels.

Finally, we hope that other researchers will improve on our definitions of model and product family. Ours are industry-based and pragmatic, mainly out of conviction, but also because we could not do any better. We hope that others will develop stronger definitions that will increase the applicability and relevance of both the variety-change framework and product competition life cycle.

23.6 REFERENCES AND NOTES

1. Peter Dickson and James Ginter, "Market Segmentation, Product Differentiation, and Marketing Strategy," *J. Marketing,* pp. 1–10, April 1987.

2. Peter Dickson, "Toward a General Theory of Competitive Rationality," *J. Marketing,* pp. 69–83, Jan. 1992.

3. Theodore Levitt, "Exploit the Product Life Cycle," *Harvard Business Review,* pp. 81–92, Nov.–Dec. 1965.

4. Nariman Dhalla and Sonia Yuspeh, "Forget the Product Life Cycle Concept," *Harvard Business Review,* pp. 102–112, Jan.–Feb. 1976.

5. Roy Rothwell and Paul Gardiner, "Re-innovation and Robust Designs: Producer and User Benefits," *J. Marketing Management,* 3(3): 372–387, 1988.

6. Susan Sanderson and Vic Uzumeri, "Industrial Design: The Leading Edge of Product Development for World Markets," *Design Management J.,* p. 28, Summer 1992.

7. K. B. Clark, "The Interaction of Design Hierarchies and Market Concepts in Technological Evolution," *Research Policy,* **14:** 235–251, 1985.

8. S. C. Wheelwright and K. B. Clark, "Creating Project Plans to Focus Product Development," *Harvard Business Review,* **70:** 70–82, March–April 1992.

9. Rothwell and Gardiner, op. cit. (Ref. 5).

10. V. A. Zeithaml, "Consumer Perceptions of Price, Quality and Value: A Means-End Model and Synthesis of Evidence," *J. Marketing,* **52:** 2–22, 1988.

11. L. G. Tornatsky and K. J. Klein, "Innovation Characteristics and Innovation Adoption-Implementation: A Meta-analysis of Findings," *IEEE Transact. Engineering Management,* **EM-29**(1): 28–45, 1982.

12. Additive and replacement change are best viewed as ends of a continuum, with most model change probably falling somewhere in between. It is to accommodate both types of change, that which increases model variety and that which does not, that we have elected to measure rate of change.

13. Tim Walsh, VP Marketing, Stanley-Proto Tools, personal communication, Norcross, Ga., Nov. 1989.

14. V. Ramakrishna and J. Harrigan, "Defect Learning Requirements," *Solid State Technol.,* **32:** 103–105, Jan. 1989.

15. These conclusions were derived from a study of variety and change in DRAMs from 1983 to 1991; model and product-family availability, variety, and rate of change were estimated from chips listed in an authoritative industry buyer's guide, *IC Master,* published annually by Hearst Business Publications, Long Island, N.Y.

16. Wilson, R. "DRAM Vendors Address Increasing Specialization," *Computer Design,* **29**:63[+], Dec. 1, 1990.

17. C.-F. Von Braun, "The Acceleration Trap," *Sloan Management Review,* pp. 49–58, Fall 1990 and "The Acceleration Trap in the Real World," *Sloan Management Review,* pp. 43–52, Summer 1991.

18. The sales data was obtained from the survey of electronic product sales that *Merchandising* and its successor magazine, *Dealerscope Merchandising,* has published annually throughout the study period.

19. L. Young, "Product Development in Japan: Evolution vs. Revolution," *Electronic Business,* pp. 75–77, June 17, 1991.

20. Philip Anderson and Michael Tushman, "Technological Discontinuities and Dominant Designs: A Cyclical Model of Technological Change," *Administrative Science Quarterly,* **35:** 31–60, 1990.

21. Michael Tushman and Philip Anderson, "Technological Discontinuities and Organizational Environments," *Administrative Science Quarterly,* **31:** 439–465, 1986.

22. Ibid.

23. Ralph Gomory, "From the 'Ladder of Science' to the Product Development Cycle," *Harvard Business Review,* Nov.–Dec. 1989.

24. George Stalk, Jr. and Thomas Hout, *Competing Against Time,* Free Press, New York, 1990.

25. R. McKenna, "Marketing in an Age of Diversity," *Harvard Business Review,* pp. 88–95, Sept.–Oct. 1988.

26. The Intel 80286 microprocessor was introduced in 1984 and was virtually eliminated from the market by 1991. The 80386 made an even faster exit.

27. Stalk and Hout, op. cit. (Ref. 24).

28. Peter Burrows, Gary McWilliams, and Paul M. Eng, "The Laptop Race: Can the U.S. Hold its Lead in Lap Four?" *Business Week,* p. 28, June 7, 1993.

29. Abernathy W., and K. B. Clark, "Innovation: Mapping the Winds of Creative Destruction," *Research Policy,* **14**:3–22, 1985.

30. Wheelwright and Clark, op. cit. (Ref. 8).

31. Douglas Williams, "MAP Reading," *Automotive Industries,* **167**: 103–104, Sept. 1987.

32. Michael A. Cusumano, Yiorgos Mylonadis, and Richard S. Rosenbloom, "Strategic Maneuvering and Mass-Market Dynamics: The Triumph of VHS over Beta," *Business History Review,* **66**: 51–94, Spring 1992.

33. Paul A. David, "Clio and the Economics of QWERTY," *Am. Economic Review,* **75**: 332–337, May 1985.

34. Japanese automakers' thrust in the 1980s toward greater variety and faster innovation is described in J. P. Womack, D. T. Jones, and D. Roos, *The Machine that Changed the World,* Rawson Associates, New York, 1990. Japanese automakers' second thoughts about that thrust were reported by a number of independent sources, including Karen L. Miller, "Overhaul in Japan," *Business Week,* pp. 80–86, Dec. 21, 1992; "Car Makers of the World: Unite," *Economist,* **324**(7770): 58–60, Aug. 1, 1992; Takeshi Sato and Shu Watanabe, "Carmakers Apply the Brakes," *Japan Times Weekly Internatl. Ed.,* **32**(24): 17, June 15–21, 1992.

35. N. D. C. Slack, "Focus on Flexibility," in R. Wild, ed., *International Handbook of Production and Operations Management,* Cassell, London, 1989, pp. 402–417.

36. Wheelwright and Clark, op. cit. (Ref. 8).

37. Gregory Basalla, *The Evolution of Technology,* Cambridge University Press, Cambridge, U.K., 1988.

38. K. B. Clark, "The Interaction of Design Hierarchies and Market Concepts in Technological Evolution," *Research Policy,* **14**: 235–251, 1985.

39. Anderson and Tushman, op. cit. (Ref. 20)

40. L. Armstrong and N. Gross, "It's a Shakier Perch for Toshiba's Laptops," *Business Week,* pp. 62 ff., Aug. 5, 1991.

41. Information about Project Teammate was obtained during conversations with individuals at three different U.S. portable computer resellers. Bitwise Computing of Troy, N.Y., which distributed the Teammate product, was particularly helpful.

42. Rick Whiting, "Notebook PC Makers Struggle to Survive at Sub-survival Prices," *Electronic Business,* **18**: 125 ff., May 18, 1992; B. G. Yovovich, "Surviving in the Treacherous Microcomputer Market," *Business Marketing,* **76**: 17, Aug. 1991; Brian Deagon, "Notebook Battles Ravaging Prices," *Electronic News,* **37**: 1 ff., Feb. 25, 1991.

43. C. K. Prahalad and G. Hamel, "The Core Competence of the Corporation," *Harvard Business Review,* **68**: 79–87, May–June 1990.

23.7 APPENDIX: MEASURING MODEL VARIETY AND RATE OF CHANGE

In historical terms, the product-family life cycle is fully defined if one knows the sales rates of the relevant models at all times during the study period. Unfortunately, future sales data are not known and historical sales data are often difficult to obtain at the model level. In this chapter we have used a simple method to arrive at an approximate description of the product-family life cycle. Our approach relies on the expected market life L_i of each model i, and the variety v_t of all the models that are available at time t. This analysis begins by assembling the history of the availability of relevant models a_{it}. From the history of model availability, the variety at a given time v_t is calculated for a given universe of n models that are known to exist during the study period:

$$V_t = \sum_{i=1}^{n} w_{it} a_{it}$$

where $a_{it} = 1$ if model i is sold at time t
$= 0$ otherwise

and w_{it} is an optional weight applied to specific models at specific times. If availability is known, the market survival life L_i for each model can be calculated:

$$L_i = \sum_{j=1}^{T} a_{ij} \Delta t$$

where Δ_t is the interval between successive measurements of availability and T is the total number of time intervals in the study period.

If model availability is estimated at intervals, new models may enter or leave the market between the times that availability is measured. If it is assumed that model arrivals and departures are uniformly distributed, the expected measurement error will be at both the entry time and the exit time of the model. Adding the two expected errors produces an extra Δ_t. Accordingly, the estimate of L_i adds an extra Δ_t to the life of the model by counting both the first and last nonzero values of a_{it}. With this estimate of average life, the average rate of design change for a given set of designs R_t can be estimated as follows:

$$R_t = \sum_{i=1}^{n} \frac{a_{it} w_i}{L_i}$$

While the measures are simple and useful, there are a number of practical concerns that arise when planning their acquisition and using their results:

1. Model availability can be measured in at least two ways. A manufacturer may think of model availability as the time when the model is in production. A customer may judge it to be available as long as it is offered for sale. These judgments often disagree. A manufacturer may produce a model prior to launching it in the marketplace in order to build inventory for the launch. Alternatively, the distribution and retailing system may warehouse unsold models for months or even years after the manufacturer has ceased production. Although we cannot argue one view over the other on any theoretical grounds, we have adopted the customer's view of model availability in the belief that outcomes of strategic significance are more likely to occur while the model is available to customers than during the period of actual production.

2. The analyst must make sure that n includes all relevant designs. The analyst has considerable freedom in defining the boundary of the product family, but must state those limits and capture all of the models or product families that fall within it.

3. The analyst must make sure that the criteria for distinguishing among the i different models are clearly stated.

The analyst must justify use of weighting factors. If the relative importance of specific models can be quantified, a weight w_{it} can be used to adjust the measures of availability and rate of change to reflect the influence of the important models. In this chapter, we have implicitly taken manufacturers' labeling of models to be accurate and have assumed that all models are of equal importance (i.e., $w_{it} = 1$). Future researchers may be able to achieve more precise distinctions. Possible weighting criteria might include experts' assessments of technological novelty, sales volumes, and contributions to profitability over time. Estimates of model lifetimes will almost certainly encounter truncated time-series data sets. The availability of individual models may be zero because the study period has ended rather than because the model has died a natural death. If the study is primarily historical in nature, the problem can be avoided by a conservative choice of study period. If the model data is current, the analyst can correct for the inherent error by conducting a survival life analysis to estimate the distribution of the remaining lifetimes for the truncated models. We examined the data using both approaches in order to ensure comfort with our characterizations and conclusions. Despite the problems and limitations, the method produces quantitative estimates of variety and rate of change that show promise in distinguishing among different products, different firms, as well as other levels of analysis.

CHAPTER 24

THE PROCESS OF MANAGING PRODUCT DEFINITIONS IN SOFTWARE PRODUCT DEVELOPMENT

Sara L. Beckman
Haas School of Business
University of California, Berkeley
Berkeley, California

24.1 INTRODUCTION

Software is one of the most rapidly growing product segments worldwide. The 1981 worldwide market for software (including software products, software services, and contract programming) was estimated at $13 billion.[1] By 1993, this number had grown to an estimated $130 billion,[2] and 15 to 20 percent annual growth is projected for the foreseeable future. Not included is the software that is rapidly becoming a critical component of the many everyday products sold in today's market. Cars now contain dozens of integrated circuits, all programmed to monitor different aspects of vehicle performance. Small home appliances incorporate intelligence, while remaining inexpensive consumer products. Companies traditionally described as hardware developers have slowly added software engineers to their staffs, shifting the balance of engineering talent away from electrical and mechanical engineering to software development.[3] These trends portend a major shift in approach to managing new-product development projects.

Much of the focus of the literature on new-product development to date has been on hardware or systems products. Little attention has been paid to software development,[4] much less to the integration of software and hardware components into systems products. While there is certainly learning to be transferred from the traditional hardware product development world to the management of software development projects, there is learning to be done in the other direction as well. Our research has focused on the personal computer and workstation packaged software industries in the United States and Japan and on understanding the processes employed in these organizations to define the products they develop.[5]

Product definition creation and management are critical aspects of the new-product development process. A well-developed product definition reflects understanding of customer and user needs, the competitive environment, technology availability, and regulatory and standards issues. A robust product definition is one that stands up to challenge and guides the product development process, allowing the development team to make tradeoffs and decisions quickly and effectively. Customer and user needs assessment is a particularly critical aspect of product definition, as is clear, widespread communication of product definition elements throughout the product development team. Change to a product definition during product development must be handled consciously and clearly, accommodating critical changes and postponing others.[6]

In this chapter, we present three case studies on product definition for products in a single software category, label-making software.[7] The three product lines we discuss cover over 90 percent of the category, thus representing a relatively complete picture of development for that market segment.

- Middle, Inc. entered the product category in early 1991 with its flagship product, AddressNow 1.0. Considered a success in the market today, Middle has recently introduced AddressNow 3.0, and considers the product an integral part of the overall company strategy.

- HiTech, Inc. focuses more on the technology aspects of label-making software and relies on its LabelIt! product line for 90 percent of the company's revenues. The products we address here include LabelIt! 2.0 and the subsequent CD-ROM version of the product incorporating clip art and on-line documentation in addition to the LabelIt! 2.0 software.

- Leader, Inc. is the market leader in the label-making software category. Here we address their most recent version of label-making software, SendQuick 4.0. As with Middle, Leader considers their label-making software an integral part of their product portfolio and emphasizes interoperability of the label-making software with other related packages in their product line.

These three companies compete fiercely with one another in what has become a commodity marketplace in recent years. They all watch one another closely, are quite familiar with one another's operations, and think explicitly about how to position themselves against one another. Much of the interaction between them will be seen in the stories told below.

The following three sections of this chapter lay out the stories of Middle, HiTech, and Leader as they engaged in the process of defining their products (AddressNow, LabelIt!, and SendQuick, respectively). Each section commences with a brief description of the timeline for defining, developing, and delivering the products. The key milestones shown are not unusual for product development projects in this industry, although there are interesting differences among the three companies. Following presentation of the timelines are descriptions of the organizational structures employed in the five project teams. Again, although there are commonalities among the projects in their use of matrix management, cross-functional team structures, there are also interesting differences. Finally, each section describes the mechanisms used by the teams to collect and analyze customer and user needs, competitive product positioning, and technology availability as they prepared their product definitions.

Typical product definitions in this segment of the software industry consist of a *marketing requirements document* (MRD) and an accompanying detailed product specification from which the development engineers (or developers) write the software code. Often these documents are summarized in a simple prioritized feature list that shows the specific features to be included in the present version of the product and is used to guide team activities and decisions. The generation and management of these documents is the focus of the product definition process.

24.2 MIDDLE, INC. AND THE ADDRESSNOW PRODUCT LINE

Middle, Inc. entered the label-making software business in 1988 through acquisition of a small software company. Middle's lack of experience with both the label-making software market and the underlying label-making software technologies placed them on a steep learning curve as they attempted to assimilate the new product line and introduce their flagship product, AddressNow 1.0. After a lengthy 2-year development process (relative to today's standards), AddressNow 1.0 was introduced on an OS/2 platform, and then ported a year later to a Windows platform. Although the OS/2 product was not successful in the marketplace, much was learned from its introduction that made the Windows version quite successful. The flagship AddressNow 1.0 product was followed rapidly by development and introduction of AddressNow 2.0 in February 1992. Middle today boasts significant market share in the label-making software category and has recently introduced AddressNow 3.0.

Figure 24.1 depicts a detailed timeline for all the products described in this chapter. The events shown on the timeline for Middle's products are quite similar to the events shown for HiTech and Leader's products, and to those for many of the projects we studied.

- *Launch.* Most product managers we interviewed defined the project launch date as the time at which they and a few of their colleagues convened to begin formal discussion of the product definition. Often, however, some product definition, and sometimes product development, work preceded the designated launch date, as market requirements and software code carried over from previous versions of the product. Middle's two products had somewhat different launch paths as the first and second entries to the market. Although Middle's acquisition had introduced a DOS version of AddressNow 1.0 to the market, it was substantially different from the OS/2 and Windows versions of the product were expected to be, and product definition started from scratch. AddressNow 2.0, however, built on the market knowledge and code generated in the 1.0 version. Thus, product definition was begun informally as the team began collecting feedback from the marketplace about AddressNow 1.0, and developers were slowly transferred to the AddressNow 2.0 team as they completed their commitments to the 1.0 product. Envision a continuous stream of new products and a (relatively) intact team moving fluidly from one version to the next, and the notion of launch dates becomes less meaningful.

- *Agreement on features.* Referred to variously as *marketing requirements document* (MRD) completion, complete specification, and concept complete milestones, this step entails "completion" of the documents that describe the major features to be included in the product. Described by most as a "fluid document" that evolves long into the development process, the MRD spells out the major features to be included in the project when it goes to market. A prioritized feature list derived from the MRD focuses and guides the development effort. The AddressNow 1.0 team spent nearly 6 months developing their MRD, as they simultaneously experimented with and learned about the underlying product technologies. The AddressNow 2.0 team also spent about 6 months preparing their detailed feature list, but developed a

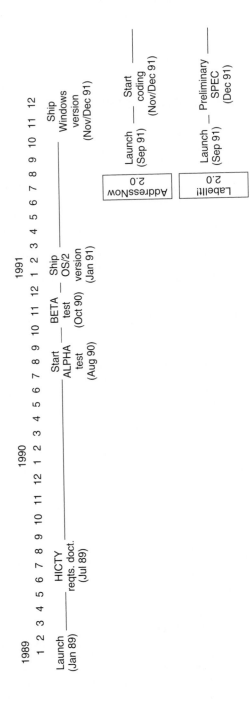

FIGURE 24.1 (a) Project timeline for Middle, Inc. (AddressNow).

24.4

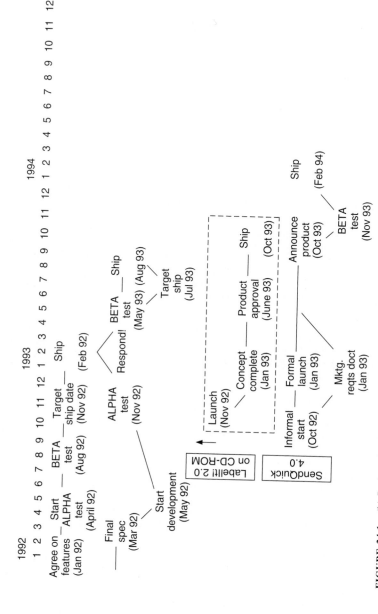

FIGURE 24.1 (b) Project timelines for HiTech, Inc. (LabelIt!) and Leader, Inc. (SendQuick).

much more detailed assessment of the costs of the features to be included and were writing code at the same time.

- *Alpha test.* Considered a major milestone in most organizations, the alpha test is typically performed when the product is "feature complete," meaning that all critical product features are fully coded and expected to work in preliminary tests, albeit with errors. Often alpha tests are conducted internal to the organization, although in the case of AddressNow 1.0, tests were performed with 12 customers in a formal design partner program begun nearly a year before product shipment. AddressNow 2.0 was alpha-tested internally, although a few external customers were invited to preview the product on Middle's computers.

- *Beta test.* Another major milestone in the process, beta tests require fully feature complete code that has no "crashing" bugs in it. Typically performed a month or two before shipment, beta tests are used to collect last minute feedback on critical bugs to be fixed, and to introduce the product early to important customers. Both versions of AddressNow were beta-tested with about 35 to 40 external customers. By version 2.0, Middle had developed a more formal program for identifying beta test candidates, allowing them to target their test activity at specific environments in which they wanted to see their product perform.

- *Shipment.* Once a determination is made that the product is ready to ship, a "golden master" is produced and the product is turned over to manufacturing for production about 2 weeks prior to the ship date. Those wondering about the ability of software companies to ship product on time will be interested to know that none of the five products described in this chapter shipped as scheduled. AddressNow 1.0 for Windows shipped 2 months later than planned, and AddressNow 2.0 shipped 3 months late. (Reasons for late shipment are provided in the sections on product definition change management.)

Note the compression of development schedules over the years. All three companies have learned to develop and deliver products more quickly, allowing them to introduce products on 12- to 18-month cycles which are relatively standard in the industry.

24.2.1 Project Organization

All the projects described in this chapter, and many in our larger study, employed cross-functional teams and matrix management structures. None of the projects was managed by a single individual; rather, project management was shared by representatives of the marketing and development organizations. A shared vision for project outcome was imposed by clear market share improvement and ship date schedule goals. Most project teams started with a small core group and added members throughout the development process as needed. Development engineers typically remained with the project throughout its duration, while other participants (e.g., documentation and help) would join the team full-time only when needed. Although consistent with these common themes, Middle, HiTech, and Leader each chose slight variations on this basic structure (shown in Fig. 24.2).

A very small group launched the AddressNow 1.0 product definition effort. One product manager (who joined Middle through the acquisition) and two developers (code writers) spent the first 6 months of the development cycle learning about OS/2[8] and the underlying product technology, and defining the product. As the product definition was formalized, the team grew and roles became more clearly defined.

Project management was shared between the product manager and the project man-

AddressNow 2.0

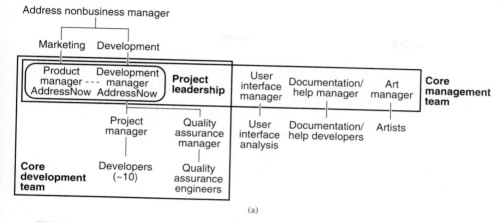

(a)

FIGURE 24.2 (a) Project organization for Middle, Inc. (AddressNow 2.0).

LabelIt! 2.0

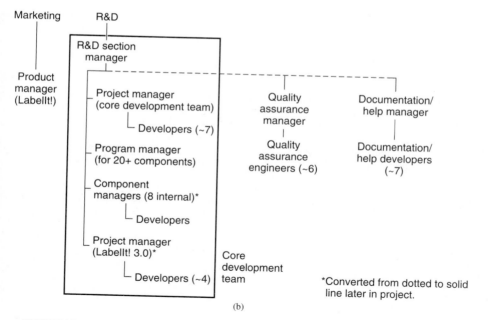

(b)

FIGURE 24.2 (b) Project organization for HiTech, Inc. (LabelIt! 2.0).

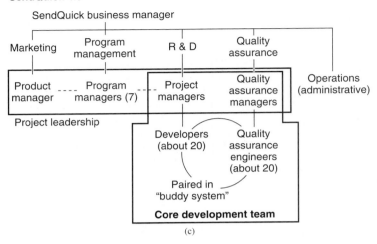

SendQuick 4.0

FIGURE 24.2   (*c*) Project organization for Leader, Inc. (SendQuick 4.0).

ager. The *product manager* measured the product's success in the marketplace and took responsibility for coordinating resources across the project and managing the product definition. The *project manager* determined whether schedules were met and directly managed the developers assigned to the project. They were joined by representatives from documentation and quality assurance. As the product developed, resources were slowly added to the team until it reached 30 to 35 people in all. The original cross-functional management team remained intact throughout.

The development team was collocated, facilitating communication among team members, and team management fostered a strong sense of stability and trust. Frequent status meetings kept the team up-to-date and allowed for rapid issue resolution. Senior management delegated most product and schedule decisions to the team for resolution, obtaining updates through periodic project reviews. When AddressNow 1.0 for Windows was shipped, both the product and project managers were promoted, making way for establishment of a new management team for AddressNow 2.0.

Definition of AddressNow 2.0 was launched when the new product and project managers convened the core team—involving, in addition to themselves, the development manager, quality-assurance manager, user interface manager, documentation-help manager, and art manager—to establish the goals for the product. A month or so later, just before shipment of AddressNow 1.0 for Windows, the full development team met and the project was formally started. As work on AddressNow 1.0 wound down, people slowly transferred to AddressNow 2.0.

On AddressNow 2.0, development was jointly managed by the product manager, who saw herself as responsible for knowing the market and coming up with a product to meet the needs of the market and a development manager who managed both the development and quality assurance organizations. The product was developed with 10 developers at peak staffing. The total head count in the label-making software business unit reached about 80 people, all focused on some aspect of the AddressNow product line.

24.2.2 Product Definition Development and Management

Product definition development and management changed significantly between versions 1.0 and 2.0 of AddressNow. With version 1.0, Middle had much less data to confirm its product definition, the market for label-making was still ill-defined with few well-developed competitors, and the underlying technologies had not been well tested in the market. By the time version 2.0 commenced development, more information on market needs and competitive offerings became available, allowing for more rigorous product definition. In this section, we describe the approaches taken to gathering customer and user needs and competitive positioning information, assessing and choosing from available technologies, and managing changes to the product definition during development.

Customer and User Needs Assessment. Customer and user needs assessment was done much more informally on AddressNow 1.0 than on the other four products we studied, as it was Middle's first major entry in the marketplace. (See Fig. 24.3 for a comparison of methods used to collect customer and user needs information.) The product manager, experienced in the industry and the underlying product technology, felt that she understood the market well. She described people at Middle as being more focused on product features, while she focused on ease of use. She spent a lot of time "thinking, reading, playing with the product, going to trade shows," and generally learning as much as she could about the new technology and its potential impact in the market. But, because she had worked in the business for so long, she felt that she knew the customers well and didn't need further formal interaction with them to characterize the product. Thus, the market research for AddressNow 1.0 was informal. It involved talking on the phone with people, asking them why they did or did not own (the DOS version) of AddressNow, how they used it, if they did own it, and what they liked or disliked about competitors' (specifically HiTech's) products. The product manager simply called people she already knew or used contacts developed through the sales organization or at tradeshows.

Introduction of the OS/2 version of AddressNow 1.0 generated significant customer input for definition of the Windows version of the product. Customer support phone lines, sources of input for all products discussed in this chapter, yielded insights on what customers liked and disliked about AddressNow 1.0. Within the customer support organization, which supported all Middle's products, there was one individual assigned to monitor call activity on AddressNow, develop statistics on the calls received, and prepare a top-10 support list. This list was updated regularly and communicated to the AddressNow 1.0 product manager. (On later versions of AddressNow, the customer support representative became a member of the development team.) In addition to gathering data from customer support, the AddressNow 1.0 product manager actively cultivated an internal customer base. Saying "we have tons of real live users right here," she launched a major effort to generate as much early feedback on AddressNow 1.0 as possible from users right at Middle. (She also suggested that if people internal to Middle did not want to use the product, then the product certainly required rethinking.)

By the time AddressNow 2.0 was launched, Middle had a well-developed external base of customers from whom to draw information. The customer support phone lines continued to be a primary source of information, and the team began to use call volume as a means of justifying investment in developing a given feature or resolving a specific bug in the program. (The number of calls was directly related to customer support costs; it was assumed that those costs would be reduced directly by inserting the feature or fixing the bug.) Large customers and high-use customers were identified

Information Access Approach/Source	Address-Now 1.0	Address-Now 2.0	LabelIt! 2.0	LabelIt! 2.0 CD-ROM	Send-Quick 4.0
Indirect:					
Thinking/reading	X				
Tradeshows	X				
Industry analysts	X		X		
Competitive analysis			X		X
Distributors				X	
Sales organization				X	X
Direct:					
Informal interaction with customers	X				
Customer support calls (feedback from installed base)*	2 (calls on OS/2 used for Windows)	X	5	4	5
E-mail with customers		X			
Customer wish lists		X			X
Internal users*	5		4	4	5
Usability analyses		X	X		5
Top customer inputs		X			
Customer council			X		
Customer focus groups*	3		3	3	3
Customer visits to discuss product needs*	3		4	5	5
Customer visits to watch products in use*	3		5	3	4
Contextual inquiry					X
Purchased market research studies*	1		3	4	5

FIGURE 24.3 Customer and user needs assessment. Product managers were asked to complete a survey rating the extent to which they used a set of techniques in assessing customer and user needs on a scale from 1 to 5. A 1 indicates "to no extent" and a 5, "to a great extent." Asterisks and numbers are shown in the table for those techniques included on the survey. Many techniques were mentioned during the interviews that were not directly captured in the survey, and those are simply listed and use of them indicated by an "X." The AddressNow 2.0 product manager did not return our survey.

and special attention paid to the "wish lists" they generated. Direct communication as well as e-mail (electronic-mail) communication were used to access information from specific customers. Extensive testing was done throughout the development process to gain early input from the customer base.

24.2.3 Prototypes

The critical role of prototyping in new-product development efforts is now being recognized in the literature on hardware product development. Software developers provide well-developed models of the use of prototypes to gather early feedback from customers, as the more advanced companies in our study employed extensive usability testing early in the development process as well as both alpha and beta testing later in the cycle. Usability testing is a time-consuming and expensive process that entails sitting a user down in front of a terminal to execute a specific task. Both the screen and the user's face are videotaped, and users are sometimes asked to describe what they are doing. This data is then analyzed and feedback provided to the development team as to changes to be made. Of the three companies reviewed here, Middle and Leader proved the most skilled in testing to elicit detailed and useful customer feedback.

Middle began usability testing for AddressNow 2.0 nearly a year before product shipment. (They planned to start much earlier in the development cycle on AddressNow 3.0.) At that time, they contracted with an external lab to perform the tests, but have now invested in their own usability testing capability internally, as they view the feedback as so critical to the development process. The product manager oversaw the usability testing efforts for AddressNow 2.0, sharing selected critical insights with the rest of the development team.

In addition, Middle conducted both alpha and beta tests (described in the section on timelines) on AddressNow 2.0. Two rounds of beta testing were done before product shipment, as some late changes were made to the product. Middle has formalized its beta site management process, putting in place a beta manager who identifies potential beta sites and catalogs them according to the hardware environment in which they work, the types of activities they perform with the software, and their ability to provide good feedback to Middle. Beta sites are then chosen in concert with the product positioning strategy to emphasize feedback from customers that best fit the target profile.

Competitive Analysis. A number of product managers we interviewed listed competitive analysis as one means of generating customer and user needs, and some started their product definition efforts with competitive product reviews. Middle's product managers did not suggest competitive analysis as a primary vehicle for eliciting customer and user needs information, but did execute very detailed analyses. (See Fig. 24.4 for a list of methodologies used in performing competitive analyses.) Large, detailed "feature tables" are developed that catalog the specific features of all major competitive products. As those products have become increasingly complex, the task has become quite daunting. Much of the information for the table comes from the purchase and use of competitive products (which all the teams described here do).

The product manager typically obtained the product and played with it first before making it available to others on the development team. Product managers for both versions of AddressNow created presentations on the competitors' products for the development team. The AddressNow 2.0 product manager even went so far as to pretend that she worked for Leader and was selling SendQuick to the Middle salesforce. She sold SendQuick to the point at which the salespeople were ready to buy it over AddressNow, and then spent the rest of the meeting working with them to tear apart her presentation and the SendQuick product itself.

The AddressNow 2.0 team also undertook usability testing with competitive products. In this case, they had representative users execute the same tasks with both AddressNow and the competitive products. They paid particular attention to the process users employed to execute their tasks, seeking to minimize the number of steps users took to complete their tasks with Middle's software relative to that of the competitors. Data gathered from these exercises proved to be an extremely valuable supplement to the basic feature table analysis they had performed previously.

Finally, in addition to the detailed analyses of competitive products, Middle's product managers stepped back and watched the big industry picture. With information from analysts, tradeshows, the press and general industry "gossip," they built a picture of the messages their competitors were delivering, the positions they were taking in the market, and directions they were heading. This information was used to think

	Address-Now 1.0	Address-Now 2.0	Labellt! 2.0	Labellt! 2.0 CD-ROM	Send-Quick 4.0
Information Access:					
Purchase and use*	5	X	5	4	5
Press reviews	X	X			X
Tradeshows	X	X	X		X
Competitive usability tests		X			
Engage customer focus groups in competitive product evaluation*	3		4	3	5
Watch use of competitors' product by customers*	3		4	3	5
Collect competitors' evaluations					X
Analysis Techniques:					
Feature comparison table	X	X			
Positioning assessment	X	X			X
Demos of competitive products		X			X

FIGURE 24.4 Competitive analysis. Product managers were asked to complete a survey rating the extent to which they used a set of techniques to assess competitive products. Asterisks indicate those techniques listed on the survey. Ratings ranged from 1 indicating "to no extent" to 5 indicating "to a great extent." Techniques mentioned during the interviews are simply listed with "X"s. The product manager for AddressNow 2.0 did not return the survey.

through positioning of the next version of Middle's product which in turn dictated some of the product's features.

Technology Assessment. In addition to customer and user needs assessment and competitive analysis, product development teams must choose from among a number of available technologies (usually software platforms in this case). Although the choice of technologies is often quite straightforward, on the basis of the installed base of a given platform, the timing of adoption is less easy to determine.

Platform choice at Middle was driven from the corporate level, and was based on potential market size and alliances Middle had formed with other players in the market. In hindsight, of course, choosing OS/2 as the first platform on which to launch AddressNow 1.0 was a mistake and critically delayed introduction on a Windows platform. At the time the decision was made, however, OS/2 was still a viable option. Subsequent decisions have been somewhat easier, although deciding when to move from one version of Windows to the next continues to be a difficult process for the AddressNow team.[9]

Technology issues were not limited to choice of operating system. Given the graphic intensive nature of the AddressNow product line, compatibility with various videodisplay technologies and printing standards had to be addressed. A decision was made, for example, to provide automatic conversion of labels designed in color to a black-and-white palette for printing on the commonly used black-and-white laser printer, thus accommodating both color display and black-and-white printing technologies. Competitors chose not to provide this capability.

Product Definition Change Management. The notion, made popular in much of the literature on product development management, that one freezes a product specification and then develops product to that specification, clearly does not hold in the software companies we studied. Instead, these companies recognize that change will happen in the course of product development and put mechanisms in place to deal with it. Major decisions, such as choice of platform, were not changed during development in any of the projects we studied. Product feature prioritization was changed on a number of projects, typically with very careful analysis and consideration. And small changes in response to feedback from usability analyses were made regularly.

The AddressNow 1.0 for Windows project went through a major reorientation late in the development cycle, as the product manager had a sudden insight as to how the product should be positioned in the marketplace. Although the team argued with her at first, not wanting to make the major changes required, she ultimately won them over and together they approached senior management to gain their approval as well. In hindsight, it is clear to the team that they made the right decision, although it cost them a couple months on their shipment schedule. The product manager justified the changes, saying "Until you begin to sell it and market it, the product concept just doesn't come into focus. You have to practice the story again and again and again."

Although the changes made on AddressNow 1.0 for Windows were more extensive than those on a typical project, both development teams remained open to the possibility of change throughout the development process. The AddressNow 2.0 team incorporated a number of small changes suggested by the usability analyses. In the end, they made a single decision to slip shipment by 3 months to accommodate the numerous small changes to be made. Changes are not undertaken lightly, as rescheduling the many, complex interdependencies among elements of the product was a sizable task. Middle employed a sophisticated schedule management system, increasing their flexibility to late changes.

24.3   HITECH, INC. AND THE LABELIT! PRODUCT LINE

HiTech relies on the LabelIt! product line for over 90 percent of the company's revenues. It considers itself a "technical" leader in the label-making software field, investing significant internal resources in the development of the underlying technologies. LabelIt! 2.0 was introduced in August 1993 after a very painful 2-year development cycle. Its introduction was followed shortly by introduction of the LabelIt! 2.0 software on CD-ROM. LabelIt! 2.0 was considered a very successful product, as its sales were more than double those predicted in its first 6 months on the market. It was introduced with a significantly lower cost of goods than LabelIt! 1.0, and received unanimously good reviews in the market. From a technical perspective, development of LabelIt! 2.0 yielded HiTech's first two patent applications on the underlying label-making technologies.

LabelIt! 2.0 followed a very difficult path through its definition and development. (See Fig. 24.1 for timeline.) The project commenced in September 1991 when an R&D section manager recruited two project managers, an architect and a marketing manager, to work on the "direction of the product, defining the main features to be emphasized in the release." In October 1991, the team received approval from the executive staff for their product concept (including revenue projections and general product description). At that time, the team size increased, and work began on the preliminary product specification, which was to describe the product's user interface, features, and underlying technologies. Review of the first preliminary specification in December 1991 resulted in a recommendation that the specification be cut in half to meet schedule requirements, and the final specification was completed in March, 1992.

Meanwhile, enough code had been written, in part through modification of LabelIt! 1.0 software, that usability testing could begin, involving about 100 customers between March and June 1992. Formally, however, the development effort began in May 1992, after the final specification was completed. Work was split up into a series of 6-week milestones which allowed separate, distinct sets of features to be coded and tested independently. The R&D section manager overseeing the project wholeheartedly believed in managing project in milestones saying, "Anything that takes longer than 6 weeks to implement—has to be broken down into something smaller." The product finally shipped in August 1993, 1 month behind schedule, the best on-time performance in HiTech's history.

The timeline for LabelIt! 2.0 on CD-ROM was driven entirely by LabelIt! 2.0's schedule. The project began formally in November 1992, when the product manager joined the team. He led the charge to develop a product requirements document that not only described the content of the CD-ROM (LabelIt! 2.0, a demonstration facility, on-line product documentation), but also the choice of technologies. The product concept was approved in April 1993, after the team sold senior management on the notion of selling product via CD-ROM. By the product approval milestone in June 1993, the team had developed a detailed sense of what the product would be, what features it would include, and what the market plan would be. Development of much of the product was dependent from that point forward on the timeline for the LabelIt! 2.0 software itself, and the CD-ROM version shipped in October 1993.

24.3.1 Project Organization

LabelIt! 2.0 was somewhat unusual in its organizational structure, as there were four different product marketing managers assigned to the project during its development, making the marketing participation in the project fairly weak. As a result, the R&D section manager retained full responsibility for running the project and ultimately delivering to customer requirements. Although the R&D section manager asserts that HiTech is "not marketing driven, we're market driven," it seemed apparent throughout the interview that HiTech is, in fact, quite engineering-driven.

About 82 people worked on the LabelIt! 2.0 development team in all, some in Kansas City at a location garnered during one of HiTech's acquisitions. The core team consisted of a project manager overseeing seven engineers, a program manager who oversaw development of the twenty or so components of the product that were being developed in other groups both internal and external to HiTech, and, near the end of the project, a project manager for LabelIt! 3.0 with four engineers dedicated to starting on the next development effort, all of whom reported directly to the R&D section manager. In addition, reporting to the R&D section manager on a dotted-line basis, were the quality-assurance manager and about six quality-assurance engineers, who were also responsible for testing the externally developed components, a documentation-help group consisting of seven people who developed documentation, on-line help, and tutorials, and, later in the project, the eight project managers for the internally developed components (see Fig. 24.2b).

Communications on this team were clearly more and complex than those on Middle's teams because of the geographic separation of the team and the number of technology components to be integrated into the product. The R&D section manager held formal update meetings on a regular basis, using videoconferencing technology with the Kansas City team members. Coordination of the work of the 21 component development teams was particularly crucial, and caused the most trouble during the development effort. In hindsight, the R&D section manager suggests that matrix management isn't such a good way of managing projects. He felt that the core engineering team concept worked well in keeping people focused on their goals, but that having various components developed in semi-independent teams was not a workable long-term solution. In fact, he argued that HiTech's strategy of developing its product in multiple components "was too much. We ended up doing more integration than we had anticipated. I actually had a full-time program manager whose job was tracking all the dead bodies...." He found that external sources were much more reliable than internal ones, as they remained committed to schedules and product specifications, while the internal teams assumed that the LabelIt! 2.0 core engineering team would accommodate any changes they wished to make.

The CD-ROM project, although tightly linked to the LabelIt! 2.0 team's outcomes, was smaller and more focused. The product manager not only took responsibility for gathering customer and user needs but also represented the voice of the customer support, international marketing, and sales organizations. (CD-ROM offers new approaches to product distribution, implying different roles for the sales organization.) The project manager, located in Kansas City, oversaw the work of three development engineers, none of whom worked full-time on this project, as well as contributions from the documentation and quality assurance organizations. Finally, there was a localization group involved to adapt the product for overseas applications. The core team remained intact throughout the product development effort, and communicated through weekly meetings called and run by the product manager at which they focused on making critical decisions about the product. (Status updates were done via informal daily communications using e-mail or phone.)

24.3.2 Product Definition Development and Management

The engineering driven nature of the HiTech culture comes through in a review of the means by which they developed their product definitions. Although thorough in the development of 200-page product specifications, these specifications seemed more focused on technical product features than they were on specific customer and user needs. Nevertheless, considerable work was done to collect external data for the product definition process.

Customer and User Needs Assessment. Customer and user needs assessment for LabelIt! 2.0 started with a thorough analysis of competitive products and a subsequent brainstorming session at which a long list of potential product features was developed. The R&D section manager led this effort, accompanied by a product marketing manager and a couple of people from the development organization. They augmented their brainstormed list with a detailed list of specific enhancements requested through the customer support phonelines. This list, prioritized by call volume, allowed them to estimate returns on investment for development of a given feature by estimating reductions in customer support call handling. (Note that Middle used a similar approach to justifying development investment.)

With the long, detailed list of possible features in hand, the team turned to determining what the primary marketing messages for the new product would be (e.g., data connectivity, usability). Although all three companies described here emphasized specific marketing objectives or goals for their products, HiTech's goals were the most technical of the three; R&D's influence on the product clearly showed. The team further narrowed the list of possible features by calling on a council of 10 strategic customer representatives to help prioritize the feature set and by engaging over 150 personnel internal to HiTech in a feature prioritization exercise. This data was integrated to finalize the feature list for LabelIt! 2.0.

While HiTech continues to use customer support phone lines as an important source of customer and user needs information, they are moving away from use of the customer council, as they have determined that the council does not represent the bulk of HiTech's customer base. Engineers are increasingly encouraged to learn more about the broader customer base by sitting in on customer support phone calls for an afternoon, making a few visits to customer sites, and attending customer visits to HiTech's facilities. The R&D section manager firmly believes that the engineers are the ones who need to know what is going on with customers, as "they will build the products," and thus measures them on making a certain number of contacts each year. Virtually all the data HiTech collects is from their installed base.

The CD-ROM team had a somewhat different objective in collecting customer and user needs information, as the media was of more interest to them its content. In contrast with the other teams (at Middle and Leader and for LabelIt! 2.0), the CD-ROM team relied heavily on purchased market research data to profile the existing and expected installed bases of CD-ROM. From acquired data, the team built various profiles of CD-ROM users in an attempt to more fully describe their own market opportunity. They also undertook extensive research with customers themselves, speaking with large HiTech users about their plans for CD-ROM—when they planned to move to it, what applications they were running or planned to run, etc. Finally, they interviewed key distributors to determine how they would handle the new-product format. The team kept a (computerized) bulletin board on which customer and user needs information was posted and thus accessible to all.

24.3.3 Prototypes

There were many prototypes built during the course of LabelIt! 2.0's development that ranged from prototypes developed by changing small parts of LabelIt! 1.0 to test different user interface options to full-blown prototypes of various program components. Some prototypes were used to elicit customer feedback. Before the final detailed product specification was finished, modified LabelIt! 1.0 and visual basic prototypes were taken out to various different daily users of the product and detailed observations made on their use. This allowed early changes to the product to be made before significant investment in code writing. Other prototypes were used more for internal R&D purposes, as they developed models of the components that would eventually be available for the product and tested their compatibility.

Following extensive usability testing early in the development cycle, formal alpha tests of LabelIt! 2.0 were conducted with 10 external customer sites. These tests entailed code that had complete functionality and a specified mean-time-between-failures (MTBF) performance level, but did not include all the components in complete form. Beta testing was postponed until very late in the process as the introduction of Middle's product (AddressNow 2.0) and the late cancellation of the data management component disrupted the process. (See section on product definition change for more detail.) By then, a fully integrated (all 20 components) version of the product was presented to about 300 external sites as well as to everyone inside HiTech. Given the late start to the beta testing cycle, it is not clear that HiTech was able to do much with the input on LabelIt! 2.0.

The CD-ROM product team ran an extended alpha test. In addition to a few early visual basic usability test prototypes, they periodically burned in a new CD-ROM with the latest version of LabelIt! 2.0 and worked with three to four sites to review the product. Rather than look for bugs in the software or usability issues (which were in the purview of the LabelIt! 2.0 team), they focused on the CD-ROM-specific aspects of their product (e.g., setup mechanism).

Competitive Analysis. HiTech considered both Middle and Leader's products to be primary competitors for LabelIt! 2.0, and learned about both products by purchasing and using them. HiTech asked customers what they thought of the competitive products, and sometimes engaged in formal testing programs with customers of competitors' products. In general, however, the R&D section manager felt that "we're probably our own best customers," and suggested that they look at competitive products primarily to gain good new product ideas. As with the other players in the industry, HiTech employees attended tradeshows and reviewed industry publications for additional information. There was no discussion, unlike that at Middle and Leader, of understanding competitors' market positions and trends in that positioning. If there had been stronger product marketing presence on this product, there likely would have been more competitive analysis.

The CD-ROM product team was interested primarily in competitive products offered on CD-ROM. At the time, Middle had not yet chosen CD-ROM as a distribution medium; Leader had, but offered several of their products in a single package. Another, smaller and more obscure competitor in the label-making software market did offer their package on CD-ROM, so was considered a primary competitor in this specific situation. Because the focus of the team was on packaging, pricing, and presentation, they spent most of their time analyzing these competitive product characteristics rather than look at the performance of the software itself.

Technology Assessment. The platform choice for LabelIt! 2.0, as the R&D section manager put it, was "pretty clear cut." Windows 3.0 had a distinct hold on the market at that time, and HiTech's customers had clearly indicated their interest in moving to the new platform.[10] HiTech's development teams also maintained close relationships with printer vendors so that they could get their software tested with as many different technologies as possible.[11]

The CD-ROM team had a slightly more difficult decision to make. They had to select from among several different CD-ROM standards. Using some basic criteria such as being able to read the CD-ROM from multiple drives and have it networked, they chose from among the various options available. Other media formats (e.g., read/write CD-ROM and tapes) were not being sold in sufficient volumes to be considered.

Product Definition Change Management. Of the five products reviewed in this chapter, LabelIt! 2.0 experienced the most change in its definition during the development phase. Two major events drove the reorientation: First, Middle introduced AddressNow 2.0 6 months before LabelIt! 2.0 was scheduled to ship, providing a new standard for the user interface provided on the product. Second, a critical data management component of LabelIt! 2.0 was not adequately developed, and a late decision was made to remove it from the product altogether.

In response to Middle's product introduction, a "hit team" of two engineers, one marketing person, one quality-assurance person, and two documentation/help people, were isolated from the rest of the team and given 4 weeks to: Improve the first 15 minutes of ease of use of the product, emphasize LabelIt!'s strengths, and stress Middle's weaknesses. Because most of the issues involved the user interface, the hit team was able to parse off their section of the code and work in isolation from the rest of the team to complete their effort as planned in one month. (Members of the core engineering team made up the hit team and thus were familiar with the product code.) At that time, it was clear what changes had to be made to the product, and that introduction of LabelIt! 2.0 would have to be postponed by one month as a result. Nevertheless, the hit-team project was called "Project Magic," and it was.

Handling the data management component was another story altogether. The data management component development team, although internal to HiTech, intended its output to be sold as a stand-alone product, and so did not accept the direction given to them by the LabelIt! 2.0 team. Despite numerous attempts to get senior management to understand the situation and deal with it, the team was allowed to operate on its own until 4 months prior to LabelIt 2.0's scheduled shipment. By then, it was too late to employ the approach used in response to AddressNow 2.0's introduction. Instead, a single development engineer was assigned the task of finding a workaround solution that ultimately eliminated the need for the data management component, but sacrificed considerably on product features. Four weeks of 16-hour days on the part of the individual engineer kept the overall project on schedule. Although the end result lacked the functionality they had hoped for, it did serve to keep LabelIt! 2.0 competitive in the market. Meanwhile, managers for the other internal components were asked to report solid line to the R&D section manager for LabelIt! 2.0.

The CD-ROM product had a much less tumultuous experience. Although there was some scaling back of requirements when the team learned that one of its technology vendors would not be able to deliver as planned, the product remained relatively unchanged throughout the development process.

24.4 LEADER, INC. AND THE SENDQUICK 4.0 PRODUCT

Leader claims lead market share—60 percent or more—with its entry in the label-making software market, SendQuick, due in part to the product's integration with other desktop management software offered by Leader. SendQuick 4.0, introduced in February 1994, is the most recent version of label-making software we studied. It was introduced after a relatively short (13-month) development cycle that commenced informally in October 1992 and was kicked off formally in January 1993. The MRD and SendQuick 3.0 code formed a strong base for the project kickoff, although the product manager described the MRD as "pretty fluid" at that point in time.

Like HiTech, Leader managed SendQuick 4.0 development in modules, each of which focused on a specific feature set and was managed on its own timeline. There were three major development milestone dates at which code was frozen and handed off to the quality assurance organization to test and debug while the development organization moved on to develop the next module of code. The overall project proceeded through a "code complete" milestone and a "visual freeze" milestone, at which time the product's user interface was frozen and pictures taken for inclusion in product documentation.

SendQuick 4.0 was announced 3 months before it was shipped, as Leader wished to position it within its broader product line. This resulted in considerable pressure on the development team to get early versions of the product out for market review and testing. Beta testing began in November 1993 and was considered a "market beta" in this case, as the team was not able to use much if any of the feedback generated that late in the process to change the product itself. Rather, the "beta" served to introduce the product early to customers eager to see what it would do. SendQuick 4.0 finally shipped in February 1994, 3 months behind its original schedule.

24.4.1 Project Organization

The project team at Leader was structured differently than those at Middle and HiTech, although historically Leader had employed a similar form. As at Middle and HiTech, there was a product manager for SendQuick 4.0 who took primary responsibility for reflecting the customer and user needs to the development team. She had no direct reports, but was responsible for overseeing progress on the project and integrating the various elements of the effort. Rather than work directly with a project manager (and the developers), however, the product manager for SendQuick 4.0 worked through a team of seven *program* managers, six of whom took responsibility for specific feature areas (e.g., user interface, clip art, templates) and one of whom was responsible for product internationalization (e.g., ensuring that dialog boxes were large enough to accommodate long German words).[12] Program managers, in turn, worked closely with the project managers to negotiate the product specification and then ensure that code was built to specifications. Product managers maintained their traditional role of managing developers to schedule and quality specifications. Product definition was a joint effort among the product manager, program managers, and project managers as well as representatives of the quality assurance function. Team members shared the common objective of defining and developing a "best of breed" product that would capture market share in a highly competitive marketplace.

There were roughly 20 developers dedicated to SendQuick 4.0 development collocated with about 20 quality-assurance people and another 30 or so people associated with other aspects of the SendQuick product line (e.g., documentation). Each quality-assurance engineer was assigned to a development engineer to work in a "buddy system" to simultaneously develop and test code throughout the development cycle. This substantially improved team spirit overall, and made the difficult decision of when to ship the product (with which remaining bugs) easier as there was a shared sense of the code's status.

Collocation of the product team facilitated frequent informal communications that were supplemented by frequent meetings. The entire 80-person SendQuick team met weekly for demonstrations of competitive products, status updates, and reviews of all products. Smaller teams were convened to review status on individual projects or on certain feature areas of a given product. Review meetings were held at each major milestone to approval passage to the next phase of the project.

24.4.2 Product Definition Development and Management

The SendQuick 4.0 development team employed many of the same techniques for collecting and evaluating customer and user needs, competitive positioning, and technology availability that Middle and HiTech's teams used. Because SendQuick 4.0 was developed more recently than the other products we have described, however, they addressed some new issues and employed some new techniques for customer analysis. Both Middle and Leader have recently undertaken major internal efforts to integrate product across their product line, making them easier for customers to use. SendQuick 4.0 was Leader's first attempt at extensive integration, forcing the team to deal with complex interoperability issues and dictated technology choices. Despite the restrictions this placed on the SendQuick 4.0 team's ability to fulfill customer requirements, they undertook experiments in "contextual inquiry" to gather very detailed information on customer environments.

Customer and User Needs Assessment. "Contextual inquiry" entailed watching customers use SendQuick software in their own environments and documenting in detail how they use the product. A cross-disciplinary team, led by the product manager, conducted the on-site research with the customers. The data they collected was later analyzed to provide insights into the tasks customers performed with SendQuick, why they chose the approaches they did, and why other approaches were not chosen. Through this process, the team learned a number of things that it was not able to learn through usability testing, focus groups, and other market research as they actually observed customers in their own environments. Contextual inquiry was thought to yield particularly useful information, and will be used more on SendQuick 5.0.

As at Middle and HiTech, customer support phone lines were a major source of data on customer and user needs for Leader, too. Phone calls were categorized when received and the number of calls in each problem area tallied. SendQuick 4.0 (and each product revision that preceded it) accounted for the top 5 to 10 customer support categories in generating the MRD and prioritized feature list. At Leader, customer support costs were charged to the product development team (although the function was housed in a central corporate organization), so the team paid particular attention to high-volume call categories. Once a week, members of the development team joined the customer support engineers, listening in on phone calls and helping answer questions. In addition, Leader maintained a database of customer "wishes" and, on SendQuick 4.0, the top four customer wishes were accommodated with new-product features.

Leader made more extensive use of broader market research survey work than Middle and HiTech, including attitude surveys of registered SendQuick users and random market research surveys to understand potential users outside their own installed base. Focus groups were employed to assess product positioning and focus advertising campaigns rather than to review product features or content.

24.4.3 Prototypes

As with its counterparts, Leader performed extensive usability testing early in the development cycle. Although they subcontracted management of the usability testing outside the organization, tests were conducted in an internal lab and observed by a range of people from the product manager to the developers themselves. As with Middle, users were asked to perform a specific task with the software and, at Leader, were asked to describe the process they are going through as they performed it. Even before actual code was available, screen mockups were shown to users (who were asked in this case *not* to touch the mouse) and were asked what they expected to see next. As reported by other organizations, the usability analysis process was described by the product manager for SendQuick 4.0 as "really, really eye-opening," particularly for the developers of the features being tested.

There was no formal alpha testing for SendQuick 4.0, and beta testing was really considered a "market beta," as the product was sent out too late to allow for any substantive changes from resulting feedback. In fact, at Leader, the costs of beta testing were charged to the sales organization, and were considered as much a sales tool as a means of collecting customer feedback during product development.

Competitive Analysis. The SendQuick product manager purchased competitive products as soon as they went on the market, played with them (using them to generate her own address list and labels), and wrote analyses on the products. She used the competitive information herself to generate the list of features that would be included in SendQuick 5.0 (the next version of the product), although she emphasized Leader's wish to lead rather than imitate or follow the competition. (All three companies made this point!) The rest of the product development team was exposed to information on competitive products at the weekly team meetings, and written analyses were sent to the salesforce.

Other competitive information was made available by the salesforce, as they collected documents left behind at customers by their competitors. These documents often showed the competitors' views of how their products (AddressNow and LabelIt!) compared to Leader's SendQuick, and were useful for thinking through future product positioning.

Technology Assessment.   Technology choices at Leader were dictated by corporate, challenging the SendQuick 4.0 development team to integrate its product with others in Leader's line. While interoperability standards were set at the beginning of the project, there was considerable iteration among the various product teams throughout the development effort to meet them. A considerable amount of development team energy was expended on interoperability concerns rather than on new-product features, which was a source of frustration for the team. Disparate goals among the various product teams at Leader caused frequent friction in identifying common technology standards.

Product Definition Change Management.   Although SendQuick 4.0 missed its intended shipment schedule by 3 months, little changed in the product's basic definition. The team did decide to add a few features late in the game, as they were concerned that they had focused too much on interoperability issues and not enough on "sizzle" that would cause customers to upgrade or switch from a competitors' product. Although the developers complained about these last-minute additions, the product manager did not believe that the schedule would have slipped as a result of the changes. Rather, the schedule slipped because of problems in developing some core software technologies that were to be shared across Leader's products. When the technology was not developed on time, several product schedules were pushed out.

The SendQuick 4.0 team showed flexibility similar to that displayed in many of the software teams we studied, as they accommodated numerous minor changes to the product late in the process. Inputs collected from the usability analysis which was conducted throughout the development process caused the team to make a few changes after the "visual freeze" milestone. The product manager justified these changes, despite the cost to reshoot much of the documentation (changing a common screen means that many of the pictures shown in the manual have to be redone), saying "You don't want the process to get in the way of making the right decision about changing something." And as for the conflict arising from the late changes, she says "...they're still friends. They hate you at the moment, but in the end the mutual respect is there."

24.5   CONCLUSIONS

Each case study presented here provides its own set of lessons, as each team did some things well and showed some room for improvement. A few common themes arise from the case studies that yield useful lessons to a broader audience.

1. These teams interacted extensively with their customers throughout the development cycle to test product acceptance and gather detailed, meaningful feedback. Product mockups and prototypes are relatively easy and inexpensive to develop in software, making extensive interaction possible. As simulation tools, such as those used to develop the Boeing 777, become more widely available, hardware product development teams will be able to efficiently engage more customers throughout their development cycles. These teams may want to learn from software development colleagues who have gone before them.

2. Each of these companies views their label-making product line as a continuous flow of products. Teams remain largely intact from revision to revision, providing continuity in knowledge from product to product. Understanding that the current revision of the product will be shortly followed by another allows the development team to more readily make decisions to postpone inclusion of a feature to the next product revision. Thus, product introductions are not perpetually postponed as developers strive to get one more feature in.

3. The teams we studied adopted a balanced view of change, allowing critical changes during the development cycle and postponing others. Major decisions, such as the platform on which these products were developed, were made prior to launching development and were not changed. Relatively minor decisions, such as the language in a dialog box, were changed quickly if feedback from usability analyses suggested they should be. Decisions in between—such as HiTech's accommodation of Middle's product introduction—were carefully considered and made by the team after extensive assessment and deliberation. None of the teams pretended that the requirements specifications they began with would remain intact throughout the development process. This realistic view allowed the teams to properly and effectively deal with changes that did occur.

While the software development project teams described here could likely learn something from their hardware development team counterparts, they have something to teach as well. As the hardware world changes, introducing more software content into their products and making greater use of simulation tools, it will almost certainly migrate to using a number of the tools and techniques that the software development world has to offer. Hopefully these case studies provide a launching point for transferring some of that learning.

REFERENCES AND NOTES

1. OECD, 1985.

2. International Data Group, 1993.

3. An informal survey of several companies in the computer industry indicated that over 70 percent of the new-product development engineers employed in these organizations are software engineers. Nevertheless, many of these organizations still think of themselves as "hardware" companies.

4. A notable exception is work done by Michael Cusamano at MIT. His book *Japan's Software Factories* describes management of software development projects in both U.S. and Japanese companies in great detail. His recent work on Microsoft's product development processes sheds light on development activities in consumer software companies.

5. Primary participants in the research include Drs. Sara Beckman and David Mowery of the Haas School of Business and Mr. Michael Dunn of Context Integration. We have collected data on over 40 new-product development projects in the United States and Japan, focusing on their approaches to collecting customer and user needs, competitive positioning and technology data, and using that data to define the product they were to develop.

6. For more on product definition and its management from earlier research projects, see Glenn Bacon, Sara Beckman, David Mowery, and Edith Wilson, "Managing Product Definition in

High-Technology Industries: A Pilot Study," *California Management Review,* **36**(3), Spring 1994.

7. Product category, company names, and product names have been disguised to maintain confidentiality.

8. An earlier DOS version of the product had been developed by the acquired company and Middle brought it to market. As with most graphically based products, label-making software evolved most rapidly and gained most acceptance in the marketplace once placed on an OS/2 or Windows platform.

9. A number of the product managers we interviewed complained about the stronghold that Microsoft has on the market, and the unreliability with which they bring new platforms to market. They were quite blunt about their frustrations in determining when the latest version of Windows would be stable enough to support new-product development and about their sense that Microsoft purposely confuses its competition in the platform introduction game.

10. The decisions for LabelIt! 3.0 are much less clear-cut. A sense that "Microsoft is trying to keep everyone dancing" came across in a number of our interviews, as product and project managers struggled to understand whether Windows NT and Chicago were ready to be used.

11. There are vendors who provide a service of testing software on various different types of hardware.

12. The introduction of the program management buffer between the product and project managers was found in other cases in our broader study. Generally, the program managers relieve the product managers of some of the day-to-day management of the project, leaving them instead to focus their attention on maximizing customer satisfaction.

CHAPTER 25

PRODUCT PLATFORM RENEWAL

CONCEPT TO COMMERCIALIZATION

Alvin P. Lehnerd

Shadowstone Group
Darlington, Maryland

Manufacturing corporations in endless pursuit of growth and profitability tend to be biased toward the relentless search for new products, new products, new products, on and on and on.

Frequently forgotten or ignored by management in their growth and profitability pursuits is the latent and untapped potential of their existing mature products and product lines. Rarely do they explore these product areas for growth and enhanced profitability that could be harvested via *product platform renewal* (PPR) programs.

A manufacturing company's management should seek the answers to this question:

> What would be the growth and profitability consequences of completely replacing our outdated product platforms with new modern product platforms designed, processed, and marketed using the state of the art approaches embracing programs of standardization, product and subsystem platforms, integrated design, process, and market development techniques using concurrent and simultaneous multifunctional teams?

In many instances, the answers would be quite remarkable and, in many instances, unbelievable. It is the author's belief that PPR techniques will produce the fastest, easiest, and most efficient approaches in boosting sales and profits; refreshing the corporation in the relaunch of renewed product platforms.

However, new-product initiatives should not be abandoned at the expense of PPR initiatives. Both initiatives should be harmoniously funded and pursued concurrently and simultaneously. The tendency seems to favor doing one initiative at a time, leaning heavily on the new-product programs.

Why are manufacturing companies averse to PPR programs? My opinion? _____. Manufacturing management finds it incomprehensible to pursue projects that (figuratively speaking) essentially bulldoze their existing products into oblivion because projects of this nature will (1) cost too much, (2) take too much time, (3) be incomprehensibly complicated, (4) require resources that do not exist in the company, and/or (5) absorb inordinate amounts of the "in-house" resources already fully loaded on today's projects that cannot be diverted to work on such radical and risky initiatives.

This chapter will share with the reader a recent example of a PPR project that will refute these five concerns. We hope to entice the readers to think seriously about their own product lines that have aged, become outdated, and have lost their sizzle, excitement, and competitiveness in the marketplace. Further we hope to provide encouragement to look inward to their organizations by examining current practices of resource allocation and assessing the current efficiency and effectiveness of these practices in the pursuit of growth and profitability.

This chapter will also overview only the highlights and the major elements of a PPR program with, apologies in advance, for the brevity demanded in the few pages allotted to this chapter. Suffice it to say that there are many steps in the product, process, and market development activities in new product and PPR programs. All the steps required did take place; the manner and method in which they were executed is the key to the speed and effectiveness of the success story to be shared, which we have entitled "project X."

25.1 PROJECT X: A PRODUCT PLATFORM RENEWAL

25.1.1 Introduction

A Midwest manufacturer was recently a benefactor of a PPR program that required only 367 days from concept to first production. The project involved a small multidisciplinary team totally dedicated to the project. The team was directed by the author, a senior VP (vice president) of the company who had previously directed PPR programs in other industries.

To encourage the readers to hang in and continue their reading, we begin with Fig. 25.1, showing the remarkable results of this product platform renewal project: project X.

Illustrated is the impact on the product-line platform with regard to part count, material cost, and labor content, comparing the old to the renewed.

The product line consisted of three product platforms: the basic line (low end), the standard line (midpriced), and the premium line (high end). As you will note, the part count was reduced in all three product platforms (basic, standard, and premium) by 67, 37, and 40 percent, respectively. The material cost ($) reduced by 10, 12, and 29 percent, respectively. The direct labor content ($) reduced by 71, 50, and 54 percent, respectively. It was noteworthy that the labor content ended up being essentially the same for all three product categories: 81 to 82¢ per unit.

The impact was pervasive throughout the entire product line. These results suggest that there must be a lesson or two to be learned by the study and review of the approaches used in this platform renewal project.

Not shown in Fig. 25.1 is another major accomplishment. The new product line was enriched by an expanded list of features and functions, heretofore not available on the old line, offering to the consumers lower costs, operating efficiencies, and more features than offered before in the old line; in addition, the dealer margins were significantly increased, providing more profit and incentives to focus on the new lines.

In the pages that follow, a synopsis is presented as to why the project was conceived, staffed, and managed; what pressures brought on the need for the project, and what initiatives, tools, rules, methodologies, etc. were brought into play to enable a product platform renewal requiring only "367 days from concept to first production."

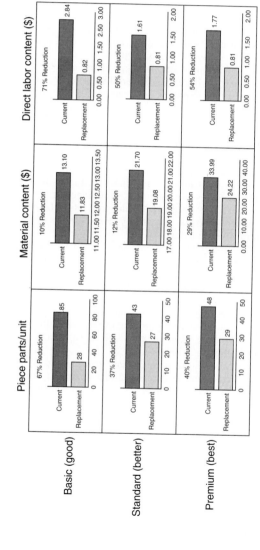

FIGURE 25.1 Product design comparison.

25.1.2 Background Hypothesis

The product line discussed in this chapter is typical of many product lines found in mature and long-established manufacturing companies. A company's product lines are the result of evolutionary developments usually taking place chronologically over the years and decades of the company's existence. Products are invariably designed one at a time, and as the lines expand, the evolving products share very few parts, components, processes, tooling, etc., and, as new materials and processes become available, only the newer designs enjoy the advantages of these new capabilities. Older products rarely are reengineered in design or process and gradually do fall behind and become outdated. Older designs just don't command the attention or resources needed to take the advantages enjoyed by the newer designs.

Companies grew by a continuous stream of products over the years, filling their "market baskets," i.e., catalogs with expanded product offerings added one at a time. Traditionally this was the way companies behaved; designing one product at a time. That's just the way things happened and, at the time, that seemed to be okay.

The consequences, however, in retrospect, are not pleasant. The company has saddled itself with product-line arrays that have become too complicated, having too many variations; too many catalog numbers; too many parts, tools, processes, and vendors; and too much labor, inventory, floor space, and pieces of capital equipment. In addition, redesign efforts took too much time and cost too much; and in doing so, the company's resources were continually stretched to unreasonable extremes. Old designs traditionally absorb excessive amounts of engineering resources, soaking up practically all available time in the redoing of old products. The development of new materials, processes, and technologies essential to enrich the company's product portfolios is deferred or diminished to trivial levels.

Over time, the product platforms become obsolete in feature, performance, and costs, eventually becoming unpopular lines to sell, promote, and excite the customers. "Keeping the line hot" work reaches a state of very little improvement at the expense of a lot of engineering efforts (time and money). The products gradually go into decline and are eventually abandoned.

As this phenomenon is happening, marketing becomes frustrated because of the decline of customer acceptance for their products in the marketplace, and has no other recourse other than to convince management to seek more attractive products offering lower costs and/or more features via outsourcing.

If this initial tactic is successful, management gets excited by these results. The natural course of events leads them to extending this strategy to the remainder of the product line, triggering the abandonment of its once-held advantages in manufacturing. Continued success leads to more abandonment of its manufacturing capabilities. Outsourcing of large numbers of its products shuts down factories and initiates tactics of chasing cheap labor markets, in essence, walking away from its once-revered manufacturing capabilities and effectiveness (thus hollowing out the corporation).

The product platform discussed in this chapter suffered from this same dilemma. The product platforms were viewed as an outdated and a trivial element of the company's product offerings. The only efforts allocated to the products was cost reduction via simple design changes and reprocessing, struggling to offset the inflationary influences of labor and materials, rarely providing the customer with new benefits or features. Marketing had already contracted an OEM to supply a low end product. Another supplier of similar products saw the opportunity and made overtures to management in the hopes of convincing them to abandon their existing line and allocate to them the complete responsibility for the design and manufacture of the entire product line, relieving the company of its need to manufacture any of the existing products in this category.

I happened to be the senior management VP involved in this case. It was my conclusion that this outsourcing strategy was not a rational strategy for the corporation in the long run. My strategy was product platform renewal, i.e., to completely redesign the entire product line all at once, abandon the old lines' designs and processes of manufacturing, and having done so, enable the company to relaunch a new product array that would be modern, cost-effective, and rich in performance and features, resulting in renewed excitement in the marketplace, leading to sales growth and increased profitability. This was the genesis of project X.

Developing the "thought architecture" (Fig. 25.2) became the essential element, the glue, the vision, the compass in orchestrating such a venture. Getting the thought architecture locked in my brain and the brains of the team to be engaged in this project was extremely vital in accomplishing this task.

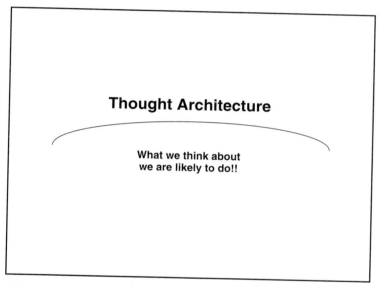

FIGURE 25.2   Thought Architecture.

25.1.3   Thought Architecture: Project X

Project X original "thought architecture" was as follows:

1. Assume that the company had *no* existing capability whatsoever in designing and manufacturing these products.

2. Assume that we were another manufacturer, i.e., new competitor, that was desirous of getting into this product business.

3. Assume that we wanted to produce a product line that was "uncontested" in cost, value, and feature.

4. Assume that our product line would have an absolute minimum of direct labor content, less than 1 percent of cost of sales.

5. Assume that our product line would have the fewest number of parts and components and would have no fasteners.

6. Assume that our product line would have the absolute minimum amount of material content.

7. Assume that our product line would have the highest utilization of material content, i.e., the customer would essentially receive all of the material we had to buy to produce the product. Material yields in the high 90 percent levels.

8. Assume that our product line could be used in our competitors' products.

9. Assume that our products could serve international markets as well as our own in North America.

10. Assume that our products could offer the greatest number of features and benefits to the end users, fulfilling the perceived as well as the unperceived needs of the customers.

11. Assume that our products could be produced in a minimum amount of factory floor space, requiring a minimum amount of capital equipment and able to be produced in small lots, i.e., a lot size of one (1), without the penalties of time-consuming setup and excessive downtime.

12. Assume that the time to produce a product, the raw material to a finished product in a box, could be minutes, not hours, days, or weeks.

13. Assume that our product-line platform could deliver a full range of products covering all the price and feature points, i.e., lowest-cost platform with minimum features to the highest-cost full-feature platform.

14. Assume that our product line would be capable of retrofit to all our existing products, enabling the culling of all the old product offerings.

15. Assume that our product platforms would be able to supply customers with special features, i.e., colors, without excessive cost, delays, or disruption of the manufacturing and fulfillment systems.

16. Assume that our product line would not require a complete overhaul of the in-place management systems of our manufacturing and customer fulfillment centers.

17. Assume that our product-line manufacturing system would be a turnkey totally integrated and contiguous packaged system to allow relocation to any manufacturing facility without major reinvestment, downtime, and disruption to the continued supply of finished product.

18. Assume that our product platform design architecture would be so elegant that future product derivatives could be added without major investments in design, processes, and time to market.

With this thought architecture, was it possible to design, develop, and manufacture a completely new product platform that could claim to be uncontested in cost, value, and feature? It was my conviction that the quest for the uncontested cost-value-feature product platform would demand consideration of all 18 assumptions as acceptable constraints and would be the thought architecture drivers throughout the project.

However, articulating this thought architecture all at once would have doomed the project and labeled it as the impossible dream. Who in their right mind would tackle such a project with such a high-risk factor? This was a problem that had to be handled very delicately, unfolding the goals one step at a time. Using the theory of the hidden hand, i.e., shielding the unenlightened from the entirety of the ultimate goals, unveiling the goals gradually as the project moved on, would give the participants confidence as the tasks were completed, one by one. Careful gradual feeding to the team in a digestible fashion avoided the demoralizing trauma that would result in the team faced with this seemingly impossible task. As confidences were built, the 18 assumptions would be unveiled sequentially during the project. The project was similar to the famous Lewis and Clark expedition, i.e., taking a trip into the unknown and uncharted regions, heretofore untried by the inexperienced.

With this thought architecture as a driver, several weeks were spent in doodling and sketching, studying the design and process architecture of the competitors' products, and mentally walking through all the hierarchical elements of the product and its major subsystems down to the elemental subsystems. In this exercise care was taken to catalog what elements and subsystems were common to all producers and were at parity on a global basis. Isolating the parity elements, the task became very easy, the design and process R&D was narrowed to only the elements and subsystems of the products that could be influenced by novel approaches. This thought process unveiled concepts of a product platform that could be designed to satisfy the majority of the 18 thought architecture assumptions listed above. The impossible dream conceptually could be reached and a new product platform could be developed producing a product line that would be uncontested in cost, value, and feature.

So project X became a reality and was launched—a project that could be managed by a very small team, fully dedicated to the "new" product platforms designs.

25.1.4 Getting Ready: Forming the "Full-Time" Team

The first team member recruited was a marketing individual who had been working some years in the existing product line and had been a player in the *original equipment manufacturer* (OEM) contacts in the outsourcing initiative. He was familiar with the product line and the marketplace, having a good working knowledge of the features and benefits the consumer required, and was familiar with the competitive scene. Once convinced and on board, he became the key individual in building the remainder of the full-time team.

The next additions were a product designer and a manufacturing engineer, both having familiarity with the products. The designer came from the product development group and the manufacturing engineer came from the factory that produced the old product line. Both had worked on many of the cost reductions and process improvements of the old line. The manufacturing management wanted to loan the manufacturing engineer as needed, but we held our ground and insisted that he be-come a permanent member of the team full-time. Later a market development specialist was recruited to the team.

The core team of only four members was now on board.

25.1.5 Collocation: An Essential Ingredient

The full-time team members were collocated, i.e., officed immediately adjacent to one another. Collocation is, in my opinion, an absolute essential element in team projects. Collocation creates the bonding, the camaraderie, the cohesiveness, the "glue," the building of interpersonal relationships, and the cross-fertilization of the members—all essential ingredients in team building.

Collocation, in many companies, is considered a needless distraction, an unnecessary expense, and a trivial element in team effectiveness. Collocation is the team's unspoken catalyst creating the bonding and commitment needed in focused, fast, high-risk projects.

25.1.6   Team Resources: A Project Arena of Their Own

The team was assigned a sequestered room, a project room, a totally dedicated conference room that was initially called the "I dunno" room. This moniker was selected because we were overwhelmed by the "I dunno's." Practically every question we could ask ourselves had the same answer, i.e., "I dunno." The project room became the repository of all the "I dunno's." The project could not meet the goals and objectives of uncontested cost-value-feature products until we could convert the "I dunno's" to "I know's." The room was to contain all the items that we did not know about the market, the competitors, the components, processes, etc., essential in meeting the goals.

The team used this room for essentially all of their meetings, and it served as the repository and display area for all the needed information to be used in the project. Listing all that we did not know about the competitors, the component suppliers, the competitive products, etc., the team could methodically, one by one, convert the "I dunno's" to the "I know's." All "I dunno's" had to be displaced by the "I know's."

After some weeks, the "I dunno" room became the "I know" room.

25.1.7   Market Research

The team engaged an industrial research consulting company for a fast-track comprehensive update of the market place for their products under study. Their charter was to determine

1. The size of the market
2. Who the top participants were
3. The market share of the competitors
4. The market share of the OEM producers
5. The pricing and discount practices
6. The growth rates past, present, and future
7. What the entire peripheral market size was

Their findings were as follows:

1. Of all the competitors in the market, only two produced their own products; all others were using OEMs.
2. In the previous 8 years, three OEMs had garnered 63 percent (Fig. 25.3) of the unit volume serving this industry. None had been in the manufacture of these products for the industry prior to their entry.
3. The profit margins of the OEMs was much thinner than the margins enjoyed by the suppliers to the end users.
4. One aggressive OEM had changed its market approach and was selling direct to the end users, garnering fatter margins than the industry had been enjoying for many years.

The report was most revealing and enlightening. This information supported the strategy to not only supply the company's customers with the new uncontested cost-value-feature product line but allowed the design of the line to fit all of the company's competitor products, enabling the salesforce to sell direct into competitors' accounts.

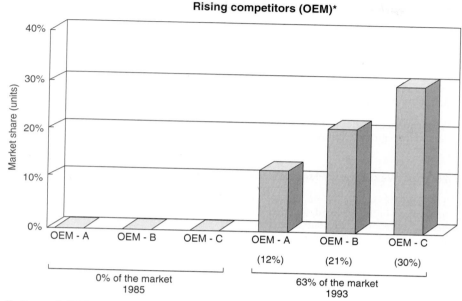

FIGURE 25.3 Rising competitors (OEM). In 8 years, three OEMs capture 63 percent of this market.

25.1.8 Competitive Product Analysis

Competitive products were obtained, disassembled, and analyzed. Reverse-engineering approaches were implemented to fully understand how these units were manufactured. Material costs, labor costs, and component costs were estimated. Feature arrays were developed in matrix form, and dimensional features were documented to enable the company's engineers to design their new products to fit all competitors' products.

The team visited several OEM factories, observing firsthand the methods and processes used by them in their manufacturing techniques. It was agreed beforehand that if we determined that the OEMs had distinctive advantages in labor rates, overhead structure, manufacturing processes, etc., that could not be significantly offset by our design and manufacturing approach, the project would be scuttled and we would settle for the strategy to source vs. manufacture. The trips reinforced our confidence that we were on the right track. Our approach could not be beaten by the labor, overhead, and material content advantages observed at the OEMs.

The team also trekked to Europe to visit an international show and exhibit where a host of manufacturers from all over the world were displaying and demonstrating their products and components common to the product lines we were working on. Here again, the team was encouraged by what was discovered; i.e., our approach was superior to the international offerings in our product line.

25.1.9 Next Steps: Conceptual Product and Process Design

The product line was laid out in a platform approach across the entirety of the product offerings:

1. Sizes
2. Features
3. Cost platforms—basic, standard, and premium

As mentioned earlier, the mechanical dimensional features were structured to fit all competitive products as well as our own.

25.1.10 Next Steps: Market Research—Conjoint Analysis

The array of features, pricing strategies, and feature groupings across the product-line hierarchy—basic, standard, and premium—presented a confusing array of options that needed validation. The team availed itself of one of the best methods and techniques for bringing logic and order to the vast array of variables: a well-designed conjoint study.

The conjoint study accessed an array of 100 respondents, viz., dealers, installers, facility managers, and end users, involving respondents of the company's products as well as the competitors'. Key information about customer preferences, pricing trade-offs, features, benefits, and costs was obtained.

This analysis provided the roadmap for the product-line offerings, arraying the platforms by sizes, features, performance characteristics, and costs. Further, the preference grid that was developed gave guidance in how the product line should be cataloged and featured. Lead products were aligned with the prevailing needs and desires of their target audiences, i.e., purchase decision makers and influencing end users. This strategy, albeit subtle, proved to have a major influence on the buying decisions, a result unanticipated before product launch.

(*Note: Conjoint analysis* is a method used to analyze product preference data and simulate consumer choice. It is also used to study factors that influence consumers' purchase decisions. Products possess attributes such as price, color, guarantee, environmental impact, and predicted reliability. Customers typically do not have the option of buying the product that is best in every attribute, particularly when one of the attributes is price. Consumers are forced to make tradeoffs as they decide which products to purchase. Conjoint analysis is used to study these tradeoffs.)

25.1.11 Next Steps: Integrated Design and Process Development

The prevailing objective, i.e., uncontested cost-value-feature, was the ever-present driving force. The integration of an optimum manufacturing process influenced the mechanical design architecture. Feature options, lot size constraints, and time to manufacture could not be ignored in the design. A lot size of one (1) could not be a constraint, change of sizes could not be a constraint, and time to manufacture could not be a constraint. The manufactured components were so designed to allow size and feature variations without penalties of major setup costs, inventory buildup, packaging changes, etc.

The results—an integrated manufacturing system that was an integrated, contiguous, continuous fabrication, assembly, test, and pack line. The step-by-step manufacturing processes were totally linked (integrated), arranged in the same area (contiguous), and structured to operate in a "flow-'n'-go" manner (continuous), i.e., as one product enters operation number one (operation 1), a finished tested product in a box exits the system, i.e., one in–one out. All steps of the process being linked eliminated batching or staging except when nonstandard products were needed, and parts were intercepted for special treatment and then reinserted when the unique feature was completed, i.e., when a nonstandard paint finish was specified. Those components were diverted to a paint system and then returned and inserted into the line for completion.

Using this integrated design and process approach, the standard products are produced, raw material to finished product in a box, in 11 minutes traveling about 160 ft in distance. Production rates are determined by the number of operators on the line; the normal rate would be about four products per minute.

25.1.12 Material Utilization: The *U* Factor

Rarely do design and process engineers in manufacturing companies focus on material utilization, i.e., how much material is required to produce the part compared to the amount of the material the customer receives in the finished product form. The ratio of these two numbers is called the *U factor.*

$$U \text{ factor} = \frac{\text{material in finished part}}{\text{material required to make part}}$$

For example, if a finished part as received by the customer weighs 6 lb and the amount of material required to make that part weighs 10 lb, the *U* factor is .60% ($\%_{10}$ = 0.6 × 100 = 60%)

In the product design/process system of this project the *U* factor exceeded .9 percent. In other words, the customer received 99$^+$ percent of all the material that the manufacturer had to buy to produce the product. This is a clear example of what it takes to have an uncontested cost-value product line.

25.1.13 Speed to Market: Concept to Initial Production (367 days)

Speed to market was another essential ingredient to project X. "How can a new product platform be conceived, developed, tooled, and launched in such a short period of time?"

Project X's answer was: "The stringent adherence to the following multiplicity of effective team principles, serious consideration of the need for action, and embracing the thought architecture's 18 issues as they were enfolded with understanding and acceptance." The team articulated, documented, and practiced as a unified group from project initiation to completion, frequently revisiting the thought architecture and fine-tuning and tweaking as the need arose.

Let us summarize:

1. A serious need for action was evident, and a major decision was about to be made concerning the future of the product line: abandon manufacturing and outsource the line.

2. A senior officer of the company was uncomfortable about the decision to outsource as a viable long-range strategy.

3. A demanding thought architecture was developed, first in the mind of the officer, then transitioning into the minds of the project team.

4. A small multidisciplined team (four or five members) was recruited to work on the project full-time under the direct supervision of the senior officer. The disciplines represented by the team were marketing, product design, manufacturing, and market development. The team leader was the marketing member.

5. The team was collocated, i.e., officed in immediate proximity of one another, enriching the team with the benefits of integrated commitment and the associated interpersonal relationships that evolve.

6. The team was provided a project room—a "war-room" facility for their exclusive use.

7. The team was provided resources to commission a market study to better and fully understand the market, the competitive situation, and the source of products, i.e., integrated producers vs. OEMs.

8. The team was charged with total responsibility for the project design, tooling, manufacturing, marketing and market development, pricing, advertising, promotion, training, etc.

9. The team had the flexibility to commission experts in specific critical technical aspects of the design and tooling, reaching outside the company to find the experts to focus on the specifics essential to meet the goals of cost and performance.

10. The team used advanced marketing techniques, i.e., conjoint analysis for guidance in feature and benefits as well as pricing and positioning of the product platforms.

11. The team developed the business plans, capital requisitions, pro forma P&Ls (profit/loss statements), strategic marketing plans, etc.

12. As the project moved from the development and tooling phase into manufacturing the product, the team was relocated en masse to the manufacturing plant to continue the intimate linkage to the product line as production commenced and the product was launched, i.e., "running the business."

13. Most importantly, the team never lost control of the project as it interfaced throughout the many other elements of the company that had to be interfaced and utilized, i.e., the development, tooling, manufacturing, sales, and distribution department's infrastructures. It was their project and is still their project and their business to run. They act like a "small company within a large company."

In closing, the author adds another simple cookbook recipe for team projects which is brief and somewhat exacting, nonetheless important for success which is entitled "cautions in managing team programs."

25.2 CAUTIONS IN MANAGING TEAM PROGRAMS

25.2.1 Coping with Barriers, Disconnects, and Fractures of Team-Driven Projects

Team-driven projects are fragile and are fraught with continuous barrages of barriers and obstacles as they progress through their assignment. Large corporations, with all their complex interdependencies, practices, structures, rules, and procedures, tend to burn out teams, leaving them with diluted final results, unreached goals, and exasperating circumstances.

Teams invariably will lose control of their projects as they interface with the seemingly endless elements that must be attended to in designing, developing, manufacturing, and selling their products.

To be successful, teams have to run the gauntlet of the establishments they work in. The stamina and tenacity of the team have to be world-class for the team to survive and succeed in their project. Teams without a top-management sponsor ever-present to provide guidance, counsel, resources, and most important, "air cover" protecting them from the land mines, the surprises, the politics, the "not invented here" (NIH) attitude, the territorial prerogatives, etc., are almost assuredly going to fail or underperform.

The following author's checklist, i.e., gauntlet, for team guidance is provided. Every item in the "gauntlet" that follows must be answered with a "yes" or "no" by the team. The team must cope with and constantly work on converting all or the majority of the "no"s to "yes"s. The team desirous of a surprise-free environment throughout their project must take this list seriously or risk failure and disappointment.

Checklist of Essential Inquiries in Team-Driven "Global Cost-Value-Feature" PPR Projects (Running the Gauntlet of Inquiry)

1. Is there a senior manager of stature and power (influence) in the organization who wants the project and will assume influence over ownership and accountability?
   Yes_____ No_____

2. Have other top-management people been convinced, and are they committed to the project? Have they assigned a "champion" of stature and power to lead, guide, and manage the endeavor? (Assume that individuals have been polled and a consensus exists.)
   Yes_____ No_____

3. Is the project important to the success of the corporation? Yes_____ No_____

4. Does or could the project have "global" potential? Yes_____ No_____

5. If successful in its development and implementation, will the project make a significant impact or difference to
   a. The product line? Yes_____ No_____
   b. The end user, the customer? Yes_____ No_____
   c. The company's profits, market share, and growth? Yes_____ No_____
   d. The company's competitors? Yes_____ No_____

6. Are new and previously unapplied technologies accessible that will enrich and expand capabilities or create new features and functions (heretofore not available or integrated into existing products or new products, ours or our competitors)?
   Yes_____ No_____

7. Will a discontinuity and/or a new standard of performance be created by this project?
   Yes_____ No_____

8. Will this project fulfill unperceived needs of the end users, the customers?
   Yes_____ No_____

9. Can this project be isolated and insulated from the established bureaucratic quagmire of the corporation and managed as a fast-track project given air cover and protection throughout the fragile fledgling inception to commercialization?
   Yes_____ No_____

10. Are these critical functions involved and committed at concept development stages of the project?
    a. R&D Yes_____ No_____

b. Marketing   Yes_____ No_____
c. Product engineering   Yes_____ No_____
d. Manufacturing   Yes_____ No_____
e. Design and styling   Yes_____ No_____

Further Advice and Counsel to the Team Participants. Replies should not be simply intuitive replies. Every "no" response must be considered a "disconnect" and a serious barrier to successful concurrent or simultaneous projects. "No" is the alert warning sign, a beacon to be reckoned with. Ignoring the "no" or pretending it is not to be an important "no" will inevitably indict the errantly cavalier team player's attitudes that so often creep into many project's team(s). Teams must constantly strive to get the "no"s eliminated or provide conclusively that the "no" is not significant and can be culled or defused in its ability to scuttle or damage the project.

Time and time again, projects will be delayed, scuttled, wander aimlessly, drift from initial goals, underperform, be fraught with surprises and failures, be redefined, exceed budgets, miss schedules, frustrate and infuriate management, get people fired, and lose customers because the "no"s were ignored and not attended to.

If the "no"s cannot be eliminated, then the project, in many cases, should be tabled or dropped. If the project is not important enough to command "yes"s from question 1 down through the list, maybe it isn't important enough to do!

Don't waste time and resources on "dry wells" and nuisance and futile projects, filling the company's skeleton closets, bone yards and "unmarked-gravestone cemeteries" with unthinkable sunk costs of misguided and mismanaged projects that can't endure and survive the critical scrutiny of the gauntlet of critical inquiry.

25.3 EPILOG

Sixteen months after first shipment of the products of project X and 6 months into its first full fiscal year of production, the project X team reported the following:

1. The annual unit volume will be nearly double the unit volume of the old product platform.

2. Year-to-date sales are tracking at 38 percent ahead of budget (plan).

3. Year-to-date pretax profits are tracking 90 percent ahead of budget (plan).

4. Manufacturing celebrated one year, i.e., 52 weeks, of meeting 100 percent of production schedules.

5. After a defect-free finished-goods audit of the first 250,000 units, the finished-goods audits were discontinued.

6. The old platform culling resulted in the elimination of 228 style numbers. The new platform requires only 74 style numbers.

7. Manufactured parts inventory turns are tracking at ≥ 60 turns per year.

8. Purchased material and component inventory turns are tracking at ≥ 40 turns per year.

9. The product platform manufacturing unit enjoys the highest sales dollar per square foot in the company, the lowest percentage of direct labor in the cost of sales of manufactured goods in the company, and the lowest fixed and variable overhead and asset ratios of all current manufactured products of the company.

10. The sales division has reported success in breaking into competitive accounts

heretofore unpenetrated and was able to pull along other company's products with the order.

11. The project X team is still intact and resides in the manufacturing plant in close proximity of the manufacturing area. At present the team is working on future product platforms and product derivatives using the cost/value design and process architectures so successful in this product platform renewal.

25.4 CONCLUSION

Product-line renewal programs properly structured and staffed with small teams given full control and guided by thought architecture developed at the concept stage and tenaciously held on to throughout the project will produce outstanding results. "Uncontested cost/value products" are still obtainable and can be produced in the United States!

MANAGING MANAGEMENT OF TECHNOLOGY

CHAPTER 26

MANAGING THE MANAGEMENT OF TECHNOLOGY PROCESS

Lynn W. Ellis

University of New Haven-Scholar in Residence
West Haven, Connecticut

26.1 INTRODUCTION

"Manage the process"—that is the key to technology management. All the planning in the world will not get new technology implemented, so the process itself has to be managed. This chapter is about how to get there, how to make it happen, and how to face up to change. These tasks are what organizations have to do after choosing a technical strategy. And these are not just managerial tasks, because people at the bench level really run the process. Managing the process yields engineering performance effectiveness, and makes the desired impact on downstream operations.

Analysis of the process of managing technology management begins with a choice of a model. The process to be managed changes with time along the innovation process. It also changes with the sector to be managed, whether this is resources, infrastructure, or functions.

26.1.1 An Innovation Process Model

Many companies now manage their innovation activities on a classic open-systems basis, in the manner shown in Fig. 26.1. This is done from inputs through processes to outputs, with minimal feedback.

Depending on the degree of R&D-marketing coordination, the inputs may come from these two organizational units. Or commercial information may come in serial form as marketing tosses specifications over the wall to R&D. In many companies

FIGURE 26.1 Classic open-systems model. (*Source: Compiled by the author.*)

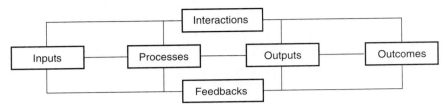

FIGURE 26.2   Partially closed systems model. (*Source: Brown and Svenson.*[1])

also, the output point is where design and manufacturing information is tossed over the wall to the downstream implementing organizations: manufacturing, divisions, or subsidiaries.

In comparison to how modern companies appear to be organizing, this simple model seems to be missing the interaction with other organizational units in the company. Today these interactions are not only in a serial interaction form with feedback but also in various parallel or reciprocal feedforward interaction forms. Outputs from R&D have to become outcomes of value to general managers and shareholders.[1] A representation of this modified model is shown in Fig. 26.2. The effect of modifying this model is that, while making it more complex, it allows a much greater variety of potential *leading* indicators to be studied. For the purposes of this chapter, this expanded model will be used to characterize the measurements of outcomes, outputs, interactions, and inputs which need to be made to make the process effective.

26.2 PROCESSES AS A FUNCTION OF TIME

The processes to be managed vary from earliest exploration to "sales-order specials" for older products. Some remain the same, some must be added as time progresses, and some are less important at the later stages of the time scale.

For convenience, this time scale for innovation has been divided into four sectors, as shown in Fig. 26.3: exploratory concepts, opportunity creation, core competency development, and product improvement. Over this range, some processes stay the same, and others gain or lose importance.

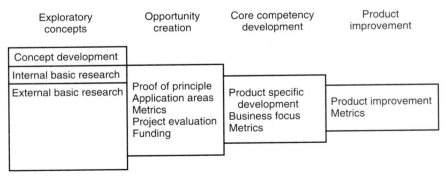

FIGURE 26.3   Processes as a function of time. (*Source: Adapted from IRI Research-on-Research Committee.*)

26.2.1 Processes That Are the Same from Exploration through Extension

Some processes to be managed do not change along the time cycle. People have to be managed. Funding has to be sought. Managing requires focusing on effectiveness, which means doing the right thing, rather than on efficiency which is measured by output divided by input. These processes are the same for all managers, not just ones of technology: plan; organize, lead, measure, and control,

Managing People.   This comes after planning, which does change along the time cycle, and is covered below. Since other chapters in this book cover managing people more extensively, only a brief overview is covered here.

Once the plan is set, organizing becomes the next task. Organization involves the division of work to make effective the specialization of individuals. It also requires integration mechanisms to make whole the divided work in the interests of the entire organization.

The old-fashioned division of work was by minifunction within the R&D organization. Research threw results over the wall to development, which threw models over the wall to engineering, which—you get the idea. This overspecialization made best use of individual talents, but took time. Chrysler used to take 5 to 6 years to create and first deliver a new car, and now does it in 2.5 years.[2] They do it by project teams, which is the increasing trend in U.S. industry. Teams bring project ownership and team-member empowerment. The use of cross-functional project teams has now reached the 70 percent level in the United States.[3]

Leadership Following Organization.   The top R&D manager used to select all department heads and project managers. Should managers continue to do this, or should teams elect a captain? Project and program championship now becomes an issue for leadership. Launching a new team requires that management provide the initial team leader. But after that, if a natural leader develops, the manager should quietly allow that individual's succession, moving the previous leader onto the next crisis.

Leadership also involves staff motivation and rewards. Morale is a consequence of motivated and rewarded team members. Rewards may be either extrinsic (money, promotion, etc.) or intrinsic (recognition, etc).[4] Leadership also involves the skills assessment of the quality of personnel, arranging training time each year of the proper quality, and conflict management.

Measuring Time.   Time measurement occurs throughout the process. It is not enough just to measure milestones. Effective metrics also include staff hours and costs on a project basis, and cycle times through the project's stages. It is wise also to measure the satisfaction of the downstream organization with R&D's timeliness. Many other measurements are needed depending on the stage along the time cycle.

Control.   This is a final people management task. What you measure is what you control to, and what you get as results. Thus, careful selection of metrics is important to send the right message to the staff. Top-down control is necessary, but not sufficient. Peer evaluation of technical accomplishments is a desirable form of control.

Funding.   Funding of the process tasks is also an activity that is carried on the entire length of the technology process. In the exploration phase, the technology manager may be a primary determinant of what gets funded. Business unit manager selection and evaluation ensures that the R&D is important to achieving the outcomes desired by management.

26.2.2   Processes Most Important in Exploration through Development

The processes most important from exploration through development are technical planning, and ensuring the quality and preservation of technical output. By product improvement time, the customers and the field salesforce mostly guide technical planning, and the amount of technical contribution needed is less than for the earlier stages.

Technical Planning.   Technical planning for exploration involves the generation of concepts, and the balancing of internal and external research efforts. Increasingly, the focus is on external alliances, as covered below in the section on managing external resources. By opportunity creation, technical planning has shifted into an alliance with business managers for the selection of projects and the concentration on application areas. Technical planning still needs to outline what is needed to reach proof of principle. Developing core competencies brings technical planning more to a product-specific mode focused on business needs.

Ensuring Quality.   Ensuring the quality and preservation of technical output is another task from exploration to development. Acquisition of technical knowledge is the technology manager's responsibility, done through training, hiring, or alliances. This and technical planning need metrics to guide the manager. The traditional R&D output metrics are number and quality of patents, copyrights, and trade secrets. These fail modern needs in two ways: general managers are interested in outcomes, not in outputs; and patent or copyright activity may be more a function of the activity of the legal department than of R&D work of value to the company. Still, a common rule is "one patent per megabuck of sales."

Outcome Metrics.   Outcome metrics are of more value to higher management. These include measuring sales protected by a proprietary position secured by R&D activity. Another useful output metric which leads into outcomes is to measure at the end of each stage the percent of started ideas which are adopted by the next phase and eventually find their way into a product. Turning this into an outcome metric, the technology manager needs to show higher management how this percentage flows through to the "bottom line" (net income). The outcome *effectiveness index* (EI) may be expressed mathematically as follows:[5]

$$EI = \frac{\% \text{ new-product revenue} \times (\% \text{ net profit} + \%R\&D)}{\%R\&D}$$

Purists among the author's accounting colleagues point out that net profit is after interest and taxes which have nothing to do with technology. Thus, the term in the parentheses may be replaced with equal effect by operating profit, or by earnings before interest, taxes, and R&D.

26.2.3   Other Subprocesses Need to be Added with Opportunity Creation

With the onset of opportunity creation, or as some call it directed applied research, new subprocesses come into play. The first of these is the need for the effort to have an alignment with the strategic direction of the firm—the "directed" part of the definition. One achieves this by bringing into the direction process those managers who have the ability to take the results and run with them, often through a group activity.[6,7] Directing opportunity creation necessitates individuals who have knowledge of cus-

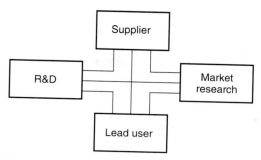

FIGURE 26.4 The lead user concept. (*Source: Herstatt and Von Hippel.*[8])

tomer needs. Most R&D staff have minimal coupling in this area. One such solution is to identify a lead user, or teaching customers with whom both the R&D staff and the marketeers interact.[8] This is shown graphically in Fig. 26.4. This solution has the advantage over just having marketing let R&D know the customers' needs in that it removes one intermediate source of attenuation and distortion. Whatever the method, customer contact time is the key metric to be used in calibrating knowledge of customer needs.

Opportunity Creation. Opportunity creation needs goal clarity. As an old New England saying goes, "If you don't know where you are going, any road will take you there." Many R&D groups don't know where they are going, and fail to deliver as a result. This happens not only because there is insufficient knowledge of customer needs but also because the direction from above is imprecise. This led one pair of authors to call it the "fuzzy front end."[9] Cutting out the fuzziness is part of managing the technology management process.

While project selection and evaluation need technical and business input in earlier stages of the technology management process, opportunity creation is the point where such input also has to be financial. A business is an organization which measures its results in money. Key higher executives live by the dollar sign—so must technology managers. What really matters is not the form of financial measurement, since it can be shown that they are all mathematically related.[10] The most important metric is the one that the firm uses customarily, since it is the one that other claimants within the firm use to present their cases for funding in competition with technology managers. These other claimants include capital projects for existing plant facilities, industrial engineering projects for cost reduction, and market development beyond existing markets. If the return on an opportunity creation project in technology does not match up to these other potential uses of that scarce commodity called "money," it will not be approved.

Still another metric needed in opportunity creation is the response time to competitor moves. This is the idea of customer cycle time composed of the "fuzzy front end" of marketing time plus the response time through R&D and implementation, as shown in Fig. 26.5. To be competitive, companies must be able to respond to competitor moves, as well as to ideas from the customer. If a company has adopted cycle time in all departments, the response time is measured as the sum of the segments. But if any one department in the chain fails to track cycle time, the overall time metric will be imprecise.

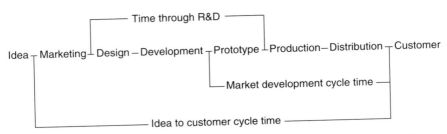

FIGURE 26.5   Three cycle time measures clock the innovation value chain. (*Source: Compiled by the author.*)

Customer Satisfaction.   This is still another metric needed from opportunity creation onward. A recent survey identified desirable forms of metrics, measuring the customer's satisfaction with your firm as well as the service or product.[11] Over 50 percent of companies use an index, market share, or a lead user to measure customer satisfaction, and often two or more of them. An overall customer satisfaction index tested had responsiveness, technology, and quality-reliability as its main components, but not cost competitiveness. For a cost-competitiveness satisfaction rating, measure market share and responsiveness, and continue use of time measurements once you have started to use them. Faster or slower cycle times affect different components of an overall index. Speedy innovation is associated with customer satisfaction with high technology and customer communication. A slower pace of innovation is associated with customer satisfaction with cost competitiveness, program-contract management, and product quality and reliability. From follow-up interviews, the customer was identified not only as the end user or consumer, but also as the downstream implementation chain and the distribution chain, as shown in Fig. 26.6. R&D managers should select measurements from the survey results to provide useful leading indicators for enhancing customer satisfaction in their own companies.

26.2.4 New Subprocesses Begin with Core Competency and Product Development

By this point in the innovation cycle, managers need to focus on their goals for implementation. Typically, those downstream in the organization place quality, time, and cost as primary objectives.[12] Quality needs to be designed into the product or service.[13] Cost of the product or service begins to become a paramount issue. Only time in the implementers' eyes is governed in the same manner as discussed earlier.

Product quality and reliability tend to be measures of customer satisfaction with the product, rather than with the company as an innovator. Defect quality and reliabili-

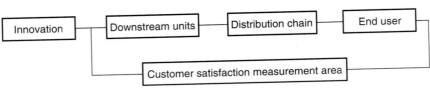

FIGURE 26.6   Steps from innovation to the end user. (*Source: Compiled by the author.*)

ty of an innovative offering are two distinct, but complementary categories. As the customer for an innovative service or product sees it, defects are occasions when performance is not as expected. Defects per million opportunities is an appropriate measure. Reliability is the time the product is free from defects, often measured as mean time between failures (MTBF).

Product costs lead directly into measuring the prospective financial return versus the development and product costs. This is the main business focus.

Another new issue is that of technology transfer. The cross-functional team metrics discussed below are one way to quantify this task. Another is the degree of formality in the transfer process.

26.2.5 By-Product Extension and Improvement: Some Subprocesses Declining in Importance

Strategic alignment tends to decline in importance, because the markets to be served are now known. The metric then becomes one of deciding whether the extension is worth it—does the incremental profit exceed the incremental costs?

The quality and preservation of technical output also decline in importance, because extensions and improvements rarely generate patents, and the skills required are now known by several people.

26.3 MANAGING THROUGH RESOURCES, FUNCTIONS, AND INFRASTRUCTURE

Three sectors need to be managed: resources, infrastructure, and functions. The resources available to the innovation manager are both external and internal. External resource management will be addressed first. Because they overlap in content, internal finance resources will be addressed with the financial function, and internal human resources will be addressed with infrastructure.

26.3.1 Managing External Resources

Managing external resources involves spanning external boundaries. Besides spanning the boundary to the financial world, the innovation manager must span the boundary from inside the company to other parts of the outside world. That boundary is, in practical terms, two different boundaries. The first is the one to the business and customer environment which is essentially commercial in nature, and the other is the one to the sources of technical knowledge. Both must be spanned equally effectively if the company is to prosper.

26.3.2 Buildup of Commercial Knowledge

Four tasks characterize the methodology of building up knowledge spanning the commercial boundary. First, a vision is needed of what the firm is trying to do. The external environment for business encompasses the three remaining tasks: the environment common to all, the industry's particular characteristics, and the competitive situation. Knowledge of each of these underlies establishing a clear picture of customer needs and opportune ideas for products and services.

Start with vision. How does one get vision? The methodology outlined below will help systematize the search, but one must have the ability to see what has not yet been seen by others. This analysis is usually called *strengths, weaknesses, opportunities, and threats* (SWOT). Vision starts with opportunities, with the question "What might we do?" This is an ideal point for brainstorming sessions. There are many choices, so that this question narrows down the list only slightly. Next, the question becomes, "What could we do?" Threats limit choices, as do internal strengths and weaknesses. It may be possible to obtain some strengths externally and to offset some weaknesses externally. At this point, the list becomes much shorter.

"What would we do?" The values of those deciding come into play. If one is ethical, criminal acts don't stay on the list, and so on. Finally, a choice has to be made: "What should we do?" If several options still remain on the list after "might," "could," and "would," some selection has to be made. Visionary people go through this cycle instinctively. For the rest of us, the discipline of going methodically through this cycle is necessary. To convert this selection into a detailed focused plan, the external environment needs to be analyzed.

26.3.3 Environmental Assessment

Opportunities and threats exist in the external environment. Since industries and competitors may be assessed separately, the first analysis should be of the part of the external environment that is common to all industries and competitors. This environment is economic, demographic, political, and social (including labor relations). It is also always changing. To establish a baseline, the existing environment should be surveyed first.

Present Factors.   What present factors now control the environment? On the basis of experience with earlier brainstorming sessions, a group tends to run out of good ideas after about 20 suggestions. These suggestions generally depend on the character of the business which has been chosen. Some businesses are more prone to the fluctuations in the economic business cycle. Others depend on political acts, such as the degree of government support for the Information Superhighway. It is important to probe deeply into the nature of this external environment to elicit as many ideas as possible from differing viewpoints.

The next task is to decide which of these factors are the most important. For example, a university needs to base its undergraduate planning on the number of 17- to 21-year-olds available for forming a student body. The number of babies born this long ago was the lowest in the prior 20 years, resulting in a recession in the present undergraduate student population. Again, on the basis of experience, about six of these factors will be seen to be most important after a vote of the brainstorming group.

Future Factors.   Which future factors will dominate in the next few years? Five years is a good number for most businesses, but some are shorter-term, such as toys; and some longer-term, such as mining. Again, the brainstormers will cite many of the present factors, but some new ones will appear, such as shifts in health-care financing and upcoming elections. Twenty or so ideas is a good base from which to develop another six most important factors.

Change.   What has changed? What impact will these changes have? A university can't change the number of babies born two decades ago, but it can see that the birth rate has been rising for a decade, now causing municipalities to build new high schools. Thus a university can plan for the time when these teenagers reach college age. By the time that a brainstorming group has gone through this analysis of the environment, a clear list of opportunities and threats will have emerged.

26.3.4 Competitive Assessment

The best method of competitive analysis is to create a matrix of competitors versus key characteristics. Which competitors should be chosen? The three to six competitors who have 60 to 80 percent of the market is a usual selection for the columns of the matrix. If the industry is more fragmented than that, some logical grouping may be required; for example, all importers as a group might fit in some industries.

The other axis of the matrix consists of about 10 rows of the principal characteristics of companies which affect competition in the chosen market. Usually, this includes the four "p"s of marketing: product, price, place (market outlet), and promotion. Up to six other characteristics may include costs, finance, technology, location of plants, or any others which describe the industry's competitive conditions. For each characteristic, your firm's rating as a competitor should be rated against each other competitor or group of competitors. The rating should be quantitative, if possible, such as 15 percent higher in price.

The prepared mind will translate the above analyses into a knowledge of customer needs. From this list of needs, good ideas will be generated readily for useful and economic service or product offerings, which take proper account of the environmental and competitive threats, and the strengths and weaknesses of your firm as a competitor. If it is difficult to generate such a list, help may be needed in the form of new people or consultants. For most companies with which the author is familiar, however, the list of prospects usually well exceeds the available funding. It does not take a large or costly organization to run such an analysis—a handful of the right people is usually sufficient.

26.3.5 Buildup of Technical Knowledge

The innovation manager is custodian for the company of the gateway to new knowledge across the company's external boundary. It is not necessary for this knowledge to be technical in nature. However, for a large fraction of companies, the knowledge needed for competitive advantage will be in science and technology. Building up the technological knowledge base thus becomes an important boundary spanning task. The choices are to do it all yourself, or to multiply your effort with outside sources. If the latter is your choice, the sources need to be identified and managed.

Technological Self-sufficiency. This is the classic model of the innovative firm of the past, such as Thomas Edison's group at Menlo Park, New Jersey. His example was followed by a number of early industrial technical laboratories, where innovation depended entirely on internal resources. This lead to the acronym NIH (not invented here) as a way of saying that imports from across the external boundary were not wanted. The facts of life are that, for all but a few firms, NIH is now too costly these days. In addition, the entrepreneurial spirit has made the sources of innovation abound in small firms, and in your own company's suppliers. Government spending has created a large base in government laboratories which is available to be tapped. The flow of advanced research funds is now largely to universities, not to industrial laboratories. The growing base of scientific and technical knowledge outside the firm's own country is also there to be accessed. Managing to span this external boundary to technology is now central to the innovation manager's task.

Managing Alliances with Others.   This may take many forms. These range from simple one-on-one relationships, usually in contractual form, to multiple company consortia focused on a common goal. Some of these are discussed in more detail below.

The boundary-spanning task begins with a fundamental commitment to multiply your company's own technical effort by some substantial factor using these external resources. There has not been much written in a scientific manner about how to manage to do this. However, there are quite a few articles and conference papers which are descriptive in nature on how companies have managed alliances with partners. These seem to all point to two common principles: trust, and agreed-on measurements.

Mutual trust between partners in alliances of all forms seems to be essential for multiplication of technical effort. As one executive put it "...the relationship we have with our suppliers is...'virtual enterprise'—it really is as if we are becoming one big, seamless value-added chain."[2] This is diametrically opposed to the adversarial relationship that most purchasing departments have adopted with vendors of all sorts. Until and unless this shift is made to being on the same team, effort enhancing alliances will not work.

Building trust is a difficult task. It begins with encouraging friendships between key executives involved in setting up the alliance. In most countries overseas, executives stay for long periods with their company, and expect the same of U.S. companies, which often is not the case. Communication between partners must be open and comprehensive as to goals for the venture. Disagreements must be resolved equitably, surprises avoided, and contractual disputes worked out amiably. The people in the trenches must work in each others' facilities and not try to work the alliance just by telecommunications.

Partners bring differing metrics to any alliance. Each partner has to take back information to its management in a form which will make the alliance appear to be an extension of its internal effort. Reconciling these metrics can be difficult in a one-on-one partnership, but may end up being a major stumbling block in multicompany consortia. Resolution of these two factors seems basic to successful external boundary spanning in the search for new technical knowledge.

26.3.6   Sources of Scientific and Technical Knowledge

The potential sources of multiplying knowledge are many. Most firms, however, find that there are five most valuable places to search. Suppliers have already been mentioned. Universities in the United States are recipients of most research grant funds, and thus fountains of new science. Government laboratories have their areas of expertise, such as material and massive computation, which may be what an individual firm lacks. Joint ventures have been used for years domestically, and are currently a leading method of tapping into research and development being done overseas. Research consortia of many companies are the newest of the potential sources, following recent favorable U.S. government rulings on antitrust issues, and the growth of similar intergovernmental sponsored groups in Europe. Each source has its particular advantages for partnering, and each also has its own disadvantages.

Suppliers.   Suppliers have many advantages as partners. They have the industrial mentality, and thus fit well with your company's values. Many have substantial research teams, such as Goodyear, a major tire manufacturer, and one of auto manufacturer Chrysler's new product team members.[2]

Universities.   As a result of many years of government funding, universities are now the home of the leading edge of science in the United States. Thus, they are the logical place to obtain new technical knowledge. This may be obtained through contracts for consulting, licensing of their technology, or applied-research contracts. Some universities have specialized facilities which may be used to supplement those in house. Finally, if you do not have close relations with universities, you will not have access to their best students when they graduate.

Barriers to using universities are many. The ownership of intellectual property rights needs to be resolved. "Publish or perish" is a university need which may run counter to company interests of keeping its technology from others.

Also in the security area is the need of the researchers to know about elements of the company's business strategy, which some firms may be reluctant to cede. Responsiveness, particularly short time to complete, is another issue. Companies need to have results at low investment costs because of the difficulties of capital approval. The termination procedure for the agreed-on program must be in place and understood, both because universities need lead times to rearrange their activities and because many research assistants graduate and want to move on, thus prejudicing the project's finish. Despite these disadvantages, universities are best used for basic knowledge, specialized facilities, and precompetitive research.

Government Laboratories.   Government laboratories are also advantageously used for specialized facilities, unique knowledge such as materials, ability to conduct massive computation, talented personnel, and product development skills. It is also worth getting to know the people who may in the future have the power to regulate your activities. The disadvantages are several. Government laboratories tend to lag behind universities in resolving intellectual property rights. Freedom-of-information laws limit their security. They are often suspicious of industry, and shift with the prevailing political winds. Access to seed money is not readily available to explore new ideas. And because of the downsizing which accompanies the current "reinventing government," they often have lots of internal turmoil.

Joint Ventures.   Joint ventures and other forms of partnering with single companies (other than suppliers) form another means of multiplying effort. This category ranges from simple licensing across to jointly owned separate companies. Such alliances have the advantage that both companies have an industrial mindset which facilitates cooperation. Partnering may be between firms with different specialties in the same markets, such as technology and distribution, respectively. Or they may be with noncompeting or smaller firms in other fields in which both partners have their own unique contribution, but have their individual reasons for preserving their independence, rather than merging. Disadvantages include NIH, reluctance to divulge confidential strategies, skepticism about effectiveness and relevance, responsiveness, intellectual property rights, inexperience with identifying partners, managing partnering, and negotiating terms.

Research Consortia.   These are the newest form of partnering in the United States. An early example is the Microelectronics and Computer Technology Corporation (MCC), a consortium of approximately 22 shareholder companies and 48 associate members. These consortia are established to share the costs of the evolution of new science and technology of strategic importance, on a large scale, far beyond what individual firms could afford. Similar efforts in Europe are driven by the need to share costs of new technology across many national boundaries.

Success factors include having a vision inspired by those collaborating. Sustaining collaboration must be learned by all partners and regularly reinforced. Boundary spanners need to be rewarded for their contributions. The problems found are those discussed above, but the multiple heterogeneous voices to be heard and satisfied make the management of consortia a complex and time-consuming task.

26.3.7 Organizing and Systematizing Technology Acquisition

Management across external boundaries needs to be organized and made systematic. Successful use of outside sources requires first an understanding of what types of programs you are willing to place externally. From among the sources listed above, which will do best to achieve your objective? This requires focusing on the source's true competencies. For science one might go to universities. For mature product lines, this might be consultants, etc. This also requires deciding what to keep inside, which for most firms is usually key core technologies, new initiatives, and strategic innovation on which new breadwinners will be based.

Next comes the need to handle interpersonal relations. Dedicated personnel must be assigned as program champions, and kept on that task until it has been completed. Colleague-to-colleague relations are essential to form trust. Some cultural change may be needed in your own organization, just as it will be needed in that of your partner. Speed to market, commitment, and urgency will help to give successful results quickly and establish the creditability of the alliance to both partner companies.

A consistent regular external review process will be needed to bring each colleague's peers and superiors into play. The innovation manager has to depend on the commitment of some downstream part of the organization to implement the finished innovation. Preference should be given to an activity council which includes representatives from the groups onto which the innovation will be grafted.[6,7] Groups benefit from the greater knowledge of a variety of members and the increased number of ideas generated. Participation in such a council builds commitment to the result and appreciation of the decisions which led up to it.

Finally, the metrics for alliance programs need to include all those each partner uses for inside R&D programs. These will include financial ones such as program costs, sales, product costs, margins, and capital. They will also have to include nonfinancial measures as discussed previously in this book, such as timeliness, customer satisfaction, and measures of strategic intent. If your prospective partner is not up to your standard of metrics, or vice versa, the conflicts over metrics may keep the alliance from achieving its goals.

26.3.8 Effect of Internal Organization on Boundary Spanning

This chapter has so far treated boundary spanning as if the organization were a single entity. In practical terms, however, the boundary spanning tasks are different for commercial and technological boundaries. About two-thirds of U.S. companies are organized by product lines or have their team organizations by products or product platforms. For these companies, the boundary spanners to the commercial world are correspondingly decentralized.

The technology boundary-spanning task is not quite as simple. Most decentralized companies have shown a need for some centralization of this task. The usual form for this is a central laboratory, even if the bulk of the technical effort is decentralized. For example, Chrysler keeps a small central scientific group effort of 69 people who work with its four platform teams.[2] To further avoid duplication of effort, they have also introduced an informal network of "Tech Clubs" formed by the engineers of like technologies from each of its platform teams and the central scientific research activity. The management of such informal groups has its own set of skills.[6,7]

26.4 MANAGING THE INTERACTION WITH FUNCTIONS

The management of interaction processes between functions or departments may be done by various integrating mechanisms. The first group of these concerns the R&D-marketing interface, including commercial development. The downward transfer of technology is a second group of interactions. The now widespread use of cross-functional teams, including concurrent or simultaneous engineering, is seen as the best overall management practice.

26.4.1 Integrating Mechanisms

Organizational theory holds that two principles apply to a firm: work is divided to gain the benefits of specialization, and the divided units must be integrated for the benefit of the company as a whole. In the smallest units, integration is done by the boss. As units grow, the general manager must find other mechanisms for integration, to ensure that all decisions referred to the boss do not create a superbottleneck. Some of these mechanisms include procedures, integrating individuals, integrating departments such as commercial development, temporary groups, and cross-functional teams.

In the growing organization, the usual integrating individual at a lower level is the project or program manager in a matrix management form. Raised to the top of the firm, this individual may become a chief operating officer. The coordinating individual also runs into limits to what an individual can handle, and the need for coordination may grow into an integrating department. These three forms are inherently top–down in structure.

The alternate approach is to structure integrating mechanisms in a bottom–up fashion. This is the logical extension of a trend that started a half-century ago to decentralizing companies "federally," like the split between Washington and the state capitals. Well over half of U.S. business is now decentralized, principally by product line, although other divisions of work are also found such as by geographic area or key customer. What has changed in recent years is that this trend has proceeded even further down to decentralization by groups and teams. First came the use of permanent or standing committees for management coordination. This was followed by the use of temporary groups, an alternate form of matrix management.[7] Names for temporary groups range from ad hoc committees through to task forces. Advantages of both types of groups are the greater knowledge base from which decisions are drawn, and the participation of lower-level people increases the understanding and acceptance of the decisions made. Since technological innovation has so increased in complexity, with the necessary knowledge more often residing in the workforce rather than in the managers, that it is not surprising that the group form of organization has grown in use.

Disadvantages of groups, which must be overcome by their leaders, are the conflicts they may generate, and the longer time taken to reach decisions.

26.4.2 Managing the Financial Boundary

Managers of innovation need to integrate their effort across to the financial organization, upstream through marketing toward the customer, and downstream through the operating chain of the firm. Working along the financial boundary, the innovation manager must establish boundary-spanning mechanisms. What these are depends first on the firm's financial culture. The successful innovation manager learns to work within this culture, and not to fight it. The empowered organization has its own needs for financial boundary spanning. This spanning must include some linkage to finance on project selection and evaluation. Finally, the boundary with finance comes to setting the budget which is needed for the effective management of innovation.

26.4.3 Managing within Your Firm's Financial Culture

Companies have strong variations in their driving cultures. Three of these are described below: the financially driven culture, the activity-based management-costing culture, and a more pliable culture in the financial sense.

The Financially Driven Culture.   This culture is one having a view of innovation as an investment, as opposed to merely adding up costs as the accounting profession practices. Now that it is clear that innovation's role does not affect current return because of the "acceleration trap," managers must focus their efforts on the growth part of total return.[14,15] All financial methods recognize the time value of money, which is the essential element separating finance from accounting.

In a decade of writing about the financial boundary, the author has clearly reflected his industrial experience with financially driven cultures. Most of these references are just as valid today for dealing with the financial boundary, when read recognizing that product management and development need to be more growth-focused, and less benefit/cost-oriented than in the environment in which these earlier views were written.

Innovation managers need to reach across the financial boundary and comprehend the investment focus of financial managers. This means learning the expectations for returns versus risks that top management have. It also means putting forth requests for project approvals in financially justified form, using the selection method most acceptable to those executives across the financial boundary.

The Activity-Based Management and Costing Culture.   This culture is based on management and financial accounting, which operate in a framework of the current income and assets statements. Thus, it does not factor in the time value of money as does the investment approach. Management accounting in recent years has come in for strong criticism and multiple calls for reform.[16] One reform movement has been to move into *activity-based management* (ABM) and *activity-based costing* (ABC).[12] In this modernized form of cost accounting, tasks are broken down into repetitive activities for costing. That which is not repetitive runs the risk of being considered a time waster. However, innovation is seldom repetitive. ABM/ABC has accepted nonfinancial measures such as quality and timeliness in addition to costs.

What should the innovation manager do if ABM/ABC is the firm's culture? Again, fighting the culture is not a good solution, although educating those across the financial boundary is a needed activity. Following the ABC ideas of activity-based analysis of repetitive tasks, the first areas to encompass should be those of a more repetitive nature. These would include naturally support activities, such as laboratory analysis. Customer order specials, contract engineering, and qualification testing are tasks which also naturally fall into a more repetitive analysis mode. It is not clear, however, that research and much of product development can be easily adapted to ABC.

26.4.4   The More Pliable Culture

In a more pliable culture, hopefully, innovation will be allowed to take place in most firms that recognize the need for a financial and accounting boundary, but are not a slave to it. The innovation manager in such a company should tend this boundary carefully. First, using a financial project method selected with the financial manager as one criterion for project selection is recommended. Other arguments for selection or not should be raised as appropriate to the occasion. Second, along the part of the boundary facing the supply or implementation part of the organization, matching the use of costing methods used on the other side of the boundary is appropriate. Nonetheless, as expressed by the author a decade ago (Ref. 10, p. 185), the innovation manager "must consider financial methods as only tools with which to defend research expenditures in a financial and political environment, rather than as a complete means to fulfillment of the overall research management task."

26.4.5   Managing the Upstream Boundary: The R&D-Marketing Interface

The R&D-marketing interface is an important internal boundary. To help managers in deciding how to create a better marketing-R&D interface, two examples are given of a group approach and an integrating department.

Product Planning Committees.   These committees are a natural bottom–up grouping of marketing and R&D professionals, with the addition of other functions such as general management and manufacturing, depending on the nature of the company. In ITT Corporation, these committees started in the 1930s, as a consequence of the need for intergeographic technical coordination due to the many national subsidiaries of that company. Renamed "task forces" in the 1970s, they obtained greater input of general and product-line managers. The primary contribution these committees made to innovation was in the screening of new ideas from the field, rather than the generation of them by committee members. This helps select a lesser number which appear attractive enough to be worth more rigorous analysis through market assessment and financial evaluation. Only once in his career has the author found such a committee to have too few ideas—they usually have far too many, and screening is desirable to devote resources to the most promising. The one exception was funded for an outside company to organize brainstorming for them, which generated 60 ideas they had not considered, thus quickly transforming their product planning committee into a screening mode.

Commercial Development of New Products.   This is a consequence of the realization that some functions are not economically able to be decentralized. Commercial development is thus the creation of an integrating department of marketing and R&D while retaining a centralized operating department. Unless a business is totally labor-intensive, the sharing of fixed assets is usually more cost-effective than letting each decentralized unit have its own plant facilities. Up to a point, the close coupling of a decentralized management to its market may offset this implementation cost penalty, but as the business becomes more capital-intensive, plant centralization economies of scale shape the organization structure.

Thus, it is not surprising that this type of integrating mechanism is found most often in heavy industry such as primary metal and chemical manufacturing.

In the 1960s, long before the author had heard of the term *commercial development,* he found that this need for centralization also applied to electronic equipment manufacture. The structural lesson in organizing the marketing-R&D interface is that some balance is needed in decentralizing completely where the cost disadvantages overwhelm the closeness to the customer and shorter communication lines of the smaller unit.

26.4.6 Downward Transfer of Technology

Equally important to innovation is the internal organization which transfers technology downward to the implementation chain of the company. As was noted in Fig. 26.5, market development cycle time is composed of the time from the start of the prototype to the completion of customer acceptance of the first delivery. This cycle time often spans several organizational units, and for many firms is handled through teams as discussed below.

In companies which have not adopted the team approach to this transfer downward, few of the research results cited give much guidance for organization. The operations side of any business is focused on repetition and process, and is thus averse to change. Current accounting assigns the costs of the intermediate technology transfer organization (such as industrial engineering) into manufacturing overheads, thus providing a disincentive for manufacturing managers to devote effort to this task. Reforming accounting treatment to allow the manufacturing department's costs for introducing new products to be separate from overheads on current production has recently been proposed.[16]

As a consequence, the trend in recent years has been to demand that more of the technology transfer be done in and charged to the R&D unit. The author's only recommendation in this case is to arrange the assignment of development engineers to the manufacturing organization for as long as it takes to make sure that the transfer has been made effectively. In a company managed by the author, this was made a standard cost estimate as part of the development proposal, at 2.5 percent of forecast annual manufacturing cost.

26.4.7 Managing Infrastructure and People

After managing resources and functional interactions, the technology manager must deal with infrastructure and people. In particular, how does that person organize? Does the manager organize with specialization so as to get the best use of a few key specialists? Or does the manager organize by task, which brings together various specialties into a single group, which produces the proper focus on what needs to be done? In recent years, the latter has won out because it permits managing time more effectively.[2] Even more, the groups or teams have broadened to include other functions and outside suppliers as well.

Cross-Functional Teams. Such teams are clearly the best overall management practice, incorporating not only the innovation chain, but also the implementation one. They incorporate R&D, engineering, and manufacturing, and usually—but not always—marketing. This use of cross-functional teams is usually called *concurrent engineering* in the United States, and often *simultaneous engineering* in Europe.

Concurrent engineering appeared on the scene a few years ago as a conceptual solution to the time delays encountered in the traditional manner of serial engineering, in which a project moved in a steady, but slow progression from marketing down through functional stages of research, development, and engineering and finally out through manufacturing to a customer.[17] Concurrent engineering was conceived as an overlapping, cross-functional process, allowing each function to begin earlier as a result, and thus reducing overall time from idea to customer. This chapter will address how some experienced companies are successfully managing this concept.

Arguments in favor of this practice are that it solves the upstream/downstream coordination problem mentioned above, lowers costs, reduces cycle time, shortens lines of communication, and brings those most knowledgeable into the decision path.

Against this practice it is argued that the will to stay with concurrent engineering had faded at many of the firms surveyed who had initially followed this practice. Where it had been enthusiastically sustained, the organizations had been transformed with shorter cycle times and fewer engineering changes. Management of human resources was the most difficult issue in two areas: balancing team preservation against the demands for staffing newly initiated projects and linking teams back to the unchanged parts of the organization, particularly functional executives. Despite these concerns, concurrent engineering is now widely used, as explained in a subsequent section.

Background Literature. Some background literature shows that the term "concurrent engineering," was put forward by the Defense Advanced Research Projects Agency's (DARPA) Defense Initiative on Concurrent Engineering (DICE) at the University of West Virginia.[17] Efforts at cycle time reduction indicated that many redesigns come because the development has not been reviewed for manufacturability, quality, serviceability, or testing before being "thrown over the wall" into manufacturing. This view held that paralleling of engineering and manufacturing would alleviate the problem, but clearly neglected marketing, as might be expected where the supplier was expected to respond to a government specification. At about the same time, the Department of Defense also adopted total quality management (TQM) as an objective, so that both initiatives interacted in companies doing defense business.[18] As an engineer in an aerospace company put it in a follow-up interview to the 1994 study:[3]

> To meet the Defense Department's requirements for TQM for a new (airframe) model, a totally new concept has been initiated called integrated product development teams, marking the first time in the company's history that a genuine "concurrent" engineering effort has been implemented.

Several issues appear from the preceding discussion. First, the currency of many of the cited sources indicates this shift to a time orientation is rather recent. Several threads relating to application of concurrent engineering seem to have been identified: there is a need to address the innovation management issue of reducing the cycle time from concept to product introduction into the implementation process; the inclusion or exclusion of marketing seems dependent on the closeness ⸢ the business to the end user; and last, the most effective method of implementation—parallel or by teams—and the management techniques are imprecise.

Use of Concurrent Engineering.   This was shown in two separate surveys in 1993 and 1994 to have median value of 70 percent of the budget for R&D projects.[3] Correlation between concurrent engineering and other factors was found in these surveys.[3] Managers who use concurrent engineering are also likely to use value engineering, and have a higher new product to sales percentage. Concurrent engineering is now so prevalent that concurrent engineering use was not correlated with earnings before interest and taxes, industry, and idea to customer or R&D cycle time measures, except marketing cycle time. Even so, the increased new-product percentage of sales fuels the engine of profit growth in the firm, making concurrent engineering a worthwhile engineering management practice. Cost competitiveness and growth through new products should be foremost in engineering managers motives for using concurrent engineering. Lower users of concurrent engineering may also see a shift away from minor projects into major ones or extensions, and find some higher use of quality function deployment (QFD), or have greater regulation.

26.5 MANAGING TIME

What is now known about measuring times of cycles, dates, and staff hours was previously summarized. Once these times are recorded, they must be further diagnosed and managed by reducing times, consolidating time, concentration, and measuring contribution.

The philosophies of time management covered below derive from about four decades of use by the author in management practice and in the business school classroom.[19] Simply expressed, "Time is inelastic—you can throw money and people at any project, but you can't make more time!"

26.5.1 Diagnosing Time

Once time has been recorded as days, dates, and labor-hours, it must be diagnosed: Where did time go? Even for projects close to schedule, answering this question will be helpful in understanding how effective the organization is. Where schedules are not being met, diagnosis is essential to understand what management steps must be taken to correct the problem.

26.5.2 Pruning Time Wasters

Once time has been recorded and diagnosed, proactive time management must begin. This author's experience has been that eliminating time wasters is the easiest place to gain free time for other needed activities. Thus, the first task is to prune out time wasters. After time wasters are pruned, the productive time remaining needs to be regrouped into larger time blocks to enable time to be used effectively. This is particularly true in a multitask environment which includes that of supervisors. The time which is made free may be in increments so small that it does not allow effective use, while regrouped schedules may not only free up larger blocks of time but also prune away the time lost in dropping one task and picking it up later.

26.5.3 Concentration

"Do first things first" is the guiding principle of concentration. Pareto's 80/20 rule is the principal guideline of how to concentrate: 80 percent of results are produced by 20 percent of effort. The corollary is also true: 80 percent of effort produces only 20 percent of results. Profitability is aligned with the proportion of effort spent on major new projects. Thus, the action needed may be summarized concisely: Feed the stars. This requires finding some effort that was not previously available, some of which can be supplied by reducing time wasters.

26.5.4 Contribution

In place of the vague term results used above, in time management contribution needs a more complex definition: time saved, at little or no impairment of financial performance; or improved financial performance needing little or no extra time.[15] It is often easy to reduce cycle time by tackling incremental projects, but this has been shown to cause worse financial results. The focus on contribution needs to be outward from the innovating organization, not inward and downward. Contribution results mostly from satisfying a customer, who then will pay a remunerative price for the service or product.

Inward to the organization is the task of making the average person productive, in the sense of working toward the definition of contribution mentioned above. Human resource contributions need also to be focused not on making individuals happy, but on what will aid their own contributions. Contribution from meetings needs also to be focused on time effectiveness, at minimum cost in time away from assigned tasks.

26.5.5 Building on Strength

The organization should make its strengths relevant to time performance of the assigned tasks. This means staffing primarily with just the number of people needed, all with their own appropriate skills and expertise. The necessary degree of empowerment needs to be given to the innovating group so that it can conduct its work rapidly, with a minimum of delays to coordinate with external interfaces. The task of the innovation manager is not to direct personally all work, but to neutralize weakness, personality, and conflict in the organization, which might lower its contribution. This does not mean suppressing conflict, but keeping it sufficient to build creativity, and not too much to impair effective cooperation. Equally, the task of the group is to make its boss productive by keeping up the flow of information needed to satisfy higher management.

26.5.6 Decision Making

In an empowered innovation organization, the manager makes few decisions. The team, which is closer to the task, makes many decisions on the spot, freeing the manager to make the most important decisions. In this environment, the managerial effort is to define effective decisions, and make decisions effectively.

The effective managerial decision is one which addresses a generic problem. Most such problems are masked by symptoms which should not be treated individually. If the project is late—a symptom—is the generic problem a bad estimate, or understaffing? The manager must realize that the required knowledge rests at the lower level of the group, and seek out the problem from the symptoms. Also, the group needs to participate in making the decision so that they will understand and respect the decision.

Making decisions effectively at the innovation manager's level is an example of the scientific method. Which alternatives are there to address the generic problem? How does each address the need, and what new risks does it entail? Which criteria should be used to make the decision? How does the choice among alternatives get made? Once made, how does the choice become adopted? Who changes? What does the manager do if they won't change? All this is easy to say, but hard to do.

26.6 MANAGING CHANGE—THE ULTIMATE PROCESS

Managing change is the ultimate process of technology management. To get there (where you want to go), you have to change from where you are now, except in the narrow case of a start-up company. Industrial psychologists use the analogy to changing the shape of an ice cube. To make such a change, it is necessary to melt the ice. In an organizational context, this means unfreezing past practices. This is done by education and training, by teams as covered above, by temporary groups, by use of a change champion, and by any means reasonably at the manager's disposal to highlight the need for a change.

Back to the ice cube, once it is melted, it becomes water and may be sloshed around. At this point, the manager begins to reshape by example. Select a group which has been most receptive to change. Chrysler chose the Viper car.[2] The "watery" group was made up of volunteers who were given the task of cutting in half the time for introduction of a new car, and they did it!

Then the manager must begin the task of refreezing the organization (ice cube) into a new shape. The best missionaries are always those who have been most recently converted. Use them to form the nuclei of new teams. Preach the gospel from the change champion. Send the top managers with much publicity to visit the pilot team, and spread the word in company newspapers or on the public address system. Put pictures of the team where everyone entering the plant can see them. Refreezing is encouraged by the setting of benchmarks and goals, revised procedures, new measurements, and rewards based on the new metrics. This leads into commitment building between the project teams and their leaders, and general management. Finally, codify the new way of doing business in standard practices, and in the orientation manuals for new hires.

26.6.1 Managing Major Cultural Change in R&D and Marketing

For those not yet into concurrent engineering, or having difficulties in handling the changes required by introducing it, a few comments on the change process are relevant here. First, this change requires a behavioral change, for which many innovation managers are poorly prepared, and texts or experts should be consulted.[20] Having an experienced facilitator available to the team leader should be a regular practice. When the change is put into place, a period of adjustment takes place, with the facilitator easing over the rough spots. These facilitators are available in many companies' human resource departments, or can be obtained through outside consultants.

The success factors in managing major change in R&D were studied recently.[21] The most dominant success factors started with top management's quick recognition of the need to change, and being fully committed to it. In U.S. companies, the use of change champion was next most highly rated as a success factor, followed by teams, while in Japanese companies the use of teams was ranked third. Since concurrent engineering is a team activity, it seems logical to emphasize teams over the use of a change champion, but the latter may be appropriate in larger firms. Piloting the change in part of the organization builds experience in making the change, and pro-

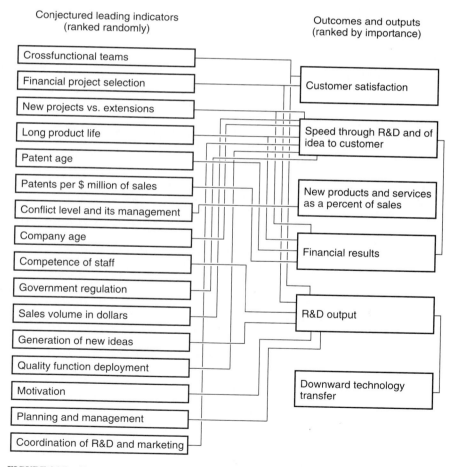

FIGURE 26.7 Simplified model of effectiveness measurements.

vides a source of missionaries to preach to those needing to be converted, as was done at Chrysler.[2]

26.7 THE CHALLENGE

It is easy to offer separate prescriptions for each process management task mentioned above. The problem is to put them all together without shooting yourself in the foot. Many interactions exist between metrics, as shown in Fig. 26.7 (page 26.23). These interactions may be positive, or they may be negative—the brief study from which the flowchart in Fig. 26.7 was made did not allow for such clarification. Many of the articles cited in this chapter go deeper into some of these factors.

As you manage the process of technology management, the challenge is to identify what you should do, and what you should measure, to get the most results from the least controls. This is industry-specific, and company-specific. All that this chapter can do is make you start thinking. What you measure is what you get. Make sure what you get is what you want.

26.8 REFERENCES

1. M. G. Brown and R. A. Svenson, "Measuring R&D Productivity," *Research Technol. Management,* **31**(4): 14–15, July–Aug. 1988.
2. R. A. Lutz, "Implementing Technological Change with Cross-Functional Teams," *Research Technol. Management,* **37**(2): 14–18, March–April 1994.
3. L. W. Ellis and C. C. Curtis, "Concurrent Engineering: Use and Related Measurements," 1995, submitted for publication to *IEEE Transactions on Engineering Management,* in press.
4. L. W. Ellis and S. Honig-Haftel, "Reward Strategies for R&D," *Research Technol. Management,* **35**(2): 16–20, March–April 1992.
5. M. E. McGrath and M. N. Romeri, "From Experience: The R&D Effectiveness Index: A Metric for Product Development Performance," *J. Product Innovation Management,* (11): 213–220, 1994.
6. L. W. Ellis, "Effective Use of Temporary Groups for New Product Development," *Research Management,* **XXII**(1): 31–34, Jan.–Feb. 1979.
7. L. W. Ellis, "Temporary Groups: An Alternative Form of Matrix Management," in *Matrix Management Systems Handbook,* David I. Cleland, ed., Van Nostrand Reinhold, New York, 1983.
8. C. Herstatt and E. Von Hippel, "From Experience: Developing New Product Concepts Via the Lead User Method: A Case Study in a 'Low Tech' Field," *J. Product Innovation Management,* **9**(3): 213–221, Sept. 1992.
9. P. G. Smith and D. G. Reinertsen, *Developing Products in Half the Time,* Van Nostrand Reinhold, New York 1991.
10. L. W. Ellis, *The Financial Side of Industrial Research Management,* Wiley, New York, 1984.
11. L. W. Ellis and C. C. Curtis, "Measuring Customer Satisfaction," *Research Technol. Management,* **38**(5): Sept.–Oct., 1995, pp. 45–48.
12. S. Hronec, *Vital Signs,* AMACOM, New York, 1993.
13. J. M. Juran, *Juran on Quality by Design: The New Steps for Planning Quality into Goods and Services,* Free Press, New York, 1992.

14. C. C. Curtis, "Nonfinancial Performance Measures in New Product Development," *J. Cost Management*, **8**(3): 18–27, Fall 1994.

15. L. W. Ellis and C. C. Curtis, "Speedy R&D: How Beneficial?," *Research Technol. Management*, **38**(4): July–Aug., 1995, pp. 42–51.

16. L. W. Ellis and R. G. McDonald, "Reforming Management Accounting to Support Today's Technology," *Research Technol. Management*, **33**(2): 30–34, March–April 1990.

17. A. Rosenblatt et al., "Special Report: Concurrent Engineering," *IEEE Spectrum*, **28**(7): 22–37, 1991.

18. *Total Quality Management Guide*, vols. I, II, United States Department of Defense, Washington, D.C., 1990.

19. P. F. Drucker, *The Effective Executive*, Harper & Row, New York, 1967.

20. M. R. Weisbord, *Productive Workplaces: Organizing and Managing for Dignity, Meaning and Community*, Jossey-Bass, San Francisco, 1990.

21. G. F. Farris and L. W. Ellis, "Managing Major Change in R&D," *Research-Technology Management*, **33**(1): 33–37, Jan.–Feb. 1990.

CHAPTER 27
MANAGING INTERFACES

K. K. Brockhoff
Institute of Research in Innovation Management
University of Kiel
Kiel, Germany

A. K. Chakrabarti
New Jersey Institute of Technology
Dean, School of Industrial Management
Newark, New Jersey

J. Hauschildt
Institute of Business Administration
University of Kiel
Kiel, Germany

A. W. Pearson
Manchester Business School
Manchester, United Kingdom

27.1 INTRODUCTION

An important area for attention in most organizations is the development and introduction of new products and processes. This involves a progression from idea to implementation, which requires the cooperation and interaction of people from many parts of the business, from different functions, and increasingly across geographic as well as organizational boundaries. It leads to more and more organizational interfaces. With time to market becoming an increasing focus for attention, this provides a real challenge to management.

Over many years researchers have studied how the quality of interface management impacts on the success or otherwise of the innovation process, and it has become clear that an area requiring serious attention is the relationship between R&D and marketing. This is so for a variety of reasons, particularly the ability to effectively link technological developments to market needs, which has been identified by many researchers as a major factor in distinguishing success from failure.[1,2] The value of good market research at an early stage has also been emphasized.[3] Good communication between the different parties is therefore essential, and yet this is one of the areas about which most concern is often expressed.

Research studies which confirm the feelings of many managers include those of Gerstenfeld et al.,[4] who concluded that the lack of a continuing, collaborative relationship between R&D and marketing was the main cause for more than half of project failures which occurred for nontechnical reasons. In another study the degree of harmony and joint involvement between R&D and marketing was identified to have a significant effect on the success of new-product development projects.[5] These results have been supported by more recent studies revealing the effectiveness of information transfer and the understanding of user needs to be major variables affecting project outcomes.[6] Practitioners as well as researchers confirm these findings and agree on the problems facing managers who are continually looking for ways of increasing the efficiency and effectiveness of the innovation process. In the following sections we therefore consider some of the basic characteristics of interfaces in general before examining more specifically those involving R&D. Factors which are important to the project level are then discussed, and the chapter concludes with an examination of the important roles which people play in the innovation process.

27.2 ORGANIZATIONAL INTERFACES

27.2.1 Interfaces Explained

In technical terms, *interfaces* are observed at the point where two or more separate entities need to be joined together in order to achieve a certain output. Most often, some amount of material, energy, or information has to be transferred from one to the other. It is of value to design the interface to ensure that there is no loss of the aforementioned flow of exchanges, and in particular that there is not a one-sided constraint of the flow. If we observe people who exchange goods, financial resources, or information between one another to achieve a common objective, we can identify an analog to the technical interface, namely, the organizational interface between these people. Again, the flows exchanged should not experience significant losses and should not be constrained in a one-sided manner. To achieve this, organizational interfaces need to be managed. However, as people are capable of developing personal objectives and producing conflict, this complicates the management problem over and above the technical design problem.[7]

Specific instruments to handle organizational interface problems therefore need to be available. Here, we are concerned with organizational interfaces that occur within one organization, more specifically, within a company. Furthermore, we consider organizational interfaces which can be observed between two or more subunits of the organization in question that are not placed in a hierarchical order such that one is superior to the other and has the power to issue orders, and that cannot resort to a common superior to solve day-to-day interface problems and conflicts.[8] This situation has become more and more common as companies have been reengineered by introducing strategies of internal decentralization or by establishing group processes, for instance, in production.

27.2.2 Approaches to Interface Management

This situation rules out two approaches that could otherwise be adopted to solve interface problems. As we consider only interfaces within an organization, there is less

TABLE 27.1 Instruments for Interface Management

Orientation	Personal	Impersonal

Structure	Team building in
	  Committees
	  New-product groups
	  Project teams
	  Task forces
	Staff work
	Matrix organization
	Readiness to integrate various
	subcultures
	Consideration of cooperation support-
	ing characteristics in choosing personnel
		Distance reduction by
		  Decentralization
		  Spatial arrangements
		Planning with decompositioning
		algorithms
		Programs
		Transfer pricing
		Incentive systems to support
		coordination

Process	Joint formation of objectives and goals Development of shared visions for the organization Avoiding extreme solutions for partial or functional objectives Learning about differences in activities, e.g., by job rotation or further education Internal information meetings	Securing up-to-date information on plans by information exchange Networking Simultaneous engineering Job rotation programs Cooperation-supportive company culture Milestone planning and monitoring Quality function deployment

possibility of applying market mechanisms in the strict sense to coordinate the activities. The other approach, essentially a command system, cannot be applied as neither subunit has hierarchical power over the other. This leaves us searching for other alternatives, including the following suggestions:

1. Problems arising from organizational interfaces may be handled as a full-time job if these problems occur frequently, if they are relatively important, or if they absorb a substantial amount of input in their managerial handling. This job can be performed by appointing an interface manager such as a coordinator, a project manager, or a liaison person, or by an individual who has specific skills to perform such a job, for instance, the so-called internal star, who has learned to acquire and to disseminate information within an organization to help it to run smoothly.[9,10]

2. Organizational interface problems may be handled by the people who are part of the interface as an activity that is accessory to their main job. Four possible categories can be identified, which might best be seen as structural-versus-procedural activities, and personalized-versus-nonpersonalized activities. In Table 27.1 we give some examples for each of these four groups.

27.2.3 Selection of Managerial Instruments

The choice of instruments depends on the conditions under which organizational interfaces occur. A few guidelines have been developed that can help to select appropriate instruments.[11] These are as follows:

1. The level at which the interface problem occurs should be considered. Interfaces may occur between *strategic business units* (SBUs), between functional areas within an SBU, and between projects within functional areas, as well as within projects. If information exchange is involved at all those levels, it is greatly facilitated by joint committees at the SBU level, and by teams at the project level.

2. The type of exchange should be considered. For example, sequential transfer, reciprocal transfer, and pooled transfer, in which a multitude of units need to share one common scarce resource, may be considered. In the first case, motivational and goal conflicts are likely to arise, and in the last case we are likely to observe distribution conflicts, while the case of reciprocal transfers may give rise to both types of conflicts. It is obvious that motivational and goal conflicts can be avoided by sharing information, which, for instance, can be encouraged by participation in a job rotation program. Conflicts over scarce resources need the development of some allocation mechanism, such as a program planning rule or a pricing rule.

3. The reason for creating an interface should be considered. Capacity limits, diseconomies of scale, and productivity gains from specialization could all be reasons for splitting a task and having it performed by more than one individual. In the first two cases standardized qualities need to be achieved to assure comparability of the outcomes. In the third case the assurance of the appropriate mix of qualities is of prime importance. Again, this can be achieved by different coordination mechanisms.

4. Task characteristics determine the choice of the instruments for interface management. Such characteristics are frequency, repetition, standardization, and complexity. It is obvious that parts of a frequently recurring task can be coordinated more effectively by the use of plans and programs, while a more-or-less unique task may require more team building and personal information transfer.

27.3 R&D INTERFACES

27.3.1 Overview

R&D interfaces may be considered at two levels:

1. R&D is a business function that arises from a specific specialization with the objective of creating new technological knowledge to be used by the company. It can be distinguished from other business functions, such as production or marketing, and if it is further specialized into a more research-oriented central laboratory and one or more development-oriented decentralized laboratories, we find more interfaces as a consequence of the extension of functional specialization. Little standardization and a small degree of repetitiveness of R&D work should be kept in mind if instruments for interface management need to be selected.

2. R&D work is split into a number of projects. This gives rise to two types of interface problems. One problem is due to the fact that the projects compete for the joint use of some scarce resource, such as finance, a special type of equipment, specialized labor, and so on. To solve the resulting conflicts, a preference order over the projects needs to be developed. This can be supported by different types of project evaluation schemes which have been well documented in the literature.[12] Another problem area results from the fact that the selection of projects and the definition of their characteristics (such as performance parameters, cost, time of completion) need to incorporate knowledge that is best available in business functions outside R&D. The incorporation and the updating of this information in the project planning may be facilitated by establishing a project steering group for large projects or a project review committee for smaller projects that determines when milestones should be reached. We shall return to this problem area later after considering some of the characteristics of particular interfaces.

27.3.2 R&D and General Management

In many companies general management is concerned with R&D only on the occasion of major organizational changes and the annual budget decision. The uncertainty inherent in R&D work seems to preclude a more specific inclusion into the strategic planning of the company, although this situation is clearly changing.[13] If it is independent of corporate objectives, it is too easy for R&D to choose its own long-term goals. Therefore, one should soon observe severe effectiveness problems. This is particularly true if the R&D function is centralized, if it is located at a distance from the strategic business units, and if its budget is taken entirely from the overhead of the company. Such conditions reduce the felt need to communicate with the other business functions and—because of the distance—the likelihood of informal as well as formal exchange of information.

Interface problems may be reduced if the head of R&D is represented at the same level as the other functional areas on the general management committee, whether this is the board in a corporation or the management committee of an SBU to which the R&D group belongs. It is interesting to observe that board representation of R&D in larger corporations is often different in Europe, Japan, and the United States.[14] Board representation may be a prerequisite for continuous involvement of R&D in strategic planning,[15-18] and it must include a discussion of the functions that the company wants R&D to perform. Thus it is important to agree on the priorities to be given, for example, to product support and to innovation through new technologies. To solve the interface problems at this level is considered as not only a strictly managerial problem but also a problem of national concern in international competition.[19]

27.3.3 R&D and Production

R&D has a considerable influence on the production cost of new products, and it is the primary supplier of new production technology when a firm develops its own process innovations, such as in the chemical industry. Furthermore, the joint development of new products and their process technology becomes imperative to survive in the competitive battle. The interest of R&D personnel in innovating may not be shared by production departments when it tends to slow production rates, and thus interferes with reaching their short-term objectives.[20,21] The work organization and the value systems of these functional areas can differ as a result of the differing levels of uncertainty involved and the timeframes to completion in the typical jobs to be performed as well as of the differences in the educational level of the average employee in either department.[22,23] Although the two areas have been called "cultural cousins,"[24] this does not necessarily mean that they feel familiar with each other.

It has been suggested that the particular interface problems may be eased by joint reward systems that tend to coordinate motives and action plans. However, questions about how to design and administer such systems remain largely unanswered.[25] Joint formulation of development goals can be considered as a first step toward harmoniousness between the functions.[26] Job rotation programs that help their participants understand differences in work conditions and subcultures and that smooth transfers of projects from R&D to production by letting R&D people move into production with the completed project or joint representation on project management teams appear to be more efficacious. The more that parallel work needs to be done in the different departments on a specific project as a means to speed up the time to market, the more demand arises for intense, real-time communication exchange. This is greatly facilitated by using CAD systems.

27.3.4   R&D and Marketing

The R&D-marketing interface could involve sequential transfers in either direction, depending on whether the company is more interested in technology-push or in market-pull innovations. There are two major issues at this interface. The first issue concerns the information exchange on new products and their development at competitively superior levels (which involve timing, cost, and quality). This aspect will be considered later at the project level.

The second issue revolves around the definition of a technology strategy and a marketing strategy per product group that support each other. A high-capacity, technology-push strategy is of little value if it is matched with a defensive, low-key, low-capacity marketing strategy. In terms of strategy, empirical research seems to indicate that most companies experience cleavage or discord more than harmony or agreement between the two functions we discuss here.[27–29] This is not always due to a lack of overlapping teams or of meetings that set out to define strategies but about differences between the underlying assumptions on the present state of the company (together with its economic environment), the planning horizons, the available resources, and the willingness to bear risks. The identification of such differences can be taken as starting points for the selection of instruments for interface management.

A number of suggestions involve improvements in the integration of marketing and technology planning.[30–32] Organizational instruments such as the ones mentioned with respect to the R&D-production interface may also be used here. It appears, however, that it is more easy to arrange job rotation programs from R&D to marketing than vice versa, and it should be observed that the results of learning by job rotation become effective only with a considerable time lag after taking a new position in a firm.[33] It is also of interest to note that the demands on the managers of both departments to bridge the interface are slightly different.[34,35] Reducing the distance between both departments serves to increase communication, which, in turn, helps alleviate interface problems.[36] This idea is relevant to overcome other interface problems as well.

27.4   INTERFACES AT THE PROJECT LEVEL

27.4.1   Communication and Commitment

Many interface problems are identifiable at the project level. They can arise for a variety of reasons, including lack of clarity of objectives, inadequate resourcing, changing priorities, ineffective planning, and poor monitoring of progress. Improvement in all these areas can be made through better communication across the different functions within the organization and between all parties who are concerned with and can influence the outcome. Such improvements inevitably result in a higher level of commitment, a focusing of effort, and an increased likelihood that project objectives will be more closely met within agreed-on time and cost estimates. In many cases such improved performance is reflected in a reduced time to market. This can have a significant impact on profitability when product life cycles are being increasingly squeezed by new technology and by agile and aggressive competitors in the marketplace.

Having identified some of the key interface problems at the project level, we are in a strong position to look for ways and means of providing solutions. Here the theory and the practice are moving in the same directions, and it is convenient to look at the approaches suggested under the headings of *planning* and *monitoring,* with the management of uncertainty being a key influence on actions and outcomes.

27.4.2 Project Planning

As stated in the previous section a key need in project management is to obtain commitment by all people involved not only to the objectives but also to the process by which these goals are most likely to be achieved. The organizational behavior literature emphasizes the importance of clarifying goals, roles, and procedures as early as possible. Doing this is, of course, a team effort and makes it clear that it is not very sensible to draw up plans for a project without the direct involvement of those who will be responsible for carrying them through. Unfortunately, this is not always possible in practice, and the results can be seen in the number of failures which occur for this reason. The "house of quality" approach is well known as an aid to promoting communication between R&D and marketing at an early stage in a project.[37] While this approach can be considered as a planning heuristic that helps to integrate information from marketing and R&D, more analytic approaches have also been developed. These approaches use multivariate statistical techniques, such as conjoint measurement, to arrive at optimal product positioning.[38] There is no shortage of techniques available to assist in the activity planning process. These are well documented in the literature and include bar or Gantt charts, various forms of network analysis, precedence diagrams, and research planning diagrams or flowcharts.[39] All of these have been used in practice with varying degrees of success but the most flexible, and the one now being increasingly used, is based on the flowchart. The reasons for this are its ability to handle uncertainty and the clear way it shows important features of projects such as the need to consider repeating activities when the results do not turn out as expected and the likelihood of different outcomes. The emphasis on decision points and on the prior identification of performance criteria to check progress against also makes flowcharts excellent vehicles for monitoring and for resource management.[40] A simple example of such a flow diagram is shown in Fig. 27.1.

A point which cannot be overemphasized is the real value gained by using such an approach as a communication device which not only clearly pinpoints issues identified at the outset of a project but also keeps these in focus at subsequent reviews. Even more benefit is obtained if a monitoring procedure is established which focuses attention on key decision points, or milestones, and the progress which is being made toward achieving these goals.

27.4.3 Project Monitoring

If we define a *milestone* as a point at which it is anticipated that an agreed-on level of performance is expected, it is not necessary to force any particular form of planning procedure on a project. In fact, one could foresee situations in which milestones could be identified from past experience with a reasonable degree of accuracy without the use of any formal analysis. Such a situation might arise if the project were very simi-

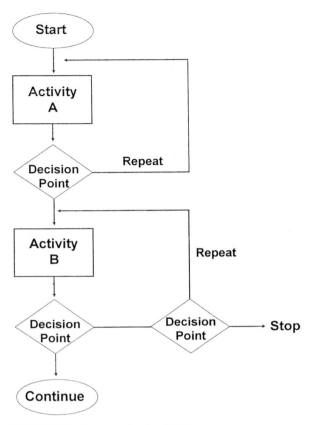

FIGURE 27.1   Research planning diagram.

lar in nature to previous work and the people involved were all highly experienced in the area and also highly motivated to achieve the goals. This type of project is, however, not very common in R&D, and therefore it can be assumed that more detailed information is required in order to establish the time and cost likely to be needed to achieve not only the ultimate goal but also the intermediate milestones. The value of the latter becomes immediately obvious when the development of a practically useful monitoring system is being considered. The identification of milestones is an important management task, as is the agreement about the interval between them. Too few intermediate milestones is often a recipe for disaster because insufficient information about progress is made available to interested and involved parties. Too many milestones—that is, too short an interval between them—can create dissatisfaction because it implies overcontrol and hence a lack of confidence in the people undertaking the work. In practice, three to five milestones per calendar year seems to be an appropriate number to aim for, although this will depend on the type of project and the experience and motivation of the people involved. The value of using milestones as the basis of monitoring becomes obvious if a form of chart such as that shown in Fig. 27.2 is used. This simple form of presentation emphasizes not only the progress made to date,

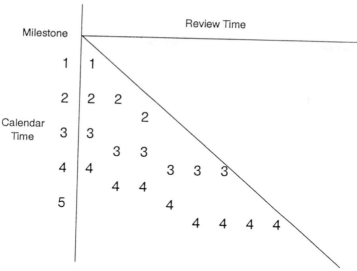

FIGURE 27.2 Progress chart.

that is, how many milestones have been reached or passed, but also what is the perceived situation regarding future progress. It is therefore both an historical and a future-oriented form of presentation. The value of the latter cannot be overemphasized, as under conditions of competitive pressure, it is more important to know that a project will not be completed for a further, say, 6 months than to know that it is 95 percent complete. Such information, signaled as far as possible in advance, can trigger a variety of actions, for example, to stop the project early and save spending scarce resources, to change priorities and accelerate progress to beat the competition in the marketplace, or to reevaluate the objectives to see whether such a change might provide competitive advantage. Any such changes must, of course, take into account the views of many different people and ensure that the output generated from the progress chart can be easily communicated to everybody in the organization, and also to parties outside, very quickly. The use of this approach has proved to be very valuable in practice and has improved both the efficiency and the effectiveness of R&D project management in many organizations.[41]

27.4.4 Uncertainty, Project Management, and Communication

The approach to project planning and monitoring outlined above is of general application, but it is important to recognize the impact which different forms of uncertainty have on the process, in particular the degree to which the objectives or goals can be clearly specified and the knowledge that is available about the means which are available to achieve these goals. The importance of these two dimensions is easily recognizable from the moment that the construction of a plan in the form of a flow diagram is attempted. Clearly, if such a diagram indicates that there are a lot of points at which feedback loops are incorporated, the uncertainty about means is relatively high. When the outputs from the completion of activities show up as many branches, then the clarity of ends is low. In extreme cases in which the construction of such a diagram would require multiple feedback loops and multiple branching at key decision points, it is obvious that planning in any degree of detail is likely to be of little, if any, value. In such situations carrying out a feasibility study prior to the decision to embark on the project is called for, and recognizing this at the earliest possible point in time can be of real value.

The influence of uncertainty on the ability to plan and progress a project is too obvious to need much debate, but unfortunately practice indicates that not enough attention is paid to such an analysis to the detriment of performance. The reasons for this lack of focus on the key dimensions of uncertainty are not clear, but a way of promoting discussion on the issues involved is to present the information that is available about activities and projects in the form of a simple diagram as shown in Fig. 27.3.

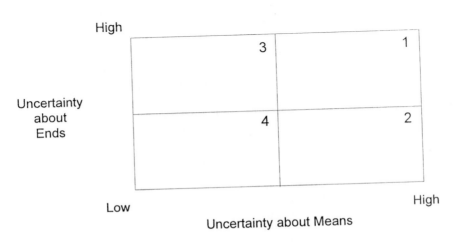

FIGURE 27.3   Uncertainty map.

Many messages for interface management can be drawn from an analysis of project characteristics presented in this form. For example, as stated earlier, there is little value in trying to formally plan activities which fall into quadrant 1, but it is also important to emphasize that unless any information generated from such activities is connected to those who can see its potential, in either technical or marketplace terms, it will have no impact on the organization's performance. At the other end of the spectrum, projects which fall into quadrant 4 can be relatively easily planned, and the need is for total commitment and adequate resourcing. Projects in quadrant 3 are unlikely to be successful because they are not focused well enough. Here clarification of goals is required in a process which must, inevitably, involve people from many functions within the organization.

Projects which fall into quadrant 2 are more difficult to handle because, by definition, they are aimed at achieving goals which have been identified as important, but on the evidence available the means cannot be clearly specified. Such projects can be real winners if successful but there are many examples of failures due to an inability to find a means to achieve the objectives or where solutions, when found, have come too late to achieve market success. In general, this type of project consumes considerable resources in time and effort, and earlier termination might have been desirable. In such cases the progress chart described earlier is helpful in indicating that there are real problems but does not take away from management, across all functions, the need to very carefully assess the costs and benefits against the anticipated future risks if work is continued on the project. The value of using this framework of analysis for identifying interface problems has been reported in the literature.[42]

Dynamics of Roles in R&D Projects and Their Implications for Interface Management. The functions and roles of innovation managers and their opponents in R&D projects are important because such projects need the cooperation of many people from various functional areas. Implementation of inventions for commercial success entails many types of activities and therefore requires many types of talents. During the life of a project different people assume different informal and semiformal roles which supplement and complement the communication channels and decision-making loci. The literature shows a plethora of roles which are important for the success of innovation. If we take a closer look, it will appear that these roles can be reduced to their specific contributions in different phases of the innovation process. Examples of this are shown in Table 27.2, which matches the roles to the phases, and in Table 27.3, which specifies the power bases for the incumbents of these roles. Both tables provide a framework of understanding and systematizing the diversity of these

TABLE 27.2 Phase-Determined Roles in Innovation Projects

Timing	Function	Role incumbents identified in the literature
Early	Stimulation of the innovation process	Initiator, catalyst, stimulator
	Solution development	Solution giver, information source, idea generator, technologist, inventor
	Process management: communication	Gatekeeper, idea facilitator, communication star
	Process management: cooperation	Champion, legitimizer, orchestrator, idea salesperson, administrative star, resource linker
	Decision making	Investor, sponsor, decision maker
	Realization within the firm	Executor, realizator
Late	Realization between firms	Relations promoter

Source: Chakrabarti and Hauschildt.[43]

TABLE 27.3 Power-Determined Roles in Innovation Projects

Power base	Participants
Expertise	"Technical innovator," "technologist," "inventor," "Fachpromoter"
Hierarchical potential	"Chief executive," "executive champion," "Machtpromoter"
Access to material resources	"Business innovator," "investor," "sponsor," "entrepreneur"
Organizational knowledge and communication intensity	"Product champion," "project champion," "entrepreneur," "Prozeßpromoter"

roles and show the commonality among the various concepts reported in the literature.[43]

The basic concept for the division of labor starts from a three-person constellation and are based on theoretical ideas which have been modified to include both product and process innovations.[44]

1. The expert (fachpromoter) has all the technical knowledge regarding the innovation and is the inventor, the idea generator, and the creative genius. The technical alternatives, internal mechanisms, and the potential of the new product or process are known by the expert.

2. The sponsor (machtpromoter) disposes resources and starts the process of decision making and the realization of the innovation. Decisions on the budget, the capacity, and the personnel in favor of the innovation must be based on a broad view and long-term perspective of the firm's strategic objectives. The sponsor is a doer whose promises can be relied on and who has the power to block or hinder opposition.

3. The champion (process promoter) acts as a linkage having the knowledge of the organization, and knowing who should be concerned with the innovation, thus connecting the sponsor with the expert. The strength here lies in the ability to translate the technical language of the innovation into one which is commonly used in the organization. By becoming a salesperson of the new idea, the champion is able to develop a plan of action and, using diplomatic talents, provides access to different people within the organization.

In the innovation process the activities of the expert are essential for bringing forth the technical knowledge for information search and processing, alternative development, and evaluation. The champion, who is the navigator of the innovation process evaluating the potential of the idea in terms of its salience for the firm, must understand the economic implications, and also needs information about long-term goals and strategies without formally defining them. The champion brings the new ideas to the notice of the sponsor, arranges for support and allocation of resources, determines the course of action by decomposition of the problem, and decides on the sequence and timing of actions. This person becomes involved in developing a "montage" of partial decisions to a working whole; steers the goal formation process; resolves conflicts between competing goals, different departments, and controversial information; and helps the top management to determine the fit between the innovation and corporate strategy. The champion "sells" the innovation to all members of the firm who are likely to be affected by it and explains, teaches and motivates, deals with opponents by getting the activities legitimized by power centers, and, if necessary, gets opposition blocked by the sponsor indirectly.

The division of labor is efficient only when the members of the innovation team work closely together to develop common goals. In successful cases they meet without formal procedures. They are not isolated from one another and maintain frequent communication. They develop and speak a language of their own. They work on mutual respect. Their communication is open without any reservation for fear of reprisal. Team members are hard-working people and engage themselves enthusiastically. The traditional view looks on promoters as members of one organization, thus conceptualizing the model as an intraorganizational phenomenon. The modern development in technology and organizational processes requires a broader view of the innovation process, involving multiple organizations in a cooperative, interorganizational mode. Thus, corresponding and interacting promoter teams will emerge in the supplier firms as well as in the user firms.[45]

27.4.5 Necessary Opponents

According to this division of labor in the promotion of the innovation, there can be a corresponding structure of opposition. We argue that opposition is not necessarily of negative influence, but it may raise the quality of the solution.

1. The opposition needs technical experts or technical opponents. They may promote a competing idea to raise technical arguments against the innovation.
2. The opposition needs power opponents. According to the principle of countervailing power, the sponsor may be blocked by means which correspond to this power basis. So the power opponent needs high status and/or control of finance or other resources. Power opponents restrict the budgets or the capacities for the innovation and/or may facilitate investments in competing projects. Both power promoters and opponents have competing points of view about strategies and long-term objectives.
3. The opposition needs counterchampions who act as overall opponents. They have to activate the networks of all those who are silent spectators and might join the opposition. They will identify and coalesce the diffused pockets of resistance. They use the informal relations to spread the spirit of opposition. Their most important contribution to the innovation process will be that the people in leadership positions will be alerted and the organization as a whole will have a better understanding of the innovation. As a result, their activities will induce management to engage in the problem.

Is promotion and opposition a question of position or action? If we look into the basis of power, it seems to be a question of position. This refers mainly to the sponsor and to the power opponent as well as to the experts. On the other hand, if we look on the champions, we will find that either promotion or opposition consists of certain activities, namely, initiating, stimulating, connecting, facilitating, and encouraging.

27.4.6 The Roles in a Dynamic Perspective

Most of the studies referred to above have concentrated on projects with narrowly defined focus: in one given project, in a given span of time, in one given organization, in one given environment. But in reality the locus of decision making must move higher up in the organization as the scope of the project is expanded. The innovation problem may start as a research question in a laboratory. But as the project moves into the developmental phase, other subunits, such as engineering, design, and production, have to participate in the project. Coordination problems will arise and will require hierarchical solutions. In a third stage the entire organization, including marketing, finance, investment, and personnel departments, will be engaged in the innovation process. The original roles of the incumbents will not be the same when the project is moving from one stage to the other. People who acted as sponsors on the laboratory base will have to champion the project in the development and construction unit. And if the project moves even higher up in the hierarchy, the same persons may have to act as experts. So in each of these stages new sponsors and champions have to be found and must meet the challenges of these roles. On the other hand, the communication lines among the people who assumed these roles in former stages must remain open to maintain the continuity for the flow of information. And moreover, the acting promoters must accept that they have to behave and to articulate themselves quite differently, but according to their respective roles. Figure 27.4 shows the development and the roles of the project.

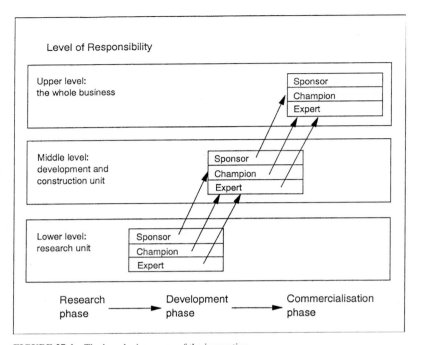

FIGURE 27.4   The broadening scope of the innovation.

Figure 27.4 describes the dynamics of the roles during the different phases of an innovation project. At the early stage of research, the project needs to be sponsored by the head of the R&D organization. The technical specialists provide scientific and technical solutions to the project. The champion nurtures the idea and influences the sponsor to help garner the necessary resources to keep the project alive. When the idea moves to the next phase, the role of the sponsor must be assumed by someone higher up in the organizational hierarchy who commands the authority to commit resources in terms of both money and people. The sponsor of the earlier stage must now assume the role of champion to nurture the idea and protect it from other competing projects. This process follows as the project grows and moves through different phases and different organizational subunits become involved in the project.

Three different organizational models for R&D and innovation projects—stage-dominant, process-dominant, and task-dominant—have been identified. The stage-dominant approach dictates that the project be divided into discrete stages and that clear transfer of responsibility for different stages must be defined. In the task-dominant model, coordination among the participants is achieved through a project team consisting of people with relevant expertise. The process-dominant model represents a more fluid approach to organizing the project teams in which the members are phased in or out depending on the need for their expertise. The different models for coordination vary in effectiveness with the varying types of factors such as environmental uncertainty, nature of the project, and organization structure.[46]

Apart from the type of organizational model, we argue here that the roles such as

TABLE 27.4 Role Requirements

Item	Sponsor	Champion	Technical expert
Technical knowledge	Not very important	Somewhat important	Critically important
Creativity	Less important	Somewhat important	Very important
Knowledge about the strategic issues	Very important	Important	Moderately important
Communication skills	Important	Critically important	Important
Political skill	Important	Critically important	Not so important
Formal authority	Critically important	Not important	Not important
Tolerance for ambiguity	Moderately high	High	Not necessarily high

sponsor, champion, and expert are independent of the organizational models and need to be fulfilled in each of these cases. These semiformal roles are meant to bridge the gap between project phases and organizational boundaries. One of the implications for this point of view is that it requires different types of skills and organizational power bases. The effectiveness of the role incumbents will be related to the appropriate mix of skills and power bases as shown in Table 27.4

Table 27.4 summarizes the various requirements for the different roles. When a person's role changes, it must be remembered that the skills needed for one role may not be ideal for the other. The behavior of the incumbent thus must change with the progress of the project.

27.5   CONCLUSIONS

Research indicates that there are different and multiple roles in innovation management. Each role has some unique contributions to make in terms of technical expertise, process facilitation, and organizational resources. Different skills and power bases may be required for subsequent phases of the project. But different people will have to assume the roles, or, in part, the same people will have to assume other roles. This requires a change in behavior, expertise, and knowledge. This change will ask for a deliberate and conscious learning process. The failure of many innovation projects may be due to a failure of this learning process.

There is also considerable evidence to confirm that the process of setting and agreeing on goals has an important part in overcoming some of the problems encountered in interface management. The organizational development literature provides useful guidance on how to do this and points to the importance of understanding the type of project and the nature of the uncertainty likely to be encountered. Finally, it is clear that there will be few projects in which progress is exactly as expected, and here we can draw on research and practice in planning and monitoring, which provides us with some relatively simple approaches to information management and communication. If this is presented openly and honestly, there is a much better chance that potential conflicts will be handled in a more constructive way and to the benefit of all partners, and if this does happen, the effectiveness of the organization will undoubtedly be improved.

27.6 REFERENCES

1. J. Langrish, M. Gibbons, W. G. Evans, and F. R. Jevons, *Wealth from Knowledge—Studies of Innovation in Industry,* Macmillan, London-Basingstoke, 1972.

2. R. Rothwell, "The Characteristics of Successful Innovators and Technically Progressive Firms (with some Comments on Innovation Research)," *R&D Management,* **7:** 191–206, 1977.

3. R. G. Cooper, "Identifying Industrial New Product Success: Project NewProd," *Industrial Marketing Management,* **8:** 124–135, 1979.

4. A. Gerstenfeld, C. D. Turk, R. L. Farow, and R. F. Spicer, "Marketing and R&D," *Research Management,* **XII:** 409–412, 1969.

5. W. E. Souder and A. K. Chakrabarti, "The R&D/Marketing Interface: Results from an Empirical Study of Innovation Projects," *IEEE Transact. Engineering Management,* **EM-25:** 88–93, 1978.

6. W. E. Souder, *Managing New Product Innovations,* Lexington Books, Lexington, Mass., 1987.

7. K. Brockhoff, *Management organisatorischer Schnittstellen—unter besonderer Berücksichtigung der Koordination von Marketingbereichen mit Forschung und Entwicklung,* Vandenhoeck & Ruprecht, Göttingen, 1994.

8. K. Brockhoff and J. Hauschildt, "Schnittstellen-Management—Koordination ohne Hierarchie," *Zeitschrift Führung + Organization,* **6:** 183–190, 1993.

9. B. Saxberg and J. W. Slocum, Jr., "The Management of Scientific Manpower," *Management Science,* **B-473–B-489,** 1968.

10. W. B. Brown and R. C. Schwab, "Boundary-Spanning Activities in Electronics Firms," *IEEE Transac. Engineering Management,* **EM-31,** 105–111, 1984.

11. Brockhoff and Hauschildt, op. cit. (Ref. 8), pp. 183–190.

12. J. Balderston, P. Birnbaum, R. Goodman, and M. Stahl, *Modern Management Techniques in Engineering and R&D,* New York, Van Nostrand Reinhold, 1984.

13. W. H. Gruber, O. H. Poensgen, and F. Prakke, "The Isolation of R&D from Corporate Strategy Management," *Research Management,* **6:** 27–32, 1973.

14. J. Travice, "How US Companies Measure Up," *Science,* **1:** 23, 1993.

15. J. Berthel, S. Herzhoff, and G. Schmitz, *Strategische Unternehmensführung und F&E—Management,* Springer, Berlin, 1990.

16. M. Domsch and J. Fischer, "Entscheidungsgremien und strategisches Forschungsmanagement," *Zeitschrift für betriebswirtschaftliche Forschung,* pp. 851–868, 1990.

17. A. L. Frohmann, "Putting Technology into Strategic Planning," *Calif. Management Review,* **2:** 48–59, 1985.

18. M. J. Liberatore and G. J. Titus, "Synthesizing R&D Planning and Business Strategy: Some Preliminary Findings," *R&D Management,* **4:** 207–218, 1983.

19. National Science Board, Committee on Industrial Support for R&D, "Why U.S. Technology Leadership is Eroding," *Research Technol. Management,* **2:** 36–42, 1991.

20. J. B. Quinn and J. A. Mueller, "Transferring Research Results to Operations," *Harvard Business Review,* **1:** 49–66, 1963.

21. M. E. Ginn and A. H. Rubenstein, "The R&D/Production Interface: A Case Study of New Product Commercialization," *Product Innovation Management,* **1:** 158–170, 1986.

22. M. Carlsson, "Aspects of Integration of Technical Functions for Efficient Product Development," *Research Management,* **1:** 56–66, 1991.

23. H. Gerpott, *F&E und Produktion,* Florentz, München, 1991.

24. A. D. Biller and E. S. Shanley, "Understanding the Conflicts between R&D and Other Groups," *Research Management,* **5:** 16–21, 1975.

25. S. A. Bergen et al., "The R&D/Production Interface in Four Developed Countries," *R&D Management,* pp. 201–216, 1988.

26. Gerpott, op. cit. (Ref. 23).

27. K. Brockhoff and A. K. Chakrabarti, "R&D/Marketing Linkage and Innovation Strategy: Some West German Experience," *IEEE Transact. Engineering Management,* **3:** 183–190, 1988.

28. K. Brockhoff and A. W. Pearson, "Technical and Marketing Aggressiveness and the Effectiveness of Research and Development," *IEEE Transact. Engineering Management,* pp. 318–324, 1993.

29. U. Weisenfeld-Schenk, *Marketing- und Technologiestrategien: Unternehmen der Biotechnologie im internationalen Vergleich,* Poeschel-Schäffer, Stuttgart, 1995.

30. N. Capon and R. Glazer, "Marketing and Technology. A Strategic Coalignment," *J. Marketing,* pp. 1–14, July 1987.

31. Frohmann, op. cit. (Ref. 17).

32. G. Specht and K. Michel, "Integrierte Technologie- und Marketing-Planung mit Technologie-Portfolios," *Zeitschrift für Betriebswirtschaft,* pp. 502–520, 1988.

33. R. K. Moenart et al., "R&D/Marketing Integration Mechanisms, Communication Flows, and Innovation Success," *J. Product Innovation Management,* **1:** 31–45, 1994.

34. A. K. Gupta, D. L. Wilemon, and S. P. Raj, "R&D and Marketing Dialogue in High-Tech Firms," *Industrial Marketing Management,* pp. 289–300, 1985.

35. A. K. Gupta and D. L. Wilemon, "Improving R&D/Marketing Relations: R&D's Perspective," *R&D Management,* **4:** 277–290, 1990.

36. T. J. Allen and A. K. Fusfeld, "Design for Communication in the Research and Development Lab," *Technol. Review,* **6:** 64–71, 1976.

37. J. R. Hauser and D. Clausing, "The House of Quality," *Harvard Business Review,* pp. 63–73, May-June 1988.

38. P. E. Green and A. M. Krieger, "Recent Contributions to Optimal Product Positioning and Buyer Segmentation," *Eur. J. Operational Research,* pp. 127–141, 1989.

39. A. W. Pearson, "Planning and Control in Research and Development," *Omega,* **6:** 573–581, 1990.

40. D. G. S. Davies, "Research Planning Diagrams," *R&D Management,* **1:** 22–29, 1970.

41. Pearson, op. cit. (Ref. 39), pp. 573–581.

42. A. W. Pearson, "Innovation Strategy," *Technovation,* **10:** 185–192, 1990.

43. A. K. Chakrabarti and J. Hauschildt, "The Division of Labour in Innovation Management," *R&D Management,* pp. 161–171, 1989.

44. E. Witte, *Organisation für Innovationsentscheidungen—Das Promotoren Modell,* Vandenhoeck & Ruprecht, Göttingen, 1973.

45. H. G. Gemünden, *The Relationship Promoter—Key Person for Interoganisational Innovation Co-operations,* Working Paper, University of Karlsruhe, 1994.

46. W. E. Souder and A. K. Chakrabarti, "Coordinating Marketing with R&D in the Innovation Process," in B. V. Dean and J. L. Goldhar, eds., *Management of Research and Innovation,* North Holland, Amsterdam, 1980, pp. 135–150.

CHAPTER 28

FACTORS INFLUENCING EFFECTIVE INTEGRATION IN TECHNICAL ORGANIZATIONS

J. Daniel Sherman
Chairman, Management and Marketing Department
University of Alabama in Huntsville
Huntsville, Alabama

William E. Souder
Alabama Eminent Scholar Endowed Chair in
Management of Technology
Founder and Director, Center for the Management of
Science and Technology (CMOST)
University of Alabama in Huntsville
Huntsville, Alabama

28.1 INTRODUCTION

In any complex new-product development project, the effective integration of diverse inputs from cross-functional groups is important in meeting schedule and budget requirements. Integration is often critical to the success of the new product or system. In the case of basic research, incremental innovations, or small-scale projects, the organization can utilize small dedicated teams (e.g., skunkworks) with lower cross-functional integration requirements. However, with large-scale projects in which there are many interdependent subsystems and coordination is required across functional areas or organizations (i.e., contractors and subcontractors), the issues of integration become more crucial.

28.2 THE BASIS FOR INTEGRATION REQUIREMENTS

Integration requirements can be partially understood through an analysis of technological and environmental uncertainty[1,2] or through the analysis of the information processing requirements of the new-product development effort.[3,4] With greater levels of uncertainty associated with a new technology or project, greater amounts of information must be processed between decision makers during development. If the technology and the market are well understood, then planning can be more certain and precise. However, if these aspects are not well understood, then information must be acquired during the development effort. This may necessitate ongoing changes in priorities, schedules, resource allocations, staffing requirements, etc. Therefore, the greater the levels of uncertainty, the greater the amount of information that must be processed among decision makers during development.[5,6]

Uncertainty can be conceptualized as the difference between the amount of information required to complete a task and the amount of information previously possessed by the organization. With generally increased levels of uncertainty, integration is affected because planning and decision making are subject to ongoing modification. This increases information-processing demands across the organization. Specific sources of uncertainty include customer uncertainty, competitive uncertainty, technological uncertainty, and resource uncertainty. *Customer uncertainty* refers to unrealized user requirements.[7,8] *Competitive uncertainty* is a function of the absence of information regarding the activities of competitors.[9,10] *Technological uncertainty* refers to the lack of knowledge regarding the solution of technical problems.[11] *Resource uncertainty* refers to the absence of information regarding the financial, technical, or human resources needed (human resource uncertainty) to successfully develop the new product.[12,13] Specifically, in this fourth category, *human resource uncertainty* refers to incomplete information regarding the staffing needed to complete the project. In some cases, this includes uncertainty regarding manufacturing capabilities. *Financial resource uncertainty* refers to the level of uncertainty regarding the costs of development. *Technical resource uncertainty* is the uncertainty that exists relative to the laboratory equipment or pilot plant facilities required.[14,15]

Figure 28.1 illustrates how these four major sources of uncertainty combine to influence the general level of uncertainty that many organizations face in starting the development of a new-product innovation.[16] Customer uncertainty and technological uncertainty are usually positively related. Greater customer and technical uncertainty generally brings concomitantly greater uncertainty regarding the competitive environment. These three factors jointly affect the organization's resource uncertainty.[17-19] As these combined levels of uncertainty increase, the requirements for integration both within R&D and between R&D and other's (e.g., marketing, manufacturing, procurement) increase. When contractors are involved, integration demands increase further across these interfaces.

28.3 ORGANIZATIONAL STRUCTURE AND INTEGRATION

28.3.1 Hierarchy

In order to achieve integration across functional departments or groups, the most basic structural mechanism is the hierarchy itself. Under conditions of low uncertainty the organizational hierarchy may be the primary mode of cross-functional integration. However, with even minimal lateral information processing needs, overreliance on hierarchical coordination becomes inefficient and problematic.

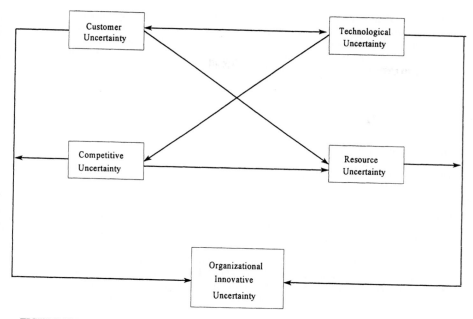

FIGURE 28.1 Sources of uncertainty. (*Source: Moenaert and Souder.*[4])

28.3.2 Informal Direct Contact

The second most basic mode of integration is the informal direct contact between managers. If a decision is limited in scope, a solution can be developed which incorporates the expertise of personnel in the respective departments or work units. As noted by Galbraith,[20] from an information processing perspective, the use of informal direct contact combined with decentralized decision making prevents upward referral and overload of the organizational hierarchy. Of course, higher levels of management may be involved in either approving the decision or in the process itself if the magnitude of the problem requires higher-level involvement. However, for design decisions or technical problems that have limited scope, informal direct contact between the relevant technical managers (with formalized approval) is the second most fundamental form of integration.

28.3.3 Liaison Positions

Galbraith[21] suggests that the next level of integration requires the creation of liaison positions. When the frequency of contacts or volume of information flows between two groups is high, the liaison role can be used to serve as a point of contact and coordination. Liaison positions are particularly useful in cross-functional (intraorganizational) interfaces, e.g., interfaces between government laboratories and project offices, between R&D and marketing divisions, or between corporate research labs and divisional product development organizations. Liaison roles may also be necessary in cross-organizational coordination, such as interfaces between government laboratories and contractors. The creation of these integrating roles naturally bypasses the lines of communication involved in the upward referral of requests for information or technical assistance. Thus, these roles overcome organizational blockages that may result in selective filtering, time delays in response, or simply the failure to initiate needed communications.[22,23]

28.3.4 Temporary Cross-Functional Teams

The next-higher level of integration involves the temporary cross-functional team or task force. Liaison positions work well for coordination between two units or functional organizations. However, when a technical problem or decision arises which requires inputs from several units or functions, a temporary cross-functional team is needed.[24] These teams may be composed of representative managers and/or technical specialists from the relevant functional groups, departments, or divisions. Some team members may be assigned temporarily, but on a full-time basis. Others may be assigned on a temporary basis. Personnel assignment decisions will naturally be a function of the task requirements. These types of teams must be distinguished from permanent cross-functional teams. The temporary teams are true temporary groups. They exist only until the technical problem is solved or design decision is reached.

28.3.5 Permanent Cross-Functional Teams

The next progressive mode of integration extends the temporary cross-functional team to the permanent (or semipermanent) cross-functional team.[25] In the permanent team, some individuals may be assigned on a full-time basis, while other personnel are assigned on a part-time basis. In this sense, the temporary and permanent teams are very similar. Again, these assignment decisions are a function of the task requirements. The use of cross-functional teams must be differentiated from a project organization, in that personnel typically continue to report to their functional managers in their functional organizations.

Current developments in the use of teams and concurrent engineering organizations are indicative of the recognition of the inadequacies of traditional cross-functional integration techniques. Overutilization of functional structures has resulted in lethargic sequential new-product development processes. Current applications in reengineering organizational structures and processes are a response to a fundamental problem of poor integration in the past. The movement away from sequential design and development processes, and toward concurrent processes and the use of cross-functional teams, will reduce cycle times and improve coordination in product development in many industries. This movement away from sequential to concurrent coordination is facilitated by the use of computer-aided design and other information system innovations.

28.3.6 The Boeing 777: An Example of Effective Integration

The development of the Boeing 777 aircraft is an excellent illustration of the effective use of cross-functional teams to achieve optimal levels of cross-functional and cross-organizational integration. In designing the 777, Boeing created approximately 240 teams which were labeled "design-build teams." These teams included cross-functional representatives from engineering design, manufacturing, finance, operations, customer support, customers, and suppliers.[26] The teams were organized around specific components, systems, or parts of the aircraft, rather than around their functional specializations.

Boeing's use of customer participation on these design-build teams was unprecedented in commercial aircraft development. Commercial airlines collaborating with Boeing on the design included United, American, Delta, British Airways, Quantas, Cathay Pacific, Japan Airlines, and All Nippon Airways.[27] In total, these customer airlines identified approximately 1200 items that they wanted in the design of the aircraft. Many of the most important customer needs were in the area of maintainability. Traditional approaches to customer input (viz., reliance on the first level of integration, i.e., hierarchical integration) would have taken the form of correspondence or meetings at an upper management level. However, by having actual airline maintenance personnel participate on the teams, improvements resulted that would otherwise be imperceptible to design engineers and airline management.[28] Typical examples are enlarging push buttons on exterior access panels so that maintenance crews working in cold climates would not need to remove gloves, redesigning the avionics bay so that workers could physically turn around, and moving light positions to improve night visibility for maintenance crews.[29]

Suppliers were also included on the design-build teams. For the Boeing 777, there were a total of 545 suppliers, with 58 of these headquartered in over 12 different countries. Integration was facilitated with the use of three-dimensional CATIA computer aided design software, and the networking of over 2000 workstations. This allowed the suppliers to have real-time interactive interface with the design data.[30] Moving to 100 percent digital design meant the virtual elimination of traditional methods of using blueprints, large mockups, and master models. It also meant that simultaneous supplier and manufacturing inputs could facilitate concurrent engineering. CATIA pre-assembly checks allowed the engineers to visualize parts and interrogate the system to determine costly misalignments and interferences. CATIA was also used to locate gaps, confirm tolerances, and analyze balances and stresses on parts and systems. This system allowed tool designers to get updated design data directly from engineering to speed the development of tooling.[31] This level of integration was made possible by the cross-functional and cross-organizational design-build teams, and by networking the suppliers with Boeing's manufacturing and engineering functions. The result was a reduction of rework and factory floor changes by more than 50 percent compared with the Boeing 767—the predecessor to the 777.[32] More importantly, because the design-build teams incorporated customer needs and manufacturing requirements, the 777 development was a cost-based design at the outset. This is a significant contrast to the classic sequential approach in which the product moves from design to engineering to manufacturing, and then finally to the customer.[33]

28.3.7 Project Manager Positions

Leadership is a central issue in the coordination of the activities of cross-functional teams. With small numbers of teams, leadership and coordination problems can be managed by the relevant functional managers and/or by the personnel on the teams. In some cases, leadership responsibilities can be centered in the department where the work (or technical problem) is centered. Autonomous (or self-managing) teamwork principles can often be used in these situations. Although intragroup coordination problems may thereby be overcome, coordination with other functional units and other cross-functional teams may still be problematic. In any case, this issue inevitably leads to the next progressive form of integration: the project or product manager. At this stage of integration, the project or product managers play integrating roles which cut across departments. The creation of this cross-functional role in effect creates a matrix organization, with significant capacity to achieve organizational integration.

28.3.8 Functional Matrix Structures

Three basic types of matrix structures exist: the functional matrix, the balanced matrix, and the project matrix. In the *functional matrix* system, a project manager with limited authority is designated to coordinate the project across different functional areas. In this model, functional managers retain primary responsibility and authority for their specific segments of the project. Some personnel may be assigned from the functional areas to the project manager on either a full-time or part-time basis. However, much of the work is conducted in the functional areas, with the cross-functional team members working with the project manager, who then coordinates with the functional managers. In the functional matrix, the project manager operates primarily on the basis of expert or referent power.[34] A major problem with this model is that the project manager's limited formal authority can severely restrict that person's ability to coordinate efforts across departments. One common solution to this problem is to give the project manager budgetary authority. This creates the so-called balanced matrix structure.

28.3.9 Balanced Matrix Structures

In the *balanced matrix,* the project manager is assigned to "oversee" the project. The project manager shares the responsibility and authority for completing the project with the functional managers. Project and functional managers jointly approve a wide range of decisions relevant to the completion of the project. Thus, this approach is sometimes referred to as the "shared authority" matrix structure.

28.3.10 Project Matrix Structures

The third form of matrix is the *project matrix.* In this form the project manager assumes the primary responsibility for completing the project and thus assumes significant authority. In this model, the role of the functional managers is to assign personnel as needed and to provide technical expertise. In this model, the functional managers serve as "landlords" and managers of the up-to-date repository of available resources.

It is important to note that each of these three types of matrix structure has equal capacity in terms of integration. Therefore, the decision regarding which matrix structure should be utilized is a function of the personnel needed in functional-versus-project tasks, the determination of where authority should be concentrated on the basis of strategic priorities and organizational size.

28.3.11 Project Organization

The highest level of integration is achieved in creating a project structure by going one step beyond the matrix and creating completely self-contained projects.[35] If one focuses exclusively on integration, this appears to be the best alternative when integration requirements are extremely high. However, it is also important to note that this form of organization is not without its disadvantages. The cost associated with the duplication of specialized functions makes this the most costly alternative. In the matrix form, specialized talent can be shared across projects. But this is not the case in the project organization. In addition, support functions must also be duplicated in the project organization, and this significantly affects costs.

28.3.12 Integration and Cost-Effectiveness

On the basis of these observations, the decision regarding the appropriate mode of integration and consequent organizational structure cannot be based on a single criterion. Rather, the issue must be resolved on a cost-effectiveness basis. With each successive mode of integration, personnel (worker-hours) costs and administrative overhead costs increase.[36] This means that one should utilize the least costly mode of integration necessary to achieve the desired or needed level of integration. For example, a project organization should only be used if the needs require this level of integration. Similarly, a matrix structure should not be utilized if less costly forms of integration will satisfy the integration requirements. Figure 28.2 illustrates the relationships between each successive mode of integration, information-processing capacity, and costs.

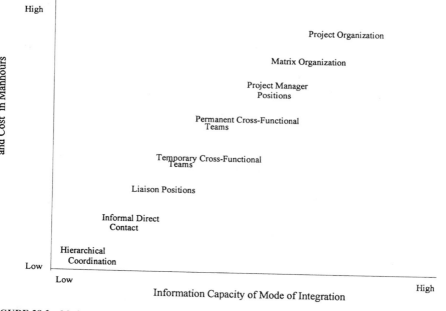

FIGURE 28.2 Modes of integration, information-processing capacity, and cost.

28.4   INTEGRATION AND COST-EFFECTIVENESS: THE CASE OF THE U.S. ARMY MISSILE COMMAND

To illustrate the issues involved in achieving the necessary levels of integration in light of the cost-effectiveness criteria, the following case example is given. This case is based on recent organizational changes at the U.S. Army Missile Command (MICOM), implemented in response to the Department of Defense budget reductions and the need to improve efficiency in managing the development of major weapon systems. MICOM is responsible for research, development, production, and world-wide support of U.S. Army missile systems, including foreign military sales.

28.4.1   Historical Background of the MICOM Problem

During the 1950s, early missile systems developed at Redstone Arsenal in Alabama relied predominantly on functional structures. In managing the development of these systems across functional units, several modes of integration were used: informal contact, liaison positions, temporary cross-functional teams, permanent cross-functional teams, and project managers with limited budget authority. Managing systems across functional areas and contractors proved to be cumbersome for MICOM and coordination problems persisted.

In 1958 the Army Ballistic Missile Agency (ABMA) was created. This was the immediate predecessor organization to MICOM. During this same time frame, the National Aeronautics and Space Administration (NASA) was created, and the Marshall Space Flight Center was born under the leadership of Werner Von Braun. With this creation of two separate organizations, from an integration perspective, what occurred next is most interesting.

Both NASA and the ABMA were experiencing a wide range of integration problems given the nature of the magnitude and complexity of their engineering projects. The solution which NASA adopted in 1961 was the use of *matrix structures,* which assigned individuals from NASA's functional science and engineering organizations to program and project offices. This system allowed flexible staffing and the optimal use of specialized technical talent. This system of project matrix structures continues to this day within NASA.

The Army adopted a very different organizational approach by continuing to use their traditional functional directorates with their technical divisions. In 1961, this structure was augmented with the *project office concept,* which involved placement of a complete project management office for missile system development under the direction of a project manager and a project office. The project manager was responsible and accountable for the successful accomplishment of all phases of the development and deployment of a particular weapon system. The decision to create a project organization occurred when a new concept or prototype had been developed in the functional laboratory directorates, or developed in conjunction with outside contractors. Once feasibility had been determined, a project office was then created to intensively manage the development, production, deployment, and support of the new system. However, the approval to create each new project office was variously restricted to the Secretary of Defense, the Secretary of the Army, the Army Chief of Staff, and the Commanding General of the Army Material command.[37] Thus, MICOM did not have the sole authority to create any project management structures that they thought were needed. This system remained in place with relatively little change from the early 1960s through the late 1980s.

28.4.2 Problems with This System and the Decision to Change It

As early as 1984, studies were being initiated to modify the project management system MICOM had been using for the previous 20 years.[38] Thus, the initial impetus for change actually preceded the Department of Defense (DOD) budget cutbacks following the decline of communism in eastern Europe and the former Soviet Union. There was concern within DOD that some project offices had become self-perpetuating entities, lingering on long after the need for intensive management had passed. In many cases, project offices had become inflexible organizations that resisted needed changes in structure, staffing level, and skill mix as their programs evolved through the system life cycle. Concern also emerged over the practice of establishing project offices as self-contained separate organizations which resulted in some duplicating of functions. This self-containment made it quite difficult to reorganize, reduce, or terminate a project organization without extensive personnel turbulence.[39] Further impetus for the modification of MICOM's project management structures was brought about by pressures stemming from the experiences of both the Air Force and the Navy, who had successfully utilized matrix structures since the early 1960s.

Despite these pressures for changes, it is important to note that MICOM's project management system worked extremely well in terms of achieving internal project integration, as well as the necessary integration with MICOM's defense contractors. As discussed previously, this is a primary strength of the project organization system. Integration is maximized; but it is maximized at significant cost. On the basis of the logic of Fig. 28.2, moving back one level of integration from self-contained project organizations to a project matrix system had the potential to reduce MICOM's overall development costs. From a cost-benefit perspective, strong arguments could be made for MICOM to move to the project matrix system.

28.4.3 Implementing the Project Matrix System at MICOM

In moving the large complex MICOM organization back one level of integration from a project organization (with functional laboratories) to a project matrix structure, project office personnel were first grouped into three categories: project core personnel, project direct-support personnel (referred to as collocated support personnel), and the project general-support personnel.

The *project office core* consisted of the project manager (and deputy), administrative staff, and secretarial positions. This core is relatively stable and changes only as the project passes major milestones in its life cycle. The *project direct-support* positions or *collocated engineers* are dedicated to full-time support for a single project office. Collocated support positions are not as stable as the core positions and can be transferred to other project offices or functional units based on workload, skill requirements, and shifting resource priorities. However, movement of collocated personnel is limited because of the learning curves involved on most missile systems. The *general-support* personnel are assigned to the functional laboratory directorates in the Research, Development and Engineering Center (RDEC). These personnel provide matrix support to the project offices either through temporary full-time assignment or temporary part-time assignment, with time split between support for a particular project office and assignments in their functional (RDEC) directorate.[40]

In this project matrix system the project office core and the collocated engineers (i.e., direct-support personnel) are supervised by the project manager. The general-support personnel are supervised by their respective functional manager. The general-support personnel receive their performance appraisals from the functional (RDEC)

TABLE 28.1   Changes in MICOM Modes of Integration Over Time

Prior to 1961	1961–1988	1989 to present
Hierarchy	Hierarchy	Hierarchy
Informal contact	Informal contact	Informal contact
Liaisons	Liaisons	Liaisons
Temporary cross-functional teams	Temporary cross-functional teams	Temporary cross-functional teams
Permanent cross-functional teams	Permanent cross-functional teams	Permanent cross-functional teams
Project managers	Project managers	Project managers
	Project organizations	Project matrix structure

manager with input from the appropriate project office. The normal work location for the project office core personnel and the collocated or direct-support engineers is the project office. The normal work location for the general-support personnel is the functional (RDEC) directorate, with possible temporary assignment to the project office location when requirements necessitate this action. The project manager has the authority for the selection of personnel for the project office core. Functional (RDEC) management has the authority for selection of personnel for the direct-support positions (i.e., collocated engineers) and for general support (i.e., laboratory directorates). Long-term training and career development are the responsibilities of the functional managers for the general-support personnel and collocated engineers, with input from the project manager in the case of the collocated direct-support engineers.[41]

Because each successive mode of integration is inclusive of the preceding modes (see Fig. 28.2), the preceding modes of integration were implicitly utilized in MICOM's matrix system. These included the organizational hierarchy, informal direct contact across groups and organizations, liaison positions, temporary cross-functional teams, and permanent cross-functional teams. In the past, these modes of integration operated effectively within the project organizations and across the interface with the contractors. The weakest area of integration existed between the functional laboratories and the project organizations. With the implementation of the project matrix system, conscious effort was applied to enhancing liaison positions and temporary cross-functional teams to facilitate integration between the project offices and the functional laboratory directorates. A chronology of those changes is presented in Table 28.1.

28.4.4   Achieving Integration under Cost-Effectiveness Criteria

In the MICOM illustration, moving back one level of integration from a project organization to a project matrix system was not accomplished without significant friction. Many of the project offices resisted the change, performance evaluations were complicated by the collocation of engineers, and the dual reporting relationships created role conflict. In addition, accountability and control of RDEC (functional) support continued to be a problem, and the budgeting and personnel approval processes were complicated by the changes.[42]

These implementation problems were addressed with varying degrees of success by MICOM military and civilian managers. However, what is important is that under conditions of major DOD budget reductions, overall efficiencies were realized. Technical expertise within the functional laboratories (RDEC) is utilized more effectively by the project offices than had been the case in the past. The change to the project matrix system also reduced duplication in some areas, improved flexibility in assigning personnel, and aided in maintaining optimal staffing based on project requirements. One primary benefit of MICOM's move to a matrix system was the avoidance of severe layoffs and major personnel disruptions caused by the DOD's budgetary reductions during the early 1990s.[43] Clearly, this was a significant benefit to the long term viability of the MICOM organization and the health of the Huntsville community.

This case clearly illustrates that achieving integration is a problem of optimization. Because of integration requirements, organizational design decisions must be made in light of overall cost-effectiveness criteria.

28.5 NONSTRUCTURAL FACTORS INFLUENCING INTEGRATION

The integration modes discussed thus far and presented in the MICOM and Boeing cases are structural in nature. It would be unrealistic to conclude that the level and quality of integration achieved is exclusively a function of these structural modes of integration. In achieving optimal integration, other factors must also be considered. These factors could be described as organizational culture or behavioral. Let us now turn our attention to these matters.

28.5.1 Supportiveness

One organizational cultural characteristic which is important for effective integration is a pervasive norm of supportiveness. An attitude of supportiveness is important to stimulating initiative in solving cross-organizational problems, and to energizing efforts that exceed specific individual or group responsibilities in resolving cross-functional problems.[44] In addition to cross-functional and cross-organizational supportiveness, management must be supportive of the efforts and cognizant of the problems encountered by engineering personnel.[45,46] This includes helping subordinates obtain the necessary resources, technical information, or assistance when needs arise. To improve internal supportiveness, each individual should meet with the manager one level above his or her immediate supervisor as part of the annual evaluation process. This could provide useful guidance to the engineer, feedback to the manager on how those supervisors who report to her or him are performing, and give the manager additional insights into problems and challenges that subordinates two levels below are encountering.

28.5.2 Management of Conflict

A second organizational cultural characteristic which affects integration is openness in the management of conflicts.[47,48] Here the assumption is that open communication regarding disagreements on technical or administrative matters is ultimately in the best interest of the organization. Research on group decision processes has repeatedly shown that superior decisions result when a group has one or more individuals assuming the role of devil's advocate as opposed to a situation where individuals are reticent and afraid to express disagreement.[49] Particularly where the higher-level modes of integration are utilized, conflict can commonly result from staffing decisions (e.g., matrix assignments resulting in dual reporting arrangements) or from the problem of different frames of reference that are common on cross-functional teams.[50] Under these conditions, it is important that higher-level managers provide clear guidance and prioritize resource allocations consistently with the organization's strategic objectives.

28.5.3 Open Communication

Closely related to the need for openness in the management of conflict is the need for more open communication in general.[51,52] Typical problems which adversely affect integration include a lack of attention to downward communication of information on the part of higher-level managers, insufficient staff meetings that facilitate two-way communication, and an insufficient number of meetings crossing functions or projects aimed at minimizing duplication of effort and maximizing the flow of technical information. In addition, as noted in the Boeing 777 case above, enhanced external communication with suppliers, customers, contractors, and manufacturing engineers in the design process will reduce later design changes, reduce the developmental cycle time, more effectively meet customer requirements, improve maintainability, and improve product quality.

28.5.4 Organizational Identity

Another organizational culture factor which can influence effective integration is a strong sense of organizational identity and identification with the mission and goals of the organization. This is to be contrasted with a situation in which the sense of identification is primarily with a division or a department of the organization. When the departmental or divisional identification prevails, "we versus them" attitudes can develop which reduce the potential for effective integration across departments or divisions. With a higher sense of organizational identity, individuals feel a sense of pride in belonging to the greater organization, loyalty and esprit de corps are high, and individual or group interests are subordinated to the overall organizational mission and goals.

A number of factors seem to influence the development of identity. This cultural characteristic will be enhanced to the degree that the leadership of an organization can create a sense of mission and articulate a specific goal or goals with which the members of the organization identify. Organizational identity, loyalty, and commitment will also be enhanced by the degree to which the organization is viewed as an elite group, pursuing technically challenging goals of significant importance. It is primarily the responsibility of leadership within the organization to create and foster these values.[53]

28.5.5 Decentralization of Responsibility

Another factor which can affect the quality of intraorganizational and interorganizational integration is the decentralization of responsibility and authority.[54–56] Decentralization allows individuals and teams at the lower levels who possess the necessary expertise and information to make critical decisions, and then obtain approvals for these decisions from higher level management. This will encourage higher order need fulfillment, which, in turn, will foster high degrees of motivation.[57] In increasing responsibility at the lower levels, that is, empowering individuals to a greater degree, it is important that accountability also be increased. In cases in which engineers both report to a functional manager and perform work for a project manager, it is important that annual performance appraisals encompass inputs from both managers.

28.6 CONCLUSIONS

Because today's organizations employ personnel with dissimilar specialties from many different disciplines who work on complex technologies, organizational integration is vitally important but difficult to achieve in modern technology-based organizations. The situation is further complicated by the existence of various organization structures for achieving integration that vary in their cost-effectiveness. Thus, the choice of the most appropriate integration structure must be viewed as an optimization problem in balancing complex tradeoffs in uncertain environments.

In general, managers should always favor the use of the least complex organization structure that has the capability to achieve the required levels of integration. Managers should always attempt to develop appropriate organizational cultures and use the best state-of-the-art management knowledge to facilitate the effectiveness of their integration structures, thereby increasing the effectiveness/cost ratios of these structures.

The current conventional wisdom favors the use of matrix structures or project organizations as cure-all medicines for organizational integration problems. However, as this chapter has shown, these approaches can be unnecessary or wasteful. Moreover, the indiscriminate use of overpowerful methods can create problems more serious than the ones they were intended to solve.

Managers will always face the dilemma between creating a highly integrated, steady-state, efficient organization versus a fast-response, environmentally adaptive organization. Gradations of each may be achieved in various combinations by blending elements of functional and project organization structures. But managers must remind themselves that these blends are compromises that may not fully realize the best features of either extreme. The best long term organization design is necessarily the one that preserves the manager's flexibility; it meets the minimum integration requirements for today but can quickly transition to tomorrow's lower or higher integration needs. Thus, tomorrow's long-term successful managers of high-technology enterprises will be ultimately familiar with all the methods shown here and in Fig. 28.2 and with the nonstructural factors discussed in this chapter. They will be competent in flexibly applying these methods and factors appropriately, in quick response to emerging environments and dynamically changing needs.

28.7 REFERENCES

1. J. D. Thompson, *Organizations in Action,* McGraw-Hill, New York, 1967.
2. P. R. Lawrence and J. Lorsch, "Differentiation and Integration in Complex Organizations," *Administrative Science Quarterly,* **12:** 1–47, 1967.
3. J. R. Galbraith, *Designing Complex Organizations,* Addison-Wesley, Reading, Mass., 1973.
4. R. Moenaert and W. E. Souder, "An Analysis of the Use of Extrafunctional Information by R&D and Marketing Personnel," *J. Product Innovation Management,* **7:** 213–229, 1990.
5. Galbraith, op. cit. (Ref. 3).
6. J. R. Galbraith, *Organization Design,* Addison-Wesley, Reading, Mass., 1977.

7. Moenaert and Souder, op. cit. (Ref. 4).

8. R. Moenaert and W. E. Souder, "An Information Transfer Model for Integrating Marketing and R&D Personnel in New Product Development Projects," *J. Product Innovation Management,* **7:** 91–107, 1990.

9. K. Clark, "The Interaction of Design Hierarchies and Market Concepts in Technological Evolution," *Research Policy,* **14:** 235–251, 1985.

10. R. Duncan, "Characteristics of Organizational Environments and Perceived Environmental Uncertainty," *Administrative Science Quarterly,* **17:** 313–327, 1972.

11. J. Utterback, "The Process of Technical Innovation within the Firm," *Acad. Management J.,* **14:** 75–88, 1971.

12. R. Cooper, *Winning at New Products,* Addison-Wesley, Reading, Mass., 1986.

13. A. Rubenstein, A. Chakrabarti, R. O'Keefe, W. Souder, and H. Young, "Factors Influencing Innovation Success at the Project Level," *Research Management,* **19:** 15–20, 1976.

14. Moenaert and Souder, op. cit. (Ref. 4).

15. Moenaert and Souder, op. cit. (Ref. 8).

16. Moenaert and Souder, op. cit. (Ref. 4).

17. F. Milliken, "Three Types of Perceived Uncertainty about the Environment: State, Effect, and Response Uncertainty," *Acad. Management Review,* **12:** 133–143, 1987.

18. Moenaert and Souder, op. cit. (Ref. 4).

19. Moenaert and Souder, op. cit. (Ref. 8).

20. Galbraith, op. cit. (Ref. 3).

21. Ibid.

22. W. E. Souder, and A. Chakrabarti, "The R&D/Marketing Interface: Results from an Empirical Study of Innovation Projects," *IEEE Transact. on Engineering Management,* **4:** 88–93, 1978.

23. W. E. Souder and J. D. Sherman, "Organizational Design and Organizational Development Solutions to the Problem of R&D/Marketing Integration," in R. Woodman and W. Pasmore, eds., *Research in Organizational Change and Development,* vol. 7, JAI Press, Greenwich, Conn., 1993, pp. 181–215.

24. Ibid.

25. Galbraith, op. cit. (Ref. 3).

26. P. M. Condit, "Focusing on the Customer: How Boeing Does It," *Research-Technol. Management,* **37**(1): 33–37, 1994.

27. J. P. Woolsey, "777: A Program of New Concepts," *Air Transport World,* pp. 60–64, April 1991.

28. Condit, op. cit. (Ref. 26).

29. P. Proctor, "Boeing Rolls out 777 to Tentative Market," *Aviation Week & Space Technol.,* pp. 36–51, April 11, 1994.

30. Ibid.

31. R. G. O'Lone, "Final Assembly of 777 Nears," *Aviation Week & Space Technol.,* pp. 48–50, Oct. 12, 1992.

32. Proctor, op. cit. (Ref. 29).

33. Condit, op. cit.

34. J. R. French and B. Raven, "The Bases of Social Power," in D. Cartwright, ed., *Studies in Social Power,* University of Michigan Press, Ann Arbor, Mich., 1959.

35. Galbraith, op. cit. (Ref. 3).

36. Ibid.

37. Department of the Army, *Army Material Command Program Management Matrix Support,* Aug. 11, 1986.

38. Ibid.

39. Ibid.

40. Ibid.

41. Department of the Army, MICOM regulation 10-9, April 3, 1989.

42. H. H. Harris, J. D. Sherman, and E. A. Olsen, *Phase II Assessment of RDEC Matrix Reorganization,* U.S. Army Missile Command, DAAHO1-91-D-R002, June 1992.

43. Ibid.

44. A. Gupta, S. Raj, and D. Wilemon, "Managing the R&D-Marketing Interface," *Research Management,* **30**(2): 38–43, 1987.

45. A. Abbey and J. Dickson, "R&D Work Climate and Innovation in Semiconductors, *Acad. Management J.,* **26**: 362–368, 1983.

46. S. Kozlowski and B. Hults, "An Exploration of Climates for Technical Updating and Performance," *Personnel Psychol.,* **40**: 539–563, 1987.

47. W. E. Souder, *Managing New-Product Innovations,* Lexington Books, Lexington, Mass., 1987.

48. Souder and Sherman, op. cit. (Ref. 23).

49. I. L. Janis, *Victims of Groupthink,* 2d ed., Houghton-Mifflin, Boston, 1982.

50. S. M. Davis and P. R. Lawrence, *Matrix,* Addison-Wesley, Reading, Mass., 1977.

51. P. Shrivastava and W. E. Souder, "Phase Transfer Models for Technological Innovation," *Advances in Strategic Management,* vol. 3, JAI Press, Greenwich, Conn., pp. 135–147, 1985.

52. D. Ford and C. Ryan, "Taking Technology to Market," *Harvard Business Review,* **59**(2): 117–126, 1981.

53. J. H. Morris and J. D. Sherman, "Generalizability of an Organizational Commitment Model," *Acad. Management J.,* **24**(3): 512–526, 1981.

54. Abbey and Dickson, op. cit. (Ref. 45).

55. M. McCarrey and S. Edwards, "Organizational Climate Conditions for Effective Research Scientist Role Performance," *Organizational Behavior and Human Performance,* **9**: 439–459, 1973.

56. W. E. Souder, "Organizing for Modern Technology and Innovation: A Review and Synthesis," *Technovation,* **2**: 27–44, 1983.

57. J. D. Sherman, "The Relationship between Factors in the Work Environment on Turnover Propensities among Engineers and Technical Support Personnel," *IEEE Transact. Engineering Management,* **33**(2): 72–78, 1986.

CHAPTER 29

BARRIERS TO IMPLEMENTATION OF A STRUCTURE FOR MANAGING TECHNOLOGY

Anil B. Jambekar and Paul A. Nelson

Michigan Technological University
Houghton, Michigan

29.1 INTRODUCTION

All too often, organizations adopt change programs, only to find that the change is short-lived, or does not measure up to expectations. When one change program fails, another is pushed in to take its place, and then another. Introduction of new technology often follows the same path. A firm incurs all the costs of adoption of the new technology, but fails to reap the full benefits. Employees, particularly salaried staff and middle managers, seem to continue to use the old technology for their important work. Whether "hard" technology such as computer-aided process planning (CAPP) or "soft" technology such as total quality management (TQM), the outcome seems to follow the same disappointing path.

This chapter explains why many present organizations are the way they are, why it is very difficult to transform them into structures which are more compatible with managing technology, and what needs to happen in order to produce lasting transformation to a compatible structure. Many of our existing organizational structures, the ones which we label *functional, mechanistic,* or *bureaucratic,* are in fact structures which evolved to solve real problems. Any new structures which we now develop to facilitate the managing of technology will degenerate to these old structures if those old underlying problems are not addressed. Organizational change involves formation of new roles, duties, relationships, information flows, evaluations, incentives, and perceptions. If we think in terms of only departmental definitions and reporting relationships, we will not deal with the fundamental processes involved in the operation of our organization, and will not have a beneficial impact on performance. Resistance to change, particularly to changes in fundamental processes, is almost certain to occur. Management tools exist to understand what type of resistance will occur, and to influence and channel this resistance in a constructive manner. Once fundamental organizational change has occurred, it is necessary to activate the new structure, and to prevent it from degenerating back to the old structure. Again, management tools exist to bring this about.

29.2 FORCES UNDERLYING OLD ORGANIZATIONAL STRUCTURES

The functional structure is perhaps the oldest and most common arrangement for an organization. Businesses, government entities, educational institutions, churches, and even sports organizations seem to use it in spite of its many shortcomings. Moreover, it seems to have evolved quite independently in many parts of the world under a variety of cultures and historic situations, from the pre-Columbian Aztecs to the hoards of Kublai Khan. Figure 29.1 shows a generic functional structure for a business firm.

The structure has a number of obvious characteristics. The first is hierarchy. One person is at the top, the CEO. The CEO has several direct reports, and each of these has several direct reports, yielding the familiar pyramid structure. The second is unity of command. Each individual in the structure has one, and only one, supervisor, and the CEO at the top has ultimate responsibility for the organization. The third is departmentalization by function. Sales and marketing people are separated from people in production, and high degrees of specialization occur toward the bottom of the pyramid. These structures evolved to solve real problems. New structures developed to facilitate the managing of technology must address these same problems effectively in order to survive.

29.3 PRESSURE FOR EFFICIENCY

29.3.1 Transactions Costs

It takes managerial time and energy to purchase materials and services. Choosing the right set of machine tools or negotiating a long-term contract can consume thousands of dollars in managerial time and effort. These "transactions costs"[1,2] may be lower in the functional structure than, for example, in a structure organized on the basis of profit centers and cost centers because of centralized purchasing from the outside and off-budget transactions within. On the other hand, if the manager of a particular profit center needed the services of one of the firm's cost centers, say, the central engineering department, the situation would be analogous to a firm contracting out a service to

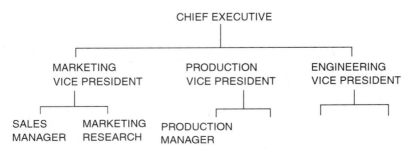

FIGURE 29.1 Functional structure.

an external vendor. The manager might not be able to readily develop a contract which deals with every possible outcome and, as a result, might become the victim of "opportunistic behavior" by the department. Part of the reason for this is the manager's "bounded rationality." The profit center manager, although quite competent as a general manager, might not be an expert on stochastic processes, and consequently might not be able to fathom exactly what ought to be involved in an engineering study to correct a vibration problem in a process. Another reason for this is that of "information impactedness," or the fact that only those engineering department employees actually working on the study get to see how bad the problem really is, and how much time and effort are needed to correct the problem. As a result, the profit center might be overcharged and never know it. In the classic functional organization, no effort is even made to impute a transfer price for the engineering study because both departments are part of the same firm.

29.3.2 Specialization

Another source of efficiency is specialization. By putting all of a specific function into one department, employees can specialize in specific activities. With specialization, employees can concentrate on a narrow range of activities, and become proficient. Time is not lost in switching from one activity to another. Moreover, employees can have their technical skills reinforced by having a manager who is technically competent in the function, and by having fellow employees in the same field for support and development. Often, there are economies from fuller utilization of specialized equipment.

29.3.3 Monitoring and Control

Other sources of efficiency are in the areas of monitoring and control.[3] In Fig. 29-1, each manager can supervise or monitor the activities of several subordinates. Cross-comparisons of performance can be effective within a narrow function. Decision making is centralized, and the higher-level manager can solicit input from lower levels.

29.3.4 Communication

Efficiency can also be achieved in communication. If a sales manager in Fig. 29.1 wishes to change the *master production schedule* which is under the jurisdiction of the production manager, the request must pass through the vice president of marketing and the vice president of production. Even though time is consumed as the request passes through channels, the relevant members of the organization are kept informed. Moreover, other members of the organization do not have to deal with the information, as they are not in the channel. When everybody in the organization is presented with every message or communication, as now can happen with e-mail systems, there is obvious information overload, and a lot of people's time is wasted wading through irrelevant information, often missing or never getting to the relevant information.

These sources of efficiency might not be present in the new organizational structures. As a consequence, the new structures might not be robust; that is, they may degenerate over a period of a year or two to the old functional structure. An example may illustrate. One university has a central photo services department that, among other things, makes slides, transparencies, and charts for presentations. Because of long lead times, and the suitability of some personal computers and workstations for graphics, many departments and individuals developed their own capabilities in these areas, at great expense to the university in terms of redundant hardware, software, and the opportunity cost of the professional time allocated to developing these capabilities. The driving force behind the duplication of effort was minimization of response time. One by one, though, the independents started to return to the central photo services department. It turns out that not all of the independents succeeded in mastering the software, or had adequate artistic aptitude to turn out acceptable work. Moreover, most discovered that the hours they were spending on graphics were hours they could not spend on their own field of specialization. For a new organizational structure to succeed, the issue of efficiency must be addressed. Thus, we must be explicit about how much efficiency we are willing to sacrifice in trade for responsiveness or wholism, or we must find ways to exploit efficiency in the new structure.

29.4   PRESSURE FOR STABILITY

Many new organizational forms arise during entrepreneurial phases in an organization's history. In the hectic burst of growth which characterizes these phases, almost any type of free form or organic structure can work because wholism is achieved through the excitement of the moment. Gradually, though, people start craving stability. Systems and procedures are established to reduce uncertainty, to make the operation of the firm more predictable. People seem to have a natural attraction to stable situations over unstable ones. For example, in spite of the realities of the 1996 job market for college graduates, most indicate preference for a *Fortune* 500 firm, should one happen to be hiring instead of downsizing, over the tens of thousands of smaller firms which actually are having problems attracting suitable talent. Excitement is nice, but even the soldier of fortune prefers the CIA.

29.5   DESIRE FOR PERSONAL GROWTH AND DEVELOPMENT

People naturally want to advance themselves through their work. In the functional organization, there generally is a clear career ladder stemming from each position. A college graduate starts in an entry-level position such as a programmer in an information systems department, and sees a sequence of positions to aspire to over a career. Each position offers opportunities for the growth and development needed to qualify for higher-level positions. In many of the newer organizational structures, there are no standard job titles nor clear career ladders. Employees are enthusiastic in the early years, but gradually this enthusiasm dissipates. Position titles start to appear and hierarchy starts to form, perhaps reflecting the relative capabilities or accomplishments of the people in the positions. Gradually, the old structure is re-created.

29.6   DESIRE FOR EQUITY

We all want to be treated fairly, and we all have our own opinions on what fairness is. Adams has advanced the notion that always and everywhere we make equity comparisons with those around us, comparing the ratio of rewards received to contributions made.[4] In the old functional organization, there was typically a *management by objectives* (MBO) program, in spite of Deming's exhortation against MBO, or some other type of formalized planning and appraisal system. It may not have been perfect, but it provided an outlet for us to sing our own praises. In the newer structures, formalized appraisal systems often are supplanted by team bonuses or other group rewards. Even though there are good reasons for this, there is a deep-seated urge by many to be rewarded individually. Moreover, there is a fear of being totally at the mercy of the system. A desire gradually emerges for some type of retrenchment policy which bases job security on seniority, and which provides safeguards against arbitrary transfers or sanctions. Over time, this can lead to a regression to the old system.

29.7 THE NATURE OF ORGANIZATIONAL CHANGE

29.7.1 Elements of Change

Chandler pointed out the need for symmetry between strategy and structure decades ago.[5] If a firm makes a major change in strategy, it needs to make accommodating changes in its organizational structure. Today it is widely believed that in addition to strategy and structure, an organization's culture, technology, and people all have to be compatible. If you introduce change in technology, you should expect to alter your corporate strategy to capitalize on the new capabilities, alter various departmental roles and relations, add personnel with new talents, and attempt to "manage" change in shared beliefs and values needed to facilitate use of the new technology.

Hall and colleagues[6] suggest that six organizational elements are involved in fundamental change: roles and responsibilities, measurements and incentives, organizational structure, information technology, shared values, and skills. Many attempts at bringing about organizational change fail because management attention is focused on organizational structure alone, or on that and only one or two other elements.

Johnson and Kaplan argue that failure to change old measurement and incentive systems, such as equipment and facility utilization measures, in manufacturing firms adopting the just-in-time (JIT) approach leads to serious implementation problems.[7] There is also general awareness that employee roles and responsibilities, shared values or culture, and skills must also change. Even the information technology change needed to support JIT is now known and accepted. But this wasn't the case in the mid-1980s, nor is it the case now with some of the more recent types of changes occurring in organizations.

29.7.2 Organizational Context

Digging deeper, Goss et al.[8] find that organizations that make fundamental changes in the way they do business, reinventing themselves in the process, must alter the underlying assumptions and invisible premises on which the firm's decisions and actions are based. This "context," they argue, is the sum of all the conclusions that members of the organization have reached, the product of members' experience and interpretations of the past, yesterday's mechanism for success. They note that context is generally hidden and requires real effort to uncover, yet if left unchanged, efforts at organizational change will be unproductive, or at best, episodic. Consider a successful all-you-can-eat restaurant which pulls up stakes and moves to a large shopping mall, expecting that the large pool of potential customers will more than offset the high rental fees. The type of menu, the assumptions about what customers value in service, and the method of preparing meals all remain unchanged. Management and staff succeed in doing what they did before, yet now the business fails because it cannot generate the high rental fees, even though the restaurant attracted more well-fed patrons than before, many grazing for hours on end at the food bar. The restaurant, instead, should have changed its assumptions and the premises on which it based decisions, recognizing that only a much higher customer throughput could generate the rental fixed cost, and that a menu closer to fast-food fare would generate that kind of throughput.

29.7.3 Resistance to Change

In order to implement lasting change, the manager needs to understand that there will be resistance to change, to any change, and that this resistance occurs as naturally as a stretched rubber band regaining its original shape after the force used to stretch it is removed. Individual and organizational resistance to change are interrelated. Individual resistance to change creates an environment which can promote organizational resistance. Organizational resistance, in turn, can fuel further individual resistance. The self-reinforcing nature of this relationship can be extremely powerful. In order to deal with the resistance, the manager needs to be able to understand its true nature, so that it might be dealt with appropriately.

Irrational Resistance. A major organizational change program almost certainly will arouse insecurity and generate fear. Salaried employees, particularly middle managers, recognize their vulnerability in this age of massive restructurings and outplacements. Insecurity and fear, however, are viewed as weaknesses; their presence arouses anxiety. We have trouble admitting to ourselves, much less to our superiors or fellow employees, that we are afraid. We protect ourselves from this anxiety by switching our focus from self-pity to a more self-acceptable position that the change itself is not in the interests of the organization, setting off the generation of a series of seemingly defensible criticisms of the proposed change. As each criticism is rebutted, another takes its place; hence, the term "irrational." This irrational resistance is triggered by more than just fear of termination. There is recognition that we will have to work much harder now to stay on top of our jobs than before, and we might dread the loss of comfort and ease. There is the fear that we will not be able to learn or master the skills needed in the new structure. There is the fear of loss of status—status which was earned over a long period of time for mastery of the old system. Under the new system, we may have a competence level at best equal to entry-level employees. We may even need to ask entry-level employees how to perform certain aspects of our job, something unimaginable under the old system. Many college professors faced this situation in the 1980s as networked PCs and workstations emerged on every part of campus, discovering that freshmen often knew more about the mechanics of the network than they did.

Political Resistance. Political resistance is almost a group version of irrational resistance. The functional unit assesses the probable consequences of the change program on its activities, and prefers the status quo. Habit and inertia, prior fixed investments and staff skills which may be rendered valueless, and the enormous amount of work which now lies ahead contribute to this. There is also the problem of relative status of the functional unit. Any change program which affects the six elements of context discussed above can affect the relative importance of organizational units. The nature and importance of quality-control departments in manufacturing firms changed during transitions to JIT, particularly as positions of inspectors were eliminated, and statistical process control (SPC) was substituted for acceptance sampling, leading to a change in fortunes for many in those departments. The relative importance of information systems departments in organizations switching to distributed computing is another case in point, particularly if a department loses control over access to information.

Cultural Resistance.   Old cultural mindsets bred in an era of market dominance are hard to shake. People are being asked to change ways of doing things which worked very well in the past, not always realizing how the success of those ways of doing things was dependent on circumstances which now no longer hold. New measures of performance which are a part of many well-conceived change programs have the potential to expose real areas of weakness in some functional units, leading to vulnerability. Even access to information, which once could be hoarded, or at least sanitized before being released, is threatening.

Ineffective Interactions.   One reason why the types of resistance to change discussed above become powerful enough to thwart attempts to succeed at change is that ineffective interactions among people prevent ideas from being discussed openly, lead to false consensus, and ensure an environment in which cooperation and coordination are either ineffective or absent.

Argyris has described defensive routines and organizational barriers which preclude effective dialog. Individuals form assumptions and beliefs which remain largely untested.[9,10] We adopt those beliefs because they are based on conclusions which were inferred from what we observe, plus the meaning we add based on our experience. What we observe is based on the data we select from a much larger pool. Our selection process is dictated by our assumptions and beliefs which act as filters. The "ladder of inference" is a common mental pathway of increasing abstraction, so natural and comfortable that we forget all but the bottom and top rungs (see Fig. 29.2). The rest of the rungs—our thought processes—are unseen, unquestioned, and extremely abstract.

Rung	
4	The theories we use to create the meanings on rung 3 (Beliefs)
3	Meanings imposed by us (Assumptions and conclusions)
2	Culturally understood meanings
1	Relatively directly observable data, such as conversations

FIGURE 29.2   The ladder of inference. (*Source: Chris Argyris.*[9])

The ladder of inference explains why most people cannot tell where their deepest attitudes come from. The data is long since lost to memory, after years of inferential leaps. Before long, our long-standing assumptions dominate our thinking and substitute for the real data. Our unchallenged assumptions are the basis of our interaction with others. We each act according to our individual conclusions about the environment we share. Our interactions, then, add to the pool of data available for us to select specific observations.

29.8 APPROACHES TO SUCCESSFULLY IMPLEMENT CHANGE

With all the organizational change which has occurred in past 15 years, some methods for success at implementing change—old, time-tested steps that work—are already widely known.

29.8.1 Explaining the Need for Change

First, the employees of the organization, and other important stakeholders such as stockholders, key customers and suppliers, and major lenders, need to be provided with a clear and honest statement of why that change is needed, focusing, where appropriate, on external forces such as heightened competition, a need to penetrate foreign markets in order to maintain minimum efficient scale, or changes in technology which threaten the viability of old-product lines.

29.8.2 Participation in Developing Plan

Participation by a broad segment of the organization in developing a plan for change may add months to the finalization of the plan, but it will save many many more months in implementation, particularly if members made real contributions to the plan. The idea is that it now is "our plan," not just the "boss's plan," and we are much more likely to fight for the success of "our plan."

29.8.3 Disarming the Resistors

It is not particularly difficult to identify those people in the organization who are going to become the most serious resistors to the plan. They often identify themselves early in the process. Often, these are the natural leaders in the organization, people who influence many others. Some firms have been successful in working with these people in an honest manner, helping them digest the background information which motivates the change. Other firms find that these are the prime candidates to send away to short courses related to the change, so that they become the "experts" who will shepherd in the new technology. Still other firms find that recalcitrant staff members are now eligible for special early-retirement programs.

29.8.4 Presentation of the Plan

Once a plan is developed, and ratified by the organization, it must be clearly explained to all. In particular, management must beat the gossip circuits to the punch by clearly, and completely, revealing how the plan will affect everyone, particularly in regard to layoffs and terminations. It is hard to give adequate hearing to a plan when you have a deep-seated fear that you will be out of a job.

29.8.5   Presenting a Path to Follow

The plan must include a path for employees to follow to learn the new way of doing things. This includes opportunities to learn needed new skills, and opportunities for practice.

29.8.6   Burning Bridges to the Past

At some point, the plan must be implemented, and the new system must be the only system. As long as it seems possible to revert to the old way, some employees, and not necessarily those at the bottom of the pyramid, will hold on to the thought that it will all go away, and we'll be back to the old system. It's like a major railroad wanting to abandon service to a region; until the tracks are ripped up, there is always the possibility of returning service, a fact not overlooked by communities petitioning government agencies to bring that about.

29.9   DEVELOPING A LONG-TERM SYSTEMS PERSPECTIVE

For reasons ranging from corporate quarterly reports of earnings, to periodic MBO appraisals, to the popularity of job hopping in U.S. industry, organizations, and the people who populate them, operate with short-term time horizons. Major change programs such as introduction of new technology or development of JIT or TQM, by contrast, take years to unfold.

The pervasiveness of organizations with the functional structure in our culture tends to produce a narrowness in people which management scholars label *functional isolation,* but which industry leaders and popular writers term the "chimney effect." Departments in functional organizations tend to be viewed by their members as fiefdoms, with the consequences that departmental goals are given more weight than organizational goals, and conflict among departments is more likely than cooperation. Unfortunately, this kind of behavior does not automatically disappear in newer structures. Our educational system which isolates mechanical engineers from accountants, and computer science majors from marketing majors, and our professional and trade journals which reinforce narrow specialization, breed functional isolation too.

Organizations must understand that commitment to technology in the 1990s requires adjustment of temporospatial orientation.[11] Most organizations can march employees by the numbers through the mechanics of improving processes and ways of doing things as in the thousands of firms which have made progress with SPC, quality circles, and some of the elementary stages of TQM. This sort of activity can be labeled *operational learning.* To get the real benefits of any of these programs, though, *conceptual learning,* or the ability to reframe problems in new contexts and timeframes, is needed. Systems thinking tools[12] and the *action science* tools[13] together have potential for enhancing conceptual learning.

29.10 FALLACY OF PROGRAMMATIC CHANGE

Beer and coworkers[14] point out that many firms have attempted to implement change by promulgating companywide mission statements, corporate culture programs, training courses, quality circles, and the like, without much success. They argue that to succeed in implementing organizational change, firms must align employee roles, responsibilities, and relationships to address the firm's most important competitive tasks. "Task alignment" focuses energy for change on the work itself, not on abstractions such as participation or culture. Their key point is that individual behavior is powerfully shaped by the organizational roles that people play, rather than the reverse. Give people new roles to play, a new organizational context, which imposes new roles, responsibilities, and relationships, and then attitudes and behaviors will change.

29.11 ACHIEVING HOLISM

Holism is the cement that binds the new organizational form together. It is the property of the systems view of the organization by which all parts and subsystems coordinate their activity toward attainment of the organizational goal much like all the anatomic parts of a human being work in concert as the human walks, talks, and goes about daily activities. To achieve holism, the members of the organization must all strive for the organization's fundamental goal such as that of providing customer satisfaction, and there must be a free flow of information and understanding among members and customers. Failure to accept the organization's goal or to provide a free flow of information and understanding will threaten achievement of holism.

29.11.1 Accepting the Organization's Fundamental Goal

In order to promote acceptance of the organization's goals, members or employees must understand why the specific goal exists, and what their department or division's role is in the achievement of that goal. A lot of dysfunctional behavior occurs when certain departments have false impressions of what their role is supposed to be. In some cases, the concept of a departmental charter, which states the department's role in attainment of organizational goals, might be a good idea. It could eliminate the possibility of employees or middle managers behaving dysfunctionally in good faith.

29.11.2 Eliminating Barriers

Barriers to the free flow of information and understanding exist for a reason, perhaps to safeguard the firm's intellectual property or marketing plans, or perhaps to preserve inequities or inefficiencies. Proprietary information needs to be identified and its security assessed. Everything else should be made available to those with a need to know. It is interesting that computer mail and distributed computing are accomplishing more in this area than most systematic management efforts. Groupware is now available to allow shared access to data on individual workstations and PCs.

29.12   SYSTEMS TOOLS

The structure of effective interactions requires that, if informed technology choice is to be converted into timely action, individuals must change the way they think, the way they act (based on their thinking), and the way they interact. Furthermore, they must be helped to understand the counterintuitive behaviors commonly exhibited by complex systems. That is possible only through taking the systems view of the technology in a long-term timeframe. The systems thinking based on system dynamics offers an opportunity to experience double-loop learning[10] and thus to learn about the systemic structure and processes which give rise to underlying symptoms. Systems thinking requires that we develop a particular way of looking at the world. It is different from most other problem-solving tools because we are a part of the system we seek to understand and influence. If we are trying to implement a technology we do not fully understand or if just doing the same thing with more intensity is argued to be the only solution, the system orientation allows us to step back and map interdependencies in the organization to generate systemic insights.

29.12.1   System Dynamics

System dynamics[11,12] helps managers select better data, interpret it better, and reach better conclusions. But the real advantage is that managers learn to question the implicit beliefs and assumptions which filter the data they select in the first place. The systems view completely changes the way they think about the problem.[15] Thus the strength of system dynamics lies in the ability to improve the way we view the problems. System dynamics tools facilitate viewing the system by moving out in space and thus understanding the interrelationships among parts, and by moving out in time to be able observe the evolution of longer-term behavior of the system. The tools also allow you to test a variety of assumptions about the future.

29.12.2   Process Consultation and Action Science

Interventions based on action science[13] focus on how we think about other people, how we act, and thus how we interact. They do not address how we think about the strategic challenges the organization faces. Standing alone, they are sound methods for achieving change in individuals, but they rely on the individual's ability to know what to change or where to move strategically.

By integrating system dynamics, process consultation, and action science with traditional strategic management and quality management tools, we have the tools needed to overcome the barriers to implementing structures to manage technologies. It should be mentioned that the tools should not be the bastion of few individuals, but should be taught to everybody to be effective.

29.13   CONCLUSION

Failure to adopt new technologies that deal with human issues often accounts for why technology initiatives do not realize the levels of success hoped for. First, ask a few questions to trigger the thought process regarding the existing management processes or the organizational structures:

1. Are your management processes vertically or horizontally integrated?

2. Where and how do your new technology and customers fit into your management structure?

3. Does your organizational structure or management process allow you to focus on the customer or simply on elements of technology?

The answers to these questions will show you how deeply you are likely to experience the physical, mental, or organizational barriers that preclude the adoption of new technologies. Horizontal focus for a management process; clearly understood interdependency of new technology, organization, and customers; and willingness by all to approach every situation with the assumption that you can learn from it and that your interpretation is only one of several possibilities are key ingredients for overcoming barriers to implementation of technology.

Now shift gears and brainstorm the following specific key questions in the context of selected value-adding technology:

1. How will the technology change jobs (tasks, time, skills needed)? What is the game plan for communicating these changes and making adjustment to accommodate the changes?

2. Are there clearly established priorities, or are people just given more to do? What is the plan for managing day-to-day impact of newly ushered-in technology on people?

3. What organizational obstacles must be overcome (existing policies, incentive systems, existing power blocs)?

4. What apprehensions, doubts, and expectations are pervasive?

5. What organizational practices exist that contradict the new-technology message?

6. Is the implementation process designed to be directly relevant to jobs, and does it make sense to the people doing the job?

7. Are managers espousing the benefits and requirements of new technology, giving a clear and consistent message?

8. What training is being offered? Is it enough and just in time? Does it improve employee value and capability?

The core message is that to achieve technological change, management must be aware that they are dealing with a pattern of employee habits, beliefs, and traditions and need to ascertain just what will be the social effects of the proposed technological change.

29.14 REFERENCES

1. Oliver E. Wiliamson and Sidney G. Winter, eds., *The Nature of the Firm: Origins, Evolution, and Development,* Oxford University Press, Oxford, 1991.

2. Oliver E. Williamson, *Markets and Hierarchies: Analysis and Antitrust Implications,* Free Press, New York, 1975.

3. Mark Casson, "Why Firms Are Hierarchical," *J. Business and Economics,* 1(1):47–76, 1994.

4. J. Stacy Adams, " Inequity in Social Exchange," in L. Berkowitz, ed., *Advances in Experimental Social Psychology,* vol. 2, Academic Press, New York, 1965, pp. 267–299.

5. Alfred D. Chandler, Jr., *Strategy and Structure: Chapters in the History of the American Enterprise,* MIT Press, Cambridge, Mass., 1962.

6. Gene Hall, Jim Rosenthal, and Judy Wade, "How to Make Reengineering Really Work," *Harvard Business Review,* pp. 119–133, Nov.–Dec. 1993.

7. H. Thomas Johnson and Robert S. Kaplan, *Relevance Lost: The Rise and Fall of Management Accounting,* Harvard Business School Press, Boston, 1987.

8. Tracy Goss, Richard Pascale, and Anthony Athos, "Risking the Present for a Powerful Future," *Harvard Business Review,* pp. 97–108, Nov.–Dec. 1993.

9. Chris Argyris, *Overcoming Organizational Defenses,* Allyn & Bacon, Needham, Mass., 1990.

10. Chris Argyris, *Reasoning, Learning, and Action,* Jossey-Bass, San Francisco, 1982.

11. Barry Richmond and Steve Peterson, *An Introduction to Systems Thinking,* High Performance Systems, Hanover, N.H., 1987, 1992.

12. George P. Richardson and Alexander Pugh III, *Introduction to System Dynamics with DYNAMO,* Productivity Press, Portland, Oreg., 1981.

13. Chris Argyris and Donal Schon, *Theory of Action,* Jossey-Bass, San Francisco, 1974.

14. Michael Beer, Russel A. Eisenstat, and Bert Spector, "Why Change Programs Don't Produce Change," *Harvard Business Review,* pp. 158–166, Nov.–Dec. 1990.

15. Peter Senge, *The Fifth Discipline: The Art and Practice of Learning Organization,* Currency Doubleday, New York, 1990.

CHAPTER 30

THE NEED FOR INTERDISCIPLINARY ENDEAVOR AND IMPROVED INTERFUNCTIONAL RELATIONSHIPS

Thomas G. Whiston
Senior Fellow, Director of Studies
(Technology and Innovation Management)
Science Policy Research Unit (SPRU)
University of Sussex
Falmer, Brighton, United Kingdom

30.1 INTRODUCTION

Nobel prizes are awarded for advances within a discipline. Promotion within academe usually depends on research success and publication within a comparatively narrow area of study. On the other hand, industry favors the pragmatist, and its area of discourse, or operation, is necessarily interdisciplinary or multidisciplinary. Therein lies the dilemma.

Expressed in more transparent terms, academic and intellectual progress has developed over the centuries by increasing reliance on specialization and endeavor; study and application within comparatively narrow confines and hence our departments or faculties of mathematics; physics; chemistry; biology; electrical, electronic, mechanical, or civil engineering; economics; psychology; sociology; or whatever. In short, until recently academe has been the domain of the specialist. In turn, this has begat students who know a lot about their own "discipline" (indeed, they are *disciplined*) but often do not know enough about other subjects.

On entry into industry and commerce, a problem emerges for, as we have just noted, industry has to *integrate* knowledge: the engineer, the accountant, the designer, the marketing and administrative individual, the materials scientist, and the R&D chemist or biologist have to work together; to communicate; to understand each other; and to see the potential, the challenge, the failings, the obstacles of each area of knowledge. In short, the various disciplines must co-operate. This is not always easy, and the greater the degree of specialization of the various functions, then (often) the greater is the difficulty of obtaining such necessary integration.[1]

While industries, or large companies, were organized on tight functional lines (the design unit or department—production planning and the accounting and marketing departments)—and where reliance was placed on hierarchical "Fordist" principles, the specialization problem received both protection and reinforcement. Increasingly, however, it is recognized that those days are long since past—at least by the aware company—and that in order to achieve (say) satisfactory design for manufacture, or "lean production," or great increases in quality of product, increased productivity, defect reduction, and so on, more functionally-integrative systems are essential.[2] This, in turn, demands much attention to improved interdisciplinary training.[3]

Indeed, as well as the recognition that much more "organic," internally communicative organizational systems—more fluid interdisciplinary systems—permit the development of more *effective* performance, is the added challenge of the new sectoral, scientific, and increasingly generic areas relating to such areas as biotechnology, new-materials science, IT (information technology)-systems application, environmentally friendly products, energy-efficient systems, integrated transport and communication systems, or whatever. All these areas (and many more could be listed) demand the cross-fertilization of numerous subdisciplines and functions. They are necessarily cross-disciplinary and multidisciplinary, thereby requiring scientist, engineer, chemist, biologist, and mathematician to not only pool their skills but now, much more than in past, to be able to enter the mindframe of their colleagues. Thus, not only is the leading edge of *intellectual* advance seen to be at the interstitial boundaries [biochemistry, biophysics, AI (artificial intelligence) and cognitive psychology, complex systems analysis],[4] the same "leading edge" pertains to industrial *organization* and commercial application.

This demands two main requirements: (1) that *academe* itself be encouraged, organized, and structured to produce much greater numbers of interdiscipline-oriented individuals and (2) that industry organizes itself to best utilize such individuals.[5] As we shall note later, there is, however, if not a third requirement, certainly an important catalyst to those two generic requirements which relates to increased and improved academic-industrial *collaboration* programs. Such collaboration pertains both to the training and research function of higher education.

In the following sections we will examine the ways in which academe and industry can best encourage and utilize interdisciplinary endeavor and provide examples of good academic-industrial collaboration which further encourages interdisciplinary training and exchange—as well as identifying problems which still must be overcome.

30.2 THE NEED FOR AND CHALLENGE OF INTERDISCIPLINARY TRAINING AND PERSPECTIVES

We can take several perspectives, some of a narrow form, others from a very broad base or platform, in order to justify and illustrate the need for interdisciplinary endeavor. We can also apply that perspective to every level of training and *retraining.*

Thus, from the point of view of *perspective,* much improved interdisciplinary training is demanded by the following:

1. At the level of the *firm or enterprise,* the transition to post-Fordist forms of work organization necessarily demands new systems organization, crosslinkage of functional areas (e.g., R&D, design, production, manufacturing, maintenance and services, internal business and economic strategy of the firm, marketing and market research) if a more smooth-flowing, organic, and intimate organizational system is to emerge.[6] When the production, or formulation, of a robust product portfolio range is aimed at, then if it is recognized, as it should be, that the temporal and strategic matching between a company's R&D portfolio and its subsequent product range, or product portfolio, is to be maintained over long time periods, it is essential that each subfunction can and should communicate with and comprehend each other[7] (see Fig. 30.1).

Perhaps nowhere is this more important than within the aegis of the design "function," or more accurately the linkage of design and innovation.[8,9] To be effective, the design function must encompass a range of functions—materials science, engineering, ergonomics, economic market analysis, production and manufacturing capability, and so on.[10] Here we see the true value of interdisciplinary training and functional coordination. We can also observe the market failure of products which can ensue if such functional integration is not attended to.

2. In some ways an emphasis on *simultaneous* or concurrent engineering recognizes the need for parallel involvement of different functions and therefore, implicitly, for greater interdisciplinary skills. However, such a position has been subjected to a

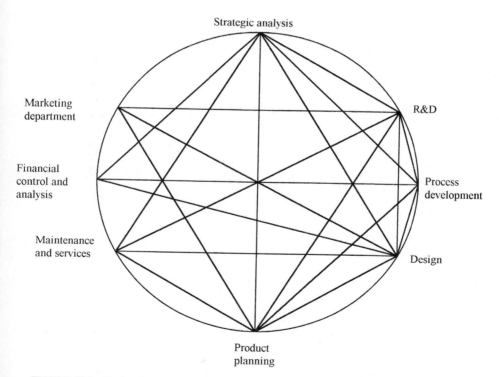

FIGURE 30.1 Further developments of multidirectional communication and involvement. (*Source: Whiston.*[1])

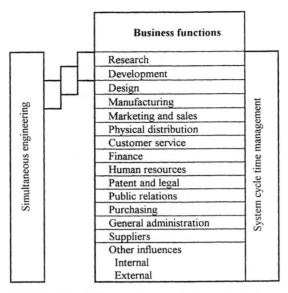

FIGURE 30.2   Differences in scope between simultaneous engineering approaches and system cycle time management. (*Source: Gaynor.*[39])

more subtle analysis and interpretation by Gaynor in recent years.[11,12] Gaynor argues that a much wider interfunctional integration is necessary to achieve the optimal training of the chain of products which a company produces (see Fig. 30.2).

In this instance we can easily see the need of much greater interdisciplinary training, experience, learning, and interfunctional cooperation than has previously been considered. A similar argument based on both detailed empirical study of numerous companies in Britain and a national questionnaire/survey of the barriers to improvement has been developed by Whiston.[13] Whiston extends the need of multidisciplinary integration to a range of areas: highlighting managerial *integration,* organizational integration, and strategic and informational integration. Such integration is highly dependent on (*a*) appropriate training and (*b*) suitable internal organizational structures (this can take many forms, e.g., reduction in reporting layers, cross-matrix linkage, location of management on shop floor—a point we shall return to in Sec. 30.3).

3. At a level above that of the firm or the enterprise, viz., at the *national* policy level, numerous countries now recognize the need to coordinate scarce national research and industrially relevant skills across or within special generic themes. This can be either *mission*-oriented (e.g., NASA space program, new alternative energy schemes, advanced robotics and sensing, etc.) or relate to *generic* areas of study (e.g., biotechnology, superconductivity, new materials, IT areas). In either case it is essential to encourage cross-functional and much interdisciplinary study—and to better train for such skills. In both the United States and the United Kingdom several approaches have been taken to facilitate and service such a need. Thus in the United States, "engineering centers" (interdisciplinary centers) and a range of academe-collaborative programs address that need.[14] In Britain the introduction of "interdisciplinary research centers" (IRCs) have been much encouraged by most of the major Research

Councils,[15] while the introduction of Directorates (e.g., the Biotechnology Directorate) has also been a means of actively and positively encouraging both interdisciplinary coordination, research perspective, and training.[16] Such a need is by no means limited to OECD or industrially advanced nations. Whiston has recently completed a senior policy analysis outlining a new research and higher-education system for the whole of South Africa.[17] Special emphasis was given to the need for improved interdisciplinary training and research.

4. If we go beyond the level of the nation (and national research, training, and sectoral policies) to the *global* level, we again come to recognize the enormous interdisciplinary challenges which face the world from a science, technology, and socioeconomic perspective. Such a topic is far too complex to be treated in any detail here but has recently been the subject of wide and detailed study by numerous researchers funded by the European Union. The work has been undertaken under a program entitled "Global Perspective 2010: Tasks for Science and Technology," and the results (the global challenge if you will) have been published in 23 volumes. An overview synthesis report is given by Whiston in vol. 1 of the studies.[18] A wide range of global challenges and needs—*new* housing, transport, energy systems, communication systems, agricultural and food production systems—are indicated. All these, it is argued, require detailed *interdisciplinary* study, planning, and in particular individuals trained in an interdisciplinary manner. The need is not only to obtain parallel economic, social, technologic, and scientific *endeavor* but also to train individuals and organizations who can *think and perceive* in such a manner. This, in turn, demands new global communication, systems linkage, and new forms of international educational systems.[19]

5. Finally, the intellectual and scientific challenges of the future, the *leading edge of research*—which ultimately informs industrial and commercial application—depends increasingly on interdisciplinary, multifunctional endeavor. Thus space research, bionics, biophysics, the science of complex systems, expert systems, artificial intelligence, pattern recognition, and so on all call for the crosslinking of disciplines and individuals or groups who can simultaneously handle concepts from seemingly disparate "disciplines." This is particularly so in the case of artificial intelligence, pattern recognition, and the computing sciences in which cognitive psychologists, neurophysiologists, information theorists, and computer-architecture combine forces. Such an area, presently employing tens of thousands of researchers, is most probably the most important scientific-conceptual platform for industrial progress over the early to middle part of the twenty-first century.

All the above establish the "need." But they do not stand alone, and all the "levels" briefly indicated above intertwine and are interdependent. They demand various forms of "response" which we shall discuss in Sec. 30.3.

However, before considering the "response and delivery" of interdisciplinary skill formation, we would note that such training, such skill formation applies at *all* levels of the educational process. Thus it is necessary to consider

- The undergraduate level of education
- The postgraduate level of education
- The research function
- The continuous learning and retraining or updating function

As we shall note below, each of these levels has in recent years been receiving increasing attention from an interdisciplinary training standpoint. To consider these levels alone is insufficient, however. For the need also is to recognize the different circumstances of industry and science from yesteryear. The new challenges of the twenty-first century[20] and the new sociotechnical paradigms that society finds itself in demand an ever more holistic approach to both study and policy response.[21]

30.3 PATTERNS OF RESPONSE TO INTERDISCIPLINARY AND INTERFUNCTIONAL NEEDS

We have indicated above the challenges and needs regarding interdisciplinary training. What dominant forms of action and response are being made? Are the academic (and industrial) barriers and obstacles being overcome? Does the ingrained specialization and reliance on disciplinary study act as a brake on the wider, more complex need? Who, or what, is acting as the driving force or catalyst toward much needed change?

The examples we provide here are merely that: examples. It is not possible to conclude as to whether there is a degree of *sufficiency* of response. Whiston[22] has undertaken several surveys, collated ongoing policy initiatives, but again this primarily tells us the sort of things that are developing; a quantitative policy analysis of the scale and appropriateness of response still awaits us. What we can say, however, is that there is much activity, numerous examples of "best practice," but the final landscape will depend on market forces and wider dissemination of need, publication, and wider knowledge of the great social commercial and scientific outcome of the endeavors.

And here we must signal one important caveat (although we also comment on this in Sec. 30.4 where we briefly consider some of the "remaining problems"). This caveat relates to the contemporary entrenchment, in many countries, of the total resource allocation given to higher education. Thus for several decades following World War II nearly every country—whether Organization for Economic Cooperation and Development (OECD), least developed countries (LDC), or newly industrialized countries (NIC)—has recognized the critical importance of their higher-education "agenda" and made ever increasing resources available.[23] This is no longer a universal trend. Budgets have come to be recognized as finite. A plateauing of funding is spoken of, and greater selectivity and concentration of resources into particular areas has resulted.[24,25] Now under such circumstances departments and faculties are forced to "defend their own turf," their own special interests. Not surprisingly, the "classical disciplines" are often better organized, more primed at so doing. Often areas of interdisciplinary study (and training) do not have a natural, well-established base or collegium.[26] As a consequence, they do not always survive so easily. Individuals who have, through their own imaginative effort, opened up new interdisciplinary ground and forged interdepartmental and new interfunctional links are often vulnerable. There is, here, a tragedy in the making. It is therefore essential that the widest publicity and awareness be given to the importance of interdisciplinary training, research, and viewpoint. There are several important actors who can play a part in the stimulation and maintenance of interdisciplinary endeavor under such circumstances: industry, government, research councils (or other national policy bodies), and social pressure groups. Each of these "actors" represents, in essence, subject matter which is inherently dependent on the *integration* of knowledge, which demands real-world multidisciplinary understanding beyond the more narrow confines of academic allegiance. It is such actors who have often been the prime movers in gainsaying the narrower, more specialized, academic tradition and thereby stimulated interdisciplinary and multidisciplinary studies—often through the example areas, which we shall discuss. The necessity is to maintain, indeed *further,* that interest, a legitimate self-interest, on a wider scale albeit at a time of financial constraint.

We indicated in the previous section that improved or enhanced interdisciplinary training (and research) is required at all levels: undergraduate, postgraduate, faculty research programs, retraining and "updating" of learning, and lifelong education programs. We also indicated that the need, in industry, is for new structures (see also Chap. 18) and also for improved academic-industrial collaboration as a stimulant to interdisciplinary study. Finally we noted that such a need—for interdisciplinary study and action—not only pertains to industry and academe but also should be viewed at the national and international level. Let us now turn to some interesting examples of such practice and, inter alia, note some of the important lessons to be learned. (Several of the examples will be taken from Britain, but similar programs are to be found in other nations.)

30.3.1 The Undergraduate Level of Education

Many undergraduate degrees are organized, essentially, at a single discipline level (chemistry, physics, electrical engineering, economics, business management, etc.). However, increasing reliance is placed on cross-disciplinary studies, "joint" degrees as a means of widening horizons. This helps the *base* on which interdisciplinary studies can later develop. But there are problems. Ironically *employers* are sometimes suspect of the relative value of such students. They sometimes fear (in an elitist fashion) that such students are not the best. This can, unfortunately, have a self-fulfilling effort on the pattern of student recruitment. Such a problem cannot be ignored but must be faced head on. Industry, in terms of perception of need and recruitment, has a big part to play here.

There is a second problem which relates (at the undergraduate level) to the degree of undergraduate specialization. Thus, for example, in the case of engineering some (many) universities aim, from the very first year of study, to encourage (say) an electronics engineering specialist, or mechanical engineering, or civil engineering, or whatever. A few universities take a broader view (e.g., Cambridge University), laying down broad foundations in the early years and "specializing" in the final year of study. These differences in approach and their implications for future policy have been the subject of much study.[27]

As the total knowledge base available to humanity increases—as it does year by year—there also emerges a problem relating to the amount and hence depth that a particular study area can be taught to in a short finite time. This is especially so in such areas as chemistry. Will the day come when an undergraduate studies (say) physical, organic, or inorganic chemistry as a specialty rather than the whole gamut of chemical theory? This has been commented on in some detail regarding future curricula challenges and indirectly has tremendous implications for interdisciplinary studies.[28] The problem of the accumulation of scientific knowledge is a fundamental one. It has to be recognized as a practical brake for many students regarding what can be covered; realistically, in a particular syllabus. The answer to this implied dilemma lies in part in improved teaching methods: teaching by principle, not by rote (unfortunately, the latter still, implicitly, dominates many college syllabuses),[29] and by life-long-learning, topping-up modules, etc.

30.3.2   The Postgraduate Level

Postgraduate students usually have two main branches of choice regarding further study: a postgraduate *research*-oriented degree or, alternatively, reliance on some form of *taught* courses. (Naturally permutations of these two approaches exist across nations with significant differences. Some countries, such as the United States, require several taught courses in parallel with the research thesis. Much also depends on the "level" of the previous undergraduate degree and the extent to which easy transition to postgraduate study can ensue.)

In relation to postgraduate *research,* national endeavors have increasingly been made to overcome the "specialty problem" of an academic supervisor (and that individual's specialized research interest) through the encouragement of interdisciplinary research programs. For example, in Britain for more than two decades now, two of the major Research Councils ESRC and SERC have, through a joint committee program (ESRC/SERC) encouraged at the national level both M.Sc. and Ph.D. interdisciplinary programs. These were reviewed in some detail by Whiston[30] because of the significant failure rate (nonsubmission of thesis or long study periods) of many students. Problems of academic supervision were seen as a major problem;[30] namely, the *single* discipline training of the previous cohort of faculty members created an "evolutionary block."

Interestingly, a comparative analysis by Whiston (privately commissioned by the U.K. SERC Research Council) of American and British views relating to interdisciplinary training revealed that the United States takes a much narrower view of what is understood by the term *interdisciplinarity* than pertains to the United Kingdom. In the United States a mechanical engineer who takes on an electronic problem is "interdisciplinary." The U.K. view of combining (say) chemistry, economics, and political regulations was viewed (by U.S. academics) as circumspect. Herein is part of the problem.

In terms of the broader reach of *taught* courses, then numerous programs or courses now exist worldwide which aim to bridge innovation studies, science and technology, business management, or whatever. Bessant has reviewed a wide number of courses within a broad international setting.[31] One of the major challenges of such M.Sc. programs is to bridge a large number of students from a single-discipline training to a wider interdisciplinary training all within a comparatively short time period. This writer is director of one such program and recognizes that fundamental difficulty. Nevertheless, the broad base ultimately provided to the graduates acts as a superb window of opportunity to industry, commerce, and governmental agencies. Such programs of study, nevertheless, are only a *basis* for the development of further skills as postgraduate students progress through their careers. Industry has much to gain from (say) an engineering student who possesses good knowledge of innovation theory, science policy, or industrial economics.

30.3.3   Collaborative Research Programs and Interdisciplinary Training

Collaborative research between academe and industry is widespread throughout the world. It is a particularly good mechanism for technology and knowledge transfer.[32] From an interdisciplinary standpoint, we noted earlier that academe tends (greatly) toward the encouragement of *specific* disciplines while industry, necessarily, has to take an interdisciplinary perspective. Joint research programs between academe and industry, particularly where research training is involved, can be an invaluable bridge across those two viewpoints. Two major programs in the United Kingdom—the Teaching Company Scheme[33] (TCS) and the Total Technology Programme, are informative in that respect. The TCS scheme encourages postgraduate researchers to address (sometimes at Ph.D. level, although not exclusively) a real-world industrial problem. There is both academic and industrial supervision. Detailed study of such research programs has demonstrated that industry, academe, the students, *and* the supervisor gain in terms of improved interdisciplinary knowledge and awareness of interdisciplinary needs.

The creation of "interdisciplinary research centers" (IRCs) and "designated research centers" (DRCs) within the United Kingdom can, and do, serve a related function and are now generating worldwide interest.[34]

In one sense the gain is in terms of not only the research topic per se but also the *interdisciplinary attitude* which is improved in the academic involved—a topic to which we now turn.

30.3.4 Retraining, Updating, and Continuous Learning with Respect to Interdisciplinary Skills

It is a perennial problem that for many industrialists (whether researchers, operations personnel, or managers) much of the knowledge they had acquired during their earlier formal academic training is now increasingly obsolete. Retraining and updating is called for, more so as new sociotechnical factors emerge—for example, with respect to IT systems, new technology, new design capabilities, and new materials.

As well as the rapid pace of scientific and technological change, which continues to accelerate, it is also recognized that new more "up-to-date" graduates will take several (sometimes many) years to rise to positions of power and authority within industry. This dual problem was recognized some time ago in Britain. In response, a national program entitled the "Integrated Graduate Development Programme" (IGD) was initiated to address the underlying dilemmas and has been reviewed in some detail by Kennedy, Rotherham, and Whiston.[35] The programs are organized at a sectoral level (computing, process industry, transport, manufacturing systems, etc.) and attract industrial workers—often at middle and senior management level—who attend on a modular basis (1 week per month for either 1 or 2 years). The largest national course is at Warwick University which involves 300 or so "students" from such companies as Rolls Royce, Rover (previously), Lucas Industries, and British Aerospace.

From an interdisciplinary perspective we would emphasize, here, several points: the courses are designed jointly by academe *and* industry, the lectures and seminars, etc., are given by *both* academics and industrialists; the courses are *pragmatic* with a strong *interdisciplinary* content; encouragement is made to crosslink individuals from different disciplinary or functional backgrounds; and common needs under real-world conditions are identified. The course participants, in being "updated" and taken for a short while from the "day-to-day pressures of commerce," develop significantly and can reenter their respective companies at senior operative, managerial, and research levels with much wider revitalized interdisciplinary perspectives.

Such programs have been extremely successful. Some companies (e.g., Rover) have instituted similar, but even more senior, programs for all their directors and most senior managers (reviewed in some detail by Whiston[36]). One of the important findings was that benefit was not only to course recipients but to *academe itself.* The academic faculties concerned found they took on, understood more fully than would otherwise be the case, and were much more sympathetic to the need for interdisciplinary perspectives: often of an economic, production, manufacturing, marketing, and technological nature. Thus the beginnings of a breakthrough of entrenched disciplinary, "academic" positions often inimical to interdisciplinary needs, ensued; the Gordian knot was broken.

30.3.5   Organizational Change within Industry and Commerce

No amount of interdisciplinary training and multifunctional awareness can prosper unless there exists within industry, or more specifically within a company, an appropriate organizational structure which encourages multifunctional collaboration. Such a structure is often absent in many companies and demands determined efforts of organizational and functional restructuring in order to obtain the necessary *integrative-functional* environment. This writer has recently devoted a text to the subject based on detailed empirical study and national surveys;[37] also a series of papers summarize salient aspects.[38] What is demanded of that organizational environment? The answer is: an organic, multifunctional, intimately crosslinked structure. It demands a much "flatter" organizational pyramid; managers who spend much time on the "shop floor" (à la the Japanese tradition); matrix managerial structures; R&D "departments" which are *not* at "green-field sites" but closely linked to all the operational, manufacturing, and marketing functions; total systems integration as outlined in some detail by Gaynor;[39] and above all an ethos of mutual involvement of all employees.

Under such conditions an interdisciplinary and multifunctional-integrative environment can thrive. Product strategies, company learning, wider diffusion of "tacit knowledge,"[40] corporate success is given a much greater opportunity of being achieved. Dynamics and systems fluidity rather than organizational sclerosis can then come to characterize company performance. In recent years 3M, DuPont, ICI (U.K.), and numerous Scandinavian and Japanese companies have increasingly followed such principles to much good market effect.

All the above illustrate the response to interdisciplinary and multifunctional needs. In some cases they are only beginnings. Much remains to be done and problems remain, which we now finally consider.

30.4   REMAINING PROBLEMS

Against the backdrop of the general case made above, we have to accept that there are numerous practical problems to be overcome or continually addressed. The more obvious ones are listed below:

1. There still exists within academe much *resistance* to true interdisciplinary training. Individual departments, disciplines "defend the turf" in many ways. Thus individual disciplines are often suspicious of wide, deep collaboration; academics themselves who have been previously trained as "specialists" can be defensive—and have true legitimate limits—regarding the form of knowledge they are either prepared or capable of transmitting. This, in turn, yields a self-perpetuating problem. Few academics constitute a Leonardo da Vinci, a Linus Pauling, or a Buckminster Fuller—although they would aspire to be Newtons or Einsteins.

2. Professional bodies (who are often responsible for accrediting degrees and approving changes in the syllabus) are often very conservative bodies. The rate of change perceived as needed (or achieved) is therefore often insufficient.

3. Industry itself is often loath to overturn the organizational structures of yesteryear. Departments, divisions, or subfunctions have their own microempires and see specialist knowledge *within* their own specialist departments as a form of power.

4. But perhaps the most important (and realistic) problem of all relates to the degree and depth that true interdisciplinarity can be encouraged and nurtured within many individuals. An individual can cover only a certain amount of conceptual material in a given time period. Specialization of knowledge *is* highly valuable. The problem then is to derive and formulate training programs which balance specificity of knowledge with wider awareness of other functional needs in a realistic but imaginative manner. Examples of such programs have been given in Sec. 30.3. Development of interdisciplinary perspective is, however, an open-ended, never-ending problem. The need, therefore, is for continuous updating of skills; close academic-industrial collaboration; emphasis on "teaching by principle," not rote learning; encouragement of creative thinking; and the sustaining of a collaborative, cooperative atmosphere *within* both academe and industry—which takes us to our final problem:

5. As noted by Faulkner et al.,[41] academe encourages in many ways a competitive outlook: competition to publish, establish reputations, be first with one's findings. Industry depends on teamwork, collaboration, and cooperation within the enterprise. These two attitudes are not complementary. For interdisciplinary work to prosper the need is for collaboration and cooperation. The final obstacle, therefore, like much in life may be more of an attitudinal, an emotional form than one due to analytic or intellectual difficulties.

True interdisciplinary endeavor, rich multifunctional linkage, and integrative organizational structures, therefore, may be both a stimulus to and a consequence of future industrial, commercial, and much wider societal challenges. Specialization of knowledge, individual prowess, has a strong history of success in both performance and motivational terms. The challenge is to balance individualism with wider cooperation, without diminishing either.[42]

30.5 REFERENCES AND NOTES

1. T. G. Whiston, *Managerial and Organisational Integration* Springer-Verlag, London, 1992.

2. Ibid., pp. 73–102.

3. Ibid., pp. 125–156.

4. Many such areas constitute the generic intellectual (academic) *and* industrial innovation pathways of future decades.

5. Jupiter, "Management of Technology: The Invisible Advantage," statement by the Jupiter Consortium Technology Management Group, Brunel University Science Park, 1988.

6. B. Twiss, *Managing Technological Innovation,* 3d ed., Pitman, London, 1990, chap. 7.

7. Whiston, op. cit. (Ref. 1), pp. 197–212.

8. R. Rothwell, P. Gardiner, and K. Schott, *Design and the Economy,* The Design Council, London, Sept. 1993.

9. R. Rothwell and T. G. Whiston, "Design, Innovation and Corporate Integration," *R&D Management,* **20**(3): 193–201, July 1990.

10. M. Oakley, ed., *Design Management,* Basil Blackwell, Oxford, 1990.

11. G. H. Gaynor, *Achieving the Competitive Edge through Integrated Technology Management,* McGraw-Hill, New York, 1991.

12. G. H. Gaynor, *Exploiting Cycle Time in Technology Management,* McGraw-Hill, New York, 1992.

13. Whiston, op. cit. (Ref. 1), pp. 125–156.

14. T. G. Whiston, *Report to the SERC on the "Workshop on Interdisciplinary Engineering Research,"* held at Purdue University (USA), Nov. 16–18 1983; SPRU Mimeo 1983, 19 pp.

15. Science and Engineering Research Council (SERC), *Interdisciplinary Research Centers* (monograph), Swindon, U.K. (undated).

16. J. Senker and M. Sharp, *The Biotechnology Directorate of SERC: Report and Evaluation of Its Achievements 1981–1987,* Science Policy Research Unit. University of Sussex, U.K., 1988, 205 pp.

17. T. G. Whiston, *Research Policy in the Higher Education Sector of South Africa,* Science and Technology Policy Series no. 3, FRD South Africa, 1994, 325 pp.

18. T. G. Whiston, *Global Perspective 2001: Tasks for Science and Technology,* vol. 1: *A Synthesis Report,* Commission of the European Communities, Brussels, 1992. (The entire range of publications encompasses 23 volumes obtainable from the European Commission, Brussels; vol. 1 summarizes the findings of the 23 volumes.)

19. This is given considerable emphasis in the EC-FAST papers; see Ref. 18.

20. T. G. Whiston, *The Training and Circumstances of the Engineer in the Twenty-First Century,* a study undertaken for the Fellowship of Engineering (U.K.) March 1987. Science Policy Research Unit, University of Sussex, U.K., 1987, 234 pp.

21. C. Perez, "Microelectronics: Long Waves and World Structural Change: New Perspectives for Developing Countries," *World Development,* **13**(3): 441–463, 1985.

22. T. G. Whitson, *Management and Assessment of Interdisciplinary Training and Research,* ICSU and UNESCO Press, Paris, 1986.

23. T. G. Whitson and R. Geiger, eds., *Research and Higher Education: the UK and the USA,* O.U. Press, Buckingham (U.K.), 1992.

24. J. Ziman, *"Science in a Steady-State":* The Research System in Transition, Science Policy Support Group, London, Concept Paper no. 1, 1987.

25. ABRC, *A Strategy for the Science Base,* Advisory Board for Research Councils, London, 1987.

26. Whiston, op cit. (Ref. 22).

27. Whiston, op. cit. (Ref. 20).

28. Ibid., chap. 2.

29. By "rote learning" we imply not a simple repetitive commitment to memory of repetitive facts but reliance on numerous examples of a "horizontal" nature which do not really take the student that much deeper into the subject matter. Teaching by "principle" emphasizes the more generic, fundamental principles which underpin a subject. The important thing is to get the balance correct.

30. Science and Engineering Research Council (SERC), U.K., *'Interdisciplinary Research: Selection, Supervision and Training—A Summary of the 'Whiston' Report,* SERC monograph, Swindon, April 1983, 29 pp.

31. Professor J. Bessant of University of Brighton, U.K. (CENTRIM Group), undertook an extensive international survey and identified several hundred courses of this type.

32. M. Dodgson, *Technological Collaboration in Industry: Strategy, Policy and Internationalisation in Innovation,* Routledge, London, 1993.

33. P. Senker and J. Senker, "The British Teaching Company Scheme: An Example of a Successful European Scheme for Technology Transfer," in *Competitiveness, Growth and Job Creation—What Contribution Can Education and Training Make?* Commission of the European Communities, Brussels, 1994.

34. An "interdisciplinary research center" aims to bring together several disciplines into a coherent focus. Usually the IRC is located at one university for the whole country. It can act as a "hotel" for expensive equipment, lead to a critical mass of expertise, and host visiting faculty. A "designated research center" (DRC) does not have to compete for research funding for several years (8 or so), thereby permitting continuity of work and building up of research or training excellence.

35. A. J. Kennedy, L. Rotherham, and T. G. Whiston, *A Review of the SERC Integrated Graduate Development Programme*, 2 vols., Report TCCR-83-005, Technical Change Centre, London, April 1983.

36. T. G. Whiston, *The Management of Technical Change, Training for New and "High" Technologies*, a report to the U.K. Manpower Services Commission on the Programme for Directors and Senior Managers of Austin-Rover held at Warwick University, 1985/86; Science Policy Research Unit, University of Sussex, 2 vols., Feb. 1986.

37. In the studies (see Ref. 1) it was found that numerous companies could not achieve satisfactory organic integration of subfunctions because of lack of interdisciplinary skills.

38. T. G. Whiston, "Managerial and Organizational Integration Needs Arising out of Technical Change and UK Commercial Structures," part 1—*Technovation*, **9:** 577–606, 1989; part 2—*Technovation*, **10**(1): 47–58, 1990; part 3—*Technovation*, **10**(2): 95–108; part 4—*Technovation*, **10**(3): 143–161, 1990.

39. Gaynor, op. cit. (Ref. 12); see especially chap. 1, "Perspectives on Managing Cycle Time," pp. 3–17.

40. P. Patel and K. Pavitt, *Technological Competencies in the World's Largest Firms: Characteristics, Constraints and Scope for Managerial Choice*, STEEP Discussion Paper no. 13, Science Policy Research Unit, University of Sussex, May 1994.

41. W. Faulkner, J. Senker, and L. Velho, *Knowledge Frontiers: Public Sector Research and Industrial Innovation in Biotechnology, Engineering Ceramics, and Parallel Computing"* Oxford University Press, Oxford, 1994.

42. T. G. Whiston, "The Last Empire: The Corporatisation of Society and the Diminution of Self," *Futures*, **23**(2): 163–178, March 1991.

CHAPTER 31

RESEARCH PERFORMANCE EFFECTIVENESS AND IMPACT

Ronald N. Kostoff*

Office of Naval Research
Arlington, Virginia

31.1 BACKGROUND

31.1.1 Definitions

Research is the pursuit and production of knowledge by the scientific method. *Research productivity* is the generation of tangible and intangible products from research. *Research efficiency* is the productivity of research per unit of input resource. *Research impact* is the change effected on society due to the research product. *Research effectiveness* is a measure of the focus of impact on desired goals.

31.1.2 Increased Interest in Research Impact Measurement

In research sponsoring organizations, the selection and continuation of research programs must be made on the basis of outstanding science and potential contribution to the organization's mission. Recently, there have been increasing pressures to link science and technology programs and goals even more closely and clearly to organizational as well as broader societal goals. This is reflected in a number of studies,[1-3] in the controversial National Institutes of Health strategic planning process, and in the controversial strategic goals proposed for the National Science Foundation.

In tandem with the pressures for more strategic research goals are motivations to increase research assessments and reporting requirements to ensure that the increasingly strategic research goals are being pursued by proposed and existing research programs. The 1992 Congressional Task Force report on the health of research[1] stated, as one of its two recommendations: "Integrate performance assessment mechanisms into the research process using legislative mandates and other measures, to help measure the effectiveness of federally funded research programs."

---

*The views expressed in this chapter are solely those of the author and do not represent the views of the Department of the Navy.

The Government Results and Performance Act of 1993 (Public Law 103-62) requires the establishment of strategic planning and performance measurement in the federal government, and for other purposes. Not only will the federal agencies be required to establish performance goals for program activities, but as the law states, they will be required to establish performance indicators to be used in measuring or assessing the relevant outputs, service levels, and outcomes of each program activity.

Because of increased world competition, and the trends toward corporate downsizing, parallel pressures exist for industrial research organizations to link research programs more closely with strategic corporate goals and to increase research performance and productivity. In tandem with the increasing governmental interests in research impact assessment (RIA) stated above, there is considerable industrial interest in RIA as well. These actions toward increased accountability in government and industry make it important that research managers and administrators in government, industry, and academia understand the assessment approaches which could be utilized to evaluate research quality and goal relevance, and that researchers gain an understanding of these evaluation approaches as well. After a brief discussion of the pros and cons of RIA, this chapter will describe the three main categories of research performance assessment techniques: retrospective, qualitative, and quantitative.

31.1.3   Underutilization of RIA

The reasons for reluctance of research sponsors and customers to implement RIA vary. The rewards in research and research management go to new discoveries, not quality assessments. Neither the costs nor the time requirements of RIA are negligible, and have to be weighed against additional research which could be performed. More immediate organizational requirements are assigned higher priority than RIA. The outcome of an RIA is not always predictable or pleasant for managers, and "pet" projects may be terminated after a rigorous evaluation. Negative results from an RIA may provide executive or legislative branch overseers, or corporate management, ammunition for budget reductions. Finally, since there is very little experience with the use of advanced evaluation techniques, there is insufficient evidence that their use will result in better payoff than use of rudimentary techniques. To many research managers and administrators, there is little to be gained from RIA, and a potential for loss.

31.1.4   Benefits of Increased Utilization of RIA

One major benefit would be to improve organizational efficiency. A properly executed RIA would target the people and the exogenous variables (management climate, funding conditions, infrastructure, etc.) necessary to increase research output relevant to the organization's goals. An RIA which increased communication among researchers and potential research customers during the conduct of research would allow a smoother conversion of the products of research to technology, through better integration of the users with the research performers.

Another major benefit would be to identify the diverse impacts of basic research. The impacts of basic research are pervasive throughout a technological society, but for the most part the impacts of basic research are indirect on technologies, systems, and end products. A major limitation of articulating the benefits of basic research has been the lack of data which could show the pathways and linkages through which the research impacts the intermediate or end products. A credible RIA of completed research would trace the dissemination of the research products through the many communication channels and would identify the multitude of near- and long-term research impacts (impact on other research fields, on technology, on systems, on education, etc.). Having this data would provide more substantive arguments for continuing to provide the necessary funds to those who control the allocation of research funds.

31.2 RESEARCH IMPACT ASSESSMENT METHODS

31.2.1 Retrospective Methods

Background and Overview. In the evaluation of research performance and impact, a spectrum of approaches may be considered. At one end of the spectrum are the subjective, essentially nonquantitative, approaches, of which peer review is the prototype.[4] At the other end of the spectrum are the mainly quantitative approaches, such as evaluative bibliometrics and cost-benefit.[5,6] In between are what can be termed *retrospective* or *case-study* approaches.[7,8]

These retrospective methods make little use of mathematical tools, but draw on documented approaches and results wherever possible. In practice, there are two major reasons why research sponsoring organizations perform retrospective studies of research. Positive research impact on the organization's mission provides evidence to the stakeholders that there is benefit in continuing sponsorship of research. Also, if the study is sufficiently comprehensive, the environmental parameters which helped the research succeed can be identified, and these lessons can be used to improve future research.

There are two major variants of retrospective studies. One type starts with a successful technology or system and works backward to identify the critical R&D events which led to the end product. The other type starts with initial research grants and traces evolution forward to identify impacts. The backward-tracing approach is favored for two reasons: (1) the data is easier to obtain, since forward tracking is essentially nonexistent for evolving research; and (2) the sponsors have little interest in examining research that may have gone nowhere.

While methods for performing retrospective and case studies may differ within and across industry and government,[8] especially concerning the research question, case selection, and analytic framework, the fundamental evaluation problems encountered are pervasive across these different methods. In the remainder of this section, a few of the more widely known case studies will be reviewed, and the key pervasive problems and findings will be identified. These retrospective studies include Project Hindsight, Project TRACES and its follow-on studies, and Accomplishments of Department of Energy (DOE) Office of Health and Environmental Research (OHER) and of the Advanced Research Projects Agency (ARPA).

Specific Retrospective Studies

Project Hindsight. Project Hindsight was a retrospective study performed by the Defense Department in the mid-1960s to identify those management factors important in assuring that research and technology programs are productive and that program results are used.[9] The evolution of the new technology represented in each of the 20 weapons systems selected was traced back in post–World War II time to critical points called "research or exploratory development (RXD) events."

Original Traces Study.   In 1967 The National Science Foundation (NSF) instituted a study[10] to trace retrospectively key events which had led to a number of major technological innovations. One goal was to provide more specific information on the role of the various mechanisms, institutions, and types of R&D activity required for successful technological innovation. Similar to Project Hindsight, key "events" in the research and development history of each innovation selected were identified, and their characteristics were examined.

Follow-on Traces Study.   In a follow-on study to TRACES, the NSF sponsored Battelle-Columbus Laboratories to perform a case-study examination of the process and mechanism of technological innovation.[11] For each innovation studied, the significant events (important activity in the history of an innovation) and decisive events (a significant event which provides a major and essential impetus to the innovation) which contributed to the innovation were identified. The influence of various exogenous factors on the decisive events was determined, and several important characteristics of the innovative process as a whole were obtained.

Recent Traces Study.   In a modern version of the TRACES study, the National Cancer Institute initiated an assessment[12] to determine whether there were certain research settings or support mechanisms which were more effective in bringing about important advances in cancer research. The approach taken was analogous in concept to the initial TRACES study, with the addition of citation analyses to provide an independent measure of the impact of the TRACE papers (papers associated with each key "event"), and by adding control sets of papers.

DARPA Accomplishments Study.   The Institute for Defense Analysis produced a document[13] describing the accomplishments of the Defense Advanced Research Projects Agency (DARPA). Of the hundreds of projects and programs funded by DARPA over its then (1988) 30-year lifetime, 49 were selected and studied in detail, and conditions for success were identified.

DOE OHER Accomplishments Book.   The approach taken by DOE was to describe the 40-year history of OHER,[14,15] and present selected accomplishments in different research areas from different points in time. This technique allowed impacts and benefits of the research to be tracked through time, and in some cases to be quantified as well.

Retrospective Studies Conclusions.   Project Hindsight, TRACES, and, to some degree, the OHER and DARPA accomplishment books had some similar themes. All these methods used a historiographic approach, looked for significant research or development events in the metamorphosis of research programs in their evolution to products, and attempted to convince the reader that (1) the significant R&D events in the development of the product or process were the ones identified; (2) typically, the organization sponsoring the study was responsible for some of the (critical) significant events; (3) the final product or process to which these events contributed was important; and (4) while the costs of the research and development were not quantified, and the benefits (typically) were not quantified, the research and development were worth the cost.

Six critical conditions for innovation were identified through analysis of these retrospective studies. The most important condition appears to be *the existence of a broad pool of knowledge which minimizes critical path obstacles* and can be exploited for development purposes. This condition is followed in importance by a *technical*

*entreprenuer who sees the technical opportunity and recognizes the need for innova-
tion,* and who is willing to champion the concept for long time periods, if necessary.
Also valuable are *strong financial and management support* coupled with *many continuing inventions in different areas* to support the innovation.

As the historiographic analyses (Hindsight and TRACES) of a technology or system have shown, if the time interval in which the antecedent critical events occur is
arbitrarily truncated, as in the two-decade-time-interval Hindsight case, the impacts of
basic research on the technology or system will not be given adequate recognition.
The number of mission-oriented research events peaks about a decade before the technology innovation. However, the number of non-mission-oriented research events
peaks about three decades before the technology innovation, and eight, nine, or more
decades may be necessary in some cases to recognize the original critical antecedent
events. Over a long time interval, the majority of key R&D events tend to be non-mission-oriented. Thus, future studies of this type should allow time intervals of many
decades to ensure that critical non-mission-oriented research events are captured.

Even in those cases when an adequate time interval was used, and critical non-mission-oriented events were identified, the cumulative indirect impacts of basic research
were not accounted for by any of the retrospective approaches published or in use
today. A recent study[16] which examined impacts of research on other research and
technology through direct and indirect paths using a network approach showed that
the indirect impacts of fundamental research can be very large in a cumulative sense.
Future retrospective studies would be more credible if they devoted more effort to
identifying indirect impacts of research. While indirect impacts of research are much
more difficult to identify than direct impacts, and the data-gathering effort is much
larger and more complex, neglect of indirect impacts reduces appreciation of the value
of basic research significantly. Use of some of the advanced computer-based technologies available today, such as the network approach mentioned above or citation analysis,[12] could identify many of the pathways of the indirect impacts of research.

While some studies concluded that the technical entrepreneur was extremely
important to the innovative process, it does not appear (to the author) to be the critical-path factor. Examination of the TRACES historiographic tracings, which display
significant events chronologically for each innovation, as well as the ARPA and
OHER case studies and accomplishments books, showed that *an advanced pool of
knowledge must be developed in many fields before synthesis leading to an innovation
can occur. The entrepreneur can be viewed as an individual or group with the ability
to assimilate this diverse information and exploit it for further development. However,
once this pool of knowledge exists, there are many persons or groups with capability
to exploit the information, and thus the real critical path to the innovation is more
likely the knowledge pool than any particular entrepreneur.* The entrepreneurs listed in
the studies undoubtedly accelerated the introduction of the innovation, but they were
at all times paced by the developmental level of the knowledge pool.

A detailed reading of those studies which attempted to incorporate economic quantification showed the difficulties of trying to identify, assign, and quantify costs and
benefits of basic research, especially at a project-investigator level. As TRACES and
other similar studies have shown, the chain of events leading to an innovation is long
and broad. Many researchers over many years have been involved in the chain, and
many funding agencies, some simultaneously with the same researchers, may have
been involved. The allocation of costs and benefits under such circumstances is a very
difficult and highly arbitrary process. The allocation problem is reduced, but not eliminated, when the analysis is applied at the macro level (integrating across individual
researchers, organizations, etc.).

One goal of all the studies presented was to identify the products of research and some of their impacts. In addition, the Hindsight, TRACES, and ARPA studies tried to identify factors which influenced the productivity and impact of research. The following general conclusions about the role and impact of basic research were reached:

- The majority of basic research events which directly impacted technologies or systems were non-mission-oriented and occurred many decades before the technology or system emerged.

- The cumulative indirect impacts of basic research were not accounted for by any of the retrospective approaches published.

- An advanced pool of knowledge must be developed in many fields before synthesis leading to an innovation can occur.

- Allocation of benefits among researchers, organizations, and funding agencies to determine economic returns from basic research is very difficult and arbitrary, especially at the micro level.

While these approaches do provide interesting information and insight into the transition process from research to development to products, processes, or systems, the *arbitrary selectivity and anecdotal nature of many of the results render any conclusions as to cost-effectiveness or generalizability suspect.* Supplementary analyses using other approaches are required for further justification of the value of the R&D.

31.2.2 Qualitative (Peer Review)

Background.   Peer review of research represents evaluation by experts in the field, and is the method of choice in practice in the United States.[4,17–20] Its objectives range from being an efficient resource allocation mechanism to a credible predictor of research impact.

Requirements for High-Quality Peer Review.   Many studies related to peer review have been reported in the literature, ranging from the mechanics of conducting a peer review, to examples of peer reviews, to detailed critiques of peer reviews and the process itself (e.g., Refs. 4, 17–19, 21–34). A nonstandard peer review approach for concept comparisons is the Science Court. As in a legal procedure, it has well-defined advocates, critics, a jury, etc. It was applied by the author to a review of alternate fusion concepts in 1977.[35] This procedure generated substantial debate and surfacing of crucial issues, but it was time-consuming compared to a standard panel assessment.

   While these reported studies present the process mechanics, the procedures followed, and the review results, the reader cannot ascertain the *quality* of the review and the results. In practice, procedure and process quality are mildly necessary, but nowhere sufficient conditions for generating a high-quality peer review. Many useful peer reviews have been conducted using a broad variety of processes, and while well-documented modern processes (e.g., Ref. 29) may contribute to the efficiency of conducting a review, more than process is needed for high quality. There are many intangible factors that enter into a high-quality review, and before examples of reviews are presented, some of the more important factors will be discussed.

   The desirable characteristics of a peer review can be summarized as[4]

- An effective resource allocation mechanism
- An efficient resource allocator
- A promoter of science accountability
- A mechanism for policy makers to direct scientific effort

- A rational process
- A fair process
- A valid and reliable measure of scientific performance

High-quality peer reviews require as a minimum the conditions summarized from Ormala:[32]

- The method, organization, and criteria for an evaluation should be chosen and adjusted to the particular evaluation situation.
- Different levels of evaluation require different evaluation methods.
- Program and project goals are important considerations when an evaluation study is carried out.
- The basic motive behind an evaluation and the relationships between an evaluation and decision making should be openly communicated to all the parties involved.
- The aims of an evaluation should be explicitly formulated.
- The credibility of an evaluation should always be carefully established.
- The prerequisites for the effective utilization of evaluation results should be taken into consideration in evaluation design.

Assuming that these considerations have been taken into account, *three of the most important intangible factors for a successful peer review are motivation, competence, and independence.* The review leader's motivation to conduct a technically credible review is the cornerstone of a successful review. The leader selects the reviewers, summarizes their comments, guides the questions and discussions in a panel review, and makes recommendations about whether the proposal should be funded. The quality of a review will never go beyond the competence of the reviewers. Two dimensions of competence which should be considered for a research review are the individual reviewer's technical competence for the subject area and the competence of the review group as a body to cover the different facets of research issues (other research impacts, technology and mission considerations and impacts, infrastructure, political and social impacts). The quality of a review is limited by the biases and conflicts of the reviewers. The biases and conflicts of the reviewers selected should be known to the leader and to one another.

The best features of different organizations' peer review practices can be combined into a heuristic protocol for the conduct of successful peer review research evaluations and impact assessments. The main aims of the protocol are to ensure that the final assessment product has the highest intrinsic quality and that the assessment process and product are perceived as having the highest possible credibility. The protocol elements are described in the following paragraphs.

Peer-Review Research Evaluations. The objectives of the assessment must be stated clearly and unambiguously at the initiation of the assessment by the highest levels of management, and the full support of top management must be given to the assessment. In turn, the objectives, importance, and urgency of the assessment must be articulated and communicated down the management hierarchy to the managers and performers whose research is to be assessed, and the cooperation of these reviewees must be enlisted at the earliest stages of the assessment:

- The final assessment product, the audience for the product, and the use to be made of the product by the audience should be considered carefully in the design of the assessment.

- One person should be assigned to manage the assessment at the earliest stage, and this person should be given full authority and responsibility for the assessment.
- The assessment manager should report to the highest organizational level possible in order to ensure maximum independence from the research units being assessed.
- The reviewers should be selected to represent a wide variety of viewpoints, in order to address the many different facets of research and its impact.[31] These would include bench-level researchers to address the impact of the proposed research on the field itself, broad research managers to address potential impact on allied research fields, technologists to address potential impact on technology and the potential of the research to transition to higher levels of development, systems specialists to address potential impact on systems and hardware, and operational personnel to address the potential impact on downstream organizational operations. The reviewers should be independent of the research units being evaluated, and independent of the assessing organization where possible. The objectives of, and constraints on (if any), the assessment should be communicated to the reviewers at the initial contact.
- Maximum background material describing the research to be assessed, related research and technology development sponsored by external organizations, the organization structure, and other factors pertinent to the assessment should be provided to the reviewers as early as possible before the review. This will allow the reviewers and presenters to use their time most productively during the review.
- Recommendations resulting from the assessment should be tracked to ensure that they are considered and implemented, where appropriate. For research programs, planning, execution, and review are linked intimately. Feedback from the review outcomes to planning for the next cycle should be tracked to ensure that the review-planning coupling is operable.

Levels of Organizational Research Evaluation.   Evaluations should be performed at three levels of resolution in the organization:

1. The highest level would be an annual *corporate-level review* of how the organization performs research. If the organization has a separate research unit, then the unit should be evaluated as an integrated whole. If research is vertically integrated with development, then the research should preferably be evaluated as part of a total organization R&D review. The charter of this highest-level assessment would be to review, at the corporate level, general policy, organization, budget, and programs (e.g., Ref. 36). Total inputs and outputs, including integrated bibliometric indicators, would be examined. Overall research management processes would be examined, such as selection, execution, review, and technology transfer of research. The overall investment strategy would be evaluated, and would include different perspectives of the program, such as technical discipline, performer, and end-use allocation. The integration of the research objectives with the larger organization objectives would be assessed. The evaluators would include, but not be limited to, representatives of the stakeholder, customer, and user community whose potential conflicts with the organization are minimal.

2. The second level would be triennial peer review of a discipline or management unit at the program level (e.g., Refs. 31, 37), in which a program is defined as an aggregation of work units (principal investigators). If the organization has a separate research unit, then the discipline should be evaluated as an integrated whole. In the nominal review, quality and relevance could be evaluated concurrently. If research is vertically integrated with development, then the research should preferably be evaluated as part of a total vertical structure R&D review. In the nominal vertical structure review, quality and relevance should preferably be evaluated separately. Thus, research evaluation must take into account how research is structured, integrated, and managed within an organization. Research quality criteria should include research merit, research approach, productivity, and team quality. Relevance criteria should include short-term impact (transitions and/or utility), long-term potential impact, and some estimate of the probability of success of attaining each type of impact. While the emphasis is on peer review, bibliometric and other types of indicators should be utilized to supplement the peer evaluation.

3. The third level would be a minimum of triennial peer review at the work unit (principal investigator) level (e.g., Ref. 29). Most of the program-level issues described above are applicable and need not be repeated here.

For each of these three levels of review, the following criteria and issues should be considered during the review as appropriate.

Criteria for Organization Reviews

- Quality and uniqueness of the work
- Scientific and technological opportunities in areas of likely organization mission importance
- Need to establish a balance between revolutionary and evolutionary work
- Position of the work relative to the forefront of other efforts
- Responsiveness to present and future organization mission requirements
- Possibilities of follow-on programs in higher R&D categories
- Appropriateness of the efforts for organization (in lieu of) other organizations
- Other organization connection (coordination) of the work

Questions to Be Asked of Organization Programs

- What is the investment strategy of the larger management unit? This would include the relative program priorities, the actual investment allocation to the different programs, and the rationale for the investment allocation. For each program being reviewed, what is the investment strategy for its thrust areas?
- What are we trying to do (in a systems concept)?
- Can specific advantage to the organization be identified if the program is successful?
- How is the system done today, and what are the limitations of the current practice?
- Would the work be supported if it were not already under way?
- Assuming success, what difference does it make to the user in a mission area content?
- What is the technical content of the program, and how does it fit with other ongoing efforts in academia, industry, organization labs, other labs, etc.?
- What are the decision milestones of the program?
- How long will the program take, how much will the program cost, and what are the midterm and final objectives of the program?

In Europe, another development line has been to commission evaluation experts either to support panels or to conduct independent assessments which may involve surveys, in-depth interviews, case studies, etc.[38] Barker[21] describes how evaluation experts coming from two main communities (civil servants and academic policy researchers) interact in evaluation of R&D in the United Kingdom. The performance of evaluations, including the synthesis of evidence and the production of conclusions and recommendations, is done by professionals, as opposed to panels of eminent persons.

Problems with Peer Review.   Peer-review problems include[4,19,39–41]

- Partiality of peers to impact the outcome for nontechnical reasons.
- An "old boy" network to protect established fields.
- A "halo" effect for higher likelihood of funding for more visible scientists, departments, and institutions.
- Reviewers differ in criteria to assess and interpret.
- The peer review process assumes agreement about what good research is, and what are promising opportunities.

These potential problems should be considered during the process of selecting research impact assessment approaches.

Another problem with peer review is cost. The true *total* costs of peer review can be considerable but tend to be ignored or understated in most reported cases. For serious panel-type peer reviews, in which sufficient expertise is represented on the panels, *total real costs will dominate direct costs* by as much as an order of magnitude or more.[37] The major contributor to total costs for either type of review is the time of all the players involved in executing the review. With high-quality performers and reviewers, time costs are high, and the total review costs can be a nonnegligible fraction of total program costs, especially for programs that are people-intensive rather than hardware-intensive.

The issue of peer review predictability affects the credibility of technological forecasting directly. A few studies have been done relating reviewers' scores on component evaluation criteria to proposal or project review outcomes. Some studies have been done in which reviewers' ratings of research papers have been compared to the numbers of citations received by these papers over time.[42,43] Correlations between reviewers' estimates of manuscript quality and impact and the number of citations received by the paper over time were relatively low. The author is not aware of reported studies, singly or in tandem, that have related peer review scores and rankings of proposals to *downstream impacts* of the research on technology, systems, and operations. This type of study would require an elaborate data-tracking system over lengthy time periods, which does not exist today. Thus, the value of peer review as a predictive tool for assessing the impact of research on an organization's mission (other than research for its own sake) rests on faith more than on hard documented evidence.

Peer-Review Conclusions.   Peer review is the most widely used and generally credible method used to assess the impact of research. Much of the criticism of peer review has arisen from misunderstandings of its accuracy resolution as a measuring instrument. While a peer review can gain consensus on the projects and proposals that are either outstanding or poor, there will be differences of opinion on the projects and proposals that cover the much wider middle range. For projects or proposals in this middle range, their fate is somewhat more sensitive to the reviewers selected. If a key purpose of a peer review is to ensure that the outstanding projects and proposals are funded or continued, and the poor projects are either terminated or modified strongly, then the capabilities of the peer review instrument are well matched to its requirements.

However, the value of peer review as a predictive tool for assessing the impact of research on an *organization's mission* (other than research for its own sake) rests on faith more than on hard documented evidence. Also, for serious panel-type peer reviews or mail-type peer reviews, in which sufficient expertise is represented on the panels, *total real costs will dominate direct costs.* The major contributor to total costs is the time of all the players involved in executing the review. With high-quality performers and reviewers, time costs are high, and the total review costs can be a nonnegligible fraction of total program costs, especially for programs that are people-intensive rather than hardware intensive.

Most methods used in practice include criteria which address the impact of research on its own and allied fields, as well as on the mission of the sponsoring organization. Nearer-term research impacts typically play a more important role in the review outcome than longer-term impacts, but do not have quite the importance of team quality, research approach, or the research merit. A minimal set of review criteria should include team quality, research merit, research approach, research productivity, and a criterion related to longer-term relevance to the organization's mission. More important than the criteria is the dedication of an organization's management to the highest-quality objective review, and the associated emplacement of rewards and incentives to encourage quality reviews.

31.2.3 Quantitative Methods

Background. Quantitative approaches to research assessment focus on the numerics associated with the performance and outcomes of research. The main approaches used are bibliometrics and econometrics such as cost-benefit and production function analysis. This section focuses on these three main approaches, briefly describes the bibliometrics-related family of approaches known as cooccurrence phenomena, briefly describes a network modeling approach to quantifying research impacts, and ends with an expert systems approach for supporting research assessment.

Bibliometrics

Foundations. Bibliometrics, especially evaluative bibliometrics, uses counts of publications, patents, citations, and other potentially informative items to develop science and technology performance indicators. The choice of important bibliometric indicators to use for research performance measurement may not be straightforward. A 1993 study surveyed about 4000 researchers to identify appropriate bibliometric indicators for their particular disciplines.[44] The respondents were grouped in major discipline categories across a broad spectrum of research areas. While the major discipline categories agreed on the importance of publications in refereed journals as a performance indicator, there was no agreement about the relative values of the remaining 19 indicators provided to the respondents. For the respondents in total, the important performance indicators were

- *Publications*—publication of research results in refereed journals
- *Peer-reviewed books*—research results published as commercial books reviewed by peers

- *Keynote addresses*—invitations to deliver keynote addresses or present refereed papers and other refereed presentations at major conferences related to one's profession
- *Conference proceedings*—publication of research results in refereed conference proceedings
- *Citation impact*—publication of research results in journals weighted by citation impact
- *Chapters in books*—research results published as chapters in commercial books reviewed by peers
- *Competitive grants*—ability to attract competitive, peer reviewed grants from the ARC, NH&MRC, rural R&D corporations, and similar government agencies

These bibliometric indicators can be used as part of an analytical process to measure scientific and technological accomplishment. Because of the volume of documented scientific and technological accomplishments being produced (5000 scientific papers published in refereed scientific journals every workday worldwide; 1000 new patent documents issued every workday worldwide), use of computerized analyses incorporating quantitative indicators is necessary to understand the implications of this technical output.[5]

Narin states three axioms that underlie the utilization and validity of bibliometric analysis. The first axiom is *activity measurement:* that counts of patents and papers provide valid indicators of R&D activity in the subject areas of those patents or papers, and at the institution from which they originate. The second axiom is *impact measurement:* that the number of times those patents or papers are cited in subsequent patents or papers provides valid indicators of the impact or importance of the cited patents and papers. However, there could be weightings applied to the raw count data, depending on the perceived importance of the journals containing the citing papers. Also, the impacts would be on allied research fields or technologies, not necessarily long-term impacts on the originating organization's mission. The third axiom is *linkage measurement:* that the citations from papers to papers, from patents to patents, and from patents to papers provide indicators of intellectual linkages between the organizations which are producing the patents and papers, and knowledge linkage between their subject areas.[5]

Use of bibliometrics can be categorized into four levels of aggregation:[5]

- *Policy*—evaluation of national or regional technical performance
- *Strategy*—evaluation of the scientific performance of universities or the technological performance of companies
- *Tactics*—tracing and tracking R&D activity in specific scientific and technological areas or problems
- *Conventional*—identifying specific activities and specific people engaged in research and development

Policy questions deal with the analysis of very large numbers of papers and patents, often hundreds of thousands at a time, to characterize the scientific and technological output of nations and regions. Strategic analyses tend to deal with thousands to tens of thousands of papers or patents at a time, numbers that characterize the publication or patent output of universities and companies. Tactical analyses tend to deal with hundreds to thousands of papers or patents, and deal typically with activity within a specific subject area. Finally, conventional information retrieval tends to deal with identifying individual papers, patents, inventors, and clusters of interest to an individual scientist or engineer or research manager working on a specific research project.

Problems with Bibliometrics. Problems with publication and citation counts include[30,40,45,46]

1. Publication counts

   - Indicate quantity of output, not quality.
   - Nonjournal methods of communication ignored.
   - Publication practices varying across fields, journals, employing institutions.
   - Choice of a suitable, inclusive database is problematical.
   - Undesirable publishing practices (artificially inflated numbers of coauthors, artificially shorter papers) increasing.

2. Citations

   - Intellectual link between citing source and reference article may not always exist.
   - Incorrect work may be highly cited.
   - Methodological papers among most highly cited.
   - Self-citation may artificially inflate citation rates.
   - Citations lost in automated searches due to spelling differences and inconsistencies.
   - *Science Citation Index* (SCI) changes over time.
   - SCI biased in favor of English-language journals.
   - Same problems as publication counts.

A major problem with bibliometrics is cross-discipline comparisons of outputs. For example, how should the paper or citation output of a program in solid-state physics be compared to that of shallow-water acoustics? What types of normalization are required to allow comparisons among these different types of programs and fields? Three types of normalization solutions are proposed by Schubert and Braun.[47]

1. **The publishing journal as reference standard.** By relating the number of citations received by a paper (or the average citation rate of a subset of papers published in the same journal—the *mean observed citation rate*) to the average citation rate of all papers in the journal (the *mean expected citation rate*), the *relative citation rate* will be obtained. This indicator shows the relative standing of the paper (or set of papers) in question among its close companions: its value is higher or lower than unity as the sample is more or less cited than the average.

2. **The set of related records as reference standard.** *Bibliographic coupling* uses the number of references a given pair of documents have in common to measure the similarity of their subject matter. Comparing a set of papers that are "similar" in this sense to a given article of the same age will yield an ideal reference standard for citation assessments.

3. **The set of cited journals as reference standard.** A promising method is based on the journal in the reference lists of the articles of the journal in question. These journals are selected by the most reliable persons, the authors of the journal, as references (in both senses of the word) and, therefore, can justly be regarded as standards of the expected citation rate.

Bibliometric Studies. Macroscale bibliometric studies characterize science activity at the national (e.g., Refs. 48, 49), international, and discipline level. The biennial *Science and Engineering Indicators* report[50] tabulates data on characteristics of personnel in science, funds spent, publications and citations by country and field, and many other bibliometric indicators. Another study at the national level was aimed at evaluating the comparative international standing of British science.[51] Using publication counts and citation counts, the authors evaluated scientific output of different countries by technical discipline as a function of time. Much more understanding is required as to which indicators are appropriate and how they should impact allocation decisions.

There have been numerous microscale bibliometric studies reported in the literature (e.g., Refs. 52–61). The NIH (National Institutes of Health) bibliometrics-based evaluations[33] included the effectiveness of various research support mechanisms and training programs, the publication performance of the different institutes, the responsiveness of the research programs to their congressional mandate, and the comparative productivity of NIH-sponsored research and similar international programs.

Two papers[12,62] described determination of whether significant relationships existed among major cancer research events, funding mechanisms, and performer locations; compared the quality of research supported by large grants and small grants from the National Institute of Dental Research; evaluated patterns of publication of the NIH intramural programs as a measure of the research performance of NIH; and evaluated quality of research as a function of size of the extramural funding institution. Most of the NIH studies focused on aggregated comparison studies (large grants vs. small grants, large schools vs. small schools, domestic vs. foreign, etc.).

Patent citation analysis has the potential to provide insight into the conversion of science to technology.[63–69] Much of the federal government support of the development of patent citation analysis was by the NSF (e.g., Refs. 70, 71). Some recent studies have focused on utilization of patent citation analysis for corporate intelligence and planning purposes (e.g., Ref. 72). Some of the data presented verify further *Lotka's productivity law,* in which relatively few people in a laboratory are producing large numbers of patents. In the example presented in Narin,[72] the patents of the most productive inventor are highly cited, further demonstrating his key importance. Narin concludes that highly productive research labs are built around a small number of highly productive, key individuals.

Despite its limitations, bibliometrics may have utility in providing insight into research product dissemination. For laboratories, these studies

- Examine distribution of disciplines in coauthored papers, to see whether the multidisciplinary strengths of the lab are being utilized fully.
- Examine distribution of organizations in coauthored papers, to determine the extent of lab collaboration with universities, industry, and other labs and countries.
- Examine the nature (basic or applied) of citing journals and other media (patents), to ascertain whether the lab's products are reaching the intended customer(s).
- Determine whether the lab has its share of high-impact (heavily cited) papers and patents, viewed by some analysts as a requirement for technical leadership.
- Determine which countries are citing the lab's papers and patents, to see whether there is foreign exploitation of technology and in which disciplines.
- Identify papers and patents cited by the lab's papers and patents, to ascertain degree of lab's exploitation of foreign and other domestic technology.

A recent comparative bibliometric analysis of 53 laboratories[73] clustered the labs into six types (regulation and control, project management, science frontier, service, devices, survey), and stated that "comparisons of scientific impacts should be made only with laboratories that are comparable in their primary task and research outputs." The report concluded further that

- Bibliometric indicators and scientific publications are not the only outputs that should be measured, but the other types of outputs differ for different laboratories.
- Bibliometric indicators are not equally valid across different types of laboratories.
- Bibliometric indicators are less useful for the evaluation of research laboratories involved in closed publication markets.

Cooccurrence Phenomena. One class of computer-based analytic techniques which tends to focus more on macroscale impacts of research exploits the use of cooccurrence phenomena. In cooccurrence analysis, *phenomena that occur together frequently in some domain are assumed to be related, and the strength of that relationship is assumed to be related to the cooccurrence frequency.* Networks of these cooccurring phenomena are constructed, and then maps of evolving scientific fields are generated using the link-node values of the networks. Using these maps of science structure and evolution, the research policy analyst can develop a deeper understanding of the interrelationships among the different research fields and the impacts of external intervention, and can recommend new directions for more desirable research portfolios. These techniques are discussed in more detail in Kostoff[74–76] and Tijssen and Van Raan.[77] The Tijssen–Van Raan paper contains an excellent exposition on mapping techniques for displaying the structure of related science and technology fields.

In particular, *cocitation analysis* has been applied to scientific fields, and cocitation clusters have been mapped to represent research-front specialties.[77] *Coword* has been utilized to map the evolution of science under European (mainly French) government support, and has the potential to supplement other research impact evaluation approaches. *Conomination,* in its different incarnations, has been used to construct social networks of researchers and has the potential, if expanded to include research and technology impacts in the network link values, for evaluating direct and indirect impacts of research. *Coclassification* is based on cooccurrences of classification codes in patents, and is used to construct maps of technology clusters.[78]

Cost-Benefit and Economic Analyses. A comprehensive survey examined the application of economic measures to the return on research and development as an investment in individual industries and at the national level.[33] This document concluded that while econometric methods have been useful for tracking private R&D investment within industries, the methods failed to produce consistent and useful results when applied to federal R&D support. A more recent analysis focused on economic and cost-benefit approaches used for research evaluation.[79] The methods involve computing impacts using market information, monetizing the impacts, and then comparing the value of the impacts with the cost of research. Principal measures described include surplus measures and productivity measures. With known benefit and cost timestreams, internal rates of return to R&D investments are then computed. The paper notes both the standard technical difficulties with these approaches and the political and organizational difficulties in implementing them.

Cost-Benefit Analyses. Cost-benefit analyses are a family of related techniques which include cost-benefit, net present value, and rate-of-return.[79–81] These approaches tend to be more widely used in industry than government. For one, or many, projects, the basic approach is similar. A starting point in time for the research is defined. The timestream of costs for product development is estimated, and the timestream of benefits from the product is estimated. Using the time value of money, the costs and benefits are discounted to the origin of time, and the net benefits are compared with the net costs. The main differences in the approaches to cost-benefit analyses are in the sophistication of the methods used to estimate the cost and benefit streams, and the time value of money.

Cost-benefit analyses have limited accuracy when applied to basic research because *of the quality of both the cost and benefit data due to the large uncertainties characteristic of the research process, as well as selection of a credible origin of time for the discounting computations.* As an illustrative example, a deterministic cost-benefit analysis was performed by the author on a fusion reactor variant.[82] Its *real problem, which pervades and limits any attempt to perform a cost-benefit analysis on a concept in the basic research stage, was the inherent uncertainty of controlling the fusion process. This translated to the inability to predict the probabilities of success and time and cost schedules for overcoming fundamental plasma research problems* (e.g., plasma stabilities and confinement times); *no credible methods were available.* Thus, the main value of the cost-benefit approach was to show that the potential existed for positive payoff from the hybrid reactor development, and that there was a credible region in parameter space in which controlled fusion development could prove cost-effective; what was missing was the likelihood of achieving that payoff.

A 1991 marginal cost-benefit study weighed the costs of academic research against the benefits realized from the earlier introduction of innovative products and processes due to the academic research.[6] The study used survey data to show a very high social rate of return resulting from academic research. While the method is innovative, future applications using more objective data sources would provide higher confidence in the computed rates of return.

Production Function. Production function approaches to evaluating research returns invoke economic theory–based assumptions relating outputs to inputs to generate an estimatable model. One only needs time-series data on output, capital, labor, and research expenditures to estimate empirically the marginal contribution of research to value-added. However, the relationship of research to value added is nonlinear and indirect. Variables such as other inputs to technology and production and marketing functions complicate the research/value-added relationship.

Much of the major recent economic work relating economic growth and productivity increases to R&D spending has been performed by three economists.[6,83–87] Mansfield's earlier study typifies the strengths and weaknesses of the production function approach. This study[83] attempted to determine whether an industry's or firm's rate of productivity change was related to the amount of basic research it performed. Mansfield developed a production function which disaggregated basic and applied research, then regressed rate of productivity increase with many different variables. The regressions showed a strong relationship between the amount of basic research carried out by an industry and the industry's rate of productivity increase during 1948–1966.

The study exemplifies the problem inherent in multiple regression analyses: determining cause and effect from what is essentially correlation. As Mansfield points out, "It is possible that industries and firms with high rates of productivity growth tend to spend relatively large amounts on basic research, but that their high rates of productivity growth are not due to these expenditures."[83] Nor does Mansfield's model specify the path(s) by which R&D investment supposedly leads to productivity improvements.

A production function approach to cost-efficiency of basic research essentially used a regression analysis between outputs and inputs.[88,89] For proposals, the method involved regressing output variables (citations per dollar, graduate students per dollar) against input variables (quality of the investigator's department, quality of the investigator, etc.). The results gave some idea of the importance of the input variables, alone or in combination, on the output variables. One obvious potential application would be prediction of proposals likely to have high productivity based on prior (input) knowledge. *Much, however, remains to be done in identifying the appropriate output measures, the appropriate input measures, and the nature of the interactions among these measures for different disciplines.*

Network Modeling for Direct and Indirect Impacts. A network-based modeling approach was devised which would allow estimation of the direct and indirect impacts of a research program or collection of research programs. The research program impacts would be multifaceted, including impacts on advancing its own field, on advancing allied fields, on advancing technology, on supporting operations and mission requirements, and so on. A major feature of the model is *inclusion of feedback from the higher development categories* (e.g., exploratory development, advanced development) *on the advancement of research.*

The model and a subsequent pilot study related to U.S. Navy R&D have been described in detail.[16] In summary, a network was constructed in which each node represented an area of research or development. The values of the links connecting each node pair represented the impact of results from the first node area on the second node area. The total impact of an area of research on other research or development was obtained by integrating over all paths from the research node to the node(s) of interest.

Expert Networks. Research impact assessment is, in essence, a diagnostic process with many diagnostic tools. In other fields of endeavor, such as medicine and machinery repair, expert systems are being used increasingly as diagnostic tools or as support to diagnostic processes. Recently, there have been efforts to develop expert-system approaches combined with artificial neural networks (expert networks) for use in R&D management, including RIA.[90–92] A brief summary of these efforts follows.

The product of these efforts is the *research-management expert network* (R-MEN), which is characterized by two complementary tools: organizational and professional development (O/PD) and expert network. The latter technology comprises an expert system (left side of brain) and an artificial neural network (right side of brain). Given a set of research, and research management, policies, and strategies, R-MEN learns concepts that hierarchically organize those policies and strategies and use them in classifying and triaging research proposals.

The R-MEN framework consists of a knowledge base and a database. Feeding into the knowledge base are four modules: a policy-strategy impartation module and a proposal data-acquisition module, both of which receive input from the O/PD process; and a research impact calculation module and a proposal review module. The knowledge base then feeds into the database through five modules: a project selection module, a resources allocation module, a project evaluation and control module, an investigator evaluation module, and an organization evaluation module.

R-MEN is implemented in three phases. Phase 1 includes the development of the strategic plan, which defines and communicates longer-term research directions, and the development of the operating plan, which specifically identifies the projects that will implement the strategic plan, taking into consideration the goals, quantifiable objectives, and development of the individual investigator and the organization.

Phase 2 represents the necessary education and management support needed to prepare the staff to participate in such an "action research" effort. This phase identifies and utilizes the critical components required to develop an environment that facilitates participative research management activities. A significant activity occurring during this phase is daily verification of individual scheduled training and development. If an individual has no recorded training and/or development within a preset period, the system will generate and send a report through e-mail directly to the office of the director for R&D. The system will be able to look at a training and/or development description(s) and compare it (them) with the background of the individual to determine whether the training and/or development is (are) suitable for that individual.

Phase 3 represents a means by which participative methods can be put into operation in developing productivity tracking systems. Significant activities occurring during this phase include project evaluation and control. This entails periodic monitoring of project milestones for applied research, and research objectives for the more basic research. If a project has no recorded fulfillment of a milestone within a preset period, the system will generate and send a report through e-mail directly to the office of the director for R&D.

If R-MEN is initially used concurrently with present research review processes, it will serve as a supplement in the form of a guide to data generation, acquisition, and processing, and a validity check. With appropriate implementation and maintenance, this knowledge technology, which utilizes demonstrated and proven approaches, methods, procedures, and techniques in an innovative and unique way, could lead to the following benefits:

- Provide a means for *effective,* policy- and strategy-oriented management through outcomes-management
- Improve management quality, reduce operation costs, and increase productivity and public trust
- Foster impact evaluation to document federally funded program and management effectiveness
- Provide short-term (3-year) program progress tracking and long-term (10-year) result(s) impact tracking
- Shield administrators, managers, and other policy makers from the complexity of the mathematics of the inference machine
- Permit the evaluation of a range of alternatives
- Permit the handling of large amounts of data
- Permit policy makers to have a better understanding of existing technical attributes of and capabilities for potential projects
- Facilitate choice of strategy compatible with agency structure and processes, and with the policy or the nature of decision making for activities scheduling and control

Quantitative Methods: Conclusions.   Bibliometric methods are valuable in quantifying the output of research. Because they do not address quality, and their numeric outputs are subject to multiple interpretations, they are not self-contained assessment methods. They are a valuable supplement to the subjective interpretative methods such as peer review.

Economic approaches have limited value when applied to assessing the potential of fundamental research, because of the uncertain nature of the data. Their validity increases as the research becomes more applied, and cost and benefit streams can be estimated more accurately.

As databases become more extensive, and computer power continually increases, data-intensive quantitative analyses will increase in use. Approaches such as cooccurrence, network modeling, and expert networks described above will become more commonplace in research assessment.

31.3 REFERENCES

1. G. E. Brown, *"Report of the Task Force on the Health of Research,"* Chairman's Report to the Committee on Science, Space, and Technology, U.S. House of Representatives, no. 56-819, U.S. Government Printing Office, Washington, D.C., 1992.

2. NAS, *"The Government Role in Civilian Technology: Building a New Alliance,"* Committee on Science, Engineering, and Public Policy, National Academy of Sciences, National Academy Press, Washington, D.C., 1992.

3. Carnegie, *"Enabling the Future: Linking Science and Technology to Societal Goals,"* Carnegie Commission on Science, Technology, and Government, Carnegie Commission, New York, 1992.

4. D. E. Chubin, "Grants Peer Review in Theory and Practice," in R. N. Kostoff, ed., *Evaluation Review,* Special Issue on Research Impact Assessment, **18:**1, Feb. 1994.

5. F. Narin, D. Olivastro, and K. A. Stevens, "Bibliometrics—Theory, Practice, and Problems," in R. N. Kostoff, ed., *Evaluation Review,* Special Issue on Research Impact Assessment, **18:**1, Feb. 1994.

6. E. Mansfield, "Academic Research and Industrial Innovation," *Research Policy,* **20,** 1991.

7. R. N. Kostoff, "Semi-Quantitative Methods for Research Impact Assessment," *Technol. Forecasting Social Change,* **44:**3, Nov. 1993.

8. G. Kingsley, "The Use of Case Studies in R&D Impact Evaluations," in B. Bozeman and J. Melkers, eds., *Assessing R&D Impacts: Method and Practice,* Kluwer Academic Publishers, Norwell, Mass., 1993.

9. DOD, *Project Hindsight,* Office of the Director of Defense Research and Engineering, Washington, D.C., DTIC no. AD495905, Oct. 1969.

10. IITRI, *"Technology in Retrospect and Critical Events in Science,"* Illinois Institute of Technology Research Institute Report, Dec. 1968.

11. Battelle, *Interactions of Science and Technology in the Innovative Process: Some Case Studies,* final report, prepared for the National Science Foundation, Contract NSF-C 667, Battelle Columbus Laboratories, March 19, 1973.

12. F. Narin, "The Impact of Different Modes of Research Funding," in David Evered and Sara Harnett, eds., *The Evaluation of Scientific Research,* Wiley, Chichester (U.K.), 1989.

13. IDA, *DARPA Technical Accomplishments,* vol. I—IDA paper P-2192, Feb. 1990; vol. II—IDA paper P-2429, April 1991; vol. III—IDA paper P-2538, July 1991, Institute for Defense Analysis.

14. DOE, *"Health and Environmental Research: Summary of Accomplishments,"* Office of Energy Research, Office of Program Analysis, Report no. DOE/ER-0194, May 1983.

15. DOE, *"Health and Environmental Research: Summary of Accomplishments,"* Office of Energy Research, Office of Program Analysis, report no. DOE/ER-0275, Aug. 1986.

16. R. N. Kostoff, "Research Impact Quantification," *R&D Management,* **24:**3, July 1994.

17. R. N. Kostoff, *The Handbook of Research Impact Assessment,* fifth e dition, Summer 1995, DTIC Report No. ADA 296021.

18. J. M. Logsdon and C. B. Rubin, *"An Overview of Federal Research Evaluation Activities,"* report, The George Washington University, Washington, D.C., April 1985; see also J. M. Logsdon and C. B. Rubin, *Federal Research Evaluation Activities,* Abt Associates, Cambridge, Mass., 1985.

19. D. E. Chubin and E. J. Hackett, *Peerless Science: Peer Review and U.S. Science Policy,* State University of New York Press, Albany, N.Y., 1990.

20. R. N. Kostoff, "Evaluating Federal R&D in the U.S.," in B. Bozeman and J. Melkers, eds., *Assessing R&D Impacts: Method and Practice,* Kluwer Academic Publishers, Norwell, Mass., 1993.

21. K. Barker, "The 'British Model'—Evaluation by Professionals," in P. Laredo and P. Mustar, eds., *EC Handbook on Evaluation,* 1992.

22. D. V. Cicchetti, "The Reliability of Peer Review for Manuscript and Grant Submissions: A Cross-Disciplinary Investigation," *Behavioral and Brain Sciences,* **14:**1, 1991.

23. S. Cole, L. Rubin, and J. Cole, *Peer Review in the National Science Foundation: Phase One of a Study,* National Research Council, NTIS acc. no. PB83-192161, 1978.

24. J. Cole and S. Cole, *Peer Review in the National Science Foundation: Phase Two of a Study,* National Research Council, NTIS acc. no. PB82-182130, 1981.

25. S. Cole, J. Cole, and G. Simon, "Chance and Consensus in Peer Review," *Science,* **214,** Nov. 1981.

26. S. E. Cozzens, "Expert Review in Evaluating Programs," *Science and Public Policy,* **14:**2, April 1987.

27. DOD, *"The Department of Defense Report on the Merit Review Process for Competitive Selection of University Research Projects and an Analysis of the Potential for Expanding the Geographic Distribution of Research,"* DTIC acc. no. 88419044, April 1987.

28. DOE, *An Assessment of the Basic Energy Sciences Program,* Office of Energy Research, Office of Program Analysis, report no. DOE/ER-0123, March 1982.

29. DOE, *Procedures for Peer Review Assessments,* Office of Energy Research, Office of Program Analysis, report no. DOE/ST-0007P, revised January 1993.

30. S. P. Frazier, *University Funding: Information on the Role of Peer Review at NSF and NIH,* U.S. General Accounting Office report no. GAO/RCED-87-87FS, March 1987.

31. R. N. Kostoff, "Evaluation of Proposed and Existing Accelerated Research Programs by the Office of Naval Research," *IEEE Trans. Engineering Management,* **35:**4, Nov. 1988.

32. E. Ormala, "Nordic Experiences of the Evaluation of Technical Research and Development," *Research Policy,* **18,** 1989.

33. OTA, *Research Funding as an Investment: Can We Measure the Returns,* U.S. Congress, Office of Technology Assessment, OTA-TM-SET-36, Washington, D.C., U.S. GPO, April 1986.

34. R. S. Nicholson, *Improving Research Through Peer Review,* National Research Council, NTIS acc. no. PB88-163571, 1987.

35. DOE, *An Evaluation of Alternate Magnetic Fusion Concepts 1977,* DOE/ET-0047, May 1978.

36. NIST, *Annual Report, 1990,* Visiting Committee on Advanced Technology, Jan. 1991.

37. R. N. Kostoff, "Quantitative/Qualitative Federal Research Impact Evaluation Practices," *Technol. Forecasting Social Change,* **45:**2, Feb. 1994.

38. E. Ormala, "Impact Assessment: European Experience of Qualitative Methods and Practices," in R. N. Kostoff, ed., *Evaluation Review,* Special Issue on Research Impact Assessment, **18:**1, Feb. 1994.

39. R. Roy, "Funding Science: The Real Defects of Peer Review and an Alternative to It," *Science, Technol. Human Values,* **10:**3, 1985.

40. J. King, "A Review of Bibliometric and Other Science Indicators and Their Role in Research Evaluation," *J. Information Science,* **13,** 1987.

41. C. Kruytbosch, "The Role and Effectiveness of Peer Review," in David Evered and Sara Harnett, eds., *The Evaluation of Scientific Research,* Wiley, Chichester, 1989.

42. R. F. Bornstein, "The Predictive Validity of Peer Review: A Neglected Issue," *Behavioral and Brain Sciences,* **14:**1, 1991.

43. R. F. Bornstein, "Manuscript Review in Psychology: Psychometrics, Demand Characteristics, and an Alternative Model," *J. Mind and Behaviour,* **12,** 1991.

44. Australia, *Research Performance Indicators Survey,* National Board of Employment, Education and Training, commissioned report no. 21, Australian Government Publishing Service, Canberra, Australia, Jan. 1993.

45. J. E. J. Oberski, "Some Statistical Aspects of Co-citation Cluster Analysis and a Judgement by Physicists," in A. F. J. Van Raan, ed., *Handbook of Quantitative Studies of Science and Technology,* North Holland, New York, 1988.

46. H. D. White and K. W. McCain, "Bibliometrics," in M. E. Williams, ed., *Annual Review of Information Science and Technology,* 1989, Journal of the American Society for Information Science, Washington, D.C., **24,** 1989.

47. A. Schubert and T. Braun, "Reference Standards for Citation Based Assessments," *Scientometrics,* **26:**1, 1993.

48. D. Hicks, B. Martin, and J. Irvine, "Bibliometric Techniques for Monitoring Performance in Technologically Oriented Research: The Case of Integrated Optics," *R&D Management,* **16**(3), 1986.

49. T. Braun, W. Glanzel, and A. Schubert, "An Alternative Quantitative Approach to the Assessment of National Performance in Basic Research," in David Evered and Sara Harnett, eds., *The Evaluation of Scientific Research,* Wiley, Chichester, 1989.

50. NSF, *Science and Engineering Indicators—1989,* National Science Board Report NSB 89-1, GPO, Washington, D.C., 1989.

51. B. R. Martin et al., "Recent Trends in the Output and Impact of British Science," *Science and Public Policy,* **17:**1, Feb. 1990.

52. J. D. Frame, "Quantitative Indicators for Evaluation of Basic Research Programs/Projects," *IEEE Trans. Engineering Management,* **EM-30**(3), Aug. 1983.

53. P. R. McAllister, F. Narin, and J. G. Corrigan, "Programmatic Evaluation and Comparison Based on Standardized Citation Scores," *IEEE Trans. on Engineering Management,* **EM-30**(4), Nov. 1983.

54. N. Mullins, "Evaluating Research Programs: Measurement and Data Sources," *Science and Public Policy,* **14**(2), April 1987.

55. N. Mullins, W. Snizek, and K. Oehler, "The Structural Analysis of a Scientific Paper," in A. F. J. Van Raan, ed., *Handbook of Quantitative Studies of Science and Technology,* North Holland, New York, 1988.

56. H. F. Moed and A. F. J. Van Raan, "Indicators of Research Performance: Applications in University Research Policy," in A. F. J. Van Raan, ed., *Handbook of Quantitative Studies of Science and Technology,* North Holland, New York, 1988.

57. J. Irvine, "Evaluation of Scientific Institutions: Lessons from a Bibliometric Study of UK Technical Universities," in David Evered and Sara Harnett, eds., *The Evaluation of Scientific Research,* Wiley, Chichester, 1989.

58. A. F. J. Van Raan, "Evaluation of Research Groups," in David Evered and Sara Harnett, eds., *The Evaluation of Scientific Research,* Wiley, Chichester, 1989.

59. T. Luukkonen, "Bibliometrics and Evaluation of Research Performance," *Annals of Medicine,* **22**(3), 1990.

60. T. Luukkonen and B. Stahle, "Quality Evaluations in the Management of Basic and Applied Research," *Research Policy,* **19,** 1990.

61. T. Luukkonen, O. Persson, and G. Sivertsen, "Understanding Patterns of International Scientific Collaboration," *Science, Technol. Human Values,* **17**(1), Jan. 1992.

62. F. Narin, "Bibliometric Techniques in the Evaluation of Research Programs," *Science and Public Policy,* **14:**2, April 1987.

63. M. P. Carpenter and F. Narin, "Validation Study: Patent Citations as Indicators of Science and Foreign Dependence," *World Patent Information,* **5**(3), 1983.

64. F. Narin, M. P. Carpenter, and P. Woolf, "Technological Performance Assessments Based on Patents and Patent Citations," *IEEE Transact. Engineering Management,* **EM-31**(4), Nov. 1984.

65. J. T. Wallmark and K. G. Sedig, "Quality of Research Measured by Citation Method and by Peer Review—A Comparison," *IEEE Transact. Engineering Management,* **EM-33**(4), Nov. 1986.

66. P. Collins and S. Wyatt, "Citations in Patents to the Basic Research Literature," *Research Policy,* **17,** 1988.

67. F. Narin and D. Olivastro, "Technology Indicators Based on Patents and Patent Citations," in A. F. J. Van Raan, ed., *Handbook of Quantitative Studies in Science and Technology,* Elsevier Science Publishers, Amsterdam, 1988.

68. B. G. Van Vianen, H. F. Moed, and A. F. J. Van Raan, "An Exploration of the Science Base of Recent Technology," *Science Policy,* **19,** 1990.

69. F. Narin and D. Olivastro, "Status Report-Linkage between Technology and Science," *Research Policy,* **21:**3, June 1992.

70. M. P. Carpenter, M. Cooper, and F. Narin, "Linkage between Basic Research Literature and Patents," *Research Management,* **13:**2, March 1980.

71. F. Narin, E. Noma, and R. Perry, "Patents as Indicators of Corporate Technological Strength," *Research Policy,* **16,** 1987.

72. F. Narin, "Technological Evaluation of Industrial Firms by Means of Patent Investigation," paper presented at VPP Professional Meeting, Nürnberg, Germany, Nov. 13, 1992.

73. R. Miller, "The Influence of Primary Task on R&D Laboratory Evaluation: A Comparative Bibliometric Analysis," *R&D Management,* **22:**1, 1992.

74. R. N. Kostoff, "Research Impact Assessment," *Proceedings: Third International Conference on Management of Technology,* Miami, Feb. 17–21, 1992. (Larger text available from the author.)

75. R. N. Kostoff, "Co-Word Analysis," in B. Bozeman and J. Melkers, eds., *Assessing R&D Impacts: Method and Practice,* Kluwer Academic Publishers, Norwell, Mass., 1993.

76. R. N. Kostoff, "Database Tomography: Origins and Applications," *Competitive Intelligence Review,* Special Issue on Technology, **5:**1, Spring 1994.

77. R. Tijssen and A. F. J. Van Raan, "Mapping Changes in Science and Technology," in R. N. Kostoff, ed., *Evaluation Review,* Special Issue on Research Impact Assessment, **18:**1, Feb. 1994.

78. E. C. Engelsman and A. F. J. Van Raan, *Mapping of Technology: A First Exploration of Knowledge Diffusion amongst Fields of Technology,* Research Report to the Ministry of Economic Affairs, CWTS-91-02, Centre for Science and Technology Studies, Leiden, March 1991.

79. H. Averch, "Economic Approaches to the Evaluation of Research," in R. N. Kostoff, ed., *Evaluation Review,* Special Issue on Research Impact Assessment, **18:**1, Feb. 1994.

80. A. Link, "Methods for Evaluating the Return on R&D Investments," in B. Bozeman and J. Melkers, eds., *Assessing R&D Impacts: Method and Practice,* Kluwer Academic Publishers, Norwell, Mass., 1993.

81. J. D. Roessner, "Use of Quantitative Methods to Support Research Decisions in Business and Government," in B. Bozeman and J. Melkers, eds., *Assessing R&D Impacts: Method and Practice,* Kluwer Academic Publishers, Norwell, Mass., 1993.

82. R. N. Kostoff, "A Cost/Benefit Analysis of Commercial Fusion-Fission Hybrid Reactor Development," *J. Fusion Energy,* **3:**2, 1983.

83. E. Mansfield, "Basic Research and Productivity Increase in Manufacturing," *Am. Economic Review,* **70**(5), Dec. 1980.

84. N. Terleckyj, *State of Science and Research: Some New Indicators,* Westview Press, Boulder, Colo., 1977.

85. N. Terleckyj, "Measuring Economic Effects of Federal R&D Expenditures: Recent History with Special Emphasis on Federal R&D Performed in Industry," paper presented at NAS Workshop on "The Federal Role in Research and Development," Nov. 1985.

86. Z. Griliches, "Issues in Assessing the Contribution of Research and Development to Productivity Growth," *Bell J. Economics,* **10,** Spring 1979.

87. Z. Griliches, "Productivity, R&D, and the Data Constraint," *Am. Economic Review,* **84:**1, March 1994.

88. H. Averch, "Measuring the Cost-Efficiency of Basic Research Investment: Input-Output Approaches," *J. Policy Analysis and Management,* **6**(3), 1987.

89. H. Averch, "Exploring the Cost-Efficiency of Basic Research Funding in Chemistry," *Research Policy,* **18,** 1989.

90. C. O. Odeyale, *Knowledge-Based Systems: Knowledge Representation and Inference Strategies for Effective and Unbiased Military Biomedical and R&D Management,* Ph.D. thesis, Walden University, Minneapolis, Minn.

91. C. O. Odeyale and R. N. Kostoff, "R&D Management Expert Networks: I. Knowledge Representation and Inference Strategies," *HEURISTICS, J. Knowledge Engineering and Technology,* **7:**1, 1994.

92. C. O. Odeyale and R. N. Kostoff, "R&D Management Expert Networks: II. Prototype Construction and Validation," *HEURISTICS, J. Knowledge Engineering and Technology,* **7:**1, 1994.

CHAPTER 32
LONG-TERM SUCCESS DIMENSIONS IN TECHNOLOGY-BASED ORGANIZATIONS

Aaron J. Shenhar
University of Minnesota
Minneapolis, Minnesota

Dov Dvir
Tel-Aviv University
Tel-Aviv, Israel

32.1 INTRODUCTION

What does *organizational success* mean? Is it the organization's profits, its sales, or its cash flow? Is it the current level of growth, or something else? And what if the organization is successful today; does this imply success in the future? According to current practice, many organizations assess their performance using immediate, and usually short-term, measures: profit, sales, returns, and growth. However, many realize that the long-term aspects of an organization's effectiveness are not less important. This realization is reflected in numerous discussions and writings published during the last 20 years. In their 1980s famous article, Hayes and Abernathy[1] claimed that one of the reasons for the decline of American industry is the pervasive myopia of what they called the "new management orthodoxy." Executives with financial and legal skills who know little about their firm's products, markets, and production processes usually rely on quantitative short-term criteria. These executives are more likely to engage in financial restructuring to boost their profits than to take risks on technological innovation. This and similar influential arguments have not produced, however, any accepted conclusion, and no significant change has been observed in American businesses so far. According to the traditional American management system, most businesses are still focused on immediate results and managed with a short-term view. Consequently, executive compensation is greatly dependent on these short-term outcomes. In contrast, as suggested by the MIT commission on industrial productivity,[2] a key factor in the managerial culture of Japanese firms is their longer time horizon.

Japanese firms are willing to take on high levels of debt in order to in-vest in new production capabilities and marketing infrastructure.

Longer time horizons are particularly important for technology-based firms, operating in rapidly growing markets and rapidly changing environments. These companies must build their technological base[3] and continuously invest in knowledge. They must develop their core technological competencies,[4] and make sure that they are preparing the necessary infrastructure for the future, and for coming generations of products. Leading-edge companies do not wait for competitors to tell them where the industry is going. They are mobilized to create and dominate emerging opportunities. Furthermore, they do not compete within the boundaries of existing industries, but make any effort to shape the structure of future industries.[5]

Even in less dynamic environments, a significant portion of present industrial results are not an outcome of recent managerial decisions. Rather, they are results of decisions taken 5, 7, and even 10 years ago. And when looking into the future, most of the strategic decisions made today will normally have an impact on the business only 5 or more years from now. Several known examples will illustrate our point. Classic as it is, the VCR industry case must still be remembered. When JVC and Sony made a lifelong commitment in the early 1960s to set specific goals on developing and marketing a home videorecorder, no American company was doing the same. Although American firms possessed the technology and were the world leaders in commercial videorecording, none of them had the vision needed to foresee or even match the introduction of the Japanese home VCRs in the mid-1970s. An entire market was lost for many years, if not forever. Similarly, Lee Iacocca admitted, in retrospect, that it would have been impossible to save Chrysler in the 1970s if it wasn't for the previous investments made by Chrysler several years before he joined the company.[6] These decisions involved some new models, such as the K car, and they served as a major component in Chrysler's successful turnaround. The second time around when Iacocca got the company out of trouble was in the late 1980s. This time, again, the solution was based to a large extent on the very successful and profitable lines of minivans whose initial investment were made in the late 1970s and early 1980s.

How, then, would companies know they are preparing for the future? Obviously, they must focus their attention on decisions and plans dedicated to tomorrow. However, they must do so while assessing their effectiveness in both the short and the long terms. The purpose of this chapter is to address these issues. It is based on a collection of studies we conducted at different organizational levels and on the experience of leading companies around the world. Our underlying premise is that "what gets measured gets done" and "you can't manage what you don't measure." As we found, technology-based organizations would benefit from adapting a multidimensional framework which would be used for scanning their entire success horizons. Such a framework will serve as a basis for the development of specific success measures for each organization at different organizational levels. The theoretical framework presented in this chapter will discuss three levels, each encompassing the previous one: the single project, the business unit, and the entire company or corporation. For each level we will describe four distinct success dimensions, each addressing a different managerial concern and directed toward a different time frame.

Although none of the companies we studied is using the entire multidimensional framework, some organizations are using various portions of it, and several come closer than others. We believe that the development of a comprehensive framework is feasible as well as practical. There is plenty of rewarding potential for any company willing to achieve better and more focused results.

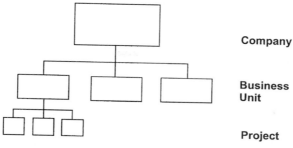

FIGURE 32.1   Three organizational levels of analysis.

32.2   THE MULTIDIMENSIONAL SUCCESS ASSESSEMENT FRAMEWORK

The three levels of analysis within an organization are shown in Fig. 32.1. The *project level* represents a temporary organization, set up to achieve a certain and finite goal. It may involve a new-product development, product improvement, process building, or the establishment of a new service. When this goal is achieved, the organization is divested, dismantled, or perhaps reshaped into a new organization for the execution of a different project. Such temporary organizations are formed as cross-functional project teams in a case of a small project, or as a multiteam project with a central project management office and many internal and external subcontractors in a case of a larger project. At the second level, a *business unit* is a more stable organization. It is usually an operating unit that sells a distinct set of products to an identifiable group of customers in competition with a well-defined set of competitors. At any given time the business unit includes several ongoing projects, each responsible for producing, building, or improving separate products. Finally, the entire *company* or corporation consists, of course, of a collection of separate business units as well as some central functions or departments such as a central research laboratory, a technology center, or a company's central test facility.

How is organizational success assessed? Studies of organizational effectiveness have been at the heart of organization theory for many years.[7] However, most empirical investigations in the field of business strategy have used only financial criteria for success—profitability, return on investment (ROI), etc. For example, typical studies at the business unit level have used the *profit improvement of marketing strategy* (PIMS) framework and the ROI criterion.[8] An important departure from the single-dimensional track (although not exactly at an organizational level) is Cooper and Kleinschmidt's study on the success of new products.[9] According to their findings, new-product success consists of three dimensions: financial performance, opportunity window, and market impact. Similarly, several studies at the project level have identified multiple criteria for measuring the success of projects, including meeting budget and schedule goals, customer satisfaction,[10] technical performance, manufacturability, and business performance.[11]

As our findings indicate, the success of an organization can be divided into four major dimensions, each addressing different concerns and each consisting of several variables (measures). The four dimensions for each organizational level are described in Table 32.1 and discussed in the following sections. While certain dimensions may

TABLE 32.1 Success Dimensions for Various Organizational Levels

Success dimension	1	2	3	4
Time horizon	Very Short	Short	Long	Very long
Organizational level				
Project*	Project efficiency Time, budget goals Other efficiency measures	Impact on Customer Performance goals Requirement goals Customer satisfaction Customer comes back	Direct business success Sales, ROI, Positive cash flow Market share	Preparing the future New technology New-product line New market
Business unit†	Profitability Sales, profit, cash flow Profitability margins	Orders and marketing Sales objectives Market share New orders (backlog) Projected cash flow	New opportunities New products New markets Customer satisfaction Customer loyalty	Future infrastructure Next-generation products Developing new technology Preparing future skills
Company	Financial performance Sales Profit Overall cash flow	Market response Stock price Growth Reputation	Strategic leverage Core competencies Strategic goals Strategic imperatives	Creating the future Defining the industry's foresight Defining unarticulated needs

*Adapted from Shenhar et al.[13]

†Adapted from Dvir and Shenhar.[19]

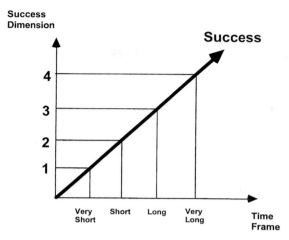

FIGURE 32.2   Time frame of success dimensions.

address different time frames and have different importance at different times, real and full success will be achieved if organizations pay attention to all four dimensions (see Fig. 32.2).

32.3   THE PROJECT LEVEL

Projects are launched today for a myriad of reasons: to develop new products, to establish manufacturing processes, to construct buildings, to build defense-related systems, or to upgrade existing products. However, no matter what the motivation for the project, the question of assessing the project success has always been controversial and dependent on the assessor's point of view. One of the most common approaches to project success is to consider a project as successful because it has come close to achieving its budget and schedule goals. Although this may seem appropriate in the short term, it is hardly sufficient. In many cases what seemed a very troubled project—suffering enormous delays and extensive cost overruns—turned out, just a few years later, to be a great success. A famous example is the Sidney Opera House, which took three times longer to build than anticipated and five times more than its original budget. However, it quickly became Australia's most famous landmark, and few tourists want to leave Australia without seeing it. Similarly, Apple's successful Macintosh computer was built on the lessons learned during the development of the nonsuccessful Lisa computer.[12] Given these lessons, was, then, the Lisa really a complete failure?

Customers' reaction to the project's outcome should also be considered. Customer satisfaction and meeting customer needs and requirements must be part of a project success assessment. Yet, ironically, a project may provide an efficient solution to customer's requirements, but be considered as a failure by the performing organization in terms of business success. Similarly, some projects seem successful in the short-term, but may turn out to be less successful in the long-run. Indeed, a long time may pass before success can be really evaluated and until initial expectations are met. To compound for these complexities, project success assessors should consider the following four major dimensions.[13]

32.3.1 Project Efficiency

The first dimension is the short-term measure expressing the efficiency with which the project has been managed. It simply tells us how the project met its resources constraints, whether it was completed on time, and whether it was finished within the specified budget. This is the immediate dimension, which is usually used for project assessment during execution and right after completion. Although meeting time and budget constraints may indicate a well-managed, efficient project, it may not prove that this project was successful in the long term and has benefited the organization in the longer term. On the other hand, with increased competition and shorter product life cycles, time to market (i.e., time from initial concept to market introduction) becomes a critical competitive component, and therefore enhanced project efficiency should be seen as adding to product competitiveness.

Some organizations may find it beneficial to consider additional measures of efficiency, such as the number of engineering changes before final design release, cost of materials and tooling, and efficiency and yield of production ramp.[14] Other measures may involve efficiency of purchasing (time to get orders out and materials in), reliability (or inverse number of prototype failure), and safety measures (number of accidents or injuries). However, one must realize that all these measures only relate to successful implementation of project execution and do not necessarily mean product success.

32.3.2 Impact on the Customer

The second dimension relates to the customer. This dimension addresses the importance one should place to the customer's requirements, and to meeting that person's needs. As was found, meeting performance measures, functional requirements, and technical specifications are all part of this second dimension, and not, as commonly assumed, part of meeting the project plan. Meeting performance has clearly a great impact on the customer. Customers, above all, are best to assess whether the product is serving their needs. Within this framework, meeting performance objectives is one of the central elements. From the developer's point of view, this dimension also includes the level of customer satisfaction, the extent to which the customer is using the product, and whether the customer is willing to come back for another project or for the next generation of the same product. Obviously, the impact on the customer is one of the most important dimensions in assessing project success.

32.3.3 Business Success

The third dimension addresses the immediate and direct impact the project may have on the developing organization. In the business context, did it provide sales, income, and profits as expected? Did it help increase business results and gain market share? Most important, did the project produce enough return on investment, and did it result in a positive cash flow?

This dimension may also apply to projects not aimed at building new products, or to internal organizational projects. For example, organizations may need to assess the success of their reengineering projects,[15] or the building of new manufacturing processes. This is the measure with which such an assessment could be done. It will include measures of performance time, cycle time, yield, and quality of the process; all of them will assess the direct impact that the project had on the organization.

32.3.4 Preparing for the Future

The fourth dimension addresses the issue of helping prepare the organizational and technological infrastructure for the future. How did the project contribute to future businesses and additional innovations? Did it produce a new technology? Did it create a new-product line? And did it create a new market? Did we also build during this project new skills that may be needed in the future? Did we develop enough core competencies to be used later across different business lines?

Project success, therefore, should be considered as a dynamic concept in which both short- and long-term implications are considered. The first dimension can be assessed only in the very short term, during project execution and right after project completion. The second dimension can be assessed after the project's product has been delivered to the customer, and the customer starts using it. Customer satisfaction (and particularly customer feedback) can therefore be assessed only after several months from the moment of delivery. The third dimension, business success, can be assessed only after reaching a significant level of sales, which usually takes between 1 and 2 years, while the fourth dimension must be assessed in the long term. Usually several years may pass before this dimension can be fully exploited and judged.

32.3.5 Discussion

The relative importance of each of these dimensions is therefore time-dependent. Different dimensions are more important at different times with respect to the moment of project completion. As mentioned, in the short term, and particularly during project execution, the project efficiency dimension is the most important. In fact, this is the only dimension that can be assessed or measured at this time. Once the project is completed, however, the importance of the first dimension declines. As time goes by, it matters less and less whether the project has met its resources constraints; and in most cases, after about 1 year, it is completely irrelevant. In contrast, after project completion, the second dimension—impact on the customer and customer satisfaction—becomes more relevant. The third dimension, business and direct success, can only be felt later. It takes usually a year or two until a new product starts to bring in profits or establish market share. And finally, preparing for the future can be recognized and assessed only much later. The long-term benefits from projects will affect the organization only after 3 or even 5 years. The relative importance of the four dimensions as a function of time is illustrated in Fig. 32.3.

The perspective of different stakeholders (developers, users, and contracting officers) with respect to these dimensions was tested in a separate study dedicated to defense development projects.[16] As the results indicate, almost all participants perceive "impact on the customer" as the most important dimension, followed by "project efficiency." The other two dimensions—"business success" and "preparing for the future"—have received much less attention and were perceived as less important.

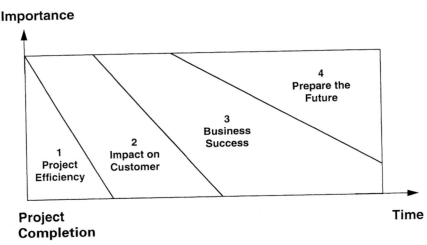

FIGURE 32.3   Relative importance of success dimensions is time-dependent.

Such findings probably indicate that most managers and other parties involved in projects are still focused on the short and medium horizons, while neglecting the long and very long implications of the project.

Finally, since different projects are initiated for different reasons, assessing the project success is also project-dependent.[17] One of the prominent factors affecting the importance of success measures is the initial level of technological uncertainty. Shenhar defined four levels of technological uncertainty in projects:[18]

> *Type A:* Low-tech projects are those projects that rely on existing and well established technologies, such as construction, road building, or "build to print" projects.
>
> *Type B:* Medium-tech projects rest mainly on existing base technologies but incorporate some new technology or feature. Examples include many industrial projects of incremental innovation, as well as improvements and modifications of existing products.
>
> *Type C:* High-tech projects are defined as projects in which most of the technologies employed are new, but existent, having been developed prior to the project's initiation. Examples are developments of new computers or many defense developments.
>
> *Type D:* Super-high-tech projects are based primarily on new, not entirely existent technologies. The Apollo Moon-landing project was an example of this kind, indicating, probably, that this type of project is relatively rare, and is usually carried out by only a few large organizations or government agencies.

For lower-uncertainty projects, meeting resources constraints may seem relevant and important, but such projects are seldom launched to create new technology or infrastructure in the long term. Their immediate success relies on meeting time and budget, and usually their expected profits are determined in advance. The importance of these

Importance

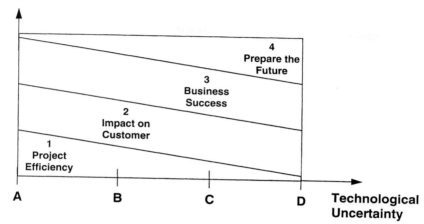

FIGURE 32.4   Relative importance of success dimensions is project-dependent.

measures changes when technological uncertainty is higher. For such projects, even poor performance in the short term and even limited business success may be compensated by long-term benefits, such as creating new markets, creating expertise in new technologies, and preparing the infrastructure for additional products for the future. Clearly, customer satisfaction and direct success are important to all types of projects. The relative importance of the four dimensions is depicted in Fig. 32.4.

32.4   THE BUSINESS-UNIT LEVEL

Organizational success at the business unit level rests on the aggregation of results achieved by different projects and products. Typically businesses are measuring their success in terms of gross sales, profit, return on investment, gross market share, etc. Some of these measures may have longer time horizons than others—for example, meeting backlog targets. While a few companies are using additional longer-term measures such as the percentage of sales from new products, a comprehensive framework of the business unit success is needed. During our studies of business units in the high-tech industry we found that business unit success should be measured using the following four dimensions.[19]

32.4.1   Profitability

The first dimension at the business-unit level addresses the issues of how well the business meets its financial and profit objectives and whether present sales generate enough cash to assure the smooth operation of the business. It reflects results of actions taken in the past that have generated recent sales. This dimension involves the traditional measures of sales and profits. However, the business should also ask how its profit margin is doing relative to similar businesses in the industry and compared to its own profitability goals.

32.4.2 Orders and Marketing

This dimension involves the next step of success in the market. It includes measures indicating prospective revenues from existing orders scheduled for delivery in the near future. It addresses the question of how successful the business is in achieving sales objectives and in creating additional orders. Is there a continuous flow of orders, and what is the current level of backlog? Of particular interest is the issue of how these orders will influence future cash requirements.

32.4.3 New Opportunities

This dimension is focused on the somewhat longer range. It measures outcomes that influence longer-return results since exploiting new opportunities may take time. It addresses the question of how successful is the organization in opening up new opportunities for new products and new services and new markets. What are the prospects of entering into these markets in the foreseeable future? This dimension also includes assessing customer satisfaction and loyalty. Are customers happy with the quality of the products and services of the business unit, and are they willing to come back for additional purchases?

32.4.4 Preparing the Infrastructure for the Future

This very long range dimension addresses the question of how well the business is prepared for future opportunities and changes. It reflects past investments that may determine the business unit's results in the years to come, and sometimes may even hurt business results in the short term. Had the business identified and made the necessary strategic decision on future technologies and future types of products? Had it invested enough in developing skills required for future markets? Does it have enough long-term programs, whose goals are beyond 2 or 3 years from now; and is it working on next generations of technology and products which are focused on 5, 7, or 10 years away?

While most organizations are using the first and second measures (profits and backlog), only a few are formally looking at longer-term dimensions. Several technology-based companies, such as Hewlett-Packard and 3M, have instituted a longer-term perspective. To focus their businesses on continuous innovation, they are assessing, among other things, the percentage of sales from new products developed within the last few years (part of the third dimension—opening the window for new opportunities).

32.4.5 An Example from the Electronics and Computer Industry

The four-dimensional framework of business-unit success was used in a study of 76 business units in the electronics and computer industry in Israel.[20] The results indicate that the success of a business unit may be divided into two parts: one encompassing the first two dimensions (profitability level and orders) and reflecting the short-run

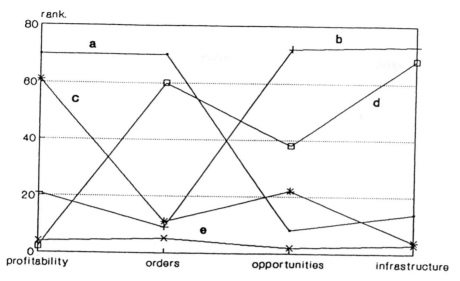

FIGURE 32.5   Ranking by success dimensions. (*Adapted with permission from Dvir and Shenhar.*[19])

economic success and the other including the remaining two dimensions and reflecting the prospects for the future.

The difference among the four dimensions and how short-term success differs from long-term success is demonstrated in Fig. 32.5. It exhibits the performance achieved by five business units and their relative ranking among the group of all businesses studied along the four success dimensions. A high rank means high performance. Unit *a* was one of the best units in terms of short- and very short range success. However, it was one of the worst in terms of longer-range success. Obviously, such a business situation requires immediate action for opening new opportunities and preparing the unit for future challenges. Without such action, the situation might rapidly deteriorate, and then affect its short-term results as well. In fact, 3 years after the study was conducted, unit *a* went into severe financial problems, and its previous sales have dropped tremendously.

In comparison to unit *a*, the position of unit *b* is quite strong regarding long-term measures. Its short-term measures, however, are rather poor. It is likely, however, that because of its good technological infrastructure and its new product and market opportunities, it will soon recover and present good performance regarding profitability and orders as well. And in fact, 3 years later this unit has proved itself as a very successful one. Intermediate cases have also been observed. Unit *c* was ranked high in terms of profitability. Regarding other dimensions, however, its ranks were quite low. Although this unit has good profits at the moment, there are no orders to provide time for reorganizing and opening new opportunities. Furthermore, it seems that management has not taken the right steps to establish the technological and human infrastructure, nor has it taken the steps for getting new orders. Obviously, this unit is on its way to decline and so indeed happened later. The data on unit d supports our theoretical framework. Its profitability level is quite low, while all its longer-term scores are higher. As can be anticipated from the data exploiting the new opportunities, using a strong technological infrastructure might improve the profitability situation. That is exactly the course that unit d has taken, and in fact, their profitability returned to an acceptable level. The data on unit e presents probably the last moment before consolidation. Prior to the study, unit e reduced its workforce by 50 percent and during the study's course it had actually taken desperate actions which proved to be too late and too little.

32.5　THE COMPANY LEVEL

The company or the corporation level represents a collection of business units and central units usually built to provide common services to the business units and the corporation (e.g., central research and development laboratory, computation center). A company or a corporation is then a diversified organization whose main goal is to create and increase value. Since competition occurs only at the business level and managing a company which consists of a collection of businesses adds costs and constraints to the business units, the corporation cannot succeed unless it truly adds value to the business units by providing tangible benefits that offset the costs of lost independence. In order to achieve such an added value, there must exist some kind of interrelationship between the business units that creates synergy. Porter identified two types of relationships:[21] the company's ability to transfer skills or expertise among similar value activities in different business units such as basic technologies, manufacturing processes, or marketing techniques and ideas; and the ability to share assets and activities such as sales force or logistics networks. The ability to transfer competitive skills across businesses and use them for gaining company advantage over competitors is expressed by the core competency concept, representing the sum of learning across individual organizational units.[22]

The corporation success of effectiveness depends, therefore, on integrative decisions and activities which cross business boundaries, some of them having an immediate impact, and others whose impact can be recognized only in the long term. The success in achieving synergy by operating as a unified company has to be measured from several angles. It has to reflect the company's short-term performance (financial performance) on one hand, and its success in creating an atmosphere of cooperation between the various business units, exploiting capabilities and core competencies across business boundaries, and establishing the vision, direction, and foundations for the future, on the other hand. Transforming these inferences into a framework of success dimensions at the company level results in the following four dimensions.

32.5.1　Financial Performance

This is the standard shortest-term dimension of organizational success. It reflects last-quarter results such as sales, profits, and overall cash. All these indicators are based on sales from existing products and services which are the result of strategic decisions made years ago. Obviously, financial performance of the entire organization represents an integration of the financial results of separate businesses and is simply an indication of total sale results of these businesses.

32.5.2 Market Response

Market response is the first dimension which aggregates various business-unit activities and the corporation's contribution to a synergistic result. This dimension reflects the cumulative effect of company and business units' strategic decisions taken in recent years and represents the integrative impact that the company has made on the markets and industries in which it is functioning. Market response may be assessed through annual growth percentage, the company's stock price, and the rate of stock growth. However, it should also include an assessment of the company's reputation and its image in the industry, among customers, and among competitors.

32.5.3 Strategic Leverage

This dimension represents the strategic strength and the competitive advantage achieved by the corporation as one entity (in contrast to a collection of separate businesses) and how well the company is prepared to compete in the market at the present time and in the near future. The most important part of this dimension is the company's set of core competencies developed during recent years in various businesses and its ability to deliver these competencies to its different collection of businesses while exploiting them as fundamental customer benefits. A core competency may be in the form of a key technology which creates the product's advantage over competing products. However, it may also be expressed in the company's developed capability in process- and manufacturing-related competencies that yield sizable cost and quality benefits, or its marketing, advertising, or distribution capabilities as a company beyond those possessed by competition.

The second part of a company's strategic leverage is the extent to which it established strategic vision, strategic goals, and strategic imperatives to achieve these goals. Within this dimension the company should assess its strategic thinking (or, for that matter, the planning that follows this thinking). Did it produce a clear vision of the company's direction in the coming years? Did it articulate this vision in a well-defined, well-understood, and well-shared set of goals that will guide development and action in the coming years? And did the company establish a collection of imperatives in the form of development projects, programs, and companywide activities that are aimed at achieving its articulated goals?

32.5.4 Creating the Future

This last and longest-range dimension is aimed at assessing the company's initiative taken to create a new future for the company, one which is beyond existing trends and predictable developments. It is its ability to see the future prior to its competitors and customers and define new needs no one has been able to articulate before.[23] Within this dimension the company should strive to write the rules in its industry and define its insight. It must have the foresight to do things that others will copy later and find new customers and uses not addressed at the moment. It should try and make a difference to customers by exceeding their expectations, by creating unimagined products, and by making the future real and tangible to customers and competitors.

Several questions should be asked for assessing this dimension. For example, does the company's opportunity horizon extend sufficiently far beyond the boundaries of existing product markets? Is there an explicit process for identifying and exploiting opportunities that lie between or transcend individual business units? Are these processes identifying new and unarticulated needs for existing and new customers which are not served at the moment? And are management and employees committed to these processes and to their level of aspiration?

32.6 IMPLICATIONS AND CONCLUSION

Success of technological organizations is a multidimensional concept. Essentially it has four major dimensions, two of them in the short term, and two in the long term. Measuring success by using only one dimension may be misleading and may not provide a complete picture of the organization's strengths or weaknesses and its future prospects. Specific measures of success should be developed for each unit and for each project, business unit, and organization at large. Such measures should match the organization's mission, strategy, and line of products, and the industry. The preceding discussion demonstrates the viability of the multidimensional approach to success, particularly in the realm of technology-based organizations. Managers at all levels must be aware of the danger in evaluating business success by short-term measures only. Profitability is merely a snapshot of a momentary situation, and it may change almost instantaneously. Good financial performance is obviously the result of good previous decisions. However, present short-term success must be accompanied by satisfactory performance assessment in other areas. Failing to do so will lead an organization into trouble. An acceptable level of orders, new business opportunities, and particularly build up of the future in terms of technology, people, facilities, and other areas, are the organization's royalty for long-term profitability and success. Organizations, in particular, must establish the long-term vision while making all businesses aware of the pitfalls of short-term thinking.

Establishing new frameworks for assessing organizational success is not easy. It requires a departure from present practices and a new way of thinking. One of the difficulties in creating a working framework for the organization is that some of the measures discussed here are subjective and could not easily be transformed into quantitative measures. Such transformation is not impossible, however, as we have seen in several of the cases we studied. It requires a careful assessment of the issues that are important to the organization at different levels. It will entail the development of specific criteria and rank assessment measures for the organization and a companywide learning process in which the organization will learn to rank itself compared to its past performance and to its competitors.

32.7 REFERENCES

1. R. H. Hayes and W. J. Abernathy, "Managing Our Way to Economic Decline," *Harvard Business Review,* **58:** 67–77, 1980.
2. M. L. Dertouzos, R. K. Lester, and R. M. Solow, *Made in America: Regaining the Productive Edge,* Harper Perennial, New York, 1989.
3. A. J. Shenhar and P. S. Adler, "The Technological Base of the Company," in G. Gaynor, ed., *Handbook of Technology Management,* McGraw-Hill, New York, 1996.
4. C. K. Prahalad and G. Hamel, "The Core Competence of the Corporation," *Harvard Business Review,* pp. 79–91, May–June 1990.

5. C. K. Prahalad and G. Hamel, *Competing for the Future,* Harvard Business School Press, Boston, 1994.

6. L. A. Iacocca, *Iacocca: An Autobiography,* Bantam Books, New York, 1984.

7. S. E. Seashore and E. Yuchtman, "Factorial Analysis of Organizational Effectiveness," *Administrative Science Quarterly,* **12:** 377–395, 1967; see also P. S. Goodman and J. M. Pennings, *New Perspectives on Organizational Effectiveness,* Jossey-Bass, San Francisco, 1977.

8. P. R. Cowely, "Market Structure and Business Performance: An Evaluation of Buyer/Seller Power in the PIMS Database," *Strategic Management J.,* **9:** 271–278, 1988.

9. R. G. Cooper and E. J. Kleinschmidt, "New Products: What Separates Winners from Losers," *J. Product Innovation Management,* **4:** 169–184, 1987.

10. J. K. Pinto and D. P. Slevin. "Project Success: Definitions and Measurements Techniques," *Project Management J.,* **19:**(1): 67–72, 1988.

11. M. Freeman and P. Beale, "Measuring Project Success," *Project Management J.,* **23**(1): 8–17, 1992.

12. F. Guterl, "Design Case History: Apple's Macintosh," *IEEE Spectrum,* pp. 34–43, Dec. 1984.

13. A. J. Shenhar, D. Dvir, and O. Levy, "Mapping the Dimensions of Project Success" *4th International Conference on Management of Technology,* Florida, Feb.–March, 1994.

14. S. C. Wheelwright and K. B. Clark, *Revolutionizing Product Development,* Free Press, New York, 1992.

15. M. Hammer and J. Champy, *Reengineering the Corporation,* HarperBusiness, New York, 1993.

16. S. Lipovetsky, A. Tishler, D. Dvir, and A. J. Shenhar, *The Relative Importance of Success Dimensions of Defense Projects,* Working Paper, Tel-Aviv University, School of Business, 1995.

17. A. J. Shenhar, D. Dvir, and O. Levy, *Project Success—a Multidimensional Strategic Concept,* Working Paper, University of Minnesota, Center for the Development of Technological Leadership, 1995.

18. A. J. Shenhar, "From Low to High-Tech Project Management," *R&D Management,* **23**(3): 199–214, 1993.

19. D. Dvir and A. J. Shenhar, "Measuring the Success of Technology-Based Strategic Business Units," *Engineering Management J.,* **4**(4): 33–38, 1992.

20. Ibid.

21. M. E. Porter, *Competitive Advantage,* Free Press, New York, 1985.

22. Prahalad and Hamel, op. cit. (Ref. 5).

23. Ibid.

CHAPTER 33

MEASURING THE BENEFITS FROM MANAGING TECHNOLOGY

Gerard H. (Gus) Gaynor

G. H. Gaynor and Associates, Inc.
Minneapolis, Minnesota

33.1 PROBLEM AND OPPORTUNITY

It may appear presumptuous to consider approaches for measuring the performance and impact of management of technology (MOT) when processes for measuring any type of technical performance seldom provide meaningful information. Macro figures provide some general information, but those figures do not suggest how the changes in performance were achieved. Those figures seldom provide adequate feedback to business-unit management and less often to the scientific and engineering communities. Measurements such as return on investment, sales and profits, investment and expense, and all other financial measurements that focus on the endpoint provide macro figures. Measurements such as amount of rework, customer returns, number of engineering changes, and late delivery provide limited information at the micro end of the spectrum. Neither of these measures establishes a basis for capitalizing on the reasons for the positive results or for resolving the problems that produced the negative results. While this chapter focuses on measuring MOT performance, I propose that MOT measurements impact organizational results far beyond what is considered as technology impact.

Measuring performance and impact of MOT enables organizations to take advantage of new business opportunities and in the process optimize the value added from investment of resources. This is true whether MOT is considered from a very narrow perspective or from an integrated and holistic perspective. There are no known processes for measuring MOT performance, but I suggest that there are measures that can be used if measurement is viewed from a new perspective. For example, measuring performance usually spans the spectrum from the war room approach with its multiplicity of charts to the macro figures that really say nothing about performance. The extreme ends of this spectrum do not provide realistic and usable cost-effective information that can be used as the basis for structuring improvement programs.

Measuring the performance of technology driven organizations and their professional talent presents special problems. The measurement of MOT must go beyond the hard macro numbers that originate in manufacturing. Managing creativity and innovation requires more detailed insight, and that measurement includes all the business-unit functions—even the firm's top management.

This chapter does not provide the "how to" for measuring MOT performance. It raises the issues that managers must face as they think about measuring performance. There are no answers today. This chapter raises some of the issues involved in the process of measuring professional performance. Toward this end the chapter considers

- Current approaches to measuring functional performance
- Doing the "up front" work
- Measuring MOT performance—the scope
- Redefining research and development
- Management of technology—point of origin
- MOT process model
- MOT measurement model
- Selecting the measurement parameters
- Scaling the parameters
- An approach to measuring MOT performance
- Questions that must be answered
- Summary and conclusions

There is no consistent definition of management of technology. Current definitions generally restrict MOT to managing research, managing engineering, and so on. I propose a broad definition of technology:

> Technology includes the tools, the techniques, the processes, and the knowledge required to accomplish a task.

Management of technology takes on many different meanings. To many, MOT means managing engineering. To others it means managing information, managing research, managing development, managing manufacturing operations, managing the activities of engineers and scientists, or managing functional activities without concern for the total spectrum of activities that encompass the business concept-to-commercialization process. Those interrelated activities must be integrated into a technology management system. There is one key word that must be emphasized in the management of technology: integration.

> MOT involves managing the system—it also involves managing the pieces. Neither the system nor the pieces can be subordinated. MOT involves integrating the "pieces" into an acceptable "whole" by focusing attention on the interdependence of the pieces.

33.2 CURRENT APPROACHES TO MEASURING FUNCTIONAL PERFORMANCE

Current approaches to measuring performance of the technology related functions center on such factors as cost, schedule, performance to requirements or specifications, amount of rework, number of drawings, number of drawing changes, and lines of code. Other activities such as hours of education; attendance at conferences; number of patents; number of confidential processes; number of new programs initiated related to quality, concurrent engineering, and so on; and many other what I term "single-issue programs." These measurements indicate level of activity but not performance.

When quality mania infected this country, many organizations began tracking cost of quality. The objective was to reduce it. Managers interpreted Crosby's "quality is free"[1] concept to mean that no up-front investment was required. These elaborate cost-of-quality reports which showed major reductions in the cost of quality in the first months of the program resulted in misplaced optimism. In reality there were few benefits. A review of manufacturing costs would often show an increase in manufacturing cost. It became a game that managers played to win the blessings of top management. Subsequently managers began counting the number of quality circles as though that was an indication of performance. Such measurements are meaningless. The numbers are not too important. The importance lies behind the numbers.

Organizations also use such figures as sales per employee, percentages of payroll allocated to various functions, and number of new products without distinguishing among the various types of new products. Sales per employee are meaningless unless inflation and capital investment are taken into account. Introducing me-too products, upgrading current products, and using the same platform for a new product are quite different from introducing a new-to-the-market or a breakthrough product that develops into a completely new business or perhaps a new industry.

Statistics of this type do not relate the performance level in the associated functions or project teams to the final macro result. Will these statistics impact future performance in any significant way? Do these statistics give us insight into the quality of the professional work? Do they tell us how professionals approach their work? Do they measure professional-level effectiveness and efficiency? Do they tell us anything about the use of the resources and the infrastructure (the percentage of high-level people doing low-level work)?

Focusing on the "project process" provides a baseline for MOT. Revisiting the fundamentals always forces an appraisal of work methods. Business activities and especially technical activities revolve around some form of project management. Research clearly demonstrates that most projects are poorly managed from the start. If there were laws against the malpractice in project management, most organizations might be forced to substitute lawyers for engineers and scientists—not only engineers and scientists but also executives and managers. Most project activity focuses on obtaining authorization for the project and then managing it to some conclusion. Recognizing that probably less than 5 percent of all projects meet the specifications and requirements, schedule, and cost commitments should force managers to review their current project management model. Research clearly shows that projects seldom have a well-developed and integrated statement of purpose, a project specification (not boiler plate), and the identified means for accomplishing the up-front work.

33.3 DOING THE UP-FRONT WORK

The *up-front work* is that work that must be performed before any major effort can be pursued to acquire the funds. During this up-front period decisions are made which impact future project performance. This up-front period requires a delineation of the essential competencies. Those competencies must go beyond the traditional approach of asking for so many electrical, computer, or mechanical engineers or so many polymer or physical chemists. Effective project management requires delineation of the specific competencies—competencies that are current. This up-front period requires determination of the adequacy of the infrastructure to support the project.

Technical and operational decisions made during this up-front work period seldom can be retracted. It is important that those decisions take into account alternative and contingency solutions. The decisions that involve implementing new technologies must be based on fact. There is little justification for attempting to implement a new technology on a tight schedule if that technology has not been reduced to practice or validated in such a way as to allow a fairly high level of confidence. Operational decisions must follow the same principles—validation in some form or at a minimum an understanding of the potential problems that may be incurred.

Doing the up-front work differs from emphasizing the project planning and programming tools. These project tools often lead project managers down the wrong path. Project tools deal with the details—essential details that come after the foundations have been laid. The emphasis must be placed on the thinking processes and the thinking that the process generates. Project planning and scheduling tools are simply tools. They are not the end. They are the means for tracking progress. They provide no intelligence into the technologies or processes.

33.4 MEASURING MOT PERFORMANCE: THE SCOPE

The word integration will be used throughout this discussion of measuring MOT performance and impact. MOT is equivalent to integration: integration of functions, integration of technologies, integration of learning, integration of competencies into capabilities, and so on—all precursors for MOT. The segmented approaches of the past must be brought together as a cohesive and effective operating system. Integration means bringing together but bringing together is a complex process. Integration goes beyond intellectual agreement. It goes beyond philosophical considerations. It goes beyond theory. MOT resides in the domain of the practitioner—the problem solver.

Integration means bringing together—bringing the pieces into the whole so that the whole is greater than the sum of the pieces. That means breaking up the fiefdoms and bringing them to focus on the objectives for the benefit of a larger entity. This does not imply breaking up the centers of excellence or the critical mass of talent so essential in a technologically oriented organization. These centers are essential, must be maintained, but must be operated effectively. These are not retirement positions for those who have made a contribution in the past.

MOT cannot grow in the command and control environment. At the same time it cannot grow in total freedom. MOT demands discipline with accommodation for many different levels of freedom. New ideas seldom emerge in the command-and-control environment. At the same time they seldom emerge in sufficient quantity and of sufficient quality in a totally undisciplined environment. Innovators need freedom, but they also need direction and acceptance of accountability to do what they said they would do.

The following question may be asked: Why measure performance and impact? The opposite may also be asked: Why not measure? Whatever activity an organization undertakes requires resources and use of the infrastructure. That implies an investment and an investment must be justified whether it applies to the factory floor, the research laboratory, the marketing organization, or any group or person. Measurement provides understanding because it deals from fundamentals. Some measurements may be quantitative; others, qualitative. There is no reason to look for third decimal point accuracy. The purpose is to determine as realistically as possible the rationalization and the justification for managing the technology issues as an integrated function. To this extent measurement of MOT focuses on measuring performance levels of business integration and the manner in which the results are achieved. The following question must be answered: What factors must be considered in developing a viable means for measuring MOT performance and impact?

The scope of issues includes

- Defining performance
- Defining impact
- Redefining research and development
- Management of technology—point of origin
- Integrating research, development, design, manufacturing, and marketing
- The business process
- MOT process model

33.4.1 Defining Performance

Performance and impact, although related, are quite different. *Performance* is basically a measure of output versus input. In essence, what benefits were received from the investment in resources and infrastructure? These performance figures do not measure what could have been accomplished. The performance may have been marginal at best because the objectives did not include any significant stretch targets. The performance may have been marginal, but bottom-line figures were enhanced through unpredictable competitive conditions that required no effort on the part of the firm. The performance may have been marginal because of emphasis on short-term results—no consideration about investing for maintaining the financial viability of the company—no analysis or measurement of the risk that may have provided greater benefits. Performance measurement must be based on fact, and not on the rationalization of extenuating circumstances—no excuses. The measurement must demonstrate the reality and not the illusion of reality.

33.4.2 Defining Impact

Impact, on the other hand, measures the value added by an activity to the business. A minor and obscure effort can make a significant impact on business performance. A modest change in a polymer may significantly enhance the performance of that polymer. A change in strategy may redirect the organization into new markets with existing technologies, production facilities, and distribution systems. A change in the measurement of process parameters often yields significant manufacturing efficiencies. A proactive effort on the part of the accounting department in relation to providing manufacturing with timely costs could yield vital information that affects future manufacturing costs. Similar opportunities exist in all business-unit functions.

33.5 REDEFINING RESEARCH AND DEVELOPMENT

The term *R&D* leads organizations to distort the value of their investment in research. After many years in integrating the technology-related functions in the real world, I concluded that R&D must be considered as two distinct but related functions and activities. Separating R&D into research and development as two distinct functions allows placing research and development in their proper position. Financial statements that show *x* percent of sales invested or expensed in R&D give a false impression—most of those funds are expended on development and not research. Obviously this depends on how research is defined but the description of research cannot be minimalized to the point where it focuses largely on development. This distinction becomes important for two major reasons: selecting people and defining the scope of the work.

33.5.1 Selecting People

Assignment of a critical mass of the essential talent determines project success. Researchers approach a project from the perspective of understanding the "how" and the "why"—they approach a solution from the input to the process. Development specialists approach the same problem from the perspective of finding a solution—a much more pragmatic approach that focuses on the output. Finding that solution may require only demonstration of the answer for the development specialist while the researcher searches for a full understanding. The distinction does not suggest greater importance of the developer or the researcher but emphasizes the different approaches used in problem solving.

Assigning the wrong type of talent to a project extends not only the total cycle time but also the total project time as well as the project timing. A researcher may go far beyond the requirements for an adequate solution and look for some new breakthrough solution. The developer working in a research-oriented program may take certain shortcuts that are bound to cause future problems. In both cases the cycle time and the total time may be significantly extended.

Consider the following example: A physicist involved in developing a specialized measurement device required certain optically sensitive materials. Instead of contacting organizations that possessed the expertise, she began her own research into the solid-state physics of the materials—a discipline totally unfamiliar to her. Eventually the project was canceled when the equipment became commercially available. This same project under a development-type person would have been completed in less time and with better results.

33.5.2 Defining Scope of Work

Defining the scope of work in research and in development requires different perspectives. Most researchers contend that research cannot be planned. Granted that some exploration of the unknown may not be planned but parts of that process can be planned. For example, the literature and information search can be planned. Some may argue that even a literature search cannot be planned. They argue that information research is an ongoing process—it is—but it is not a full-time activity. There comes a time when the "searching" must stop and the "doing" begins. The new competitive paradigm does not allow the extravagances of the past. The old model is no longer applicable. The work effort may also be planned. The contingency approaches can be developed. Failures can be anticipated.

Development work, in contrast, can be planned on a more controlled schedule after the required resources and infrastructure are identified and synthesized, and the priorities established. While some research may be required, that research most likely involves a relatively small segment of the project. This does not imply that the research effort can be ignored. The issues that require research can be identified with a high degree of specificity very early in a project.

Distinguishing between the needs of research and those of development allows differentiation of the two functions and their activities. There is no doubt that research and development must be integrated into a whole, but the difference in processes used to achieve results must be resolved. There are no formulas for this process, and each organization must determine the extent to which the management processes are differentiated. Those processes will be quite different for a firm that is technology driven and a firm that is product driven.

33.6 MANAGEMENT OF TECHNOLOGY: POINT OF ORIGIN

Figure 33.1 illustrates the various phases of MOT. The level of integration is divided into five arbitrary phases for convenience only. The number of phases and the functions to be included in a phase must be determined by the organization's needs. What should be measured depends on where MOT begins. Phase 1 of Fig. 33.1, if it includes an integrated research and development department that focuses its efforts on system requirements, provides a starting point for managing technology.

Phase 2 adds design and manufacturing which expands the scope, but it is not until phase 3 that an organization assembles the critical mass necessary to implement an MOT process management approach. Although the integration begins in phase 1, MOT does not begin until phase 3. Phase 3 includes the operating groups in the concept to commercialization process. Anything less than a phase 3 effort is equivalent to managing functions with the hope that some level of integration will be achieved.

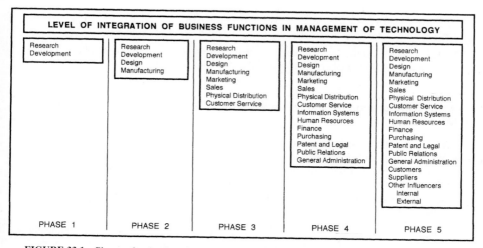

FIGURE 33.1   Phases of technology integration for management of technology.

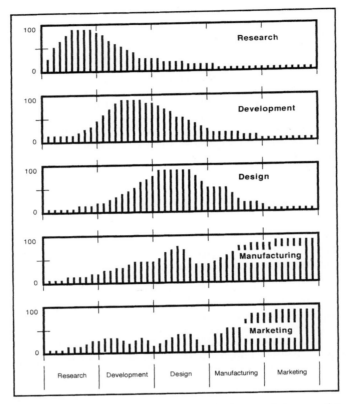

FIGURE 33.2   Percent of effort by research, development, design, manufacturing, and marketing people during the project phases of research, development, design, manufacturing, and marketing.

Phases 4 and 5 add the service functions that are essential to support the phase 3 functional activities. For further discussion, refer to Chap. 1.

33.6.1 Integrating Research, Development, Design, Manufacturing, and Marketing

Figure 33.2 illustrates the levels of involvement by research, development, design, manufacturing, and marketing as an organization begins to manage according to the principles of MOT. I have intentionally shown that each function has some involvement throughout the concept to commercialization cycle. Each function must be involved from the concept stage and cannot relinquish responsibility as long as the product is in the marketplace. Levels of intensity of involvement vary for each discipline depending on the product stage.

This involvement goes beyond a cursory acquiescence—it must go beyond intellectualizing. It must be real. It must be active. It must be directed and managed. Each discipline must not only bring the issues on the table but also actively engage in the resolution of any inconsistencies. Early involvement does not mean listening. It means doing what is necessary to meet the objectives. This approach—an integrated approach—eliminates the throw-it-over-the-wall approach so commonly used between business-unit functions. Even as we approach the end of the twentieth century, most organizations fail to involve the required people resources in the early stages. MOT cannot be implemented without this minimum level of functional integration. Obviously the level of involvement depends on the type of project and the particular industry.

33.6.2 The Business Process

Integration of the activities of phase 3 and the measurement of the results of this integration begin with an understanding of the business process model. Having spent a major part of my professional engineering career in process measurement, instrumentation, and control systems as well as management of research, development, engineering, and manufacturing, I have concluded that each of these functions operates within the limits of a generic process. On further investigation I concluded that the process model can be used as a model for managing any activity or organization. The model also extends to what are generally classed as service businesses.[2]

Figure 33.3 is a simplified model of an open-loop control system. Certain inputs, such as resources and infrastructure, are applied to the system to be transformed through a process into some desirable output. This is an open-loop system without any feedback loop. This model assumes that the system will function in spite of variations in the inputs as well as the inherent limitations of the system to maintain the desired output.

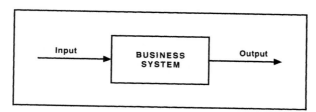

FIGURE 33.3   Simplified process model.

The model of Fig. 33.4 shows a simplified closed-loop feedback system for controlling manufacturing processes. It is similar to the model of Fig. 33.3 but provides a feedback system to maintain the system in control. The model includes the means for establishing the preset requirements for the system, accommodating inputs 1 and 2, measuring the output, comparing that output to the requirements through a comparator, measuring the difference between the output and the requirements, correcting the system's operation on the basis of difference between the output and the output specification, and sending that difference to a control system which then brings the system under control. Control systems of this type are common.

Input 1 includes the known, the controlled, and the predictable elements of the process. Input 2 includes the unknown, the uncontrolled, and the unpredictable. These unknowns of input 2 force the system into various states of unpredictability and loss of control. A manufacturing system, for example, must be able to manage the impact of the unknown, the uncontrolled, and the unpredictable. The variables in the operating system often number several hundred or more control points that must be controlled in such a way as to keep the system under control.

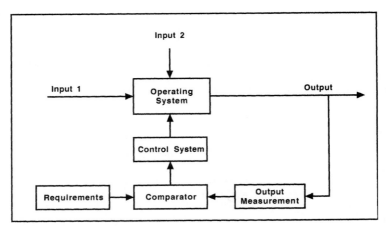

FIGURE 33.4 Process control mode.

These control points could include temperature; pressure; speed of operation; air and liquid flow measurement of different types; volumetric measurement; raw-materials quality; specialized instrumentation for measuring thickness, flaws, gloss, surface, and other conditions; various environmental measurements; and so on. All these measurements provide information to a control system that alters various activating devices to keep the process in control. This control system is managed through a set of algorithms that make the decisions relative to the controls that must be adjusted. They make the if/then decisions automatically. If any of the process parameters exceed their limits, visual or audible alarms indicate an out-of-control state. The most sophisticated systems may include a built-in automatic shutdown program which brings the process to a controlled stop. While the control system in this example deals with quantitative inputs, the application to human performance requires a much more sophisticated— and at the same time judgmental—means. Human behavior cannot be reduced to a mathematical equation. Human behavior belies prediction.

How does this process model relate to MOT? A business or firm constitutes a system. That system is composed of certain known, controlled, and predictable inputs and some unknown, uncontrolled, and unpredictable inputs that affect the management process and the output. The business resources, infrastructure, and activities with their many subdivisions are equivalent to the hundreds of variables that must be controlled in the mechanical process control model illustrated in Fig. 33.4. All the elements in the management process interact, and that interaction constitutes the basics of MOT. No algorithm exists for measuring the thousands of interactions of the resources, infrastructure, and activities elements and then controlling those interactions to optimize performance.

33.7 MOT PROCESS MODEL

The description of the process model in Fig. 33.4 can be applied to the MOT system model shown in Fig. 33.5. You may argue that the application of the system process model attempts to transfer the characteristics of mechanical systems to people. The comparison does not in any way discount the appropriateness of the model. After all,

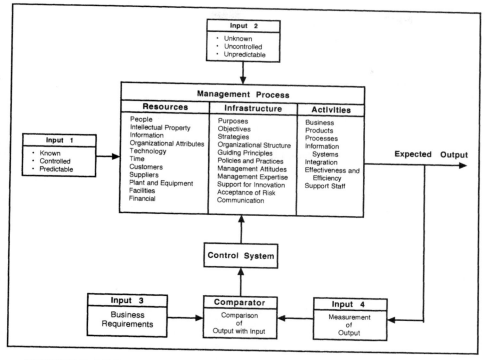

FIGURE 33.5   MOT process model derived from a typical production process.

human beings are systems. Mechanical systems display different levels of inertia, and people also display levels of inertia whenever any change is proposed. The model demonstrates the complexity of managing technology, the necessity to control the process that manages the system, and the benefit to be derived from a managed process. The discussion associated with Figs. 33.3 and 33.4 applies to Fig. 33.5.

What was described as the *operating system* in Fig. 33.4 becomes the *management process* in Fig. 33.5. The management process includes three components: resources, infrastructure, and activities. The integration of these three components forms the basis for MOT. They are interdependent. A review of these broad-based components recognizes that each can be subdivided into many subcomponents. Each of the components affects the other, and each is affected by the other. This interaction can be either positive or negative.

33.7.1   Resources

The elements listed under *resources* are affected by all the elements of the infrastructure. Those infrastructure elements receive little attention not only from managers but from other participants as well. There is no argument that people are the major resource and the most complicated of all variables. People represent the most intangible or subjective element. Most other resources can be quantified to a lesser or greater degree. The number of variables that can be applied to a single person limit our ability to neatly quantify not only skills, experience, and competence but all those other human characteristics that determine competency in social interaction.

33.7.2 Infrastructure

People work within an infrastructure that must be managed. If no support exists for innovation and acceptance of risk, innovators—people—most likely receive little support. A continuous flow of new ideas and concepts and subsequent innovations to the marketplace cannot be expected in such a situation. Support for innovation requires degrees of risk. Support for innovation without acceptance of the inherent risk yields nothing but antagonism toward management. If the CEO in the annual liturgy emphasizes the need for innovation and then fails to follow through with resources, perhaps greater prudence should have been exercised. If the firm lacks management expertise, the resources may be poorly allocated. Directing the activities of an organization may be difficult if the purposes, objectives, and strategies exist only on paper—they must be communicated in meaningful language. What do they mean to the bench chemist or the designer or the salesperson? Are they platitudes, or are they guidelines for improving performance? Similar comments apply to all the other elements of the infrastructure.

33.7.3 Activities

The classification of "activities" in Fig. 33.5 under "management processes" includes seven broad categories which interact with the resources and infrastructure. In essence, the management process evolves to a three-dimensional model that includes resources, infrastructure, and activities. The "activities" classes are arbitrary, provide a reference point, and span the continuum that embodies every business activity. These activities can be reclassified into many subcategories. Organizations must classify these activities in a way that best serves their purposes. That classification must include the system perspective.

33.8 *MOT MEASUREMENT MODEL*

The MOT measurement model describes the relationship between resources, infrastructure, and activities. Figure 33.6 illustrates this relationship on a typical x, y, z coordinate system. The three-dimensional model provides a more realistic approach since each of these three axes interacts with the other two. For example, activities depend on adequate resources and infrastructure. In turn, those activities influence the resources and infrastructure either through what has been learned during the specific activity or what has not been learned. Figure 33.6 illustrates this relationship in a unit cube in which resources, infrastructure, and activities can each have a maximum value of one (the value of 1 is arbitrary—it could be 10, 100, 1000, etc.). This relationship can be expanded or subdivided as necessary. In some instances it may be reasonable to use different scaling of the cube. For example, resources could be valued at 10, infrastructure at 5, and activities at 7. The objective however would be the same: Optimize this relationship. The optimum solution lies at endpoint B of the three-dimensional vector $A–B$ as shown in Fig. 33.6. This may be utopia and may not actually exist, but it provides a direction—it provides a target.

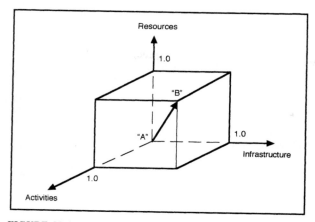

FIGURE 33.6   Three-dimensional model relating resources, infrastructure, and activities.

The first step toward implementing MOT involves an evaluation of the current resources, the infrastructure, and the activities. This evaluation will clearly demonstrate where corrective action must take place. Developing such "as is" scenarios requires an understanding of the organization and a realistic analysis of the parameters being considered. Generally the analysis shows that such as-is studies overstate the reality. An organization is rarely as good as the analysis indicates. Inflation of information constitutes a major problem which must be resolved.

33.9 SELECTING THE MEASUREMENT PARAMETERS

The major difficulty in measuring MOT performance arises from determining what should be measured and the significance of the measurement. Why should performance be measured? Is the purpose punitive? Is it for department public relations? Is it an opportunity for upward mobility? Is the purpose to win some company-sponsored award? Or is the purpose to provide guidance for future action and improve performance and impact? While the purpose may be some percentage of each of these, the major emphasis must be placed on the guidance that measurement provides for future activities and the opportunities that measurement provides for improving performance. People and projects become the critical measurement parameters.

33.9.1 People

People present the greatest difficulty in this evaluation process. It is not difficult to quantify education/learning and experience but much more difficult to quantify them in terms of their current value. Education/learning, after formalized academic education, involves many different sources for learning. Experience is one of these sources. But the significance of that experience becomes important: Is it relevant to the current situation? Because a person understands the fundamentals does not assure ability to apply those fundamentals.

There are two types of measurement for human performance: quantitative and qualitative. The quantitative presents no problem. There is no difficulty in measuring formal learning and the experiential results of that learning that are transformed into usable skills. Knowledge of calculating even the most complex, simulating, modeling both physical and mathematical, understanding of work processes, and using management by objectives intelligently—not only computer literacy but computer usage and application—and so on are examples of measurements that can be quantified. If no other means exist, normal testing procedures are available. This is not a recommendation for massive testing programs, but only recognition that sources for securing such information are available.

The qualitative issues include the intellectual and the interpersonal. The intellectual include

- Creativity and innovation
- Conceptualization
- Entrepreneurship
- Usable subject or disciplinary knowledge
- Multidisciplinary understanding and experience
- Process of knowing—thought and thinking
- Analysis and synthesis
- Problem/opportunity finding
- Problem/opportunity resolving

These intellectual characteristics cannot be assigned a number but can be identified and rated. Establishing the baseline begins with describing in simple terms each characteristic. Creativity does not imply genius. It must be defined in its context. Every creative act does not involve reaching the moon.

Qualitative scales of measurement also apply to interpersonal characteristics. These characteristics or skills include

- General people skills
- Participation
- Acceptance of ambiguity and uncertainty
- Acceptance of risk—personal and business
- Search for objectivity
- Observation
- General attitude—proactive, reactive, or inactive

These interpersonal characteristics impact performance and cannot be ignored. However, their meanings are not universal. For instance, the characteristic "aggressive" must be defined in a workable context. What does "aggressive" mean? Does it imply laying the issues on the table, or does it imply berating an individual, demanding unrealistic performance without adequate resources, or generally harassing people? Unfortunately the word "aggressive" as often used does not imply aggressive; it may just mean that a question was asked that demanded an answer.

Measuring people performance in its many aspects demands a great amount of sensitivity. The process of appraisal can destroy people and that process can also attribute knowledge, skills, and performance that do not exist. Appraisal of the "as is" demands a great amount of integrity; otherwise it is a useless process. This applies not only to people but to all other resources and the infrastructure and activities elements.

People function as individual participants as well as in teams. Use of teams is a highly overworked and over publicized concept that is promoted without sufficient data to quantify the benefits. Nevertheless, teams will continue to be used as they have been for millennia. The gurus attempting to spread the gospel of self-selected teams and self-disciplined teams have never been forced to meet a deadline, meet some undefined specifications, and deliver at the projected cost. Teams not only have their place but also are an absolute essential ingredient in managing an organization. But teams are composed of individuals, and people produce results. The touchy-feely approaches and the camaraderie are not necessary. If it is there, fine; if not, do not become concerned. An organization cannot wait for 6 months to determine the compatibility index of a team. The team cannot spend an excessive amount of time arguing minutiae. Doing must accompany philosophizing. Asking the difficult questions requires developing answers. Differences that affect the use of specific technologies must be resolved.

Teams probably provide the best organizational approach, but—like any grouping that involves two or more people—some one individual must be accountable. Problems cannot be considered as "our" problems. Problems are solved only when the rose has been pinned on some one individual. The problem is either "yours" or "mine," but not ours. The "our problem" approach probably means no one's problem. Individual and team performance must be differentiated. Both are essential.

Much has been written about appraising people performance. Perhaps too much. Multipage appraisal forms provide little, if any, useful value. Appraisal processes would be more beneficial if human resource departments eliminated the annual appraisal ritual and substituted an appraisal based on project performance when it occurs and not on some arbitrary calendar. People appraisal must be more realistic. Inflation of individual performance ratings has followed the same disastrous policies that have infected all academic institutions. Inflating appraisals only continues to lower the standards of performance.

33.9.2 Projects

People and projects provide a base for measuring MOT—people for their individual and group contributions and projects as the base around which all work can be organized. Figure 33.7 shows some possible "project" subcategories. Project type includes product, process, technology, and management. Each of these four types can be small or large. They can be simple or complex. They can be routine or creative. In reality, projects fall on a continuum from small to large, from simple to complex, and from routine to creative. Within this classification a product project could be small, complex, and routine. It could be both large and simple, yet require high levels of creativity. A product project could be small, simple, and routine or large, complex, and creative. Managers must recognize the significance of these combinations because each must be managed with different types of managers as well as different management tools. A major project that deals with known technologies and known tools requires an approach different from that for a project whose success depends on creativity. The major critical issue: Understand the project requirements and assign the people who possess the necessary competencies.

This vast array of project configurations forces managers to view "people" with a

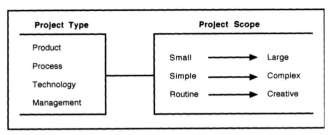

FIGURE 33.7   Simplified classification of project type and project scope.

new perspective. The category "individual" requires attention since the often misguided emphasis on teams as the new management paradigm tends to minimize individual contributions. No argument about the necessity for teams. Teams are essential, and certainly most technology related work depends on teamwork—teams are not a new discovery. But the individual cannot be subordinated to the team (group). Teams do not innovate. Individuals—people—innovate. Some one individual must develop an idea or a concept. Teams add value after an idea or concept has been communicated, and many different competencies and expertise must be introduced to add value and advance the original idea or concept to a timely conclusion. Teams often also become the greatest sources of lost time.

Project performance can be measured very easily. The project met the requirements, met the time schedule, and met the estimated budgetary costs or it did not—no rationalization—no excuses. But that measurement tells us nothing as to how the project was managed and whether any long-term benefit will prevail. Was anything useful learned in the process?

Project management demands integration, and as stated previously, integration is synonymous with MOT. Integration does not imply acquiescence or agreement; it may mean the exact opposite—dissonance and resolution of conflict. Not personal conflict, but conflict of ideas. Figure 33.2 illustrated the percentage of involvement by research, development, design, manufacturing, and marketing over the project life cycle. This relation is measurable. The functions and designated people were either involved or not involved, either contributed or did not contribute, throughout the project life cycle. No rationalization. No excuses.

Elements of the project management process can be measured, some quantitatively and others qualitatively. Measuring does not require a new bureaucracy. The information is available. Consider the following measurements:

1. Level of integration, involvement, and participation by all functions during the project life cycle.
2. Quality of a written statement of purpose. Is there a documented statement of purpose, and is it based on reality?
3. Written specifications that fully describe the project.
4. Identifying the known from the unknown—recognizing the importance of the unknown.
5. Doing the up-front work—conceptual design considerations prior to securing project funds.
6. Project process flow diagrams: Do they exist, and do they provide the details?

7. Evaluation of resources and infrastructure: Does a realistic evaluation exist?

8. Assignment of resources—not just people but all applicable resources: Do people have the required knowledge, experience, and skills, or are they learning the basics?

9. Problem-solving sessions—problem solving and integration or justification of actions and finger pointing.

10. Project reviews—detailed or cursory—informing or concealing.

These measurements are real, and most can be evaluated with a yes/no response. If a project team responded positively to each of these 10 requirements and failed in meeting their objectives, there was simply a lack of integrity among the participants during the project process.

33.10 SCALING THE PARAMETERS

Figure 33.8 shows relative values assigned to resources, infrastructure, and activities in which each is assigned a maximum value of one. The assigned numbers, relative importance, will vary among organizations and also among groups within an organization. MOT emphasizes people, support for innovation, and integration. The remaining must also be viewed as important. Even though facilities in the resource column is rated as .03, they must be available at some acceptable level. Organizational structure in the infrastructure column, while rated only at .05, also impacts the effective use of resources. Support staff and business activities in the activities column, although rated the lowest at .05, are nevertheless important for business success. Selecting these values requires judgment and a complete understanding of the firm and its business activities.

Management Process		
Resources	**Infrastructure**	**Activities**
People .25	Purposes .08	Business .05
Intellectual Property .10	Objectives .08	Products .10
Information .10	Strategies .08	Processes .10
Organizational .10	Organizational Structure .05	Information Systems .15
Attributes	Guiding Principles .05	Integration .35
Technology .10	Policies and Practices .05	Effectiveness and Efficiency .20
Time .10	Management Attitudes .08	Support Staff .05
Customers .06	Management Expertise .08	
Suppliers .06	Support for Innovation .20	
Plant and Equipment .05	Acceptance of Risk .15	
Facilities .03	Communication .10	
Financial .05		

FIGURE 33.8 Ranking the relative values of the elements of resources, infrastructure, and activities model using "one" as the unit of measure for each element.

Using the three-dimensional model illustrated in Fig. 33.6, in conjunction with the ranking of the relative values in Fig. 33.8, reveals more clearly the state of the business unit. Each of the elements in the management process must be evaluated and a value assigned for the resources, infrastructure, and activities. As was mentioned previously, utopia does not exist. Do not expect to reach the optimum of 1,1,1 on the x,y,z

axes of Fig. 33.6. Managing by the MOT principles requires at least recognizing the amount of space occupied within the cube. This type of evaluation may cause managers to disengage themselves from such a process. The results force a realistic look at the situation and the conclusions demonstrate that although the bottom-line figures may have reached the expected targets, greater opportunities exist for optimizing the investments from technology.

Consider the following conditions if resources, infrastructure, and activities are rated equally at

- 0.5, the overall rating is 12.5 percent: 87.5 percent of the cube is empty
- 0.6, the overall rating is 21.6 percent: 78.4 percent of the cube is empty
- 0.8, the overall rating is 51.2 percent: 48.8 percent of the cube is empty
- 0.9, the overall rating is 72.9 percent: 27.1 percent of the cube is empty
- 0.95, the overall rating is 85.7 percent: 14.3 percent of the cube is empty

The three-dimensional model provides a more realistic approach. What appears as 25 percent effort in a two-dimensional model translates into 12.5 percent in the three-dimensional model. Reaching a 1,1,1 solution may not be possible. Like all other measurements, the objective focuses on improvement.

33.11   AN APPROACH TO MEASURING MOT PERFORMANCE

The objective of this chapter was and is to raise the issues that might lead to some measures that provide meaning in measuring MOT performance. The preliminary discussion has provided some fundamentals. The conclusion is that the "process approach" may provide some insight. Measurement of any kind provides many dilemmas that generally lead to questionable and inconsistent numerics. Those numbers are also subject to misinterpretation. They often lack integrity because of some underlying agenda. They often attempt to quantify what cannot be quantified. Evaluation, assessment, and subsequent measurement derive from judgment and the environment in which that judgment is made. A judgment made in a profit year may be quite different from a judgment, using the same set of information, made under the influence of negative results.

Effective management of technology depends on resources, infrastructure, and activities. This trilogy of issues must be satisfied at some minimum acceptable level. The process begins with an evaluation of all the elements suggested in Fig. 33.8. The evaluation involves people and projects. People represent only 1 resource of 11, but people determine the value-adding power of the other 10. At the same time, people define the organizational infrastructure. People also determine the success level for activities, the third member of this trilogy. In the final analysis, MOT performance can be reduced to measuring people performance, project performance, and the level of integration of people in project activities.

Implementing MOT principles requires making a decision as to the starting point. As shown in Fig. 33.1, phase 3 is the starting point for MOT. Phase 3 begins the concept to commercialization approach without including the staff functions. While integrating phases 1 and 2 is essential to varying levels in all organizations, these phases do not include the necessary functions for MOT. Phase 3 begins by integrating research, development, design, manufacturing, marketing, sales, physical distribution, and customer service.

The proposed approach for measuring MOT performance focuses attention on Fig. 33.5, the MOT process. The measurement process begins with understanding input 1—the known, the controlled, and the predictable information.

Input 1. Identify, define, and understand input 1. Input 1 provides the management process with the known, the controlled, and the predictable inputs. In the mechanical process model these inputs would include materials and their characteristics, the processes used, the process parameters that must be measured and controlled, and the allowable tolerances. In the MOT application of the process model, input 1 provides the known, the controlled, and the predictable parameters about the resources and the infrastructure.

Input 2. Identify, define, and understand the unknown, the uncontrolled, and the unpredictable elements of input 2. In the mechanical process these "un-"s include such characteristics as out of specification materials, uncontrolled process fluctuations for many reasons, unpredictable environmental conditions, and unanticipated process interactions. How can the "un-"s be quantified? Anticipation, speculation, breadth of knowledge, projection of data, observation, statistical tools, and similar efforts present possible approaches. Decisions can be made with assigned allowable percentages of error and followed by if/then questions and responses.

This measurement process may involve many projects: in some situations more than 1000 projects. Those projects will vary in scope and size, as noted in this chapter. The output is merely the sum of the individual projects. But not every project must be measured. Priorities must be established not on the basis of size but on the basis of impact on business performance. In a multi-billion-dollar company a million-dollar project may severely impact future performance.

Similar factors apply to the MOT process model. For example, loss of critical talent can be predicted and followed with a contingency plan, and industry and business performance as a function of economic conditions can be predicted within acceptable limits. Competitive pressures can be predicted. While major introduction of a new technology may not be predictable, it is possible through knowledge and observation to project, anticipate, or predict future directions. Such speculation may not provide definitive answers, but the deliberate speculation forces gaining knowledge about the possible. Such assessments fall into the if/then category. Was the Japanese entrance and domination of the automotive and electronic industries predictable? Definitely, yes. Was the decline of companies like IBM, Kodak, and other well-respected corporations predictable? Definitely, yes. It is only necessary to look at the hiring practices and the "responsible for" philosophy that pushed their managements down the wrong path.

33.11.1 Management Process

Inputs 1 and 2 impact the management process in Fig. 33.5, which includes the resources, infrastructure, and activities. Your knowledge of the relation of this trilogy shown in Fig. 33.6 determines the expected output. This management process node is the organization. Figure 33.8 shows the relative ranking of the resources, infrastructure, and activities. The vector relating these elements of the trilogy shows the relative percent output of the trilogy. The numbers allocated to the elements of Fig. 33.8 may or may not meet your organization's needs. If so, create your own starting point. Then compare the information of the as-is study with the information of Fig. 33.8 or your own ranking of importance. Your numbers will obviously by qualitative.

Do not discount qualitative information. Information about people, as an example, cannot be quantified—it is always qualitative. Information about people is always subject to some form of bias. Numerics assigned to qualitative information must be viewed with some skepticism when high levels of accuracy are suggested. Qualitative measurement is not a seat-of-the-pants process. People present the greatest problem. Qualitative measurements regarding people must be based on some guidelines. What do you mean by competence, breadth of knowledge, meeting objectives, individual contributor, team member, creative, innovative, and so on? What do you mean by people person, productive, reactive, proactive, and so on? These responses cannot occur with the shrug of the shoulder that says this person is a great performer. At the same time it does not take hours to make an evaluation.

33.11.2 Expected Output

This expected output results from input 1 and input 2 acting on the "management process." Input 3 describes all the information that defines the business requirements. Just as the manufacturing process requires a statement of the requirements, our application of the process approach to MOT requires a clear statement of the expectations from the resources, the infrastructure that affects the utilization and output of the resources, and the type of activity to which the resources and the infrastructure are applied. Input 4 provides a means for measuring the expected output; input 3 and input 4 must be compared. In a manufacturing process this comparison is made automatically without human intervention. In our management process MOT model this comparison is made through human intervention. At some time in the future it may be made through some form of expert system using the principles of fuzzy logic. This comparison identifies the differences between inputs 3 and 4 that must be satisfied by some means—the control system.

33.11.3 Control System

While the control system in the manufacturing example comprises various combinations of electronic, pneumatic, hydraulic, and mechanical actuating devices, the control system for the "management process" involves people. People—individuals and teams—provide the means for control. The control system for the mechanical example requires a set of equations that links all the variables and their interactions and tolerance limits. When hundreds of variables are involved with wide ranges in tolerance, the equations (algorithm) become very complex. Keep in mind that this is a mechanical system in which input 1 can be clearly defined and even input 2 (the "un-"s) can be defined within the required operational limits. If those limits are exceeded, the system can be shut down in a systematic manner.

The control system for the MOT process resides in people. That control system must link the 11 resources, the 11 elements of the infrastructure, and the 7 major types of activities. People as individuals and in teams must decide what parameters must be modified, how those parameters can be modified, and the possible impact of the modified parameters on the system. There is no perfect answer, but there are possibilities from which the best choice can be selected. As mentioned previously no expert system exists to aid in this process. But developing a baseline does not require inordinate talent. It only requires a very high level of integrity. The evaluation of the resources and infrastructure elements must be realistic. Everything will not be a 10 on a scale of 1 to 10.

33.11.4 Controlling the Process

Recycle, recycle, recycle, and recycle. The mechanical example does not make the process corrections in one single adjustment. The adjustment is continuous and is made in steps. A machine cannot reach its optimum performance in zero time, and neither can a people-dominated process. An electric motor that is rated at 1800 revolutions per minute (r/min) does not reach 1800 r/min at the time (instantly—zero time) it is activated. The same applies to the MOT model. Many iterations will take place in order to bring the system in control. Some of those iterations may even deteriorate the process before gaining a benefit.

33.11.5 Developing the "As Is" Situational Analysis

Developing an as-is situational analysis is fraught with many problems. Third-decimal-point accuracy provides no benefit and consumes time that results in volumes of reports but not much usable information. People resources provide the greatest challenge. Evaluation of the people resources involves understanding the usable competencies. The fact that an individual made a major contribution in the past does not predict the future. What research has been done does not provide any guidance. The Leonardo DaVincis, the Edisons, and other prolific creative people represent a discontinuity in human behavior. How many are there? The majority of people devote and dedicate their talent and energy to resolving problems and in the process creating new knowledge. So this appraisal of human capability must be realistic. Evaluating the other resources can be performed by known methods and once again without time-consuming teams that focus on the minutiae.

Evaluating the infrastructure requires a look at past history. Most of the responses to the infrastructure elements are almost yes/no-type decisions. At most they require categorization in a format such as

- Do not meet minimal requirements
- Need major improvement
- Marginal but acceptable
- Meet requirements
- Exceed requirements

33.11.6 Measuring Integration

The MOT process model and the three-dimensional model determine levels of integration. The process involves integrating the three-dimensional model that synthesizes resources, infrastructure, and activities within the MOT process model. Gaining an understanding of the inputs, confirming the expected output, specifying the business requirements, and then managing those interrelationships in a holistic manner provides a workable approach. This may appear to be an insurmountable task, but most of the information is available. As stated, there is no need for third-decimal-place accuracy. Pareto's 20/80 rule applies—20 percent of the information provides 80 percent of the information. A word of caution: Make sure you focus on the 80 percent of the information that is important for your particular situation.

33.12 QUESTIONS THAT MUST BE ANSWERED

Raising questions provides a means for evaluating. This section considers each of the elements of the management process as a prompt for gaining insight into the state of the resources, infrastructure, and activities and their linkage and interdependence.

33.12.1 Resources

People

1. Do you know your people—their skills, their competencies, their interpersonal qualifications? Do you understand the interpersonal characteristics of your people in order to optimize the output of the system?
2. Are people qualifications documented in such a way to provide input for project assignment?
3. Does the classification show up-to-date skills and competencies—specific rather than general? What is the basis for your conclusions? What do you do when skills are no longer required?
4. On a scale of 1 to 10, how would you rate the performance of every employee? Can you be realistic in this appraisal? What is your approach to appraisal—once a year panic button or continuous appraisal?
5. Do you have a system for meeting individual requirements? Are the doors open for easy access? Is there sufficient flexibility to take into account individual differences?
6. Management by objectives has had minimal success. Do you know why it has failed? Do you use it? If you do practice management by objectives, do you consider yourself as part of the objective? Is it a paper mill or a management tool?
7. What attempts are made to focus the organization? Are objectives clearly delineated? Do people understand their role in the system?
8. What opportunities are provided for discussing or presenting unsolicited proposals? Is freedom of expression encouraged? Are people asked to raise unpopular issues? In project reviews, what is the policy on telling all? Do you tend to shoot the messenger?
9. Specialization is important, but has it gone too far? What programs allow for increasing breadth of knowledge and experience? What programs promote interfunctional and interdisciplinary understanding?
10. What do your people know about the company, the management, the immediate department, and so on? Purposes, objectives, and strategies form the basis of an organization: Are these clearly communicated to people at all levels in meaningful terms so that they can respond accordingly?

Intellectual Property

1. Does the organization understand the meaning of *intellectual property?* Is that understanding comprehensive, or is it limited to the technical functions? Has the organization defined what intellectual property includes?
2. Are processes available for recording, saving, and protecting the intellectual property of the organization? Are those processes used and monitored? Are those processes reviewed periodically?

3. How is the intellectual property managed? Has the company attempted to instruct about intellectual property?

4. Does the organization record use of intellectual property and publish internally how that intellectual property was used to the benefit of the organization?

5. Documentation of all types is an asset of the organization. In functions such as research, development, design, manufacturing, and marketing, what type of documentation is used? Are research notebooks a part of the system? Are design drawings updated when changes are made?

6. Document security classification presents special problems. Is there a realistic document classification process? Is that process managed? If a policy exists not to patent certain processes, how is that policy enforced?

Information

1. Does the organization differentiate between data, information, and knowledge? What is the basic philosophy regarding information? Does the process result in reams of meaningless data? Is information provided in such detail that it loses its significance?

2. Managing information plays a major role in business performance. What is the competence level of this group of professionals? Can they discriminate between what is essential and what is wanted? Do these professionals bring the users into the process? What is the value added of information in relation to its acquisition cost?

3. Most surveys show that organizations seldom provide sufficient information. Do people have sufficient information to perform their jobs? Is that information used? Does the availability or nonavailability affect job performance?

4. In a technology-oriented organization, technology transfer presents a major challenge. How is technology transferred in your organization? Is there a formal procedure? Are the necessary contributors engaged in the early stages of a project? Do they remain connected to the project to completion?

5. Feedback is a major factor in business processes. This was demonstrated in Fig. 33.4. How is feedback communicated? How is feedback used? Are the results of feedback tracked and communicated when necessary? Is the feedback timely so that it can be used effectively?

6. There is no doubt that decision processes require information. But how is information extracted from that mass of data? Does the accuracy fit the need, or is third-decimal-place accuracy provided when plus or minus 10 percent is sufficient? Achieving that accuracy may cost valuable time. How do you manage the price of information and its accuracy? Are you even aware of the cost of information?

7. Have you bought into the concept that information is a competitive weapon? Has it made you competitive? Has it allowed you to overtake your competitors? How long did the competitive advantage last? Were you able to isolate information as a competitive weapon? Did the information system add value at any point in the concept-to-commercialization process?

8. What is the organization's policy relative to information management? Is information measured by gross weight or timely contribution to decision processes? Is it used to promote understanding? Is information used to guide the future activities? Or is information part of the organization's historical society? Is the information relevant, or does it obscure the important? Is acquiring more information used as a delaying tactic? Is it used to obscure reality?

9. Information is of value only if it is timely and accurate within specific needs. Is information tailor made to the needs of the group? Must manufacturing wait for 15 days after the close of the month to know its costs? Is the right kind of information delivered to the factory floor, research laboratory, marketing and sales, and other functions? How about the feedback of information in the reverse direction and across functions?

10. Benchmarking may be an overstated concept, and it certainly is not a competitive weapon, but is the information system and its transfer function benchmarked against some standard? This does not necessarily mean against the best. Is it benchmarked at least against an internal standard? Is that standard clearly defined? Is the benchmark suitable for your organization? Do you think it is necessary to benchmark against some external standard?

Organizational Attributes

1. Acceptance of change continues as a major business issue. How does the organization manage change processes? Is sufficient in-depth thinking included to determine the consequences of the change ("in-depth thinking" does not mean excessive time for decision)? Are the precursors for implementing changes given due consideration?

2. What approaches are used for implementing change? The peace-and-quiet approach or the cognitive dissonance approach? What are the potential sources of change? How is diversity in thinking managed? Are the concepts of complete buyin hindering progress? How does the organization educate participants in the change process?

3. Openness and the open-door policy are good public relations, but are the principles practiced? What is the attitude of the organization for stimulating this openness among all levels of participants? Classification of information is critical, but does it prevent people from understanding the objectives? Is business-unit strategy communicated effectively at all levels?

4. The call for freedom of action sounds good, but do employees understand the implications of that freedom? Are responsibility and accountability part of the freedom equation? Is freedom of thought encouraged? Are the free thinkers allowed to voice their ideas and concepts? Have you established the guidelines for that freedom? What are your expectations from the person who presents diverse ideas? Do you encourage thought before action?

5. Organizations spend large sums of money on what is loosely termed "education and training." How much is education? How much is training? Is there any support for intellectual pursuit—intellectual pursuit in all matters? Does that intellectual pursuit extend to executive literacy? Does that support include understanding the basics? Is that support limited to what is generally referred to as the "tools of the trade"?

6. People are a major contributor to organizational attributes—culture is not dictated from the top—it can be stimulated from the top but develops from the bottom up. Are the differences that people bring to the table encouraged? Are those differences reinforced? Are they seriously considered? Are they allowed to emerge without recrimination?

7. Organizational commitment and loyalty, according to some researchers, is on the wane. Was it ever there in the first place—loyalty to the corporation and from the corporation, and what were the sources of the loyalty? If so, for what percentage of the people? Do people generally do what they say they will do? Are the sights

set high enough? Does the word "excellence" imply 98 out of 100, or has that term been inflated as in our academic institutions?

8. Why is it that a sense of excitement can be perceived by visitors in some groups and not in others? Which factors determine the spirit that can coalesce around resolving a problem or taking advantage of an opportunity? Do managers expect 8:00-to-5:00 performance and think they can develop a creative and challenging community?

9. Integration is not a choice; it is an imperative. How do different levels of integration impact the group's performance? Do people understand what is meant by "integration"? Is integration ever discussed? Does anyone ever consider the simple mathematical approaches to integration (remember integral calculus)?

10. Does the organization support innovation and entrepreneurship? Is this demonstrated by management's actions? Can you describe where management recently supported an innovation or entrepreneurial program? What was the payoff of such efforts not only on the bottom line but also in excitement and human performance?

Technology

1. How is technology linked to business strategy? Is technology considered in the strategy development process? Is there a technology strategy? Does that strategy direct the entrance into new technologies? Does that strategy tend to focus on technology platforms (technologies that can be applied across a broad range of products or processes)? Does the technology strategy take into account the limitations of other resources and the infrastructure? Is any consideration given to developing technology through a virtual organization concept?

2. Is the importance of technology understood at all levels beginning with the board of directors? Do any board members have the competency to make decisions that involve technology? Does management know at what level technology decisions are being made? Are there multiple levels of management approval without understanding the significance of investing in technology?

3. Is technology a technology issue or a major business issue? Does the organization differentiate between technology and business decisions? What percentage of the sales value of production involves technology? Does management take an integrated approach to technology?

4. What is the role of technology in the organization? Are technologies classified relative to their importance? How are new technologies implemented? Is there any technology scouting activity? Who tracks the latest technology directions? How is the latest information transmitted to those who have a need to know? Are technologies limited to those in research, development, design, and manufacturing, or do they encompass all technologies?

5. How are investments in technology justified? Are investments in technology justified, or are those investments evaluated? Is the investment in technology an expense or an investment? Does the justification of the investment include the rationalization of competitive technologies? Is the process systems-based?

6. Technologies are associated with products and processes. Are products and processes considered as one activity or as two separate and distinct activities independent of each other? Is sufficient consideration given to situations that involve new technologies in new products? How about new technologies in new products in existing markets? How about new products with new technologies in new markets? Are these relations understood?

7. Technologies come in many different types and sizes. Does the firm's technology allow building a technology platform? Does that platform provide synergy with other existing technologies? What is the value of such a platform for the future? Does management understand the benefits of building technology platforms?

8. Has the organization defined its critical technologies? Has it defined its technology base in meaningful terms? Are technologies classified according to any designated system? Do those definitions provide direction in technology development? Is the appraisal of technologies performed effectively? Does that appraisal provide a benefit?

9. What is management's approach to technology? Inactive? Reactive? Active? Proactive? Does management push not for the latest technology but for the most adaptive technology for the short and long terms? What percent of general management understands the related technologies? What percentage of research, development, design, and manufacturing managers understand their technologies? Are they capable of making intelligent technology decisions? What have they done in the past to keep current in the important related technologies (not as experts but from the breadth of understanding)?

10. Does the organization benchmark its technologies? If so, against whom? Does management consider benchmarking an important issue? If you are at the leading edge, do you need to benchmark? Does management understand what it takes to do intelligent benchmarking?

Time

1. Does management differentiate between cycle time, total time, and timing? Is cycle time being emphasized? Is it important in your business and your particular industry? Has your organization quantified the impact of cycle time on business performance? Are the cycle time data relevant? What does the data tell you? Are you comparing apples and apples?

2. Has your organization developed a program to focus attention on the negative impact of extended cycle time? Does your top management consider it necessary to focus attention on cycle time? If so, what has it done to focus attention? Is it action or rhetoric? Are the fundamentals understood? Do people understand the ramifications of cycle time management? Does it cost more or less?

3. Do the business unit managers accept cycle time management as an important issue? Do the functional managers accept it as an important issue? Have they done anything to reduce cycle time in real terms? If they made significant strides in cycle time management, how were the results quantified? Was it possible to isolate the benefits from cycle time management?

4. Does your organization provide any special education regarding the issues related to managing cycle time? For executives? For managers at all levels and in all disciplines or functions? For the professional staff? For administrative personnel? For production workers? If education in cycle time management is a priority, can the benefits be quantified?

5. Extended cycle time represents an added cost to the organization. Have the following issues extended cycle time? Undefined business strategy? Undefined marketing strategy? Undefined technology strategy? Insufficient integration of research, development, design, manufacturing, and marketing? Overspecialization in all professional areas? Segmentation of the organization?

6. Have the following issues extended cycle time? Lack of clearly defined and communicated objectives? Lack of integrated business planning? Lack of integrated project plans? A nonsupportive attitude by management? Inability of management to think of the business as a system? Unresolved differences in business priorities?

7. Have you identified where the major cycle time losses occur in your organization? Research-development interface? Development-design interface? Research-manufacturing interface? Research-marketing interface? Linkage between research, development, and design? Information transfer throughout the whole organization? Indecision on the part of the originator? System-imposed lost time? Boss-imposed lost time?

8. Can you identify the cycle time losses due to lack of literacy by executives and managers about technology and cycle time? Literacy related to the business by all employees? Literacy about marketing? Literacy about technology? Literacy about integration? Literacy about what makes the organization function? Managing by wandering around is a great idea but what happens during that wandering around is far more important.

9. What is your organization's attitude about managing cycle time? Is it accepted as part of a requirement for effective and efficient management? Does it require specific top–down programs directed toward improving cycle time? Does cycle time management take care of itself through normal management processes? Is cycle time management an important factor in the current competitive climate?

10. Does cycle time management relate principally to the functional groups and those involved in products and processes? Does it permeate the total organization? Does it also include total time and timing? What is your organization's philosophy regarding cycle time? Can you list any specific situations where managing cycle time provided a benefit to your organization? What factors affect cycle time in your organization (positive and negative)? What are the consequences of long cycle time for your business?

Customers

1. Are customers your partners? How do you define "partners"? Are those partnerships real or imaginary? How many customers do you have that fall in the class of partners? How many can you rely on? How many understand your products and your business?

2. Disregarding the hype in the business press, are you trying to work with your customers? Have you become so attached to voice mail that there is no way to contact you directly without spending time listening to babble about "press 1," "press 2," and so on? What kind of service do you provide your customers? Is it on time? Does it meet the customer's requirements?

3. Customers are often used as beta sites. Do you work with customers in the early stages of product development to ascertain the value of selected new features? What is your confidence level in the feedback that you receive? What are the risks in using customers as beta sites? Do you plan these beta sites in such a way as to avoid disclosing your proposed products to the competition?

4. Have you taken the time to critically evaluate the information that customers are providing? Do they understand the basics of your products? Is their system for providing you feedback verifiable? Do you try to corroborate their information with other sources of expertise?

5. Information from your customers is only as good as their integrity. What are the possibilities of customers leading you down the wrong track? Are you placing too much confidence in customer feedback? Do they provide you with a wish list without recognizing what that wish list may cost? Have you made a decision that a specific customer can contribute to your effort?

6. Customer input is beneficial, but have you considered the geographic differences that might influence product design? How about the demographics? What is the difference in input based on the organization's culture? What about market differences? What about the ability of the customer to use the new technologies? Are they a mystery, or are they usable and provide an added value?

7. Do you fully understand the customer's use of your product? Have you taken the time to determine whether the product is used according to specifications? Have you taken the time to work with the customer to optimize the output of the equipment or service? Do you know the competence level of the people using the product? Are you capable of determining the competence level of the user?

8. How do you use customer input? How broadly do you survey your customers? Is one concern or complaint sufficient reason to modify designs? Do you systematically track field problems and the circumstances under which they occur so as to make appropriate decisions? Do you push the panic button every time a field problem surfaces?

9. Do customers want to buy oranges when you are selling apples? Do they want to buy Washington State red delicious apples when you are selling crab apples? Do you come back with orders for oranges when you have only apples to sell? These are not facetious questions. They are part of daily life—always being able to sell something that is not available. How do you manage such inputs from the field? Do you ignore them? Do you consider them seriously for future activities?

10. Is there a written policy relative to customers? Is that policy reinforced by management? How much flexibility is embodied in the policy to account for different situations? What is the policy and practice from time of order entry to time of delivery and customer satisfaction? Are customer's needs taken seriously and expedited, or does the bureaucracy prevail?

Suppliers

1. What is the policy regarding suppliers? Are suppliers respected as participants? Are estimates requested without consideration of cost of preparation? Do you accept estimates from vendors who are not financially responsible? Or is the financial determination made after estimates are received?

2. Do you qualify suppliers? How is that qualification performed? Does it include people competencies, facilities, processes, and other operational issues? Does this qualification take place in the office or on the shop floor? Does it involve hands-on assessment?

3. Who makes the assessment? Is this a purchasing function? Is it a team effort by people with the required competencies? Does it follow a strict question-and-answer process, or is there freedom to adjust the process as required?

4. Can you describe the working relations with suppliers? What is the past history? What has been done to improve those relations? Are the relations long-term? At what levels does the interchange take place? Is there a mutual support structure?

5. What does the organization do to transfer knowledge to the vendor's organization?

Do you support your suppliers with education and training? Is that education and training effective?

6. How much freedom do your suppliers have? Do you accept ideas from your suppliers? How do you manage the proprietary issues? Is the process formalized?

Plant and Equipment

1. At what level do plant and equipment meet current needs? When was the plant built? How old is the equipment? Is the plant and equipment state-of-the-art? Has plant and equipment been upgraded or replaced over the years?

2. Will the plant and equipment provide for the needs during the next 5 years? Should plant and equipment be upgraded or replaced in the future?

3. Do you benchmark your plant and equipment periodically? If you do, what has been the result during the last 5 years? If you do not benchmark, why not? Can you provide a good reason?

4. What has been the automation protocol? Is automation accepted as a means for improving productivity? Are management and employees aware of the implications of automation? If you have automated, has that automation been effective?

5. Environmental and safety issues are vital for many reasons. What is your accident record? What are you doing about it? What is your approach to environmental issues? Is there a continuous improvement program?

Facilities. *Facilities* include all the other physical resources that are required to operate the business effectively. Since these facilities requirements (buildings, computers, communication equipment, automobiles, etc.) vary considerably depending on the size and type of industry, we raise only one issue: Define those facilities and determine whether they are adequate—an emphasis on adequate and not luxurious.

Financial. The financial requirements vary considerably among organizations. Some operate with significant debt, and others operate from a strong cash position. The financial resources must be available in one form or another. It is important that organizations provide the financial resources.

33.12.2 Infrastructure

The business infrastructure determines the effective and efficient use of resources. Its importance cannot be overemphasized. Infrastructure begins with clearly delineated purposes (mission), objectives, and strategy followed by organizational structure, guiding principles, policies and practices, management attitudes, management expertise, support for innovation, acceptance of risk, and communication.

Purposes. The purpose of an organization cannot be limited to making money. Money, as profit from operations, is a result from effective use of resources. But the purpose must be carefully defined. The purposes of business units, of project teams, of functional groups, and any combination of people must operate within some defined context. It is important that purposes from the top to the bottom of the organization build on one another. The question that must be answered: Why does the organizational unit exist?

Objectives.   Objectives are a self-evident fact of life—not just in business but in our personal lives as well. There is no use in attempting to justify the need for defined objectives. But those objectives must focus on the purposes of the organization, and they must be clearly and concisely defined. Does your organization establish objectives at all levels beginning with the CEO and the board of directors? Those objectives must relate directly to the purposes of the organization. Where does your organization rate on a scale of 1 to 10?

Strategies.   Strategies are important in order to determine how objectives will be met. This does not mean strategic planning. *Strategy* points the direction as to how (the "how") some objective will be accomplished. But recognize that there are different kinds of strategy that impact performance. There is a corporate strategy, a business-unit strategy, a marketing strategy, a technology strategy, an innovation strategy, and so on. Corporate strategy, while important, may be the least important. It is macro strategy, but business operates at the micro level. What is the business unit's strategy in these multiple areas? Is technology part of each of these strategies?

Organizational Structure.   The importance of organizational structure is often overstated. Personal experience has demonstrated that if you have the right people, organizational structure is not that important; and if you do not have the right people, no organizational structure will make a difference. What is your organizational structure? How many coordinators do you have? Why do you need coordinators? They are redundant. They give you an out to rationalize the malpractice in management.

Guiding Principles.   Guiding principles impact performance. Guiding principles must be differentiated from policies and practices. Does the organization practice what it preaches? This may appear as simplistic psychology, but guiding principles start with the behavior patterns of top management. If integrity is a guiding principle, it cannot be subordinated when the situation becomes critical. It cannot be rationalized at the upper levels and then be expected to flourish at the operations levels.

Policies and Practices.   Policies and practices can be a blessing or a curse—too often a curse. They can provide guidance and at the same time can destroy initiative. Shelves of policies and practices remove all decision-making responsibility and delegate it to a set of impersonal books. If you have a problem, go to section so and so and look for the answer. Judgment has been eliminated.

Management Attitudes.   Attitude determines success. Attitude at the management level permeates the whole organization. Managers who keep one foot in the office and the other foot in a personal business, on multiple corporate boards, on presidential committees, or on the golf course or some unrelated activity send the wrong message to the organization. After all, executives and managers are employees of the corporation. They are not the owners. Excessive participation in extracurricular activity also sends the wrong message. Social responsibility is essential but in the proper context. What is the value-added component of such activity?

Management Expertise.   How do you evaluate your management's expertise? How many of them have mentally retired? Management expertise is a serious issue. How many executives and managers are technologically literate? How can they make decisions regarding investments in technology without some minimal understanding? Granted that they have faith in people and in their track record of past performance, but as technology becomes more complex, some minimum level of understanding is essential. Information systems would provide greater benefit today if CEOs had some knowledge of the monsters that were being built. Management expertise is important.

Support for Innovation. Innovation provides opportunities for growth. But innovation must be supported by management and rewarded. That support implies allowing and capitalizing on mistakes—what (the "what") was learned from the failure. Innovation does not come from people who work from 8:00 to 5:00. Innovation comes from the 5 to 10 percent of people in the organization. These are dedicated people who are driven to accomplishment usually by their own ideas. Innovation does not come about through strategic plans. Innovation comes from individual competence that searches for new ways of doing things.

Acceptance of Risk. Innovation involves risk, and acceptance of calculated risk can be traced to successful organizations. Accepting risk does not mean rolling the dice. It is common to talk about people who are risk takers. Fifty years of involvement in various aspects of technology have convinced me that the competent risk takers usually evaluate their probability of success. They probably are not serious risk takers. They know their business and did their homework. How much risk do your managers accept? How do they manage risk takers? Do they encourage risk taking?

Communication. It seems almost trite to talk about communication as we approach the end of the twentieth century. But communication is a very major issue. It involves one-on-one, small groups, and larger groups. It is also important to understand that different styles and types of communication are essential. Too often communication really is nothing but double-talk.

33.12.3 Activities

Activities are classified as business, product, process, information, integration, effectiveness and efficiency, and support-staff projects. Projects provide the means for defining objectives, assigning resources, and establishing the start and finish dates. If an organization would institute the project approach from the top, including the CEO, most of the non-value-adding activities would be eliminated.

Business. Business projects relate to activities such as managing the business, identifying and communicating a business strategy, planning the future, making acquisitions, providing the resources and the business, managing the finances, integrating the activities of business subgroups, guiding the operating and staff groups toward common objectives, and the general issues involved in providing for stakeholder interests.

Product. *Product projects* are probably the best examples of the project process. The term is self-explanatory and encompasses the range of projects described previously. The product project process is used primarily in its many forms by research, development, design, and manufacturing. Functions such as marketing, sales, and the staff functions such as patent and legal and purchasing must be included in this process.

Process. The designation *process* must go beyond the traditional use in manufacturing. Process has often been denigrated because of the "by the numbers" approach generally applied. This denigration began when strategic-planning gurus focused on the process. But process—and more particularly, delineation of the process and managing that process with substance—cannot be ignored. Processes provide direction. Some may be very rigid; others may provide various levels of flexibility. Delineation of processes must be expanded throughout the organization. Consider for a moment the benefits to be gained from defined processes in all the administrative and support functions—not the bureaucratic approach but a continuous effort to minimize the number of steps in the process. Process flow is important in every function. It must be documented but cannot be at the same time a deterrent to meeting objectives. It takes balance of flexibility and rigidity.

Information Systems.   Information, while not necessarily a competitive weapon, plays a major role in business performance. But much of the investment in information systems resources does not provide any value added to the user. Management information often operates in isolation and continues to provide more data rather than information. How often have you heard "I can't depend on those computer printouts"? The project approach focuses on specifying the benefits and justifying the investment in information systems according to the same rules that govern investments in other capital equipment—that means within specifications, on schedule, and at cost. When was the last time you heard of a new management information system that met the specifications, schedule, and cost estimates?

Integration.   The integration of MOT does not come about by espousing the cause. Integration requires disciplined decision making followed by providing the resources and the infrastructure. There are no 10 easy lessons on the road to integration of any type. Integration requires education, intelligence, understanding, tolerance, intellectual integrity, observation, discrimination, creativity, dedication, and a "think and do" approach. It needs management agreement and know-how.

Effectiveness and Efficiency.   Many projects focus on effectiveness, efficiency, and the economic use of resources. Projects of this type form the fundamentals of continuous improvement and must be implemented throughout the organization. They are the source for not only financial benefit to the organization but also keeping an organization challenged and healthy for sustained business performance. Measuring the performance of these projects can be easily quantified.

Support Staff.   Support-staff projects provide the greatest challenge to management. Organizations seldom require the same analysis and justification for these activities that they require for the more tangible projects. Is there any reason why these activities should be treated differently? I propose that organizations apply the project process to all these activities and eliminate the "responsible for" type of mentality. Think how differently the introduction of a new personal appraisal system, by human resources, might be considered if the project would be bound by the same guidelines as other investments. Think how differently financial data would be collected and communicated if such activities were developed into projects and treated accordingly. Think how differently a purchasing department might function if activities were divided by projects.

33.13 SUMMARY

This chapter can best be summarized by Fig. 33.9, which links the "what," the "why," the "how," the "who," the "when," and the "where" of MOT. It is self-explanatory. Managing MOT provides benefits even though no specific theoretical proof is provid-

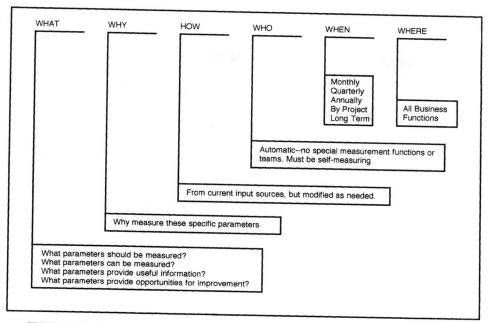

FIGURE 33.9   Inputs required for developing a model for measuring MOT performance.

ed—it is self-evident and does not require research to demonstrate its practicality. If you doubt this statement, observe how people interact and work as individuals and teams. As stated earlier, my purpose was to raise some issues, not provide a definitive solution. The process model with the three-dimensional model linking resources, infrastructure, and activities may be a starting point.

33.14   REFERENCES

1. Philip B. Crosby, *Quality Is Free: The Art of Making Quality Certain,* McGraw-Hill, New York, 1979.
2. G. H. Gaynor, *Achieving the Competitive Edge through Integrated Technology Management,* McGraw-Hill, New York, 1991.

P · A · R · T · 6

CASE HISTORIES AND STUDIES

CHAPTER 34
NUTRASWEET®—*EUREKA!*

Gary S. Lynn
Associate Professor
Management and Engineering Management
Stevens Institute of Technology
Hoboken, New Jersey

34.1 COMPANY BACKGROUND

G. D. Searle and Company was started by Gideon Daniel Searle in 1888 when he owned a single drugstore in Fortville, Indiana (Ref. 1, p. 20). The company invested in new pharmaceutical drugs, and the research paid off with several significant new drugs in the 1920s and 1930s. Searle continued to grow and prosper, and in 1950 the company issued an initial public stock offering. In 1960 for the first time G. D. Searle and Company made the *Fortune* 500 list. A transition in management occurred in 1966 when John Searle was replaced by his son Dan as president and John was appointed CEO. During this time, a corporate mission was being solidified:[2]

> Daily we must be patient with the demands of science and impatient in the struggle to make the product of our technology available to mankind at the earliest possible moment.

After Dan assumed control of Searle, the company underwent a transition. In 1967, the Searle company was reorganized into autonomous operating divisions. From 1967 to 1968 there were 46 major managerial changes in the United States alone.[3] By 1969 the strategic focus of the company had changed, and was expanded to an organization that "can pursue virtually any promising idea to a conclusion far beyond the scope of yesteryear."[4] Searle embarked on a diversification effort that included acquiring: an optical business, Pearle Vision, a retail chain of eyeglass studios; Curtiss Breeding, a company that was involved with artificial insemination of dairy and beef cows; Medidata Sciences, a computer company that offered systems for use in multiphasic clinics; Nuclear-Chicago, a firm that developed peaceful applications of nuclear radiation; and several international businesses. Searle's sales reached $163.9 million in 1969—up from $36.5 million in 1958, and net earnings escalated to $28.5 million from $7.4 million in the same period. See Table 34.1 for a detailed sales and net income breakdown. Table 34.1 shows sales and net income from 1958 to 1969.

Searle's effort to diversify, coupled with a new-product pipeline that was drying up, helped pave the way for the development of aspartame.

TABLE 34.1   Searle's Sales and Net Income Summary

Year	Net sales(thousands)	Net income (thousands)
1958	36,518	7,414
1959	41,004	7,865
1960	45,339	8,047
1961	55,677	10,881
1962	69,808	14,629
1963	86,134	19,278
1964	104,293	25,591
1965	108,314	24,051
1966	113,465	22,854
1967	132,707	26,625
1968	147,724	27,366
1969	163,936	28,491

Source: G. D. Searle *Annual Reports,* 1968, 1969.[3,4]

34.2   THE ASPARTAME OPPORTUNITY

34.2.1   The Artificial-Sweetener Market

The artificial-sweetener market was born in 1953, with the introduction of cyclamate-sweetened No-Cal cola by the No-Cal Corporation, when the diet industry was born.[5] Other companies followed No-Cal's lead: Royal Crown Cola Company introduced Diet Rite Cola in 1962, Coca Cola entered the market with Tab in 1963, and Pepsi Cola launched Diet Pepsi in 1964 (Ref. 1, p. 82). The popularity of these products increased, and by 1969 soft drinks sweetened with artificial sweeteners (cyclamate and saccharin) were found in over 70 percent of American homes.[6] The major market for artificial sweeteners—or "synthetic sweeteners," as they were known in the early 1960s—was primarily in one of two areas: tabletop sweeteners and diet carbonated soft drinks. Approximately 60 to 70 percent of cyclamates and close to 20 percent of the saccharin consumed in the late 1960s were used in the carbonated soft-drink market.[6,7]

Demand for artificial sweeteners, primarily cyclamate and saccharin, more than tripled from 1962 to 1966.[7] Cyclamate was the preferred artificial sweetener because it did not suffer from the bitter aftertaste that plagued saccharin. Du Pont held the patent for cyclamate and licensed it to Abbott Laboratories. Abbott introduced Sucaryl, a cyclamate-based tabletop sweetener, and after the company redesigned the packaging for the product (to be more of a mass-marketed consumer product and less medicinelike), sales exploded. Potential seemed almost unlimited.[8] Du Pont's patent on cyclamate expired in 1959, and with the expiration competition quickly entered. Other companies wanted to offer an artificial sweetener because of the high profit margins. (See Table 34.2 for market size.) The preferred artificial-sweetener mixture consisted of 10 parts of cyclamate to one part of saccharin, and a 13-oz package of artificial sweetener retailed for 37¢, a 32-oz package sold for 88¢, while a 5-lb bag of sugar sold for 56¢.[7,9] Market size for cyclamate and saccharin is shown in Table 34.2.

By the late 1960s, many companies were competing in the artificial-sweetener market. Adolph Food Product's sugar substitute (a tabletop sweetener made with glycine, saccharin, and gum arabic) had achieved a 25-fold increase in the consumer market and 10-fold increase in industrial markets. The future looked bright for the entire artificial-sweetener category, and in 1967 growth rates were estimated at 7 to 10

TABLE 34.2   Market Size: Cyclamate and Saccharin[6,7,10,11]

Year	Volume (in millions of pounds per year)	Dollar sales (in millions)
	Cyclamate	
1963	5	$2.5
1966	12	$6
1969	18–19	$15
	Saccharin	
1966	3.5	$5
1969	4.5–5.0	$6.5

percent per year for saccharin and 6 to 10 percent for cyclamates.[7] But then everything changed. In 1969 concerns arose regarding the safety of these two compounds which created a window of opportunity for a new artificial sugar substitute.

34.2.2   Searle and Aspartame

In 1965 three groups of scientists in the chemistry division at Searle were involved with identifying molecules which facilitated or interfered with human body chemistry. Other groups in the company would then try to figure out what to do with the substances. James Schlatter, a research chemist, was part of the chemistry group, and his responsibility was to make compounds for Robert Mazur, a senior chemist, who was using dipeptides (a particular chemical compound structure) in ulcer treatment. One of the projects in the chemical research lab was to find an inhibitor of the gastrointestinal secretory hormone, gastrin, as a possible treatment for ulcers. The chemical compound, aspartame, was an intermediate in the synthesis of the gastrin. In December 1965, James Schlatter, while working on the ulcer treatment, noticed something unusual (Ref. 12, p. 4):

> In December 1965, I was working with Dr. Mazur on the synthesis of the C-terminal tetrapeptide of gastrin. I was heating the aspartame in a flask with methanol when the mixture bumped onto the outside of the flask. As a result, some of the powder got onto my fingers. At a slightly later stage, when licking my finger to pick up a piece of paper, I noticed a very strong, sweet taste. Initially, I thought that I must have still had some sugar on my hands from earlier in the day. However, I quickly realized this could not be so, since I had washed my hands in the meantime. I, therefore, traced the powder on my hands back to the container into which I had placed the crystallized aspartyl-phenylalanine methyl ester. I felt that this dipeptide ester was not likely to be toxic and I therefore tasted a little of it and found that it was the substance which I had previously tasted on my finger.

Searle had discovered something; now the question was what to do with it.

The discovery of the sweet taste of L-aspartyL-L-phenylalanine methyl ester, also known as *aspartame* (or NutraSweet), was accidental and surprising. It was accidental because the researchers at Searle were not looking to find what they found. They were looking for a new ulcer treatment. It was surprising because the sweetness could not have been predicted from the taste of the constituent amino acids that made up aspartame; aspartic acid is tasteless and phenylalanine is bitter (Ref. 12, p. 4). The taste of aspartame was remarkably similar to regular sugar—without an aftertaste. Accidental detection was not unfamiliar to substances found to be sweet. Most previous chemical compounds that were found to have a sweet taste were accidentally discovered (Ref. 12, p. 4). Aspartame, for a still unknown reason, binds to the sweetener receptors of the tongue to produce a sweet sensation.

Ironically, aspartame was developed prior to Searle's discovery of its sweet taste by Imperial Chemical, a British company, in the early 1960s. Imperial did not seek to patent the compound because they thought it did not have any significant commercial or scientific merit. After the researchers at Searle discovered it was sweet and the results of their efforts were reported in the industry, the scientists at Imperial tasted the compound and were amazed that it, in fact, did taste sweet.[13]

Searle began to explore aspartame purely because it was scientifically interesting and not because it was perceived to have commercial potential.[13] Mazur secured funding for the project because he felt he should do so in the interest of science. As Mazur explained,[13]

> This was a novel discovery and this structure had not been known before to be sweet. We were certain what we discovered was really new. And what you do in a situation like that, if you have the time and people, is to synthesize compounds which are closely related to the original substance and see what it does to the taste. We did this for about four years. What we were exploring is called structure activity relationships which means if you change the chemical structure what does it do to the activity—in this case "activity" was taste.
>
> We were chemists and our interest was in research as opposed to development—*definitely* as opposed to production and marketing. I decided that this was really important; not commercially important—it was a really important scientific discovery, and I decided that this is something we should work on. It was strictly science, strictly curiosity and chemical research. I really and truly never ever had one thought about market size. Also there was no history that would have encouraged us to look at the marketing aspect [of aspartame], because Searle was a pharmaceutical company. They'd never even seen a food additive; they didn't have the vaguest idea what to do with a food additive—they were not interested in food additives—they were interested in drugs.
>
> I felt we should stop what we were doing, which was looking for an ulcer drug, like ICI [Imperial Chemical], and explore the sweet-taste aspect of this series [of compounds]. And that is what we did. I made that decision within a month of Jim's discovery, because it was so obvious [that the discovery was novel].

Searle continued to investigate aspartame, and the company filed a patent on the compound on April 18, 1966 (issued Jan. 27, 1970,—U.S. patent 3,492,131). The patent reflected the company's emphasis on the research of the product and not on the development or market applications. The patent covered different compounds and formulations that could produce results similar to those of aspartame.

Searle spent a considerable sum of money in the early research of aspartame. From 1966 to 1968, there were approximately two researchers working on the project, and in 1969 there were four people. This translated to a cost, in labor and laboratory overhead, of approximately one million dollars spent on aspartame research before any commercial interest was established.[13] The aspartame research was conducted as a course of normal business and not on a "bootleg" basis—Mazur was adamant about this point.[13]

> There was no bootlegging. The company was structured in such a way that there was room for this kind of research. We had approval for everything we did from the director of chemical research. He said that this discovery [the sweet taste of aspartame] was the best Christmas present he ever had.
>
> The research department was organized in such a way that it was possible for people like me to make fairly autonomous decisions. You had to get approval from the department head, but if you had some rationale it wasn't too difficult. *Absolutely this wasn't bootlegging!*

During 1969 and 1970, several outside events occurred which changed Searle's perception of the aspartame opportunity from a *scientific curiosity* to a *significant business opportunity.*

34.2.3   Safety Concerns—The FDA

The diet food industry—a $750 million/year market and growing at 10 percent annually—was stopped cold in October 1969 when Health, Education, and Welfare Secretary Robert Finch announced a ban on the sale of cyclamates.[14] With cyclamates being removed from the market, saccharin was left as the only sugar substitute. Unfortunately, saccharin was also suspected of causing health problems. As far back as 1957, scientists have shown that certain forms of saccharin could cause cancer in mice. Saccharin's dubious future combined with the cyclamate ban prompted significant questions regarding the future of all artificial sweeteners.

34.2.4   Life After FDA Cyclamate Ban

Companies reacted quickly to the FDA ban. One week after the ban on cyclamates, Coke, Pepsi, Royal Crown, and Seven-Up announced reformulation of their diet drinks; they indicated that new formulations would be on grocers' shelves in "1 to 8 weeks."[14] Most of these new mixtures used saccharin with a "shot of sugar," to help mask saccharin's aftertaste.[14] The tabletop artificial sweetener manufacturers were also panicking. In October 1969, Alberto-Culver announced that it would reformulate its SugarTwin to remove any trace of cyclamates.[15] Most other tabletop sweetener manufacturers announced similar intentions.

Kraft and General Foods had some foods containing cyclamates, but they were not overly concerned. General Foods announced that it would reformulate its D-Zerta gelatin dessert, and Kraft was not sure if it would reformulate or continue its diet line. Diet products represented a small portion of each of the company's sales—except for Kool Aid at General Foods, which represented one of its major products.[14]

Many companies and scientists were racing to find a new artificial sweetener to fill the void left by the cyclamate ban and the opening that saccharin was likely to present because of its aftertaste. Several compounds were rumored to hold promise; examples were neohesperidin dihydrochalcone (NHD), Monellin, and xylitol, but they either had an unpleasant aftertaste, did not reduce the caloric content, or else were prohibitively costly.[16–18]

34.2.5   Searle Decides to Enter the Sweetener Market

The flurry of activity, after cyclamate was banned and saccharin rumored to be removed from the market, caused Searle to view the aspartame opportunity in a new light:[13]

In 1969 the management of the company got seriously interested in this [aspartame] because cyclamate was taken off the market—out of processed foods. There was a big hole [market opening], and saccharin tasted so bad that it couldn't truly be used alone, so the market really disappeared. This [the opportunity] was so obvious, and Searle had something that tasted very good. There was no argument about this—anybody who tasted this [aspartame] would say, "It tastes great." It began to view this as an opportunity. Even the management of a pharmaceutical company couldn't fail to recognize the opportunity.

Although the cyclamate ban was a major motivator for Searle's commercial interest in aspartame, it was not the sole reason for the company's excitement about the artificial-sweetener business. As Joseph McCann, the author of *Sweet Success,* noted[19]

> They [Searle] were compelled to [enter the market]. They had an empty product pipeline and this [aspartame] was one of the more promising, although scary products. It was scary because there seemed that there was some potential even though it was in an area that was pretty strange for them to be moving into. They didn't know how to market a consumer product.

During 1968–1970, Searle had a narrow product line and relatively few products accounted for a disproportionate share of sales. The company had not introduced a significant new product for the past 5 to 6 years. Eleven of Searle's products accounted for 92 percent of 1968 sales volume.[4] These products are listed in Table 34.3.

TABLE 34.3   Searle's Product Profile

Product	Year of introduction	Use
Metamucil	1934	Treatment of constipation
Dramamine	1949	Relief of motion sickness
Banthine	1950	Treatment of ulcers and gastritis
Pro-Banthine	1953	Treatment of ulcers and gastritis
Enovid	1957	Oral contraception
Ovulen	1963	Oral contraception
Aldactone	1959	Treatment of edema and ascites in heart failure
Aldactazide	1961	Treatment of edema and ascites in heart failure
Lomotil	1960	Treatment of diarrhea
Serenace	1961	Treatment of psychotic illness
Flagyl	1963	Treatment of vaginitis

Source: G. D. Searle *Annual Report,* 1968.[3]

Because of the dearth of new products, Searle began an aggressive new-ventures effort to diversify its business. Tom Carney was appointed the head of the new-ventures group, and he had six ventures that he was evaluating—aspartame was one of these. When *cyclamate was banned,* aspartame became the top priority of all the ventures. The combination of the cyclamate ban and Searle's need to find a new business opportunity caused the company to move into an area in which they had no prior experience—the food additive business.[20]

Although aspartame seemed to suggest an opportunity, the question became, how much of an opportunity? Searle did not seem to know what they had—they had no point of reference, nor could they see clearly the complexities needed to bring a food additive from concept to market. The aspartame project represented a sizable risk in terms of development, marketing, and manufacturing. Major capital investments in new plant and equipment would be necessary. Phil Worley, who became head of marketing for aspartame, explained[20]

> Searle's management had some questions as to the potential for the product. One of the requirements for the project was to build a very large and costly plant—something

that was beyond anything that Searle had tackled before and there was real concern by the Searle family whether they should proceed with this.

John Searle, who was the chairman of Searle at the time, was not convinced that the business potential was worth the cost. He wanted to sell the aspartame project totally—to license it and use the royalties to fund drug-related research and development projects (Ref. 1, p. 33).

However, after senior management talked to the core group of people involved in the project, who had been investigating the potential of aspartame with some prospective customers, Searle decided to continue developing aspartame and preparing it for commercial launch. Several of these food companies were trying and testing sample batches of aspartame in their labs and were becoming increasingly more enthusiastic about the product.[20,21] These companies helped give Searle a better understanding of the market. The feedback from the food companies reassured Searle that there was some potential for aspartame. As Worley explained, "The reason to go ahead with this was the enthusiasm of the food and soft drink companies."[20]

Although the feedback was not quantitative, the information was instrumental in shaping Searle's view of the market. The input from the food companies helped alleviate some of Searle's apprehension, because Searle itself was having great difficulty estimating the market potential of this innovation. As Barry Homler, manager of technical services for Searle, said, "We tried to look at it [the potential of aspartame] because nobody could—it was like we had a dart board and a single dart to figure out what was actually going to happen."[22]

Robert Mazur, the head of research for aspartame, voiced similar concerns. "There was a realization that there was a commercial value to the product. There were questions being raised in terms of how large the opportunity was, and also the cost and risk of getting into it."[13]

34.2.6 Searle Aggressively Pursues Aspartame

Searle decided to actively pursue aspartame in 1970 and appointed Dr. Robert Chin as head of the aspartame project. Dr. Chin had previously directed the development and commercialization of the oral contraceptive project (the "Pill") for Searle, which was tremendously successful for the company, and he had the support and trust of Searle's management. His job was to "do it again."[23] However, exactly just what the "it" was for aspartame was an unknown.

In an attempt to define the opportunity and create a more coherent marketing strategy, Searle hired Phil Worley, who had worked in marketing in the 1960s for Abbott on Sucaryl, a cyclamate sweetener. When Worley joined Searle, its management still had doubts about aspartame's potential. To quantify demand, Worley began calling on preferred customers (General Foods, Lipton, etc.) to see if these companies would be willing to place initial orders for the product, pending FDA approval. The results of his efforts were encouraging; these early customers were excited about aspartame and were willing to place advanced orders for the product. One interesting and unexpected observation was that in some cases it was the number 2 brand in the category rather than the number 1 brand that was more receptive to trying aspartame. Worley explained:[20]

> In some cases we found that the top guy [company] was conservative, and in some cases it was the second or third position that was more aggressive in looking for growth or market share increase who was willing to go along with it [aspartame].

The results of Worley's preselling effort were seminal for Searle, and the company became more convinced of the innovation's potential and decided it must turn its attention to producing aspartame in larger quantities.

34.2.7   Aspartame Manufacturing

Until this time (early 1970s), the company was producing aspartame in sample batch quantities. However, as the project progressed, the company decided to build its own production facility and planned to spend $15 million on manufacturing. Progress was halted in January 1974 because the plant would have to be approximately two times larger than any plant that Searle had previously built, and the company was unsure of the demand and was not able to firmly establish the optimum manufacturing procedure for producing aspartame.[20,24,25] Adding to the uncertainties of the manufacturability of aspartame was the oil embargo that was making manufacturing more and more expensive. The cost to build the plant kept increasing. In 1969 the plant was estimated to cost $5 million; the cost then escalated to $10, then to $25, to $50, and by 1976 the cost estimate had reached close to $100 million. The company was reluctant to spend such a significant amount of money on an unproven and highly uncertain process. The manufacturing process in the early 1970s required 100 lb of raw material to produce approximately 3 lb of aspartame. The company did not know what it was going to do with the 97 lb of waste for every 3 lb of aspartame produced except to dump it in the ocean, and Searle was concerned about the environmental impact of dumping 97 percent of the material used in its aspartame process.[21,23]

As a result of all the manufacturing unknowns, Searle entered into a joint-venture agreement with a Japanese company, Ajinomoto, a worldwide leader in amino acid research, as its manufacturing partner. Ajinomoto had a long history in the foods and chemical manufacturing arena. In the early 1900s Ajinomoto was the largest supplier of MSG (monosodium glutamate) in the world. In fact, *ajinomoto* means MSG in Japanese.[26] Ajinomoto learned about aspartame at a scientific conference where Searle presented a paper on the sweetener. Searle barely beat Ajinomoto to the patent office with its patent application. Ajinomoto subsequently contacted Searle to see if the two companies could work together, and a relationship was established. Ajinomoto agreed to provide the first complete manufacturing process for aspartame in return for Searle's technology to produce the compound.

Ajinomoto's original plant had the capability to produce 20,000 lb of aspartame per month, but the manufacturing process was quite crude because the company was using a plant that was designed to produce another drug—L-dopa (levo-dihydroxyphenylalanine), a pharmaceutical drug for treating certain types of neurological and mental illness.[27] In the early 1970s, Ajinomoto had a monopoly on L-dopa until Monsanto became a competitor and cut the manufacturing cost of L-dopa to 10 percent of Ajinomoto's cost. Ajinomoto, in the typical Japanese fashion at that time, would not fire any workers, which left them with a large investment and little return—a fully equipped but idle L-dopa plant. When Searle discussed the aspartame process with Ajinomoto, the Japanese firm was receptive and enthusiastic. Ajinomoto produced initial runs of aspartame using the currently available equipment in their L-dopa plant. The early manufacturing plant was extremely unrefined using "pots and kettles."[23] Aspartame was important to Ajinomoto because it represented an opportunity for them to use a plant that was being underutilized.

Ajinomoto was to be paid royalties on sales of aspartame, and each company would share in the other's research. The two companies agreed to jointly market aspartame in Europe and some North African countries; Ajinomoto had exclusive rights to the Japanese market, and Searle had exclusive rights in the United States and Canada. The remaining markets would be free for each to pursue individually (Ref. 1, p. 35).

34.2.8 General Foods Joint Commercialization

A critical juncture in the commercial development and marketing of aspartame came in 1974 when General Foods decided to become more actively involved with Searle in the commercialization of aspartame. Searle needed assistance developing an understanding of the food market since its expertise was limited to drugs. For example, it needed to address questions such as[23]

- What should the pricing structure be?
- What other uses can aspartame be used for?
- What should we call the product?
- How good is the product, and how does it compare with competing products?
- How big is the market?
- Who is the target market?

Searle concluded that it needed to form an alliance with a company that had experience in bringing new food products to the market and that knew how to run focus groups, taste tests, concept tests, preference analysis, and brand-loyalty assessments. General Foods seemed a likely candidate because of its presence in the dry-packaged-food market such as Jello, Kool Aid, and cereal. Searle believed these foods would offer a stable environment for aspartame.

The relationship between Searle and General Foods solidified, and in 1975 Searle and General Foods formally formed a joint venture (or as they called it a "joint commercialization assessment team") to analyze aspartame's potential.[23]

General Foods was interested in aspartame because of its experience in the 1960s with Sugar Free Kool Aid, an exceptionally successful product that represented approximately 20 percent of the Kool Aid business. When cyclamates were banned, General Foods was forced to remove Sugar Free Kool Aid from the market. The only cost-effective alternative at the time was saccharin, and children did not like the taste.

As a result of General Foods' experience with sugar-free products such as Kool Aid and D-Zerta (a diet gelatin dessert), the company knew how to proceed to assess the market for aspartame. General Foods began conducting tests, even before the FDA approved aspartame. These tests were one of two kinds: the first was actual taste tests with their own employees (who had to sign taste-testing agreements) designed to verify that aspartame, indeed, tasted like sugar. The second type was concept tests: customers would read advertisements describing the benefits of aspartame, and then taste Kool Aid or cereal sweetened with sugar, although the participants were told that the food or beverage was sweetened with aspartame. The concept tests indicated that consumers loved the product. In fact, the result "was off the chart it was so good."[23] General Foods conducted thousands of taste and concept tests during these early years to learn what customers wanted.

Unfortunately, giving customers what they wanted proved difficult, and in some cases impossible, from a manufacturing perspective. In the manufacturing of Kool Aid, there was a dust problem—aspartame was difficult to keep contained in the dry pouring operation. Because of the high cost of aspartame (approximately $55.00/lb), the dust which was lost or wasted was expensive. This problem was remedied, but the problems associated with delivering an aspartame-sweetened cereal were not. What General Foods found in the tests was that customers loved the concept of aspartame-sweetened cereal. But a cereal sweetened with aspartame would have had to sell for approximately $5.00 per box, compared to sugar-sweetened cereals selling for $1.50 to $1.70. So neither General Foods nor Quaker, which had also completed concept tests with cereal, were able to deliver an aspartame-sweetened cereal to the market, despite the promising test results.

While the concept and taste tests were being conducted in the United States, General Foods and several other food and beverage companies performed actual commercial taste testing of aspartame outside the United States to learn the intensity of consumer interest in the product. These tests demonstrated that there was a market for the product. Worley described the international tests:[20]

> We started working outside the U.S. where some taste testing was permissible and we started getting experience with the product, and other companies experience also. As part of the agreements we had at that time was that they [aspartame customers] would provide us with information they learned from a commercial aspect, technical aspect as well as from a consumer response.

34.2.9 Aspartame Applications

As a result of the domestic and international tests, Searle forecasted the target market for aspartame to be weight watchers and diabetics. This market represented $10 to $20 million.[21] Searle, with the input from General Foods, determined that a spoon-for-spoon product (one teaspoon of sugar would equal one teaspoon of aspartame) would represent 60 percent of aspartame's volume, a packet (two teaspoons of sugar equivalency) would be 30 percent, and a tablet would be 10 percent.[23] These sales were to come from various markets:[20]

Carbonated soft drinks	60 percent
Powdered drink mixes	25 percent
Chewing gum and candy	10 percent
Tabletop sweetener	5–10 percent
Dairy products (ice cream)	5 percent

The spoon-for-spoon product was the number 1 priority for Searle, because the company believed that it offered the greatest market potential because of its ease of customer use; the packet was second; and the tablet was third.

The tablet probably would not have been introduced at all had Searle not been a pharmaceutical company who had a preference for pills and tablets.[20] The spoon-for-spoon product and the tablet required separate development work and had to be manufactured at three different locations. Each was viewed as a separate opportunity.[20]

As the market tests continued, Searle's executives developed more experience with marketing aspartame and their requirements became more exact. Trying to make a product that marketing wanted to sell continued to pose problems from a development and manufacturing perspective. Worley explained about trying to perfect the tablet:[20]

> One of the forms of Equal was a tablet. The fellow that was in charge of developing the tablet had worked on developing pills and pharmaceutical tablets for close to 20 years.

I set all this criteria up on what the tablet should do. It should be the right color, the right sweetness and so on. And he would bring these tablets over (to test) and the first ones I put in my cup of coffee—they'd sit there and they'd never dissolve. He'd look at the checklist, and say, "Well there is nothing on here about dissolving. Ok back to the lab." Then I got a call from this individual and he said, "Boy I've got it. This thing dissolves in under ten seconds. The taste was right." He was real excited. He brought the tablet over, lined up some cups of coffee, dropped it in, sure enough, it dissolved, but on the top of the coffee cup after it dissolved there was this sort of green scum. I said, "What is this green scum?" And he said, "You didn't say anything about green scum specifications."

34.2.10 FDA Approval

Searle began human testing in February 1972, and on March 1, 1973, the company filed a petition with the FDA (FR DOC 73-4260, petition FAP 3A-2885) seeking approval for aspartame for use in foods and selected beverages. The FDA approved aspartame on July 26, 1974 for several categories:[28]

- Free-flowing sugar substitute (in packets equivalent to less than two teaspoons of sugar)
- Tablets for hot beverages such as coffee and tea
- Cold breakfast cereals
- Chewing gum
- Dry bases for beverages such as coffee, tea, gelatins puddings, and dairy product substitutes (ice cream and imitation whipped toppings)

There were two stipulations in the approval: (1) aspartame must not be used in cooking (it breaks down and loses its sweetness) and (2) a warning must appear in capital letters: "PHENYLKETONURICS: CONTAINS PHENYLALANINE" (*phenylketonuria* is a rare disease that reacts adversely with aspartame). The warning label required and the limitation for nonbaking applications were foreseen. However, what was unforeseen was that Searle's spoon-for-spoon equivalent product, which was the company's number 1 priority, was not approved. This posed a significant problem to the company because the spoon-for-spoon product was the preferred embodiment for aspartame. Also the number 1 use, carbonated soft drinks, was not approved.

Another problem occurred shortly before Searle launched aspartame. Several external studies were performed by outside scientists that raised questions about the safety of aspartame. John Olney, a psychiatry professor at Washington University Medical School, and a consumer group, Consumer Action for Improved Food and Drug, found that aspartame, when used in conjunction with MSG, could cause brain damage in children.[29] Following Olney's study, the FDA on December 5, 1975 issued a stay (a restraining order) on the sale of aspartame.[30] Further complicating matters, in 1975 the FDA called into question Searle's testing and data-reporting procedure on several of the company's products, one of which was aspartame.[24]

As a result of the stay and the FDA investigation, the aspartame project all but died. Prior to FDA approval there were 100 people working on the project at Searle; after the stay, there were only 10 (Ref. 1, p. 54). The few employees who were left focused their attention on aspartame's international market applications. Aspartame floundered for the next several years until Searle changed its top management; with this change a more focused strategic direction emerged.

Don Rumsfeld, the former Secretary of Defense and Ambassador to NATO, became CEO of Searle in 1977. He changed the company's mission to stay closer to its core business and led the divestiture of over 20 of Searle's businesses.

During the late 1970s, the company's new product pipeline was drying up and Searle remained dependent on a small number of pharmaceutical products for the majority of its revenue. Sales of 15 products accounted for 89 percent of sales in 1977, and of the 15 products only four were new since 1968. Two of the company's products, Aldactone and Aldactazide represented 21 percent of sales, and these were introduced in 1959 and 1961, respectively.[31] From a new-product standpoint, the early 1980s were no kinder to Searle than were the late 1970s. The same two products, Aldactone and Aldactazide, represented 48.2 percent of the company's sales in 1980, and no significant new products were in sight.

34.2.11 The Changing Sweetener Market

While Searle was grappling with its future corporate innovation strategy, the sugar market was rapidly changing. Because of an increased demand and a decreased supply of available alternatives, sugar prices had risen to their highest level in 5 years—30 to 36¢/lb in 1980 up from 21¢/lb in 1979. And prices were expected to increase to over 45¢/lb in the near future.[32] Because of sugar's high prices, soft-drink bottlers were experimenting with new alternative sweeteners, primarily high-fructose corn syrup, which cost one-third less than sugar.[32] A cost comparison of the various sweeteners is shown in Table 34.4.

Searle decided to launch aspartame for several reasons: the high cost of sugar, the limited availability of high-quality safe alternatives, a lack of other new products, and a readily obtainable sugar substitute in the company's own labs. Because aspartame had not yet been FDA approved, management was skeptical about its potential. Sales of aspartame through 1981 had totaled only $13 million and were primarily in Canada, where aspartame had been approved as an ingredient for carbonated soft-drink use (Ref. 1, p. 47).

Because of aspartame uncertainties with FDA approval, manufacturing, and marketing, and because the product did not fit the strategic mission of the company, Rumsfeld charged Marcel Durot, the head of the Consumer Products Division, with the duty to sell the aspartame business entirely and take the proceeds to fund activities more closely aligned with Searle's core business. Unfortunately, the company could

TABLE 34.4 Sweetening Equivalence

Sweetener	Calories	Sweetness (compared to sugar)	Cost per pound	1 lb of sugar sweetening equivalence	Description
Aspartame	0.4—tablet 2—powder	200 times	$90.00	45¢	No baking
Sugar	16	1	29¢	29¢	All uses
Saccharin	1.36—powder 0—tablet	300 times	$4.00	0.013¢	All uses
Cyclamates	2—powder 0—tablet	300 times	$1.93	0.064¢	Banned
HFCS	15	1	18¢	0.018¢	Baking and beverages

Source: Bylinsky.[33]

find no buyers—not a single one.[34] Rumsfeld had to make a decision—either commit to commercializing aspartame or drop it. Many of the largest food and beverage companies were constantly inquiring about aspartame, asking for samples and testing the product in their labs. The total investment in the aspartame project up to the early 1980s totaled close to $100 million (Ref. 1, p. 52), and since Searle did not have another new product that offered more potential, the company decided in 1981 to make a commitment to aggressively pursue commercializing aspartame. Searle then sued the FDA for excessive delay in approving aspartame.

On July 15, 1981 the FDA's Commissioner, Arthur Hull, issued final FDA approval for aspartame (Ref. 1, p. 45). Aspartame was legalized for the following applications:[30]

- Tabletop sweetener—40 mg (packets of two teaspoons equivalent of sugar)
- Tablets—20 mg per tablet
- Dry beverage mixes—120 mg in an 8-oz glass
- Gelatins or puddings—32 mg per serving ($\frac{1}{2}$ cup)
- Whipped toppings—10 mg per serving (2 heaping teaspoons)
- Cold breakfast cereals—90 mg per serving
- Chewing gum—8 mg per stick

Similar to the 1974 decision, aspartame was forbidden to be used in cooking or baking and the warning label about PHENYLKETONURICS would also be required. After FDA approved aspartame, Rumsfeld became a supporter of the project. In late 1981 he said, "We're going to sell this (aspartame), but we don't know what 'it' is."[5]

The applications approved by the FDA were classified into two broad categories: using aspartame as an ingredient (dry mixes, puddings, whipped toppings, cereals, and chewing gum), and using it as an end product (as a tabletop sweetener in packets and tablets). Because of the different applications, two strategies emerged for marketing aspartame: selling the product as a tabletop sweetener under the name Equal (and Canderel in some countries) and offering it as an ingredient under the name NutraSweet. Searle would sell Equal via the typical methods—in grocery and drug stores. NutraSweet would be sold directly to firms such as General Foods, Pillsbury, Pepsi, and Coca-Cola (when approved for carbonated soft drinks) for use in their products.

As part of the company's commercialization plan for aspartame, management planned to use a "branded ingredient strategy" (similar to selling yellow dye no. 6 and calling it *molar*). The origin of the branded ingredient strategy is not clear. Some credit Robert Shapiro (who eventually became president of the NutraSweet group) with the idea, and others credit John Mullendore, who was head of the aspartame project in the late 1970s. Whoever's idea it was, the branded ingredient strategy was pivotal in NutraSweet's positioning. Consumers did not have to read a label and see "aspartylphenylalanine methyl ester"; they only saw "aspartame" and "sweetened with NutraSweet." In addition to greater perceived safety by consumers, simply seeing "aspartame" and that "little red swirl" secured for NutraSweet high brand-name recognition and customer loyalty.

Searle's pricing strategy for NutraSweet developed and became more sophisticated over time. In the early 1980s the price per pound of NutraSweet was based on a sliding scale depending on how much aspartame decreased the caloric content of the product in which it was used. However, the original $90.00 price point was determined without any formal market analysis; it was decided by Don Phillips in the late 1970s. Phillips was in charge of Searle's optical division. Rumsfeld felt that because Phillips had retail experience, he would be capable of devising a retail strategy for NutraSweet. Phillips was placed in charge of the NutraSweet project. When Phillips took control, NutraSweet was selling for $60/lb and was being rationed to customers who wanted to test it. Demand far exceeded supply. Phillips approached Annette Ripper (now Mullendore), who was in charge of marketing aspartame at the time, and after talking to Ripper about customer's wanting far more aspartame than the company could supply, told her, from then on the price would be increased from $60.00 to $90.00. The new price of $90.00 per pound was established.[21]

34.2.12 Market Launch

The strategy for the marketing mix of aspartame seemed solidified. As a check, prior to launch of aspartame, many marketing techniques were used in an attempt to quantify the market potential. Market research for Equal, based on the opinion of experts inside and outside the company, and taste and concept tests conducted in the early 1980s, indicated that aspartame could achieve 60 percent customer awareness of the product in 90 days, a 10 percent trial, with a 50 percent repeat purchase rate. Searle set its goals on these numbers.[34]

In October 1981, Searle introduced Equal in the United States,[35] and by the end of 1981, Equal had been approved in 13 countries. Also in 1981, John Robson, a senior executive with the company, succeeded in getting the patent life of aspartame extended for an additional $5\frac{1}{2}$ years, which allowed Searle to retain its monopoly on aspartame (Ref. 1, pp. 52–53).

In late 1981 and early 1982, Equal was released in selected test markets. What was surprising was that the trial rate was double what was expected (close to 21 percent) with the same awareness (60 percent) and repeat purchase (50 percent) levels as had been predicted in the early 1980s. This meant that the market for NutraSweet would be twice the original estimates.[34] The target market for Equal as late as 1981 was still saccharin users. Executives forecasted that the market for Equal would reach a 50 to 60 percent share for a spoon-for-spoon product or a 40 percent share for the packet product and the market for the ingredient NutraSweet could reach 15 percent of the sugar market with FDA approval of its use in carbonated soft drinks, or 7 to 8 percent without carbonated soft drinks.[20] The actual results showed that the ingredient estimates came pretty close to hitting the mark, but Equal actually achieved a 25 percent market share instead of the 40 percent that was forecasted.

One reason for the varying levels of accuracy in the forecasts was that companies such as General Foods, and other food and beverage companies, were helping Searle calculate the market size for the ingredient business. These food and beverage companies based their estimates on two factors: thousands of concept and taste tests with the product and past experience with their own artificially sweetened products. On the other hand, the Equal forecasts were based on Searle's own internal marketing research, and Searle did not have a solid foundation of experience on which to base the forecasts. Interestingly, it was not until the product was actually launched in regional test markets in 1981 and 1982 that the true picture came into focus. Until 1982, the company still thought the market for Equal would be saccharin users, but follow-up studies in shopping malls during the test market indicated that the market was not saccharin users at all, but rather dissatisfied sugar users. Although saccharin had a bitter aftertaste, it was discovered that saccharin users had become used to and, in fact, preferred the aftertaste of saccharin. Searle had to change its marketing and positioning of the product to reach this new target segment.

34.2.13 The NutraSweet Group

As the project began to gain momentum, aspartame needed additional company resources, and in December 1982, Rumsfeld formed the NutraSweet group to coordinate and run the aspartame operations. He appointed Robert Shapiro as its new manager. Shapiro had been Searle's corporate counsel and was one of the key executives responsible for selling many of Searle's businesses in the rash of divestitures in the late 1970s.[35]

Once Shapiro took control, he began a build-up of manufacturing and marketing. Searle was now firmly behind the project. Gone were the days of staging dog and pony shows to get people to give their approval of the program. The NutraSweet group was staffed with young and very aggressive people. By late 1982, the company's strategy was firmly established: Searle would market NutraSweet (now the generic name for aspartame) as an ingredient and not as a finished product. The only exception would be the Equal-brand tabletop sweetener packets to be sold directly to end consumers.[5]

NutraSweet sales increased to $74 million in 1982, and estimates indicated that sales volumes would increase.[5] Shapiro arranged for $25 million to expand Ajinomoto facilities, and $30 million to expand Searle's own plant in Michigan that was producing aspartame and their pilot manufacturing plant in Illinois. The cost to produce NutraSweet in the early years was approximately $55 to $60 per pound, and the company was selling the product for $90 per pound.[21]

With the success of NutraSweet imminent, a new company strategy began to emerge: expand capacity and reduce costs. Customers were being rationed NutraSweet and Searle was trying to decide how to increase capacity. One alternative was to build a dedicated state-of-the-art NutraSweet manufacturing plant to meet demand and reduce costs. In March 1983, the board approved $100 million to build a manufacturing facility, but concerns remained. As Shapiro said, "We didn't know if we could manufacture it [NutraSweet] ourselves on a large scale."[36] In 1984 the cost to build the facility that was to be located in Augusta, Georgia, had risen to $130 million, and was scheduled to be completed in 1985.[37]

34.2.14 Indications of Success

Sales of aspartame in 1982 totaled $74 million, but it was not until 1983 that Searle began to get an indication of the true demand. Two events occurred that changed the way Searle viewed NutraSweet: FDA approved aspartame for use in carbonated beverages, and a new advertising and promotion campaign was implemented—the famous gumball program. The results of these two events transformed NutraSweet from being perceived as an innovative product to a significant business opportunity for Searle.

To educate consumers about the availability of NutraSweet in soft drinks and other products, the NutraSweet group invested heavily in advertising. In the first 3 years after product introduction (1982), the company spent close to $120 million in advertising promoting the NutraSweet name.[35] This was in contrast to other artificial-sweetener companies who were spending approximately $3 to $4 million on advertising.

Ogilvy & Mather was hired to direct the NutraSweet campaign. Ogilvy launched a gumball promotion program that had a profound impact on the estimate of the magnitude of NutraSweet's potential. In March 1983, print advertisements appeared in bold-face explaining NutraSweet as a breakthrough sweetening agent. A coupon campaign was initiated. Anyone returning a coupon would receive a free NutraSweet gumball. Over 3,000,000 coupons were returned.[5] The advertising strategy worked. Within only a few years of introduction, 98 percent of all American adults could identify NutraSweet from a list of sweeteners, and 70 percent could name NutraSweet as a sweetener unprompted.[5]

The second and more significant event for Searle was FDA approval for use of NutraSweet in carbonated beverages. Searle submitted an FDA petition in 1982 for carbonated beverages and was granted approval on August 1, 1983. The carbonated beverage market was the true gem of the entire sweetener market, and the company had been trying to access this market since the mid-1970s, when there were estimates indicating that this segment could represent 70 to 80 percent of the total artificial-sweetener market. Also, control of this market was dominated by only a few players: Coke and Pepsi's combined market share was 64 percent, making this a potentially attractive market. Furthermore, consumption of carbonated beverages was increasing and the per capita growth of this segment had increased over 70-fold from the time they were first introduced. Recent growth of the soft-drink segment was impressive. In 1969 per capita consumption of soft drinks totaled 23.4 gal, and in 1985, consumption had grown to over 40 gal (Ref 1, p. 80; Ref 35). The diet soda market enjoyed similar success. In 1980 diet soft drinks penetrated 12.9 percent of the entire soft-drink industry, and by 1984 the figure reached 21.3 percent.[35]

In 1983, both Coke and Pepsi finally became NutraSweet customers, and the impact on the sales of NutraSweet was substantial. NutraSweet's sales went from $74 million in 1982, to $336 million in 1983, to $585 million in 1984.[11,38] NutraSweet had captured 86 percent of the total U.S. sugar-substitute market.[38] In Canada, NutraSweet had achieved 40 to 45 percent of the entire tabletop sweetener market. The NutraSweet Group was "relishing its success."[5] After NutraSweet was approved for carbonated beverages, Shapiro was quoted as saying, "About all we have to do is hang our shingle and say, 'we're open for business.'"[39]

This story was well on its way to a happy ending. By the end of 1984, NutraSweet had been approved in 40 countries and over 100 million consumers had tried foods sweetened with NutraSweet.[37] During this same period, the FDA had received fewer than 600 complaints on aspartame out of 70 million Americans who used it.[40] The question became: What was the company going to do for an encore?

Although NutraSweet's future seemed bright, there were many competitors trying to get a piece of the artificial sweetener market. Some companies readying products to compete with NutraSweet are listed in Table 34.5.

TABLE 34.5　　Potential NutraSweet Competitors[1,35,41,42]

Company	Product	Description
Biospherics	Lev-O-Cal	Same taste as sugar without the calories, and can be used in baking, browning, and for bulk; it also does not promote tooth decay
Hoechst AG (Germany)	Acesulfame-K (*Sunette* brand name)	Slight aftertaste and must be blended with other sweeteners; directly competing with aspartame—no baked applications
Pfizer	Alitame	Stable in baked goods, 2000 times sweeter than sugar; shelf life 2–4 times that of aspartame
Coca-Cola	Family of sweeteners	High-performance sweeteners that are 1900 times sweeter than sugar
Ajinomoto	?	Rumored to be thinking about entering the market once NutraSweet's patent expires

Several other companies were reported to be positioning themselves to enter the market once the NutraSweet patent expired on December 14, 1992; these included Beatrice, General Foods, Johnson & Johnson, Procter & Gamble, Schering AG, Adeda Chemical, Tanabe Seiyaku, Tate & Lyle, and Unilever.[41,43]

34.3 SEARLE'S NEW STRATEGIC FOCUS

In light of the imminent influx of competitors, Searle and NutraSweet needed to formulate a new plan to remain competitive. Searle's management was faced with three options:

- Maintain their present situation.
- Merge with or acquire another company to give them greater critical mass.
- Sell the firm.

Searle decided to sell the entire company to a larger corporation that had the resources to enable both Searle and NutraSweet to remain competitive in the turbulent times ahead. Monsanto decided it was interested in purchasing Searle's pharmaceutical business, but Monsanto was not interested in the NutraSweet business. Monsanto was apprehensive about aspartame soon coming off patent and was uncertain what its future potential would be. The Searle family rejected Monsanto's offer to buy part of the company—the family insisted on an all-or-nothing deal.

Searle's own executives who expressed an interest in buying the NutraSweet business were unable to come up with a plausible selling price. These executives used standard ROI (return on investment)-type calculations but could not justify the $300 million the company was asking for only the Equal tabletop sweetener business; the executives felt $150 million was closer to the correct evaluation which left a safety margin for error.[34] However, Monsanto changed its mind because of active interest from several food companies in buying the line, and decided to purchase the entire company. In August 1985 Monsanto bought Searle and NutraSweet for $2.754 billion.[44]

34.4 LIFE AFTER MONSANTO

After Monsanto purchased Searle, NutraSweet's strategic focus changed to a four-pronged action plan (Ref 1, pp. 101–102; Ref. 35):

- Increase the use of NutraSweet in existing markets.
- Develop new applications for NutraSweet.
- Create new food products by using advanced technology to solve customers' problems.
- Lower costs.

The company realized that it had to change. As George Logan, the Managing Director for Middle East and Africa Business, said in an interview: "This company was built upon 90 yard passes in the Superbowl—really big plays. We're now having to learn 'first-down football' and the running game" (Ref. 1, p. 94).

To meet these goals, NutraSweet spent $31 million on R&D in 1987. This represented 4.2 percent of sales, compared to the 1 percent that traditional food companies were spending. The strategic mission of NutraSweet evolved into one of not simply lowering cost, but of achieving *the* low-cost position. Also, the company began to place more emphasis on customer service. And finally, the third area of direction was to be the dominant international supplier of aspartame. In 1989, the company doubled manufacturing capacity—investing close to $100 million (Ref. 1, p. 96). NutraSweet enlarged its applications support department, calling it the Technology and Development Group. The charge of this group was to work with customers on possible applications for NutraSweet. By the late 1980s, this group had grown to over 200 people—190 professionals, 65 percent with Ph.D.s. Approximately one-third of all Searle's R&D scientists were in support roles for aspartame (Ref. 1, pp. 102–103; Ref. 35).

The company became the world's lowest-cost producer of aspartame and the components of aspartame. Even Ajinomoto, the original supplier, began buying product from NutraSweet because of the lower cost. NutraSweet built manufacturing plants in foreign markets where sugar was imported in an attempt to compete locally with sugar. Manufacturing costs were cut by a factor of 5 from 1983 to 1989, which allowed the company to post a 33 percent gross margin from 1985 to 1986, even though the market price had decreased by 10 percent. Unit volume had increased by 50 percent.[45] By 1988, the price of aspartame had fallen to $27/lb in Europe because of increased competition, but NutraSweet still claimed that they were able to make a profit at that price. In the United States, NutraSweet was selling for $65/lb, and the company's profit margin on 1987 sales before depreciation and R&D expense was 53 percent.[35]

If NutraSweet had been an independent company during the 1980s, it would have been number 392 in sales revenue of all U.S. companies, number 152 in net income, number 22 in sales per employee, and number 4 in net income per employee.[45] In 1986, the picture looked better than ever, as one observer noted: "The hill had been taken by the troops and now it was time to secure and defend their gains" (Ref. 1, p. 90). The company had reduced manufacturing costs, and it dominated the artificial-sweetener market. In 1988, NutraSweet accounted for 90 percent of the U.S. sugar-substitute market and 80 percent of entire worldwide sales of artificial sweeteners.[41] By 1989, NutraSweet was in over 2700 products, and by 1991 it was in over 4000 products.[46] The company had become an international manufacturer and marketer. As of 1989, 22 percent of the company's sales were outside the United States, and by 1990, NutraSweet was approved in 79 countries (Ref. 1, p. 111). NutraSweet's sales increased from $722 million in 1987, to $736 million in 1988, to $869 million in 1989, to $933 million in 1990 (Ref. 1, p. 2; Ref. 47).

The cost to manufacture NutraSweet was being reduced and the market for NutraSweet-sweetened products was becoming saturated. The company had squeezed almost all the costs and markets out of the product that it could and needed to find the next big money maker.

34.5   DISCUSSION

This case illustrates six implicit, and at times explicit, phases through which Searle passed when it marketed its breakthrough: breakthrough mentality, call to action, learning from experience, experimental marketing, fine tuning, and opportunity exploitation.

34.5.1 Breakthrough Mentality

Searle was a drug company in which long lead times for new products were common. A new pharmaceutical agent can easily require 10 to 15 years to be commercialized. This type of business recognizes the need for prolonged research, patience, and persistence in pursuing many avenues on the path to a new medical treatment. Drug research in many aspects is similar to a river flowing and branching into different tributaries. It would be difficult to predict ahead of time where the river would finally end. Pharmaceutical researchers must be given a relatively free hand to pursue the tributaries of their work, because it is difficult to know, a priori, what effort will yield a desired result.

During the initial stages of the development of aspartame this discovery was pursued because it was "scientifically interesting." Neither Mazur nor Schlatter[48] knew or cared about the market size or commercial potential of aspartame. These compounds should not have tasted sweet, but they did, and this observation was scientifically significant. Because of Searle's organizational structure, research managers were authorized to pursue these types of opportunities without much trouble. Mazur funded the aspartame project *not* covertly or on a shoestring basis, but rather as part of the normal course of business.

Aspartame fit nicely into the ilk of Searle, an organization that believed "Daily we must be patient with the demands of science and impatient in the struggle to make the product of our technology available to mankind at the earliest possible moment."[2]

Searle seemed to be the ideal type of company to pursue the early development of aspartame because this company was conditioned in the pharmaceutical mentality that breakthroughs take time and patience. So a question arises: How can a nonpharmaceutical company develop a breakthrough mentality? Perhaps the following phases will help illuminate this idea.

34.5.2 Call to Action

Breakthrough mentality allows the research of an idea to continue, but what causes a company to act on a radical innovation that is outside the company's normal scope of business in a concerted and aggressive fashion? Several events occurred simultaneously to cause Searle to view the aspartame opportunity seriously. These events consisted of the cyclamate ban in 1969—"If cyclamate had not been banned I do not think Searle would have pursued the commercial aspects of aspartame"[20]—and the drying up of the company's new-product pipeline. Without both of these, Searle may not have pursued the project as fervently as it had.

A similar call to action inspired Ajinomoto to invest heavily in the manufacturing of aspartame. Because of the disruption in its business caused by Monsanto's new L-dopa production process that dramatically reduced the cost of manufacturing the drug, Ajinomoto found itself with a plant operating at far-from-full capacity. The combination of excess capacity and the Japanese no-firing ethic led to Ajinomoto's call to action. The company had a plant, and it needed to find something to produce; aspartame seemed to offer potential at the time, and the uncertain potential of aspartame was far better than their other alternatives.

Would each company have gambled as they had without their own respective call to action? Perhaps, but the probability favors a languishing product development effort because aspartame was seven times more expensive than cyclamate, its nearest competitor.

34.5.3 Learning from Experience

One observation that became clear with the development and commercialization of aspartame was the reliance on experience. Phil Worley, the head of marketing on aspartame, drew on his experiences from working on Sucaryl at Abbott. Ajinomoto benefited from its extensive knowledge of amino acid production, which was similar to the process needed for aspartame production. General Foods provided critical direction and guidance to Searle because General Foods recognized the opportunity aspartame represented. General Food's cyclamate-sweetened Kool Aid drink was a big hit prior to its removal because of the cyclamate ban in 1969. General Foods knew that if aspartame could pass FDA scrutiny, its company could once again introduce a sugar-free Kool Aid. General Foods as well as many other food companies conducted thousands of taste and concept tests, many before FDA approval, in an attempt at defining the opportunity. These tests reinforced the notion that NutraSweet had potential. However, even the food companies, with substantial experience in the food area, encountered several critical problems, many of which were unable to be successfully resolved. One good example was aspartame-sweetened cereal. Neither General Foods nor Quaker Oats could deliver what people wanted—a low-cost, reduced-calorie, good-tasting, and sugar-free cereal.

The importance of building on experience surfaced again when Searle introduced Simplesse. One year after Kraft signed an exclusive agreement with Searle for the use of Simplesse in its line of mayonnaise, Kraft merged with General Foods. General Foods learned of the Kraft-Searle joint commercialization arrangement, and General Foods explained to Kraft the dilemma it faced when one ingredient is withdrawn from a food product (sugar from cereal) and another more expensive ingredient must be added (NutraSweet) to add bulk or texture; the result is a far more costly end product. Kraft, building on General Foods' experience, decided to cancel its contract with Searle, but Searle went ahead on its own and introduced a line of ice creams with Simplesse (Simple Pleasures). Searle, however, did not have much experience with this line of business—and Simple Pleasures flopped.

34.5.4 Experimental Marketing

Searle tried several different approaches in its search to exploit the aspartame opportunity. These tactics were the result of working with people and companies with experience, and these experiences allowed Searle to narrow the scope of product, offering choices to a manageable number, but market experimentation of the product was needed. It was as though the company were approaching the different products and markets in a venture capital-like fashion—try many investments with the hope the few that succeed will be large enough to more than compensate for the others that fail.

In the early years, Searle thought it would market several different embodiments of aspartame: spoon-for-spoon, packets, and tablets. However the FDA did not approve the company's number 1 priority product—the spoon-for-spoon product, and the company had to introduce its number 2 and number 3 products: packets and tablets. Similarly, original applications included carbonated soft drinks, powdered drink mixes, chewing gum, tabletop sweeteners, dairy products, and tablets for use in coffee, but the number 1 application, carbonated soft drinks, which was estimated to be 60 percent of the potential, was not allowed by the FDA. The company tried many product combinations: Equal; packaged drink mixes, such as Kool Aid; food mixes, like Jello and D-Zerta; packaged cereals; as well as a host of other products. The concept tests and taste tests showed that several applications offered potential; however, the in-depth market tests indicated that a few applications did not offer the possibilities that were expected. During taste tests it was found that NutraSweet tablets, for use in coffee, did *not* offer better taste than Sweet 'n Low, and aspartame would have to be sold at a substantial premium price. It was also determined that even though some concepts offered potential, such as aspartame-sweetened cereal, it could not be produced cost-effectively. As a result of the uncertain product and uncertain manufacturing technology, Searle tried several avenues with the hope that one or more would prove possible and profitable.

34.5.5 Fine Tuning

Aspartame changed drastically over time. The initial concept—the spoon-for-spoon product—was not introduced in the United States, and what was ultimately offered for sale were packets and tablets. The market changed from its originally envisioned target—saccharin and cyclamate users—to dissatisfied sugar users. The manufacturing changed radically from Ajinomoto's first production process with "pots and kettles" to a state-of-the-art facility.[23] The product, market, and manufacturing changed drastically over the course of aspartame's development. Because Searle desperately needed a new business, the company stuck with the aspartame opportunity and remained flexible, enabling Searle to change and adapt to new twists and turns as they arose. Searle changed the product and marketed its second choice—aspartame in packet form. The company learned that the market was, in fact, dissatisfied sugar users and not the saccharin market, and consequently changed its advertising and focused and fine-tuned its message to this new market. Searle continued to refine and improve the manufacturing of the product until it was even better than Ajinomoto. Over time the picture of the opportunity became clearer and the company realized where it needed to direct its efforts.

34.5.6 Opportunity Exploitation

This final stage, opportunity exploitation, was characterized by expanding the base of products that could use NutraSweet, concentrating on lowering manufacturing costs, and growing internationally in terms of both marketing and manufacturing. In this phase, Searle tried to squeeze all the potential it could from NutraSweet; this included product acceptance in over 4000 products and 79 countries by 1991, and leadership as the worldwide low-cost producer of aspartame.

34.6 LESSONS LEARNED

We can glean several lessons from this case:

  1. Because commercializing a radically new product can take 20 years or longer, companies wishing to develop and market these types of products must have a *breakthrough mentality*—a patience with the belief that there is a light at the end of the tunnel.

  2. Companies may wish to impose an urgency to the project—*a call to action*—that will cause them to stop talking and act. It can be easy to rest on past laurels, but if companies want to develop and commercialize truly innovative innovations, a proactive-versus-reactive strategy should be cultivated.

3. Companies need to *build on their experience* because many problems will arise when developing and commercializing these types of products and a company's past experience can help guide it around some of the obstacles. This third lesson does limit companies to markets they are already competing in. How, then, can companies successfully diversify into new, uncharted areas? For companies that desire to compete in areas outside their "comfort zone," one technique may assist: Live with your customer. This concept goes beyond adages of "knowing your market," or "getting close to your customer." Living with your customer means just that. For a period of several weeks, live at your customer's facility. Tag along when their salespeople make sales calls, attend their staff meetings, in short, for that several week period, be a part of their organization. The intimate knowledge you will gain should give you a springboard of knowledge to help understand customers' needs and wants, and the dynamics of the market. You will have developed a storehouse of information upon which to build the new venture.

4. Companies must *experimentally market* the radical innovation—testing different products and markets—because as Sophocles said in 400 B.C.: "One must learn by doing the *thing,* for though you think you know it—you have no certainty until you try."

Experimental marketing allows companies to remain flexible while trying various products and markets. This technique may seem disconcerting to young (and some old) marketers who have been schooled using quantitative tools and computer software products (like Lotus and Excel) to analyze various market scenarios and then select the course of action that offers the greatest discounted cash flow. This case demonstrates that when developing breakthroughs, performing these standard-types of analyses can be fraught with problems. First, when dealing with a breakthrough, customers may know, themselves, if they need the item. For example, if, in 1975, I would have asked you if you would be interested in having a "personal" computer on your desk, you most likely would have said, "No!" Frequently, with discontinuous innovation, customers cannot verbalize needs and wants that they may not yet know that they have. Also, even if customers express a desire for a product, the company may not be able to produce it—as was the case with NutraSweetened ice cream and cereal. As a result, companies wanting to commercialize breakthroughs must be more flexible in trying different product ideas on various market segments and to determining the optimal product/market combination.

5. Once companies have identified a product and a market, they will need to *fine-tune* and adjust their product offering and innovation strategy as well as the implementation of the strategy as new developments arise. Also, companies must then *exploit the opportunity,* to squeeze all it can from the new product, by improving the manufacturing process, reducing manufacturing costs, and expanding into niche as well as international markets. These factors enabled Searle to define more clearly the opportunity over time and to plan better for the successful commercialization of aspartame, and these steps should assist your company when trying to grapple with the uncertainties of developing and commercializing these types of complex, time-consuming, and resource-driven innovations.

34.7 CONCLUSION

Marketers like to think of aiming at a specific, predefined target, called the *target market,* with a well-defined marketing strategy, known as the *marketing mix* (pricing, product offering, channel of distribution, and promotional campaign). When marketing misses the "target," we feel that, as marketing professionals, we have failed. To spend thousands or millions of dollars on a pricing strategy or on an advertising campaign only to find that the pricing was too high or that the advertising was directed at the wrong segment, can send the best marketer looking for future employment. However, when dealing with radical innovation, several questions and unforeseen problems surface. *Discontinuous innovation* is a process that requires a long lead time and is highly dynamic and uncertain. It is difficult to predict the exact product that will be offered for sale. As an example, if manufacturing can figure out an easy, cost-effective way to produce the innovation, the price could be $1.00; however, if manufacturing is unable to devise an efficient process, the cost could reach $5.00, as was the case with aspartame-sweetened cereal. This scenario would change even the best-planned marketing strategy. Furthermore, because radical innovation can require 10 years or longer to develop and commercialize, the market may look entirely different than it did when the market was first examined. Consequently, the product is changing, the market is changing, and the manufacturing is highly uncertain.

So what is the message of this case? I think that the message of this case is one of *experimental marketing*, or management by muddling around. It is almost as though you are trying to navigate through a heavily misty and foggy swamp, where you can see only 5 ft in front of you. To map out a specific and exact navigational course when you are on one shore trying to get to the other would be foolish. However, a global goal is needed: "Get to the other side." The specifics will be worked out during the journey.

The strategy that is being presented from this case is similar to the story that was rumored when Seymour Cray was trying to develop the world's fastest computer. He hired a newly minted M.B.A. to develop a business and marketing plan, which the M.B.A. completed. The plan was inordinately long with a quantitative analysis on various scenarios. When Cray saw the plan, he elected to write one himself to present to prospective investors—it was a one-page plan that included both a 5-year and a 1-year plan. It read similar to the following:

- *Five-year plan:* Build the world's fastest computer.
- *One-year plan:* Be one-fifth of the way there.

Perhaps this story is an oversimplification of the complexities when trying to understand the market and manufacturing difficulties for commercializing radical innovation. However, it does capture the essence of the inherently dynamic and complicated environment that exists with discontinuous new products. Management must try to refrain from its need to precisely quantify opportunities that may not be able to be bounded so tightly in the early stages. To pursue this type of precision may be comforting to management, but may unduly restrict and possibly impede the development and successful commercialization of radical innovation.

34.8 SUMMARY

This chapter explores the development and marketing of aspartame, also known as NutraSweet. Specifically, the case described in this chapter details how Searle, the innovators of NutraSweet, went about trying to understand the market dynamics when developing and commercializing a radical innovation. The case depicts a company that was originally looking for one product—an ulcer treatment, and found another—a revolutionary artificial low-calorie sweetener. It is a story about how a company missed what it was shooting for (in terms of the originally envisioned product and market) but still managed to hit a target. The case also depicts how a company gradually developed an understanding of the potential for its innovation, as well as an understanding of the market and how this definition changed over time.

How can a company manage an uncertain product in an uncertain market with an unproven manufacturing process? This is the story of NutraSweet.

In developing this case, 16 personal interviews were conducted with executives who represented a wide range of responsibilities within Searle (now Monsanto) and outside the company. The executives interviewed ranged from Dan Searle, who was the CEO, to James Schlatter, the scientist who originally discovered the sweet taste of NutraSweet.

34.9 REFERENCES

1. Joseph McCann, *Sweet Success: How NutraSweet Created a Billion Dollar Business,* Business One Irwin, Homewood, Ill., 1990.

2. G. D. Searle *Annual Report,* 1967.

3. G. D. Searle *Annual Report,* 1968.

4. G. D. Searle *Annual Report,* 1969.

5. "NutraSweet," Harvard Business School, case 9-389-142, 1989.

6. "Diet Drinks Take a Lump," *Business Week,* p. 69, April 12, 1969.

7. "Calorie Cutters Weigh in Light," *Chemical Week,* pp. 125–127, April 22, 1967.

8. "Sucaryl," *Modern Packaging,* pp. 118–119, July 1963.

9. "April New Products Star Is Low-Cal Sugar," *Chemical Week,* p. 108, April 1968.

10. "Ban on Sweetener Sours Flavor Outlook Too," *Chemical Week,* pp. 16–17, Nov. 5, 1969.

11. Robert Teitelman, "Bittersweet," *Forbes,* pp. 36–37, Aug. 27, 1984.

12. Lewis Stegink and L. J. Filer, Jr., eds., *Aspartame,* Marcel Dekker, New York, 1984.

13. Robert Mazur, Research Fellow, Searle, Inc., interview, July 23, 1992.

14. "Diet Industry Has a Hungry Look," *Business Week,* pp. 41–42, Oct. 25, 1969.

15. "Cyclamates: How Sweet It Isn't," *Chemical Week,* pp. 30–31, Oct. 29, 1969.

16. "Can Citric Chemicals Fill the Sugar Bowls?" *Chemical Week,* pp. 21–24, July 21, 1971.

17. "Turning Sour on a Sweet Berry," *Business Week,* pp. 30 ff., March 25, 1972.

18. "Marketers Are Sweet on Birch Sugar," *Chemical Week,* p. 37, July 14, 1976.

19. Joseph McCann, author of *Sweet Success* (see Ref. 1), interview, July 21, 1992.

20. Phil Worley, interview, July 25, 27, 28, 1992.

21. Annette Mullendore, former director of marketing on the aspartame project, interview, July 27, 1992.

22. Barry Homler, Manager of Technical Services, Searle, Inc., interview, July 23, 1992.

23. Bob Ganger, Director of Development at General Foods, interview, July 27, 29, 1992.

24. "The FDA Gives Searle a Series of Shocks," *Business Week,* p. 20, Dec. 22, 1975.

25. G. D. Searle *Annual Report,* 1975.

26. Kelichi Koseki, "Marketing Strategies as Adopted by Ajinomoto in Southeast Asia," *J. Advertising Research,* pp. 31–34, April–May 1990.

27. "Searle Sweetener, Cleared by FDA Is Expected to Hit US Market Shortly," *Chemical Marketing Reporter,* p. 4, Aug. 5, 1974.

28. FDA Ruling, *Federal Register,* **39**(145): 27317, Washington, D.C., July 26, 1974.

29. "Team-up on Sweetener," *Chemical Week,* p. 16, April 16, 1975.

30. "Aspartame," *Federal Register,* **46**(141), Washington, D.C., July 24, 1981.

31. G. D. Searle *Annual Report,* 1977.

32. "High Prices May Sap Sugar's Market Share," *Business Week,* p. 70, June 16, 1980.

33. Gene Bylinsky, "The Battle for America's Sweet Tooth," *Fortune,* (July 26) p. 28 + 3.

34. Marcel Durot, former head of the Consumer Products Group, Searle, Inc., interview, July 24, 1992.

35. "The NutraSweet Company: Technology to Tailor-Made Foods," Harvard Business School, case 9-589-050, 1988.

36. Robert Shapiro, former president of Searle NutraSweet group, interview, April 14, 1992.

37. G. D. Searle *Annual Report,* 1984.

38. Julie Franz, "NutraSweet Ads Ready to Swirl in Three Countries," *Advertising Age,* Feb. 11, 1985.

39. "Aspartame," *Business Week,* p. 62, July 18, 1983.

40. *1986 Medical and Health Annual,* Encyclopedia Britannica, Chicago, 1986.

41. Susan J. Ainsworth, "New Sweeteners Crowd Sugar out of the Bowl," *Chemical Week,* pp. ; c6–8, Aug. 1988.

42. Susan Irving-Monshaw, "Sugar Substitutes Are Poised to Hit the Table," *Chemical Engineering,* pp. 47–49, July 1989.

43. Monsanto *Annual Report,* 1990.

44. Monsanto Company, *Moody's Industrial Manual,* 1988, pp. 4105–4106.

45. Robert Shapiro, presentation to securities analysts in New York, Sept. 20, 1989.

46. Joseph McCann, "Design Principles for an Innovating Company," *Acad. Management Executive,* **5**(2): 76–93, 1991.

47. Louis Therrien, "NutraSweet Tries Being More of a Sweetie," *Business Week,* p. 88, April 8, 1991.

48. James Schlatter, Scientist, Searle, Inc., interview, July 28, 1992.

CHAPTER 35

COMMUNICATION IN AN INTERNATIONAL ENVIRONMENT

LINKING THE TECHNOLOGY RESOURCES FROM FOUR COUNTRIES

Richard McNichols

NUM Corporation
Naperville, Illinois

35.1 INTRODUCTION

A model gives us a special way of thinking about things. By definition in the model, we make fixed assumptions and focus on the interplay of input, process, and output variables. This special way of thinking leads to more effective thinking. Think about your written and spoken communication, across different departments in your company, different regions of your country, different countries that still speak the same language, and finally, different countries with different languages. In a competitive world it is absolutely essential to remove the blinders and think about our communication models.

Reducing project cycle time when managing across borders is the case in point. Clarity in communication is where the rubber meets the road. The knowledge, the skills, and the technology as well as all other related elements must be in place. Projects are doomed to failure without interactive and understandable communication. Time after time, projects fail to meet the projections related to performance, cost, and schedule because of language differences, lack of translation competency and integrity, and the biases inherent in different cultures.

The following case history regarding a multinational project on a short time schedule sheds some light on these interactions and demonstrates the issues involved and approaches used to assist the communication process. This case history, while anecdotal, presents the communication issues in a multinational project through a discussion of the

- Sources of information
- International dimensions

- Brief history of the organization
- Communication process variables
- Guidelines of implementation

35.1.1 Factory Automation

A tour of manufacturers reveals a wide variety of physical environments and cultural attitudes. My trips through factories in the United States and Canada have been varied, and subject to extremes. Perhaps the worst were in the industrial Northeast, where the strong and sweaty wielded giant wrenches, positioned a replacement fitting, and pushed levers with all their might, in a dangerous and smoking industrial hell just this side of Dante. Framed by the faint illumination of control panel lights, a welder issued Fourth of July sparks to seal the final connection. In a garish choreograph, machinists nurtured balking and recalcitrant machinery back up to the rhythmic clatter of production with tense urgency to defray the cost of lost production time at a rate of $100,000 an hour.

Other factories I saw in the Southeast were cleaner and more organized, but still a lot of waste could be seen in materials handling and plant layout.

Yet, my tour of the production locations in France showed spotlessly clean plants, with most of the parts production automated, attended by women fashionably attired and in heels to boot. Telemecanique led industry with a special program attracting women with reduced hours so that they could work a little and be home for the children after school. From about 9:00 until 3:00, with time for lunch and coffee breaks, a largely female workforce manufactured products exported all over the world and then went home in time to check the children's homework. It does not get any better than this. Or does it?

35.1.2 Parent-Company History

In 1924 Telemecanique Electrique was founded in France by four engineers who developed a technology and manufactured the resultant components to start and stop large motors over long distances. In essence, they began as a system control business. By centralizing controls, they felt that people would be safer working on large machines, and could control more of them from one central location. Over the years, the company thrived, expanding into all types of electrical and electronic controls.

The company had some interesting credos. No plant should have more than 200 people at work. No plant should have a clock to punch. Limiting the plant size developed specialization, expertise, and individual recognition. Without the tyranny of the punch (time) clock, the production worker's self-esteem grew to the exacting standards of professional commitment. By building factories across France (and later the world) in small and large towns, the company took advantage of regional variations in location costs and selling prices, and reduced the susceptibility to business risks that plagued the centralized firm.

35.1.3 Unique Strengths

Dedicated workers continued to modify existing products, miniaturize, and combine functions. They also carried these efforts upstream to how they made the product. Advancements were made in factory automation, assembly lines, transfer lines, and motor controls. The resulting stream of new products and new production technologies created *phenomenal opportunities*. Each plant expanded production lines and capacity with *remarkable improvements* in quality using their own Telemecanique product and Telemecanique production technologies.

The company developed elaborate photoelectric devices and proximity detectors. Most of you who enter stores with automatic doors know intuitively what a photoelectric device is. It projects a beam of light toward a photoelectric cell. When the beam is broken, a switch is tripped, opening a door or sounding a buzzer. A proximity detector uses a different technology to trip a switch given the presence of a material under wide-ranging conditions, magnetic, plastic, hot, or cold. For example, the motor controls were soon interfaced with photoelectric cells and proximity detectors on the assembly line eliminating elaborate manual efforts to synchronize work up and down the assembly line. Work throughout increased dramatically. The company's motto was "Innovative solutions to control problems," and no one knew it more thoroughly than their engineers, as the product that was sold, made the product. The proof was in the pudding.

This is not unique to Telemecanique. Actually, many manufacturers make products that are used in their own factories. The differences in degree are startling, however. Some Japanese firms have duplicated their existing factories in the United States right down to the wallpaper. U.S. manufacturers field-test emerging technologies on production lines in their own factories before marketing. The feedback from production personnel is important. The vendor factory sells to the customer factory at an intercompany transfer price. As intercompany politics are involved in the bargaining process, the price may be unreasonably high, creating financial distortions and vociferous objections. Yet the deal is done "for the good of the team." This is not always the case; it is a matter of company culture. Another Japanese concern in the same industry departs from this approach. If the internal transfer price is excessive, plant managers are encouraged to use other sources, even their competitors, as long as they get the best price adjusted for quality and productivity.

35.1.4 Subsidiary History

In 1962 the *computer numerical control* (hereafter CNC) group was formed as a department within Telemecanique. A CNC coordinates the motors in a machine to accomplish work in a highly precise and repeatable manner. The computer itself is useful in computing mathematical equations that transcribe arcs and motions from drawings to three dimensions. Stated simply, a CNC is to parts manufacture as a word processor is to documents. Both have the abilities to catalog, produce, and "print out" tangible results.

35.1.5 The Market

A CNC is better appreciated in terms of the products it produces. One area of specialty is the lowly and often unseen gear, quietly and surely moving civilization about its business. There are gears for watches, drills, eggbeaters, and cement mixers, automotive transmissions, locomotives, and even turbine blades for jet engines and hydroelectric power generators. A gear is an interesting study in circles, arcs, and tangents. Calculating the motions is trigonometrically and numerically intensive. For example, to cut the circle for the gear, grind the teeth along the radial, allowing for the continuous rate of change needed along the teeth and make its smoothly fitting mate requires mathematical and dimensional precision. The CNC is wired to the motors that operate the machine in the same way the brain operates the muscles to move the skeleton.

The market for machine tools is a lively and diverse one. Machine tools and their manufacturers are from Germany, Japan, Switzerland, and the United States. Some machine tools fetch a price of $1.5 million, run for 30 years, and produce $1.5 million in annual revenues. This is certainly not the domain of a thinly capitalized player, nor one faint of heart. A malfunction from bad design can kill, but smooth operation can print money faster than a banana republic. The machines are often unique, one-of-a-kind, highly engineered products. The producers of machine tools are a maverick bunch, faced with the task of unusual nonrepetitive problems to solve, and a high degree of experience required. A small miscalculation can set a project back months; a middle-sized mistake can be fatal.

35.2 INTERNATIONAL DIMENSIONS OF MANAGING TECHNOLOGY AND PROJECTS

The following experiences on a recent project framed my thinking on the subject. We had a virtual monopoly on several emerging technologies developed by our international subsidiaries. This would be demonstrated at the International Machine Tool Show, which was widely attended by affluent buyers. From the back of an envelope in May to assembly in August and a live functioning exhibition starting September 5th, life could have been a risky battle of minutes. Schedules were set tightly. One slipup could jeopardize millions of dollars in ruined machinery, lost customers, and future business.

35.2.1. Project Scope

The exhibit was a remanufactured Cincinnati lathe, with a twist. Along conventional designs, a machinist manually turned the handwheels along horizontal and lateral axes to shape the parts. Outfitted with a flat-panel LCD, servo motors, and a CNC, the operator could twist the handwheels, and even feel the tension while cutting into the part. The flat-panel LCD display ergonomically gave operators feedback on the parts they machined. Where an operator's skills fell short, the CNC could be used to design the advanced geometries, and then produce quantities using fully optimized high-speed machining techniques.

The motors and servo drives came from our Italian subsidiary. Powerful like a Ferrari, delivery would take in excess of 8 weeks. The computer numeric control came from France. The flat-panel LCD display, the most visible and central ergonomic feature of the product, came from Germany. Software development that was assigned to the United States posed a specific problem. The hardware was not available during the software development, test, and debug phases. Additional up-front effort was required to precisely define the variables and control system configuration. This intercountry interaction involved close scrutiny of the system's operational parameters. Table 35.1 shows the origin of parts, assemblies, and products.

Against these pressing and immediate requirements, and in direct opposition to any apparent progress, stood formidable barriers. The barriers related to

- Language—reducing uncertainty

    The second language

TABLE 35.1 Products by Company and Language

Country	Product	Language
France	CNC	French
Germany	LCD	German
Italy	Servo motors and drives	Italian
USA	Machine	English
USA	Software	English

The translator

Clarity in communication

- Cultural issues
- Documentation
- Technical
- Financial
- Transfers between countries

35.2.2 Languages: Reducing Uncertainty

Sometimes I think that the whole purpose of human existence is to reduce uncertainty. We wonder and question perhaps without even knowing that these musings follow very predictable patterns. When we question, we anticipate a menu of simple answers. On the basis of that, we may ask more questions about the consequences of each of those answers. And finally, we need to know the costs or performance characteristics of those choices, in order to rank the answers and pass judgment. Our ability to communicate quickly and accurately *without distortion* helps reduce uncertainties.

Regardless of nationality, people differ markedly in communication skills within their own native language. Many do not possess a sophisticated ability to make a request; make that request understood; in turn, understand the reply, modify their request; and then, on the basis of what was heard, congenially formulate an action plan. Subject to the foregoing, they therefore do not differ in their ability to reduce uncertainty. Many such efforts end not in productive inquiry, but in frustration. Advanced education does not seem to help. None of you would be surprised to be told that our colleges and universities do not turn out very good writers, skilled readers, or attentive listeners. Even armed with omnipotent computers and bowed before the church of latter-day spelling and grammar checkers, the results are pretty scary. While humorous stories and anecdotes about miscommunication abound, comedy is not financially tolerable in business. A typographical error on a work order can mean product built to the wrong specification, time delays, and money wasted. Sloppy writing can open vistas of litigation on obligations and performance. Understanding problems, even those without ready answers, is vital. Understanding—who does what to whom, and when—is critical.

The Second Language. Communication skills in a second language are usually worse. The language skills taught even at the college level are enough to get one through customs, restaurants, restrooms, and some classic eighteenth-century literature but not through street slang and complex negotiations. Unless the person thoroughly knows basic grammatical rules, it is doubtful that grammatical rules will be mastered in another language to truly perceive every nuance, especially in a highly technical and pressing environment.

At this point we meet the double standard. People can be highly literate in another language but have such poor pronunciation, or a native accent, that they are reluctant to speak, although they comprehend every word spoken. At the same time, those participating in the latest "10 easy lessons approach" to a foreign language often exaggerate their level of understanding.

The Translator.   Ah! you might say, I have the answer, use a translator! They do not work out well as company slang and technical jargon are not taught, and again, the commercial worlds in revolutionary turmoil spin around urgent missives and arcane kant. If a translator is not highly technical, the results can be disastrous. Instead of a meeting of the minds, you have the befuddled brainless, one who knows what one wants to show and tell, and the other knowing full well all the possibilities of what could be said, but without any technical understanding, what probably should be said.

Translators vary in the level of their skills and most certainly their endurance. At a recent European sales meeting the translator arrived 2 hours late, because of heavy traffic. Traffic is horrendous in Paris. In the meantime, one of the European financial executives tried to translate, but did not know the product line. When the discussions covered company finances, the translation was superb, but as every chart looked the same, nobody paid any attention. Financial details bore engineers, and you, or anyone else, could not make them pay attention. The room was hot, and the speakers, lacking really dynamic presentation skills, took their toll on even the most curious, reducing the audience to a confused slumber. My favorite moment was when one of the speakers was giving his speech in English and the translator had lost her bearings and was rephrasing ever so slightly what he said, in English! I let it go on for a while and then gave her a coffee break. I could not blame her.

Management meetings which on the surface seem straightforward can be highly cloaked negotiations, employing all sorts of duplicitous strategies to close a deal. The question arises as to whether the translator functions as a translator or an interpreter. Research shows that translators who interpret often inject their own biases. These occur in technical translation as well as in management communications.

35.2.3 Clarity in Communications

If you think people's communication skills vary, just test their filing skills. We needed a method that got our message across, was countersigned by someone responsible, and was scheduled for delivery. Put only one subject in a memo, never two or three, as people need to be able to integrate your action plans with their ticklers.

35.2.4 Cultural Variables

Regrettably, if superb communication skills are not enough, the international environment introduces cultural variables that are difficult to control. The term *world class* implies universal standards, but few understand them or even know about them. Things you take for granted and count on are nonexistent or worse. An American might say "do or die," start early, work through lunch, stay late, and even reschedule vacation without complaint. A European may start late, take a long lunch, and miss an important deadline without so much as an apology. If it is an accepted company norm, there may be little you can do, except schedule your way around it. But a multinational project needs an agreement on the ground rules. These rules should take into account the diverse ways of working. In essence, the rules accommodate the idiosyncrasies of the culture without necessarily submitting to those idiosyncrasies. So, get your schedule declared and get it countersigned. Production planning is a poor place for a hidden agenda.

Technical skills, organizational skills, and swift follow-up are vital in dealing with cultures for whom time is not a series of consecutive moments, but perhaps a spaced-out trail to another dimension. You also need a sixth sense that detects when something is amiss. It could be a meeting that went too well, or a memo that is cheerfully acknowledged, or promises made, or amazing product specifications. Often questions and answers need to be repeated so that every person understands the impact of the decision, the price of success, and the penalty for failure.

35.2.5 Documentation

Technical people understand the physics of the project and the quantification of the specifications, and grasp the nuts and bolts of how things go together. Fortunately, mathematical disciplines and the related symbologies are taught congruently in European and American schools. Yet there are very troublesome fixed assumptions that people take for granted and part of our communications model that are missing or poorly positioned. These differences make the difference between success and failure. Chief differences appear in documentation. Europeans like their tables of contents at the end. Americans may find this charming, or so objectionable as to be out of the question. Europeans use paper of a different size. Size A4 paper, for instance, will not fit in standard filing cabinets—another objection. Finally, the American market has been much closer to the significant advances in technical writing pulled along by superb software development. Poorly illustrated technical manuals developed on antiquated publishing systems that are long on hyperbole and short on savoir-faire are just not going to cut it.

35.2.6 Technical Issues

Technical issues to resolve are myriad. Foreign companies may be viewed as having only a fleeting presence, spotty local distribution, and limited, if any, on-site service. Backup hardware may have to come from thousands of miles away with long delays. From there, we may encounter inch/metric measurement differences, different voltages, and different current requirements. The software menus and data-collection schemes may not be modified in mission critical applications to the customer's satisfaction. The software may not be capable of being networked.

35.2.7 Financial Issues

The devil is often in the financial details. Since the bulk of the financial details are often developed in negotiations, there are very few immutable facts. Payment terms may be negotiated backward from a standard American net 30 all the way to 120 days or more as there is no European standard, and they are used to tolerating longer collection times. A progress payment schedule is advisable with dates, performance criteria, and amounts clearly delineated. Anticipate the most common problems and incorporate them into an agreement.

35.2.8 Transfers between Countries

Transfers of materials, people, or technology need to be clearly conceived and planned in advance to predetermine the rights and obligations. Visualize the transfer to minimize logistical problems with carriers and customs. Pave the way for a low-cost shipment by getting bids on transportation rates. Making it up as you go along from FOB (freight-on-board) origin through several trucking, railroad, marine, and air freight concerns is prohibitively expensive. Make sure you have the capability to handle heavy cargo at the destination and at any point en route. Customs regulations are onerous and require great expertise. Transferring technology can fall under many different duty rates depending on legislative incentives. Even a simple thing like a warranty shipment should be a no-charge nonevent. But is it incumbent on the party to pay customs, duty, brokerage, and transport fees all the way back to the manufacturer? That no-charge warranty may cost thousands. Better check the fine print and change it before it is too late.

Maintaining your intellectual property rights is very problematic. The perception may not be the reality. What about special software written for the project? Presume that your highly secret manufacturing technology is machine-coded to perfection. Is it truly secret any more? Can the manufacturer sell it to your competitor? How you communicate these issues will depend on what you and your reader assume as important. Beyond the United States, intellectual property is not sacrosanct.

What about future purchases of spare parts or service? How will they be priced? Are the prices reasonable? What will happen if the price of the foreign currency rapidly escalates?

35.3 GUIDELINES FOR IMPLEMENTATION

Most problems in this vein are communication problems. If you follow communication from physical conception, verbalization, translation, transmission, reception, interpretation, and response, you can usually pinpoint what items need further elaboration and commitment.

Put clear translations immediately adjacent on the same page. Provide a dotted line for a countersigned commitment for follow-up. Take a minute to study Table 35.2. The

TABLE 35.2 How to Provide a Clear Translation of Your Action Plans and Deadlines for Compliance

Objective:	Proposed	Due
Part number:		
Description:		
Documentation:		
Software:		
Working prototype:		
Beta test:		
Client acceptance:		

concept is clear. You can cross-reference your communiqués to your project management software, or your Gantt chart, whatever the case may be. In practice we use Italian and German as well.

Observe the following pointers. Write simply to your international colleagues, telling them exactly what they must do and by when to meet the project commitments. It takes a great deal of effort to simplify, but that is the only way. Avoid using clichés, colloquialisms, humor, vernacular, idioms, professional mumbo jumbo, or anything vague. Whatever they do not understand, they will try to look up in a dictionary, and the predictable result is confusion and lost time.

35.3.1   Business Meetings in Virtual Reality

Much has been written about proper conduct in international business meetings. But what if you do not have to meet? How do you structure protocol? We did not want to travel anywhere for the project. All we needed was the fax machine; it was preferable to talking, and certainly to travel. All we needed was a method for getting people to buy into the project, and an effective method of reminding them that their promises were tightly scheduled and part of a highly visible project.

35.3.2   The Bulletin Board as the Lingua Franca

While passing bilingual faxes resolves simple problems, we needed to send larger reports, program listings, graphic files, and so on. Here, we found bulletin-board software that met our requirements: simplicity, file security, and—best of all—the ability to use different languages. One unintended but beneficial result was that other members of the corporation not involved with running the project made helpful suggestions.

35.3.3   Virtual Logistics

Do you think that the best way to win in foreign markets is to source and manufacture as closely as possible to the customer? We think not. By standardizing communications, from the global strategic level to the individual, across companies, and through governmental regulations, a reduction of communication problems can reengineer the manufacturing process and save time. Communiqués from faraway places look immediately familiar. Since the ambiguities have been hashed out, they can be immediately acted on. By closely managing each link of this supply chain, you could produce the same tangible results as if all the related elements were right next door.

35.3.4   The Shrinking World

As the world shrinks, work that was previously done at the manufacturing plant can be now be done more economically en route, subject to successfully resolving the language barriers above. It is a pivotal transformation in manufacturing technology.

For example, a manufacturer in Germany can ship a machine carcass by boat to New York; ship by rail to Rockford, Ill.; add controls, engineering, and software; continue on by rail to Seattle; and then ship the machine by boat again to Japan. This is called a *sea land–sea bridge.* Sea land–sea bridges combined with vastly rapidly improving airports are rapidly shrinking transit times. Travel times from the Far East through Seattle, across the states by rail, and overseas to Europe are routinely done at a price that is a bargain.

Air freight terminals have been renovated to handle large loads expeditiously; 8000 mi and 4000 lb can be routed into a domestic plant on a 48-hour basis as if it were a simple domestic shipment. There is no reason why you cannot apply just-in-time (JIT) shipment across borders *if you standardize the process of communication.*

You should be able to procure parts from Europe and Asia, and assemble the finished goods here in the United States as efficiently as if the whole operation from fabrication to final assembly were taking place in the United States. If you think that the idea is novel, get familiar with it. Global competition is heating up, and those who fail to master strong communication skills will soon be left behind. Superb communication skills will have you adeptly manufacturing en route and navigating the wayward flows of product to customers taking every advantage of fluctuating currencies and production efficiencies in the same way that Magellan used the stars.

35.4 SUMMARY

If you think about the way we communicate, much of what we can say and need to hear benefits from a formalized structure and reciprocal commitment. We have explored several ways at getting acknowledgments and commitments from our international colleagues. By standardizing on communication and commitment, we can "shrink the world" and shrink cycle time. It is important to observe these ideas as they are rapidly becoming a competitive standard.

CHAPTER 36

PRODUCT DEFINITION

KEYS TO SUCCESSFUL PRODUCT DESIGN AND MARKETPLACE ACCEPTANCE

Edith Wilson

Hewlett-Packard Company
Palo Alto, California

Until the late 1980s, high-technology industries did not need to focus on the product definition step of product development to the degree that they do today (Fig. 36.1). Many reasons underlie this. First, user and customer needs are more precisely defined in the 1990s than they were in the 1970s and 1980s. Then, there were fewer competitive products and global competition was not as intense as it is today. Consequently, firms could release a new product, complete with its deficiencies, and customers

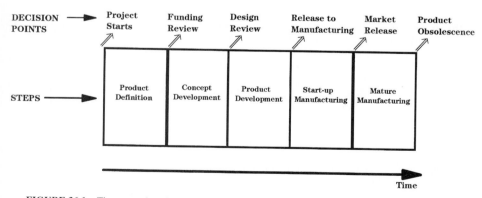

FIGURE 36.1   The steps of product development.

would buy it because there was no reasonable alternative. As technology-based products became more abundant in the 1980s and 1990s, customers became more savvy and products had to solve more of their needs to gain market acceptance. In parallel, competitors became more ubiquitous so designs had to really meet the customers' needs for a sale to be secured. Consequently, in the late 1980s, successful companies quickly developed techniques to define not only how they would develop and launch products but also how they would define what customers truly need from products.

Mastering the product definition phase of product development is an emerging capability. A number of authors have explored the subject of needing to do product definition well,[1,2] but none have outlined the steps necessary to define and screen a successful product. Here, 14 key steps of product definition will be outlined. These steps are based on action research conducted at Hewlett-Packard,[3–5] General Motors, General Electric, Motorola, Xerox, and IBM[6] in which the product definitions of products that were successful in the marketplace were compared with those of products that were unsuccessful in the marketplace. Follow-up research conducted by the Product Development Consulting Firm in Cambridge, Mass. at 43 companies verified the efficacy of the model.[7]

The 14 steps of product definition are

1. Strategic alignment
2. Understanding user and customer needs
3. Competitive analysis
4. Compliance issues
5. Localization
6. Product positioning and value proposition
7. Decision priority list
8. Risk analysis
9. Core competencies
10. Marketing channel and support
11. Strategic dependencies
12. Leadership
13. Resources
14. Project and business plan

From the author's action research and subsequent consulting work with product development teams at Hewlett-Packard, successful projects were found to have adequately researched all 14 of these steps and to have based their subsequent decisions on fact, not hunches or intuition. This critical differentiation of the methods employed by the successful versus the unsuccessful teams led to the development of a product definition model which illustrates the logical cause and effect relationships between each of the 14 steps (Fig. 36.2). Notably, both successful and unsuccessful teams use a similar product definition process; the difference between the teams continues to be the consistently higher level of knowledge the successful teams have about each factor prior to making key project decisions. This points to teams not needing more process training, but rather more coaching and guidance on adequately completing each of the product definition factors. This preparation for the product specification, plan, and financial analysis culminates in the product definition.

The product definition step which most unsuccessful projects have problems with is understanding user and customer needs. This was noted in three retrospective stud-

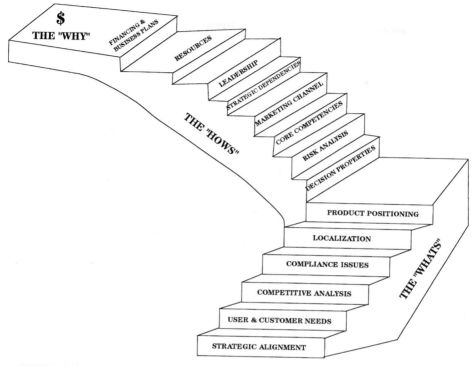

FIGURE 36.2 The steps of product definition.

ies in 1973, the first conducted by the University of Sussex, England's Science and Policy Research Unit, which was called Project SAPPHO,[8–10] the second of which was conducted at Hewlett-Packard[3] and the third conducted at General Electric, General Motors, IBM, Motorola, Xerox, and Hewlett-Packard with a grant from the Sloan Foundation on U.S. Competitiveness.[6] Ongoing consulting work with product development teams at Hewlett-Packard and by John Carter and Sheila Mello of Product Development Consulting with their clients at more than 43 companies[7] also supports the fact that difficulties with projects in this phase are rife. In analyzing projects even more thoroughly, the product definition problems most development teams struggle with are in the first steps of product definition, which we call the "what" steps since they are used to identify what the marketplace needs from a product. The "what" steps are followed by "how" and "why" steps (Fig. 36.3).

36.1 THE "WHAT" STEPS

The goal in completing the "what" steps, which include strategic alignment, user and customer needs, competitive analysis, localization, compliances, and product positioning, is to develop a compelling product positioning statement which mutually optimizes the product's contribution to the sponsoring organization and to the target mar-

The "whats"

- Strategic alignment
- User and customer needs
- Competitive analysis
- Localization
- Product positioning

The "hows"

- Project priorities
- Risk assessment
- Core competencies
- Strategic dependencies
- Project leadership
- Project resources

The "why"

- Project and business plan

FIGURE 36.3   The "whats," "hows," and "why" of product definition.

ketplace. In this section, the goal and relevance of each of these steps is explored to outline the work teams need to complete to identify what the marketplace needs.

36.1.1   Strategic Alignment

The first "what" step, strategic alignment, sets the guidelines and objectives for the team. The often observed problem with strategic alignment is that organizations devote too little effort to creating strategies. This results in plans that provide little meaningful direction to the project team and results in the organization being misguided since there is no strategic backdrop against which to make decisions. Time delays occur regularly since the projects' product definitions and expected contributions to the organization must be reexamined on a regular basis. This problem can be corrected if management spends sufficient time creating and updating the organization's strategic plans and communicating the expectation of what the project needs to contribute to the organization and what boundary conditions the project needs to operate within to the project team.

Several key points need to included. With respect to the project's contribution, the team needs to clarify what they are expected to deliver to the organization in terms of revenue vs. time, profitability, cash flow, timing, and return on investments and assets. Additionally, the team needs to identify the level of financial and human investment with which they are to complete the project. Finally, the strategy needs to guide the team with regard to how the project fits with the rest of the product line by determining whether the project is an extension of an existing line, a competitive reaction, or a revolutionary product. Additionally, the strategy should provide guidance on what segment of the marketplace the product is targeted for. The segmentation guidance is a necessary supplement to managements' strategic goals in order for the team to have boundaries within which the investigation and development of the product is to be conducted. Additional guidelines should include financial expectations for the investment, the timing of key project reviews, and how progress will be tracked.

These points are straightforward, yet many teams are not provided appropriate guidance. This results in the time to market being lengthened to address each unclear topic. Once the strategic stage has been set, the team is ready to move to the next set of "what" factors: "user and customer needs," "competitive analysis," "compliance," and "localization." These four factors need to be concurrently researched since information from each one will affect the other three. You can think of this as trying to cooptimize your understanding of this group of steps.

36.1.2   User and Customer Needs

In the author's work, understanding user and customer needs is the most elusive and most expensive part of the product definition process effort. In the United States alone, billions of dollars are spent conducting both quantitative and qualitative research to identify the needs of the marketplace each year, yet as many as half of the projects initiated either never make it to market or result in a product that is unsuccessful in the market as measured by revenues, profit, net present value, return on investment, and return on assets. Robert Cooper, in *Winning at New Products,*[11,12] states that less than 40 percent of North America's investment in new-product development yields successful products. What goes wrong? The question that is not answered is, "What is the fundamental problem that this product needs to solve for this set of people?" While this question sounds trite, it is truly difficult for many teams to arrive at a succinct answer. Why do so many teams have difficulty answering this question? First, there are many places that they can go wrong. A number of the pitfalls are explored and a number of methods are presented that can be used to answer this question. In Fig. 36.4, the process of identifying user and customer needs is diagrammed.

To begin answering the question, the team must understand which segment of the marketplace they are targeting their product for. Oftentimes, teams confound segments and homogenize the needs of many segments. This results in a product definition that does something for everyone and either does too much for one segment at too high a price or does not meet the needs of any segment at any price. Just think, if you were designing laser printers and thought that the corporate user and the sole-proprietor user were in the same market segment. You then might conclude that having a printer that could operate on a network and print extremely fast were needs of the whole segment. Well, we know that the sole proprietor would love the faster speed but is unlikely to need network printing capability. However, we also know that there is less price sensitivity in the corporate user segment than in the solo user segment. By separating the segments, we are better able to first understand and identify the specific needs of the segment and then clarify our understanding of those needs. Another reason for identifying the segments of the market are that you need to quantify the size and growth rate of each market segment and the share of that market that both you and your competitors have.

So, how does one understand the needs of the users and the customers? The author's favorite starting point is visiting customers. Caution is needed in making customer visits; do not just go with a laundry list of questions because the customers will just answer the questions you ask. The worst question you can ever ask is, "What would you like for our new product to do?" Customers invariably will tell you about their most recent problem and suggest a solution that is probably one that a competitor has already proposed to them. If you base your design on this information, it is frequently outdated prior to market release by a competitor's release of a similarly designed product.

The best customer visit is one in which a cross-functional team visits the customer site and moves the meeting out of the conference room and spends time observing or, even better, performing the customer's work. It is important to understand the whole

Attribute	Your Co.	Co. A	Co. B	Co. C
Print Quality	800 dpi	600dpi	1200 dpi	700 dpi
Price	$325	$500	$1.0K	$800

FIGURE 36.4   Customer perception of product attributes.

process of how the work is performed. For example, if your team is researching the use of printers at law offices, it will be necessary to observe how the lawyers generate their documents:

Do they dictate the documents and does a secretary then type a draft?

Do they type the drafts themselves, or do they handwrite the originals?

After the draft is completed, how are corrections made?

Does a team of administrative assistants simultaneously work on a large document, and do they each edit and print sections, or does one assistant do the whole job?

Questions about the equipment that they are using are necessary as they might not have all the same make and model, which could cause the documents to look different. This line of observation and questioning may seem absurd, but a study similar to this identified that large law offices were frustrated with Hewlett-Packard printers because they had several different models and each had a different print resolution, which resulted in complex documents having a varied appearance when different sections were printed on different models of printers.

The goal of observation is to identify what discontinuities occur while customers are doing their work. These are the opportunity points for your design and the reason a cross-functional team is necessary is that each member will bring a very different perspective about how each of them observes what needs to be addressed to appropriately meet the users' and customers' needs. However, teams need to be prepared to accept that the best that one can really hope to get out of a series of customer visits are a number of ideas of what the product ought to embody to meet the needs of the users and customers in this segment.

Customer visits are an example of qualitative methods to identify users' and customers' needs. This is not statistically valid, but Griffin found that teams need to visit at least 20 different customers to be able to statistically project their observations.[13] A couple of caveats are in order: (1) visit customers who both are and are not currently buying your product and (2) include customers in your sample who are purchasing the competitors' product or not buying at all. If you choose just one of these four subcategories, you will probably distort the data and will not be able to project your observations on the other subcategories.

Other methods of qualitative research are available. Among the most popular are focus groups,[14-18] in which a sample population (usually less than 12 people) are brought together to discuss their needs. The benefits of this method are that the identity of your company can be left unknown to the interviewees and that you eliminate bias by having an outside facilitator guiding the discussion. The team is usually able to observe the focus group from behind a one-way mirror, and the discussion can be videotaped.

The next step is to develop and validate a hypothetical product definition. To develop the product definition, the team will usually meet and brainstorm about what to embody in the new product. This is often difficult because each functional area will have a different opinion about what they heard from the customers in the qualitative interviews, and these differences need to be rectified. In this rectification, perfection should not be strived for because the next step of understanding user and customer needs is to go out and test the hypotheses that have been developed using quantitative methods. This is the step in which data is collected about what customers truly prefer and need as opposed to what they do not prefer or need. There are a variety of ways to do this, including surveys[19-24] and conjoint analysis.[25-29] A favorite is to use a $100 test in which a number of product attributes are listed and the customer is then asked to assign portions of a $100 budget to each attribute on the list in order of importance to them. Caution always needs to be taken to ensure that the data is from only one segment of the marketplace, which can be determined by looking at the standard deviations of the data; a large deviation would provide a hint that there were subtleties in marketplace preferences that the segmentation did not capture. If this method is used on a sufficiently large sample size, the preferences and needs of the segment will be determined. These methods will not be described in detail here as they are covered in the many references devoted to the subject and by consulting firms that specialize in select methods. The purpose of this quantitative revisiting of user and customer needs is to validate the hypothesis that you have developed. Through working with numerous organizations, validation of this sort has been found to shorten the development cycle because clarity about user and customer priorities exists. To get more information on needs, organizations such as the International Society of Business Marketing are useful to direct you to appropriate sources for methods.

36.1.3 Competitive Analysis

This is the next step of the "whats," and it is important to analyze each of the top three companies in your product area. The way to select which products and companies are on this short list is to identify which three companies have the most market share in your product area; these companies will be called the "giants." One caveat is to also look for companies that have significantly higher growth in market share than do any of the main competitors and to include them on your list as a "sprite" that may soon be a market share leader (Fig. 36.4). In the recent past, many people have thought of competitive analysis as the process of doing only a product component and manufacturing process breakdown. Unfortunately, this is insufficient, and much more needs to be understood if we are to be able to understand why competition has made the moves they have and to forecast what moves they will make with future products. The goal of competitive analysis is to understand the whole set of interactions a customer will have with a competitive firm and its product vs. your own firm and your product. This is a necessary prerequisite to enable your project team to identify each competitor's strategy and what areas your product needs to be strong in and where it can be weak. Researching this will provide the basis on which to identify what it will take to beat the competition in meeting the needs of the users and the customers. A useful method for cataloging this information is a QFD (quality function deployment) chart using the methods outlined by Hauser in the *Harvard Business Review*[30] and Clausing in his treatise on total quality development.[31]

Completing competitive analysis thoroughly entails understanding how a customer is initially exposed to a company and its products. Exploring whether the company is using word of mouth, advertising in journals, magazines, and newspapers, has a toll-free 800 customer service phone number, or is relying on presenting its products attractively in retail distributors is the initial goal. The critical issue then is how the users and the customers like the experience they have with the competitor and to compare the competitor's market share and growth rates with your own to fully determine whether your marketing communications are effective. Above all, you need to understand whether the competitor or you offers the best information and experience. If there is room to improve this experience, is there something unique that your project team can do to improve the experience your firm provides the customer?

Second, it is important to understand what is entailed in purchasing the product and how the customer goes about doing this. Does the product have to be ordered through a third party, or is there a mode of distribution different from the method your company is using? An example is how more and more sophisticated products such as computers, printers, and facsimile (fax) machines are being purchased at warehouse stores rather than from computer-specialty stores or through company-specific salespeople. If changes such as this affect how customers wish to purchase products similar to the ones you will be manufacturing, it is critical for you to identify this trend so that your firm can change their distribution process time for the release of the product.

One of the most critical areas to investigate about the competitive products is its set of attributes and the pricing of their products. The major attributes of your product and of each of the top three competitors' products need to be listed. Next, compare these attributes to the customers and users needs and rate them as equal (=), plus (+), or minus (−) in their ability to meet these needs. The third step is to assign a weight to each of these attributes using the customer's and user's perspective as to whether these are important (+), neutral (=), or unimportant (−). By comparing the scores of the competitors, the users, the customers, and your project, areas of opportunity for your project will emerge. See Fig. 36.4 for an example.

Next, your team will learn what is it like for the customer to receive and install the product. For many products, this is a trivial part of the interchange between the customer and the company; however, for others, this can be horrible. The first consideration is whether or not all the necessary components are delivered concurrently and in a timely manner. Next, understanding what is entailed to install the equipment is studied. For the late 1980s and early 1990s, Apple Computer Company had a distinct advantage over the personal computers because their equipment was supplied with preloaded software and printer drivers and all one had to do was plug the computer in and begin using it. Compared with spending hours installing various software applications and installing an appropriate printer driver for a personal computer, Apple won the approval of many customers for the ease with which setup could be done. Once again, the project team's goal is to understand what method would be the most satisfactory from the users' and customers' perspectives.

Next, do the users and customers of your product need any special training to use that product? Does the customer need to attend classes, is there on-line self-help, or are there manuals which customers need to read themselves? How do customers learn to use the product?

Once the customers have been trained in the use of the product, next is the step of actually using the product. Two aspects of use need to be determined: (1) how easy the product is to use and (2) whether the product really meets the customers' and users' needs and solves their problems and objectives while facilitating productivity or whether the product is difficult to use although it provides useful information. Thoroughly understanding deficiencies in this area is difficult because customers may not be able to articulate their frustrations, but, when they use something that is better, they can immediately identify the earlier problems. Using the product includes not only day-to-day use but also the support the customers will need for the product. Support may include telephone hotlines or on-line help, as well as how customers can get repairs done and purchase refills and secure upgrades for the product.

Ultimately, how the product will be discarded at the end of its life needs to be addressed. In many European countries, obsolescence is now the responsibility of the original manufacturer, so design for reuse, teardown, and recycling needs to be considered during the product definition. Now that there are an increasing number of environmental issues which prevent disposal of many subcomponents and hazardous substances, how the product is discarded is a major concern. This is to assure that the end-of-product-life experience is a smooth one for the customer.

In summary, taking the whole-life approach of looking at the entire set of interactions that a customer must have with a company to buy, use, and dispose of the product offers the product development team a variety of opportunities to identify the competitors' strengths and weaknesses and, thus, opportunities for your project team to differentiate your product. Asking "Do we need to be better, the same, or worse than each of the competitor's attributes?" is critical if you are to gain market share and accurately position your product to win market share. It is important to remember that the competition you should measure yourselves against includes not only the direct competitors but also the "sprites," which are often the companies that use alternative techniques to solve the customers' problems. An example is in the test and measurement industry, in which the giant companies were also building expensive test equipment which was used at the end of the manufacturing line to test the quality of the product. The project teams repeatedly did not consider companies that were designing software simulation tools as competitors until too late, when simulation had displaced much of the need for testing on the manufacturing line.

Additionally, if you look at the "spritely" and "giant" competitors over time, you can begin to determine what their strategies are. Some major strategy themes include the price/performance leader, the budget product, the service provider, and the performance/quality leader. It is important to identify what all competitors offer that satisfies their customers. By tracking the competitive information over several years and several product generations, you will observe trends of how competitors invest, and, if you augment this with your current intelligence, you will be able to project what moves the competitors might make in the future.

It is critical to look for novel methods to secure information about the competitors and their products. There is a great deal of information about the competition in product data sheets and brochures, but there are other ways to gain access to how the competition might be moving from other sources. These include securing a clipping service that will make copies of all articles that occur in a selection of newspapers, journals, and magazines. Additionally, having the service keep track of the classified ("want") ads in the major newspapers and in local newspapers serving the region where your competitor is located can provide extremely useful information such as hiring emphases. Other sources can be attending tradeshows, interviewing past employees of the companies, and talking to common customers and vendors. Gaining access to using the competitor's equipment can be difficult at times, but leasing the equipment from leasing companies, using the equipment at customer sites, and working with a friendly distributor can be very helpful in providing access to the products. If worse comes to worse, or the product is inexpensive, or if the information is critical to your success and the product is relatively inexpensive, just purchase the product.

The final question you need to ask is, "If we want more market share, what do we need to do?" Using the methods described above will help you identify your competitor's strategies and what moves you expect them to make in their product offerings. This is critical if you are to design your product to gain market share.

36.1.4 Compliance Analysis

One should think of compliances as unspoken user and customer needs which can exclude you from the market if they are not met. The critical issue about compliances is to understand which compliances govern the sale and use of your product type and whether they vary from country to country. The international component of compliances will be discussed under the section on localization. The basic set of compliances that each project group should investigate include the question of intellectual property and what patents you might need to secure for your product or to secure a license from another company. Next, you must determine what standards govern the design and use of the product. For example, are there IEEE standards or governmental standards which need to be adhered to? If you are designing disk drives, a common interface is an SCSI (small-computer-system interface) which is defined by an IEEE standard. In another example of designing an electrocardiograph, passing the Food, Drug and Alcohol Bureau's testing for medical product safety is a precursor to being able to sell that product.

Each of these standards must be identified prior to the design of the product because usually something needs to be designed into the product to ensure that the product meets these standards. Often, customers will not be specifying what standards they need a product to meet but will be very vocal with their dissatisfaction if the product does not meet one or more standards. For example, all personal computers are now expected to have both a parallel port and a serial port, that they will use the IEEE protocol for modems, and that the emissions requirements will meet the minimums set by over 50 nations.

At a more complex level, there may be some industry-specific testing procedures that your product needs to go through. These might include a self-inflicted standard such as assuring that toys manufactured by your company contain no toxic substances and cannot be broken into small pieces that might cause choking. Standards of this sort are often made to protect a company against liability suits.

One method to assure that standards work goes smoothly for a company is to invest some of your company's people time into the participation on these standards committees. This is important to influence the development of the standards to make use of the skills and competencies of your firm rather than those of the competitors' firms.

36.1.5 Localization

Localization not only involves making your product usable by customers in many foreign countries but also means assuring that your product meets the standards and government regulations of the different countries in which you wish to sell. Localizing to the regional language in both your written and on-line support materials and having local language selling and support processes is just the beginning of the process. Additionally, meeting the local compliances and intellectual property practices will be necessary. These protocols can be quite expensive, and you need to ascertain whether the size of a given local market warrants the investment to localize the product or whether there is a way to design the product so that it meets the needs of multiple markets with the same design and configuration. Understanding these differences should drive the design process. In one firm that sells laser printers, all written materials have no words in them; the communication is done solely with pictures. This strategy eliminated the need for translating the materials into different languages and resulted in significant savings for the company.

One simple difference throughout the world is voltage supplies. They vary from 100 to 240 V and from 50 to 60 Hz. Additionally, they require a variety of different plugs to power the product. Just this simple difference can lead to dozens of variations in a product configuration or just a few depending on the choices that are made in the design. If you choose to use a universal power supply that self-adjusts to the local power configuration, you will need only four different plug versions. However, if you do not choose the universal power supply, you may need dozens of configurations.

Additional localization challenges include the management of inventory. As the demand for products varies over both time and region, early considerations of how you might design your product architecture so that you can localize the configuration of the product as late as possible will result in large savings for your organization by not necessitating inventories tailored for specific regions. In the example of printers, this was achieved by varying only the power cord of the device for any particular country as there were no written materials in the packaging and the rest of the product design was suitable for use in any part of the world.

36.1.6 Product Positioning

After completion of the first five product definition steps—strategic alignment, user and customer needs, competitive analysis, compliances, and localization—the product positioning of the product needs to be developed. This step marks the completion of the "what" steps and outlines the point in the product development effort at which the team has identified what their company and the marketplace need from the product to satisfactorily solve their problems by some. Product positioning is sometimes called a *value chain,*[32] which is another way of stating how the product solves the target market segment's problems as well as how the product causes the customer to get more use out of the product than what the product initially costs. To develop a product positioning that will result in a product that is successful in the marketplace, the team needs to base this part of the product definition on a thorough understanding of the answers to the first five steps.

First, a summary of the business strategy and the objectives for the project must be listed. The critical questions that should be answered about the project's contribution to the strategy include, but are not limited to

- Revenue objectives for the life of the product
- Profit objectives for the life of the product
- Investment level available for developing and launching the product
- Other operational boundary conditions within which the team must operate, including staffing levels, limitations regarding cannibalizing sales of existing products, and the use of current or new vendors or distribution channels

Next, the team needs to consider the results of the research that they did in the steps of user and customer needs, competitive analysis, compliances, and localization. The team needs to keep in mind that the localization and compliance issues are, indeed, subsets of the users and customers needs but that the users and customers are very unlikely to ever articulate these needs but are also very unlikely to purchase the product if it does not adhere to the local compliances of the type of product you hope to sell. A number of methods can be used to organize the research that the team has used.[33–43] One way has been to use the QFD method, in which user needs are tabulated against the competitive solution to distinguish the areas of opportunity for your product. Teams should compare their product's ability to meet the needs of the target customer relative to the competitors' abilities and summarize the specific attributes that their product needs to embody. In this tabulation, it is critical to remember to include not only the technical specifications of the product but also less tangible needs of the customers such as how they wish to secure information regarding the product, to order the product, to install and begin using the product, to support the product, to maintain the product and, finally, to dispose of the product.

In the creation of the product positioning statement, the essence of the differentiation that your team will be embodying in your product offering is summarized. The creation of this short sentence often results in many agonizing meetings, but some hints on the appropriate basis of differentiating your product are outlined in Fig. 36.4. There, you will see that there are different phases in the life of a product and that differentiation needs to be based differently depending on which phase you are in. An additional complication to product positioning is to decide whether your new product is attempting to create a new competitive paradigm in the marketplace. The question to consider for any positioning statement is how this product will benefit target users and customers in a better fashion than any of the competition can, while contributing to the business success of your organization.

Next, with all these points in mind, the team completes the following positioning statement:

The____(X company's)____(description of the product)____product is the solution of choice for____(target market segment)____because it____(points of differentiation better than any competitive solution)____.

The XYZ company's new version Q laser printer product is the solution of choice for university students because it prints any Apple, Microsoft Windows, or Internet document or graphic at a resolution of 800 dots per inch (dpi) while costing less than $200 to purchase and less than 1¢ per page to print while occupying only 30 in^2 of desk space.

36.2   THE "HOW" STEPS

Product positioning was the last of the "what" steps and, subsequently, the team needs to change their focus to the "how" steps. In this section of the product definition effort the team will research each of the following steps:

- Project priorities
- Core competencies
- Project risks
- Strategic dependencies
- Market channel
- Project leadership
- Project resources

Each of these steps is briefly described below, although not in as much depth as the "what" steps were above. The reason for this is that, historically, market success depends more on how well the "what" steps have been completed than on how well the "how" steps have been completed.

36.2.1   Project Priorities

The author's experience has shown that teams often encounter difficulties because they and/or their management do not agree on the priorities for their project. There are several reasons for this, which will not be explored here, but a simple template for communicating the appropriate priorities for a project is described below. In Fig. 36.5, the team is given three topics to either constrain, optimize, or let float. These are the date of market release, the factory cost of the product, and the elective features that exceed the "must include" set. Notably, the project investment is not included on this project priority list. The reason for this is that we have found that using this as a boundary condition does not affect the market response to the product, which is what we are trying to optimize. In completing the grid, it is important to remember that no row or column can have two tickmarks in it. In fact, the reason many projects do get into trouble is that the team or management tries to constrain all three dimensions—time to market, factory cost, and the "want" features—simultaneously.

The major question to ask of a project team is which of the priorities is constrained, which is optimized, and which is left to float. The team will invariably respond with the question, "How do we decide which priority to place where?" The answer lies in the "what" steps and, in particular, the steps of strategic alignment, user and customer needs, and competitive analysis. Most teams have found that, if they are developing a very novel product that has little or no competition, constraining the elective features, and thus the performance of the product, is critical and that, depending on the situation, factory cost or market release date is next. The second case is a product being developed for a very competitive market that witnesses frequent new-product releases. This case usually demands that the date of market release be constrained, the factory cost optimized, and the elective features be left to float. One note of caution for the reader here: Deciding on critical changes in project priorities but not actually making these changes is the same as having no project priorities to begin with.

PROJECT PRIORITIES GRID

	Constrain	Optimize	Float
Time-to-Market (Date of Release to Market)			
Factory Cost			
Elective Features Beyond Must-Have Features			

FIGURE 36.5 Project priorities grid. To use the project priority grid, the team must decide on which three boxes to tick remembering that the project will be overconstrained if any one row or column contains more than one tick. This grid should be an embodiment of the strategic direction chosen by the team to optimize the company's success with the product.

36.2.2 Project Work Breakdown and Risk Assessment

The next step is to create a project work breakdown and to identify how the team will complete the project and which areas of the project are risky. There are several software-based project management tools, but these will probably be very cumbersome to use at this early phase of the project. A technique that has been successfully used by many teams is to assemble the multifunction project team, review the results of the research into the "what" steps and the agreed-to project priorities, and then discuss what needs to be done to complete and release the project. Next, provide each member of the team a pad of Post-it (3M Company trademark) notes and ask each individual to note each activity they think is necessary to complete the project on a separate Post-it. Additionally, they should note who should be assigned the task, the amount of time and investment required to complete the task, and the certainty that the task will take the amount of effort and investment they assigned.

The second portion of this step is to arrange all the Post-its in a chronological manner and note any overlaps or gaps and linkages in the plan. Once this is completed, the effort of creating a project plan using software-based tools is simplified.

The last portion of this step is to highlight any substeps in which there is not at least a 90 percent level of certainty that the task will be completed within the available resources. These are the risks for your project and will lead to delay and overspending if they are not contained early in the project.

36.2.3 Core Competencies

The next step is to look at the work breakdown and the risks and to identify which risk areas require the organization to have new core competencies to complete the project. The competencies can be for any functional area and are not limited to the technology area as described by Prahalad and Hamel in the *Harvard Business Review*.[44] The reason for listing these at this time is that the project team needs to decide how they will acquire the competencies. The available techniques are to develop the competency, buy it from a vendor, or ally with another company or institution who can provide the competency. If the competency is not currently available, the team needs to decide whether they should proceed with the whole project or delay the project while they resolve the issues of acquiring the necessary core competencies and explore alternative solutions. It is critical to note that misjudging the difficulty of acquiring the necessary core competencies in sufficient time is second only to not understanding the users' and customers' needs in leading to marketplace failure of products.

36.2.4 Strategic Dependencies

In the strategic dependency step, each project partner is identified. These are the other divisions in the organization and outside companies, consultants, and organizations that are supplying something to the project. It is key to think of every portion of the project effort in identifying the strategic dependencies so that you do not overlook some critical aspect of the effort. Some of the common dependencies are integrated circuit and other component suppliers, the distribution companies handling the overseas distribution of your product, independent software suppliers that you need to influence to assure compatibility of your product with theirs, standards committees which will set guidelines to which your project must adhere, and governmental agencies, to name only a few.

36.2.5 Project Leadership

The leadership of the project needs to be excellent and stable for a project to be successful. Most evident is that the people guiding the project must be capable of motivating the rest of the team so that they will enthusiastically work hard during the difficult periods and will seek the guidance of the leader to address any difficulties that might be stymieing the progress of the project at all times. In high-technology projects there are a few traps that management often falls into when choosing a project leader. The first is to select a person who is technically superior to the rest of the project team members. This choice is often burdened with difficulties because, often, technically superior people lack people skills and are not adept at facilitating tradeoffs between different departments' needs or managing the softer people issues encountered in most projects. Additional difficulties with using extremely technically talented people are that they sometimes micromanage the technical details at the expense of the other function and the business goals of the project. The author's guidance is to choose the project leader for the ability to listen and to facilitate problem solving. Additional skills the project leadership will need to have are the abilities to communicate the project's problems and successes with their management in written and verbal forms, to be capable of managing the project schedule and budget, and to be cognizant of when they need help on a particular issue and how to secure the help they need to resolve the issue. Some excellent references on project leadership include Wheelwright and Clark[45] as well as others.[46-49]

36.2.6 Project Resources

Simply put, project resources are money and people. Yet, there are some subtleties to having the right people and money at the right time. In the 1980s many firms moved to thinking of people as being interchangeable so long as they had the same job description. In the 1990s, we have recognized that not all people are created equal and that for some of the specialty jobs on a project, it will be necessary to have specific people. In jobs such as designing the firmware for electronic test instrumentation at one division at Hewlett-Packard, each of seven projects needed a firmware engineer and had included this in their budgeted staffing plans. When the overall department staffing was analyzed, it was found that each of the projects needed a specific engineer, L. W., yet there was no possibility for him to work on each of the seven projects. In hindsight, this division lacked a core resource competency, a number of firmware engineers of talent equal to that of L. W.

On the funding side of the resources, the team needs to generate an accurate estimate of the use of people and the capital and expense that they will need from the end of the product definition phase until the market release of the product. Additionally, the staffing, capital, and expense run rates for manufacturing and marketing after market release need to be accurately estimated at this time. The resource planning exercise is enormously simplified when the team leverages data from previous projects. Obvious inputs are the past staffing levels for similar activities, actual labor and yield information from production, staffing levels from marketing, and material and yield rates from procurement and production, respectively. This step marks the last of the "how" steps and is a critical one, because in it you flag the difficulties you might encounter if the right resources are not in place and it is the creation at a critical input to the determination of (the next step of product definition: the project and business plan) why your organization should invest in this project.

36.3 THE "WHY" OF PRODUCT DEFINITION: THE PROJECT AND BUSINESS PLAN

In this final step of product definition, the product specification is created as are the supportive staffing and funding plans. Additionally, a financial pro forma is developed which outlines the return on investment the product will secure given a number of possible market scenarios.

The product specification can be in a number of different formats. There is no one right product specification, but good ones are based on the product positioning required for the product to be successful in the marketplace. Most specifications include, at the highest level, a data sheet listing what the product will do for the users and customers; at a deeper level, a detailed specification which is often called an *external specification*; and, finally, an internal specification which will include detailed information about the mechanical, electrical, firmware, software, and other components of the product as well as their interactions. Smith and Reinertsen[1] do an excellent job describing how to specify a product and suggest that effort put into defining the boundaries and the interaction of the boundaries of the different subcomponents will make the design and the integration processes much simpler.

The staffing levels are next calculated and are based on what you are going to do. If there are shortages or excesses of people, the scope of the project needs to be commensurately adjusted. Then the financial side of the expenditures is investigated; funding for the people, staff overhead, and manufacture and sales of the product is calculated.

Next, the market projections of sales volumes and sales price vs. feature are integrated, and the team calculates a number of financial variables to determine whether the project is a good investment. Included are net present value (NPV), internal rate of return (IRR), return on investment (ROI), cash flow (CF), and return on assets (ROA), to name only a few. Each of these tells you something a little different about how well or poorly you might do with the investment. The caution that is offered here is that you should not rely on any one of these metrics but look at the whole picture and ask if the investment makes sense for your organization given the objectives you need to meet.

The final effort in this step is to analyze which, if any, of the possible scenarios would be best for your organization to invest in. The primary criterion for selection will be some measure of profitability, as that is the primary objective of most publicly traded companies. Sometimes, there will be other reasons for pursuing a project even if it is not profitable. Some commonly cited reasons include making a competitive move and augmenting an existing product line to have a fuller range of products.

36.3.1 The Product Definition

The "why" step marked the last of the 14 product definition steps, and the product definition phase of the project is best concluded with a review meeting of these 14 steps, a sharing of the uncertainties and risks the team believes the project will encounter, a review of the financial forecasts for revenues and expenditures, and a recommendation on the best path forward. Additionally, it can be very helpful for the team to test their product definition using the methods outlined in Refs. 50–56. This method was developed by the author to identify any project weaknesses which could cause delays in the project or a marketplace failure of the project. By querying every person associated with the project, from general management to individual contributors from each functional area, on each factor, any weakness that might exist on the project, and any differences in opinion about the status of the project will be highlighted. The test mechanism to discover this information is a multiple-choice question on each of the 14 steps. A score of 1 means there is a problem; a score of 5 means that the step has been excellently researched and no problems have been encountered. After having each member of the extended management and team complete the test separately, means and standard deviations are calculated. Any means less than 3.0 or standard deviation greater than 1.0 is noted as being indicative of project weaknesses and discrepancies in perception on those individual steps. This has proved to be a satisfactory method to note product definition problems at several companies.[7] The likelihood of the project meeting the business goals is correlated to the scores: a mean of 1 indicates a low likelihood; a mean of 5, a very high likelihood. For example, for the Hewlett-Packard Laserjet 2P and 3 Series products, the assessment test score means were all between 4 and 5 and the standard deviations were below 1. These products marked the beginning of Hewlett-Packard's dominance in the laser printer market. Teams need to remember that the purpose of doing this exercise at the end of the product definition phase is to highlight where they need to focus to quickly address any project problems, successfully reduce the likelihood of project failure, and maximize the likelihood of market success. If a project has the majority of the mean scores at less than 3, management and the team should hold critical discussions on how to redeploy the project. If there are one or two low means, action plans to improve the score are in order, and if all means are very high, it is unlikely that the project will encounter weaknesses in the future. In interpreting the standard deviations, scores greater than 1 require attention. Usually, discussion between management and the team will rectify the differences in opinion. If this has not occurred, an action plan to resolve the outstanding issues must be developed. However, if there are critical issues that cannot be improved, canceling the project should be considered.

In summary, product definition can be broken into a series of interdependent steps which project teams need to answer. One of the keys to a successful product in the marketplace is completing each step thoroughly so that the product definition can be based on fact and not supposition. By questioning each step individually, product definition problems can be pinpointed and resolved, or a decision to redeploy the project can be made.

36.4 REFERENCES

1. Preston G. Smith and Donald G. Reinertsen, *Developing Products in Half the Time,* Van Nostrand-Reinhold, New York, 1991.

2. W. I. Zangwill, *Lightning Strategies for Innovation: How the World's Best Firms Create New Products,* Lexington Books, New York, 1993.

3. Edith Wilson, *Product Definition Factors for Successful Designs,* unpublished thesis, Stanford University, Stanford, Calif., Department of Mechanical Engineering, Dec. 1990.

4. Edith Wilson, "Improving Market Success Rates through Better Product Definition," *World Class Design to Manufacture J.,* **1**(4): 13–15, 1994.

5. Edith Wilson, "Product Definition: Assorted Techniques and Their Marketplace Impact," *Proceedings of the IEEE International Engineering Management Conference,* 1990.

6. G. Bacon, S. Beckman, D. Mowery, and E. Wilson, "Managing Product Definition in High Technology Industries," *Calif. Management Review,* **3**(3): 32–56, Spring 1994.

7. J. Carter and S. Mello, *Best Practices Survey 1994: Product Definition Executive Summary,* Product Development Consulting, Inc. and The Management Roundtable, Cambridge, Mass., 1995.

8. Science and Policy Research Unit, *Success and Failure in Industrial Innovation,* Center for the Study of Industrial Innovation, London, 1972.

9. C. Freeman, *The Economics of Industrial Innovation,* 2d ed., MIT Press, Cambridge, Mass., 1982 (summarizes SAPPHO's findings).

10. R. Rothwell, C. Freeman, A. Horley, V. I. P. Jervis, Z. B. Robertson, and J. Townsend, "SAPPHO Updated: Project SAPPHO Phase II," *Research Policy,* **3**: 258–291, 1974.

11. Robert G. Cooper, *Winning at New Products,* Addison-Wesley, Reading, Mass., 1988.

12. Robert G. Cooper, *Winning at New Products: Accelerating the Process from Idea to Launch,* Addison-Wesley, Reading, Mass., 1993.

13. Abbie Griffin, *Functionally Integrating New Product Development,* MIT Sloan School of Management doctoral thesis, Cambridge, Mass., 1989.

14. David L. Morgan, *Focus Groups as Qualitative Research,* Sage, Newbury Park, Calif., 1988.

15. Richard A. Krueger, *Focus Groups: A Practical Guide for Applied Research,* Sage, Newbury Park, Calif., 1988.

16. Bobby J. Calder, "Focus Groups and the Nature of Qualitative Marketing Research," *J. Marketing Research,* **14:** 353–364, Aug. 1977.

17. Abbie Griffin and John R. Hauser, "The Voice of the Customer," *Marketing Science,* pp. 1–27, Winter, 1993.

18. Glen L. Urban, John R. Hauser, and Nikhilesh Dholakia, *Essentials of New Product Management,* Prentice-Hall, Englewood Cliffs, N.J., 1987, pp. 105–110.

19. Jerome Kirk, *Reliability and Validity in Qualitative Research,* Sage, Beverly Hills, Calif., 1986.

20. Abbie Griffin and John R. Hauser, "The Voice of the Customer," *Marketing Science,* pp. 1–27, Winter, 1993.

21. Nicholas Bateson, *Data Construction in Social Surveys,* George, Allen, & Unwin, London, 1984.

22. Gary W. Burchill, *Concept Engineering: The Key to Operationally Defining Your Customers' Requirements,* Center for Quality Management, Cambridge, Mass., 1992, p 155.

23. Eric Von Hippel, *Novel Product Concepts from Lead Users: Segmenting Users by Experience,* Working Paper, Sloan School of Management, MIT, Cambridge, 1984.

24. Paul E. Green, Donald S. Tull, and Gerald Albaum, *Research for Marketing Decisions,* 5th ed., Prentice-Hall, Englewood Cliffs, N.J., 1988, pp. 712–715.

25. Paul E. Green and Yoram Wind, "New Way to Measure Consumers' Judgment," in *New-Product Forecasting,* Yoram Wind, Vijay Mahajan, and Richard N. Cardozo, eds., Heath, Lexington, Mass., 1981, pp, 89–108.

26. Paul E. Green and V. Seenu Srinivasan, "Conjoint Analysis in Marketing: New Developments with Implications for Research and Practice," *J. Marketing,* pp. 3–19, Oct., 1990.

27. V. Srinivasan and Gordon A. Wyner, "CASEMAP: Computer-Assisted Self-Explication of Multiattributed Preferences," in *New-Product Development and Testing,* Walter Henry, Michael Menasco, and Hirokazu Takada, eds., Lexington Books, Lexington, Mass., 1989, pp. 91–112.

28. Paul E. Green, Donald S. Tull, and Gerald Albaum, *Research for Marketing Decisions,* 5th ed., Prentice-Hall, Englewood Cliffs, N.J., 1988, pp. 715–730.

29. Paul E. Green and V. Srinivasan, *Conjoint Analysis in Marketing Research: A Review of New Methods,* Graduate School of Business, Research Paper 1071, Stanford University, Stanford, Calif., Oct. 1989.

30. J. R. Hauser and D. Clausing, "The House of Quality," *Harvard Business Review,* **88**(3): 63–73.

31. Don Clausing, *Total Quality Development,* ASME Press, New York, 1994, pp. 121–124.

32. Michael E. Porter, *Competitive Advantage,* Free Press, New York, 1985.

33. George A. Kelly, *The Psychology of Personal Constructs,* Norton, New York, 1955.

34. Paul E. Green, Donald S. Tull, and Gerald Albaum, *Research for Marketing Decisions,* 5th ed., Prentice-Hall, Englewood Cliffs, N.J., 1988, pp. 712–715.

35. C. Merle Crawford, *New Products Management,* 3d ed., Irwin, Homewood, Ill., 1991, pp. 123–129.

36. Jonathan Gutman and T. J. Reynolds, "A Pilot Test of a Logic Model for Investigating Attitude Structure," in *Attitude Research under the Sun,* J. Eighmey, ed., American Marketing Association, Chicago, 1979, pp. 128–150.

37. Dorothy E. Leonard-Barton, Edith Wilson, and J. Doyle, "Commercializing Technology: Imaginative Understanding of User Needs," paper presented at the Sloan Foundation Conference on the Future of Research and Development, Harvard University, Cambridge, Mass., Feb. 1993.

38. John Case, "Customer Service: The Last Word," *Inc. Magazine,* pp. 88–92, April 1991.

39. Karen Holtzblatt and Hugh Beyer, "Making Customer-Centered Design Work for Teams," *Communications of the ACM*, pp. 93–103, Oct. 1993.

40. Paul Eisenstein, "Your True Color," *Family Circle*, p. 40, Nov. 1, 1994.

41. Gary Levin, "Anthropologists in Adland," *Advertising Age*, pp. 3, 42, 49, Feb. 2, 1992.

42. Yoji Akao (author and editor in chief), *Hinshitsu Tenkai Katsuyo No Jissai* (English translation: *QFD: Integrating Customer Requirements into Product Design*), translated by Glenn H. Mazor, Productivity Press, Cambridge, Mass., 1990.

43. Lawrence R. Guinta and N. C. Praizler, *The QFD Book: The Team Approach to Solving Problems and Satisfying Customers through QFD*, Amacom, New York, 1993.

44. C. K. Prahalad and G. Hamel, "Core Competence of the Corporation," *Harvard Business Review*, **90**(3): 79–89, 1991.

45. Steven C. Wheelwright and Kim B. Clark, *Revolutionizing Product Development*, Free Press, New York, 1992.

46. S. Mizuno, *Management for Quality Improvement*, Productivity Press, Cambridge, Mass., 1988, pp. 116–128.

47. Michael F. Wolff, "Building Teams—What Works (Sometimes)," *Research Technology Management*, Nov.–Dec. 1989, pp. 9–10.

48. Therese R. Welter, "How to Build and Operate a Product Design Team," *Industry Week*, pp. 35–50, April 16, 1990.

49. Japan Human Relations Association, eds., *Kaizen Teian 1: Developing Systems for Continuous Improvement through Employee Suggestions*, Productivity Press, Cambridge, Mass., 1992.

50. Paul Green, "Concept Testing Approaches," in *Marketing Science Institute Conference Summary: Roles for Research and Models in Improving New Product Development*, Report 90-120, Dec., 1990, p. 7.

51. M. J. Rokeach, *Beliefs, Attitudes, and Values*, Free Press, New York, 1973.

52. Glen Urban, "Preliminary New Product Forecasting," in *Marketing Science Institute Conference Summary: Roles for Research and Models in Improving New Product Development*, Report 90-120, Dec., 1990, pp. 8–9.

53. Jonathan Gutman, "A Means-End Chain Model Based on Customer Categorization Processes," *J. Marketing*, **46**: 60–72, Spring 1982.

54. Paul Eisenstein, "Your True Color," *Family Circle*, p. 40, Nov. 1, 1994.

55. Gary Levin, "Anthropologists in Adland," *Advertising Age*, pp. 3, 49, Feb. 2, 1992.

56. Allan D. Shocker and William G. Hall, "Pretest Market Models: A Critical Evaluation," *Marketing Science Institute Research Program Special Report*, Report 86-107, Sept. 1986, pp. 86–107.

CHAPTER 37

BENCHMARKING MANUFACTURING TECHNOLOGY USE IN THE UNITED STATES

Paul M. Swamidass, Ph.D.

Associate Director, Thomas Walter Center
for Technology Management
Auburn University
Auburn, Alabama

37.1 INTRODUCTION

Manufacturers frequently want to know the extent of manufacturing technology use by competitors or by plants in industries similar to their own. Further, manufacturers would like to know the benefits, if any, associated with manufacturing technology use. For example, manufacturers are keen to know whether productivity and quality are enhanced by investments in manufacturing technologies. What are the most common benefits attributable to manufacturing technology? Unfortunately, the information on the subject is very limited. To acquire such information we must develop industry norms or benchmarks on manufacturing technology use. The process of developing such norms and benchmarks can be very tedious and expensive.

This study benchmarks the use of 15 more frequently used technologies and benefits associated with them. This large study was made possible by the valuable support of the National Association of Manufacturers and the National Science Foundation.

The 15 technologies benchmarked here are identified in Table 37.1, and each technology is described in detail in the glossary in Sec. 37.9 (App. 2). Further, Sec. 37.8 (App. 1) explains the data collection procedures and validation.

37.1.1 Soft and Hard Technologies

This study has categorized technology on the factory floor into two kinds: hard technology and soft technology. *Hard technologies* are hardware- (and associated software-) based technologies such as FMS, CAD, and CAM. *Soft technologies,* on the

TABLE 37.1   List of Technologies Covered by the Study

	Hard technologies investigated
1.* AGVs	Automated guided vehicles
2.* Automated inspection	
3. CAD	Computer-aided design
4. CAM	Computer-aided manufacturing, including programmable automation of single- or multimachine systems
5. CIM	Computer-integrated manufacturing; extensive use of computer hardware and software to link all aspects of a manufacturing plant from order entry to shipping for real-time planning, scheduling, and control; CIM may integrate across plants, states, and countries
6.* CNC	Machines with computerized numerical control
7.* LANs	Local area networks
8. FMS	Flexible manufacturing systems; automated multimachine systems linked by an automated materials-handling system
9. Robots	All kinds
	Soft technologies investigated
10. JIT	Just-in-time manufacturing
11. Manufacturing cells	See glossary in App. B
12. MRP I	Material requirements planning
13. MRP II	Manufacturing resource planning
14. SQC	Statistical quality control
15.* TQM	Total quality management

*Technologies not included in the last study.

other hand, are techniques such as statistical quality control (SQC/SPC), just-in-time production (JIT), and manufacturing resources planning (MRP II).

37.1.2   Technology Use in Small versus Larger Plants

There is an across-the-board difference in technology usage between small (less than 100 employees) and large plants (100 or more employees). It is notable, however, that small manufacturers use all the technologies investigated in this study. See Table 37.2 for the contrasting characteristics of small and large plants in the sample. Throughout the report, we provide selected information by plant size.

37.2   MANUFACTURING TECHNOLOGY USE

Figure 37.1 assesses the penetration of manufacturing technologies in all plants surveyed, but does not distinguish between extremely skilled users and others. According to Fig. 37.1, CAD (84 percent), TQM (72 percent), JIT (71 percent), and CNC (71 percent) have the most widespread use in U.S. manufacturing. We did not investigate TQM and CNC in an earlier study, but on the basis of the widespread use of the technologies reported in Fig. 37.1, their inclusion in this study has been beneficial. CAD is the most widely used technology (84percent) according to Fig. 37.1, as it was in the

TABLE 37.2 A Comparison of Averages for Small and Larger Plants

	Small Plants (employment < 100)	Large plants (employment ≥ 100)
Sample size, *n*	551	463
1. Sales, $ million	5.35	87.2
2. Employment	46	443
3. Sales per employee, $000*	114	144
4. Rejections, %	3.48	4.54
5. Inventory turns	8.3	7.8
6. Cost of goods sold	58.9% of sales	62.50% of sales
7. Product lines	23.7	23.51
8. Models	71.5	143.00
9. Average lead time, weeks	6.25	7.97
10. Direct-labor costs	21.5% of sales	14.67% of sales

*Thousands of U.S. dollars.

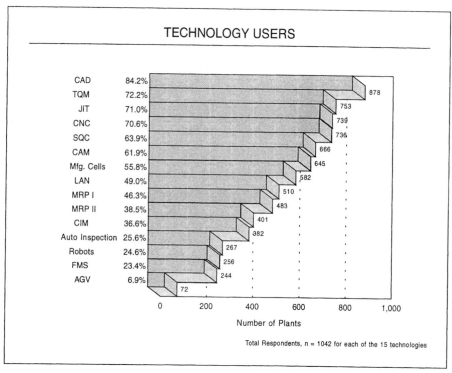

TECHNOLOGY USERS

Technology	%	Number of Plants
CAD	84.2%	878
TQM	72.2%	753
JIT	71.0%	739
CNC	70.6%	736
SQC	63.9%	666
CAM	61.9%	645
Mfg. Cells	55.8%	582
LAN	49.0%	510
MRP I	46.3%	483
MRP II	38.5%	401
CIM	36.6%	382
Auto Inspection	25.6%	267
Robots	24.6%	256
FMS	23.4%	244
AGV	6.9%	72

Total Respondents, n = 1042 for each of the 15 technologies

FIGURE 37.1 Technology users.

previous study, and it is well on its way toward universal use. Automated guided vehicles (AGVs) are not used widely because they either are not relevant to most operations or are not cost-efficient in most situations.

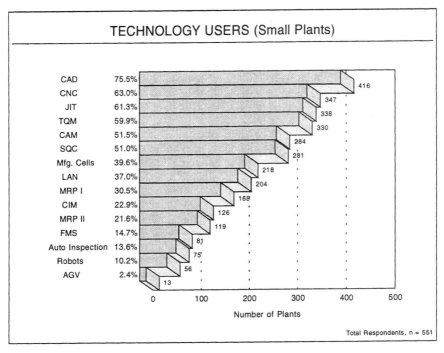

FIGURE 37.2   Technology users (small plants).

FMS use, at 23.4 percent, is notable because these are very expensive systems, which build flexibility in small-volume production through automation. Additionally, CIM use at 36.6 percent should be considered a major progress because CIM is expensive and time-consuming to implement.

Figures 37.2 and 37.3 show how technology use differs with size. While 75.5 percent of small plants use CAD, 94.8 percent of the larger plants do so. JIT use is more common among larger plants by nearly a 20 percent margin. The proportion of larger plants using manufacturing cells is nearly twice that of small plants. In the case of some technologies in the figures, the proportion of larger plants using the technologies is many times that for small plants; *thus, larger plants more frequently employ advanced technologies than do their smaller counterparts.* Some of the accepted explanations are affordability, know-how and risk-taking capability of larger plants.

In Fig. 37.4 technology use by skill levels is displayed. Figure 37.4 shows that, while the use of JIT and TQM is widespread, fewer users are extremely skilled in the use of these technologies than in the use of CAD and CNC.

37.2.1   Relevant-Technology Use

Figure 37.5 shows the extent of technology use by those who consider the technology as relevant to their operation. Thus, CAD penetration is very high, at 97.4 percent among those who consider it to be relevant to their operation. In the case of larger plants, CAD use is 100 percent when the technology is relevant (91.6 percent in the case of small plants).

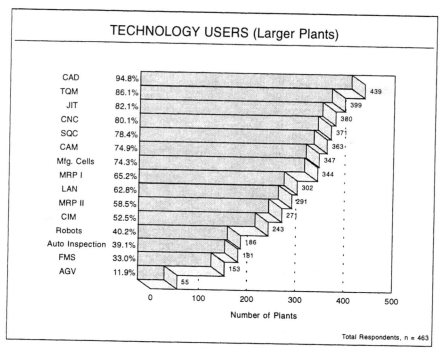

FIGURE 37.3 Technology users (larger plants).

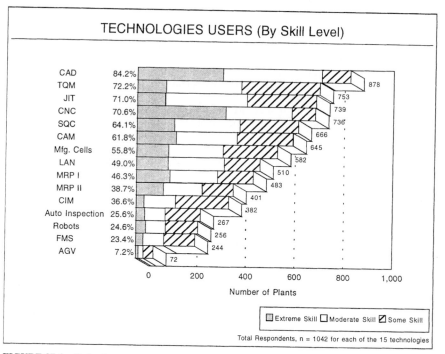

FIGURE 37.4 Technology users (by skill level).

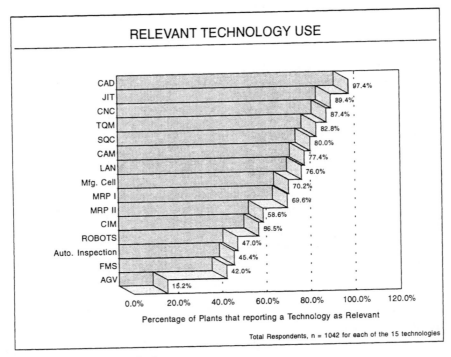

FIGURE 37.5 Relevant technology use.

It is most remarkable that JIT is used by nearly 90 percent of plants that consider it to be relevant. This technology, which was imported from Japan only about 15 years ago, has reached an extremely high level of acceptance in U.S. shop floors. Those who understand JIT know that, once a plant begins to use JIT, it is virtually imperative for its competitors to adopt JIT if they are to remain in competition. This technology shows how a powerful new idea can be transplanted in a relatively short time from one country to another.

Other relevant technologies used by more than 75 percent of plants are: CNC, TQM, SQC, CAM, and LAN. Manufacturing cells at 70.2 percent shows a very strong penetration as well. Manufacturing cells are catching on because of their overall value as a flexibility-, quality-, and productivity-enhancing tool, which can provide nice returns for a relatively small investment.

37.2.2 Skilled Users of Technologies

Figure 37.6 shows that a third of the manufacturers are extremely skilled in the use of CNC and CAD. Figure 37.7 reports that, when moderately skilled users are also considered, CAD is used by nearly three-fourths of manufacturers, considerably ahead of the penetration of CNC. The high level of penetration of both these technologies indicates that the combination of the two is at the leading edge of the transformation taking place on the factory floor. Figures 37.8 and 37.9 show how small and larger plants differ in the extremely skilled use of technologies. While only 25.5 percent of small plants are extremely skilled in the use of CAD, the corresponding percentage for larg-

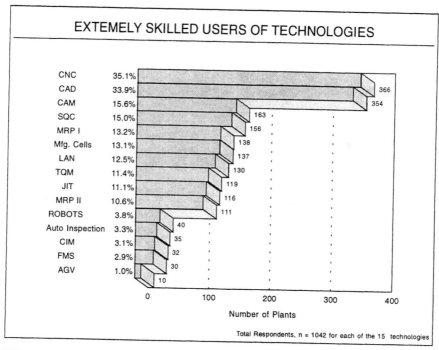

FIGURE 37.6 Extremely skilled users of technologies.

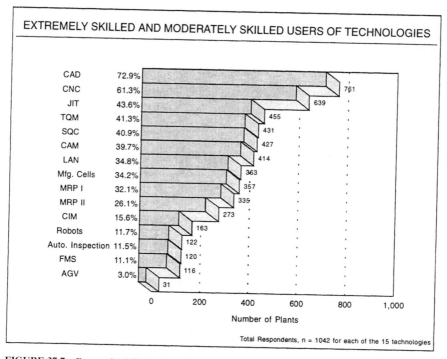

FIGURE 37.7 Extremely skilled and moderately skilled users of technologies.

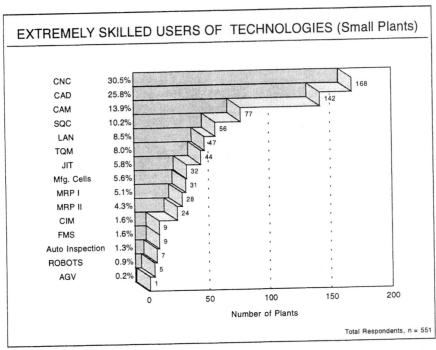

FIGURE 37.8 Extremely skilled users of technologies (small plants).

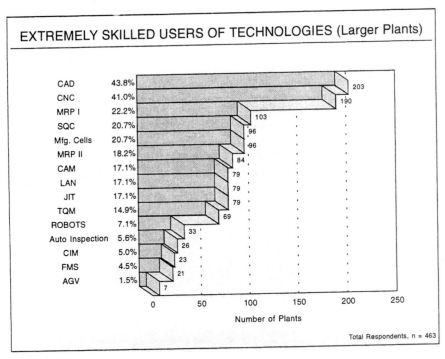

FIGURE 37.9 Extremely skilled users of technologies (larger plants).

er plants is 43.8. Figures 37.8 and 37.9 show, by comparison, that the training needs for small and larger plants can be very different.

In Fig. 37.7, extremely and moderately skilled users are combined. This figure illustrates that the rate of penetration for JIT (43.6 percent), TQM (41.3 percent), and SQC (40.9 percent) are all very similar, which suggests that some technologies are adopted simultaneously.

37.2.3 Technology Clusters

One of the most important findings of the study is that technology use occurs in clusters. That is, the use of one technology is often closely associated with the use of one or more related technologies.

There are at least two reasons for this occurrence. First, some technologies prepare the way for the adoption of other technologies: The use of CAD, for example, appears to pave the way for the adoption of CAM. Second, there are overlapping principles across several technologies that make them practical to implement synergistically together. The adoption of TQM, JIT, and SQC illustrates this point. Tables 37.3 and 37.4 show the joint use of several technologies.

TABLE 37.3 Joint use of CAD, CAM, and CNC

		Secondary use, %		
		CAD	CAM	CNC
Primary use	CAD		70.9*	76.5
	CAM	96.41		86.1
	CNC	75.41	91.2	

*Read as: 70.9% of CAD users use CAM.

General note: according to this table, 96.41 percent of CAM users use CAD, 91.2 percent of CNC users use CAM, and 86.1 percent of CAM users use CNC.

TABLE 37.4 Joint Use of JIT, TQM, and SQC

		Secondary use, %			
		JIT	TQM	SQC	Cells
Primary use	JIT		83.0*	74.5	68.7
	TQM	81.2		78.7	67.1
	SQC	82.5	88.8		69.3
	Cells	87.5	86.8	79.5	

*Read as: 83% of JIT users use TQM.

We infer that these three technologies occur in a cluster. As CAD is the most commonly used technology, it appears that CAD use is a precursor to CNC and CAM use. Thus, a delay in the use of CAD delays the entry of the firm into the world of advanced-technology manufacturing because it delays the use of complementary technologies such as CAM and CNC. This finding reinforces the importance of promoting CAD skills in the country. In this context, the CADD Skill Standards Project undertaken by the Foundation for Industrial Modernization (a research arm of the National Coalition for Advanced Manufacturing) is very timely.

There is a synergistic match between CAD and CNC technologies as the output of CAD can be used as inputs to CNC and CAM, thereby reducing the setup time to almost zero. We conclude the following: *The use of CAM and CNC is tied to the use of CAD. The three form a core of hard technologies at the heart of automation on the factory floor.*

Table 37.4 shows that the three soft technologies, JIT, TQM, and SQC, appear jointly in manufacturing plants. According to the table:

1. More than 80 percent of TQM, SQC, and cell users use JIT (column 1).

2. More than 80 percent of JIT, SQC and cell users use TQM (column 2).

Thus, the three soft technologies are frequently used in conjunction. It appears that cells and SQC are used to implement JIT and TQM. We conclude that *the use of soft technologies (TQM, JIT, cells, and SQC) occurs in a cluster, which is at the heart of effective and efficient factory operation today.*

Other technologies occurring jointly are MRP I and MRP II (65 percent of MRP I users use MRP II). Thus, MRP II coexists with MRP I in most plants.

37.2.4 Technology Use Since the Last Study

A similar study was conducted by the author in 1990. Out of the 1042 respondents participating in the current iteration of the study, 216 indicated that they participated in the first study also. In Fig. 37.10, we see the number of plants reporting the most significant improvements since the first study. CAD (47 percent) and TQM (36 percent) are the two technologies in which plants report most significant improvements since the last study. Figure 37.10 is helpful in identifying the technologies in which plants have experienced the most and the least improvements over the last few years. We see that a negligible percentage of manufacturers are investing in automated inspection and AGV.

37.3 BENEFITS OF INVESTMENT IN TECHNOLOGY

Manufacturing technologies, when selected wisely and used, enable manufacturing firms to compete better through reduced cost, increased quality, reduced manufacturing lead time, reduced time to market, and increased product variety. In spite of the many benefits of manufacturing technology use, however, the adoption of manufacturing technologies is often slow for reasons such as poor and confusing equipment justification methods, poor understanding of the benefits of modern technologies, and/or lack of capital. This study strives to correct the lack of information about the benefits attributable to manufacturing technology.

Figure 37.11 reports the claims of respondents concerning benefits attained from the use of advanced manufacturing technologies. Two out of three respondents report that reduced cycle time was a direct benefit of their investment in advanced manufacturing technologies—far and away the most frequently cited benefit. The overwhelming frequency with which this benefit is mentioned (66 percent) was unexpected. From Fig. 37.11, one could conclude that the reduction of cycle time is perhaps one of the most significant reasons for investing in new technologies. Prior to this study there was no definite link between technology usage and cycle time reduction. In Fig. 37.11,

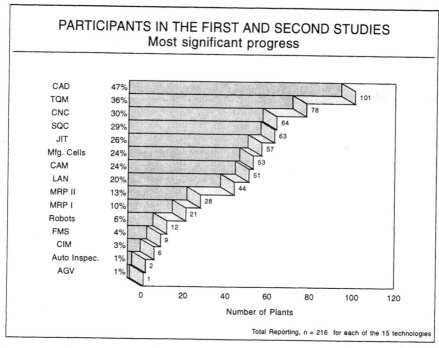

FIGURE 37.10 Participants in the first and second studies (most significant progress).

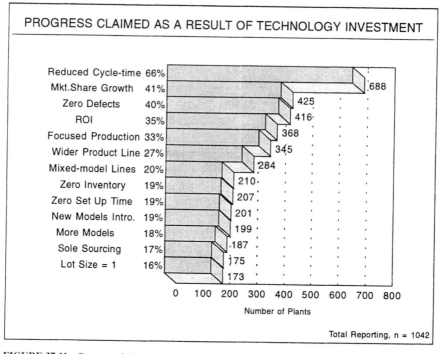

FIGURE 37.11 Progress claimed as a result of technology investment.

other top-rated benefits of advanced technology use are market share growth, zero defects, ROI, focused production, and wider product lines.

The average plant reported 3.7 benefits from the list in Fig. 37.11 as a direct result of technology use. More than 85 percent of the respondents reported one or more benefits. About a fourth of the respondents reported six or more benefits.

37.3.1 Benefits Increase as Skill Level Increases

Figure 37.11 lists 13 different direct benefits reported by manufacturers. Often a single technology is associated with more than one benefit. Out of all the technologies considered, the skilled use of manufacturing cells is associated with the most sizable increase in the number of benefits. Therefore, *for manufacturers looking for multiple benefits from investments in technology, the first and foremost choice should be manufacturing cells.* Further, as mentioned earlier, we found that the use of cells is closely associated with the use of JIT and TQM because cells seem to facilitate the implementation of these technologies; this is an added benefit.

37.3.2 Benefits of Extremely Skilled Technology Use

Inventory Turns. Figure 37.12 shows that extremely skilled users of AGV, SQC, JIT, and automated inspection outperform others on inventory turns. Since all these

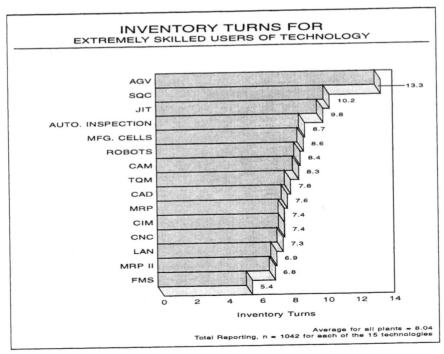

FIGURE 37.12   Inventory turns for extremely skilled users of technology.

technologies are generally implemented in repetitive manufacturing environments, the high inventory turns associated with these systems are consistent with expectations.

Lead Time from Order to Shipment.   Now, in the 1990s, when time-based competition is a major competitive asset, it is useful to know which technologies are associated with reduced lead time. Further, the most commonly mentioned benefit associated with advanced-technology use was lead-time reduction (see Fig. 37.11). According to Fig. 37.13, MRP II and JIT are associated with the shortest average lead time at 6.5 and 7.2 weeks, respectively. Hard technologies such as AGV, automated inspection, and robots are associated with longest average lead times. The average lead time for all plants is 7.19 weeks (6.25 weeks for small plants and 7.9 weeks for larger plants).

Rejection and Rework Rates.   Extremely skilled use of FMS and JIT, according to Fig. 37.14, offers the lowest rejection and rework rates. On the opposite end of the spectrum, CIM, CAD, and SQC are associated with higher rejection and rework rates. The apparent paradox could be explained by several possible reasons. High average rework and rejection rates for SQC users could be due to the very nature of the manufacturing environment. That is, rejection rates could be even higher if not for SQC. Also, SQC is often implemented after quality levels have deteriorated to a seriously low level. Still another possible explanation is that SQC identifies the defects that are rejected or reworked before shipping to customers. Therefore, non-SQC environments may be hiding defects that are not detected before shipping to customers. Additional

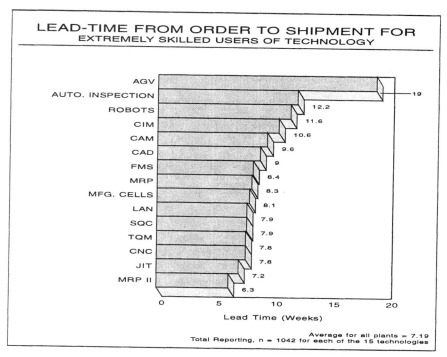

FIGURE 37.13   Lead time from order to shipment for extremely skilled users of technology.

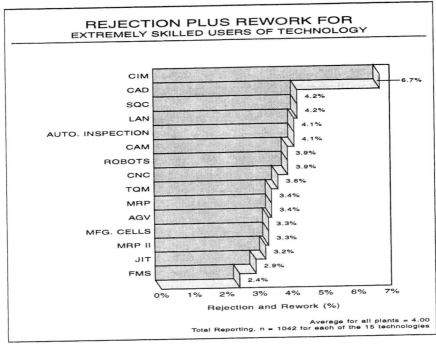

FIGURE 37.14   Rejection plus rework for extremely skilled users of technology.

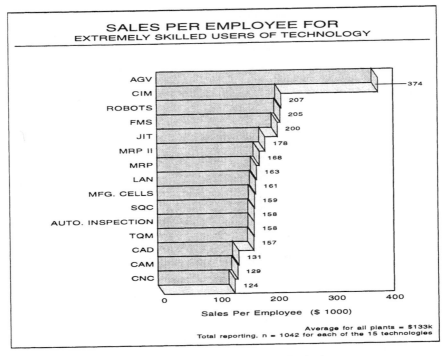

FIGURE 37.15   Sales per employee for extremely skilled users of technology.

investigation is needed to understand this more fully. The average rate for all plants is 4.0 percent; small plants average 3.47 percent, and larger plants average 4.54 percent.

Sales per Employee. According to Fig. 37.15, hard technologies such as AGV, CIM, robots, and FMS are associated with higher sales per employee, which usually improves with automation.

Return on Investment. According to Fig. 37.16, extremely skilled use of JIT, MRP II, and AGV is associated with superior return on investment. The appearance of JIT at the top of the list should not surprise anyone. It is significant that, while the average ROI for all plants is 12.99 percent (11.5 percent for small plants and 14.7 percent for large plants), extremely skilled users of all technologies report much better than average ROI in every case except automated inspection (12.7%). *The evidence here is very strong that extremely skilled users of all technologies considered here enjoy superior return on investment.*

Figure 37.16 shows how powerful JIT can be; it leads all other technologies on ROI (17.6 percent). The emergence of JIT as an effective and efficient technology is notable. Extremely skilled use of JIT is associated with the best performance or close to the best performance in each of the five performance measures in Figs. 37.12 to 37.16, i.e., JIT is associated with the best performance regardless of the performance measure used. Therefore, we conclude that *JIT use is found in the best performing firms, or the best performing firms use JIT. Either way, for the practitioner, the conclusion is easy to see,* ***use JIT!***

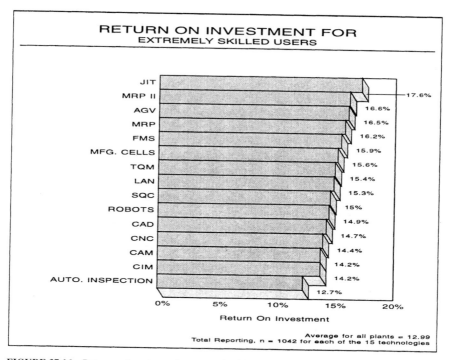

FIGURE 37.16 Return on investment for extremely skilled users.

37.3.3   Characteristics of Extremely Skilled Users

Direct-Labor Costs.   We studied how direct-labor costs as a percent of sales were affected by the extremely skilled use of various technologies. In soft technology environments such as MRP I and II, direct-labor costs are the lowest, at 12.3 percent of sales (see Fig. 37.17). CAM, automated inspection, robots, and CIM are associated with very high labor costs as a percent of sales. The high labor costs in these plants appear to justify the investment in these hard technologies. Alternatively, plants with higher labor costs may be attracted to these technologies.

Number of Product Lines.   According to Fig. 37.18, JIT environments are associated with the maximum number of product lines (43.8), while automated inspection and AGV exist in environments with the least number of product lines at 13.4 (about a third of what JIT environments handle). Given that JIT environments are associated with high profitability, the large number of product lines seem to be beneficial to JIT manufacturers; JIT perhaps helps manufacturers to broaden their market base and further helps them enter new markets quickly. Alternatively, plants with wider product lines may be attracted to JIT. The finding concerning AGV and AI is consistent with what we would expect; that is, they are suitable for high-volume, low-variety manufacturing environments.

Making to Stock.   We studied how the skilled use of various technologies varied with make-to-stock production, which is typified by repetitive, large-volume production. Figure 37.19 shows that, where SQC is used with extreme skill, average make-to-stock production as a percentage of total production is the lowest. The opposite is true

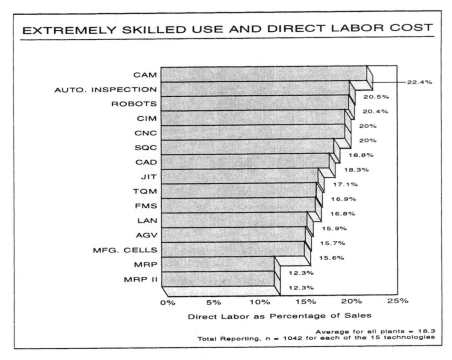

FIGURE 37.17   Extremely skilled use and direct-labor cost.

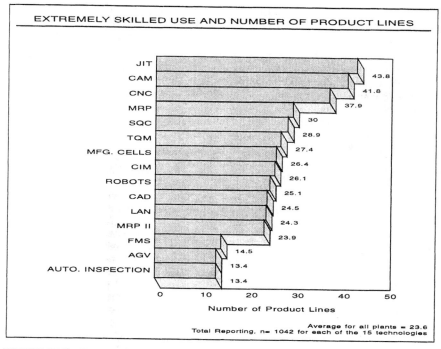

FIGURE 37.18 Extremely skilled use and number of product lines.

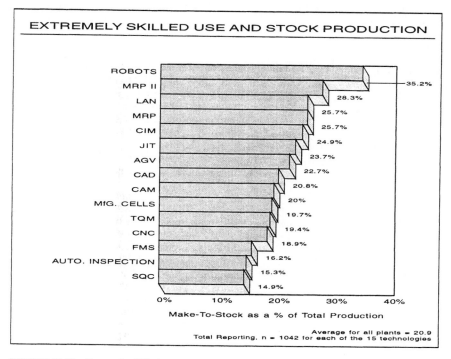

FIGURE 37.19 Extremely skilled use and stock production.

where robots are used with extreme skill. This means that extremely skilled use of robots is found more often in make-to-stock environments (high-volume production) and extremely skilled use of SQC is more prevalent in make-to-order environments (low-volume production); this is consistent with expectations because SQC can be labor-intensive and works well in low-volume environments. In high-volume environments, some form of automation of the inspection process is called for. Further, manufacturers are reporting that extremely skilled use of robots, LAN, and MRP I and II are more prevalent in make-to-stock environments.

37.4   *INVESTMENT DECISIONS*

37.4.1   Reasons for Not Investing in Technologies

Manufacturers expressed three reasons for not investing in advanced manufacturing technologies: (1) irrelevance of the technology; (2) lack of know-how; and (3) lack of capital. The results are presented in Figs. 37.20 to 37.22.

Not Relevant.   According to the Fig. 37.20, specialized hard technologies such as AGV, FMS, and robots are not relevant to about half the manufacturers; this is understandable given factory size, batch size, manufacturing processes used, etc. The findings are consistent with common wisdom.

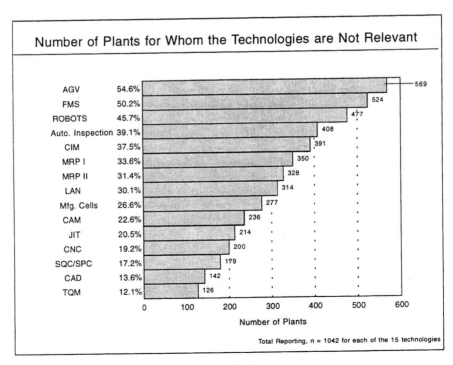

FIGURE 37.20   Number of plants for whom the technologies are not relevant.

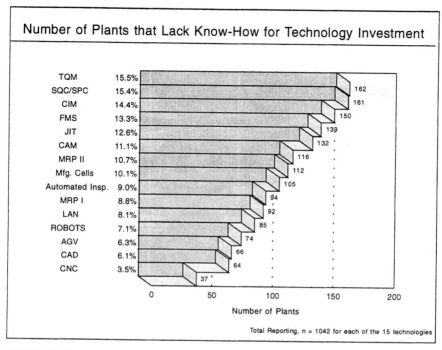

FIGURE 37.21   Number of plants that lack know-how for technology investment.

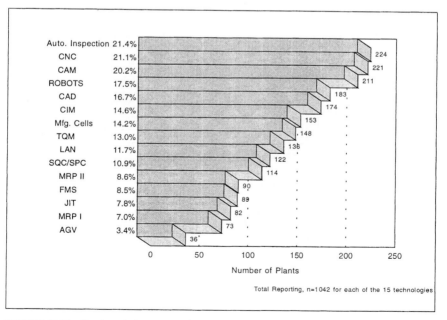

FIGURE 37.22   Number of plants that lack capital for investment.

Lack of Know-How.   Figure 37.21 is remarkable for the very low percentage of firms reporting that they lack know-how for investment in each of the 15 technologies. For example, in the worst case (TQM), only about 15 percent report that they lack know-how. It is interesting that the technologies most frequently hindered by lack of know-how are TQM and SQC, which are not capital-intensive technologies. Rather, they are training- and people-intensive technologies, which require considerable discipline on the part of users. These technologies further require the involvement of employees at all levels. In other words, these are not technologies that one could buy off the shelf and install on the shop floor for instant use. It takes time to gain extreme skill in the use of certain technologies such as TQM, SQC, CIM, FMS, and JIT, which are at the top of the list in Fig. 37.21.

Lack of Capital.   Figure 37.22 shows that hard technologies such as automated inspection, CNC, CAM, robots, and CAD are the technologies most likely to be slowed by lack of capital. According to the figure, hundreds of respondents would invest in these technologies if sufficient capital were available to them.

37.4.2   Investment Recovery Time

To determine how investment practices of manufacturers may have a bearing on capital spending on manufacturing technologies, we investigated the recovery time allowed for investments. Figure 37.23 shows that nearly 55 percent allow less than 2

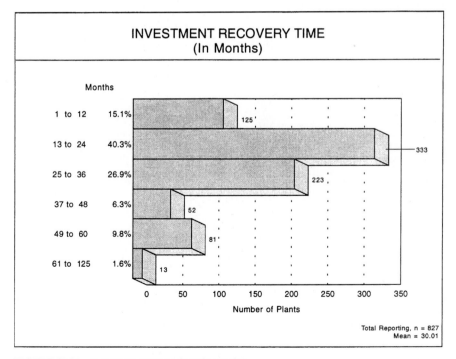

FIGURE 37.23   Investment recovery time (in months).

years for capital recovery, with about 40 percent allowing between 1 and 2 years for capital recovery. This may be too short a time for recovering investments in manufacturing technologies. Only about 18 percent allow more than 3 years for capital recovery. Plants that are entirely U.S.-owned reported a recovery time of 30 months as opposed to 26.6 months reported by foreign plants (foreign ownership in excess of 50 percent).

It is notable that while 50 percent of small plants allow less than 2 years for recovery, nearly 60 percent of larger plants do so. Further, while 22 percent of small plants allow 3 to 5 years for capital recovery, only 11 percent of the larger plants do so. This indicates that smaller plants are unable to recover investments in technology as fast as larger plants. This may partially explain the less frequent use of manufacturing technologies in small plants when compared to larger plants.

37.4.3 Sources of Assistance

More than 70 percent of respondents credit their technology vendors and customers for providing the assistance needed to invest in advanced technologies (Fig. 37.24). At nearly 43 percent, customers' role in providing assistance to manufacturers in technology investment decisions is by far the single most commonly reported source. The rise of TQM and JIT, which include a heavy dose of supplier development, may explain part of the finding. This finding also reveals a healthy cooperation among industrial suppliers and customers.

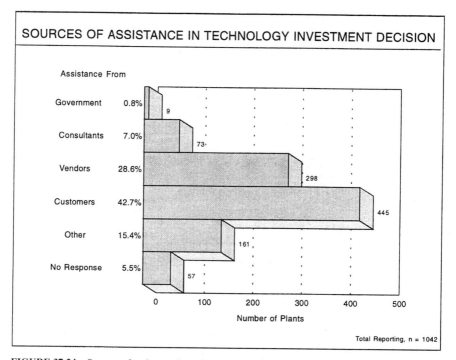

FIGURE 37.24 Sources of assistance in technology investment decision.

37.5 *PLANS TO BECOME SKILLED IN TECHNOLOGY USE*

The desire to become extremely skilled in TQM, CAD, and JIT appear at the top of manufacturers' plans for the next 2 years. According to Fig. 37.6, CNC, CAD, and CAM are the top three technologies in which manufacturers are now extremely skilled. Therefore, it appears that Fig. 37.25 records a shift in manufacturers' interests. Figure 37.25 should be of interest to those who supply production technologies to manufacturers.

Figures 37.26 and 37.27 show that, for small plants, CAD, TQM, and CNC (biased toward hard technologies) are the top items in their plans while, for larger plants, TQM, JIT, and manufacturing cells are at the top of their plans (biased toward soft technologies). Thus, small plants are intending to master certain hard technologies while larger plants are targeting certain soft technologies.

Figure 37.28 takes Fig. 37.25 one step further by capturing the number of plants that would most likely invest in the technologies shown if *funds were available.* The value of the two figures is that, while 52.8 percent of the manufacturers (Fig. 37.25) have plans to become extremely skilled in TQM in the next 2 years, only 34.8 percent (Fig. 37.28) would invest in TQM if funds were available. However, TQM is the most frequently mentioned item in both figures. This could be interpreted to mean that 18 percent (i.e., 52.8−34.8) plan to become extremely skilled without the need for any additional dollar investment in TQM. It is an indication that at least a portion of the manufacturers who intend to become extremely skilled will need no more additional investment either because they will become adept at using the technology in which

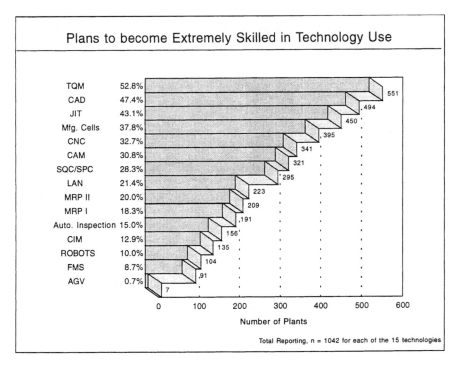

FIGURE 37.25   Plans to become extremely skilled in technology use.

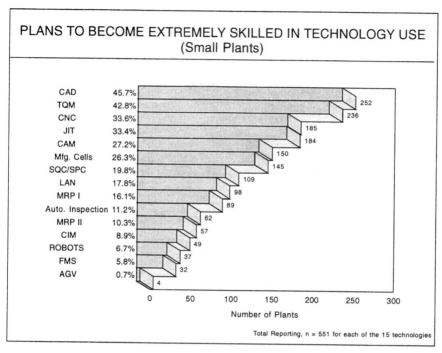

FIGURE 37.26 Plans to become extremely skilled in technology use (small plants).

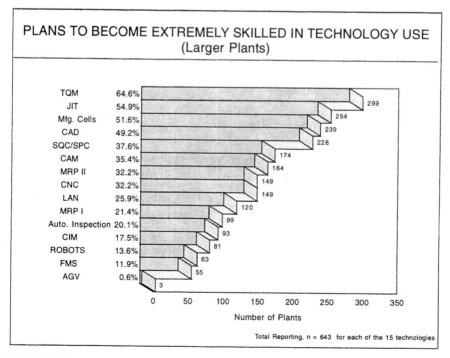

FIGURE 37.27 Plans to become extremely skilled in technology use (larger plants).

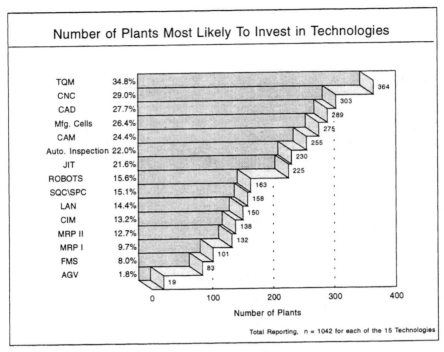

FIGURE 37.28 Number of plants most likely to invest in technologies.

they have already made an investment, or because TQM can be implemented without cash investments.

In comparing Figs. 37.25 and 37.28, we find that, where hard technologies are involved, fewer firms report plans to become extremely skilled without additional investment. If funds were available, plants would invest in robots and automated inspection more than any other technologies.

By implication, adoption of these technologies is hampered by lack of investment more than other technologies. By the same token, manufacturers report plans to become extremely skilled in the use of JIT, TQM, SQC, and manufacturing cells even when funds are not anticipated. This is a good example of the fact that funds are not always the determining factor in the implementation of certain soft manufacturing technologies.

37.6 AN OVERVIEW OF THE RESPONDENTS

This section describes the plants responding to this study. The purpose of this section is to give readers a sense of the industries and manufacturing facilities under investigation and to enable the readers to identify more readily with the respondents.

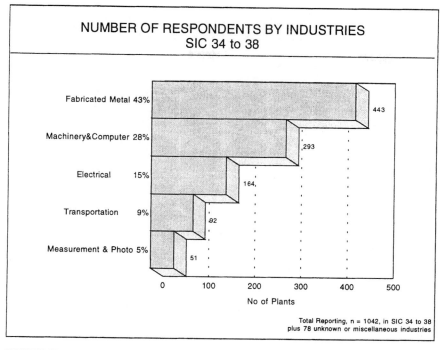

FIGURE 37.29 Number of respondents by industries, SIC 34 to 38.

37.6.1 Industries Covered by the Study

The descriptions of the industries participating in this study and the number of participants from each Standard Industrial Classification (SIC) code are given below (see also Fig. 37.29):

		Participants	Percent
SIC 34	Fabricated metal, except machinery and transport equipment	442	42.5
SIC 35	Industrial, commercial machinery, computer equipment	293	28.1
SIC 36	Electrical, other electrical equipment	164	15.7
SIC 37	Transportation equipment	92	8.8
SIC 38	Measurement instruments, photo goods, watches	51	4.9
	Total	1042	100.0
Unknown	Other manufacturing or unknown (excluded from all subsequent analysis)	79	—
	Total response	1121	—

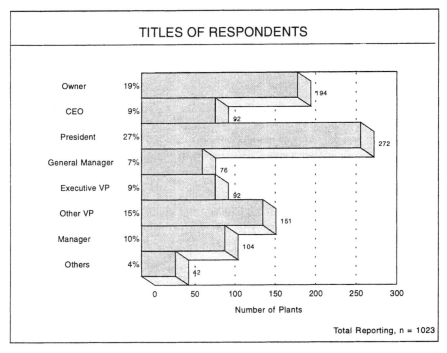

FIGURE 37.30   Titles of respondents.

37.6.2 Respondents

The questionnaires were completed by top management in the manufacturing facilities participating in this study. As Fig. 37.30 shows, 86 percent of respondents were vice presidents or higher.

37.6.3 Employees

Figure 37.31 shows the distribution of plants participating in this study by size of employment. The average plant has 228 employees. Plant employment was used as the metric for classifying plants as "small" (less than 100 employees) or "large" (100 or more employees). And, because differences between responses from the two groups were significant enough, some findings in this report are differentiated by plant size.

37.6.4 Annual Sales

The average annual sales for a plant is $47.2 million. The combined total sales for all the plants in this study is in excess of $49 billion. The distribution of participants by annual plant sales is presented in Fig. 37.32. According to the figure, 71 percent of the respondents have annual sales of less than $20 million, and 91.7 percent have annual

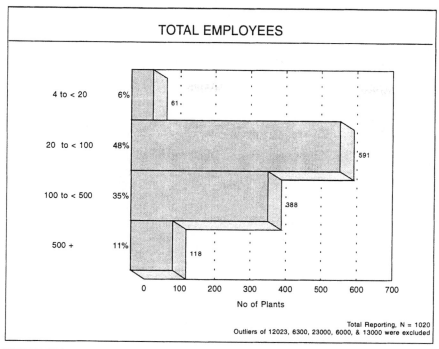

FIGURE 37.31   Total employees.

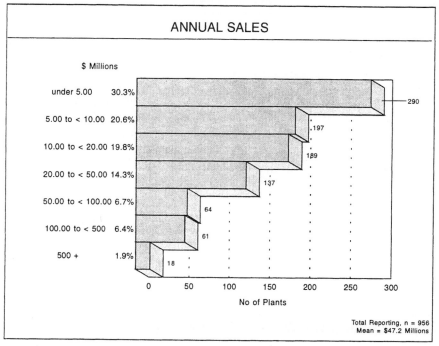

FIGURE 37.32   Annual sales.

sales of less than $100 million. Small plants average $5.35 million in sales, with 87 percent reporting sales less than $10 million. Large plants average $87.2 million in sales.

37.6.5 Export Sales

According to Fig. 37.33, only 15.5 percent of the manufacturers in these industries do not export, or in other words, nearly 85 percent have export sales. About half (44 percent) export less than 10 percent, while one-fifth (20.2 percent) export more than 20 percent of their sales. Among small plants, 22 percent had no export sales, whereas among the larger plants, only 6.5 percent reported no export sales.

37.6.6 Defense-Related Sales

Figure 37.34 shows that more than one-half of the manufacturers do not make defense-related products, while another third send less than 10 percent of their output to the Defense Department. Less than 8 percent of manufacturers sell more than 25 percent of their output to defense.

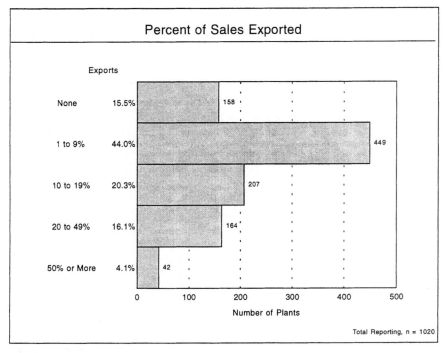

FIGURE 37.33   Percent of sales exported.

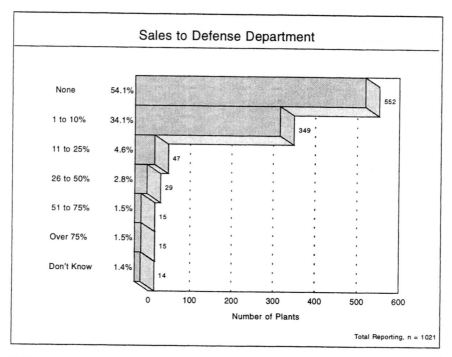

FIGURE 37.34   Sales to Defense Department.

37.6.7   ISO 9000 Certification

The ISO 9000 series is an internationally accepted quality management standard developed by the International Organization for Standardization (ISO). Figure 37.35 shows that only 4 percent of manufacturers in the industries covered by this study are already certified in ISO 9000 series. However, a healthy 23 percent (about one-fourth) are in the process of certification. Only 13 percent are not interested in ISO 9000 certification. Among small plants, 1 percent are certified, whereas 7.8 percent of larger plants are certified. While 15.6 percent of the small plants are in the process of certification, 33.2 percent of larger plants are in the process of doing so.

37.6.8   Dominant Manufacturing Processes

Figure 37.36 shows that job shops are the most common in the survey sample. This is explained by the fact that 54 percent of manufacturers employ fewer than 100 employees (see Fig. 37.3). One-fifth (20.7 percent) of all plants predominantly use manufacturing cells. While 13.8 percent of small manufacturers use cells predominantly, notably 28.2 percent of larger manufacturers do so; that is, larger plants use cells twice as often as smaller plants. Manufacturing cells provide an important avenue for capturing cost efficiencies as well as flexibility. [See the glossary in App. 2 (Sec. 37.9) for a definition of manufacturing cells.]

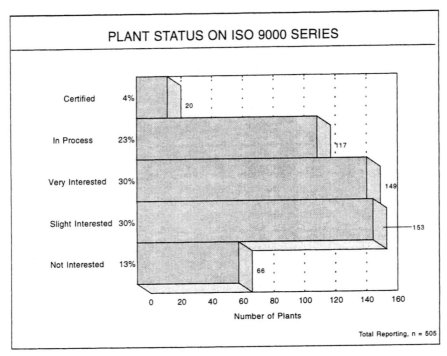

FIGURE 37.35 Plant status on ISO 9000 series.

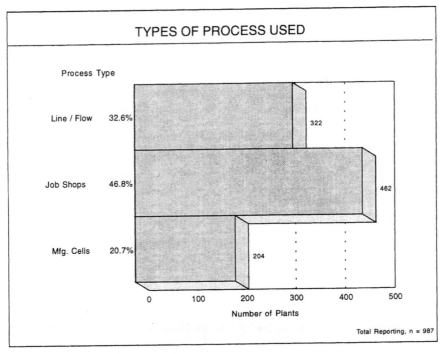

FIGURE 37.36 Types of process used.

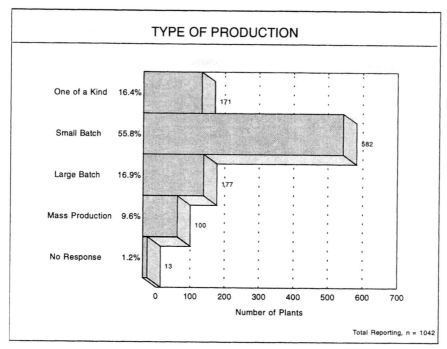

FIGURE 37.37 Type of production.

37.6.9 Batch Size

According to Fig. 37.37, *small batch production* (55.8 percent) dominates the industries studied. Part of this could be explained by the abundance of smaller manufacturers and the changing emphasis on small batches imposed by the use of JIT methods, flexible machines, reduced setup times, etc. Only 10 percent engage in mass production. While 18.8 percent of small manufacturers use large-batch or mass production, nearly twice the proportion (i.e., 35.8 percent) of the larger manufacturers use large-batch or mass production.

37.7 SUMMARY

This study benchmarks the usage of 15 selected technologies in U.S. manufacturing plants; all 15 technologies investigated are described in the glossary in Sec. 37.9 (App. 2). Notable findings on technology usage are (1) CAD, TQM, and JIT are the most widely used technologies regardless of skill levels; (2) among extremely skilled users, CNC and CAD are the most widely used technologies; (3) use of technologies occurs in clusters; (4) 100 percent of larger plants with 100 or more employees, who consider CAD to be a relevant technology, use CAD; and (5) *most important,* the extensive use of seven technologies, CAD, CAM, CNC, TQM, JIT, SQC, and manufacturing cells is setting the pace of competitive manufacturing in American factory floors.

37.7.1 Nature of Plants Studied

This study benchmarks manufacturing plants, *not* entire manufacturing firms with multiple plants. We studied plants as opposed to entire companies because technology use practices vary significantly among the various plants of a given company. The average characteristics of benchmarked *plants* are:

- Total employment is 228 workers.
- Plant sales is $47.2 million.
- Sales per employee is $133,000.
- Pretax return on investment (ROI) for 3 years is 12.99 percent.
- Average plant has 24 product lines.
- Inventory turns is 8.04.
- The lead time from order to shipping is 7.19 weeks.
- Rejection and rework rate is 4.00 percent.
- More than 84 percent of the plants participating in the study have export sales.
- Forty-five percent have some sales to the U.S. Department of Defense.
- About 30 percent of the respondents use line or flow production.
- Nearly 45 percent use job shops.
- About 20 percent use manufacturing cells predominantly for production.
- Over 55 percent of the respondents use small batch production.
- The percentage of foreign-owned plants is 3.4.

Number of Technologies Used. The average plant uses 7 different technologies; only 30 plants (about 3 percent) use none of the 15 technologies investigated. Eighteen percent of the plants use 10 or more technologies; the maximum number in use is 14 different technologies. The correlation between the number of technologies used and the number of benefits claimed is .49. The moderate correlation means that the indiscriminate use of several technologies does not translate into benefits.

The average plant is an extremely skilled user of 1.8 technologies, with 63.3 percent reporting extreme skill in the use of at least one technology. About 20 percent of the plants are extremely skilled in the use of four or more technologies. On the negative side, 37.3 percent of the plants are *not* extremely skilled in *any* of the technologies surveyed.

37.7.2 Benefits of Technology Use

Most Frequently Cited Benefits. Reduced cycle time, market share growth, progress toward zero defects, and ROI are the top four direct benefits of technology use reported by our respondents. *The most frequently mentioned direct benefit of technology use was reduced cycle time, cited by 66 percent of respondents.* One can infer from the finding that manufacturers are most often investing in manufacturing technologies to reduce the cycle-time from order to shipping.

Benefits of Extremely Skilled Technology Use. Just-in-time manufacturing (JIT) took the top honors as the one technology whose extremely skilled use was unfailingly associated with superior performance, regardless of the performance metric used. Specifically, extremely skilled users of JIT reported a superior inventory turn at 10 (average = 8.04), the second lowest manufacturing lead time at 7.2 weeks, the lowest rejection and rework rate at 2.9 percent (average = 4.00 percent), superior sales per employee at $178,000 (average = $133,000), and the best ROI at 17.6 percent (average 12.99 percent). With these rewards associated with JIT, it is no wonder JIT is used by nearly 90 percent of those who consider it relevant to their operations.

Number of Benefits Reported. The average plant reported 3.7 benefits directly attributable to the use of one or more technologies. More than 85 percent of the plants reported one or more benefits attributable to technology use, which also means that 15 percent reported no benefits. About 25 percent of the facilities reported six or more benefits attributable to technology use. About 3 percent (37 plants) reported 10 or more benefits attributable to technology use.

37.7.3 Future Plans to Become Extremely Skilled Users

TQM, CAD, and JIT, in that order, are the most frequently cited technologies in which U.S. manufacturers plan to become extremely skilled users in the next 2 years.

37.8 APPENDIX 1: DATA COLLECTION

37.8.1 The Sample

This is a study of individual manufacturing *plants, not* a study of manufacturing *firms.* A questionnaire was sent to 4453 member firms of the National Association of Manufacturers (NAM) in the SIC industrial classifications 3400 through 3899 (followed by one reminder 3 weeks later, in July 1993).

Split Sample. To examine the validity of the study, we developed a split sample. After mailing the questionnaire and one reminder, we received 556 usable responses; this formed the first "half" of the split sample. To increase the responses and to acquire the second "half" of the sample, instead of sending a mere reminder in the form of a card, we again sent the entire questionnaire to those firms that did not respond to the first mailing. We also followed this with a reminder card. The second "half" of the sample yielded 565 usable responses. Thus, the total usable response was 556 + 565 = 1121; 25 responses were unusable and 3 responses came after February 15, 1994, the cutoff date; the resulting response rate was 25.8 percent. Our estimate shows that the response from larger plants was much higher.

In Table 37.5, we present the averages for nine major variables from the two samples for comparison. The similarity of the averages is an indication of the lack of significant bias in the total sample. All subsequent analyses were performed by pooling the two split samples into one pooled sample of 1121.

Distribution by Plant Size (Employment). Table 37.6 compares the distribution of plants by size in the split samples. The two samples are very similar on the basis of this comparison; no particular bias is evident.

TABLE 37.5 Statistics for Split Samples

	Sample 1	Sample 2	Total
Sample size (*n*)	556	565	1121
Sales, $000,000*	38.4 (*n* = 487)†	56.4 (*n* = 465)	47.2 (*n* = 952)
Employment*	251.4 (*n* = 513)	203.5 (*n* = 502)	227.7 (*n* = 1015)
Sales per employee, $000	130.5 (*n* = 484)	136 (*n* = 460)	133 (*n* = 944)
Rejection, %	3.92	4.07	4.00
Inventory turns	7.61	8.16	7.89
Cost of goods sold, % of sales	0.609	0.605	0.606
Product lines	22	25	24
Average lead time, weeks	7.2	7.2	7.2
Direct labor, hours*	18.31	18.27	18.3

*Averages exclude outliers.

†Averages based on the number of firms (*n*) reporting.

TABLE 37.6 Distribution of Respondents by Size in the Split Samples (Including Outliers and Miscellaneous Manufacturing)

Employment	Sample 1		Sample 2		Total	
	Respondents	%	Respondents	%	Respondents	%
<49	159	29.18	178	32.60	337	30.89
50–99	121	22.20	122	22.34	243	22.27
100–499	196	35.96	192	35.16	388	35.56
≥500	69	12.66	54	9.89	123	11.27
Total providing employment data	545	100.0	546	99.99	1091	99.99
Total responding	556		565		1121	

37.8.2 Data Validation

Size. The industries covered by this study are identical to those covered by a Bureau of Census (BOC) study (U.S. Bureau of Census, *Manufacturing Technology: Factors Affecting Adoption 1991*, AMT/91-2, Current Industrial Reports, Government Printing Office, Washington, D.C., 1993). The BOC study estimated the total number of plants in the United States with 20 or more employees in 1991 in each industrial classification covered by the study. The BOC estimate predates this study by at least 2 years, but the changes in the distribution in the national population during the 2 years may not be substantial.

Table 37.7 compares the distribution of plants in this study with the distribution of plants in the BOC study on the basis of size (employment). The NAM study is slightly biased toward larger plants. Yet, the two samples are similar in that plants with 0 to 99 employees are the largest subgroup in both samples, and plants with 550 or more employees form the smallest subgroup in both samples. The slight bias toward larger plants in the NAM sample is because the BOC sample covered all plants with 20 or more employees. Fewer firms with 20 to 50 employees tend to be members of NAM. Yet, the sample is reasonably comparable to the BOC estimate.

TABLE 37.7 Comparison of the Distribution of Plants by Employment Size

Employment, n	BOC estimate, %	NAM sample, %
0–99	71	53.2
100–499	24.1	35.6
500+	4.9	11.3

Industry. In Table 37.8 below, we compare the distribution of plants by SIC classification with the BOC study serving as the reference. Table 37.8 shows that the distribution of manufacturing establishments in the United States is roughly comparable to the distribution of the respondents to this study with a slight bias toward SIC 34 (metal fabrication industry) in the NAM sample. This slight skewing toward one industry may be due to the slight bias toward larger plants in the NAM sample.

TABLE 37.8 Distribution by Industry

	BOC estimate for the USA (firms with 20 + employees), %	1994 NAM respondents, %
SIC 34	31.6	42.5
SIC 35	33.2	28.1
SIC 36	16.6	15.7
SIC 37	9.5	8.8
SIC 38	9.1	4.9
	100.1	100

37.8.3 Differences in Response Rates for Small and Large Plants

The last study was directed entirely at larger firms that paid $300 or more in dues to NAM. This time, we contacted all firms in the appropriate SIC classifications without regard for size or dues. In the last study, we contacted 2015 firms, whereas in this study, we contacted 4453 firms. That is, 2438 more firms were contacted this time. In columns 3 and 4 of Table 37.9, we show our estimate of small and larger firms contacted in the first study and the additional mailing in the second study to small and larger firms. In the table, for the sake of making an estimate of response rates by plant size, we are making an assumption (see note below) that larger plants among the respondents belong to larger firms contacted by us, and small plants among the respondents belong to small firms contacted by us. Using this assumption, we find in the last column of Table 37.9 that the response rate for larger plants is nearly 43 percent and the response rate for smaller plants is nearly 20 percent.

(*Note:* There is a small chance that some of the small plants participating in the study belong to larger firms. Therefore, the response rate for larger plants in Table 37.9 may be a slight underestimate. *Overall, larger plants are better represented in our sample.* The chance of larger plants from small firms participating in this study is either negligible or nil; if a small firm had larger plants, the firm would have been classified as a larger firm. *Throughout this report we mention separate figures for small and larger plants for the benefit of readers from both groups.* Small plants use

TABLE 37.9 Increase in Response by Size When Mailing Increased

Firm employees	First study	Additional* mailings in second study	First study response by plant size	Second study response by plant size	Increase in response	Increase in response for 1 additional mailing	Response* rate First study, %	Response* rate Second study, %
<100	1089	1840	101	580	479	0.26	9.3	19.8
≥100	811	383	262	511	359	0.94	32.3	42.9
Unknown	115	215	22	52	—	—	—	—
Total	**2015**	**2438**	**385**	**1191**	**838**	—	—	—

*Estimates.

technologies less often than do larger plants but *small plants do use most technologies, and therefore it is worth studying technology use and trends in small plants.*)

37.9 APPENDIX 2: GLOSSARY OF MANUFACTURING TECHNOLOGY TERMS*

37.9.1 Automated Guided Vehicles (AGVs)

Automated guided vehicles (AGVs) are unmanned carriers or platforms that are controlled by a central computer that dispatches, tracks, and governs their movements on guided loops. AGV systems utilize infrared, optical, inertial, embedded wire, or ultrasonic methods for guidance. AGVs are useful primarily for materials handling, in which they deliver inventory from holding to production areas, or between workstations as a replacement for conventional forklifts and rigid transfer lines. Some AGVs are used in assembly systems, while others provide production platforms that support products such as automobiles and engines while work is performed.

37.9.2 Automated Inspection (AI)

Automated inspection (AI) is defined as the automation of one or more steps involved in the inspection procedure. Automated inspection (AI) takes advantage of highly advanced sensor technologies to perform inspection functions once performed by humans (or not performed at all). It can reduce manufacturing lead times and product cost associated with manual inspection. Further, automated inspection allows for 100 percent inspection to be integrated into the manufacturing process.

37.9.3 Computer-Aided Design (CAD)

Computer-aided design (CAD) is a computer software-hardware combination used in conjunction with computer graphics to allow engineers and designers to create, draft, manipulate, and change designs on a computer without the use of conventional drafting. CAD systems allow for tremendous speed, precision, and flexibility over traditional drafting systems.

37.9.4 Computer-Aided Manufacturing (CAM)

Computer-aided manufacturing (CAM) incorporates the use of computers to control and monitor several manufacturing elements such as robots, CNC machines, storage and retrieval systems, and AGVs. CAM implementation is often classified into several levels. At the lowest level, it includes programmable machines controlled by a centralized computer. At the highest level, large-scale systems integration includes control and supervisory systems.

---

*These definitions were compiled by Michael Hickman, Graduate Research Assistant at the Thomas Walter Center under the guidance of the author.

37.9.5   Computer-Integrated Manufacturing (CIM)

Computer-integrated manufacturing (CIM) involves the total integration of all computer systems in a manufacturing facility; the integration may extend beyond one factory into multiple manufacturing facilities in one or more countries and into the facilities of vendors and customers. CIM integrates all computer systems that handle everything from order to shipment of final product. The integration involves accounting, finance, management, engineering, design, production, manufacturing, and equipment. The idea is to form one large system that connects all activities so that common information is shared on a real-time basis. While the scope of CAM is generally limited to the factory floor, the scope of CIM can extend far beyond the factory floor.

37.9.6   Computer Numerical Control (CNC) Machines

Computer numerical control (CNC) machines are locally programmable machines with dedicated micro- or minicomputers. CNC provides great flexibility by allowing the machine to be controlled and programmed on the floor by the machine operator. Further, CNC allows machines to be integrated with other complementary technologies such as CAD, CIM, and CAM. CNC also serves as the building block for flexible manufacturing systems.

37.9.7   Flexible Manufacturing System (FMS)

A *flexible manufacturing system* (FMS) is a group of reprogrammable machines linked by an automated materials-handling system and a central computer. The intent of such a system is to produce a variety of parts that have similar processing requirements with low setup costs. The typical FMS system is designed around and dedicated to the manufacture of a family of parts. The FMS is typically designed to run for long period, with little or no operator attention and fills the need for machining in a batch environment. FMS, unlike the old dedicated production lines, can react quickly to product and design changes.

37.9.8   Just-in-Time (JIT) Manufacturing

The concept of *just-in-time* (JIT) *manufacturing* is a philosophy that requires materials and goods to arrive "just in time" to be used in production or by the customer. Embedded in the philosophy of JIT is a "continuous habit of improving" and the "elimination of wasteful practices." The latter means not only to looking for ways to cut cost but also to continually eliminate everything that does not directly contribute to the company's objectives or add value to the product. One of the most recognizable aspects of JIT is the low levels of inventory with which it is associated. In the JIT system, inventory is seen as a necessary evil.

37.9.9   Local Area Network (LAN)

Local area networks (LANs) are the backbone of communication systems that connect various devices in a factory to a central control center. The LAN, through the control center, allows for the various devices connected to the network to communicate with one another for exchange of information and control. The types of devices that can be attached to the network include computers, programmable controllers, CNC machines, robots, data-collection devices, bar-code readers, vision systems, and automated inspection systems.

37.9.10   Manufacturing Cell (MC)

A *manufacturing cell* (MC) is composed of a small group of workers and machines in a production flow layout, frequently a U-shaped configuration, to produce a group of similar items called *part families* in dedicated production areas. Proponents of cellular manufacturing have claimed several benefits for this type of production system, including less inventory, less materials handling, improved productivity and quality, improved worker job satisfaction, smoother flow, and improved scheduling and control. Cells can be used for machining, fabrication, and assembly, as well as combinations of the three. Cells permit a degree of automation, improved flow, and reduction of direct labor, and provide the vehicle for implementing manufacturing innovations such as FMS and JIT. Flow patterns in cells vary considerably. Some resemble flow shops, while others are more like job shops. Because manufacturing cells are so adaptable, almost any manufacturer could gainfully use cellular manufacturing.

37.9.11   Materials Requirements Planning (MRP or MRP I)

Materials requirements planning (MRP I) is primarily a scheduling technique, a method for establishing and maintaining valid due dates or priorities for orders using bills of material, inventory and order data, and master production schedule information as inputs. MRP I has been around since the mid-1960s.

37.9.12   Manufacturing Resource Planning (MRP II)

Manufacturing resource planning (MRP II) is a direct outgrowth and extension of closed-loop materials requirements planning (MRP I) through the integration of business plan, purchase commitment reports, sales objectives, manufacturing capabilities, and cash-flow constraints. MRP II reports may include dollar value of shipments, product cost, overhead allocations, inventories, backlogs, cash-flow projection, and profits.

37.9.13   Robots

The Robotics Institute of America defines the *industrial robot* as "a programmable, multifunctional manipulator designed to move material, parts, tools, or specialized devices through various programmed motions for the performance of a variety of tasks." The basic purpose of the industrial robot is to replace human labor under certain conditions. The programmable nature of robots provides the flexibility to manufacture a variety of products. The industrial robot was developed to generate higher output at a lower cost in situations that require high-repetition, high-precision, large-capacity workloads, and working in hazardous environments (e.g., paint booths, chemical processing, and welding).

37.9.14 Statistical Quality/Process Control (SQC/SPC)

Statistical quality or process control (SQC/SPC) applies the laws of probability and statistical techniques for monitoring and controlling the quality of a process and its output. SQC/SPC can be used to reduce variability in the process and output quality. It contributes to the implementation of JIT and TQM.

37.9.15 Total Quality Management (TQM)

Total quality management (TQM) is built on the principle of continuous quality improvement in manufacturing, as well as in the entire organization. It works well with frequent feedback of performance measures to various system elements empowered to make changes in their operation such that the system moves closer and closer to its stated goals, which keep up with customer needs.

CHAPTER 38

NEO-OPERATIONS STRATEGY

LINKING CAPABILITIES-BASED
COMPETITION TO TECHNOLOGY*

Aleda V. Roth

Kenan-Flagler Business School
University of North Carolina
Chapel Hill, North Carolina

38.1 INTRODUCTION

Globalization and technological advancement are transforming the fundamental nature of competition. Today manufacturers are struggling to define what it means to be "global." Each plant, division, and line of business must understand how it fits into the whole; each must understand its contribution to competitive advantage. As the various entities place their pieces within the strategic jigsaw puzzle of global competition, entirely new forms of manufacturing systems are emerging. Away from stand-alone electronic technologies, integrated advanced technologies are transforming centuries-old industrial organizations. Especially within the industrialized nations, the picture many see is termed "agile manufacturing" (Burgess, 1994; Goldman et al., 1995; Goldman and Nagel, 1993; Maskell, 1994; Roth, 1996). Roth and Giffi (1995) push this vision more broadly toward capabilities-based competition on strategic agility. *Strategic agility* is "the ability to produce the right products at the right time at the right place at the right cost" (Giffi, 1994, p. 6). Accordingly, strategic agility is operationalized by the set of *combinative competitive capabilities*—on quality, delivery, flexibility, and price leadership.

This research shows that world-class manufacturers exhibit more strategic agility than do their non-world-class counterparts. From a theoretical perspective, the simultaneous acquisition of multiple competitive capabilities adds to the firm's "absorptive capacity" to deploy its operations resources as a source of formidable enterprisewide advantage. [See Cohen and Levinthal (1990) for a discussion of absorptive capacity theory in R&D management; also see Kogut and Zander (1992) for a discussion of combinative capabilities and knowledge.] Roth and Jackson (1995) apply the theory to capabilities-based competition in service operations. *Absorptive capacity* represents a

---

*The author is most greatful to each of the executives who shared their valuable experiences in the conduct of the field research, and especially to Mr. Robert Badelt for Nortel's support of the theory development phase of this research in 1992–1993, and the Craig Giffi of Deloitte & Touche Consulting Group for his valuable insights. An earlier version of theoretical framework was presented at the Executive Forum on Global Business Reengineering, CATO Center for Applied Business Research, University of North Carolina at Chapel Hill.

synthesis of accumulated organizational know-how, technological prowess, and social values that enables an entity to identify, assimilate, and exploit new knowledge; hence, there are path dependencies in knowledge accumulation. Simply put, the level and types of prior knowledge largely determine the rate of new-knowledge assimilation. Thus, it is not coincidental that the world-class group takes a holistic approach to their operations strategy. The empirical evidence presented here supports the hypotheses that strategic agility is correlated with technological leadership and best manufacturing practices.

Absorptive capacity theory requires a fundamentally new and expanded view of operations—a neo-operations strategy. *Neo-operations* strategy is defined as a path-dependent pattern of integration, infrastructural, and structural choices that create dynamic value-chain processes critical to order winning, qualifying, and keeping. These processes build combinative capabilities, accelerate the rate of organizational learning, and leverage knowledge-based competencies for competitive advantage in the global marketplace. Neo-operations strategy requires an ongoing assessment of strategic alignment of operations with activities within the firm and along the value chain. Without question, the conceptualization of neo-operations strategy requires a revolutionary—not evolutionary—change in executive mindsets and practice. "Under the mantra of neo-operations strategy, manufacturing (firms) must exploit information-intensive, knowledge-based competencies to create a seamless value chain and competitive capabilities" (Roth and Giffi, 1995, p. 3).

Thus, to complete the strategic jigsaw puzzle, world-class manufacturing firms have redefined operations strategy because traditional definitions of manufacturing strategy are woefully incomplete. Neo-operations strategy replaces the logic behind traditional manufacturing strategy that is oriented toward the shop floor. The traditional logic is functionally defined as the adoption of a portfolio of manufacturing improvement programs, tools, and activities that coincide with a specific manufacturing task (Adam and Swamidass, 1989; Hayes et al., 1988; Krajewski and Ritzman, 1993; Skinner, 1978; Schonberger, 1986). Hill (1989) classifies manufacturing tasks as "order winners" and "order qualifiers." Neo-operations strategy differs from conventional wisdom in seven important ways, as delineated in Table 38.1.

The extant manufacturing strategy literature says nothing about information age logic; it is rooted in industrial-age wisdom. It says little about boundary-spanning, such as cross-functional integration. The few exceptions are directed at bifunctional interfaces such as Hill (1989); Fitzsimmons et al. (1991); Wheelwright and Clark (1992); Kim et al. (1992); Vickery et al. (1993); and Berry et al. (1991). Yet, Roth and Giffi (1994, 1995) have empirically shown that world-class manufacturers have a broader, dynamic view of manufacturing—one that emphasizes the criticality of boundary spanning to increase strategic agility. The world-class group consistently differentiates itself by integrating production processes with other parts of the company, with parts of different companies through alliances and partnerships, and with customers and suppliers along the value chain. Such boundary spanning provides unique opportunities for customer value integration (Roth et al., 1995). For some manufacturers, like Hewlett-Packard, Xerox, GE, or Nortel, the deployment of neo-operations strategies literally means turning their organizations topsy-turvy along business process lines that span the global supply chains. One executive interviewed in this research said, "Manufacturers must now be ready to 'turn on a dime.'" Moreover, Corbett and Van Wassenhove (1993) question the trade-off logic of traditional manufacturing strategy.

In the following pages, a backdrop for neo-operations strategy with an overview of the macro forces that are influencing manufacturing firms' strategies and behavior is given; next, an overview of the evolution of management perspectives establishes the

TABLE 38.1 Contrasting Traditional Manufacturing and Neo-Operations Strategies

Strategic elements	Traditional manufacturing strategy	Neo-operations strategy
Production paradigm	Industrial-age, mass-production logic (economies of scale and scope)	Postindustrial, information-age logic (economies of integration and knowledge)
Manufacturing task	Trade-off on "order winning" capabilities	Combinative capabilities for delivering customer value and customer retention in a global marketplace; strategic agility
Strategic contents	Structural and infrastructural choices aligned to a single manufacturing task	Portfolio of path dependent structural, infrastructural, and integration choices aligned with changing customer value functions and business strategies (DMS)
Technological prowess	Functional manufacturing process employing stand-alone or linked shop-floor technologies	Boundaryless value/supply chain operations enabled by cross-functional and cross-organizational integrating technologies
Process choices	Volume and life cycle dependent	Agile and market dependent, fast changing
Product attributes	Tangibles (product quality)	Tangibles plus intangibles, including embedded and value-added services, speed to market, etc.

trajectory of capabilities-based competition. Under the assumptions of absorptive capacity theory, a series of hypotheses are generated around the technological prowess of world-class manufacturing. Third, this chapter seeks to advance both theory and practice with a theory of how combinative competitive capabilities are acquired in the first place (Ferdows and DeMeyer, 1990; Noble, 1995; Noble and Schmenner, 1994; Miller and Roth, 1994). Competitive progression theory casts doubts on the tradeoff model of capability development. The Global Manufacturing Technology and Vision Project database was used to develop operational indexes as proxies for the constructs presented here and to test five hypotheses generated from this research (Roth et al., 1993). The chapter concludes with a discussion of the results and future directions.

38.2 DRIVERS OF CHANGE

Why the hubbub about neo-operations strategy? Quite simply: The rapidity of change has outstripped traditional methods and approaches to manufacturing. Increasingly, the world-class pacesetters are "out of the box" thinkers who have succeeded by catalyzing their combinative competitive capabilities to tie production to customer orders, scrap forecasts, and use production to squeeze out working-capital inventories. Globally competitive manufacturers attract, maintain, and grow market share by thoroughly "unlearning" old production rules and contextualizing their technological know-how for competitive advantage. Coming to grips with how their business processes and product lines synergistically fit together often calls for satisfying global customers with galloping speed—without packing warehouses. This is the essence of strategic agility. Obviously, manufacturers cannot negate the key macro forces which are fundamentally driving change and altering the basis of global competition: virtual markets, savvy customers, speed to market, technological choice, and global information availability.

38.2.1 "Virtual" Markets

Global markets are seemingly "virtual." Globalization is causing markets to continually change and swap boundaries, niches, and individual customers. Virtual markets and customers are significantly impacting operations strategies. On a regional basis, emerging markets in Eastern Europe, China, South America, and Southeast Asia are stressing localized capacity development to meet needs. In other global regions, countervailing forces seek to "lock" out competition. Consider trade barriers posed by Japan's policies, NAFTA, and the European Common Market. Taking an industry slice, the retail giants like Wal-Mart and K-Mart, the increasing number of consumers who are entering purchasing cooperatives, and those who are buying from mail order and electronic catalogs are fundamentally altering the suppliers' order fulfillment processes from purchasing to manufacturing to distribution to after-sales services. Even industrial markets are in constant flux. For example, hospitals are forming huge purchasing blocks; global firms are consolidating purchases worldwide under blanket orders; and more firms, like Baxter and Johnson and Johnson or IBM and Apple, are simultaneously suppliers and competitors—depending on products, global region, and customers. Like amoebas engulfing floating prey in water droplets, markets will change forms and shapes as long as it is opportunistic to do so. From the supplier side, the plethora of emergent technologies and customer information bases also provide infinite ways of carving up markets.

38.2.2   Age of Customer Choice

Today's customers want it all! Savvy global customers have high expectation levels—and they, too, have better access to product information and databases on which to make more informed choices. Consequently, global customers are demanding more uniformity in product quality and service standards around the world. Global service firms can provide some insights for manufacturers. Take McDonald's, for example. McDonald's global success is based on its operational ability to simultaneously add "local" content to its product line, and at the same time, deliver services consistently around the globe. Ironically, however, in the age of customer choice, there are fewer prospects for sustained loyalty. As current and emerging rivals fight fiercely for the same customers, the customers become more fickle. Consider the continuous jousts among long-distance carriers AT&T, MCI, and Sprint and the cola battle raging anew around the world between Pepsi and Coke with niche players like Snapple nipping at their share.

38.2.3   Speed to Market

Speed to market is the rallying cry of global manufacturers seeking to reduce their new-product development cycles (Merrills, 1989; Stalk and Hout, 1974; Giffi et al., 1990). Estimates ranging from one-third to two-thirds of a firm's annual sales revenues are derived from new products developed within the past 2 years. Put another way, speed, or lack of it, may account for up to two-thirds of a company's profits or losses. On the darker side, processes that enable rapid new product introduction also increase product proliferation and process complexity. Research and development, product design, manufacture, distribution, and servicing must be managed using globally distributed assets and resources—whether owned or outsourced. No part of a global business remains unscathed with heightened speed-to-market goals.

38.2.4   Myriad Technological Choices

The array of technological choices available today is exploding exponentially. New manufacturing technologies such as CAD/CAM, FMS, and CIM are changing the rules of production and extending the bounds of operations capacity (Goldhar and Jelinek, 1985; Goldhar et al., 1991; Maskell, 1994). The richness of the options in an integrated resource planning hierarchy influences how goods are being designed and produced (Gray et al., 1993). Moreover, technological advancement is reshaping product definitions. The so-called strategic bill of products now includes both physical goods and intangibles, such as value-added services and information (Giffi et al., 1990). Estimates suggest that as much as 80 percent of the U.S. $6 trillion GNP (gross national product) is information-based. While the array of technological choices are expanding at lightning speed, the ability of individuals, organizations, and even society at large to exploit those options cannot keep pace with the rate of technological development. As technological advancement encroaches on every aspect of daily work, the technical skills and creativity of people and their leaders are the limiting resources—not capital and equipment (Roth and Marucheck, 1994).

38.2.5 Information as a Global Commodity

Advancements in electronics and telecommunications make information a global commodity. Global access to the information superhighway, Internet, is a prime example. Access to the Internet and other on-line services (e.g., America On-line, Prodigy, Netscape, Compuserve, and Gopher) provides new sources of value-added. Computer-based commerce speeds the order fulfillment process to gather orders electronically and simultaneously drive EDI purchase orders and enterprise scheduling of resources to fulfill them, anywhere in the world, 24 hours per day. Even more importantly, it provides the firm with unprecedented market and process information. Those firms that can assimilate, translate, and organize bits of information into products and services faster and better than the competition will have a distinct advantage.

In summary, each of the aforementioned drivers of change is creating heightened enterprisewide process complexity and confusion. Coping mechanisms for managing parts of the value-chain network must extend beyond the realm of the traditional manufacturing function—and even beyond the firm in some cases. Prevailing macro forces are causing leading manufacturers to innovate and reinvent themselves. Anecdotal evidence suggests successful coping mechanisms are melting away icy bureaucracies. Product simplification and process reengineering are two widely touted approaches. By breaking down the walls between product design and manufacturing, for example, leading firms are "designing simpler products, making them in fewer permutations, and with fewer parts, and getting them to market with fewer delays" (Henkoff, 1995, p. 136). In other cases, competitors, like the U.S. "big three" automakers, are cooperating by sharing designs for electrical connectors for efficiency. Automotive engineers at the big three have concluded that design complexities on electrical gadgets make little sense, cost too much, and create too many opportunities for things to go wrong.

As leading manufacturers struggle to make sense out of the new competitive dynamics, adding to their turmoil is the conclusion that both the magnitude and the rate of change in the macro forces have hastened the pace at which competitive capabilities must be improved. Even though a vast number of organizational experiments are under way, it is inefficient and even risky for executives to misunderstand what lies ahead. Operations decisions made today impact tomorrow's competitive capabilities (Hirasawa et al., 1995). A macro view of the neo-operations strategy provides a trajectory of capabilities-based competitiveness.

38.3 STRATEGIC MANUFACTURING EPOCHS

A first-order questions is: What are the strategic directions of manufacturing pacesettrs? Second, what are the best paths for getting there? In this section, based on 10 years of intensive field and systematic survey research on world-class organizations, I propose a "strategic map" that outlines the broad contours of the competitive terrain; in the next section, a path is indicated. Business historians speak of the great industrial divides as punctuated periods where radical changes in technology occur within organizations. Without question, manufacturing firms are in the midst of the next major divide as society moves from the electromechanical industrial age to the digital information age. One only has to look back to the sweeping economic transformations at the last turn of the century when society plowed out of the agrarian age to see the impact of a major technological discontinuity. The past merely hints at the revolutionary upheavals. It says little about what challenges and opportunities the information age brings.

Paul Allaire, Chairman and CEO of Xerox, once said, "To do things differently,

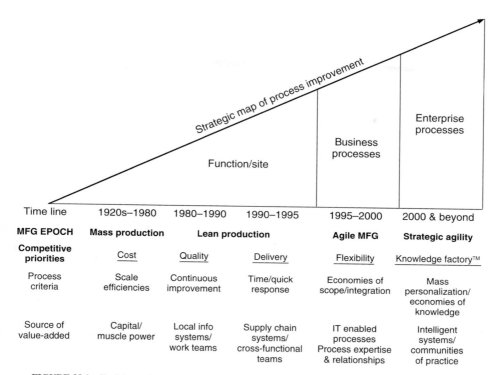

FIGURE 38.1 Evolving management perspectives: neo-operations view of capabilities-based competition.

you must see things differently." To see operations differently, I propose a juxtapositioning of the macro economic view of the industrial divides into a meta view of capabilities-based competition. Coincident with each industrial divide are *strategic manufacturing epochs*—those strategic moments in time where top management perspectives of what competitive capabilities are required to compete dramatically shift. Given that the manufacturing pacesetters are the arbiters of the near future, Fig. 38.1 represents a strategic map of their evolutionary trajectory. Notably, many of those outstanding performers, dubbed as "world class," are approaching the twenty-first century practices today. Each strategic epoch requires manufacturing executives to see and do things differently from the past to achieve stellar performance; each dramatically changes management perspectives along three important dimensions:

- *Competitive priorities* capture how capabilities-based competition will evolve.
- *Process performance criteria* indicate how operations should be evaluated.
- *Sources of value-added* signify where the greatest paybacks from investments reside.

These dimensions are intimately bound up with the firm's operations strategies, of which people, R&D, markets, products, and process and information technologies are critically linked. Herein lies the problem. To the extent to which it is possible to move seamlessly between epochs, executives must orchestrate requisite investments in people and technology today to prepare for tomorrow. The strategic map points to where in the organization investments will typically garner the greatest strategic paybacks. For example, time-based competition generally requires process reengineering at the business-unit level, whereas for strategic agility, innovations are required at the enterprise level. Each epoch calls for increasingly higher levels of process integration and coordination. Forging ahead mandates fundamentally new competitive capabilities that I argue in the next section are built cumulatively, and hence, are path-dependent [see Hirasawa et al. (1995) for a detailed discussion of path dependencies in manufacturing operations].

Unfortunately, the exact timing of the beginning of a new epoch for any particular industry is unknown. In contrast, the end of an epoch is typically represented by a discernible convergence of underlying technological bases, know-how, and industry cultures—the bundle of values, assumptions, and beliefs that delimit the bounds of management practices. Smaller firms have been observed either to become entrenched in an epoch or to seemingly leap-frog epochs (Fiegenbaum and Karnani 1994). As expected from population ecology theory, small firms are at the greatest risk. Those that grow typically fall back into the industry norms. Epochal change is marked by confusion about how best to invest resources. Products and processes become quickly obsolete, and niche competitors nip on the firm's heels. There are also discontinuities in practice as firms "experiment" with alternative modes of technology and organizational structures. One only has to consider the pervasive renewed interest in integrating new-product development and manufacturing process designs, redeploying and upgrading worker skills, reconfiguring distribution systems, adopting advanced process and information technologies, and benchmarking best practices.

Each new epoch unglues the shared values and shatters the technological traditions that preceded it, and, most importantly, each unravels the basis of industry competition. There is frequently a false sense of comfort by taking shelter in the past—or even denial. Dysfunctional management behavior abounds (Roth, 1996). Well-rehearsed approaches that built executives in the past won't succeed in the future. Using the metaphor, putting a dinosaur on a crash diet to reengineer it, won't save it! In Darwinian terms—it's the wrong species! No wonder that over 70 percent of reengineering projects fail. Only taking costs out of a dying operation won't bring in new customers, but it will make products a commodity faster. Killing off the organization's knowledge base with unscrupulous across-the-board downsizing is the equivalent of an organizational lobotomy that will only hasten its demise. Shark competitors will sweep good people up and use them to penetrate the most vulnerable spots. Examination of the evolutionary trajectory of competitive priorities indicates how operations strategies were reinvented during each epoch to meet the gaps.

38.3.1 The Industrial Age Heritage

Prior to the industrial revolution, production was based on individual skills—those of artisans and domestics (Piore and Sabel, 1984). In the craft epoch (not shown on Fig. 38.1), physical product attributes were highly variable; no two products were exactly alike, and many goods and services were in short supply. The modern factory system was introduced by Adam Smith in *The Wealth of Nations*. Smith's conceptualization of the division of labor seeded the mass-production epoch. The massive business transformation to the industrial age was not fully realized, however, until the beginning of the twentieth century in North America. Frederick Taylor formalized the principles of "scientific management," and Henry Ford and his engineers put the stake into the ground with the production of the Model-T.

Mass production symbolized the dominant management philosophy during the majority of the industrial age. Producing large volumes of standard products demanded by mass markets at affordable prices was the primary manufacturing objective. Fueled by the post–World War II economy, the power wielded by its military industrial complex, and little global competition, the dominant U.S. manufacturing strategies typically followed product life-cycle theory which gave rise to the product-process matrix (Hayes and Wheelwright, 1984). Problem solving was linear and mechanistic. Managers believed they could derive significant operational efficiencies by increasing production volumes. Consequently, by adopting dedicated technologies and realizing the learning curve phenomena, they could dramatically lower unit costs. These manufacturing strategies myopically focused the first-order benefits of production: machine utilization, productivity of direct labor (the operators of electromechanical devices), and inventory savings. The mass-production theme reflects the Western industrial heritage.

38.3.2 Beyond Mass Production

During the 1970s and 1980s, the Toyota Motor Company, now a business legend, pioneered a new manufacturing perspective. Toyota's "lean production" system changed the strategic directions of manufacturing (Womack et al., 1991). Using a highly skilled, flexible workforce, Toyota, and subsequently other leading practitioners, changed production arithmetic and shattered the sacrosanct walls between manufacturing direct and indirect labor. Lean production capitalized on continuous flows of small lots by minimizing changeover costs and managing bottleneck operations, and employed the principles of continuous improvement (*kaizen*) to reduce waste. Throughput improvement programs, such as just-in-time (JIT) production, simultaneously attacked manufacturing overhead, speeded up delivery processes, and offered customers better value

Continuous improvement and quick response to customers became the criteria for time-based competition in manufacturing. In North America, many companies like Northern Telecom developed deliberate manufacturing strategies around the construct of "time-based competition" (Merrills, 1989). While lean production was relegated primarily to happenings on the factory floor, value-chain activities that impacted continuous flows began to spill over into broader-based business unit practices. Moreover, the role of direct workers expanded from carrying out time-consuming, routine tasks to tackling the more challenging activities of redesigning work and work processes. Worker goals now emulated those of traditional knowledge workers: eliminated quality defects, improved delivery performance and customer service, and lowered costs. Best practices of high-performance, cross-functional work teams and total employee involvement seized top executive attention, as shifts away from localized information systems and toward more integrated systems. At the same time, employee empowerment created knee-jerk reactions and generated significant tensions in traditional manufacturing organizations.

These strategic movements from craft production to mass production to lean production were accompanied by changing definitions of competitive priorities. With each epoch, the competitive ante was ratcheted up. Product quality, originally defined as "buyer beware" in the craft epoch, evolved total conformance to customer specifications. Xerox's quality leadership or Motorola's "six sigma" (6σ) quality, Japanese *pokeyokes* (fail-safing processes), and robust product designs are examples. Manufacturers are recognizing that customers demand "best" value; however, the bar is continually raised for what is deemed "best." For example, value has evolved from sheer "availability" in the craft epoch to combined conformance, reliability, performance, features, and moderate price in the lean production epoch. In the 1980s, Toyota, for instance, introduced cars with more features than their U.S. counterparts offered in the same price brackets; the same was true for GE Motors' business.

Value-chain relationships have also evolved. Characteristic of the mass-production epoch are cut-throat purchasing tactics, large supplier bases, antagonistic relationships, and rock-bottom pricing. GM's infamous Lopez legacy is an exemplar. Axiomatic to lean production are cross-firm cooperation and strategic alliances with suppliers. Lean production brought fewer suppliers accompanied by more "hard wiring" in supplier-manufacturer relationships. Suppliers won orders by delivering the right quality, on time; cost became an "order qualifier" (Giffi et al., 1990). Leanness in manufacturing, too, has proved insufficient to keep pace with evolving global markets and technological forces (Stewart, 1992).

An industry-led group is advancing the next generation of manufacturing under the aegis of "agile manufacturing" (Iacocca Institute, 1991; Goldman et al., 1995). Agile manufacturing is accompanied by new paradigms from scale to scope, from hardware to integrating software, from stand-alone production to integrated manufacturing technologies, from supply-chain information systems to virtual value-chain relationships among customers and suppliers, and from a continuous improvement to radical redesign of business unit processes. If these industry leaders are correct, then at the dawn of the agile manufacturing epoch, the neo-operations strategies of world-class manufacturers will leverage advanced technologies and best practices toward broader, enterprisewide agility. The goal of customization, which is the mass production of customized goods and services (Davis, 1987; Pine, 1993), will prioritize mass personalization of products and services.

38.3.3 Propositions on World-Class Manufacturing

Manufacturing's radical transformation from the industrial age to the information age is perhaps no more apparent than in its evolving technology adoption. As a component of neo-operations strategy, here we explore three issues concerning the deployment of advanced manufacturing technologies by world-class manufacturers. For semantic clarity, *manufacturing technology* is defined here as hardware, physical equipment, software, tools, techniques, and methods of production or making which extend human capabilities (Schon, 1967). More simply, it is a "physical structure or knowledge embodied in an artifact (software, hardware, methodology) that aids in accomplishing a task" (Leonard-Barton, 1990, p. 3).

Technological advancement creates opportunities for *economies of scope.* Economies of scope "exists when the total cost of production given a quantity of two or more products in one facility is less than or equal to the total costs of producing the same quantity of those products in a set of facilities each of which is dedicated to a single product" (Noori and Radford, 1995, p. 264). In practice, this indicates that a high degree of process flexibility is required to produce more variety, faster, at similar quality and costs. With the appropriate infrastructure, Goldhar et al. (1991) suggest that *integrating* flexible production and information technologies can translate into fundamentally different product designs, production methods, and business processes. Flexible processes, especially software-driven, computer-integrated systems, afford *economies of integration* (Noori, 1990; Noori and Radford, 1995). "Economies of integration refers to the simultaneous presence of economies of scale and economies of scope" (Noori and Radford, 1995, p. 264). "Intelligent" integrated production—including flexible computer-numerically controlled (CNC) machine tool centers, robot assemblers and materials handlers, and expert-system diagnostics and process control software—are the hallmark of modern computer-integrated manufacturing (CIM) systems.

For more than two decades, however, manufacturers have been struggling with

CIM, but few have been truly successful in their attempts to fully automate the shop floor and link it to other functions in the organization (Giffi et al., 1990; Roth, 1986; Roth et al., 1993). Under the assumption that outstanding business performance is enhanced by *economies of scope and integration,* experience with multiple advanced technologies is a necessary, but not sufficient, condition of superior performance. This leads to the following hypothesis:

> H1:   World-class manufacturers have more experience with multiple advanced manufacturing technologies, including integrating and information technologies, in contrast to their non-world-class counterparts, ceteris paribus.

The second assumption is that necessary prerequisites for technological advancement toward strategic agility are *combinative* "hard," engineering-based sciences and "soft," social sciences. This is radically different from the separation of "craft" and "science," as exemplified by job shop and mass-production manufacturing, respectively. Experimentation, statistical process control, expert systems controls, process simulation, computer modeling, global task coordination via satellite communications and wide area and local PC networks, and advanced materials science are commonplace in the best manufacturing entities. Equally important, but less frequently explored, are the social systems designed to exercise the creative possibilities afforded by scientific knowledge and technology (Leonard-Barton, 1995; Galbraith et al., 1993). The "new" combinative sciences behind neo-operations require executives to also debunk Tayloristic notions of functional "separatism" (Roth, 1996, to appear). As operations becomes an integrated part of enterprisewide global processes, greater proportions of manufacturing work will be spent on knowledge work, especially communication and information handling (Roth and Marucheck, 1994; Badelt and Roth, 1995). Designing agile global business processes is more complex than simulating total product design and processes of manufacturers with supercomputers. Ethnographers and other social scientists are being deployed by companies like Xerox to obtain extraordinarily detailed understandings of the fabric of the firm's tightly interwoven sociotechnical systems (Seely-Brown, 1993).

To leverage combinative hard and soft sciences, managers must know more than how each individual part of the system performs in isolation but how the parts interact synergistically as a whole. Uncovering and recognizing these integrating *patterns* is important to extrapolating a portfolio of management choices. Consider the traditional "over the wall" transfer from engineering to manufacturing. Prototypes worked in the laboratory, but not in full production. Today, systems thinking has bolstered the application of simultaneous or concurrent engineering (Giffi et al., 1990; Wheelwright and Clark, 1992; Leonard-Barton, 1995). This systems approach is quite different from the single-faceted, slice-and-dice approach of mass production. Moreover, top management must always be mindful that each part of the system works in conjunction with all others. In this research, I assume that best practices are a reasonable proxy for the application of the state of the art in the social sciences in manufacturing. Best practices applies to the infrastructure, including development of employees as information processors, decision makers, and facilitators; integrated manufacturing strategies; cross-functional teams, and so on. Lack of attention to the whole can be more lethal than any individual failure. Thus I hypothesize:

> H2:   World-class manufacturers are more likely to combine best practices with the application of advanced technologies than their non-world-class counterparts, ceteris paribus.

Third, customer requirements for increased value-added will catapult agile manufacturing into an epoch of strategic agility, which I propose is necessary to satisfy knowledge-era customers. "Mass personalization"—the ability to discover each customer's unique needs and then deliver them beyond their expectations—must extend beyond the adoption of advanced technologies, per se. Roth and Jackson (1995) empirically show that a firm's absorptive capacity is limited by the technological competencies of employees. Thus, the term *technological leadership* is applied to encompass the entities' overall level of technologically competent knowledge workers, state-of-the-art manufacturing processes, organizational knowledge, and unique process capabilities. Technological leadership produces *economies of knowledge.* "Economies of knowledge means that the firm is able to use its business acumen, combined with skilled people and experience with advanced technologies, to create an organization that consistently identifies, assimilates, and exploits new knowledge more effectively and efficiently than the competition" (Roth, 1996, p. 30). Generative organizational learning and know-how are created by enhancing people skills with the application of advanced process and information technologies. Economies of knowledge enable new blendings of integration, infrastructure, and structure to generate entirely new business economics. "The ultimate logic is that each employee is potentially a business" (Davis and Davidson, 1991). Notably, technological leaders define the rules of competition; they are not bound by the rigid, hard-wired facilities, technologies, rules, and locations that belong to the industrial age. New questions arise, however, concerning how to organize manufacturing, where to locate production and control, and what technologies should be used under which circumstances.

Clearly, technological leadership extends operations beyond the manufacturing function toward synergistic, enterprisewide processes. Many of the basic ideas and skills developed for agile manufacturing can be expanded in moving the firm toward strategic agility. Strategic agility represents the metamorphosis of the production metaphor from that of a mechanistic, "working" machine to that of "The Knowledge Factory™"—an organic, learning organization that produces knowledge as a key by-product (Roth et al., 1994). As operations evolve from physical production of tangible goods to the manufacture of invisible assets—service, knowledge, and ideas—the new work is largely cognitive, organic, and virtual. The term *cognitive* suggests that superior people skills and organizational knowledge are limiting resources, as opposed to capital and equipment; *organic* means that operations is constantly changing and growing; and *virtual* implies that work can take place anywhere at any time. Technological leadership stresses intelligent connectiveness of invisible assets, such are the marrying of communities of practices and natural workgroups with intelligent support and control systems.

Knowledge workers, including managers, engineers, technicians, and workers who are technically equipped to skillfully maneuver the organization in the knowledge era must adeptly use increasingly powerful, versatile, and user-friendly advanced technologies to enhance their cognitive productivity. Thus, finding ways to leverage knowledge workers such as the development of communities of practice is of paramount importance for strategic agility. Communities of practices are the formal or informal workgroups who get the work done (Seely-Brown, 1993). Take, for instance, Oticon, a Danish manufacturer of hearing aids. Oticon has recreated its business to create strategic agility (LaBarre, 1994). Lars Kolind, Oticon's CEO, envisions the relationships among value-chain networks to create his vision of a holistic, "organic" business. Oticon's knowledge workers bring "clusters" of skills to the job. They bring the intelligence that uniquely integrates the new tools of production—computers, telecommunications, and automated systems; and they dynamically reconfigure themselves into "virtual communities of practice" in order to move seamlessly between projects and tasks. More formally:

H3: World-class manufacturers have higher levels of technological leadership than their non-world-class counterparts, ceteris paribus.

38.4 COMPETITIVE PROGRESSION THEORY

As strategic thinkers, executives know that they must first have a clear understanding of factors critical to success—those that will win orders to grow the business and that will retain customers to sustain the business in an increasingly turbulent global marketplace. Critical success factors comprise the manufacturer's current portfolio of competitive capabilities and those requisite capabilities prioritized for meeting the competition in the future. Understanding how capabilities are acquired most effectively and efficiently becomes a first order question. Stemming from the Japan 2000 Research Project, Jinicho Nakane first proposed that building manufacturing capabilities must follow a specific sequence, regardless of what competitive priorities were planned. Nakane (1986) rationalized distinctions between competitive priorities and capabilities like this: "In general, if some companies want to offer 'flexibility' as a competitive priority, it is necessary that at least they have already qualified for a minimum level of abilities on quality, dependability, and cost improvement. If they have not such ability, they get a chaos condition and end tragically." Ferdows and De Meyer (1990) subsequently reported the empirical observations that generic competitive capabilities seemed to accumulate in an orderly fashion from quality to delivery to flexibility to cost. They used the metaphor of a sandcone to explain a phenomena they observed in their European sample of manufacturers. Later in my own empirical research with various coauthors, I repeatedly observed this "sandcone" effect in aggregate U.S. samples of manufacturers, cross-sectionally, and over time (Roth and Miller, 1992; Roth and Giffi, 1994) and within various industries including the automotive (Roth and Marucheck, 1992), electronics (Roth and Chapman, 1993), and services (Roth, 1994).

Prior empirical research describing the "sandcone" effect was not a theory, per se, but rather a series of observations taken at different points in time over different samples. These observations lay the foundation for a competing theory; i.e., the *competitive progression theory* of knowledge-based competencies. Recall that strategic agility is operationally defined by the simultaneous accumulation of multiple competitive capabilities (see also Fig. 38.2). This new theory posits:

Sustainable combinative competitive capabilities accumulate in a sequential progression forward—from quality to delivery to flexibility to price leadership—over an innovation cycle, leading to strategic agility. Despite any intended generic competitive priorities to the contrary, higher level organizational knowledge-based competencies are required to move efficiently between successive stages within a cycle; and that over successive innovation cycles, increasingly higher levels of combinative capabilities are attained by repeating the progression. Because innovation cycles are shortening due to the rapidity of technological progress and globalization, other sequences of generic capability-seeking behaviors produce system entropy, forcing firms either to recycle back to acquire first order capabilities (e.g., quality or delivery) or to make capability trade-offs that are less endurable over the long run.

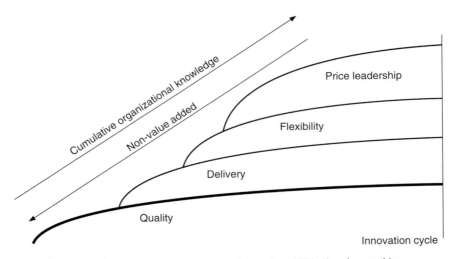

FIGURE 38.2 The competitive progression: a new theory of capabilities-based competition.

Two knowledge-based competencies explain the observed competivive progression. First, is what I call, the laws of *operations physics*; the second, pertains to the firm's overall *capacity* for managing complexity and stimulating organizational learning. Operations physics offers a theoretical basis for the observed simultaneity of benefits accruing in the progression. All organizational work processes have input, transformation, and output variability associated with differences in materials, changing customer demands, hard and soft technologies, procedures and systems, worker skills, and so forth. From statistical theory on process control, we know that process variance can be partitioned into common and special cause variation. Each process has natural tolerance limits which specify its inherent capability to conform to external requirements (Deming, 1987). Imagine now that generic capabilities tend to share overlapping work processes. This property implies "physical" process co-variance.

By analogy, a basic tenet of competitive progression theory is this: the work processes associated with creating any single generic capability can be partitioned into one of two categories: (1) unique, or capability-specific process properties (a_i, i = 1 to 4, for quality (Q), delivery (D), flexibility (F), or price leadership (PL), respectively); (2) interdependent properties, or process commonalities. Clearly, changing one capability will automatically impact the others. The cumulative magnitude of the impact will vary by the relative degrees of process commonality, or overlap; and the direction is related to process synergy. Suppose a firm is working toward six sigma product quality. A portion of this effort translates into better delivery. There is less rework, better process predictability, and more residual organizational energy to devote to delivery. Therefore, $\Delta D = x (a_2, \Delta Q)$. Similarly, work processes that influence Q also impacts other upstream capabilities F and PL, but so does improved delivery. Thus, $\Delta F = y (a_3, \Delta Q, \Delta D)$ and PL $= z(a_4, \Delta Q, \Delta D, \Delta F)$. Each newly acquired, or enhanced generic capability, can act synergistically to modify other capabilities with which it shares common processes. Under certain conditions, however, diminishing returns are possible (Hirasawa, Menor, and Roth 1996).

Absorptive capacity (Cohen and Levinthal 1990) offers the rationale for the precedence constraints and sequencing observed in the progression. Absorptive capacity represents a synthesis of organizational know-how, technological prowess, and social values that enable the firm to identify, assimilate, and exploit new knowledge. Over the past decade, I have observed literally hundreds of manufacturing facilities and practices in action. I noticed that as manufacturers moved through each successive stage in the progression, the proverbial "bar" was raised. Managers needed to tackle increasingly greater process compexities and higher level organizational obstacles along the way. These hurdles took a multiplicity of forms: technological constraints, deficiencies in people skills, bureaucracy, communications, general management abilities, and/or other types of organizational slack. Because forward progression necessitates passing more stringent hurdles, prior manufacturing process choices may inhibit progress.

Extrapolating from absorptive capacity, organizational knowledge is *path dependent* on the level of prior accumulated knowledge. Theoretically speaking, increasingly higher levels of capabilities translate into requirements for heightened knowledge-based competencies. Consider the following: Each innovation cycle coincides with spanning a tree of processes, wherein the relative organizational knowledge required for process improvements branch outward from the site/function level (Q) to the supplier/customer distribution system processes (D), to crossing functions and business units (F), and finally to enterprise processes that span the global value network (PL). Thus, the precedence constraints can be gleaned from the perspective of rapidly expanding process complexities. The following further illustrates these path dependencies.

38.4.1 Why Quality First?

From a pragmatic perspective, quality alone is no longer good enough for competitive advantage. More and more empirical data supports the assertion: Quality is an "order qualifier." Even in service organizations, the quality of "tangibles" used in the delivery system affects customers' perceptions of value and satisfaction. The most successful quality practitioners, such as Xerox, Motorola, and Toyota, continue to reset the standards expected by customers. No company is immune from the pressure to serve its customers more effectively. To do so, they must first acquire quality capabilities. Competitive progression theory posits that quality is a prerequisite for the long-term capability development. I argue that quality capabilties are much more than offering order qualifiers to the marketplace. The pursuit of quality affords effective and efficient approaches to process variance reduction and organizational learning [e.g., *kaizen* (continuous improvement)].

Senior executives in high-performing organizations have long recognized that the Deming-Shewhart plan-do-check-act wheel is a primary building block of process knowledge, and that increasing the velocity of rotations through the wheel accelerates learning. When quality is viewed as only an "outcome," and not an ongoing process for learning, research indicates that it typically becomes a vehicle for downsizing and cost cutting. In these instances, quality is not sustainable; it quickly becomes a fad, and benefits are short-term. Executives will look for the next panacea after the troops discover the ploy. Thus, quality capabilities are predicated on an understanding of production process capability afforded by what the grass-roots organization has learned about making "good" products and services. It is the "knowing" that is important to building organizational absorptive capacity. Basic knowledge competencies, derived from the mastery of total quality management (TQM) principles, form the foundation for tackling more higher-level, more process-complex capabilities. Quality capabilities include:

1. Understaning customer requirements now and those anticipated in the future.
2. Mastering "science," technology, and/or the subject matter behind products and production processes.
3. Knowing processes: mapping, common-cause/special-cause variation, process capabilities, and systems thinking.
4. Developing proactive "solution" search behaviors; e.g., vision, benchmarking, and intrinsic motivation.
5. Understanding of psychology and social sciences, including leadership, communications, conflict management, and change management.
6. Using tools for making process improvement a seamless part of daily work, including sharing of individual and group tacit knowledge and basic quality methods.

38.4.2 Delivery Reliability Expands Process Capabilities

Delivery reliability means filling customer orders completely by the dates promised 100 percent of the time. For consistent, on-time delivery capabilities, more extensive process predictability is required, crossing over manufacturing boundaries to the distribution system. Clearly, delivery reliability cannot be cost-effective without product quality. In world-class companies, superior customer relationships thrust the enterprise into a state of extraordinary customer focus (Giffi et al., 1990). Consequently, at the beginning of the second step of the competitive progression, more information and process knowledge sharing must occur cross-functionally and along the value chain. Customer-supplier relationships add depth to organizational knowledge.

Adding to the delivery process complexity, products and relationships must be local and global simultaneously. Manufacturing companies are increasingly challenged to deliver on locally imposed constraints on globally similar products. A global supplier must offer a standard product but be able to customize it for local consumption or use. A simple example: Hewlett-Packard's DeskJet printers, although globally sourced at a single point, must have manuals written in the domestic language and components that can run on local electrical currents. HP customers demand a globally, uniform base price with cost-plus to account for freight and other uncontrollable regional expense differences. Thus, distribution processes must be capable of delivering the same levels of support and service around the world, and at the same time, be able to reliably provide "local" contents in the offering (e.g., environmental, safety, energy, cultural issues).

38.4.3 Flexibility Shatters Boundaries

Quality and delivery pave the way for flexibility. Gerwin (1993) discusses the creation of manufacturing flexibility. Stewart (1992) illustrates flexibility in practice. "Closeness to customers," enabled by delivery capabilities, heralds intimate customer knowledge which is a prerequisite for flexibility. In relationships, process interdependencies are a business norm. Integration provides a competitive edge for process flexibility, defined as the ability to produce varying sizes, use various materials with quick changeover capability, or manufacture varying product mixes and volumes. At a higher level, customer relationships enable the organization to bring new products to market quickly and other innovations. One highly profitable *Fortune* 500 manufacturer routinely uses its product development engineers as customer service agents and field service technicians. Relationships are efficient means to incorporate customer knowledge and information within core business processes to improve overall responsiveness, increase speed to market, and expand market penetration. Customers are also one of the organization's best sources of competitor intelligence and new ideas.

38.4.4 Price Leadership Leverages the Enterprise

Next, sustainable price leadership occurs only when product innovations and enterprise processes add to the firm's competitive arsenal by reducing total costs over a product-family life cycle. Understanding product and process variation facilitates the use of flexible technology to reduce costs. One leading Japanese manufacturer said during an interview, "We push process improvements until people become the limiting resource, then only do we introduce automation." Under these circumstances, technological advancements can be effectively used to substitute for direct labor, enlarge people competencies, streamline processes, or expand the product functionality, definition, and/or ease of use for the customer. It is for this reason that leading manufacturers will be better able to attain economies of knowledge.

Thus, manufacturers who prioritize price leadership must understand the sources of product and process variation, establish deliberate strategies for accelerating learning, and leverage enterprise processes. Hewlett-Packard's global supply-chain approach to its printer business (Lee and Billington, 1995), Baxter I. V. Systems Divisions Operations Strategies (Roth et al., 1994), and Nortel's Switching Global Business Process Reengineering (Badelt and Roth, 1995) are good examples. Otherwise, total cost reduction tactics may actually *hollow out* long-term competitiveness rather than enabling firms to cumulatively "have it all!" With each step on the progression, enterprise nonvalue-added costs are driven out.

In summary, competitive capabilites are interdependent both in "physical" ways since they share common work processes, and in "intangible" ways, since increases in knowledge-based competencies are necessary (Roth, 1996; Marucheck, Kemp, and Trimble, 1994; Roth and Giffi, 1994). Movement on the progression impacts the whole progression—like the Ferdows–De Meyer (1990) metaphor of building a sand-cone. To the extent that the cumulative model of capability development is built on the notions of operations physics and absorptive capacity, the phenomena should hold broadly. I hypothesize:

> H4:   Generic combinative competitive capabilities are built cumulatively, from quality to delivery to flexibility to price leadership, ceteris paribus.

Taking a holistic view, each step upward simultaneously increases the firm's knowledge base—its absorptive capacity to exploit and use external knowledge (Cohen and Levinthal, 1990, pp. 137–138):

> A technologically progressive environment is not simply determined by past performance....It also depends on the firm's absorptive capacity....The greater the organization's expertise and associated absorptive capacity, the more sensitive it is likely to be to emerging technological opportunities...some organizations have the requisite technical knowledge to respond proactively to the opportunities present in the environment.

Roth and Jackson (1995) empirically show that the absorptive capacity of people enhances capability development, which underpins market performance. Experience with multiple technologies, especially integrating, enabling, and information-based technologies, should increase the firm's absorptive capacity to leverage those technologies for building multiple capabilities. Recall that strategic agility was defined in terms of possessing combinative competitive capabilities. Technological leadership, which matches people skills and experiences with advanced technologies, is expected to enable significant levels strategic agility. Thus, the following hypothesis:

> H5: A holistic approach to technology will coincide with strategic agility, ceteris paribus.

38.5 COMPETITIVE CAPABILITIES AND TECHNOLOGICAL EXPERIENCE

Determining how to "best" run tomorrow's manufacturing organization is a recurrent issue for top management. Strategic benchmarking of critical success factors prevents "blind" benchmarking because it identifies the current competitive strengths and weaknesses of a broad array of global players. Strategic manufacturing epochs and competitive progression theory provide a context for crafting new strategy, but empirical tests of the propositions are important: Are capabilities developed cumulatively on a global basis? Have world-class manufacturers made more headway along the competitive progression, and hence, can they be defined as "more agile"? Is capabilities-based competition associated with technological progress? Are best manufacturing practices also linked to agility? To address these questions, this research subjects to empirical scrutiny the current capabilities of manufacturing leaders. Manufacturing technologies, particularly those which have been proposed as enablers of strategic agility, are also considered.

38.5.1 Research Database

Since the mid-1980s, trends and developments in the strategies and performance of manufacturers have been benchmarked annually through the DTT-UNC Global Manufacturing Technology and Strategy Vision Project, a collaborative survey research effort between Deloitte Touche Tohmatsu International and researchers from the Kenan-Flagler Business School at The University of North Carolina at Chapel Hill (hereafter DTT-UNC GMTSP). The original survey was initiated in 1987 as a means to triangulate the findings of one of the most comprehensive field studies to assess the state of the art in world-class manufacturing in North America sponsored by the National Center for Manufacturing Sciences (Giffi et al., 1990). The sampling frame was the *Fortune* 1000 manufacturers supplemented by additional lists from directories of manufacturing firms reporting at least $50 million in annual revenues. Manufacturing divisions or business units from larger parents may have less than $50 million in annual revenue. By 1993, the survey included Japan, Mexico, Brazil, South Africa, and the European Common Market countries of France, Germany, Italy, Spain, and the United Kingdom. Survey research was complemented by case-based field research in five global corporations. Senior executives, line staff, and employees were interviewed regarding their perspectives of evolving manufacturing strategies and concerns. The GMTSP is a strategic benchmarking study designed to gauge broad-scale manufacturing strategies and strategic directions, in contrast, the more tactical benchmarking at the plant or functional unit level.

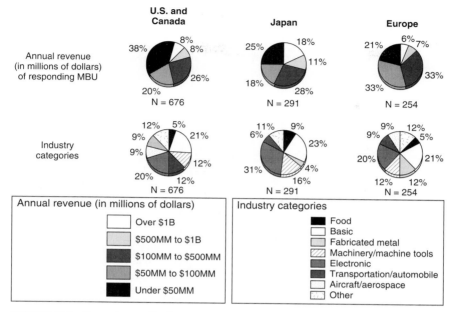

FIGURE 38.3   Respondent profiles by annual revenue and industrial sector.

In the 1992/93 study, senior manufacturing executives from more than 1300 manufacturing business units (MBUs) completed the survey. For purposes of the analyses presented in this report, 254 MBUs from Europe, 291 from Japan, and 676 from North America were included. The breakdown of industries and revenues of responding MBUs are given in Fig. 38.3. A follow-up study on a small sample of nonrespondents, combined with industry statistics, indicates the findings are biased toward manufacturing leaders. Over 40 percent had parent organizations with greater than $500 million in annual sales revenues; and 45 percent are market leaders with greater than 25 percent market share for their primary products.

38.5.2 Manufacturing Business-Unit Performance

Of the total 1221 respondents, almost one-third (32 percent) indicated that their manufacturing organization was "world-class or approaching (WC); the remainder tended to be "competitive" (NWC). Respondents were separated by objective measures of business-unit performance in order to validate their own designation of world-class status and as not to create a tautology in examining their capability and technology profiles (Hambrick, 1984; Venkatraman and Ramanujam, 1987). Objective measures were collected on average annual sales, return on assets, inventory turns, market share of primary products, and pretax profits. Table 38.2 shows that the world-class group exhibits stellar performance on traditional business-unit criteria In addition, we used additional perceptual measures of business performance and outside criteria to validate the world-class status, such as reputation and expert opinions.

TABLE 38.2 Manufacturing Business-Unit Performance* by World-Class Status[†]

Performance criteria	WC	NWC	F	P
Annual sales (range: 1–5: 1 = < $50B to 5 = > $1B)	3.41	2.83	45.23	≤.0001
Return on assets (range: 1–6: 1 = < 5% to 6 = > 30%)	2.72	2.52	4.64	≤.0314
Inventory turns (range: 1–6: 1 = < 2 ×/year to 6 = > 15 ×/year)	3.65	3.34	13.50	≤.0002
Market share (primary products) (range: 1–7: 1 = < 5% to 7 = > 60%)	4.32	3.83	19.63	≤.0001
Pretax profits (range: 1–7: 1 = loss for year, 2 = break-even to 7 = > 25%	3.65	3.39	7.41	≤.007

*Actual absolute performance measures were captured on rank-ordered categories, e.g., annual sales volume "under $50 million," "$50–$100 million", F statistics were based on ranked data.

[†]*Abbreviations:* WC, world class; NWC, non-world-class; B, billion.

38.5.3 Operational Indexes

Indexes are data reduction devices wherein a respondent's responses to several related items may be summarized into a single score. In general, composites of several items taken together provide a more comprehensive and accurate measure of a latent construct in comparison with single questionnaire items. Exploratory factor analyses and reliability analyses were performed separately for the construction of indexes over the three conceptual domains of theoretical interest: competitive capabilities, technological experience, and best manufacturing practices. Note that there were no published instruments available which were shown to have sufficient reliability and validity across global regions. Each of the multiattribute indexes here had sufficient reliability for newly developed scales (Churchill, 1979; Nunnally, 1978). Each item displays construct validity consistent with the extant literature. The North American data was used for index calibration. The Japanese and European subsamples were used as validation samples. Importantly, the psychometric properties hold for each global region, suggesting that the results are not an artifact of the set of firms in the calibration sample and making cross-cultural comparisons possible.

Content and construct validity was established for the indexes. Content validity indicated that the items included in the survey adequately represent the concepts in the domain in which the generalizations are made. Content validity is largely determined during the design and pilot phases of instrument construction. To maximize content validity, the extant literature on world-class manufacturing was examined and related field research incorporated expert judgments of executives. The instrument has been continuously refined from 1989 to 1993. The survey instrument contains a large pool of representative items on manufacturing strategy contents, competitive capabilities, and performance. Regarding construct validity, each index was examined to determine whether it actually corresponded to the theoretical concept of interest and shared variance. Each index possesses the property of unidimensionality (Carmines and Zeller, 1979).

Generic Competitive Capabilities. A list of 42 capability attributes found in the literature was presented to the respondents as follows: "Listed below are the typical critical success factors for competing in an industry. Scan over the entire list. Please circle the number that best corresponds to your business unit's current competitive strength on each factor, relative to your primary competitors in the same markets." The instruments used to operationalize the capabilities employed five-point self-anchoring scales. Of the 42 capability attributes, 20 captured the five generic dimensions of interest in this research. The remaining capability indexes (not shown here) will be used to assess unique, as opposed to generic, competitive strength in later research. The generic capabilities of quality, delivery, flexibility, and price leadership were operationalized as multiattribute indexes from the set. Table 38.3 displays the items which make up each index, the coefficient alphas by global region, and the item-to-total index correlations. Similar information is given for an index of process-focused technological leadership.

The generic capability indexes are

- *Quality:* The quality index consists of five attributes generally associated with superior product quality: product reliability, customer perceived quality, the degree to which the product conforms to standards, durability, and the general functionality of design. These five items represent the "tangible" attributes indicated by Garvin's (1987) dimensions of quality.
- *Delivery:* Three attributes signify delivery reliability. The first is manufacturing's

TABLE 38.3 Operational Indexes of Generic Competitive Capabilities

Cronbach alphas and item-to-total index correlations by global region

Mnemonic	Capability indexes Representative index items*	Overall	Europe	Japan	North America
			Standardized alphas		
	Quality index	.77	.70	.86	.75
			Item-to-total correlations		
RELIABILITY	Reduced probability of product breakdowns	.60	.49	.72	.61
PERCEIVED QUALITY	Customer-perceived quality	.55	.45	.68	.51
CONFORMANCE	Degree to which product meets stds.	.55	.43	.70	.51
DURABILITY	Product longevity	.52	.44	.64	.51
PERFORMANCE	Functionality of design	.48	.49	.60	.44
			Standardized alphas		
	Delivery index	.75	.78	.79	.74
			Item-to-total correlations		
DELIVERY RELIABILITY	Ship on time	.65	.68	.68	.63
ORDER CONFIRMATION	Rapidly confirm customer order delivery date	.63	.68	.66	.61
COMPLAINT HANDLING	Promptness in handling customer complaints	.47	.49	.54	.45
			Standardized alphas		
	Flexibility index	.76	.76	.80	.76
			Item-to-total correlations		
MIX FLEXIBILITY	Rapidly change product mix	.68	.68	.74	.67
VARIETY	Manufacture broad product mix in same facilities	.49	.51	.59	.46
VOLUME FLEXIBILITY	Rapidly change production volumes	.60	.57	.66	.61
CUSTOMIZATION	Rapidly handle "custom" orders	.38	.33	.39	.42
CHANGEOVER TIMES	Reduce product changeover time	.51	.55	.58	.50

Capability indexes

Mnemonic	Representative index items*	Overall	Europe	Japan	North America
				Standardized alphas	
	Price leader index	.82	.74	.85	.83
			Item-to-total correlations		
LOW PRICE	Offer lower priced products than competitors	.70	.60	.76	.72
LOW MFG COSTS	Manufacture similar products at lower cost	.66	.54	.68	.68
MEET COMPETITION	Meet competitors' prices	.65	.54	.74	.68
				Standardized alphas	
	Process technological leadership index	.82	.78	.85	.83
			Item-to-total correlations		
STATE-OF-THE-ART	Develop state-of-the-art manufacturing processes	.66	.59	.71	.68
SKILLED WORKERS	Workforce with superior technological skills	.65	.64	.64	.67
UNIQUE MFG PROCESS	Develop unique manufacturing process capabilities	.62	.51	.69	.65
ORG KNOWLEDGE	Superior technological know-how in organization	.65	.62	.73	.62

*For each item average respondents were asked to rate their current business unit's competitive strength, relative to their primary competitors in the same markets. Items were measured on a five-point self-anchoring scale, from "1 = relatively weak competitive strength" to "3 = strength" to "5 = market leader." Indexes for each competitive capability were constructed by averaging the items in the representative set.

ability to ship on schedule. This requires a high degree of internal coordination. The remaining two attributes indicate information flows with customers, either by having the process capability to rapidly confirm orders or to be able to promptly resolve complaints.

- *Flexibility:* To be able to respond quickly to the macro forces surrounding markets and customers, manufacturers must have responsive business processes. This degree of responsiveness requires a significant degree of coordination and process integration cross-functionally and cross-organizationally with customers and suppliers. Five items that connote process flexibility are the ability to rapidly change product mix, to produce a wide variety within the same facilities, to rapidly change production volumes, rapidly handle custom orders, and reduce product changeover times. Economies of scope, speed and economies of integration are implied by the aforementioned attributes.

- *Price leadership:* Price leadership construct is represented by three items: the ability to offer lower priced products than competitors, the ability to manufacturer similar products at a lower cost, and the ability to meet competitors' prices. Price leaders are able to garner higher margins and can be more price responsive than the competition. Therefore, they will have more resources for investment and can simultaneously satisfy customers and stakeholders.

- *Technological leadership:* Technological leadership is a sociotechnical measure of three critical resources: "hardware," "orgware," and "thoughtware." The four-item index consists of superior manufacturing processes in terms of the state-of-the-art applications and their unique capabilities. The third attribute reflects the firm's overall technical know-how in the organization. This reflects, in part, the firm's ability to leverage its organizational knowledge. The fourth item is a workforce with superior technological skills. Taken together, they are manifestations of the firm's potential absorptive capacity.

The ability to compete on the first four generic capabilities—quality, delivery, flexibility, and price leadership—together implies that the firm has the strategic agility to use different combinations of its resources to meet the diverse needs of its customers at different times. *Technological leadership* refers to a combination of resources which are used to build competitive capabilities and core competencies (Prahalad and Hamel, 1991). A measure of process technological leadership was operationalized in this study as described in the fifth entry above.

Advanced Manufacturing Technologies.   One measure of technological advancement is the firm's relative degree of experience with advanced technologies. Although literally hundreds of different types of advanced technologies are available to manufacturers, this research is restricted to advanced hardware, software, and methods technologies which can be broadly applied to manufacturing operations, including those for product and process design, automating production, enhancing production planning and control, and information technologies. [See Noori and Radford (1995, pp. 279–284) and Giffi et al. (1990, chaps. 7 and 10) for detailed descriptions of the advanced technologies covered here.] Executives were asked to respond to each of 31 technology types as follows: "Listed are a set of specific technologies and techniques that are being applied in manufacturing. Please indicate the degree of experience your MBU has with each, from no experience to 'state-of-the-art' experience."

Out of the technologies presented, 27 were found through item analyses to represent five broad dimensions of advanced operations technology: process enablers, computer numerical control, design interfaces, production planning and control, and information systems. Multiattribute technology indexes, the items making up each index, and their corresponding reliabilities are given in Table 38.4. Of special note is the high degree of internal reliability of each index across global regions. These classifications provide strong support for the implementations suggested in the manufacturing technology literature.

Each of the technology indexes is now summarized:

- *Process Enablers:* This index comprises eight items which provide advanced support functions for production. One set of support activities is used to monitor the production activities by gathering information about the process. These include computer-aided inspection, computer-aided testing, vision systems, and fail-safing processes. If adjustments are necessary, humans can intervene almost on an exception basis. A second category of enablers supports materials handling and storage. Technologies in this group are automated storage and retrieval systems, automated materials handling, and bar coding to identify the correct items to be picked, inventoried, and retrieved. Robots are multifunctional machines which can perform both direct production and production support activities. Because robots were included here, it is likely that their primary usage is for manufacturing support tasks, such as materials handling, machine loading, and pick-and-place maneuvers.

- *Information Technology:* The information system index represents eight basic information technologies (ITs) for communications, integration and coordination, intelligence, and control. Information technologies are used to automate administrative information processing tasks internally, such as electronic mail (e-mail), and externally by integrating information cross-functions and electronic data interchange with suppliers and customers. Integration of manufacturing internally relies on technologies that link together information from the various work centers, such as local area networks, personal computers, relational databases, and artificial intelligence and expert-systems software and hardware that enables machines to intelligently sense and manipulate their environments.

- *Design Interfaces:* Six items constitute an index of product and process design tools that are used to enhance the productivity of engineers and facilitate the effective coupling of engineering and manufacturing activities. Product design tools included in the index are computer-aided design (CAD), which creates, stores, and retrieves product and process information; computer-aided engineering (CAE), which ensures that design specifications are met; and value analysis and value engineering, which are methods for product and process simplification. Design for manufacture and simultaneous or concurrent engineering provide a new framework and approach for designers and manufacturing to work collaboratively in order to develop products and processes concurrently.

- *Production Planning and Control:* This index comprises four disciplines and control mechanisms that enable the organization to plan, schedule, and execute production and materials requirements effectively. Just-in-time (JIT) is a management philosophy of throughput improvement, lead-time reduction, and waste reduction. Kanban systems are the manual information control systems that are used to physically "pull" products through production. MRP-II is a companywide system for planning and controlling resources. Many companies operate MRP-II and JIT concurrently, or a hybrid of each. *Total productive maintenance* (TPM) is a method for enabling production employees to become responsible for setting up and adjusting their own equipment and for performing minor routine maintenance.

- *Flexible Automation:* This two-item index represents production equipment that enables the production of a variety of products with little time for changeovers.

TABLE 38.4 Operational Indexes of Advanced Operations Technologies

Cronbach alphas and item-to-total index correlations by global region

Technology indexes					
Mnemonic	Representative index items*	Overall	Europe	Japan	North America
				Standardized alphas	
Process enablers index		.87	.85	.92	.86
				Item-to-total correlations	
ROBOTICS	Robotics	.60	.57	.62	.61
CAI	Computer-aided inspection	.67	.63	.78	.64
CAT	Computer-aided testing	.66	.60	.81	.64
BC-SF	Bar coding for shop floor	.52	.48	.67	.52
AS/AR	Automated storage and retrieval	.66	.63	.80	.64
AMH	Automated materials handling	.64	.59	.73	.63
VISION	Vision systems	.65	.61	.77	.64
KAEYOKE	Fail-safing production processes	.56	.53	.71	.54
				Standardized alphas	
Information technology index		.86	.85	.90	.85
				Item-to-total correlations	
RDMS	Relational database management systems	.57	.58	.70	.54
EDI	Electronic data interchange	.61	.63	.74	.55
LAN	Local area networks	.67	.63	.76	.67
PC	Personal computers	.52	.49	.54	.55
E-MAIL	Electronic mail	.59	.58	.67	.56
ORG INFO INTG	Integrated information systems across functions	.66	.65	.75	.63
MFG INTEG	Integration of manufacturing and engineering systems	.66	.66	.73	.62
AI	Artificial intelligence and expert systems	.57	.43	.68	.59

	Standardized alphas			
	.86	.81	.91	.86
Design-manufacturing interfaces index	*Item-to-total correlations*			
CAD — Computer-aided design	.60	.63	.65	.57
CAPP — Computer-aided process planning	.57	.42	.75	.57
CAE — Computer-aided engineering	.73	.66	.86	.72
S_ENG — Simultaneous or concurrent engineering	.70	.61	.79	.72
DFM — Design for manufacturing	.67	.55	.80	.70
VA/VE — Value analysis and value engineering	.60	.54	.65	.60
	Standardized alphas			
	.80	.80	.87	.79
Production planning and control (PPC) index	*Item-to-total correlations*			
JIT — Just-in-time	.62	.60	.77	.59
MRP-II — Manufacturing resource planning	.48	.55	.60	.45
KANBAN — Kanban	.64	.69	.66	.61
TPM — Total productive maintenance	.60	.54	.72	.61
TMS — Tool management systems	.59	.55	.73	.59
	Standardized alphas			
	.84	.84	.90	.83
Flexible automation	*Item-to-total correlations*			
CNC — Computer numerical control	.72	.73	.83	.71
DCN — Direct numerical control	.72	.73	.83	.71

*For each item, respondents were asked to rate their manufacturing business unit's relative experience with the technology on a five-point self-anchoring scale, from "1 = no experience" to "3 = moderate experience" to "5 = state-of-the-art experience." Each technology index was constructed by averaging the item responses over each set of representative items.

Each technology contains its own microcomputer and is controlled by software programs. The computer numerical control (CNC) machines are not dependent on a host computer, whereas the distributed numerical control (DNC) machines are connected to a central computer and can share programs and information.

38.6   RESULTS AND DISCUSSION

Hypothesis 1 states that world-class manufacturers have experience with multiple advanced manufacturing technologies. This follows directly from the economies of scope and economies of integration required in postindustrial manufacturing (Goldhar et al., 1991; Noori, 1990; Noori and Radford, 1995). Using the five operational indexes of experience with advanced manufacturing technologies, a two-way analysis of variance was performed with world-class status and global regions as the main effects (Table 38.5). The results show that there are significant interaction effects between world-class status and global region ($p < .001$) for each of the five categories of technology. There is some juxtapositioning of the relative magnitudes of technological experience among global manufacturers.

The North American manufacturers tended to have the same pattern of technological experience buildup between the world-class and non-world-class groups, from adoption of information technologies (ITs) to the design and manufacturing interfaces types to production planning and control to flexible automation to process enablers. Their counterparts in Europe and Japan did not show uniform adoption patterns of technology between the world-class and non-world-class groups. For example, experience with flexible automation ranked highest for the world-class Japanese manufacturers, in contrast to the non-world-class Japanese group, whose experience with flexible automation was dismally low. There has been a tremendous drive toward IT-enabled business process reengineering over the past 5 years, especially in the Western regions; therefore, it is not surprising that IT was the top-rated technology group in North America and Europe, controlling for world-class status. It was notable that in both regions world-class manufacturers had the same experience profiles with multiple advanced technologies, but not their world-class counterparts.

The technology profiles depicted in Fig. 38.4 support H5. The world-class groups within each global region have relative magnitudes of experience with advanced technology that are significantly above their non-world-class counterparts. These data suggest that the world-class groups are more likely to be actively engaged in pursuing a broad array of technologies, especially those that have boundary-spanning potential such as the IT and design interfaces. The theory of absorptive capacity suggests that synergistic benefits can be derived from these cumulative experiences. In other words, the world-class groups aspire to build a greater capacity to leverage advanced manufacturing technologies in order to gain added flexibility due to economies of scope and economies of integration.

Hypothesis 2 posits that world-class manufacturers are more likely to combine best practices with applications of advanced technology. To test H2, a multiattribute index was derived from an exploratory factor analysis of manufacturing programs believed to represent the content of best manufacturing practices (Ward et al., 1994; Giffi et al., 1990). The index reflects the relative degree of managerial commitment and resources placed on nine key action programs to improve the effectiveness of manufacturing over the past 3 years. Items included are: (1) "work teams," (2) "broadened job responsibilities," (3) "continuous improvement (*kaizen*)," (4) "total employee involvement," (5) "worker training for new skills," (6) "manufacturing cells," (7) "statistical

TABLE 38.5 Relative Experience with Advanced Technologies by Global Region

Ranked in order of group means and two-way analyses of variance*

World-class group

WC Europe	Mean†	WC Japan	Mean	WC North America	Mean
IT	2.94	Flexible automation	2.94	IT	2.96
Design-manufacturing interfaces	2.71	Design-manufacturing interfaces	2.85	Design-manufacturing interfaces	2.85
Planning and control	2.62	IT	2.84	Planning and control	2.84
Flexible automation	2.57	Process enablers	2.82	Flexible automation	2.73
Process enablers	2.53	Planning and control	2.71	Process enablers	2.36

Non-world-class group

NWC Europe	Mean	NWC Japan	Mean	NWC North America	Mean
IT	2.48	IT	2.20	IT	2.58
Flexible automation	2.43	Design-manufacturing interfaces	2.07	Design-manufacturing interfaces	2.36
Design-manufacturing interfaces	2.37	Planning and control	2.03	Planning and control	2.35
Planning and control	2.32	Process enablers	2.02	Flexible automation	2.30
Process enablers	2.03	Flexible automation	1.95	Process enablers	1.90

Two-Way ANOVAs*

	World-class effects		Regional effects		Interaction WC status/region	
	F	P	F	P	F	P
IT	71.87	.0001	6.41	.0017	7.07	.0009
Design-manufacturing interfaces	78.92	.0001	1.11	.3285	6.05	.0024
Planning and control	75.69	.0001	3.42	.0331	6.73	.0012
Flexible automation	39.65	.0001	.49	.6103	7.75	.0005
Process enablers	119.98	.0001	15.08	.0001	.00	1.0000

*All overall F statistics were significant at .0001 level. Only F statistics for main and interaction effects are given.

†Means range from "1 = no experience" to "5 = state-of-the-art experience."

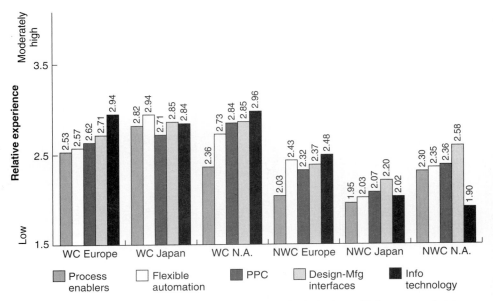

FIGURE 38.4   Experience with advanced manufacturing technologies by world-class status and global region.

process control," (8) "integrated manufacturing strategy with the business strategy," and (9) "worker-cross-training programs." Each item was measured on five-point self-anchoring scales, where 1 = "no commitment," 3 = "moderate commitment," and 5 = "significant commitment."

Each manufacturing business unit's best-practices index score is computed as the average of its responses over the nine scale items. The reliability coefficient, Cronbach's alpha, for the index was .89. Moreover, the relative emphasis on best practices was significantly greater for the world-class group in contrast to the non-world-class ($p<.0001$). This best practices scale is similar to the infrastructural capability programs reported by Ward et al. (1994, p. 343) and includes an integrated strategy item which is an aspect of their manufacturing involvement construct. My best-practices index is a broader infrastructural composite than theirs because it also includes process improvement tactics of cellular manufacturing, statistical process control, and other continuous improvement programs. The best practices index is a proxy for the application of the social sciences to operations by linking the "know what" with the "know how."

Table 38.6 contains the correlations between the relative degree of management emphasis on best manufacturing practices and the MBU's experience with various types of advanced technologies. Hypothesis 2 is sustained by these data overall and by global region. While statistically significant, the association between experience with flexible automation and best practices is the weakest. This may reflect the mistaken belief by many executives that computer-controlled technologies can substitute for labor without additional investments in the manufacturing infrastructure (Roth, 1986). It was commonplace in the 1980s to make hefty investments in automation without taking a holistic perspective. These results support the thesis that for technological

TABLE 38.6 Linking Experience with Advanced Technology and Best Manufacturing Practices
Pearson correlation coefficients and probabilities by global region

Type of technology	Overall sample	Europe	Japan	North America
Process enablers	.46 ($p < .0001$)	.47 ($p < .0001$)	.46 ($p < .0001$)	.51 ($p < .0001$)
Information technology (IT)	.46 ($p < .0001$)	.50 ($p < .0001$)	.45 ($p < .0001$)	.48 ($p < .0001$)
Design-manufacturing interfaces	.44 ($p < .0001$)	.39 ($p < .0001$)	.46 ($p < .0001$)	.47 ($p < .0001$)
Production planning and control	.51 ($p < .0001$)	.58 ($p < .0001$)	.42 ($p < .0001$)	.55 ($p < .0001$)
Flexible automation	.28 ($p < .0001$)	.18 ($p < .006$)	.39 ($p < .0001$)	.32 ($p < .0001$)

advancement, both hard and soft sciences are required (Adler, 1993; Seely-Brown, 1993). Cross-functional and cross-organizational integration, enabled by technology, must be accompanied by capable people and processes.

The third hypothesis addresses the technological leadership as a competitive weapon. Technological leadership is manifest by a high degree of organizational know-how, combining people with superior technological skills, and state-of-the-art process technology. (See Table 38.2 for technological leadership index definition.) To validate the operational measure, the results in Table 38.7 show that the technological leadership index is indeed associated with both the adoption of best practices and experience with advanced manufacturing technologies. Most importantly, technological leadership extends technical competence beyond the manufacturing function to the organization. For example, it is one thing to have experience with advanced manufacturing technology; it is quite another to be at the leading edge. General management must know how to direct it and the organization must have the absorptive capacity to use it.

Figure 38.5 highlights technological leadership capabilities of the global players. In support of Hypothesis 3, as predicted, the world-class group exhibits higher levels of technological prowess in comparison to the non-world-class group in each global region. Advanced technology is a necessary but not sufficient condition for performance. These findings corroborate those of Ward et al. (1994, p. 351), who state "Technology alone is not enough. At least one other dimension of proactiveness must be vigorously pursued." World-class manufacturers are more apt to be proactive in their multifaceted approach to technological leadership. World-class players may indeed be gaining competitive advantage through economies of knowledge. Future research is required to determine exactly how they are exploiting their "invisible" knowledge assets.

Hypothesis 4 states that the cumulative model of capability development is ubiquitous. Figures 38.6 and 38.7 offer the most substantive support for competitive progression theory to date. On average, the levels of capability development in a broadly segmented sample of global manufacturers follow hypothesized progression. For each competitive capability, the differences between world-class and non-world-class groups are significant at the .001 level (Table 38.8). The average competitive capabilities of the non-world-class group are not only relatively low but—contrary to conventional wisdom—also follow the theoretical progression. Competitive progression predicts that the logic of statistical variation, process capabilities, and value-chain process knowledge increases a manufacturing entity's inherent capabilities to deploy its resources most effectively. Consider, for example, a non-world-class manufacturer attempting to compete on price, without understanding the progression logic concerning quality or delivery competencies. Traditional cost-cutting methods in manufacturing have been shown to only exacerbate quality and delivery problems, which subsequently drive costs up. Ample empirical evidence exists. The reader has only to refer to Crosby's *Quality Is Free* or the classic works of quality gurus Deming or Juran. [See Giffi et al. (1990, chap. 2) for an overview of the gurus and a comprehensive picture of the "quality revolution" and world-class manufacturing.]

Holding world-class status constant, the competitive progression exists within each global region (Figure 38.7). More compelling evidence that competitive progression is rooted in basic operations physics is suggested by the observed progression in the non-world-class group. To the extent that the competitive progression theory is robust, future research must address how to accelerate progress in any innovation cycle. Moreover, practitioners need to think long and hard in not pursuing the logic of the theory in their manufacturing strategies. Quality must be addressed first, but not just

TABLE 38.7 Technological Leadership Capabilities and Experience with Advanced Manufacturing Technologies

Pearson correlation coefficients and probabilities by global region

	Overall sample	Europe	Japan	North America
Type of technology				
Process enablers	.31 ($p < .0001$)	.30 ($p < .0001$)	.36 ($p < .0001$)	.32 ($p < .0001$)
Information technology	.29 ($p < .0001$)	.36 ($p < .0001$)	.37 ($p < .0001$)	.25 ($p < .0001$)
Design-manufacturing interfaces	.32 ($p < .0001$)	.36 ($p < .0001$)	.37 ($p < .0001$)	.32 ($p < .0001$)
Production planning and control	.30 ($p < .0001$)	.34 ($p < .0001$)	.30 ($p < .0001$)	.31 ($p < .0001$)
Flexible automation	.26 ($p < .0001$)	.25 ($p < .0001$)	.32 ($p < .0001$)	.25 ($p < .0001$)
Manufacturing infrastructure				
Best manufacturing practices	.36 ($p < .0001$)	.42 ($p < .0001$)	.35 ($p < .0001$)	.36 ($p < .0001$)

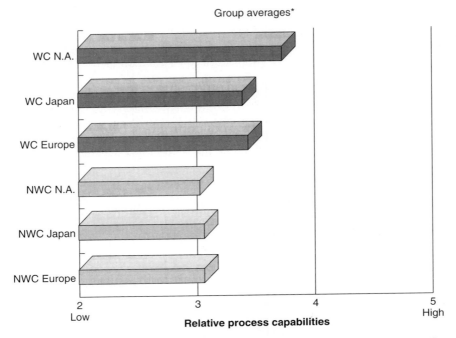

Group averages*

FIGURE 38.5 Technological leadership capabilities by world-class status and global region. (See Table 38.8 for group means and tests of significance.)

to enhance product quality. Even more important are organizational learning and development of in-depth process knowledge.

A two-way ANOVA (analysis of variance) indicates that there is a significant interaction between world-class status and global region for each capability, except price leadership. As seen in Table 38.8, the relative rank ordering of the capabilities within each world-class status varies by region for quality, delivery, and flexibility. Even though technological leadership is a means to an end, it is interesting to see where it maps onto the competitive progression. One important anomaly may guide future research. Technological leadership is closely aligned with process flexibility. In fact, for the world-class group, technological leadership uniformly *precedes* flexibility, whereas in the non-world-class group, it *follows* flexibility. Could it be that the world-class group is "pulling" out the organizational and technological barriers before taking a giant step from delivery to flexibility? In contrast, the non-world-class group appears to be feeble-mindedly "pushing flexibility" without having the necessary technological prowess. Controlling for global region, technological leadership capabilities in the non-world-class group consistently lag behind their world-class counterparts.

Finally, Hypothesis 5 states that a holistic approach to technology will be related to combinative competitive capability development. Possessing high levels of multiple capabilities has been linked to better business performance (Roth and Miller, 1992; Vickery et al., 1993; Noble, 1995; Cleveland et al., 1989) and is tantamount to the

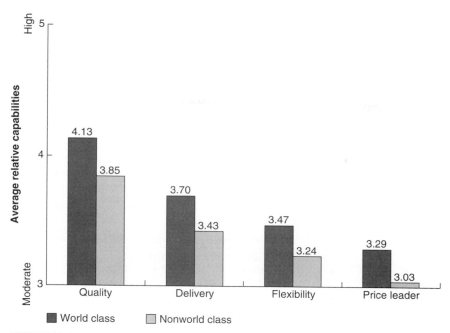

FIGURE 38.6 Competitive progression by world-class status.

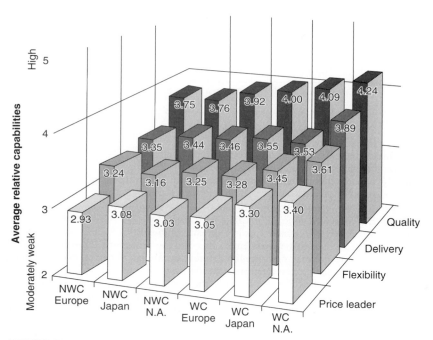

FIGURE 38.7 Competitive progression by world-class status and global region.

TABLE 38.8 Relative Competitive Capabilities by Global Region Controlling for World-Class Status

Ranked in order of group means and two-way analyses of variance

World-class group					
WC Europe	Mean[†]	WC Japan	Mean[†]	WC No. Amer.	Mean[†]
Quality	4.00	Quality	4.09	Quality	4.24
Delivery	3.55	Delivery	3.53	Delivery	3.89
Technology leader	3.46	Technology leader	3.41	Technology leader	3.74
Flexibility	3.28	Flexibility	3.45	Flexibility	3.61
Price leader	3.05	Price leader	3.30	Price leader	3.40

Non-world-class group					
NWC Europe	Mean[†]	NWC Japan	Mean[†]	NWC No. Amer.	Mean[†]
Quality	3.75	Quality	3.76	Quality	3.92
Delivery	3.35	Delivery	3.44	Delivery	3.46
Flexibility	3.24	Flexibility	3.16	Flexibility	3.25
Technology leader	3.08	Technology leader	3.08	Technology leader	3.04
Price leader	2.93	Price leader	3.08	Price leader	3.03

Two-way ANOVAs*					
	World-class effects		Regional effects		Interaction WC status/region	
	F	P	F	P	F	P
Quality	59.27	.0001	8.60	.0002	4.67	.0095
Delivery	34.46	.0001	4.79	.0085	7.98	.0004
Flexibility	37.76	.0001	2.21	.1097	6.47	.0016
Price leader	30.67	.0001	4.75	.0088	2.45	.0865
Technology leader	146.46	.0001	.07	.9322	10.04	.0001

*All overall F statistics were significant at .0001 level. Only F statistics for main and interaction effects are given.
†Means range from "1 = relatively weak competitive strength" to "5 = market leader."

firm being strategically agile. To test H5, two competing models were formulated and tested through canonical correlation analyses.

Briefly, canonical correlation is the most general of the linear models in statistics. *Canonical correlation analysis* is a technique for extracting a linear combination of a set of criterion (or dependent) variables such that the linear composites are maximally correlated. Each linear composite is called a *canonical variate*. Orthogonal to the first pair of canonical variates, a second pair of linear combinations is then extracted which again are maximally correlated with each other and the procedure is repeated until the maximum possible number of orthogonal linear combinations have been extracted. The maximum number of canonical pairs in the first model tested is four; in the second model, it is two. For parsimony, only the statistically significant sets of linear composites and their respective canonical structure loadings are reported here. Structure loadings are like factor loadings in that they represent the correlation of the canonical variate with each variable in the respective predictor or criterion set. For the purpose of drawing inferences about the meaning of the underlying dimensions, only loadings with an absolute value of .30 or greater are regarded as sufficient.

Correlations between the two canonical variates on each function are assessed by the significance of canonical correlation. The squared canonical correlation represents the maximum percent of variance accounted for by the canonical relationship.

The first model represents a "disjunctive" technology strategy in which the set of broad-based technological experiences is the predictor of combinative competitive capabilities as the criterion set. The second model is formulated with a "holistic" technology strategy, combining technological leadership and best manufacturing practices as the predictor set [see Giffi et al. (1990, pp. 141–145) for related empirical research on holistic and disjunctive manufacturing strategies].

The first model tests the multivariate relationships between the set of generic competitive capabilities and experience with advanced manufacturing technologies taken as a set. The results in Table 38.9 are quite interesting. Overall there was a statistically significant canonical correlation between a high level of competitive strength on the criterion set of quality, delivery, flexibility, price leadership, and technological experience. The canonical function 1 criterion set, highly descriptive of strategic agility, was related to a set of four of the five broad-based technological experiences types in the predictor set. In Model 1, the adoption of flexible automation was not related to strategic agility. This finding presented a quagmire since flexible automation was expected to be a forerunner of flexibility and price leadership. A second canonical function emerged in the overall set whereby process enabler technologies, flexible automation, and design interfaces were positively related to quality, but negatively correlated with delivery and flexibility. At first this second function also appeared puzzling. On inspection, although weak, canonical function 2 in the overall sample is indicative of strategies for highly capitalized, standardized production.

When viewed by global region, several disparities emerged in contrast to the overall sample. The Japanese canonical structure relationships mimicked those of the overall sample. In contrast, the European canonical criterion set did not equate to our definition of strategic agility, while the North American one did. Each multivariate relationship revealed quite different technology strategies. Within the European group, high levels of experience with process enablers, information systems, and production planning and control, devoid of experience with flexible automation and design-manufacturing interfaces, were associated with high delivery, flexibility, and price leadership capabilities. This appears to be due to a traditional shop-floor focus on delivery and flexibility allowing for a huge variation in product quality. In North American canonical function 1, relatively high levels of experience with all technology groups, except flexible automation, were related to increased strategic agility, whereas in the North American canonical function 2, high levels of technological experience produced high quality but disabled process flexibility. The percent of variance accounted for by the canonical relationships between experience with technology and experience with multiple technologies was relatively low, ranging from 3 to 15 percent. These Model 1 results provide no consistent explanations between the role of advanced manufacturing and competitive strength.

In the competing Model 2, technological leadership which represents superior people and process technologies combined with best manufacturing practices was chosen to predict strategic agility (Table 38.10). The criterion set in canonical function 1 in each region connotes strategic agility which is highly correlated with the combination of technological leadership and best practices. Importantly, the holistic approach to technology explained significant percentages of the variance accounted for, from 21 percent in Europe to 46 percent each in North America and Japan. The first canonical function in each region corroborates the findings of Roth and Jackson (1995) that the firm's absorptive capacity is limited by the expertise of people who must have the requisite knowledge to leverage the technology for competitive advantage. Technological

TABLE 38.9 Model 1: Linking Experience with Technology and Competitive Capabilities by Global Region

| | Canonical structure matrix* | | | | | |
| | Overall sample | | Europe | Japan | North America | |
	Function 1	Function 2	Function 1	Function 1	Function 1	Function 2
Predictor set						
Process enablers	.66	.40	.46	.83	.44	.60
Information technology	.76	.15	.55	.81	.53	.36
Production planning and control	.91	.12	.60	.91	.85	.41
Flexible automation	.28	.87	−.33	.73	−.05	.94
Design-manufacturer interfaces	.70	.62	−.06	.91	.49	.75
Criterion set						
Quality	.69	.67	−.02	.81	.53	.78
Delivery	.76	−.38	.85	.78	.73	−.13
Flexibility	.72	−.31	.70	.70	.80	−.30
Price	.57	−.12	.52	.69	.62	−.03
Strength of canonical relationships						
Canonical correlation R	.27	.17	.31	.39	.25	.22
Canonical correlation squared	.07	.03	.10	.15	.06	.05
Significance canonical R	p<.0001	p<.0001	p<.05	p<.0002	p<.0001	p<.0009

*Only significant canonical functions displayed overall and by global region. Coefficients represent correlation of the index with the corresponding canonical variate.

TABLE 38.10   Model 2: Linking "Holistic" Strategies and Competitive Capabilities by Global Region

| | Canonical structure matrix | | | |
| | Europe | | Japan | North America |
	Function 1	Function 2	Function 1	Function 1
Predictor set				
Technology leader	.94	−.34	.99	.99
Best practices	.67	.72	.40	.43
Criterion set				
Quality	.72	−.07	.79	.79
Delivery	.69	.31	.81	.74
Flexibility	.77	−.39	.76	.70
Price leader	.35	.63	.64	.35
Strength of canonical relationships				
Canonical correlations R	.46	.18	.67	.67
Canonical correlation squared	.21	.03	.46	.46
Significance canonical R	$p<.00001$	$p<.05$	$p<.0001$	$p<.0001$

*Only significant canonical functions displayed overall and by global region. Coefficients represent correlation of the index with the corresponding canonical variate.

leadership is a primary lever, and best practices, a secondary lever for strategic agility. Model 2 results are highly consistent with the emerging practice-based research on agility (Goldman et al., 1995). Arguably, a holistic approach to technology choices is a key component for the neo-operations strategies of firms seeking to compete on strategic agility. European manufacturers' strategic choices incorporated a modestly higher level of emphasis on best practices in contrast to their global counterparts. At the same time, their strategic payoffs in agility were no better than those of the North Americans and slightly worse than the Japanese.

Considering the findings reported in Tables 38.8 and 38.9 together, the implications are striking. Experience with technologies, regardless of type, many be a necessary but insufficient condition for competing. Interestingly, in both Japanese samples, the various types' experience with multiple advanced technologies seemed be highly correlated with one another, as did the criterion sets. Yet, the holistic strategy of Model 2 showed about three times the explained variance of Model 1. A higher level of confidence can be generated by explicitly developing a technology strategy in which people and the manufacturing infrastructure are consistent with the business strategy. Aligning requisite people skills and technological skills is perhaps the greatest challenge for acquiring strategic agility and gaining economies of knowledge.

38.7 CONCLUSIONS

This exploratory research brings forth a number of new concepts and ideas for managing postindustrial manufacturing enterprises under the mantra of neo-operations strategy. In particular, the research indicates that winning in global markets requires the technological leadership to leverage new knowledge in the form of combinative competitive capabilities. Executives need an expanded vision of combinative capabilities and an understanding of how lasting capabilities are acquired. This chapter proposes a new competitive progression theory of capability development. A series of hypotheses has been developed to test alternative strategies that counter the mass-production logic. Empirical tests of the hypotheses were made using a global database of more than 1200 manufacturing business units, spanning North America, Europe, and Japan. Stellar performers, called the "world-class group," have distinctive approaches to technology that are associated with higher levels of strategic agility, operationalized as combinative competitive capabilities on quality, delivery, flexibility, and price leadership. Contrary to conventional wisdom, the world-class manufacturers around the globe share similar strategic profiles. This fact, coupled with competitive progression theory, offers evidence of a convergence hypothesis in operations management—namely, that through economies of knowledge, world-class players are more apt to capitalize on global resources, and hence are less dependent on local resources for competitive advantage. The subject of a unifying theory of operations is ripe for future research.

A number of intriguing questions remain to be explored. For example, how does the concept of factory focus fit with progression theory? Paradoxically, focus capabilities are directly linked to process choices, which seemingly supports tradeoff theory and runs counter to competitive progression theory (Skinner, 1978; Berry et al., 1991). The notion behind factory focus is that simplicity drives out complexity, leading to better performance. One could extrapolate from competitive progression theory: A process focus contributes to accelerated movement through the progression for an innovation cycle because variance and complexity are reduced. Ironically, focus limits the firm's absorptive capacity to change production technologies, as described in the "great nuclear fizzle" case (Hill, 1989). In other words, is a process-focused factory less agile? Under what circumstances does it matter?

Another important question is: Given the competitive progression theory, why do some firms, even some of the most successful, stop making progress? The answers to this question are very complex and are the subject of future research. Early findings suggest that a combination of factors are culprits, including attempts to bypass steps in the progression, organizational "setpoint" or inertia to change, management complacency, technological constraints in the body of scientific knowledge, and future physical and attitudinal barriers imposed by the set of today's operations choices. A related question is: Are competitive progression theory and resource-based theory (Ulritch and Barney, 1984) compatible or contradictory theories? What do these theories imply for the direction in which the progression is pursued? Under what conditions is it more effective to work the progression backward (e.g., reengineer) versus always pushing forward toward strategic agility? More research is required to provide alternative theories and to test them.

What do these findings mean for technological progress? Briefly, capabilities-based competition encapsulates the organization's ability to learn, to assimilate, and to apply knowledge (Kogut and Zander, 1992). Strategic agility, however, does not mean that the firm must manufacture in every country in the world. Rather, it requires that manufacturing enterprise possess the organizational intelligence and operational know-how to seize opportunities afforded by the global marketplace. Future neo-operations management research must incorporate the study of human and organizational barriers to technological advancement. Operations managers must know how to create an environment for accelerated learning (Roth et. al., 1994), truly understand how people really work best, and how technology can expand their competencies.

A growing body of evidence suggests that new management perspectives are necessary (Drucker, 1992). Agile enterprises will be less concerned with routinization and scale economies—and more with acquiring and deploying organizational intelligence to compete. Over and over again, winners in global markets are accumulating generic competitive capabilities that allow them to play out multiple options simultaneously.

For example, speed-to-market may be the order winner in one market whereas delivery may be more critical in another. Interestingly, Porter-like (Kotha and Orne, 1989) static notions of strategic positioning are seemingly at odds with the neo-operations strategy. Dynamic leveraging of a portfolio of integration, infrastructural, and structural choices with changing customers demands is what will potentially create exciting levels of "customerization" (Hirasawa et al., 1995). Unfortunately, many manufacturers are stuck in the monotony of the industrial age; their anachronistic, mechanistic manufacturing perspectives limit their ability to become savvy global players.

38.8 SUMMARY

Confronted by quantum shifts in global markets and a rapidity of movement in technological progress, firms require unprecedented levels of strategic agility. Strategic agility is operationally defined as combinative competitive capabilities on quality, delivery, flexibility, and price leadership. The traditional logic of the industrial era does not equip manufacturers for competing on strategic agility. As a result, the game plans of leading manufacturers around the world dynamically changing to build capabilities-based competitiveness. In this chapter, I argue that twenty-first century competition requires a fundamentally new strategy: Neo-operations strategy links combinative competitive capabilities and advanced manufacturing technology strategies under the rubric of economies of knowledge. The stage for building strategic agility through neo-operations strategy is set. Hypotheses about how sustainable competitive capabilities are accumulated support a new competitive progression theory.

38.9 REFERENCES

Adam, E. E., Jr., and P. M. Swamidass, "Assessing Operations Management from a Strategic Perspective," *J. Management,* **15**: 181–203, 1989.

Adler, P. S., "Managing Flexible Automation," *California Management Review,* **71**(1): 97–108, 1993.

Badelt, R. and A. V. Roth, "Business Process and Reengineering and Knowledge Imperatives," The Knowledge Advantage II Conference Proceedings, Strategic Leadership Forum, November 16–17, Chicago, 1995.

Berry, W. L., C. C. Bozarth, T. J. Hill, and J. E. Klompmaker, "Factory Focus: Segmenting Markets from an Operations Perspective," *J. Operations Management,* **10**: 363–387, 1991.

Burgess, T. F., "Making the Leap to Agility," *International Journal of Operations and Production Management,* **14**(11):23–34.

Carmines, E. G., and R. A. Zeller, *Reliability and Validity Assessment,* Sage, Beverly Hills, Calif., 1979.

Churchill, G. A., "A Paradigm for Developing Better Measures of Marketing Constructs," *J. Marketing Research,* **16**(2): 64–73, 1979.

Cleveland, G., R. Schroeder, and J. Anderson, "A Theory of Production Competence," *Decision Sciences,* **20**(4): 655–668, 1989.

Cohen, W. M., and D. A. Levinthal, "Absorptive Capacity: A New Perspective on Learning and Innovation," *Administrative Science Quarterly,* **35**: 128–152, 1990.

Corbett, C. and L. Van Wassenhove, "Trade-Offs? What Trade-Offs? Competence and Competitiveness in Manufacturing Strategy," *California Management Review,* **35**(4):107–122, 1993.

Davis, S., *Future Perfect,* Addison-Wesley, Reading, Mass., 1987.

Davis, S., and B. Davidson, *2020 Vision,* Simon & Schuster, New York, 1991.

Deming, W. E., *Out of the Crisis,* MIT Center for Advanced Engineering, Cambridge, Mass., 1987.

Drucker, P. F., *Managing for the Future: The 1990's and Beyond,* Truman Talley Books, New York, 1992.

Ettlie, J., "Product-Process Development Integration in Manufacturing," *Management Science,* **41**(7): 1224–1238, 1995.

Ferdows, K., and A. De Meyer, "Lasting Improvements in Manufacturing Performance: In Search of New Theory," *J. Operations Management,* **9:** 168–184, 1990.

Fiegenbaum, A., and A. Karnani, "Output Flexibility—A Competitive Advantage for Small Firms," *Management Science,* **40:** 1444–1454, 1994.

Fitzsimmons, J. A., P. Kouvalis, and D. N. Mallick, "Design Strategy and Its Interface with Manufacturing and Marketing: A Conceptual Framework," *J. Operations Management,* **10**(3): 398–415, 1991.

Galbraith, J. R., and E. E. Lawler, III and Associates, *Organizing for the Future: The New Logic for Managing Complex Organizations,* Jossey-Bass, San Francisco, 1993.

Garvin, D., "Competing on the Eight Dimensions of Quality," *Harvard Business Review,* **65**(6): 101–109, 1987.

Gerwin, D. "Manufacturing Flexibility: A Strategic Management Perspective," *Management Science,* **39:** 395–410, 1993.

Giffi, C.A., "The New Manufacturing Strategy," *Harvard Business Review,* pp. 154–155, May–June 1994.

Giffi, C. A., A. V. Roth, and G. Seal, *Competing In World Class Manufacturing: America's 21st Century Challenge,* Business One Irwin, Homewood Ill., 1990.

Goldhar, J., and M. Jelinek, "Computer Integrated Flexible Manufacturing: Organizational, Economic and Strategic Implications," *Interfaces,* **15**(3): 94–105, 1985.

Goldhar, J. D., and M. Jelinek, "Plan for Economies of Scope," *Harvard Business Review,* **61**(6): 141–148, 1983.

Goldhar, J. D., M. Jelinek, and T. W. Schlie, "Competitive Advantage in Manufacturing through Information Technology," *Internatl. J. Technol. Management,* Special Publication on the Role of Technology in Corporate Policy, 162–180, 1991.

Goldman, S. L., and R. N. Nagel, "Management Technology and Agility: The Emergency of a New Era in Manufacturing," *Internat. J. Tech. Management,* **8:** 18–38, 1993.

Goldman, S. L., R. N. Nagel, and K. Preiss, *Agile Competitors and Virtual Organization,* Van Nostrand-Reinhold, New York, 1995.

Gray, A. E., A. Seidmann, and K. E. Stecke, "A Synthesis of Decision Models for Tool Management in Automated Manufacturing," *Management Science,* **39**(5): 549–567, 1993.

Hambrick, D. C. "Taxonomic Approaches to Studying Strategy: Some Conceptual and Methodological Issues," *J. Management,* **10**(1): 27–41, 1984.

Hayes, R., and S. Wheelwright, *Restoring our Competitive Edge: Competing through Manufacturing,* Wiley, New York, 1984.

Hayes, R. H., S. C. Wheelwright, and K. Clark, *Dynamic Manufacturing,* Free Press, New York, 1988.

Henkoff, R., "New Management Secrets from Japan—Really," pp. 135–146, Nov. 27, 1995.

Hill, T., *Manufacturing Strategy: Text and Cases,* Irwin, Homewood, Ill., 1989.

Hirasawa, Y., L. Menor, and A. V. Roth, "Dynamic Manufacturing Strategy: Implications for Global Agility," UNC-Kenan-Flagler Business School Working Paper presented at the Institute for Operations Research and Management Sciences Fall Meeting, New Orleans, October 29–November 1, 1995.

Iacocca Institute, *21st Century Manufacturing Enterprise Strategy,* Lehigh University, Bethlehem, Pa., 1991.

Kim, J. S., L. P. Ritzman, W. C. Benton, and D. l. Snyder, "Linking Product Planning and Process

Design Decisions," *Decision Sciences,* **23**: 44–60, 1992.

Kogut, B., and U. Zander, "Knowledge of the Firm, Combinative Capabilities, and Replication of Technology," *Organization Science,* **3**(3): 383–397, 1992.

Kotha, S., and D. Orne, "Generic Manufacturing Strategies: A Conceptual Synthesis," *Strategic Management J.,* **10**: 211–231, 1989.

Krajewski, L. J., and L. P. Ritzman, *Operations Management: Strategy and Analysis,* 3d ed., Addison-Wesley, Reading, Mass., 1993.

LaBarre, Polly, "The Dis-Organization of Oticon," *Industry Week,* July 18, 1994, pp. 23–28.

Lee, H., and C. Billington, "The Evolution of Supply Chain Management Models and Practices at Hewlett-Packard," *Interfaces,* **25**(5), Sept.–Oct. 1995.

Leonard-Barton, D., "Modes of Technology Transfer with Organizations: Point-to-Point Versus Diffusion," Harvard University Working Paper 90-060, 1990.

Leonard-Barton, D., *Wellsprings of Knowledge,* Harvard Business School Press, 1995.

Maskell, B. H., *Software and the Agile Manufacturer: Computer Systems and World Class Manufacturing,* Productivity Press, Portland, Oreg., 1994.

Merrills, R., "How Northern Telecom Competes on Time," *Harvard Business Review,* pp. 108–114, July–Aug. 1989.

Miller, J. G., and A. V. Roth, "A Taxonomy of Manufacturing Strategies," *Management Science,* **40**(3): 285–304, 1994.

Nakane, J., "Manufacturing Futures Survey in Japan: A Comparative Survey 1983–1986." Waseda Univ., Tokyo, 1986.

Noble, M. A., "Manufacturing Strategy: Testing the Cumulative Model in a Multiple Country Context," *Decision Sciences,* **26**(5): 693–720, 1995.

Noori, H., *Managing the Dynamics of New Technology: Issues in Manufacturing Management,* Prentice-Hall, Englewood Cliffs, N.J., 1990.

Noori, H., and R. Radford, *Production and Operations Management: Total Quality and Responsiveness,* McGraw-Hill, New York, 1995.

Nunnally, J., *Psychometric Theory,* McGraw-Hill, New York, 1978.

Pine, J. B., *Mass Customization: The New Frontier in Business Competition,* Harvard Business School Press, Boston, 1993.

Piore, M. J., and C. F. Sabel, *The Second Industrial Divide: Possibilities for Prosperity,* Basic Books, New York, 1984.

Prahalad, C. K., and G. Hamel, "The Core Competence of the Corporation," *Harvard Business Review,* pp. 79–91, May–June 1991.

Roth, A. V., "Achieving Strategic Agility Through Economies of Knowledge," *Strategy Leadership, (formerly Planning Review),* **24**(2): 30–37, Mar.–Apr. 1996.

Roth, A. V., *Strategic Planning for the Optimal Acquisition of Flexible Manufacturing Systems Technology,* Doctoral Dissertation, Ohio State University, Columbus, Ohio, 1986.

Roth, A. V., and R. J. Chapman, "Competing in the Electronic Industry: Benchmarking World-Class Performers," *J. Electronics Manufacturing,* **3**: 39–57, 1993.

Roth, A. V., and C. A. Giffi, "Winning in Global Markets: Neo-Operations Strategies in U.S. and Japanese Manufacturing," *Operations Management Review,* **10**(4): 1–35, 1995.

Roth, A. V., and C. A. Giffi, "Critical Factors for Achieving World Class Manufacturing: Benchmarking North American Manufacturing Strategies," *Operations Management Review,* **10**(2): 1–29, 1994.

Roth, A. V., C. A. Giffi, D. Shinsato, and M. J. Fradette, *Vision in Manufacturing: Planning for the Future,* Deloitte Touche Tohmatsu International Research Report, Cleveland, Ohio, 1993.

Roth, A. V., and W. Jackson, "Strategic Determinants of Service Quality and Market Performance," *Management Science,* **41**(11): 1720–1733, Nov. 1995.

Roth, A. V., J. Julian, and M. Malhotra, "Assessing Customer Value for Reengineering: Narcissistic Practices and Parameters from the Next Generation," in V. Grover and W. Kettinger, eds., *Business Process Change: Reengineering Concepts, Methods and Technologies,*

Idea Group Publishing, Harrisburg, Pa., pp. 453–474, 1995.

Roth, A. V., and A. S. Marucheck, "Defining World Class Manufacturing," *Proceedings of the Annual Meeting of the Decision Sciences Institute,* 1992.

Roth, A. V., A. S. Marucheck, A. Kemp, and D. Trimble, "The Knowledge Factory for Accelerated Learning Practices," *Planning Review,* pp. 26–33,46, May–June 1994.

Roth, A. V., and J. G. Miller, "Success Factors in Manufacturing," *Business Horizons,* **35**(4): 1973–81, 1992.

Schon, D., *Technology and Change: The New Heraclitus,* Delacorte Press, New York, 1967.

Schonberger, R. J., *World Class Manufacturing: The Lessons of Simplicity Applied,* Free Press, New York, 1986.

Seely-Brown, J., "Research that Reinvents the Corporation," in R. Howard, ed., *The Learning Imperative,* A Harvard Business Review Book, Boston, 1993.

Skinner, W., *Manufacturing in the Corporate Strategy,* Free Press, New York, 1978.

Stalk, G., and T. M. Hout, *Competing against Time,* Free Press, New York, 1974.

Stewart, T., 'Brace for Japan's Hot New Strategy," *Fortune,* Sept. 21, 1992, 62–74.

Ulritch, D., and J. B. Barney, "Perspectives in Organizations: Resource Dependence, Efficiency, and Population," *Acad. Management Review,* **9**(3): 471–481, 1984.

Venkatraman, N., and V. Ramanujam, "Measurement of Business Economic Performance: An Examination of Method Convergence," *J. Management,* **13L1:** 109–122, 1987.

Vickery, S. K., C. Droge, and R. R. Markland, "Production Competence and Business Strategy: Do They Affect Business Performance," *Decision Sciences,* **24**(2): 435–456, 1993.

Ward, P. T., G. K. Leong, and K. K. Boyer, "Manufacturing Proactiveness and Performance," *Decision Sciences,* **25**(3): 337–358, 1994.

Wheelwright, S. C., and K. B. Clark, *Revolutionizing Product Development: Quantum Leaps in Speed, Efficiency, and Quality,* Free Press, New York, 1992.

Womack, J. P., D. T. Jones, and D. Roos, *The Machine that Changed the World,* Rawson Associates, New York, 1991.

APPENDIX A

INSTITUTIONS OFFERING COURSES OR ADVANCED DEGREES IN MANAGEMENT OF TECHNOLOGY

This appendix includes a listing of Universities offering courses or advanced degree programs in the management of technology (MOT). The selection comprises various geographic areas as well as educational institutions with different perspectives on the requirements for advanced degrees in MOT. The degree programs span the continuum from special courses tacked onto an M.B.A., an M.B.A. with specialization in some aspect of MOT, and what by today's standards might be considered as an advanced degree in MOT sponsored jointly by the school of engineering and the school of management. Some universities also offer single courses in MOT.

While MOT is an evolving and multidisciplinary field that is practitioner-oriented, the associated advanced degree programs often fall short of meeting practitioner requirements. Chapter 1 of this handbook outlines the many issues involved in MOT. But few of the suggested advanced degree programs would meet those requirements.

From this editor's perspective, one who has spent almost 50 years in technology-related operations, the requirements at a minimum require all the basic courses in an M.B.A. plus those courses related specifically to managing and integrating the technology functions within the business unit. If approached from a one-time education process, the work would require the equivalent of three full academic years. The challenge to academia and industry: Design the educational delivery system for MOT based on a continuous (lifetime) learning process. Management of technology requires people with not only depth but breadth of knowledge and experience as well.

University of Alabama in Huntsville

Complete address:	MSM Program Administrative Science Building, Room 102 The University of Alabama in Huntsville Huntsville, AL 35899
Phone:	(205)895-6024 *or* UAH Admissions Office at (800)UAH-CALL
E-mail:	evansd@email.uah.edu
Principal contact:	Dr. Dorla Evans
Degree offered:	Master of Science in Management (M.S.M.)

The leaders of Huntsville's dynamic business and government sectors urged the University of Alabama in Huntsville (UAH) to provide leadership in a business area neglected by traditional business programs: managing technology. The future of American business lies in technology. That's why the heart of our Master of Science in Management (M.S.M.) is in the management of technology (MOT). MOT links management with engineering and science to help organizations meet the challenges of fast-changing technology. It is relevant in both service and manufacturing organizations. Our MOT program goes beyond the traditional M.B.A. program to prepare individuals to make decisions in environments of greater uncertainty and global competition.

Huntsville provides the ideal environment for an MOT program, surrounded as we are by many high-technology organizations. The campus is part of the second-largest research park in the United States. The park houses more than 50 *Fortune* 500 firms. NASA and the U.S. Army Missile Command are also in Huntsville. Students interact with managers in these organizations, who daily face the special struggles associated with technology. We instill in our students a commitment to teamwork, an essential ingredient to solving highly complex problems. UAH and its College of Administrative Science are nationally recognized. Our master's program has been recognized by the National Research Council as 1 of 19 programs in the nation with a major thrust in the management of technology.

UAH's M.S.M. curriculum includes an exciting mixture of business and technology. You will take courses in specific disciplines and other courses which integrate the material you have learned. At the beginning of your program, you will develop an overview of MOT by exploring emerging technologies and their business potentials. You will interact with instructors in the graduate program, who will link their courses with MOT. Late in the program, you will study with our MOT Eminent Scholar, Dr. William E. Souder. He will acquaint you with critical aspects of the new-product development process, from idea to marketing analysis, manufacturing, and delivery. Your last course will prepare you to develop strategies to use technology to your firm's competitive advantage. In the remaining courses, you will develop specific advanced business skills and their application to MOT, such as forecasting technology development, marketing new products, and managing information technology.

University of Brighton Business School
Center for Management Development

Complete address:	Mithras House
	Lewes Road
	Brighton East Sussex BN2 4AT, U.K.
Phone:	01273-642947
Fax:	01273-642980
E-mail:	
Principal contact:	Ms. Sheena McCann
Degrees offered:	M.S. Technology Management, M.B.A.
	Technology Management

Aims of the programs:

- To provide an advanced understanding of the functions of management
- To develop a strategic orientation to the application of technology
- To provide students with a challenging opportunity to explore and develop their

ideas in depth by focusing on an issue on the use of technology within a strategic context as the subject for a dissertation

- To develop an understanding of the key issues in the effective management of developing and implementing new technologies
- To provide opportunities for personal development
- To provide an awareness of the functions of management and their interaction with technology decisions
- To develop competence in the execution of projects involving the implementation of new technology through exposure to a variety of approaches and techniques

The program provides two award streams: M.Sc. Technology Management and M.B.A. (Technology Management). The program consists of intensive self-contained 5-day modules (Monday to Friday), 2-day workshops, plus a dissertation.

M.Sc. technology management	Core modules	M.B.A. technology management
10 modules	Technology management and policy	11 modules
1 workshop plus dissertation	Systems and organizations Design and product innovation Process innovation	2 workshops plus dissertation
Modules	Project implementation Managing change	Modules
Managing technical professionals Technology assessment plus one option	Corporate strategy	Accounting and finance Markets and competition Information systems management
	Quantitative methods workshop Dissertation	Business-context case study workshop

The M.B.A. (Technology Management) is awarded by the University of Brighton; the M.Sc. Technology Management is awarded jointly by the University of Brighton and the University of Sussex.

Carleton University

Complete address:	Department of Systems and Computer Engineering Mackenzie Building, Room 4462 Carleton University 1125 Colonel By Drive Ottawa, Ontario, K1S 5B6
Phone:	(613)788-5740 *or* (613)788-2386
Fax:	(613)788-5727 *or* (613)788-2532
E-mail:	ttm@sce.carleton.ca *or* donald_gerwin@carleton.ca
Principal contact:	The Graduate Secretary *or* Prof. Donald Gerwin

Degrees offered: Master of Engineering in Telecommunications Technology Management

The objective of the program is to develop engineers and computer scientists into managers who can deliver innovative telecommunications systems and services. The program focuses on a synthesis of communications systems engineering and management of engineering processes. Major areas of attention include network design, protocols and performance, software engineering, wireless and satellite communications, and manufacturing systems analysis. Students take required courses and electives in communications systems engineering and in the management of engineering processes, and complete a research-based thesis or a problem investigation project. Admission requirements: Applicants must have a bachelor's degree in electrical engineering or computer science or related discipline, with honors standing, and at least 2 years of work experience in the telecommunications industry.

Georgia Institute of Technology

Complete address:	School of Management
	Management of Technology Program
	Georgia Institute of Technology
	Atlanta, GA 30332-0520
Phone:	(404)894-1464
Fax:	(404)894-1552
E-mail:	william.rigg@mgt.gatech.edu
Principal contact:	William M. Riggs
Degrees offered:	Graduate Certificate in Management of Technology, Master's Degree in Management of Technology

The Georgia Tech MOT program is an institutewide, interdisciplinary effort, centered in the School of Management. The program has three major components: executive/professional education, programs for full-time students, and research. Executive/professional education is unusually important in MOT, since so many engineers and scientists eventually need further education to enhance their ability to deal with the business and policy issues that are in the context of technology development and commercialization. In response to this need, the Georgia Tech MOT program offers a master's degree program in MOT delivered on the "executive" format—alternate weekends, interspersed with two residency periods of 1-week duration, and a European residency lasting 2 weeks. This program attracts technical professionals and managers with 5 to 15 years of relevant work experience, who are sponsored by their employers. In addition to the master's program, short courses on MOT topics are offered several times per year. An MOT Certificate is offered for full-time students, with enrollment open to any graduate student with a technical background. The Certificate requires 21 credit hours of study, including three specialized MOT courses. The keystone of the program is an MOT project course in which interdisciplinary teams of students work in local companies to solve real technology management problems. A key objective of the MOT program is to help focus attention on the ongoing results of the considerable MOT-related research under way at Georgia Tech, and to help shape the future research agenda. One element of this is a working paper series on innovation, entrepreneurship, and the management of technology. The program has a strong international component, consistent with the rapid internationalization of business activity and the accompanying flow of technology across national boundaries for commercial purposes. The international activities of the program include the international residency and accompanying coursework in international technology management as part of the M.S. in MOT, and research and programmatic collaboration with various leading universities in Europe.

Harvard Business School

Complete address:	MBA Admissions Soldiers Field Road Boston, MA 02163
Phone:	(617)495-6127
Fax:	(617)496-9272
E-mail:	
Principal contact:	Admissions Office
Degree offered:	M.B.A.

The Harvard Business school provides a series of elective courses in technology and operations management in its traditional M.B.A. program. These courses include

- Business logistics: from product supply to after-sales service
- Designing, managing, and improving operations
- Managing innovation
- Managing product development
- Operations strategy

Janice Hammond, Kim Clark, and Steven Wheelwright are all faculty teaching in the technology and operations management area:

> In our area we deal with innovation, and it is important to remember that not every idea flies. The dynamic of our teaching group and the quality of class discussions help us refine ideas, gain feedback, and build insight. We learn from each other's experiences and from our students.
> The relationship between research and teaching is nowhere more integrated than in the area of technology and manufacturing. On-going relationships with the world's leading manufacturing firms allow faculty members to bring into the classroom the issues that will determine competitive success in the coming decades. Managers of the future must understand the power of technology in business. We are trying to help students see this fundamental link. It is exciting to be part of shaping a new generation of business leaders.

Stuart School of Business
Illinois Institute of Technology

Complete address:	565 West Adams Street Chicago, IL 60661
Phone:	(312)906-6526
Fax:	
E-mail:	
Principal contact:	Dr. Joel D. Goldhar
Degree offered:	M.S. in Operations and Technology Management

Successful organizations of all types—business, government, banks, health care, and the military—need highly skilled people and sophisticated technology in order to develop and deliver high-quality products and services more quickly and at low cost. These organizations also need a new kind of manager who can understand technology, work in teams, manage complex situations, supervise highly trained professionals, and initiate change—managers who know how to meet their goals and objectives. Illinois Institute of Technology's M.S. in Operations and Technology Management will help you become such a manager. The program is designed for people who get things done and want to learn how to get them done better, faster, and at lower cost. This advanced degree program, the only one of its kind in the Chicago area, emphasizes problem solving, people skills, and the practical realities of team building and project management. It focuses on complex operations that depend on the deployment of advanced technology and highly educated people. It concentrates on systems that demand high quality and rapid reaction times. The program is designed so that you and the students who enter with you will move through as a group. As you progress in the program, you and your classmates will have a common base of knowledge. This lets your professors maintain better control of the course content and, ultimately, it means your learning time will be more productive. The faculty, full-time academics, with extensive teaching, research, and consulting experience, are supplemented by expert practitioners from industry. This dual perspective, which integrates theory and practical experience, will prepare you for problem solving in a future where business challenges will differ vastly from those of the present. The program consists of 12 courses, each of four quarter hours, for a total of 48 quarter hours. Courses are offered on Saturday mornings and afternoons in a lock-step sequence, so the courses will be offered just when you need them. Students whose reimbursement options are limited may complete the program on a 30-month schedule. The program is designed for individuals who have a minimum of 6 years' work experience, a demonstrated leadership ability, and a strong commitment to an intellectually demanding academic experience.

INSEAD (The European Institute of Business Administration)

Complete address:	INSEAD Boulevard de Constance 77305 Fountainebleau Cedex France
Phone:	(33) 1 60 72 42 90
Fax:	(33) 1 60 74 55 13
E-mail:	execed@insead.fr
Principal contact:	Executive Education, Janet Burdillat
Degree offered:	INSEAD Diploma, certifying attendance

INSEAD's Strategic R&D Management Program is a 1-week program running once a year in April. The program is designed for senior managers involved in R&D or with responsibility for technical organizations who need a creative business perspective in order to manage imaginatively the interfaces with other functions in the organization. It does not review the basics of R&D management. The aim is to broaden outlook by improving management techniques and remaining attentive to the market, while maintaining technological competence. The participants are international and of widely varied experience. In about half of the sessions, cases based on real companies are used, many of which have been specifically developed by INSEAD faculty for this program. They are usually discussed in small groups before a lecture or class demonstration. A special feature of this program is the Management of Technology and Innovation project (MTI), which has been running for 10 years. This is a research partnership between INSEAD, other academic institutions, and a group of large international corporations, where research has been carried out to determine the strategic and organizational capabilities needed to transform technology into competitive commercial products. The director of the program, Arnoud de Meyer, is Associate Dean at INSEAD, Professor of Technology Management, and has a Ph.D. from the University of Ghent. His research interests concern the management of technical innovation and the use of new technologies in manufacturing. He has published extensively about communication in R&D departments, the use of information systems in manufacturing, and the internationalization of the R&D function.

INSEAD's International Manufacturing Program is a 2-week program running once a year in June. Its objective is to show that manufacturing, properly managed, can be a formidable competitive weapon in this age of increasing global competition. The program is designed for senior managers in manufacturing. These may be from corporate manufacturing or other technical functions such as engineering, R&D, or information technology. On past programs, participants have represented as many as 16 different nations. The faculty teaching on past programs have been of seven different nationalities. The contents of the program are influenced by and designed on the basis of original INSEAD research in the field of manufacturing management. Cases on real companies are used in about half of the sessions to illustrate the points being taught. This takes the form of participants reading, then working in small groups, and finally coming together in class for discussion. Special features of this program include benefiting from the research of the Global Manufacturing Futures Project, which has been ongoing for the past 14 years. This is research carried out by INSEAD, plus an American and a Japanese university, looking at the comparative techniques, approaches, and manufacturing performance resulting from the strategies of major manufacturing firms. Faculty research is presented to participants, and a computer simulation is used as part of the action learning process. The director of the program, Luk Van Wassenhove, is professor of Operations Management and Operations Research and Doctor in de Toegepaste Wetenschappen Katholieke Universiteit te Leuven. His current research focuses on supply-chain management, time-based competition, business process reengineering, and quality and process improvement. He has published extensively on these subjects. Before joining INSEAD, he was professor of operations research at the Econometric Institute of the Erasmus University, Rotterdam. He also served on the Faculty of Engineering at the Katholieke Universiteit te Leuven.

Haas School of Business
University of California

Complete address:	Haas School of Business University of California, Berkeley S545 Student Services Building Berkeley, CA 94720-1900
Phone:	(510)642-4993
Fax:	(510)642-2826
E-mail:	aberbook@euler.me.berkeley.edu
Principal contact:	Susan Aberbook
Degree offered:	Certificate offered to master's students in Engineering and Business on completion of Management of Technology course requirements

The Management of Technology (MOT) program was started in 1988 to bring together Haas School of Business and College of Engineering faculty in joint research and teaching efforts that address various aspects of the technology commercialization and adoption process. The industrial partners who helped establish the program emphasized that it should cover the entire range of functional disciplines—manufacturing, marketing, and R&D (engineering)—as well as all aspects of the technology development process—from product definition and technology development resource allocation through process development, manufacturing rampup, and delivery of the new technology to the marketplace. The MOT program supports a certificate program open to master's and Ph.D. students in both the College of Engineering and the Haas School of Business. Certificates are awarded to students, along with their degrees, when they complete four elective courses in the management of technology. Well-received MOT courses include Professor Paul K. Wright (Mechanical Engineering) course on Intelligent Manufacturing Systems, Visiting Professor David McKendrick (Haas School) course on Managing Innovation and Change, and a new course cotaught by Professor Alice Agogino (Mechanical Engineering) and Dr. Sara L. Beckman (Haas School) on Managing the New Product Development Process. Many engineering students also participate in the MOT Joint Learning Seminar, coled by the former Vice President, Manufacturing at Hewlett-Packard, Hal Edmondson, and Professor David Hodges, Dean of the College of Engineering, in which the students complete technology-oriented projects with local technology companies. In addition to basic courses on management of technology subjects, MOT seminars have been added that expose students to current technologies and technology management issues. The Emerging Technologies Seminar, hosted through the Center for Information and Technology Management, invites speakers to address various technology topics such as mobile computing and multimedia software development. The MOT Seminar on Project Management, coled by Dr. Sara Beckman (Haas School) and Professor Bill Ibbs (Civil Engineering—Construction Management), invited project managers from biotechnology, software, and electronics manufacturing companies to talk about the basics of project management.

Krannert Graduate School of Management

Complete address:	Krannert Graduate School of Management
	Krannert Building
	Purdue University
	West Lafayette, IN 47907-1310
Phone:	(317)494-4413
Fax:	(317)494-9658
E-mail:	herbm@mgmt.purdue.edu
	brady@mgmt.purdue.edu
Principal contact:	Dr. Herbert Moskowitz
Degree offered:	Master of Science in Management

The Manufacturing/Technology Management option represents a very flexible, cross-functional degree program in the Krannert Master of Science in Management program. This option was developed and is administered through the Center for the Management of Manufacturing Enterprises (CMME). The option has an interdisciplinary focus that takes full advantage of the strengths of the Krannert culture, most notably cross-functional interaction and cooperation. It is predicated on the belief that in addition to understanding the flow of materials and products through the manufac-turing process, a successful manufacturing/technology manager must understand and appreciate the nature and importance of innovation and product development, manag-ing quality and costs, developing human resources, understanding information tech-nology and its strategic importance, understanding the global marketplace, and creat-ing strategic value through technology. The option has six core modules in which students must complete one course from a set of electives: Human Resources in Manufacturing, Product Design and Development, Production Planning and Control, Systems Integration and Information Management, Total Quality Management, and Manufacturing Strategy. In addition, electives from the Purdue University Schools of Engineering are strongly recommended. Extensive fellowship and internship support are also available. Significant feedback is incorporated into the option through the industrial advisory board of CMME.

Center for Innovation Management Studies
Lehigh University

Complete address:	621 Taylor Street
	Bethlehem, PA 18015
Phone:	(610)758-3427
Fax:	(610)758-3655
Principal contact:	Dr. Alden S. Bean
Degree offered:	Master of Science in Management of Technology (M.S.-MoT)

Lehigh University's M.S.-MoT program is designed to meet the needs of technologi-cally driven companies—companies whose competitive advantage depends on the generation and implementation of technology. MoT is a relatively new field that is on the cutting edge of advanced management training. It integrates business and engi-neering interests in a number of new and exciting ways, focusing on technology-relat-ed issues dealing with strategic and tactical decisions in technology-intensive firms and industries. The program prepares graduates to effectively deal with

- Internal accountants, financial managers, marketing and sales executives, multi-functional team members, economists, etc.
- External sources of technology, strategic alliance partners, customers and vendors, cooperators and competitors, regulators and public policy makers, etc.
- Organization transformation and change; creating, acquiring, commercializing, and implementing technology; strategy formulation and implementation to achieve competitive advantage; etc.

The M.S.-MoT degree is a 36-credit program including an M.S. thesis. The program consists of nine required courses and three electives which can be completed in 1 year or part time in 2 years. The MoT curriculum is designed specifically to achieve the above objectives and offers application to real-world situations through case studies and a thesis. The master's thesis is a capstone educational experience in which the knowledge students have acquired in the program is applied in a real-world research setting. Ideally, this will involve field research in the form of a case study related to an aspect of managing the generation of or implementation of innovative technology in the student's organization or other field site.

Massachusetts Institute of Technology
Sloan School of Management
MIT Management of Technology Program

Complete address:	Suite E52-101
	50 Memorial Drive
	Cambridge, MA 02142-1347
Phone:	(617)253-3733
Fax:	(617)253-3154
E-mail:	mitmot@sloan.mit.edu
Principal contact:	Ms. Rochelle Weichman
Degree offered:	Master of Science in the Management of Technology

The MIT Management of Technology Program is an educational experience designed to develop strategic leaders for organizations. To be effective, technology-based leaders must develop a special managerial perspective—one that is different from that offered by a general management or M.B.A. program. The MIT Management of Technology program was developed to offer this perspective. The program is dedicated to developing leaders who

- Understand the critical linkages between technology and the overall business strategy
- Lead high-performance, cross-functional teams to produce the right product or process for the right application
- Develop and implement strategies for leading the organization's technological advances and innovative changes
- Are skilled in working across international and cultural boundaries
- Work effectively with technical and nontechnical professionals

The MIT Management of Technology program has been offered jointly by the Sloan School of Management and the MIT School of Engineering since 1981. It was the first master's degree program in the world to focus directly on the management of technology. The Management of Technology program is education, not training. Participants complete the program with an expanded vision with respect to their functional roles, their entire organization, their responsibilities to society and their country, as well as their personal and professional responsibilities in the world. The curriculum consists of a carefully balanced mixture of formal coursework by management and engineering faculty, as well as formal and informal seminars with senior managers responsible for the technical resources and technology strategy of their organizations. The Management of Technology program is divided into five segments with several courses in each segment:

- Strategic Aspects of Technology Management
- Managerial Decision Making
- Human and Organizational Factors in Technology Management
- Managing Product/Process Development
- Applied Research

The basic curriculum is fixed but requires three graduate-level electives.

Center for the Development of Technological Leadership
Institute of Technology, University of Minnesota

Mailing address:	Center for the Development of Technological Leadership 107 Lind Hall 207 Church Street SE Minneapolis, MN 55455
Phone:	(612)624-5747
Fax:	(612)624-7510
E-mail:	MOT@cdtl.umn.edu
Principal contact:	Dr. Y. Shulman, Director (612)624-9807 shulman@cdtl.umn.edu http://www.cdtl.umn.edu/
Degree offered:	Master of Science in Management of Technology (M.S./MOT)

CDTL Description. The Center for the Development of Technological Leadership (CDTL) was established at the University of Minnesota in 1987 with an endowment from the Honeywell Foundation. The mission of CDTL is to promote leadership in technology by supporting appropriate research and by providing Institute of Technology students and technical professionals in industry with educational opportunities for increased breadth and depth in technical management, business, and liberal arts.

CDTL is an interdisciplinary center with participation by the Institute of Technology, the Curtis L. Carlson School of Management, the College of Liberal Arts, the Hubert H. Humphrey Institute of Public Affairs, and the College of Agriculture.

MOT Program Description. Technical professionals must remain current in their fields of expertise. Those who seek to manage and lead technological activities must also be able to competently manage innovative projects; motivate and challenge specialized staff; strive toward corporate goals; understand foreign cultures; manage conflicts; communicate persuasively and effectively; and integrate technical functions with other parts of the organization, including marketing, manufacturing, and strategic planning.

The University of Minnesota's Master of Science in the Management of Technology Program (M.S./MOT) provides technical professionals with the knowledge, skills, tools, and values that will enable them to assume leadership roles within their organizations and bring technical advances to the global market.

Developed in 1989, this intensive, practitioner-oriented graduate program is designed for experienced engineers and scientists who have achieved springboard positions in their organizations and have demonstrated a potential for technological leadership. It is a 2-year, executive-format program with classes held on alternating Fridays and Saturdays. Two off-campus residencies are held each year. The 2-year integrated curriculum develops additional technical expertise and specific managerial capabilities required of those who direct groups of technical professionals or who oversee technological activities. International technology management issues are addressed during a 10-day international residency held in the Asia-Pacific region during the second year.

The Management of Technology program is the first of its kind to be offered in the upper Midwest. Because of its diversity, the availability and support of eminent faculty, and its strong, longstanding relationship with the corporate community, the University of Minnesota is uniquely qualified and positioned to offer this program.

Polytechnic University

Complete address:	6 Metrotech Center Brooklyn, NY 11201
Phone:	(718)260-3175
Fax:	(718)260-3874
E-mail:	gschilli@duke.poly.edu
Principal contact:	George A. Schillinger
Degree offered:	Master of Science in the Management of Technology

As one of the first accredited universities to offer an advanced degree in the Management of Technology, Polytechnic is a leader in developing the basic concepts and approaches involved in MOT. The curriculum melds the content of an MBA program with new subjects that meet the requirements of changing, innovative, and technology-intensive corporations. But rather than grafting a few courses onto a traditional management program, Polytechnic has developed a thoroughly innovative, integrated curriculum. The course of study incorporates a range of new and important technological subjects and issues in the framework of traditional management education. The emphasis throughout is on practical applications, and on the effective management of the enterprise. Technological subjects and issues include the following: the process of innovation in the firm; the strategic management of research and development, engineering, and operations; the reduction of new-product development times; the effective use of information systems and technologies; the effect of new technologies on strategies of the firm; moving into new technologies: timing and choice; internal technology venturing; strategic alliances for technology acquisition and product development; high-tech marketing; risk management of technological projects; development of core competencies; and technology development in the "boundaryless" corporation with customers, suppliers, and even competitors. In this new curriculum, even traditional subjects such as finance are taught with technological issues in mind. For instance, calculations on return on investment must now take into account the risks associated with obsolescence of existing capital equipment and technical personnel. Similarly, instruction in conventional methods of operations management considers the new knowledge and skills needed to manage continually changing products and processes and the interaction of manufacturing with product development, marketing, and strategy. The curriculum as a whole reflects an intensive reappraisal of what modern managers need to know to compete in complex volatile, technology-intensive, and highly competitive global markets. They include the most recent insights, theories, and lessons learned by technology managers, academic researchers, and management consultants from recent U.S., European, and Japanese experience.

National Technological University

Complete address:	700 Centre Avenue Fort Collins, CO 80526
Phone:	(970)495-6400
Fax:	(970)498-0601
E-mail:	GerryJ@mail.ntu.edu *or* Tina@mail.ntu.edu

Principal contact:	Dr. Gearold Johnson, Academic Vice President
Degree offered:	Master of Science Degree

The NTU curriculum in Management of Technology (MOT) allows technical professionals and managers to complete a 2-year interdisciplinary program of instruction leading to a Master of Science Degree in MOT. This program links engineering, science, and management disciplines and addresses the planning, development, and implementation of an organization's strategic and operational objectives in bringing today's technologies to the marketplace. MOT deals with the issues and methodologies involved in innovation, technology transfer, integration of technologies into strategic objectives of the firm, managing technical resources, evaluation of obsolescence, replacement criteria, managing large and complex projects, and technological assessment and evaluation. Candidates for a degree in Management of Technology complete 45 credits over the course of the 2-year program. NTU has brought together some of the nation's leading academic instructors and business experts in the formulation, development, and implementation of this program. The degree is pursued on a part-time basis; each trimester the students take 2 courses at their work location via satellite for a total of 12 courses over the 2 years. The courses include Technology and Economic Analysis, Managerial Accounting and Technical Systems, Manufacturing Systems and Technology Strategy, Technology and Financial Decision Making, Managing and Leading Technical People, R&D Management, Business Unit Strategy and Operations, Taking Technology to Market, Strategic Management of Technology and Innovation, Quality Management, Scientific and Technical Institutions, and Analysis of Emerging Technologies. The students work together as a group; during the course of the program, the students are brought together for seven 1-week residencies at different locations. At these residencies, the MOT faculty, as well as top experts from industry, education, and government, participate as lecturers. Additionally, all students complete individual field projects at their companies.

School of Postgraduate Management Studies (SPMS)

Complete address:	Faculty of Business Administration National University of Singapore Singapore 119260
Phone:	(65) 772-6324
Fax:	(65) 775-3955
E-mail:	FBAWPK@nus.sg or FBAKS@nus.sg
Principal contact:	Dr. Wong Poh Kam, Deputy Director, SPMS, Director, Centre for Management of Technology
	Dr. Kulwant Singh, Deputy Director, Centre for Management of Technology
Degree offered:	Master of Science

The Master of Science in Management of Technology program is offered in collaboration with the Postgraduate School of Engineering (PSE), National University of Singapore. It addresses both the managerial as well as the technical aspects of technology-intensive businesses in order to assist technical professionals to assume increasing managerial responsibilities as their careers progress. The program is particularly relevant for those involved in R&D management, engineering project management, and manufacturing management as well as product development and marketing in high-tech industries. Technical entrepreneurs, government technology planners, and other technical personnel who are expecting increasing management responsibilities will also benefit from the program. The M.Sc. program has a strong link with the Faculty's Centre for Management of Technology (CMT), which serves as a focal point to conduct industry-related research and consulting work in technology management. This ensures that the program has a good mix of theoretical and practical approaches to the management of technology. All M.Sc. (Mgt. of Tech.) candidates must complete a total of 12 courses from core management and technology streams of study. Four technology streams of study are available: systems engineering, manufacturing and electronics, civil engineering, and information technology. Candidates can therefore design a study plan that emphasizes specialization in domain-specific technology management skills or covers broad technology management disciplines. Besides providing a foundation in core management skills, the curriculum is designed to provide management training in technical functions, such as management of R&D, product innovation and commercialization, management of technical professionals, and technology strategy. Courses are offered on both part-time and full-time bases. Admission requirements are a good undergraduate degree and relevant work experience. An acceptable GMAT or GRE score, while not required, is recommended. The term runs from July to November and January to April. The program is into its fourth year and has an average intake of about 60 students per year. About 10 percent of the students come from overseas.

Rochester Institute of Technology

Complete address:	104 Lomb Memorial Drive Rochester, NY 14623-5608
Phone:	(716)475-6221
Fax:	(716)475-7450
E-mail:	rjbbbu@rit.edu
Principal contact:	Dr. Robert Barbato, Associate Dean
Degree offered:	Technology Management M.B.A.

The major benefit of earning a Rochester Institute of Technology M.B.A. is the number of technological resources available on campus, from advanced laboratories for statistics to courses for highly specialized technologies. The student gains the specialized knowledge to translate technological advances into new products and services through the Technology Management M.B.A. concentration. Concentration courses include Introduction to Technology Management and Managing High-Tech Organizations and two electives. Students may elect to have a second concentration area giving them cross-functional expertise. The RIT M.B.A. degree program requires 18 graduate credit hours. RIT is on the quarter system and has rolling admissions. The RIT College of Business is accredited by ASCSB.

Rutgers, the State University of New Jersey
Faculty of Management

Address:	Department of Organization Management 180 University Avenue Newark, NJ 07102
Phone:	(201)648-1650, 648-5984, 648-5982
Fax:	(201)648-1664

Principal contact:	Dr. George F. Farris, Director Technology Management Research Center
Degrees offered:	M.B.A. with concentration in MOT, Ph.D. Management of Innovation and Technology

Description of MOT Program: We offer a concentration in the Management of Innovation and Technology for our M.B.A. students. Students must take four elective courses beyond their other M.B.A. requirements, including two courses which are required for the concentration and two others which can be selected from a list of possible options. The required courses include Management of Science and Technology (the process of innovation, stimulating creativity and translating it to organizational output, and developing an innovative climate) and Organizations of the Future (managing knowledge-based organizations, with permeable boundaries, networked relationships, and the transformation processes under way in such organizations). Additional courses are Management of Technical Professionals (in cooperation with the National Technical University), High Technology Entrepreneurship, Management of Information Service Organizations, and a number of other courses. Our school also offers courses on the management of innovation and technology at both the undergraduate and Ph.D. levels.

George Washington University

Complete address:	2130 H Street, NW, Suite 632 Washington, DC 20052
Phone:	(202)994-7541
Fax:	(202)994-4606
Principal contact:	Ms. Patty Rice
Degree offered:	Master's in Engineering Management (emphasis on marketing of technology), D.Sc.

This concentration prepares students to market technical products and services in the domestic and international markets. Marketing is the business function that identifies unfulfilled needs and wants and defines and measures their magnitude, determines which target markets the organization can serve, decides on appropriate products, services and prepares to service those markets, and calls on everyone in the organization to "think and serve the customer." We stress the fundamentals of marketing and their application in the technical field of engineering.

This concentration also provides a systematic treatment of global-scale marketing, with an emphasis on marketing management and strategy, and deals with how technical marketing managers can analyze, understand, and be more effective in the global marketing environment. Issues of global marketing are treated within the broad context of U.S. competitiveness, i.e., the ability to compete in national and global markets. The primary objective is to explore and manage the specific array of strategic issues involved in entering overseas markets and in conducting marketing operations on a global, as opposed to a domestic scale.

University of Western Sydney, Macarthur

Complete address:	Faculty of Business and Technology University of Western Sydney, Macarthur P.O. Box 555 Campbelltown NSW 2560, Australia

Phone:	046 203699
Fax:	046 266683
E-mail:	l.delacy@uws.edu.au
Principal contact:	Lizette DeLacy
Degree offered:	Master of Technology Management

The Master of Technology Management is an innovative program designed to assist managers to plan, develop, and implement technological and managerial innovations to shape and accomplish the strategic and operational objectives of an organization. This course is designed for managers who are seeking to improve their understanding of the issues involved in economic development based on technological innovation and the application of technology to achieve continuous improvement. It will improve managerial effectiveness by developing a knowledge of operations management in a technology based enterprise. The multidisciplinary nature of managing technology requires managers to be able to understand the many interlinking facets of organizations and systems both large and small. The master's program has been designed to enable and encourage interaction between participants to enhance their skills as communicators, leaders, and team members. The Master of Technology Management is designed, developed, and presented by a team of academics and business people who are focused on making technology work to optimize the management and performance of business organizations. Entry to the Master of Technology Management requires successful completion of a degree in a technical or business-commerce area plus a minimum of three (3) years of postgraduate work experience or Graduate Diploma or Graduate Certificate with a satisfactory grade point average and appropriate work experience. In exceptional circumstances, the University may recommend an entry into the program when the applicant does not have these formal requirements, but can demonstrate equivalent knowledge and experience. The program consists of five core subjects and three elective subjects offered over two semesters in the full-time mode. The core subjects provide interdisciplinary skills in quality management, innovation, technology management strategies, performance measurement, and decision support systems. The elective streams allow participants to focus on a particular technological environment such as manufacturing, energy, transport, or water resources.

The University of Utah

Complete address:	Management of Technology Program 106 KDGB The University of Utah Salt Lake City, Utah 84112
Phone:	(801)581-7758
Fax:	(801)581-7214
E-mail:	mot@business.utah.edu
Principal contact:	Dr. Kevin W. Willoughby Director, Management of Technology Program
Degrees currently offered:	M.B.A. (Management of Technology), M.S. (Management of Technology)

The Management of Technology Program at the University of Utah was established in 1995. It is a comprehensive universitywide program, based on an axis between the

College of Engineering and the David Eccles School of Business, with cooperation of other parts of the University such as the College of Science, the Technology Transfer Office, the Office of the Vice President for Research, and the Office of the Vice President for Academic Affairs. The program is designed with "management of technology" as an integrative theme which links various disciplines and departments, rather than as a separate academic field; but a core group of specialized courses in the field of management of technology has been established by both the College of Engineering and the School of Business as the centerpiece of the program. The program has a number of special features, including

1. An international subtheme (emphasizing international technology ventures and global business)
2. Courses tailored to the distinctive strategic context of "middle-sized" economies (e.g., the Rocky Mountain states in North America, or the smaller nations of Latin America and Asia)
3. Educational experiences for students in the program emphasize "learning by doing" through interdisciplinary teams working on "real-world" problems of technology enterprises
4. MOT aspects of the program designed to complement existing engineering and scientific programs at the University of Utah

Students engaged in specialized graduate study in the sciences, medicine, or engineering may enroll in the Management of Technology Program while remaining within their core area of study; they would graduate with an M.S. or Ph.D degree in their core scientific or technical field, with an uij¥   MOT emphasis integrated into their program. Alternatively, students engaged primarily in business studies may complete an M.B.A. or Ph.D. degree with an emphasis in MOT technology. Modules in the MOT are presently being developed for undergraduate programs (in business and engineering) and for the Executive M.B.A. degree. The University of Utah's Management of Technology Program also includes an interdisciplinary universitywide seminar series and cross-departmental research activities.

APPENDIX B
ACRONYMS AND ABBREVIATIONS

AACSB	American Assembly of Collegiate Schools of Business
ABMA	Army Ballistic Missile Agency
AHP	analytical hierarchical process
AMHT	advanced manufacturing hardware technology
AMMT	advanced manufacturing management technology
ARPA	Advanced Research Project Agency
ASEE	American Society for Engineering Education
ASEM	American Society for Engineering Management
BDA	behavioral decision aids
BPR	business process engineering
CAD	computer-aided design
CAM	computer-aided manufacturing
CAPP	computer-aided process planning
CC	communications and computer
CE	concurrent engineering
CEO	chief executive officer
CIM	computer integrated manufacturing
CNC	computer numerical control
CPM	critical path method
CS	corporate strategy
CT	computerized tomography
DARPA	Defense Advanced Research Projects Agency
DICE	Defense Initiative on Concurrent Engineering

DOD	Department of Defense
DFT	distance from target
DHM	decentralized hierarchical modeling
DRAM	dynamic random access memory
DRC	designated research center
DSS	decision support system
EE	envision and enact
EEC	European Economic Community
EMS	Engineering Management Society
EMSA	enterprise modeling by structural analysis
EPA	Environmental Protection Agency
ES	expert system
ETMI	European Technology Management Initiative
FCC	Federal Communications Commission
FDA	Food and Drug Administration
FMS	flexible manufacturing systems
GERT	Graphical Evaluation and Review Technique
GMTSP	Global Manufacturing Technology and Strategy Vision Project
GNP	gross national product
GPSS	General-Purpose Systems Simulation
IAMOT	International Association for Management of Technology
IEEE	Institute of Electrical and Electronics Engineers
IMD	Institute for Management Development
INFORMS	Institute for Organization and Management Sciences
IPM	integrated process management
IRC	Interdisciplinary Research Centers
IT	information technology
IURC	Industry University Research Centers
JIT	just-in-time
KBDSS	Knowledge-based Decision Support Systems
KSA	knowledge, skills, and abilities
LCD	liquid crystal displays
LDC	least developed countries
LED	light emitting diodes
MAUA	Multiattribute Utility Analysis

MAUF	Multiattribute Utility Function
MAUT	Multiattribute Utility Theory
MBA	Master of Business Administration
MBU	manufacturing business unit
MRD	material requirements document
MRP	material resource planning
MTBF	mean time between failures
NASA	National Aeronautics and Space Administration
NC	numerical control
NGT	next generation technology
NIC	newly industrialized nations
NIH	National Institutes of Health
NIH	not invented here
NSF	National Science Foundation
OECD	Organization for Economic Cooperation and Development
OEM	original equipment manufacturer
O/PD	organizational and professional development
OSHA	Occupational Safety and Health Administration
PC	personal computer
PDSA	plan-do-study-act
PERT	program evaluation and review technique
PIMS	profit improvement of marketing strategy
PPM	Project Portfolio Management
PPP	phased project planning
QFD	quality function deployment
R&D	research and development
RDEC	Research, Development, and Engineering Center
RIA	research impact assessment
R-MEN	research-management expert network
RR	review and react
RTC	resistance to change
SBU	strategic business unit
SCI	science citation index
SGC	stage-gate concepts
SME	small manufacturing enterprise
SPC	statistical process control

SSF	Space Station Freedom
STEP	standard for the exchange of product model data
SWOT	strengths, weaknesses, opportunities, threats
TAI	technology advantage index
TAS	technology advantage score
TC	technology cycle
TCS	teaching company scheme
TG	technology gradient
TGN	technology gradient network
TIM	technology and innovation management
TOP	integration of technology, organization, and people
TPM	total productivity model
TQM	total quality management
TSTM	total systems approach to management
TTM	time to market
VGA	video graphics array

AUTHOR INDEX*

---

*This index includes authors whose work is discussed in text. It is not a listing of contributors to the *Handbook*.

SUBJECT INDEX